Seabuckthorn (*Hippophae* L.)
A Multipurpose Wonder Plant

About Editor-in-Chief

Prof. Virendra Singh, a world-renowned seabuckthorn scientist, is well known for his valuable contribution in the development of science of seabuckthorn in India. He has been working on seabuckthorn in dry temperate Indian Himalayas, since 1993 and worked on various aspects like genetic diversity and improvement, biochemical characterization, selection and introduction of high yielding local and exotic forms and cultivation technologies of seabuckthorn. He is also instrumental in motivating Indian scientists to start research work on clinical and experimental studies and standardization of various health products of seabuckthorn in his own university and about 35 other research organizations of India. Presently, he is Consortium Principal Investigator of Indian Council of Agricultural Research's funded scheme under National Agricultural Innovation Project (NAIP) entitled *"A value chain on seabuckthorn (Hippophae* L.*)"* (1 million USD, during 2008-14), a multi-organizational seabuckthorn R&D programme in dry temperate Himalayas. He is credited with several research papers on seabuckthorn published in reputed international and national journals. He has also edited four books and authored two books on seabuckthorn in collaboration with international experts.Dr.Virendra Singh is a Member of Board of Directors of International Seabuckthorn Association (ISA) and also a Member of Scientific Committee of ISA. In order to sensitize the Indian scientists, policy makers, farmers and private companies, Dr. Singh, along with his team, organized the International Workshop on Seabuckthorn, held on February 18-21, 2001 at India International Center, New Delhi. As Organizing Secretary, he has also organized two National Conferences on Seabuckthorn in 2010 and 2011 at Palampur. Dr. Singh is also Secretary of Seabuckthorn Association of India. He has also chaired sessions in ISA's Conferences held at Berlin, (2003), Beijing (2005) and Xining (2011). He is widely traveled and visited several countries of Europe, Asia and North America. For his contribution on seabuckthorn, Dr. Singh was awarded "Outstanding Contribution Award on Seabuckthorn" by "International Seabuckthorn Association" during ISA's Conference on 4th September 2011 at Xining, China. He has also been honoured with "Golden Jubilee Endowment Visiting Fellowship Award" by Institute of Chemical Technology, Mumbai on March, 8, 2013.

Seabuckthorn (*Hippophae* L.)
A Multipurpose Wonder Plant

Vol. IV: Emerging Trends in Research and Technologies

Editor-in-Chief
Virendra Singh

Associate Editors
Baoru Yang (Finland), Sonika Choudhary (India), Jörg-Thomas Mörsel (Germany), Yury A. Zubarev (Russia), K. Mohini, Sonika Singh, V.K. Sharma, R.K. Rana and Manohar Lal (India)

2014
Daya Publishing House®
A Division of
Astral International Pvt. Ltd.
New Delhi – 110 002

Cataloging in Publication Data—DK
Courtesy: D.K. Agencies (P) Ltd. <docinfo@dkagencies.com>

Seabuckthorn (Hippophae L.) : a multipurpose wonder plant / editor-in-chief, Virendra Singh ; associate editors, Baoru Yang ... [et al.].
v. 4 cm.
Includes bibliographical references and index.
Contents: v. 4. Emerging trends in research and technologies
ISBN 9789351242666 (Hardbound)
ISBN 9789351301066 (International Edition)

1. Sea buckthorn. I. Singh, Virendra, 1961- II. Yang, B. (Baoru)

DDC 634.74 23

Published by : **Daya Publishing House®**
A Division of
Astral International Pvt. Ltd.
– ISO 9001:2008 Certified Company –
81, Darya Ganj, Near Hindi Park,
Delhi Medical Association Road,
New Delhi - 110 002
Phone: 011-4354 9197, 2327 8134
Fax: +91-11-2324 3060
E-mail: info@astralint.com
Website: www.astralint.com

Laser Typesetting : **Classic Computer Services**
Delhi - 110 035

Printed at : **Salasar Imaging Systems**
Delhi - 110 035

PRINTED IN INDIA

Dedicated to

Professor Heikki Kallio

by

Dr. Baoru Yang, Professor in Food Development

Department of Biochemistry and Food Chemistry, University of Turku, FI-20014 Turku, Finland

PROF. HEIKKI KALLIO AND HIS SEABUCKTHORN

The seabuckthorn research at University of Turku started close to half a century ago, marked by the pioneering work of Prof. Arne Rousi and his group on the botanical characterization and description of the global natural populations of *Hippophae rhamnoides* L. This botanical approach was extended to a much wider scope covering chemistry, biochemistry, plant physiology, and effects on human physiology and health by the passionate work of Prof. Heikki Kallio, the Chair of Food Chemistry at the Department of Biochemistry and Food Chemistry, University of Turku. His work has not only made seabuckthorn research an important part of the strategy of Food Sciences at the University of Turku, but also formed significant contribution and promotion to global progress on research and development of seabuckthorn. Therefore, it has been a long journey on seabuckthorn with a vision to popularize the plant in his country and globally.

Prof. Heikki Kallio in Seabuckthorn Forest.

THE FIRST VISIT TO CHINA

As a member of an official delegation, representing Finland, Prof. Heikki Kallio participated in the First International Conference on Seabuckthorn held in Xian, China, in 1989. The other member of the Finnish delegate was Dr. Kauko Salo from the Finnish Forest Research Institute. At the conference, Kallio presented a research on supercritical fluid chromatographic-mass spectrometric analysis of carotenoids in seabuckthorn. He also presented the paper "Evolutionary relationships and systematics in *Hippophae*" on the behalf of Prof. Arne Rousi, who was the Rector of the University of Turku and was not able to participate in the conference because of some other engagements. Prof. Kallio always recalls the Xian meeting as one of the best seabuckthorn conferences ever organized. He was deeply impressed by both the successful work of the organizers and the research presented by Chinese and Russian scientists covering multiple aspects of seabuckthorn. After the meeting, he visited natural growth sites of seabuckthorn in the mountainous areas and several seabuckthorn factories in Shaanxi Province. The Xian meeting was a turning point that ignited his true passion on seabuckthorn and started his cooperation and friendship with his colleagues in China, Russia and India. At the time, in Finland, seabuckthorn was just a tiny berry plant on the coasts of the Baltic Sea, hardly known by any other people than the fishermen on their islands.

After his first visit to China in 1989, China and seabuckthorn have become very important key words in the life and work of Prof. Heikki Kallio. He has visited China many times in the following years as a seabuckthorn expert and had interaction meetings with several Ministers of Water Resources.

Prof. Heikki Kallio and Prof. Lü Rongsen, Two Seabuckthorn Legends in a Field Study of Wild Seabuckthorn Resources in Datong County, Qinghai Province, China, in 2011.

SCIENCE ON SEABUCKTHORN FROM FATTY ACID COMPOSITION AND HEALTH EFFECTS TO GENE EXPRESSION AND METABOLOMICS

The wide spectrum of health benefits of seabuckthorn shown by its use in traditional medicine and by work of Chinese and Russian scientists was fascinating. As a biochemist, Kallio wanted to know the composition and biochemical mechanisms behind these health effects. However, the compositional information on seabuckthorn was quite limited in the literature. In Turku, the work started with gas chromatographic analysis of fatty acids of seeds and berries of seabuckthorn of different subspecies. Samples were collected from different parts of the world ranging from the mountains in North-Western China and the coasts of the Baltic Sea to the Altai Region in Siberia. After comparison of the total fatty acids, different lipid classes including triacylglycerols and phospholipids were also analyzed. Later, the chemical analyses covered systematic comparison of vitamin C, tocopherols, tocotrienols, sterols, flavonoids, lignans, sugars and fruit acids in seabuckthorn of different subspecies, growth locations, harvesting dates and years. The results were among the earliest systematic compositional data published on seabuckthorn in peer-reviewed international journals.

Another major area of seabuckthorn research at Turku is the biological activities of the berries and different fractions of the berries and their effects on human physiology. Human clinical studies were carried out together with partners in Finland and other countries to investigate the effects of seabuckthorn berries and oils on atopic eczema, platelet aggregation, cholesterol levels, inflammation, liver function, diabetes, and expression of adhesion molecules on endothelial cells of blood vessels. In addition, a number of studies have been carried out in cooperation with colleagues in China and Finland to investigate the bioactivities of seabuckthorn using *in vitro* and other experimental models, such as effects of seabuckthorn oils on gastric ulcer, wound healing as well as on DNA and lipid oxidation.

The scientific path of seabuckthorn research of Prof. Heikki Kallios group in Turku proceeds along with and reflects the progress in the science of chemistry, biochemistry and biology. While the major methods and analytical tools used in the earlier research were chromatography and mass spectrometry, the strategic approaches of the most recent work are characterized by NMR analysis for studying the effects of seabuckthorn on human metabolomics. NMR is also used for finger-printing seabuckthorn of different origins and gene expression of enzymes for biosyntheses of key metabolites as a result of interaction between genetic background and environmental factors. This progress and broadening of scope is reflected in the list of seabuckthorn publications of the research group of Heikki Kallio (Appendix II).

Along the path, there has been a continuous effort of raising research funds in many small pieces. Seabuckthorn is not yet a strategically important plant in Finland. The Finnish government has not so far put much financial resources in seabuckthorn research and development despite the high public awareness of the health benefits of the berry. Regardless of the challenges, the science progresses in its own pace.

THE LEGEND OF WORLD'S MOST NORTHERN SEABUCKTHORN RESEARCH STATION AND SISU

The idea sounded absurd and crazy to many people when Prof. Heikki Kallio and his friend Hannu Lappalainen planned to transport seabuckthorn bushes pollinated in Sammalmäki in Southern Finland to Kittilä 185 km further north from the Arctic Circle. In Kittilä, the temperature drops to as low as -40 C and the sun does not rise at all for almost two months in winter. But they were crazy and stubborn enough to do this. The bushes produced berries in the first year. In the following years, the Russian cultivars died because of the extremely harsh winter. Two Finnish cultivars "Tytti and Terhi", transplanted as small shrubs, survived. During the first few years, the crop was nothing to be mentioned. After some years of growth and acclimation, ripening of berries in bushes started in late autumn, when the night temperatures are constantly below 0 °C, producing in the end beautiful yellow-orange berries each year. The ripe berries may remain in bushes over the freezing cold winter. This is another example showing how the nature makes miracles.

The seabuckthorn bushes in Kittilä were originally the properties of the Sammalmäki Seabuckthorn Society. A part of them were later donated to the University of Turku as the world's most northern seabuckthorn experimental site. A range of research has been carried out and is still on-going using berries from these bushes. Examples are comparison of composition and physiological effects of berries of "Tytti and Terhi" grown in Canada, Sammalmaki, and Kittilä as well as influence of growth environment on composition and gene expression of seabuckthorn enzymes.

Pictures (Left): The World's Most Northern Seabuckthorn Experimental Plot in Kittilä Belonging to the University of Turku. (Right): Seabuckthorn Berries Covered by Frost in Winter in Kittilä.

When foreigners get to know Finland and the Finnish people, they often come across the expression SISU. I have been wondering what SISU really means. The story of Heikki, Hannu, "Tytti and Terhi" gives me the best interpretation of SISU as a character of the small nation in the far north on the planet.

INTERNATIONAL COOPERATION AND GLOBAL RESEARCH AND DEVELOPMENT ON SEABUCKTHORN

To promote the global seabuckthorn development, the International Center for Research and Development of Seabuckthorn (ICRTS) was established in 1995 in China as an organization under the Ministry of Water Resources. Prof. Heikki Kallio and Mr. Veli-Markku Korteniemi as representatives of Finland were the members of the Coordinating Committee of ICRTS. Since then, ICRTS had been the active body organizing various international cooperation and conferences on seabuckthorn until the establishment of the International Seabuckthorn Association during the International Seabuckthorn Workshop held at New Delhi in 2001. Prof. Heikki Kallio and his colleagues in Finland hosted the meeting of the Coordinating Committee of ICRTS in 1997 and the annual meeting of the Board of International Seabuckthorn Association in June 2012 in Turku.

As the Chairman of the Scientific Committee of the International Seabuckthorn Association, Heikki Kallio is committed to the research and development of seabuckthorn at the global scale to promote the health and well-being of humans and especially the less privileged populations as well as the environment. In his own research, he has been cooperating with scientists from China, Russia, India, Germany, England, Canada, and Romania. But his vision, commitment and effort on seabuckthorn extends far beyond these countries and his own field of research.

SEABUCKTHORN FACTORY IN THE FAR NORTH

Being a scientist with seven years of experience as research director in food industry, Prof. Kallio has never limited his interest and activities to his laboratory at the University. In early 1990s, Prof.

Dr. A.W.Jasra (Pakistan), Ms. Xia Jingfang (China), Prof. Heikki Kallio (Finland), Prof. J. Thomas Moersel (Germany), Prof. Virendra Singh (India), Prof. Yuanlin Tai (China), Prof. Natalia Demidova (Russia), Prof. Utioh Alphonsous (Canada) and Prof. Lu Rongsen (China), Members of Board of Directors, International Seabuckthorn Association (ISA), in the meeting of ISA at University of Turku, Turku, Finland on 05.06.12.

Heikki Kallio and Mr. Veli-Markku Korteniemi established Aromtech Ltd. Later, the first supercritical fluid extraction factory in Scandinavia was built in Tornio (Finland), a small town close to the Arctic Circle. They had a vision of industrial utilization of seabuckthorn and other northern plant species for promoting human health, in an environment friendly manner. The flagship product of Aromtech is Omega7 Seabuckthorn Oil extracted with CO_2, a product representing one of the first international brands of seabuckthorn and being marketed in over 20 countries. Symbolizing high-tech industry of the food and natural product sector in Finland, Aromtech is a fast growing small enterprise with global business covering sectors of food supplements and bioactive ingredients of food and personal care products.

PRIVATE SEABUCKTHORN PLANTATION ON MASKINNAMO ISLAND IN THE ARCHIPELAGO OF SOUTH-WESTERN FINLAND

Family and friends are very important to Prof. Heikki Kallio and Sinikka Kallio. They spend their summer holidays in the summer place on the Maskinnamo Island in the Archipelago of South-Western

Prof. Heikki Kallio (Finland) honoured with International Seabuckthorn Association's Life Time Achievement Award, along with other ISA Awardees Prof. Natalalia Demidova (Russia), Prof. Virendra Singh (India) and two others (China) during ISA's Conference held on 04.09.2011 at Xining, China.

Finland. About ten years ago, they planted over 40 seabuckthorn bushes of Russian cultivars as well as wild Finnish and Chinese origins on seashore of the island. Each year, Heikki and Sinikka hand-pick hundreds of liters of berries and deliver them to family members, relatives and friends. Grand children are their top priorities. Watering bushes, cutting the grass, picking the berries, fishing and sailing with grand children are important elements of their lives on the island.

THE FUTURE

What is still to come? Who knows? Maybe seabuckthorn plantation at the Kevo Subarctic Research Station close to the coast of the Arctic Ocean, or maybe planting seabuckthorn in Peru besides the field of Andean grains and potatoes, or maybe new piers and saunas on the islands. Time shows. One thing for sure, work continues as life goes on. The next scientific generation is ready to take over the responsibility.

CV OF PROF. HEIKKI KALLIO

Kallio, Heikki Paavo Tapio, born at Turku, Finland, May 22, 1947

Common Education

- Matriculation at Turku 1965, Master of Science (Biochemistry), University of Turku 1971, Doctor of Philosophy (Food Chemistry), University of Turku, 1976.

Academic Employments

- First job in the area of Sciences, Research Assistant (microbiology) of the British Antarctic Survey, 1970
- Research Assistant (analysis of food volatiles) at the Technical Research Centre of Finland 1971-1975
- Associate Professor (Food Chemistry) at the University of Turku 1980-1990
- Visiting Professor (Food Chemistry) at the University of California Davis, Department of Food Science and Technology, and the Department of Environmental Toxicology 1985-1986. (Aroma Chemistry and Environmental Chemistry with Professors Walter Jennings, Rose-Marie Pangborn and Takayuki Shibamoto)
- Visiting Professor (Food Chemistry) in the Australian Wine Research Institute, University of Adelaide, Adelaide Australia 1990-1991 (Chemistry of wine volatiles and chemistry of lipid structures together with Dr. Patrick Williams and Dr. Graeme Currie)
- Professor of Food Chemistry, University of Turku, Finland 1991-current
- Visiting Professor (Food Chemistry and Functional Foods) at Jinan University, Guangzhou, China 2008-present

Employments in Industry

- Research and Development Director of Saarionen Ltd (Agriculture and Food Production), Tampere, Finland 1975-1980
- Research and Technology Director of Turun Muna Ltd (Food and Feed) 1980-1982

Selected Academic and Scientific Duties at the University of Turku

- Head of the Department of Biochemistry and Food Chemistry (2009-2012)
- Vice Dean/Dean of the Faculty of Mathematics and Natural Sciences (1994-1996)
- Member of the University Council, University of Turku (1994-1996)
- Chairman of the Board of Functional Foods Development Centre, University of Turku (2002-2007)

Selected Societal Duties

- Finnish Food Research Foundation, Cabinet Scientific Board, 1987-2012
- Finnish Food Research Foundation, Chairman of the Scientific Board, 1999-2003
- Federation of the European Chemical Societies, Food Chemistry Division, 1991-
- Pyhäjärvi-Institute Foundation, Board, 1989-99, 2003- (vice-memb. 99-2002)
- Member of the Coordinating Committee of International Center of Research and Development on Sea Buckthorn (ICRTS, Beijing, Ministry of Water resources, P.R. China) (1995-2007)
- Chairman of the Scientific Committee of the International Sea Buckthorn Association (2008-)

Scientific and Teaching Activities

- Food chemistry and biochemistry
- Chromatographic-mass spectrometric research on natural bioactive components

- Lipid and aroma chemistry
- Industrial supercritical fluid technologies
- Northern natural food resources
- Together with the research group over 300 original scientific publications of which over 220 in peer-reviewed Scientific Journals. Bioactive compounds and properties of sea buckthorn have been the most common targets with more than eighty publications. 25 PhD Theses and > 80 MSc Theses supervised.

Other Activities

- Northern frontiers
- Finnish Archipelago
- Photography

Research Papers

1. Ecclestone, C., Yang, B., Tahvonen, R., Kallio, H., Rimbach, G.H. and Minihane, A.M.2002. Effects of an antioxidant-rich juice (seabuckthorn) on risk factors of coronary heart diseases in humans. *J. Nutr. Biochem.* 13 (6): 346-354.
2. Järvinen, R., Kaimainen, M. and Kallio, H. 2010. Cutin composition of selected northern berries and seeds. *Food Chem.* 122: 137-144.
3. Järvinen, R., Silvestre, A.J.D., Gil, A.M. and Kallio, H. 2011. Solid state ^{13}C CPMAS NMR and FT-IR spectroscopic analysis of cuticular fractions of berries and suberized membranes of potato. *J. Food Compos. Anal.* 24: 334-345.
4. Järvinen, R.L., Larmo, P.S., Setala, N.L., Yang, B., Engblom, J.R.K., Viitanen, M.H. and Kallio, H. 2011. Effects of oral seabuckthorn oil on tear film fatty acids in individuals with dry eye. *Cornea* 30: 1013-1019.
5. Johansson, A., Korte, H., Yang, B., Stanley, J.C. and Kallio, H. 2000. Seabuckthorn berry oil lowers platelet aggregation. *J. Nutr. Biochem.* 11: 491-495.
6. Johansson, A., Laakso, P. and Kallio, H. 1997. Characterization of seed oils of wild, edible Finnish berries. *Z. Lembensm. Unters. Forsch.* A204: 300-307.
7. Johansson, A., Laakso, P. and Kallio, H. 1997. Molecular weight distribution of the triacylglycerols of berry seed oils analysed by negative-ion chemical ionization mass spectrometry. *Z. Lembensm. Unters. Forsch.* A204: 308-315.
8. Kallio, H. and Yang, B. 2002. Seabuckthorn: A unique berry for natural health. *Innov. Food Technol.* 15: 81-82.
9. Kallio, H., Lassila, M., Järvenpää, E., Haraldsson, G., Jonsdottir, S. and Yang, B. 2009. Inositols and methylinositols in seabuckthorn (*Hippophaë rhamnoides*) berries. *J. Chrom. B* 877: 1426-1432.
10. Kallio, H., Nieminen, R., Tuomasjukka, S. and Hakala, M. 2006. Cutin composition of five Finnish berries. *J. Agric. Food Chem.* 54: 457-462.
11. Kallio, H., Yang, B. and Peippo, P. 2002. Effects of different origins and harvesting time on vitamin C, tocopherols, and tocotrienols in seabuckthorn (*Hippophaë rhamnoides*) berries. *J. Agric. Food Chem.* 50: 6136-6142.

12. Kallio, H., Yang, B., Peippo, P., Tahvonen, R. and Pan, R. 2002. Triacylglycerols, glycerophospholipids, tocopherols and tocotrienols in berries and seeds of two subspecies (ssp. *sinensis* and ssp. *mongolica*) seabuckthorn (*Hippophaë rhamnoides* L.). *J. Agric. Food Chem.* 50: 3004-3009.

13. Larmo, P., Alin, J.A., Salminen, E.K., Kallio, H. and Tahvonen, R.L. 2008. Effects of seabuckthorn berries on infections and inflammations: A double-blind, randomized, placebo-controlled trial. *Eur. J. Clin. Nutr.* 62 (9): 1123-1130.

14. Larmo, P.S., Järvinen, R.L., Setälä, N.L., Yang, B., Viitanen, M.H., Engblom, J.R.K., Tahvonen, R.L. and Kallio, H.P. 2010. Oral seabuckthorn attenuates tear film osmolarity and symptoms in individuals with dry eye. *J. Nutr.* 140: 1462-1468.

15. Larmo, P.S., Yang, B., Hurme, S.A.M., Alin, J.A., Kallio, H., Salminen, E.K. and Tahvonen, R.L. 2009. Effect of a low dose of seabuckthorn berries on circulating concentrations of cholesterol, triacylglycerols, and flavonols in healthy adults. *Eur. J. Nutr.* 48: 277-282.

16. Lehtonen, H.M., Järvinen, R., Linderborg, K., Viitanen, M., Venojärvi, M., Alanko, H. and Kallio, H. 2010. Postprandial hyperglycemia and insulin response are affected by seabuckthorn (*Hippophaë rhamnoides* ssp. *turkestanica*) berry and its ethanol-soluble metabolites. *Eur. J. Clin. Nutr.* 64: 1465-1471.

17. Lehtonen, H.-M., Lehtinen, O., Suomela, J.P., Viitanen, M. and Kallio, H. 2010. Flavonol glycosides of seabuckthorn (*Hippophaë rhamnoides* ssp. *sinensis*) and lingonberry (*Vaccinium vitis-idaea*) are bioavailable in humans and monoglucuronidated for excretion. *J. Agric. Food Chem.* 58: 620-627.

18. Lehtonen, H.M., Suomela, J.P., Tahvonen, R., Vaarno, J., Venojarvi, M., Viikari, J. and Kallio, H. 2010. Berry meals and risk factors associated with metabolic syndrome. *Eur. J. Clin. Nutr.* 64: 614-621.

19. Lehtonen, H.M., Suomela, J.P., Tahvonen, R., Yang, B., Venojärvi, M., Viikari, J. and Kallio, H. 2011. Different berries and berry fractions have various but slightly positive effects on the associated variables of metabolic diseases on overweight and obese women. *Eur. J. Clin. Nutr.* 65: 394-401.

20. Leskinen, H.M., Suomela, J.P., Yang, B. and Kallio, H. 2010. Regioisomer composition of vaccenic and oleic acid containing triacylglycerols in seabuckthorn (*Hippophaë rhamnoides*) pulp oils: Influence of origin and weather conditions *J. Agric. Food Chem.* 58: 537-545.

21. Linderborg, K.M., Lehtonen, H.M., Järvinen, R., Viitanen, M. and Kallio, H. 2012. The fibres and polyphenols in seabuckthorn (*Hippophaë rhamnoides*) extraction residues delay postparandial lipemia. *Int. J. Food Sci. Nutr.* 63 (4): 483-490.

22. Manninen, P., Häivälä, E., Sarimo, S. and Kallio, H. 1997. Distribution of microbes in supercritical CO_2 extraction of seabuckthorn (*Hippophaë rhamnoides*) oils. *Z. Lebensm. Unters. Forsch.* A204: 202-205.

23. Partanen, R., Hakala, P., Sjovall, O., Kallio, H. and Forssell, P. 2005. Effect of relative humidity on the oxidative stability of microencapsulated seabuckthorn seed oil. *J. Food Sci.* 70 (1): 37-43.

24. Partanen, R., Yoshii, H., Kallio, H., Yang, B. and Forssell, P.2002. Encapsulation of seabuckthorn kernel oil in modified starches. *JAOCS* 3: 219-223.

25. Suomela, J.P., Ahotupa, M., Yang, B., Vasankari, T. and Kallio, H. 2006. Absorption of flavonols derived from seabuckthorn (*Hippophaë rhamnoides* L.) and their effect on emerging risk factors for cardiovascular disease in humans. *J. Agric. Food Chem.* 54: 7364-7369.

26. Suomela, J.P., Sandell, M., Vaarno, J., Lehtonen, H.M., Tahvonen, R., Viikari, J. and Kallio, H. 2012. Children's hedonic response to berry products: Effect of chemical composition of berries and *hTAS2R38* genotype on liking. *Food Chem.* 153 (3): 1210-1219.

27. Tiitinen, K., Hakala, M. and Kallio, H. 2005. Quality components of seabuckthorn (*Hippophaë rhamnoides*) varieties. *J. Agric. Food Chem.* 53: 1692-1699.

28. Tiitinen, K., Hakala, M. and Kallio, H. 2006. Headspace volatiles from frozen berries of seabuckthorn (*Hippophaë rhamnoides* L.) varieties. *Eur. Food Res. Technol.* 223 (4): 455-460.

29. Tiitinen, K., Vahvaselkä, M., Hakala, M., Laakso, S. and Kallio, H. 2006. Malolactic fermentation in seabuckthorn (*Hippophaë rhamnoides)* juice processing. *Eur. Food Res. Technol.* 222: 686-681.

30. Tiitinen, K., Yang B., Haraldsson, G., Jonsdottir S. and Kallio, H. 2006. Fast analysis of sugars, fruit acids and vitamin C in seabuckthorn (*Hippophaë rhamnoides* L.*)* varieties. *J. Agric. Food Chem.* 54: 2508-2513.

31. Wang, Y., Wu, Y., Kallio, H., Yang, B., Wang, J., Wang, B. and Cui, G. 2008. Effects of seabuckthorn pulp oil and seabuckthorn seed oil on radiation skin lesions in rats. *Journal of Xian Jiaotong University (Medical Sciences)* 29 (6): 705-709.

32. Xing, J., Yang, B., Dong, Y., Wang, B., Wang, J. and Kallio, H. 2002. Effects of seabuckthorn (*Hippophaë rhamnoides* L.) seed and pulp oils on experimental models of gastric ulcer in rats. *Fitoterapia* 73: 644-650.

33. Yang, B. and Kallio, H. 2001. Fatty acid composition of lipids in seabuckthorn (*Hippophaë rhamnoides* L.) berries of different origins. *J. Agric. Food Chem.* 49 (4): 1939-1947.

34. Yang, B. and Kallio, H. 2002. Composition and physiological effects of seabuckthorn (*Hippophaë*) lipids. *Trends Food Sci. Technol.* 13: 160-167.

35. Yang, B. and Kallio, H. 2002. Effects of harvesting time on triacylglycerols and glycreophospholipids of seabuckthorn (*Hippophaë rhamnoides* L.) berries of different origins. *J. Food Compos. Anal.* 15: 143-157.

36. Yang, B. and Kallio, H. 2006. Analysis of triacylglycerols of seeds and berries of seabuckthorn (*Hippophaë rhamnoides*) of different origins by mass spectrometry and tandem mass spectrometry. *Lipids* 41 (4): 381-392.

37. Yang, B., Ahotupa, M., Määttä, P. and Kallio, H. 2011. Composition and antioxidative activities of supercritical CO_2 extracted oils from seeds and soft parts of northern berries. *Food Res. Int.* 44: 2009-2017.

38. Yang, B., Halttunen, T., Raimo, O., Price, K. and Kallio, H. 2009. Flavonol glycosides in wild and cultivated berries of three major subspecies of *Hippophaë rhamnoides* and changes during harvesting period. *Food Chem.* 115: 657-664.

39. Yang, B., Kalimo, K., Mattila, L., Kallio, S., Katajisto, J., Peltola, O. and Kallio, H. 1999. Effect of dietary supplementation with seabuckthorn (*Hippophaë rhamnoides*) seed and pulp oils on atopic dermatitis. *J. Nutr. Biochem.* 10: 622-630.

40. Yang, B., Kalimo, K.O., Tahvonen, R.L., Mattila, L.M. and Kallio, H.P. 2004. Dietary supplementation with seabuckthorn (*Hippophaë rhamnoides*) oils and skin surface roughness of patients with atopic dermatitis. *Global Seabuckthorn Research and Development* 1 (2): 44-47.

41. Yang, B., Kalimo, K.O., Tahvonen, R.L., Mattila, L.M., Katajisto, J.K. and Kallio, H. 2000. Effect of dietary supplementation with seabuckthorn (*Hippophaë rhamnoides*) seed and pulp oils on the fatty acid composition of skin glycerophospholipids of patients with atopic dermatitis. *J. Nutr. Biochem.* 11: 338-340.

42. Yang, B., Karlsson, R., Oksman, P. and Kallio, H. 2001. Phytosterols in seabuckthorn (*Hippophaë rhamnoides* L.) berries: Identification and effects of different origins and harvesting times. *J. Agric. Food Chem.* 49 (11): 5620-5629.

43. Yang, B., Linko, A.M., Adlecreutz, H. and Kallio, H. 2006. Secoisolariciresinol and matairesinol of seabuckthorn (*Hippophaë rhamnoides* L.) berries of different subspecies and harvesting times. *J. Agric. Food Chem.* 54: 8065-8070.

44. Yang, B., Zheng, J. and Kallio, H. 2011. Influence of origin, harvesting time and weather conditions on content of inositols and methylinositols in seabuckthorn (*Hippophaë rhamnoides*) berries. *Food Chem.* 125: 388-396.

45. Zheng, J., Kallio, H., Linderborg, K. and Yang, B. 2011. Sugars, sugar alcohols, fruit acids, and ascorbic acid in wild Chinese seabuckthorn (*Hippophaë rhamnoides* ssp. *sinensis*) with special reference to influence of latitude and altitude. *Food Res. Int.* 44: 2018-2026.

46. Zheng, J., Yang, B., Trepanier, M. and Kallio, H.2012. Effects of genotype, weather and latitude conditions on the composition of sugars, sugar alcohols, fruit acids and ascorbic acid in seabuckthorn (*Hippophaë rhamnoides* ssp. *mongolica*) berry juice. *J. Agric. Food Chem.* 60 (12): 3180-3189.

Book Chapters and Papers in Conference Proceedings

1. Kallio, H. and Yang, B. 2012. Tyrnin monipuolinen koostumus tuottaa erityisiä terveysvaikutuksia. *Kehittyvä Elintarvike* 2: 68-69.

2. Larmo, P.S.; Yang, B.; Järvinen, R.L.; Setälä, N.L.; Viitanen, M.H. and Kallio, H.P. 2012. Nutrition in the treatment of dry eye with special attention on sea buckthorn oil. In: *Emerging Trends in Dietary Components for Preventing and Combating Disease*. In: B.S., Patil, G.K. Jayaprakasha, K.N.C. Murthy, N.P. Seeram (eds.). *ACS Books Series* Vol 1093, pp. 533-543.

3. Lundén, S.; Tiitinen, K. and Kallio, H. 2010. Aroma analysis of sea buckthorn berries by sensory evaluation, headspace SPME and GC-olfactometry. In: Blank, I., Wüst, M., Yeretzian, C. (eds.): In: *Expression of Multidisciplinary Flavour Science*. Proceedings of the 12th Weurman Symposium. Institute of Chemistry and Biological Chemistry, Zürich University of Applied Sciences, Switzerland, pp. 490-493.

4. Sandell, M.A.; Tiitinen, K.M.; Pohjanheimo, T.A.; Kallio, H.P. and Breslin, P.A.S. 2008. Why naturally healthy berry may be seen as unpleasant and non-appetitive. In. *Food Flavour: Chemistry, Sensory Avaluation and Biological Activity*. Eds. Takeoka, G.; Ebeler, S.; Kubota, K. and Tamura, H. American Chemical Society, Oxford University Press, pp. 219-228.

5. Yang, B.; Wu, Y.; Liu, Q.; Wang, B.; Kang, J.; Wang, J. and Kallio, H. Supercritical CO_2 extracted seabuckthorn pulp oil and seed oil improve blood microcirculation. In: V. Singh *et al.* (Eds.) 2008.

Seabuckthorn (Hippophaë L.)-A Multipurpose Wonder Plant. Vol. III: Advances in Research and Development. Dya Publishing House, New Delhi, India. pp. 268-271.

6. Kallio, H.; Yang, B. and Kallio, S. 2008. Tyrnin koostumus, ekologia ja terveellisyys ovat tasapainossa. http://www.arctic-flavours.fi/nettilehti/artikkeli.php?aid=18&lid=6

7. Tiitinen, K. and Kallio, H. 2007. Compounds related to sensory properties of sea buckthorn (*Hippophaë rhamnoides* L.). In: V. Singh *et al.* (Eds.) 2008. *Seabuckthorn (Hippophae L.). A Multipurpose Wonder Plant. Vol. III: Advances in Research and Development.* Dya Publishing House, New Delhi, India. pp. 187-204.

8. Tiitinen, K; Vahvaselkä, M., Hakala, M., Laakso, S. and Kallio, H. 2006. Effect of malolactic fermentation on the volatile aroma compounds in four sea buckthorn varieties. In: *Developments in Food Science 43: Flavour Science: Recent Advances and Trends.* W.L.P. Bredie and M.A. Petersen (Editors), Elsevier B.V., Oxford, UK, pp. 101-104.

9. Yang, B. and Kallio, H. 2006. Physiological effects of seabuckthorn (*Hippophaë rhamnoides*) fruit pulp and seed oils). In: Singh, V.; Yang, B.; Kallio, H.; Bala, M.; Sawhney, R.C.; Gupta, Möersel, J.-T.; R.K.; Lu, R.; Tolkachev, O.N. (Eds.) 2006. In: *Seabuckthorn (Hippophae L.). A Multipurpose Wonder Plant. Vol. II: Biochemistry and Pharmacology.* Daya Publishing House, New Delhi, India. pp 363-389.

10. Yang, B.; Kallio, H. Lipophilic components of seabuckthorn. In: Singh, V.; Yang, B.; Kallio, H.; Bala, M.; Sawhney, R.C.; Gupta, Mörsel, J.-T.; R.K.; Lu, R. and Tolkachev, O.N. (Eds.) 2006. In: *Seabuckthorn (Hippophaë L.). A Multipurpose Wonder Plant. Vol. II: Biochemistry and Pharmacology.* Daya Publishing House, New Delhi, India. pp 70-97.

11. Yang, B. and Kallio, H. 2005. Sea buckthorn (*Hippophaë rhamnoides*): A review on composition of berries and health effects of berry products. *Proc. 1st International Conference of Food Choice and Healthy Eating Focusing on Vegetables, Fruits and Berries*, pp. 15-22, September 1st-3rd, 2005, Kauhajoki, Finland.

12. Yang, B and Kallio, H. 2003. Bioactive components of berries of three subspecies of sea buckthorn (*Hippophaë rhamnoides*). In: Mörsel, J.-T., Thies, S. 2003. Congress of the International Seabuckthorn Association: *Seabuckthorn–A Resource of Health, a Challenge to Modern Technology.* Berlin, Germany, September 14-18, 2003. pp. 70-74.

13. Kallio, H. and Yang, B. 2003. Health effects of sea buckthorn (*Hippophaë rhamnoides* L.) oils. In: Mörsel, J.-T., Thies, S. 2003. Chinese and Russian knowledge and claims supported by Finnish research. In: Congress of the International Seabuckthorn Association: *Seabuckthorn–A Resource of Health, a Challenge to Modern Technology.* Berlin, Germany, September 14-18, 2003. pp 82-89.

14. Yang, B. and Kallio, H. 2003. Effect of sea Buckthorn oil (*Hippophaë rhamnoides*) on skin; Eastern tradition and modern research. *Personal Care*, Nov., pp. 46-49.

15. Kallio, H. 2003. Professor Arne Rousi – A Cultural Messenger of Science. In: Singh, V.; Kallio, H.; Sawhney, R.C.; Gupta, R.K.; Lu, R.; Eliseev, I.P.; Khabarov, S.N., Korovina, M.A.; Skuridin, G.M.; Shchapov, N.S. (Eds.) 2003. *Seabuckthorn (Hippophaë L.)- A Multipurpose Wonder Plant.* Indus Publishing Company, New delhi, India. pp. 9-11.

16. Yang, B., Gunstone, F. and Kallio, H. 2003. Oils containing oleic, palmitoleic, γ-linolenic and stearidonic acids. In: *Lipids for Functional Foods and Nutraceuticals*. Gunstone, F-D. (ed.). The Oily Press, Bridgwater, UK. pp. 263-290.

17. Yang, B.; Kallio and H. 2002. Supercritical CO_2-extracted sea buckthorn (*Hippophaë rhamnoides*) oils as new food ingredients for cardiovascular health. In: *Proc. Health Ingred. Europe 2002*. Paris-Nord Ville Pinte, Paris, France, Sep 17th-19th 2002. 7 pages.

18. Yang, B.; Kallio, H.; Peippo, P.; Tahvonen, R.; Pan, R.; Zhou, X.; Fu, Y. and He, Y. 2001. Tocopherols in seeds and berries of two subspecies of seabuckthorn (*Hippophaë rhamnoides* L. ssp. *sinensis* and ssp. *mongolica*). In: *International Workshop on Seabuckthorn; A Resource for Health and Environment for the 21st Century*, February 18-21, 2001, New Delhi, India, p.151-153.

19. Kallio, H.; Yang, B.; Peippo, P.; Tahvonen, R.; Pan, R.; Zhou, X.; Fu, Y.; Zhang, S. and Goncharov, P.L. 2001. Triacylglycerols and phospholipids in the seeds of sea buckthorn (*Hippophaë rhamnoides* ssp. *sinensis* and ssp. *mongolica*). In: *International Workshop on Seabuckthorn: A Resource for Health and Environment for the 21st Century*, February 18-21, 2001, New Delhi, India, p. 136-139.

20. Yang, B.; Kallio, H.; Koponen, J. and Tahvonen, R. 2001. Free and esterified sterols in seed oil and pulp/peel oil of sea buckthorn (*Hippophaë rhamnoides* L.). In: *Biologically-active Phytochemicals in Food*. Eds. R. Fenwick *et al.*, Eurofoodchem XI, Norwich Research Park, Norwich, UK, 26-28 September 2001. pp.24-27.

21. Kallio, H.; Yang, B.; Wang, B.; Wang, H.; Wang, J.; Song, J. Meng, H. and Zhao, H. 2001. Animal experiments on the anti-inflammatory and analgesic effects of sea buckthorn (*Hippophaë rhamnoides* L.) oils. In: *Biologically-active Phytochemicals in Food*. Eds. R. Fenwick *et al.*, Eurofoodchem XI, Norwich Research Park, Norwich, UK, 26-28 September 2001, pp. 69-73.

22. Kallio, H.; Yang, B.; Tahvonen, R.; Hakala, M. 1999. Composition of sea buckthorn berries of various origins. In: *International Workshop on Seabuckthorn* (*IWS-99*), August 29-September 2, 1999, Beijing, P.R. China. pp 17-23

23. Yang. B; Kallio, H.; Kalimo, K.; Mattila, L.; Kallio, S,; Tahvonen, R. and Katajisto, J. 1999. Effects of dietary supplementation of seabuckthorn (*Hippophaë rhamnoides*) oils on fatty acids in patients with atopic dermatitis. In: *International Workshop on Seabuckthorn* (*IWS-99*), August 29-September 2, 1999, Beijing, P.R. China. pp. 24-28.

24. Yang, B.; Kallio, H.P.; Kalimo, K.O.; Mattila, L.M.; Tahvonen, R.L.; Kallio, S.E. and Katajisto, J.K. 1999. Effects of dietary supplementation with sea buckthorn (*Hippophaë rhamnoides*) seed and pulp oils on fatty acid composition of plasma lipids in patients with atopic dermatitis and measurement of skin surface roughness. In: *Functional Foods–A New challenge for the food chemists*. Eds. R. Lásztity, W. Pfannhauser, L. Simon-Sarkadi, S. Tömösközi. Publishing Company of TUB, Budapest, Hungary. pp. 124-131.

25. Kallio, H.; Manninen, P.; Häivälä, E.; Sarimo, S. and Korteniemi, V.-M. 1997. Aseptic production of seabuckthorn oil capsules based on supercritical fluid extraction. In: *Worldwide Research and Development of Seabuckthorn*. Eds: Lu Shunguang and Li Min, China Science and Technology Press, Beijing, P.R. China, pp. 157-158.

26. Kallio, H.; Manninen, P.; Häivälä, E. and Sarimo, S.S. 1996. Production of sea buckthorn oil by supercritical fluid extraction. In: The 7th International Symposium on Supercritical Fluid Chromatography and Extraction, March 31 -April 4, 1996, Indianapolis, Indiana. p.15-16.

27. Kallio, H.; Malm, H.; Kahala, M. and Oksman, P. 1989. Analysis of sea buckthorn carotenoids by supercritical fluid chromatography/mass spectrometry. In: International Symposium on sea buckthorn (*H. rhamnoides* L.), Xian, China, October 19-23, pp. 123-127.

Books on Seabuckthorn

1. Singh, V., Kallio, H., Sawhney, R.C., Gupta, R.K., Lu, R., Eliseev, I.P., Khabarov, S.N., Korovina, M.A., Skuridin, G.M. and Shchapov, N.S. 2003. *Seabuckthorn (Hippophaë L.)-A Multipurpose Wonder Plant. Vol. I: Botany, Harvesting and Processing Technologies* (V. Singh, Ed. in Chief, 2003), Indus Publishing Company, New Delhi, 518p.

2. Singh, V., Kallio, H., Yang, B., Morsel, J.T., Sawhney, R.C., Bala, M., Dwevedi, S., Geetha, S., Tyagi, S.P., Li, T.S.C., Lu, R., Lu, S. and Zubarev, Y.A. 2008. *Seabuckthorn (Hippophaë L.)-A Multipurpose Wonder Plant. Vol. III: Advances in Research and Development* (V. Singh, Ed. in Chief, 2008), Daya Publishing House, New Delhi, 566p.

3. Singh, V., Yang, B., Kallio, H., Bala, M., Sawhney, R.C., Gupta, R.K., Morsel, J.T., Lu, R. and Tolkachev, O.N. 2006. *Seabuckthorn (Hippophaë L.)-A Multipurpose Wonder Plant. Vol. II: Biochemistry and Pharmacology* (V. Singh, Ed. in Chief, 2006), Daya Publishing House, New Delhi, 600p.

Dedicated to

Dr. Ramesh Chand Sawhney

by

Prof. Virendra Singh

Department of Biology and Environmental Sciences,
CSK Himachal Pradesh Agricultural University, Palampur – 176 062, H.P., India

DR. RAMESH CHAND SAWHNEY AND HIS PASSION FOR SEABUCKTHORN

Dr. R.C. Sawhney, a renowned scientist of India and former Director of Life Sciences, Defence Research and Development Organization (DRDO), New Delhi, presently is Registrar of Indian Institute of Technology, Mandi, Himachal Pradesh. Born in Himachal Pradesh in 1949, he joined Defence Institute of Physiology and Allied Sciences (DIPAS), a lab of DRDO at New Delhi in 1978, after his doctorate in Experimental Medicine from Postgraduate Institute of Medical Education and Research, Chandigarh with a brief stint at All India Institute of Medical Sciences, New Delhi. During the last 29 years of his scientific career with DRDO he has made significant contributions in optimizing performance of the soldiers under environmental vagaries in the areas related to a. Endocrine adaptation to environmental stress, b. Physiological glucocorticoids in prevention of acute mountain sickness, c. Application of Yoga in improvement of quality of life of HIV positive cases, and d. Coronary artery disease regression through life style intervention and Herbs for the health of armed forces

He has developed a modality for quicker induction of troops to high altitude using physiological glucocorticoids for emergency army missions. He also initiated and steered Programme Charak 'Herbs for the health of armed forces' of DRDO in which 27 products based on seabuckthorn are being transformed into nutraceuticals/herbal medicines to improve operational efficiency of the troops at high altitude and provide protection against radiation exposure, chemical and biological warfare agents. He also developed a lifestyle intervention modality for regression of coronary artery disease.

The seabuckthorn research in various laboratories of Defence Research and Development Oorganization (DRDO) of India started, when I met Dr. R.C. Sawhney in 2000, which became a turning point of Indian Seabuckthorn Programme. He organized my meeting with senior scientists of Life Science of DRDO at New Delhi. He was then Head of Human Physiology Division at Defence Institute of Physiology and Allied Sciences (DIPAS), a DRDO lab at New Delhi. The description of the plant that it is able to grow under very harsh climatic conditions in the high altitude environment at subzero temperature was very fascinating to Dr. Sawhney. At this time, Dr. Sawhney was working in the area of high altitude physiology and was looking for a nutraceutical to be used as an adaptogen for the armed forces personnel deployed at high altitude and being exposed to not only to low barometric pressure but also to extreme cold and higher radiation under low humidity. When evaluated in an animal model of cold-hypoxia-restrain stress, the seabuckthorn fruit extract was able to provide a fair degree of protection against these multiple stresses suggesting that the berries of this plant has biomolecules, which can be used as adaptogens.

This prompted Dr. Sawhney to visit high altitude areas in Western Himalayas to understand natural behavior of the golden bush. It was quite surprising to note that plant grows wildly in Western Himalayas upto the high altitude, which human habitation is existing, suggesting that there may be a fair degree of symbiosis between human habitation and seabuckthorn. These observations prompted Dr. Sawhney to join hands with our University to further explore health potential of this plant and co-organized the International Seabuckthorn Workshop in New Delhi in 2001, which culminated into establishment of International Seabuckthorn Association. A multi-Institutional programme to explore health potentials of this plant was started in which nine life sciences laboratories of DRDO worked together with an aim to develop products for human use. Both fruits and leaves of seabuckthorn (*Hipophone rhamnoides*) growing under hypoxic environment at a high altitude of more than 3500 metres were used using several *in vitro* cell lines and animal models.

His investigations on seabuckthorn (SBT) leaves are unique, reported for the first time and has added significant value to the potential of this wonder plant. Studies carried out by him and his team have demonstrated that seabuckthorn fruits and leaves have significant cytoprotective, immunomodulatory, antioxidant and anti-inflammatory activities. The seabuckthorn leaf extract alone stimulated IL-2 and γ-IFN production, but did not alter IL-4 secretion suggesting that SBT leaf extract specifically activates the cell mediated immune response. Based on these observations, a cheap and most potent Herbal Adjuvant for immunization against different diseases is being developed for the human application.

The leaf extract was also found to have significant wound healing activity, when tested in incision wound model and was found to stimulate protein, hydroxyproline and DNA synthesis besides activating angiogenesis. The leaf extract also has significant hepato-protective activity, when tested against CCl_4 induced acute hepatotoxicity in rats. When administered before exposure to hypobaric hypoxia both leaf extract and seed oil, were found to have significant adaptogenic activity and enhanced hypoxic tolerance by inhibiting hypoxia induced transvascular fluid leakage in the rat lungs and brain. The leaf extract was able to curtail hypoxia induced single strand DNA breaks in C6-glioma cells and restored mitochondrial integrity suggesting that SBT leaves have significant neuroprotective activity under hypoxic environment. The seed oil and leaf extract were found to have significant vaso-relaxant activity when tested in aortic ring model using polygraphic recording system. The seed oil was also found to have significant antiatherogenic and cardioprotective activity, when evaluated in hypercholesteremic rabbits. These studies have suggested that seabuckthorn leaf extract and seed oil can be developed as plant drugs or nutraceuticals against a variety of diseases including those at high altitude areas.

His present passion has been to augment seabuckthorn cultivation in the high altitude areas of Himalayas and help the industries in product development.

AWARDS

In recognition of his outstanding research contributions, he has been honoured with Defence Technology Group Award 2006; Distinguished Service Award, World Congress on Clinical and Preventive Cardiology, 2006; P. N. Shah Oration Award, 2003; Best Presentation Award International Conference on Seabuckthorn, 2003; Laboratory Scientist of the Year Award, 2003; Surg Rear Admiral M.S. Malhotra Research Prize, 1985; 1987; 1998; 2002; 2003 and 2006; Defence Research and Development Organization Silicon Research Medal, 2002; S.S. Misra Gold Medal, 1977; Indian Council of Medical Research; Gold Medal for first position in M.Sc. besides many scientific orations in national and international forums.

PAPERS ON SEABUCKTHORN IN JOURNALS

1. Geetha, S., Sai Ram, M., Singh, V., Illavazhagan, G. and Sawhney, R. C. 2002. Antioxidant and immunomodulatory properties of seabuckthorn-an *in vitro* study. *J. Ethnopharm.* 79: 373-378.

2. Geetha S., Sai Ram, M., Singh, V., Illavazhagan, G. and Sawhney, R. C. 2002. Effect of seabuckthorn on sodium nitroprusside-induced cytotoxicity in murine macrophages. *J. Biomed. Pharm.* 56: 463-467.

3. Geetha, S., Sai Ram, M., Mongia, S.S., Singh, V., Illavazhagan, G. and Sawhney, R. C. 2003. Evaluation of antioxidant activity of leaf extract of seabuckthorn (*Hippophae rhamnoides* L) on chromium (VI) induced oxidative stress in albino rats. *J. Ethnopharmacol.* 87:247-251.

4. Geetha, S., Singh, V., Sai Ram, M., Ilavazhagan,G., Banerjee, P. K. and Sawhney, R. C. 2005. Immunomodulatory effects of seabuckthorn (*Hippophae rhamonoides L.*) against chromium (VI) induced immunosuppression. *Mol. Cell Biochem.* 278: 101-109.

5. Lilly, G., Karan, D., Chanda, S., Srivastava, K. K., Sawhney, R. C., and Selvamurthy, W. 2003. Immunomodulatory effects of agents of plant origin: *Journal Biomed. Pharmacother.* 57: 296-300.

6. Narayanan, S., Ruma, D., Gitika, B., Sharma, S.K., Pauline, T., Sai Ram, M., llavazhagan, G., Sawhney, R.C., Kumar, D. and Banerjee, P.K. 2005. Antioxidant activities of Seabuckthorn (*Hippophae rhamnoides*) during hypoxia induced oxidative stress in Glial cells. *Mol. Cell. Biochem.* 278: 9-14.

7. Ganju, L., Padwad, Y., Singh, R., Karan, D., Chanda, S., Chopra, M. K., Bhatnagar, P., Kashyap, R. and Sawhney, R. C. 2005. Anti-inflammatory activity of seabuckthorn (*Hippophae rhamnoides*) leaves. *International Immunopharmocology* 5: 1675-1684.

8. Gupta, A., Kumar, R., Pall, K., Banerjee, P. K. and Sawhney, R. C. 2005. A preclinical study of the effects of seabuckthorn (*Hippophae rhamnoides* L.) leaf extract on cutaneous wound healing in albino rats. *Int. J. Lower Extremity Wound* 4: 88-92.

9. Padwad, Y., Ganju, L., Jain, M., Chanda, S., Karan, D., Banerjee, P. K and Sawhney, R. C. 2006. Effect of leaf extract of seabuckthorn on lipopolysaccharide induced inflammatory response in murine macrophages. *International Immunopharmocology* 6: 46-52.

10. Kanupriya, D. P., Sharma, S.K., Sai Ram, M., Ilavazhagan, G., Sawhney, R. C. and Banerjee, P. K. 2006. Flavonoids protect U-937 macrophages against tert—butyl hydroperoxide induced oxidative injury. *Food and Chemical Toxicology* 44:1024.

11. Ranjith, A., Kumar, K. S, Venugophalan, V. V., Arumughan, C., Sawhney, R. C. and Singh, V. 2006. Fatty acids, tocols, and carotenoids in pulp oil of three Seabuckthorn species (*Hippophe rhamnoides, H salicifolia,* and *H. tibetana*) grown in the Indian Himalayas. *Journal of the American Oil Chemists' Society* 83: 359-364.

12. Vijayaraghavan, R., Gautam, A., Kumar, O., Pant, S. C., Sharma, M., Singh, S., Kumar, S., Singh, A.K., Nivsakar, M., Kaushik, M.P., Sawhney, R. C., Chaurasia, O.P. and Prasad, G.B.K.S. 2006. Protective effect of ethanolic and water extracts of sea buckthorn (*Hippophae rhamnoides* L.) against the toxic effects of mustard gas. *Indian J. Expt. Biol.* 44: 821-831.

13. Gupta, A., Kumar, R., Pal, K., Singh, V., Banerjee, P. K. and Sawhney, R. C. 2006. Influence of seabuckthorn (*Hipphophae rhamnoides* L.) flavone on dermal wound healing in rats. *Mol. Cell. Biochem* 290: 193-198.

14. Saggu, S., Divekar, H. M., Gupta, V., Sawhney, R. C., Banerjee, P. K. and Kumar R. 2007. Adaptogenic and safety evaluation of Sea buckthorn (*Hipphophae rhamnoides*) leaf extract: A dose dependent study. *Food Chemical Toxicology* 45: 609-617.

15. Basu, M., Prasad, R., Jayamurthy, P., Pal, K., Arumughan, C and Sawhney, R. C. 2007. Anti-atherogenic effects of seabuckthorn seed oil. *Phytomedicine* 14: 770-777.

16. Geetha, S., Jayamurthy, P., Pal, K., Shweta, P., Ratan, K. and Sawhney, R. C. 2008. Hepatoprotective effects of seabuckthorn (*Hippophae rhamnoides L.*) against carbon tetrachloride induced liver injury in rats. *J. Sci. Food Agric.* 88: 1592-1597,.

17. Geetha, S., Sairam, M., Kumar, R., Iiavazhagan, G. and Sawhney, R. C. 2008. Cytoprotective and antioxidant activity of seabuckthorn (*Hippophe rhamnoides* L) against sodium nitroprusside induced oxidative damage in Iymphocytes. *Biomedicine* 28: 16-23.

18. Jayamurthy, P., Geetha, S., Dhananjay, S., Malhotra, S. A., Harinath, K., Ratan K., Sawhney, R. C. and Arumughan, C. 2008. Modulatory effects of seabuckthorn (*Hippophae rhamnoides L.*) in hypobaric hypoxia induced cerebral vascular injury. *Brain Res. Bull.* 77: 246-252.

19. Jain, M., Padwad, Y., Mishra, K. P., Chanda, S., Karan, D., Katiyal, A., Sawhney R. C. and Lilly, G. 2008. Effect of *Hippophae rhamnoides* leaf extract against Dengue virus infection in U937 cells. *Phytomedicine* 15: 793-799,.

20. Mishra, K. P., Chanda, S., Karan, D., Ganju, Lilly and Sawhney, R. C 2008. Effect of seabuckthorn flavone on immune system: An in-vitro approach. *Phytotherapy Research* 22: 1490-1495.

21. Gupta, A., Upadhyay, N. K., Sawhney, R. C. and Kumar R. 2008. A poly-herbal formulation accelerates normal and impaired diabetic wound healing. *Wound Repair and Regeneration* 16: 784-790.

22. Upadhya, N. K., Kumar, R., Mandotra, S. K., Meena, R.N., Siddique, M.S. and Sawhney, R. C. and Gupta, A. 2009. Safety and healing efficacy of seabuckthorn (*Hippophae rhamnodes L.*) seed oil on burn wounds in rats. *Food and Chemical Toxicology* 47: 2009 1146-1153.

23. Madhu, B., Prasad, J., Singh, S., Tiwari, S. and Sawhney, R. C. 2009. Whole-body radioprotective effects of seabuckthorn leaves. *Herbs, Spices and Medicinal Plants* 15: 204-216.

PAPERS IN PROCEEDINGS OF INTERNATIONAL CONFERENCES AND BOOKS ETC.

1. Geetha, S., Sai Ram, M., Singh, V., Illavazhagan, G. and Sawhney, R. C. 2001. Cytoprotective properties of seabuckthorn in murine macrophages. In: *Proceeding of International workshop on Seabuckthorn* (Ed Singh, V. and Khosla, P. K.), pp. 221-224.

2. Geetha, S., Sairam, M., Singh, V., Ilavazhagan, G. and Sawhney, R. C. 2004. Cytoprotective, immunomodulatory and anti-stress potential of seabuckthorn. In: *Proceeding of 1st Conference of the International Seabuckthorn Association* (Eds. Morsel, J.T. and Thiese, S.), pp. 108-124. Berllin, Germany.

3. Gupta, A., Pal, K., Singh, V. and Sawhney, R. C. 2001. Healing potential of Seabuckthorn (*Hippophae rhamnoides*) on cutaneous wounds in albino rats. In: *Proceeding of International workshop on Seabuckthorn* (Ed. Singh, V. and Khosla, P.K.), pp. 231-234, 2001.

4. Gupta, A., Kumar, R., Upadhyay, N. K., Sawhney, R. C. 2007. Healing potential of seabuckthorn (*Hippophae rhamnoides* L.) in normal and impaired wounds: A review. In: *Seabuckthorn-A Multipurpose Wonder Plant* (Ed.-in- Chief, V. Singh), Vol. III. Daya Publishing, N. Delhi, pp. 284-293.

5. Geetha, S., Basu, M., Jayamurthy, P., Malhotra, A. S., Pal, K., Prasad, R., Kumar, R., and Sawhney, R. C. 2007. Protective and therapeutic potentials of seabuckthorn (*Hippophae rhamnoides* L.). In: *Seabuckthorn-A Multipurpose Wonder Plant* (Ed. V. K. Singh), Daya Publishing, N. Delhi, pp. 245-253.

6. Jayamurthy, P., Geetha, S., Jayamurthy, H. S., Dhananjay, S., Harinath, R. K. and Sawhney, R. C. 2011. Modulation of hypoxia induced pulmonary vascular leakage in rats by seabuckthorn (*Hippophae rhamnoides* L.). In: *Evidence Based Complementary and Alternative Medicine*, Vol., Article ID 574524, 13p. doi: 1093/ecam/nep 199.

BOOKS ON SEABUCKTHORN

1. Singh, V., Kallio, H., Sawhney, R.C., Gupta, R.K., Lu, R., Eliseev, I.P., Khabarov, S.N., Korovina, M.A., Skuridin, G.M. and Shchapov, N.S. 2003. *Seabuckthorn (Hippophaë L.)-A Multipurpose Wonder Plant. Vol. I: Botany, Harvesting and Processing Technologies* (V. Singh, Ed. in Chief, 2003), Indus Publishing Company, New Delhi, 518p.

2. Singh, V., Kallio, H., Yang, B., Morsel, J.T., Sawhney, R.C., Bala, M., Dwevedi, S., Geetha, S., Tyagi, S.P., Li, T.S.C., Lu, R., Lu, S. and Zubarev, Y.A. 2008. *Seabuckthorn (Hippophaë L.)-A Multipurpose Wonder Plant. Vol. III: Advances in Research and Development* (V. Singh, Ed. in Chief, 2008), Daya Publishing House, New Delhi, 566p.

3. Singh, V., Yang, B., Kallio, H., Bala, M., Sawhney, R.C., Gupta, R.K., Morsel, J.T., Lu, R. and Tolkachev, O.N. 2006. *Seabuckthorn (Hippophaë L.)-A Multipurpose Wonder Plant. Vol. II: Biochemistry and Pharmacology* (V. Singh, Ed. in Chief, 2006), Daya Publishing House, New Delhi, 600p.

Dedicated to

Prof. Lu Rongsen

by

Prof. Pei Shengji

President, Chinese Association of Ethnobotany,
Kunmin Institute of Botany, Chinese Academy of Sciences, Helingtan, Kunmin, China

LIFE AND THE CONTRIBUTION OF PROF. LU RONGSEN-A PROMOTER OF GLOBAL SEABUCKTHORN DEVELOPMENT

In autumn 1984, as a junior researcher, Prof. Lu Rongsen took part in a field expedition in the West Sichuan, which is considered as one of the richest regions in biodiversity in China. He collected a specimen of a berry that proved as a famous wild plant-seabuckthorn in following years. The berry was analyzed in the laboratory of Chengdu Institute of Biology (CIB), Chinese Academy of Sciences (CAS). Lu Rongsen was surprised with the analysis result,which shown the high vitamin C content to 1200 mg/100g, which was higher than any fruit and vegetable. Since then, Prof. Lu Rongsen started his profession on seabuckthorn research for life time. There are many aspects of Prof. Lu Rongsen's contribution in the field of seabuckthorn development in China and the world.

ACHIEVEMENT IN SCIENTIFIC RESEARCH ON SEAKUCKTHORN

In 1985-1995, encouraged by Russian achievement on seabuckthorn development, Chinese Government paid more attention on seabuckthorn development in whole China, Prof. Lu Rongsen involved in all activities from the beginning. He made a lot of field surveys and collected plenty of samples from all the species and subspecies of *Hippohae* in China. Based on analysis of all these samples, he published a series of papers and gave many suggestions to Chinese government for guiding seabuckthorn development. From 1997 to 2007, Prof. Lu Rongsen extended his study from China to Europe and other Asian Countries, because *Hippophae* is a genus that contains many species and subspecies and widely distributed in Eurasia continent. The main chemical composition of *Hippophae* have been systematically analyzed based on the reports of surveys on the distribution and survival status of the species and subspecies of *Hippophae* in Eurasian continent. Therefore, Prof. Lu Rongsen might be the first person who made systematical studies on distribution, chemical characterization, cultivation and integrated utilization of *Hippophae.* His studies on *Hippophse* have been reported in several international and national journals and conferences and his research contribution was highly appreciated. The results of these studies have been used in seabuckthorn breeding and commercial cultivation. Prof. Lu Rongsen engaged in seabuckthorn research for the last 28 years (1984-2012) and more than 30 papers and 2 books on seabuckthorn were published. The main research field is development of *Hippophae* genetic resources and sustainable uses.

CONTRIBUTION TO CHINA'S SOIL AND WATER CONSERVATION

Soil and water erosion is the biggest environmental problem in China. China's government was making a lot of efforts to control the soil-water erosion for more than decades, but the efforts were partially successful. In 1998, China suffered a very big flood over large portion of China that made the soil erosion situation even worse. Knowing the ecological value and economic benefit of seabuckthorn, Professor Lu Rongsen submitted a report to Prime Minister Mr. Zhu Rongji and Vice-Prime Minister Mr. Wen Jiabao. The main points of the report were to request central government pay more attention on development of seabuckthorn resources and use its strong bio-ecological function to control the soil erosion and water loss. Mr. Zhu and Mr. Wen quickly gave written instructions to State Development and Reforming Committee, The Ministry of Water Resource and State Forestry Bureau, which became aware of the potential of seabuckthorn and decided to give more support on it. After that, a "National Programmme on Seabuckthorn" with yearly budget of 7 million US$ was approved and implemented by National Seabuckthorn Administration Center, Beijing. Now, the artificial seabuckthorn plantations in northern China have been planted in over 2 million hectares, which have played great role in controlling soil erosion and water loss, particularly in NW China and descent areas of China.

PROMOTING GLOBAL SEABUCKTHORN DEVELOPMENT

Prof. Lu Rongsen was the first person who introduced the achievement of seabuckthorn development in China to the world through a book: *"Sea buckthorn: A Multipurpose Plant Species for Fragile Mountains"* (published by ICIMOD in 1992) that played a crucial guiding role and set a trend in global seabuckthorn development. Prof. Lu Rongsen has actively promoted international cooperation on seabuckthorn development. Since 1990s, he traveled to many countries (India, Pakistan, Nepal, Bolivia, Russia and other countries) to disseminate the knowledge and research progress of seabuckthorn and help many countries to establish their seabuckthorn projects. Consequently, the development, utilization and exploitation of seabuckthorn have been spread from Russia and China to many other counties all over the world. With Prof. Lu Rongsen's help and promotion, International

Center for Research and Training of Seabuckthorn (ICRTS) was set up at Beijing in 1995. In 2001, International Seabuckthorn Association (ISA) was established that receives more than 30 member countries and hundreds of scientists, technicians, and enterprisers. Seabuckthorn development is really becoming a research hot point and seabuckthorn industry is well established in the world. Based on his outstanding work in sea buckthorn research and development, Prof. Lu Rongsen has been conferred "The Lifetime Award" by International Seabuckthorn Association at the 5th Conference of ISA on 4th September, 2011 at Xining, China.

At the age of 70 years, Prof. Lu Rongsen still continues to contribute on seabuckthorn development for the benefit of humanity and environmental conservation and world scientific community, farmers and policy makers recognizes his great contribution on seabuckthorn.

RESEARCH PAPERS

1. Lu, R. 1985. A study on exploitation potentialities of subtropical mountain land of Sichuan, China. In: *Proceedings of the International Symposium on Ecology of Tropical And Subtropical Mountain Areas*, October, 1985, China Academic Publishers, Chengdu, China.
2. Lu, R. 1988. A study of seabuckthorn resources in Sichuan, China. *Hippophae* 1 (1). (In Chinese)
3. Lu, R. 1988. The potentiality of growing deciduous fruit trees in arid valley of the Hengduan Mountains, China. *Mountain Research* 3 (4). (In Chinese)
4. Lu, R. 1990. A study on seabuckthorn (*Hippophae* L.) resources in China. *Acta Horticulture Sinica* 17 (3).
5. Lu, R. 1990. Application of chromatographic analysis of flavonoids races to taxonomy of seabuckthorn (*Hippophae* L.). In: *Abstracts of Books of the XXIII International Horticultural Congress*, August 27-September 1, 1990, Firenze, Italy.
6. Lu, R. 1990. Seabuckthorn resources and it's underexploited potential in the Himalayan region. In: *International Centre for Integrated Mountain Development*, MFS Series No. 12, Kathmandu, Nepal.
7. Lu, R. 1991. The nutrients in the leaves of *Hippophae* plants and their potential utilization. *Hippophae* 4 (4): 43-45.
8. Lu, R. 1992. The bioresources in the dry valleys. In: *The Dry Valleys of The Hengduan Mountains Region*, p.83-97, Science Press. (In Chinese)
9. Lu, R. 1993. Seabuckthorn resources and its underexploited potential in the Himalayan region. In: *Horticultural Development in the Hindu Kush-Himalayan Region*, Oxford and IBH Publishing Co. Pvt. Ltd.
10. Lu, R. 1993. The chemical composition of *Hippophae* fruits in China. In: *International Symposium on Seabuckthorn (Hippophae rhamnoides L.), Synthesis of Report*, August 23-25, 1993, Barnaul, Russia
11. Lu, R. 1993. Utilization of biological resources (Seabuckthorn). In: *A Serious Book of Comprehensive Survey in the South-west China*, p.115-120, Science Press. (In Chinese)
12. Lu, R. 1994. A technical report of plantation establishment in Nepal. In: *Proceedings of an International Workshop on Rehabilitation of Degraded Lands in Mountain Ecosystems*, December, 19-22, 1994, Baoshan, Yunnan, China.
13. Lu, R. 1994. The application of plastic film technology in China. ICIMOD Publications, Kathmandu, Nepal.

14. Lu, R. 1995. Eco-geographical distribution of seabuckthorn and prospects of International cooperation. In: *Worldwide Research and Development of Seabuckthorn*. Proceedings of International Workshop on Seabuckthorn, Beijing, China, p.11-22, Science and Technology Press.

15. Lu, R. 1998. *Enterprise in mountain-specific products in western Sichuan, China*, ICIMOD Publications, Kathmandu, Nepal.

16. Lu, R. 1998. Nutritive components of the leaves of *Hippophae* in China and their possible utilization. *Seabuckthorn Research*, ICRTS 1 (1): 24-27.

17. Lu, R. 1999. A new seabuckthorn resources-*Hippophae goniocarpa*. In: *Proceedings of International Workshop on Seabuckthorn*, China Science and Technology Press, Beijing, China.

18. Lu, R. 1999. Combating desertification with seabuckthorn. In: *Combating Desertification with Plants*, p. 291-298, Kluwer Academic/Plenum Publishers, New York.

19. Lu, R. 2001. Introduction of seabuckthorn (*H.salicifolia*) from high altitude to low altitude area. In: *Proceedings of International Workshop on Seabuckthorn*, p. 89-91, February 18-21, 2001, New Delhi, India.

20. Lu, R. 2003. A study on hybrids crossbred Russian varieties and Chinese varieties of seabuckthorn. In: *Proceedings of 1st Congress of the International Seabuckthorn Association*, September 14-18, 2003, Berlin, Germany.

21. Lu, R. 2003. *Hippophae* and its general chemical compositions. In: *Proceedings of 1st Congress of the International Seabuckthorn Association*, p. 20-35, September 14-18, 2003, Berlin, Germany.

22. Lu, R. 2003. Introduction and cultivation of *Hippophae salicifolia*. *The Global Seabuckthorn Research and Development* 1 (1): 21-23. (In Chinese)

23. Lu, R. 2003. *The Role of Seabuckthorn in Improving Ecological Environment in West China*, p. 58-60, Science and Technology Review, Beijing, China. (In Chinese)

24. Lu, R. 2004. Preliminary report on systematic chemical analysis of *Hippophae* L. *The Global Seabuckthorn Research and Development* 1 (1): 21-23. (In Chinese)

25. Lu, R. 2004. Seabuckthorn: A multipurpose plant for mountain people. In: *Workshop on Strategies for Improving Livelihoods through Income generating Activities in Mountain Regions*, FAO, p. 44-60, February 26-27, 2004, Shambery, France.

26. Lu, R. 2006. Seabuckthorn research. In: *The Nutritional Value of Seabuckthorn*, Science Press, Beijing, China. (In Chinese)

27. Lu, R. 2007. The correlation between seabuckthorn berry quality and altitudes of its growing location. In: *Proceedings of the 3rd International Seabuckthorn Association*, p. 5-8, August 12-16, 2007, Quebec City, Canada.

28. Lu, R. 2009. A comparison study on seabuckthorn oil and other edible oils in their main components. In: *Proceedings of the 4th International Seabuckthorn Association Conference*, p. 115-116, September 1-6, 2009, Belokuriha (Altai Region), Russia.

29. Lu, R. 2011. The flavonoid contents in the berries and leaves of *Hippophae*. In: *Abstract Proceedings of the 5th International Seabuckthorn Association Conference*, p. 59, September 3-8, 2011, Xining, Qinghai, China.

BOOKS ON SEABUCKTHORN

1. Lu, R. 1992. *Seabuckthorn-A Multipurpose Plant for Fragile Mountains*. Occasional Paper No. 20, ICIMOD, Kathmandu, Nepal, 62p.

2. Singh, V., Kallio, H., Yang, B., Morsel, J.T., Sawhney, R.C., Bala, M., Dwevedi, S., Geetha, S., Tyagi, S.P., Li, T.S.C., Lu Rongsen, Lu, S. and Zubarev, Y.A. 2008. *Seabuckthorn (Hippophae L.)-A Multipurpose Wonder Plant. Vol. III: Advances in Research and Development* (V. Singh, Ed. in Chief, 2008), Daya Publishing House, New Delhi, 566p.

3. Singh, V., Yang, B., Kallio, H., Bala, M., Sawhney, R.C., Gupta, R.K., Morsel, J.T., Lu Rongsen, and Tolkachev, O.N. 2006. *Seabuckthorn (Hippophae L.)-A Multipurpose Wonder Plant. Vol. II: Biochemistry and Pharmacology* (V. Singh, Ed. in Chief, 2006), Daya Publishing House, New Delhi, 600p.

Dedicated to

Prof. Elizaveta I. Panteleeva

by
Prof. Yury A. Zubarev

Lisavenko Research Institute of Horticulture for Siberia
49 Zmeinogorskiy Tract, Barnaul, Russia

PROF. ELIZAVETA I. PANTELEEVA- A GREAT RUSSIAN SEABUCKTHORN BREEDER

The legendary scientist Prof. Elizaveta I. Panteleeva was born in village Konstantinovka in Altai kray in 1939. In her childhood she worked at farm helping her parents and then she started dreaming to become a horticulturist. In early 1930s, a new branch of Agriculture was started in Altai and first gardens of various crops were established by worldwide known academician Mikhail Lisavenko. First time, seeds of seabuckthorn were collected from the wild bushes and sowed at selection plots. As we will know later they were sowed for Prof. Elizaveta Panteleeva.

Elizaveta's dream to become a horticulturist came true in 1958, when after finishing her secondary school, she entered the School of Horticulturist, which belongs to Altai Experimental Station (later known as Lisavenko Research Institute of Horticulture for Siberia, Barnaul). There, she met Prof. Jivko Gatin, who was then Head of the Division of Seabuckthorn Selection. A bit later in 1959, she was invited to his laboratory as young and prospective researcher on seabuckthorn. From that period of time, the real story of future seabuckthorn programme of Russia had been started.

In 1964, Prof. Elizaveta Panteleeva became the Head of Seabuckthorn Division, and for the next 40 years, she was the world leader in selection and breeding of seabuckthorn. She started to work on seabuckthorn, when seabuckthorn collection were represented by only 21 specimens of freely pollinated hybrids from the bank of Katun river in Altai Mountain. Because of her great enthusiasm for seabuckthorn, dedication and hard work by her and her team, the seabuckthorn collection rapidly grew up and first three varieties of seabuckthorn, *i.e.* "Novost Altaia" (Novelty of Altai), Dar Katuni (Gift of the Katun river) and Zolotoy Pochatok (Golden branch) were released by Lisavenko Institute in 1963. These three early varieties of seabuckthorn became a basis for many other varieties all over the world.

For the period of more than 50 years Prof. Panteleeva developed 46 varieties of seabuckthorn, which are distinguished for different valuable features like complete thornlessness, palatability (one of the best in the world), large berries size (upto 1.5 gram/berry), high productivity (up to 35 tons per hectare in irrigated areas), different periods of ripening (from July to October.) etc. Presently, the total collection of seabuckthorn at Lisavenko Institute comprises more than 57000 hybrid specimens.

She is the author of program and methods for seabuckthorn selection and breeding, which is officially used in Russia.

For distinguished career and research contribution on seabuckthorn breeding, Prof. Elizaveta Panteleeva was honoured as a Laureate of USSR State Prize (1981), the greatest award in former Soviet Union. She was also awarded by three government's medals, including Order "*For Great Services for Homeland*", as well as numerous Local and International awards.

Prof. Elizaveta Panteleeva stills works at the Lisavenko Institute because of her zeal, dedication and of course for her emotional attachment to seabuckthorn. She gives lessons to students. She is working with next generation of seabuckthorn breeders and continues to select the best seabuckthorn varieties in the world.

PUBLICATIONS

1. Lobatcheva, I.I., Letchamo, W., Huszar, L., Muchortov, S.A., Malkova, N.N. and Panteleeva, E.I. 2002. Evaluation of Siberian seabuckthorn cultivars in Washington State. In: *Trends in New Crops and New Uses* (J. Janick and A. Whipkey, Eds.), p. 402–404. ASHS Press, Alexandria, VA, 133p.
2. Panteleeva, E.I. 1998. Seabuckthorn breeding on lengthening of pedicle and diminishing of a breakaway (taking off effort). In: *Material of International Symposium on Seabuckthorn* (*Hippophae rhamnoides*), p. 144-145, Russia, 248p.
3. Imamaliev, G.N. and Panteleeva, E.I. 1980. Characteristics of some forms of seabuckthorn (*Hippophae rhamnoides*) in the foothills of the great Caucasus Azerbaijan-SSR USSR. *Rastitel"nye Resursy* 16 (4): 547-549. (In Russian)
4. Kaninina, I.P. and Panteleeva, E.I. 1987. Breeding of seabuckthorn in the Altai. Advances in Agricultural Scinces, p. 76-87. (In Russian)

5. Khabarov, S.N., Mikhailova, N.V. and Panteleeva, E.I. 1993. Seabuckthorn plantation as improving system of influence on soil. In: *Proceedings of the 2nd International Symposium on Seabuckthorn*, p. 308-310.

6. Khabarov, S.N., Mikhailova, N.V., Panteleeva, E.I. and Bartenev, V.D. 1993. Seabuckthorn plantation regeneration after machine harvesting. In: *Proceedings of the 2nd International Symposium on Seabuckthorn*, p. 383-386.

डा. एस. अय्यप्पन
सचिव एवं महानिदेशक
Dr. S. AYYAPPAN
SECRETARY & DIRECTOR GENERAL

भारत सरकार
कृषि अनुसंधान और शिक्षा विभाग एवं
भारतीय कृषि अनुसंधान परिषद
कृषि मंत्रालय, कृषि भवन, नई दिल्ली 110 001

GOVERNMENT OF INDIA
DEPARTMENT OF AGRICULTURAL RESEARCH & EDUCATION
AND
INDIAN COUNCIL OF AGRICULTURAL RESEARCH
MINISTRY OF AGRICULTURE, KRISHI BHAVAN, NEW DELHI 110 001
Tel.: 23382629; 23386711 Fax: 91-11-23384773
E-mail: dg.icar@nic.in

Foreword

Seabuckthorn (*Hippophae rhamnoides* L.), a nitrogen fixing multipurpose plant (family *Elaeagnaceae*) of cold regions of Himalayas, has been recognized as a useful source of anti-oxidants (vitamin C, E, carotenoids, flavonoids etc.), therefore, being harnessed in food, cosmetic and health industries in China, Russia, Germany, Finland, India and other countries. It is also a useful fuel wood and fodder in the cold deserts of Himalayas. China has exploited its potential in ecological rehabilitation of degraded mountainous lands and overall improvement of agro-ecosystem and economic upliftment of rural economy of Loess Plateau and other provinces. While developmental agencies are engaged in afforestation of seabuckthorn for environmental conservation of degraded mountainous lands of Himalayas, CSK Himachal Pradesh Agricultural University, Palampur, DRDO, other R&D institution of India have developed appropriate technologies on seabuckthorn, which can be harnessed for the commercial utilization of this wonder plant for the economic upliftment of rural economy in Himalayan states of farmers and health benefits to domestic animals and human beings.

I appreciate the efforts of Prof. Virendra Singh of CSK Himachal Pradesh Agricultural University, Palampur, who in collaboration with global seabuckthron experts prepared this very valuable book, which will make it easier for those, who are interested to explore seabuckthorn further. The book entitled "*Seabuckthorn (Hippophae L.)-A Multipurpose Wonder Plant Vol. IV: Emerging Trends in Research and Technologies*" has covered various aspects of seabuckthorn plant, which make it a useful source on the subject to scientists, students, entrepreneurs and policy makers etc. I congratulate Prof. Singh and his co-editors in producing a useful and quality book on seabuckthorn.

Dr. S. Ayyappan

Secretary, Department of Agricultural Research and Education
and
Director General, Indian Council of Agricultural Research
Dr. Rajendra Prasad Road, Krishi Bhawan, New Delhi – 110 001, India

Preface

Global Seabuckthorn Programme, which was initiated by Russians during 1930s, truly led by Chinese since 1980s, has been joined by many others like Germany, Finland, India, Canada, Sweden and many other countries during the last one decade. Therefore, seabuckthorn programme has come of age, a global success story. The potential of seabuckthorn in environmental conservation of mountains and health protection has been realized globally by the vision, guidance and hard work of seabuckthorn legends like Prof. H. Kallio (Finland), Prof. R.C.Sawhney (India), Prof. Lu Rongsen (China) and Prof. E. I. Panteleeva (Russia) and many others. Therefore, the present book on seabuckthorn is dedicated to these great inspiring seabuckthorn scientists, who have shaped the global seabuckthorn programme.

Seabuckthorn, one of the greatest gifts of Mother Nature to humanity, was discovered as early as 8th century during Tang dynasty, as mentioned in Tibetan medical classic "*rGyud bzhi*". Russians discovered its nutritional and medicinal values and used it in its space programme. Development of many high yielding thornless varieties of ssp. *mongolica* by Russian workers made it easier to introduce in gardens of Russia, but, several countries have introduced Russian varieties due to their large berries, few thorns, high oil and rich nutrients. Seabuckthorn subspecies like ssp. *sinensis* (China) and ssp. *turkestanica* (Himalayan countries) have great potential in ecological rehabilitation of fragile mountains, as China has shown.

The discovery of a number of anti-oxidants, like vitamin C, E, carotenoids and polyphenols and sterols led to development of a number of seabuckthorn based health food and cosmetics and some medicines, globally. The first seabuckthorn drug "Hipporhamin", an anti-viral product of leaf tannin extract, was developed by the team led by Prof. O.N. Tolkachev in Russia. Preclinical research on seabuckthorn oil and leaf extract has been done on several aspects like anti-oxidant, anti-bacterial, healing of wounds and gastric ulcer, cardiovascular and skin diseases etc. globally. A number of clinical trials on seabuckthorn have also been carried out in several countries like China, Finland, India, Germany and Russia. Seabuckthorn propagation and cultivation and processing technologies have also been improved. Consequently, seabuckthorn is being cultivated globally by the farmers and a number of health food, skin care and food supplements are in the global market. However, harnessing seabuckthorn as cheap, high quality and effective health product still remains a big challenge before us. Seabuckthorn is an excellent soil binder and it has been planted in over 2.5 million ha marginal mountainous lands, which help in ecological rehabilitation of fragile mountain ecosystems in difficult mountain zones like Loess Plateau, China. Germany has increased seabuckthorn cultivation area from 300 ha in 2000 to 1500 ha in 2012. Inspired by the success stories of China, Russia and Germany, similarly, other countries like India and Mongolia have launched national programmes on seabuckthorn.

Some of us joined hands and decided to compile the global literature on seabuckthorn for use of researchers and students, as there was little information on seabuckthorn in English, most of the information being in Russian and Chinese languages, when we started the work on the book in 1999. Since then, 4 volumes of the book entitled *"Seabuckthorn: A Multipurpose Wonder Plant"* containing about 200 chapters on various subjects of seabuckthorn have been published. The books covered almost all aspects of seabuckthorn, *i.e.*, propagation, cultivation, breeding, diseases and pests, post harvest management, chemistry and biochemistry, pharmacology and environmental conservation etc. These 4 volumes of the book proved very useful, as almost all the laboratories globally engaged on research on seabuckthorn have them and it is being used as reference book and many experts are taking help of the research findings and new technologies on seabuckthorn given in the book. We can say almost whole knowledge on seabuckthorn in the form of 4 volumes of the book is on your table, as a seabuckthorn library.

This 4th volume of the book covers the new research findings and emerging advanced technologies on various aspects of seabuckthorn, which will prove useful for seabuckthorn resource development and processing of various seabuckthorn health products.

We recognize and appreciate the cooperation, assistance and guidance of all seabuckthorn experts working globally, who contributed their chapters and made it possible to produce this quality book on seabuckthorn for use of global scientists, students, policy makers, entrepreneurs etc.

Prof. Virendra Singh
Editor- in-Chief
and
Editors

Contents

PART II: GENETICS AND BREEDING

PART III: MOLECULAR BIOLOGY AND BIOTECHNOLOGY

PART IV: PHYSIOLOGY

PART V: CHEMISTRY AND BIOCHEMISTRY

PART VI: PHARMACOLOGY

PART I:
PROPAGATION AND CULTIVATION

Chapter 1

Propagation Technologies of Seabuckthorn (*Hippophae rhamnoides* L.) from Softwood Cuttings

Virendra Singh, Sonika Choudhary, R.K. Rana* and L.K. Sharma**

Department of Biology and Environmental Sciences,
CSK Himachal Pradesh Agricultural University, Palampur – 176 062, H.P., India
*Krishi Vigyan Kendra (CSK HPAU), Bajaura (Kullu), H.P., India
**Krishi Vigyan Kendra (CSK HPAU), Kukumseri, Lahaul and Spiti, H.P., India

SUMMARY

In view of increasing demand of seabuckthorn in commercial cultivation and afforestation for ecological rehabilitation of degraded mountainous lands, efforts have been made by workers to standardize the fast method of propagation of seabuckthorn from softwood cuttings, which need to be reviewed for their adaptation in cultivation programme. Global studies have found that for propagation of seabuckthorn from softwood cuttings from one-year old shoots with length generally in the range of 15-40 cm, collected during late June to August were beneficial. Cuttings with few leaves on top were found most useful in rooting. Growth hormones like IBA, NAA, IAA, etc. have been used by various workers for accelerating root formation in seabuckthorn. The most usable hormones were IBA and NAA for root development. The basal part of cuttings are placed first in water and then in solution of growth hormone at lower concentration at room temperature for 14-16 hours or at high concentration for few seconds, then planted in soil media of mist system with 90 per cent humidity. The soil medium generally comprises of sand on the top of soil layer in the mist bed, which provide conditions for root growth. Saplings may be transplanted same season in nursery or field or next year.

Keywords: *Seabuckthorn, Propagation from softwood cuttings, Hormone treatments, Soil media and Humidity.*

Seabuckthorn (Hippophae L.): A Multipurpose Wonder Plant, Vol. 4: Emerging Trends in Research and Technologies (Virendra Singh, *Editor-in-Chief*, 2014), p. 3–21, Daya Publishing House, New Delhi, India.

INTRODUCTION

Seabuckthorn has emerged as a useful plant for health food, cosmetic and drug industries globally, due to its rich medicinal properties and its role in environmental conservation of degraded mountainous lands. On one hand, seabuckthorn is being planted for raising orchards for commercial purpose in countries like China, Russia, India, Germany and on other hand, it is being used for afforestation programmes for greening of hilly regions (Rongsen, 1992). In order to meet the requirement of plantation programmes, propagation methods of seabuckthorn needs to be reviewed, as many scientists in several countries have done a lot of work on this aspect of seabuckthorn. Seabuckthorn can be propagated from the seeds, hard or softwood cuttings, grafted seedlings and root turion seedlings (Singh and Gupta, 2003). However, there is no control on number of male and female plants borne from the seeds. This method may be useful for introduction of exotics and meeting nursery targets of afforestation of marginal lands. Therefore, in order to meet the demand of development of adequate commercial seabuckthorn resources with a regulated ratio of male and female plants, propagation from cuttings should be preferred. Plants raised from the stem or branch cuttings maintain the genetic properties of mother plants and female plants can bear fruit earlier than the female plants raised from the seeds. Therefore, propagation from cutting is an important technology for raising artificial plantations, and introducing and acclimating new and improved varieties of seabuckthorn.

Propagation from hardwood cuttings is also quite useful, as not much infrastructure is required. However, the hardwood cuttings are prone to environmental changes in nursery field conditions. The most common propagation method is raising plants from softwood, as saplings are raised under controlled conditions and they are not prone to any environmental changes during growth season which ensures the high success rate, as standardized by several workers (Avdeev, 1976; Wahlberg and Jeppsson, 1992; Wu *et al.*, 2010). Propagation from softwood cuttings are widely practiced in Europe, Russia and China. Massive nurseries can be raised from the green cuttings under controlled temperature and humidity conditions in mist chambers. The mist system ensures vigourous root formation and growth in the cuttings. The method needs the artificial mist sprays and a plastic film house with other equipments.

SOFTWOOD PROPAGATION TECHNOLOGIES

There are a number of aspects of seabuckthorn which we need to understand, while propagating seabuckthorn from softwood cuttings. Among them, time of collection of softwood cuttings, size of cuttings, hormone treatment, treatment with minerals, composition of soil medium, structure of mist system, overwintering and shifting of saplings to nursery or gardens, etc.

Time of Collection of Cuttings

The maturity level of the seabuckthorn shoots and physiological condition of the softwood cuttings are important for the development of roots. Timing of collection of cuttings affects the root formation. Softwood cuttings may be collected from selected plants in late spring when leaves are no longer growing but prior to lignification of shoots, *i.e.* the end of June to the beginning of July (Li, 2008). Avdeev (1976) reported a higher rooting rate of 96-100 per cent in the softwood cuttings collected during the beginning of growth of plants in early spring (June) in Russia. The cuttings collected during intensive growth of mother plants gave low rooting percentage, but again at the termination of active growth gave higher rate of 86-100 per cent rooting. Osipov and Morozova (1983) estimated 65-

70 per cent rooting in the cuttings collected during early June, which is the time of vigorous shoot growth. Potapov (1978) observed that semi-hardwood cuttings, collected in late June or early July from Altai region of Siberia, Russia, were beneficial. The day timing for cuttings collection did not affect the rooting. Buonous *et al.* (1992) observed the better rooting in cuttings, when they were collected in late June. Kniga (1989) reported the optimum time of cutting collection was late May in Kiev region of Ukrain. Kondrashov and Kuimov (1987) observed that cuttings taken in late June from severely pruned branches (pruning was conducted in early spring before bud break) successfully rooted (95-98 per cent) under a mist in the green house. In Balsgaird, Sweden, cuttings are collected during the end of June or beginning of July (Wahlberg and Jeppsson, 1992). According to Shuhua *et al.* (1995), the green (semi-wood) cuttings should be taken from the healthy plants (without pest) during middle of July to the beginning of August. Therefore, the best time for collection of softwood cuttings is from end of June to August. Cuttings should be collected during the early morning, when the leaves are covered with dew or during late afternoon, preferably on cool day (Li, 2008) (Figure 1.1). Wu *et al.* (2010) showed that the most significant method was to collect the semi-lignified cuttings from the middle part of branches between the end of June and the end of July.

Figure 1.1: Visiting Seabuckthorn Stand, Siberia.

Storage of Shoots/Cuttings

Shoots/cuttings are immediately put into the polythene bags (Figure 1.3), or poly or moist burlap sacks to protect them from wind drying and withering. These cuttings can remain fresh for 3 days in this condition or 7 days in water without losing their rooting capacity (Rongsen, 1992). In Russia, cuttings are placed in poly or moist burlap sacks immediately after cutting in a shaded and cool place. The cuttings are collected in the polybags and then taken to cool and shady place for further preparation (Rana, 2011). Water is sprayed over these cuttings to keep them moist.

Figure 1.2: Top and Side Shoots of Seabuckthorn Plant.

Characteristics of Shoots

The cuttings from small or thin young shoots lack the ability to form the roots when planted in soil. Cuttings from highly lignified branches also have a poor capacity to form the roots. Better results of rooting are obtained when the cuttings with some lignifications are collected from the terminal or side shoots (Figure 1.2). Studies have shown

Figure 1.3: Softwood Cuttings Stored in a Polybags in Shaded Place.

that the shoot with medium growing power in the same layer of the crown, of the same age (1 year), exposed to sunshine, are most suitable for raising the nurseries of seabuckthorn. Best results occur when cuttings are collected from terminal or side shoots. Wahlberg and Jeppsson (1992) observed that the cuttings collected from the top part of a branch gave better results than the cuttings collected from the middle and basal parts of branches. The rooting rate of cuttings with top bud is much higher (90 per cent) than without top bud. It is presumed that the top bud produces hormones to stimulate rooting (Shuhua *et al.*, 1995). Rooting rate of cuttings from 1-year old branches is higher (100 per cent) than 2-year old (90 per cent) and 3-year old branches (73 per cent). Cuttings collected from the upper part of the same plant had higher rooting (80.0 per cent) than the lower part (63.3 per cent). Green or soft cuttings of seabuckthorn have stronger ability to produce several adventitious roots within a short period of plantation. Yang *et al.* (2011) observed the rooting effects of stem cuttings from 1-year old shoots were better than from 2-year old shoots. Stanys *et al.* (2010) observed that cuttings with lowest lignification level, taken from terminal shoot part, regenerated roots more intensively. Cuttings taken from juvenile plants lag behind the cuttings from bearings plants according to the rhizogenesis speed. The length of cuttings is crucial for the rhizogenesis of seabuckthorn cuttings. As per guideline of the Lisavenko Institute of Horticulture for Siberia (LIHS), Barnaul, Russia (Rana, 2011), the cuttings are collected from the mother plants maintained in the mother orchards. The mother orchard of the different varieties to be propagated should be planted separately at closer distance. In the mother orchard, the male plants should not be planted to avoid fruiting in the female plants and to get more vegetative growth every year. Every year, the mother plants are pruned hard to get maximum number of current year shoot and its growth. The cuttings are collected in the month of June to July when the new growth starts maturing. Salikhov (1986) found IBA favourably affected rooting in softwood cuttings taken from the base of the shoots but had no marked effect on the apical cuttings.

Preparation of Cuttings

Usually cutting should have more than 2 nodes and foliage is removed from the lower portion of the cutting. Cuttings are collected by sharp knife or by the blade of a rajar on 3-4 mm below bud. The lower leaves are removed, leaving 2-4 leaves at the tip of cutting (Figure 1.4). Before the plantation in soil medium, the cuttings are again cut and sufficient leaves are removed so that the lower leaves are not in contact with the soil medium after plantation. Cuttings should be planted in a way that atleast 2 leaf nodes are below the surface of the soil media.

The cutting length should be 10 to 15 cm (Lebeda, 2008; Li, 2008). Papp (1982) found that the cuttings of 10-12 cm length were most suitable for propagation, whereas Pletneva (1983) recommended 15-20 cm length of the cuttings. Garonovich (2003) used cuttings of 12-15 cm length with a single bud on the top. Shuhua *et al.* (1995) observed the best results of rooting in 7-12 cm long cuttings treated with NAA. In an experiment in pots ($7 \times 7 \times 8$ cm^3), Wahlberg and Jeppsson (1992) observed that the cuttings of 10-12 cm length collected from the top part of a branch gave better results after treatment with 50 ppm IBA for 24 hours than the cuttings collected from the middle and base parts of branches.

At LIHS (Rana, 2011), softwood cuttings are prepared by removing 2/3rd of lower leaves carefully without damaging the bark. Salikhov (1986) observed that presence of the apical points in different types of cuttings improved rooting, and their removal reduced rooting and subsequently growth. Cuttings with some wood or with a heel, rooted and developed well. Using cuttings with a heel increased labour productivity and gave up to 88-92 per cent rooting under commercial conditions.

The key to high rate of success or rooting is to maintain the cuttings turgidity during the period from collection and well rooted in the propagation facility. Roots are borne out at the part of 0.5-2.5 cm

Figure 1.4: Preparation of Seabuckthorn Softwood Cuttings in Russia and China.

above the cut surface of the cuttings, belonging to the type of epidermal rooting. Roots are always many and concentrated, and generally arranged in the form of brush on the crevice of epidermis (Shuhua *et al.*, 1995). It was recommended that cuttings of 7-12 cm length with top bud and 2-3 pairs of leaves (Figure 1.4), selected from the lower part of maternal plants, lower part treated with 500 ppm NAA for 10 seconds and planted in the soil matrix of sand:saw dust:soil under seabuckthorn vegetation with a ratio of 10:3:1 or 10:7:0. Seedlings can be transplanted to nursery after 25-30 days in cuttings bed and 7 days in shed for training. Yao and Wei (2011) showed that when cuttings of seabuckthorn were treated with GGR6, the survival rates were 50 per cent, 66.7 per cent, 83.3 per cent and 93.3 per cent with 1 leaf and 1 bud, 2 leaves and 2 buds, 3 leaves and 3 buds and 4 leaves and 4 buds. The survival rates of the last one (4 leaves) was obviously higher than the others.

However, the cuttings are of the length of 30 to 50 cm (40 cm ideal) at LIHS, Barnaul (Rana, 2011).

Treatment of Cuttings with Hormones

Treatment of cuttings with hormones promotes root formation and healthy growth of plants. Growth hormones like IBA, NAA, IAA, IPA and ATP have been used by various workers for accelerating root formation in seabuckthorn. The lower end of cutting is dipped into water and then in rooting hormone (*e.g.* indole butyric acid, 0.1 per cent), before planting in media, like sand or perlite (Li, 2008). The basal part of cuttings are placed at 2-3 cm depth of solution of growth hormone at room temperature for 14-16 hours, then washed with water and planted. Avdeev (1976) observed that softwood cutting treated with IBA at 50 mg/1 which were planted in 2:1 peat/sand substrate, started rooting on the 9-13th day. Potapov (1978) observed that IAA at 200-400 mg/l or IBA at 100-200 ml/l improved rooting in the cuttings. Spraying the leaves with hormone required less material and was simpler to apply than treating the base of the cuttings. Superphosphate applied to the cutting bed enlarged the root system.

Papp (1982) observed that the cut end could be dipped in any of the solutions of NAA, IAA and IBA. However, Balabushka (1990) observed that IAA effect on rooting was better than IBA and chlorophenoxyacetic acid. According to Ivanicka (1988), the semi-lignified cuttings should be treated IBA (0.1-0.3 per cent). Garonovich (2003) recommended treatment of cuttings with IBA (50 mg/litre), IAA (100 mg/litre) and NAA (50 mg/litre) during 16 hours.

Figure 1.5: Treatment of Cuttings with IBA, Russia.

Figure 1.6: Quick Dip in NAA Solution, China.

2,4-D at low concentrations was used by Polikarpov *et al*. (1988) to increase the rooting of softwood cuttings in cultivar "Malinovka" of seabuckthorn and the apple rootstock M.26 but, at the concentration tested (not specified), 2,4-D reduced the percentage of rooting in cuttings of readily rooting *Hippophae* cultivars. The stimulation of root development in M.26 by 2,4-D was comparable to the stimulation by IBA.

Bounous *et al*. (1992) collected and treated seabuckthorn cuttings in late June with 3000 ppm IBA. Balabushka (1990) compared the effect of IAA, IBA and chlorophenoxyacetic acid on rooting of seabuckthorn and found that IAA showed better results as compared to other hormones. Wahlberg and Jeppsson (1992) found better results of propagation, when the cuttings were treated with 50 ppm IBA for 24 hours.

Lu Rongsen (personal comm.) suggested that before the treatment by hormone, cuttings should be placed in water to keep them fresh. Then, these cuttings are taken out of the water and bundles of 40-50 cuttings, kept together by a rubber band, are placed into a box filled with solution of 50 ppm IBA at a depth of 3-4 mm. All the cuttings soaked in the IBA solution are covered with a plastic film to maintain a temperature of 20-25 °C for 16 hours. Before planting, these cuttings should be taken out of the IBA solution and washed to remove the hormone. Shuhua *et al*. (1995) concluded that dip of a cutting in NAA solution at 300-1000 ppm for 10 seconds accelerated root formation up to 80-100 per cent. It was observed that for European seabuckthorn, IBA has been given a preference for cutting treatment (Figure 1.5), whereas in China, NAA has been widely used (Figure 1.6).

Lebeda (2008) observed that if the treatment of cuttings with hormone is not performed, then they are placed into basin with running water for few hours. For seabuckthorn, the most effective growth-promoting hormone is Indole acetic acid (IAA), which is applied at the following concentrations: For cuttings from upper unripe part of sprout at 10-25 mg/l; for cuttings from middle ripened part 25-50 mg/l; and cuttings from the lower mature part at 50-100 mg/l. The cuttings prepared for planting are placed at 2-3 cm depth into the solution of growth-promoting hormone at room temperature for 14-16 hours and washed in water and then, they are planted.

Stanys *et al*. (2010) observed 40-50 mg/litre IBA reliably quickens rhizogenesis in cuttings. He also found that IAA does not accelerate process of rhizogenesis, but 40-50 mg/l IAA increases the percentage of rooted cuttings.

The rooting effects of donor plant age, plant growth regulators and their concentrations on stem cuttings in Russian seabuckthorn were investigated by Yang *et al.* (2011), who found NAA was more effective than IBA and ABT2 among plant growth regulators and the 200 mg/l concentration of plant growth regulator was the best. Overall, applying 200 mg/l concentration of NAA to treat 1-year old shoot cuttings for 30 min could achieve the best rooting with the rooting rate, root number and average root length being 88.87 per cent, 10.8 item/panicle and 10.47 cm, respectively.

At LIHS (Rana, 2011), the cuttings are arranged in bundles of 50 cutting in such a way that the lower portion/base of cuttings are at the same level (Figure 1.5). A fresh cut is made at the base of the cuttings and dipped in water. Then, the cuttings are treated with 40 ppm IBA solution for 10-12 hours or 500 ppm IBA as quick dip for 5-10 seconds.

Wu *et al.* (2010) showed that the most significant method was cuttings presoaked in NAA 50 mg/kg solution or IBA 100 mg/kg solution for 16 hours, with the suitable proportion of transplanting medium with soil and sand. The rooting rate reached 80 per cent.

Magnetized Water Treatment

Magnetized water has also been tried by some workers to affect root formation in the softwood cuttings of seabuckthorn. Tsarkova and Polikarpova (1980) dipped seabuckthorn cuttings in magnetized water at 18-20 °C for 18 hours. The magnetized water was obtained by passing it through a magnetic field (2400 oersted at 0.33 m/sec.). Dipping of cuttings had no beneficial effect on rooting but propagation under MV mist gave up to 100 per cent rooting. Subsequent shoot growth was 33-40 per cent greater than in the control (ordinary mist).

Radiation Treatment

Some scientists have tried gamma radiation for higher rooting in the cuttings (Potapov and Borodachev, 1979). The cuttings taken from the shoot apices and irradiated at 500 rad showed the best rooting rate (97 per cent) and those taken from shoot bases showed only 76 per cent rooting, as compared to 70-71 per cent rooting in control. Higher irradiation doses decreased rooting percentages, especially in cuttings taken from the shoot bases. The 2000 rad dose was found lethal. Laser technique has also been found useful in propagation of seabuckthorn. Budagovskii *et al.* (1993) treated softwood and hardwood cuttings of seabuckthorn, apple, black currant and red currants before plantations. In this method, 20,000-150,000 cuttings could be treated with laser rays per hour. Laser treatment stimulated rooting in all the species studied, which ranged from 89 to 93 per cent, as compared with 14 to 66 per cent in the control.

Mist Chamber

Different types of polyhouses with provision of mist systems are used in Russia (Figure 1.7), Mongolia (Figure 1.8) and China (Figure 1.9). In a Russian polyhouse of size 90 × 10 m (900 m^2) (Figure 1.7), one can propagate 1,00,000 saplings of seabuckthorn per year (Rana, 2011). In order to obtain high rates of success, we need to keep the cuttings turgid during the rooting period. Cuttings require 80-90 per cent humidity to remain turgid during the rooting process. Therefore, the propagation from the softwood cuttings is carried

Figure 1.7: Structure of Polyhouses and Mist System at LIHS, Barnaul, Siberia, Russia.

out in the plastic film house, equipped with an artificial mist sprayers and drainage system. Different activities like preparation of soil media, treatment of cuttings with hormone, plantation of cuttings, weeding, etc. are done in the mist system (Table 1.1).

Table 1.1: Findings from Global Studies on Propagation of Seabuckthorn from Softwood Cuttings

Parameter	*Observations*	*Source*
Time of cuttings	Since June	Avdeev (1976), Russia
Collection	End of June to the beginning of July	Li (2008), Canada
	Since early June	Osipov and Morozova (1983), Russia
	Since late June	Buonous *et al.* (1992), Russia
	Since late May	Kniga (1989), Ukrain
	Since late June	Kondrashov and Kuimov (1987), Russia
	June to July	Rana (2011), Russia
	End of June - beginning of July	Wahlberg and Jeppsson (1992), Sweden
	Middle of July to the beginning of August	Shuhua *et al.* (1995), China
	End of June and the end of July	Wu *et al.* (2010), China
Location of shoots in canopy	Cuttings collected from the top part of a 1-yr old branch	Wahlberg and Jeppsson (1992), Sweden
	Top shoots of the 1-year old branches	Shuhua *et al.* (1995), China
	Top shoots	Stanys *et al.* (2010)
	Top and base shoots	Salikhov (1986), Russia
	Middle part of branches	Wu *et al.* (2010), China
Size of cuttings	30 to 40 cm	Rana (2011), Russia
	15-20 cm	Pletneva (1983), Russia
	1215 cm	Garonovich (2003), Belarus
	10 to 15 cm	Lebeda (2008), Ukrain
	10 to 15 cm	Li (2008), Canada
	10-12 cm	Papp (1982), Russia
	10-12 cm	Wahlberg and Jeppsson (1992), Sweden
	7-12 cm	Shuhua *et al.* (1995), China
Hormone treatment	50 mg/l IBA	Avdeev (1976), Russia
	200-400 mg/l IAA	Potapov (1978), Russia
	100-200 ml/l IBA	Potapov (1978), Russia
	40 ppm IBA solution for 10-12 hours or 500 ppm IBA as quick dip for 5-10 seconds	Rana (2011), Russia
	0.1 per cent IBA	Li (2008), Canada
	0.1-0.3 per cent IBA	Ivanicka (1988)

Contd...

Table 1.1–*Contd...*

Parameter	*Observations*	*Source*
	IBA (50 mg/litre), IAA (100 mg/litre), NAA (50 mg/litre) during 16 hours	Garonovich (2003), Belarus
	2,4-D	Polikarpov *et al.* (1988), Russia
	3000 ppm IBA	Bounous *et al.* (1992)
	50 ppm IBA for 24 hours	Wahlberg and Jeppsson (1992), Sweden
	IAA at 10-25 mg/l for cuttings collected from upper part of branch, 25-50 mg at middle part and 50-100 mg at basal part of branch for 14-16 hours	Lebeda (2008), Ukrain
	50 ppm IBA for 16 hours	Rongsen (1992), China
	500 ppm NAA for few seconds	Shuhua *et al.* (1995), China
	50 mg/kg NAA or IBA 100 mg/kg for 16 hours	Wu *et al.* (2010), China
	40-50 mg/l IBA	Stanys *et al.* (2010)
	200 mg/l NAA	Yang *et al.* (2011), China
	100-200 mg/kg GGR	Yao and Li (2011), China
Soil media	Peat:sand:: 2:1 mixture	Avdeev (1976), Russia
	Peat, polystyrene granule and sand (1-2:1:0.05)	Ivanicka (1988), Russia
	Two layered substrate, composed manure, sand and soil in equal proportions, spread on the floor of the green house, a sand layer made on the top of the substrate	Solovyeva (1998), Russia
	One layer of soil and then FYM @20-40 tone and then the layer of sand @ 50-60 tone in the polyhouse of the size 90 × 10 m (900 m^2)	Rana (2011), Russia
	Bottom parts are filled with layers of gravel or crushed bricks (15 cm) and layer of coarse-grained sand (5 cm), then layer of humus:sand::1:1 soil (15 cm). Upper most layer of substrate of pure coarse-grained sand or mixture of sand with peat with a ratio of 3:1 (5 cm)	Lebeda (2008), Ukrain
	Media mixes, such as a 2:1:1 by volume peat: vermiculite:perlite or 3:1 peat:perlite moss mixtures have provided good results	Li (2008), Canada
	Matrix of sand:saw dust (and soil under seabuckthorn vegetation) is prepared with a ratio of 10:7	Shuhua *et al.* (1995), China

Soil Media

The seed bed is prepared flat and the medium of the seed bed may be a mixture of sand and soil from seabuckthorn vegetation for appropriate *Frankia* and *mycorhizae*. Avdeev (1976) used peat:sand :: 2:1 mixture for the plantation of cuttings. Cuttings are planted in the media with atleast 2 leaf nodes are in the media below the surface. Ivanicka (1988) used mixture of peat, polystyrene granule and sand (1-2:1:0.05), which proved to be useful for healthy growth of roots and plants. Shuhua *et al.* (1995) observed that the rooting rate and the root biomass were the maximum, when the matrix of sand:saw

Figure 1.8: Structure of Polyhouses and Mist System in Mongolia.

Figure 1.9: Chinese Circular Polyhouse with Misting from a Pipe.

dust (and soil under seabuckthorn vegetation) is prepared with a ratio of 10:7 in China. It is sterilized with 0.5 per cent $KMnO_4$ solution. The seed bed is watered to keep the medium wet enough. The light permeability ratio of the shed is maintained at 50-60 per cent, 2 m above the ground.

Lebeda (2008) recommended that the bottom parts are filled with layers of gravel or crushed bricks (15 cm) and layer of coarse-grained sand (5 cm), then layer of humus: sand :: 1:1 soil (15 cm). Upper most layer of substrate of pure coarse-grained sand or mixture of sand with peat with a ratio of 3:1 (5 cm). Solovyeva (1998) used two layered substrate, which consisted of well composed manure, sand and soil in equal proportions, spread on the floor of the green house, a sand layer made on the top of the substrate.

Cuttings can be rooted in any substrate providing good air/water relationships. Media mixes, such as a 2:1:1 by volume peat: vermiculite: perlite or 3:1 peat: perlite moss mixtures have provided good results (Li, 2008). Wu *et al.* (2010) showed that suitable proportion of transplanting medium with soil and sand was 26"1. The rooting rate reached to 80 per cent. Yao and Wei (2011) observed that rooting of cuttings was better in the soil matrix with sand and turfy soil than the others.

At LIHS (Rana, 2011), the polyhouse is prepared for propagation from softwood cutting by putting one layer of soil and then FYM @20-40 tone, and then the layer of sand @ 50-60 tone in the polyhouse of the size 90 × 10 m (900 m^2). Every year, the sand is removed from the polyhouse and then the fresh sand is added in the greenhouse. The best medium for the rooting of softwood cutting is 90 per cent sand and rest soil and FYM. The sand is spread evenly in the polyhouse and raised beds of convenient length are made and leveled (Figure 1.7).

Plantation of Cuttings

In China, before planting the cuttings, a marker is used to make small shallow furrows in the seedbed. The distance between the furrows should be 7 cm and cuttings 3 cm. Then, the cuttings are inserted into the soil mixture at a depth of 1.5-2.0 cm. Plantation of cuttings more closely may be spoiled by mould. After the planting of cuttings, seed bed is again watered to make the medium around the cuttings settled down (Rongsen, 1992).

According to Lebeda (2008), before the plantation of cuttings in media, the substrate is slightly compacted and moistened, then marked. The planting space of cuttings is kept at 3-5 cm × 7-10 cm. Cuttings are planted into targeted points upright at 4-5 cm of depth, soil around cuttings compacted and after the completion of planting, beds are amply watered with mist spray.

At LIHS (Rana, 2011), after IBA treatment, the cuttings are planted in the polyhouse or in mist propagation chamber. The cuttings are planted at close spacing of about 5 × 5 cm. Over the leveled beds, a line marking machine or tool, *i.e.* marker is moved to make the lines, where the cutting has to be planted. Then cuttings are planted in the marked position (Figures 1.10 and 1.12). After every 15 minutes, the misting is done to keep the cuttings moist during completion of planting of cuttings.

Figure 1.10: Plantation of Cuttings in Sand, Russia.

Figure 1.11: Plantation of Cuttings in Polybags, China.

Figure 1.12: Plantation of Cuttings in a Polyhouse, China (A), Seabuckthorn Cuttings in a Polyhouse, Russia (B).

In China, cuttings are planted in vermiculite containing polybags in the trays placed on brick walls in the circular polyhouse (Figures 1.11 and 1.12A).

Management of Humidity and Temperature

Aftercare improves the rooting rate and survival of cuttings and plants. Cuttings can be rooted in any substrate or media, providing good air and water relationship. Humidity can be controlled in a number of ways, but the best way is with automatic misting units, such as brass mist nozzles, controlled by electronic time clocks and a solenoid.

Lebeda (2008) found that for better rooting in cuttings, during the first few days after planting, the cuttings are watered freauently (3-5 times per day), but not amply, using a watering can or hose pipe with atomized nozzle. In hot sunny weather, the cuttings should shaded by whitening glass or film by chalk or pulling gauze over them. Weeds should be immediately eliminated. In a month after planting, it is necessary to begin conducting the training of cuttings and preparing them for winter. At first, the careful aeration of structures should be carried out (not for long to open up plastic-film covers or slightly raise hot-bed panes), and later, approximately in the mid of August, plastic-film covers and panes are taken off permanently.

Garonovich (2003) is of the view that optimal air temperature for taking roots should be 22-27 °C. Relative humidity should not be lower than 90 per cent. After 15 days, watering is reduced to 1-2 times a day, but adequate. The rooting of cuttings depends upon the temperature and moisture in the matrix (soil and air around the cuttings). At the time of planting of cuttings, the temperature of the soil should be higher by 1-3 °C than the air. Generally, when the daily average temperature of open land is up to 18.5 °C, the average temperature of the soil is up to 19 °C and the relative humidity of the air is 40-50 per cent. However, within the plastic film, the average temperature of day and night should be 19 °C, the average temperature of the soil should be 24 °C and the relative humidity of the air should be 88-100 per cent by using spray-mist in time. This high level of humidity is indicated by the appearance of water layer on the leaf surface of the cuttings.

The mist system may be kept running on day and night for 30 days. Initially, 10 seconds of mist is supplied at every 30 minutes. After the root start developing, we gradually reduce the duration of misting. The cuttings develop roots in 4-6 weeks.

In case, the water layer disappear, it is necessary to spray mist until the leaf surface retains the water layer again. After the cuttings have already produced sufficient roots, the humidity should be gradually decreased and the film house should be well ventilated.

At LIHS (Rana, 2011), after planting, the misting is done after every 5 minutes to keep them wet. After 15-20 days, the rooting will start then the misting is done after every 15 minutes and with increase in rooting, the interval of irrigation is increased. At the end of the season, the cuttings become ready for planting in the orchards.

Application of Fertilizers

Simonav *et al.* (1986) observed that growth hormones (humic acid and Agrostilin) and minor elements (Mn and Zn) markedly improved the planting material produced during the first year and this reduced the cost of growing on of non-standard plants during the second year in the nursery. Treatment of softwood cuttings with humic acid at 150 g/10 litre water and application of Mn and Zn to the rooting substrate have been recommended. Krylova (1991) treated the cuttings with IBA at 25-50 mg/litre and planted in an unheated peat/sand substrate under mist. Seabuckthorn responded readily to high K rates.

Garonovich (2003) recommended the use of fertilizers of N80, P150, K100 in soil media before plantation of cuttings, top dressing of one-year old seedlings with N100, P250, K100 and liquid top dressing with ammonium nitrate (20 g/m^2).

Li (2008) reported that to ensure the vigourous plant growth under mist system, apply fertilizers in the root zone with a weekly application of 20-20-20 fertilizers at 2000 ppm (2 g/l) applying them immediately after the first roots are formed and continue during the growing season.

Micro-cuttage Method of Propagation

Li Wei of Beijing University, Beijing, China has perfected the new method of propagation of seabuckthorn from small piece of shoot, known as "Micro-cuttage method of propagation" (Wei, 2011). Generally, the normal length of softwood cuttings for propagation is 20-40 cm. Li Wei reduced the length of shoot and conducted the trial on different length of shoots for rooting (Figure 1.13). The cuttings of 2-3 cm length with 2 leaves or cuttings of 4-5 cm with 3 leaves gave above 95 per cent survival and rooting rate (Figure 1.14) when treated with 100 mg/l of GGR before plantation in media mixture sand:peat soil in ratio of 3:2 in a mist system.

Figure 1.13: A. 1-2 cm length and 1 leaf, B. 2-3 cm and 2 leaves, C. 4-5 cm and 3 leaves and D. 20-40 cm length and few leaves.

Figure 1.14A: Plantation in Media in Mist System.

Figure 1.14B: Growth Under Mist.

OVERWINTERING

With the onset of autumn, the rooting cuttings should be gradually tempered. First, the plastic film is removed for a short period everyday and this is increased day by day. After 15-20 days, the plastic film can be removed permanently. After the onset of winter, for overwintering and to save them from frost and snow damage, mulches of 5-6 cm thick layer of needles, leaves, dry humus or peat are spread over the beds (Ermakov, 1985; Rongsen, 1992). During growth period, shoots may be trained to prepare them for winter. For the same, careful aeration of whole structure is carried out (but not for long, open up plastic film cover or slightly raise hotbed panes). In mid of August, plastic film covers and panes are removed completely.

They must be acclimated and hardened off prior to planting outside the propagation bed (Lebeda, 2008). The best method is to over-winter the rooted cuttings in the propagation bed (covered with micro foam or mulched with peat moss or wood chips) or remove them and store in polybags at -2 °C (Li, 2008). Under the mulches, poisoning baits are placed against the mice. Before the initiation of bud burst in next spring, rooted cuttings are dug out and transplanted in nursery for better growth or directly in orchard.

TRANSPLANTATION IN NURSERY OR FIELD

The cuttings may also be transplanted to the field in late August or September when roots are well developed. This method works well if cuttings have been rooted directly into containers. Transplants should be kept moderately well irrigated. New site is prepared in autumn by ploughing the soil and

Figure 1.15: Transplantation of Saplings in Nursery.

Figure 1.16: Transplantation in Orchard.

mixing fertilizers (10 kg/m^2 humus, 30 g/m^2 double super-phosphate, 300-500 g/m^2 wood ash). Plants are planted at a spacing of 20 cm × 15 cm in the nursery site (Figure 1.15). Weeding and tillage operations are carried out timely, whereas moisture in the soil is maintained at 75-80 per cent. After one year of growth, they may be transplanted into the plantation field (Figure 1.16).

CONCLUSION

For fast multiplication, particularly of improved cultivars and varieties of seabuckthorn, few months or one-year old softwood cuttings are used. As global studies have found, the cuttings should be collected from late June to August months and stored in polybags to avoid their drying. Cuttings of proper size are prepared by removing most of the leaves from lower side of the cuttings, except few in top of shoot. They are treated with low concentration, mostly IBA or NAA for long duration or high concentration for few seconds before plantation in soil media in a mist system. Cuttings are planted in soil media comprising in most cases, and soil layer at base followed by FYM and sand on the top. Humidity need to be maintained at high level in mist system to maintain turgidity of the cuttings and better rooting. Saplings, well rooted can be transplanted in the nursery fields or orchards in same season or next year, depending on the growth of saplings.

REFERENCES

1. Avdeev, V.I. 1976. Propagation of *Hippophae rhamnoides* by softwood cuttings under mist. *Kratkie tezisy Dokl 2-i Vses Konf Molodykh Uchenykh po Sadovodstvu*, p. 72-74. (In Russian)

2. Balabushka, V.K. 1990. Results of the trials of growth regulators on summer cuttings of introduced woody plants. *Byulletin Glavnogo Botanicheskogo Sada* 156: 65-67.

3. Bounous, G., Bullano, F. and Peano, C. 1992. Softwood cuttings of *Amelanchier candensis, Cornus mas, Elaeagnus umbellate* and *Hippophae rhamnoides. Monti-e-Boschi* 43 (4): 51-57. (In Italian)

4. Budagovskii, A.V., Budagovskaya, O.N., Gudi, G.A., Mokrousova, G.I. and Gulshina, E.V. 1993. Laser technology in horticulture. *Sadovodstvo-I-Vinogradarstovo* 3: 6-7. (In Russian)

5. Ermakov, B.S. 1985. Introduction of low-growing seabuckthorn into the Moscow area. *Biologicheskie Aspekty Introduktsii, Selektsii i Agrotekhniki Oblepikhi*: 58-63. (In Russian and English)

6. Garonovich, I.M. 2003. Introduction of *Hippophae* L. in Belarus. In: *Seabuckthorn (Hippophae L.)-A Multipurpose Wonder Plant. Vol. I: Botany, Harvesting and Processing Technologies* (V. Singh, Ed. in Chief, 2003), p. 137-151, Indus Publishing Company, New Delhi, 518p.

7. Ivanicka, J. 1988. Propagation of unusual fruit crops from softwood cuttings under mist. *Vedecke-Prace–Vyskumncho-Ustavu-Ovocnych-a-Okrasnych-Drevin-v-Bojniciach* 7: 163-170.

8. Kniga, N.M. 1989. Characteristics of rooting of softwood cuttings of top and small fruit species in relation to natural photoperiod. *Fiziologiya I Biokhimiya Kulturnykh Rastenii* 21: 403-409. (In Russian)

9. Kondrashov, V.T. and Kuimov, V.N. 1987.Vegetative propagation of *Hippophae rhamnoides. Sadovodstvo* 6: 13-16. (In Russian)

10. Krylova, I.I. 1991. Mineral nutrition of softwood cuttings of top and soft fruit crops. *Sadvodstvo-I-Vinogradarstvo* 4: 16-18.

11. Lebeda, A.F. 2008. Propagation of seabuckthorn (*Hippophae rhamnoides* L.) in Ukrain. In: *Seabuckthorn (Hippophae L.)-A Multipurpose Wonder Plant. Vol. III. Advances in Research and Development* (V. Singh, Ed. in Chief, 2008), p. 22-26, Daya Publishing House, New Delhi, 566p.

12. Li, T.S.C. 2008. Propagation methods of seabuckthorn in Canada. In: *Seabuckthorn (Hippophae L.)-A Multipurpose Wonder Plant. Vol. III. Advances in Research and Development* (V. Singh, Ed. in Chief, 2008), p. 14-21, Daya Publishing House, New Delhi, 566p.

13. Osipov, Y.V. and Morozova, G.M. 1983. Propagation of seabuckthorn by cuttings. *Sadovodstvo* 12: 20-21. (In Russian)

14. Papp, L. 1982. Importance and propagation of seabuckthorn. *Erdo* 31 (7): 309-312. (In Russian)

15. Pletneva, T.M. 1983. Intensification of seabuckthorn nurseries. *Agrotekhnika i Selektsiya Sadovykh Kul'tur*: 149-154. (In Russian)

16. Polikarpova, F.Y., Protchev, A., Kov, A.V. and Tikhomirov, I.A. 1988. The effectiveness of applying 2,4-dichlorophenoxyacetic acid for root development in softwood cuttings. *Plodovodstvo V Nechernozemnoi Polose*, Moscow, USSR, p. 13-20.

17. Potapov, F.F. 1978. New methods of seabuckthorn propagation from green cuttings. *Rastitel'nye Resursy* 14 (2): 215-220. (In Russian)

18. Potapov, S.P. and Borodachev, M.N. 1979. Treatment of rose and seabuckthorn softwood cuttings with gamma irradiation 6 OCD. *Subtropicheskie-kul'tury* 3: 103-105. (In Russian)

19. Rana, R.K. 2011. *Training Report on Breeding and Production of Seabuckthorn in Russia at Barnaul*, 27 June-11 July, 2011, CSK HPAU, Palampur, HP, India, 16p.

20. Rongsen, L. 1992. *Seabuckthorn-A Multipurpose Plant Species for Fragile Mountains*, Occasional paper No. 20, ICIMOD, Kathmandu, 62p.

21. Salikhov, M.M. 1986. Rooting capacity of different types of *Hippophae* softwood cuttings. *Sovershenstvovanie Vyrashchivaniya Plodovykh Kul'tur v Nechernozemnoi Polose*: 61-68. (In Russian)

22. Shuhua, H., Xin, Z. and Zhixiang, Z. 1995. Studies on cutting propagation techniques of seabuckthorn. In: *World Wide Research and Development of Seabuckthorn*. Proceedings of International Workshop on Seabuckthorn, p. 118-132, Beijing, China, 206p.

23. Simonav, I.N, Poludennyi, L.V., Baikalov, P.A. and Volkova, T.V. 1986. Effect of manganese, zinc and growth regulators on the growth of *Hippophae rhamnoides* transplants. *Pitanie-Plodovykh-Rastenii*: 105-109.

24. Singh, V. and Gupta, R.K. 2003. Propagation of seabuckthorn (*Hippophae rhamnoides* L.). In: *Seabuckthorn (Hippophae L.)-A Multipurpose Wonder Plant. Vol. I. Botany, Harvesting and Processing Technologies* (V. Singh, Ed. in Chief, 2003), p. 315-333, Indus Publishing Company, New Delhi, 518p.

25. Solovyeva, A.E. 1998. Features of propagation of new seabuckthorn varieties. In: *Proceedings of International Symposium on Seabuckthorn* (*Hippophae rhamnoides* L.), p. 187-188, Russia, 248p.

26. Stanys, V., Treikauskiene, J. and Staniene, G. 2010. Propagation of seabuckthorn using softwood cuttings. *Sodinnkyste ir Darzininkyste* 29 (1): 29-39.

27. Tsarkova, T.F. and Polikarpova, F.Y. 1980. Effect of magnetized water on the development of seabuckthorn softwood cuttings. *Lesnoe-Khozyaistvo* 2: 53-54.

28. Wahlberg, K. and Jeppsson, N. 1992. Development of cultivars and growing techniques for seabuckthorn, blackberry, honeysuckle and rowan. *Verksamhetsberaettelse Balsgaard* (Sweden), p. 86-100. (In Swedish)

29. Wei, L. 2011. Rapid propagation of seabuckthorn from micro-cottage. In: *5th International Seabuckthorn Association's Conference*, 4-8th September 2011, Xining, China.

30. Wu, X., Liu, F., Wang, R.H. and Yang, J.F. 2010. Study on softwood cutting propagation techniques of *Hippophae rhamnoides* L. *Northern Horticulture* (1).

31. Yang, L.X., Wang, H.N. and Fan, J. 2011. Effects of donor tree ages and plant growth regulators on the softwood cutting propagation of *Hippophae rhamnoides*. *International Journal of Automation and Computing* 33 (6): 107-111.

32. Yao, J. and Wei, L. 2011. Study on rapid propagation technology for seabuckthorn by micro-cuttage. *Global Seabuckthorn Research and Development* 2.

Chapter 2

Standardization of Agro-techniques of *Hippophae salicifolia* D. Don (Seabuckthorn) of Sikkim Himalayas

B.C. Basistha, N.P. Sharma, L. Lepcha and A. Sen*

Sikkim State Council of Science and Technology, Development Area,
Gangtok – 737 101, Sikkim, India
*Department of Botany, University of North Bengal, Raja Ramohanpur, Distt. Darjeeling, India

SUMMARY

From the present study on *Hippophae salicifolia* D. Don growing in Sikkim and north-east Indian Himalayas, it was observed that the plant is not very big in size, but trees up to 15 meters have been recorded. The ripe fruit is 4-6 mm length and 3-6 mm in width with orange red colour contained various compounds like vitamin C (300-1600 mg/100 g), vitamin E (200 mg/100 g), lycopene, flavonoids (0.2 per cent in juice), amino acid, etc. The study on germination of seeds in different soil media was done; pertile and brown soil media gave good result. Rooting and callus formation using growth regulators like IBA, NAA and IAA in different concentrations (ppm) showed interesting growth results, in which IBA (55-60 ppm) showed more callus formation and rooting within three months. Studies carried out in different chemicals for germination of seeds of seabuckthorn in which GA_3 (700 ppm) treated seeds germinated in 8 days as compared to 17 days in water alone (control). Further response of hormones in root growth showed that IAA at 100 ppm showed mean root length of 9.4 cm. This type of study for plant with high economic potential, can lead to the better understanding and management in the natural condition.

Keywords: *Hippophae salicifolia, Dalle-chuk, Lepcha, Nepali, Bhutia, Seabuckthorn, Hormones, Soil, Modulated, Riverine, IBA, NAA and IAA.*

INTRODUCTION

Sikkim, Himalayan state of India is situated in north-east India at 28°07′48" and 27°04′46" North latitude and 88°00′58" and 88°55′25" East longitude with a total area of 7096 sq km (Anonymous,

Seabuckthorn (*Hippophae* L.): *A Multipurpose Wonder Plant, Vol. 4: Emerging Trends in Research and Technologies* (Virendra Singh, *Editor-in-Chief*, 2014), p. 22–29, Daya Publishing House, New Delhi, India.

1982) and has an average distance of 64 km from east-west and 114 km from north-south. This tiny state has immense biological diversity without any of its parallels elsewhere in the world (Sharma *et al.*, 1995), with elevation ranging from 300 m to 8500 m asl (Basistha *et al.*, 2001).

Hippophae salicifolia (Seabuckthorn) locally known as *Tarobo* (Bhutia), *Ree-singri* (Lepcha) and *Ghanguru / Achuk or Dale-chuk* (Nepali) is a fascinating plant species known and used by humans for centuries (Dhakal, 2001).

Hippophae salicifolia belongs to family *Elaeagnaceae*, is a dioecious, wind pollinated plant with more than 60 years of lifespan (Rongsen, 1992; Jeppson *et al.*, 1999; Nepal *et al.*, 2001). Seabuckthorn (*Hippophae* L.) is a multipurpose and economic plant of the dry temperate Himalayas and is a native of European countries, China and India (Singh, 1998). The species is known for fixing atmospheric nitrogen into the soil through symbiotic relationships with root inhabiting micro-organism the *Frankia* by formation of root nodules (Andreeva *et al.*, 1982). The roots of plant are also known for better ecological advantages, such as soil binding and erosion control at fragile hilly slopes (Rongsen, 1992; Singh *et al.*, 1995; Singh, 1998). Apart from its ecological value, it has medicinal value also (Dhakal, 2001).

In cold deserts of India, the plant starts its foliage in April and bears flowers in monsoon and fruiting starts in August and retains till April (Singh, 1998). But in Sikkim, it was observed that flowering started in March and fruiting in April which ripens in October-November and retained till March. Since, its root system is well developed; the plant could hold soil even in fragile condition.

Agro-techniques of seabuckthorn through different methods like cuttings, seed propagation (Auauzato *et al.*, 1987), layering, etc. have also been done in the past which along with natural habitat shall help to understand the genus closely in the local environment.

In Sikkim, the plants are observed to be growing in the areas of about 500 ha at different slopes and north east aspects in colonies or lone stand of trees (6-10 meters), on the either side riverbanks (riverine) and non river side (non-riverine) areas in and around Zema I, Zema II, Zema III, Puchi and Chaten of Lachen valley and adjoining areas Lachung, in the north district of Sikkim. The plant has been found growing at the altitudes ranging from 2377 m to 3093 m asl (Table 2.1). The plant for its faster regeneration favoured the landslide areas (new soil) of sun facing slopes and riverbanks with

Table 2.1: Geophysical Parameters at Natural Habitat of Seabuckthorn in Sikkim

Parameters	*Natural Habitat*			*Experimental Field (at Tadong) (1424 m)*
	Low Altitude (2377 m)	*Medium Altitude (2600 m)*	*High Altitude (3093 m)*	
Aerial temperature	23°C	19°C	19°C	26°C
Humidity	89 per cent	92 per cent	17 per cent	92 per cent
Soil pH	6.6	6.6	6.8	6.6
Soil temp.	17.8°C	18°C	13.9°C	19.3°C
Soil moisture	65 per cent	70 per cent	35 per cent	70 per cent
Soil colour	Brown	Brown	Brown	Brownish-Black
Aspect	North	South-East	South-East	South-East
Management	Natural	Natural	Natural	–
Site cover	20-40	20-40	20-40	–

sandy to course textured soil (Basistha, 2001; Basistha *et al.*, 2001). Seabuckthorn has good association with other floral species like *Poplus* sp., bamboos, *Daphnie*, *Rhododendron* sp., conifers, etc. Its association with *Alnus nepalansis* (also actinorhizal plant) at the lower altitude where seabuckthorn started northwards from that point at an altitude of 2377 m. This reveals variation in climatic features of soil and air at natural orchard of seabuckthorn.

The local people use the plant as cheap source of firewood. It was learnt from the locals that the wood of seabuckthorn burns with almost blue flame and gives a good heat with minimum of smoke and ash. People prepare jam and pickles out of the fruits, which, they said, is good for digestion. Local women weavers, craftsmen use the juice from the fruit as dying ink for their traditional blankets, bedcovers and sweaters, etc.

MATERIALS AND METHODS

Frequent surveys on occurrence of the plant and morphological studies were carried out in 1999-2000 in different seasons, *i.e.* flowering season (February-March), fruit ripening season (November-February) and vegetative growth season (April-July) and different physico-climatic conditions and data were recorded. In the month of June 2000, the specimens of the plant parts were collected from different areas and herbarium was prepared for further studies. The area was divided into three categories based on altitude *viz.*, low (2377 m), medium (2600 m) and high (3093 m) altitude of the *Hippophae salicifolia* growing area. A study on propagation of the plant was carried out at natural habitat and fields of Sikkim State Council of Science and Technology at 1424 m.

The fruits were also collected, both pulp and dried fruits were sent to Central Food Technological and Research Institute (CFTRI), an ISO 9001 and 14001 organization, which is a premier Institute in India, located at Mysore, Karnataka.

In-situ geophysical chemical studies analyzing different parameters were recorded from the field of natural orchard of *Hippophae salicifolia* at Lachen and Lachung valley using digital thermometer (Multi thermometer, CE make, humidity meter (Model LR_6 Mignon AA, England make), soil pH and moisture were determined with on the spot tester soil pH and moisture meter (Takemura, Japan make, model DM-15) and the soil temperature with a soil thermometer. The recorded data was interpreted with that of the trial field and other similar habitats of the species (Table 2.1).

Both hardwood and softwood cuttings were taken from the matured trees (approximately 5-10 years old) from the natural orchard in an appropriate manner in the month of January and were later waxed and pretreated with Indole butyric acid (IBA), Naphthalene acetic acid (NAA) and Indole acetic acid (IAA) at different concentrations for overnight. The treated cuttings were later transferred and planted into separate nursery bed of 0.5 × 0.5 sq m with in 15 cm spacing. The soil composition was maintained as sand : soil : compost at 70 : 20 : 10. Watering and manuring were done as per the climatic situations. Twenty cuttings were used for the treatment with each concentration of 50 ppm, 55 ppm and 60 ppm of IBA; 300 ppm, 350 ppm and 400 ppm of NAA; and 50 ppm, 100 ppm and 200 ppm of IAA. The result was recorded and tabulated accordingly.

Air layering was carried out simultaneously by taking about 20 numbers of 1-2 years old branches at sun facing side amongst 5-10 years old trees and using local moss and soil mixture supported by transparent polythene strips during the month of April, and collected in the month of late July and planted in the experimental field at 1424 m.

Propagation was also studied using freshly collected seeds from the well ripened fruits collected from the natural habitat in the month of December. These seeds were dried for 5 days and washed properly with water and soaked for 5 hours. The seeds were then imbibed. The germination was done in raised beds with different soil types/soil compositions. The imbibed seed germination was also done in Petridishes treated with varying concentrations of 50 ppm, 55 ppm and 60 ppm of IBA; 300 ppm, 350 ppm and 400 ppm of NAA; and 50 ppm, 100 ppm and 200 ppm of IAA and were studied in the laboratory (Tadong, Gangtok) at room temperature. These were later transferred to 10" polybags for further growth. The mean root length data related to germination in each of the concentrations pertaining to growth, rooting and nodulations, etc., were recorded and analyzed.

RESULTS

Morphology

Hippophae salicifolia usually a bushy tree measured up to 15-18 m in height, in some area of Zema II and in the adjoining river bank with presence of heavy branches and a big foliage canopy. In Lachen near the monastery, male and female trees nearly 15-20 m measuring with girth of 2.5 m has been standing.

The leaves measured up to 7.2-9.7 cm in length and 1.1-2.0 cm width with 3-5 mm stalk length, linear, lanceolate, dorsi-ventral with and lower side covered with silvery satellite scales and pubescent hairs.

The male and female flowers were markedly different in their appearance, as male floral buds consisted of 4 to 6 flowers in cluster, while the female floral buds possess only one flower and rarely, two in the current season growth and in the older branches. Both male and female flowers had no petals, and possess four stamens and one pistil with conspicuous single ovary in male and female flowers, respectively. Often, floral buds were found mixed with the vegetative buds and were rarely in pure.

The roots of the plant were seen mostly confined around to 1-1.5 m of the soil with plant height of 1-2 m. A long taproot was found in most cases, penetrating down the surface up to 0.5 to 1m in the non-riverine areas. Approximately, two-year old plant was measured with root length of 1m at Zema III (3093 m asl).

The ripe fruits of seabuckthorn in its habitat were orange or reddish yellow in colour and were measured 4 to 6 mm in length and 3 to 6 mm in width with small black pustules at maturity, over the outer covering of the fruit with a notch. The seeds of the plant were brown to dark-brown in colour, 3 to 4 mm in size and the test weight of 0.97 g. The roots were found restricted to comparatively less depth in the soil with maximum percentage of humus and black soil.

Habitat

From Table 2.1, it is revealed that *Hippophae salicifolia* grows naturally in fragile slopes with high risks of landslides and flash washes, ranging from 2377 m asl to 3093 m asl. The plant survived and grew well in the experimental plot situated at Tadong, Gangtok at an altitude of 1424 m asl. During the season, not much variation was observed in the temperature of natural habitat at medium and high altitudes (19 °C), which may be due to the gentle rise in slopes of the study area. The soil pH at the natural habitat found to be varying from 6.6 to 6.8 and the moisture level was in the range of 35 to 70 per cent. The plant was happily growing in almost open areas in brown soils facing south east aspect with plenty of natural light.

Germination Performance

The Table 2.2 presents the details on germination and growth pattern of seedlings in different soil based media. It shows that germination in artificial media (Kel-perlite) and black soil took 17 and 18 days, respectively, as compared to faster germination in sandy soil (13 days), which may be due to more aeration as compared to the other two soils used. 50 per cent germination was seen better in brown soil which took 21 days. 85-90 per cent germination was seen in perlite soil with better average growth (1.9 cm/month). Perlite being an artificial media may not be advisable for the rural conditions and it can be concluded that brown soil may be better for the germination and growth of *Hippophae salicifolia* seeds. Sandy soils may be better for initial growth but may not suffice the nutrients for later growth and development.

Table 2.2: Germination Performance of *Hippophae* Seeds in different Soil Based Media at Laboratory Field, 4700 ft asl

Soil Type	*Days to Initial Germination*	*Days to 50 per cent Germination*	*Total Germination (per cent)*	*Growth per Month (Average, cm)*
Black soil	18	26	40	1.9
Brown soil	17	21	80-85	1.7
Perlite	17	21	85-90	1.9
Sandy soil	13	22	80-85	1.8

Vegetative Propagation with Growth Treatment

Table 2.3 shows that IBA at 60 ppm is better for root development (45 per cent) with a significant root length (31.5 per cent) in comparison to the control and other treatments. The callus formation (60 per cent) and mean root length (9.4 cm) was observed to be higher in IAA at 100 ppm treated cuttings. Among the different concentrations of NAA, 300 ppm NAA gave the good results showing 35 per cent success in callus formation and 15 per cent success in root formation with mean root length of 2.5 cm.

Table 2.3: Response of Growth Hormones in Successful Formation of Callus and Rooting in Seabuckthorn Cuttings (90 days after planting)

Treatments (ppm)	*Total Rooting*		*Total Callus Formation*		*Root Length (cm)*	
	(Nos.)	*(per cent)*	*(Nos.)*	*(per cent)*	*Total*	*Mean*
IBA 50	7	35	8	40	43	4.8
IBA 55	8	40	9	45	14.6	1.8
IBA 60	9	45	9	45	31.5	3.93
NAA 300	3	15	7	35	7.5	2.5
NAA 350	0	04	20	–	–	–
NAA 400	0	0	4	20	–	–
IAA 50	2	10	10	50	2.2	1.1
IAA100	5	25	12	60	4.7	9.4
IAA 200	3	15	9	45	25.5	8.5
Control (H_2O)	5	25	1	5	19.5	3.9

The result from the Table 2.3 shows that cuttings may show good results if treated with 60 ppm IBA in terms of rooting, callus formation and root development as compared to NAA and IAA at the above concentrations.

The average numbers of nodes on the cuttings of *Hippophae salicifolia* were 40 measuring 50 cm and the required length of cuttings for propagation remains 25 cm with 20 buds in each, having average weight of the cuttings as 75 g, the required quantity of the prepared cuttings for the vegetative propagation of seabuckthorn as follows:

- Average weight of cutting-75 g.
- Crop geomentry-2 × 2 m.
- Cutting position while planting-vertical.
- Cutting for replacement of damage/dead cuttings-20 per cent.
- Average number of nodes on the cutting of 50 cm=40.
- Type of set/cutting to be used for planting=25 cm.
- Average number of damage buds per cuttings=10.
- Spacing (row to rows) on planting=200 cm.
- Number of cuttings required= 10,000 × 2 × 2 = 2500
- Number of cuttings required for replacement=20 per cent of 2500
- Total cuttings required =3000
- The weight of cuttings required/hectare is 3000 × 75/1000 × 100=225 kg/ha.

Table 2.4: Response of Growth Regulators in Germination of Seeds in Petridish at Room Temperature

Particulars	*Control*	*IBA (ppm)*			*NAA (ppm)*			*IAA (ppm)*		
–	H_2O	50	55	60	300	350	400	50	100	200
Mean root length (cm)	3.9	4.8	1.8	3.9	2.5	00	00	1.1	9.4	8.5
Mean root numbers	4.0	4.2	2.9	3.9	3.3	00	00	2.5	5.6	5.3

From the Table 2.4, it was shown that the seeds grown in petridish containing IBA 50 ppm solution soaked filter paper showed better germination performance in both rooting and shooting.

The findings of Tables 2.3 and 2.4 were related with the findings of Donald (1997) who reported that treatment of cuttings of *Pinus ellitti* with 1 per cent IBA for successful root initiation. Similar results were also reported by Tsar'kova (1988) and Rongsen (1992) that 50 ppm IBA presented better initiation of rooting in softwood cuttings of *Hippophae*.

The observations of layering showed that 13 out of total 20 (65 per cent) layerages with successful rooting and foliage growth were collected from Lachen area, and are transplanted in field conditions (1424 m). It was, however, noted that some of the rooted branches were already nodulated with 2-6 lobed nodules, somewhat whitish in colour but intact with the root mass. The layerages had good success rate in the experimental plot without any treatments.

Fruit Analysis

As per the studies carried out from the sample sent to CFTRI, Mysore, the composition of fruits and leaves of seabuckthorn from Sikkim were: Leaves of seabuckthorn contained 11 to 22 per cent of crude protein, 3 to 6 per cent of crude fat and some flavonoids; fruit contains 60-80 per cent juice in which sugar, organic acids and vitamins were rich. The vitamin C content was 200 to 500 mg/100 g, fruits contained 3-5 per cent of pulp oil and 8-18 per cent of seed oil that were rich in unsaturated acids, β-carotene and vitamin E. In addition, certain content of flavonoids was also found in the residues of fruit; and the fruit may be used for making soft drinks, wines, healthy food, medicines and cosmetics.

DISCUSSION

The present study was conducted in Lachen and Lachung valley of North Sikkim revealed some interesting facts. Seabuckthorn plants in riverbank-sandy soil produced higher number of nodules irrespective of the plant size as compared to that from the non-riverine slopes.

In the trial plot, it was observed that the seedlings of even smaller size produced sufficient root mass and formation of nodules were even higher as compared to the plant from natural habitat. Nodulation was even better in the sandy soil with sufficient gravel and organic matters added with cow dung manure. The average number of nodules from 100 seedlings (5-15 cm) was recorded at 12/ plant. The root mass was observed to be poor in case of waterlogged soil and found to be developing better in sandy and coarse sandy soil, which is comparatively lower than the riverbank areas of natural habitat.

It was also noted that the seabuckthorn seedlings grew at greater heights in non-riverine areas of lower altitude and higher altitude but in the case of nodulation, it was reversed. In non-riverine areas of lower altitude, there were fewer nodules per plant as compared to the plants in non-riverine areas of higher altitudes. The roots were found restricted to comparatively lesser depth in the soil with maximum humus and black soil. The heights of the plant were observed to be reducing with the increase of altitude both at riverine and non-riverine areas. The trees up to 14-20 meters can be seen in non-riverine areas of Lachen and seemed to be old natural orchards. The old trees are slowly being replaced by new bushy plants.

CONCLUSION

The present study on agro-techniques for *Hippophae salicifolia* in Sikkim Himalayas reveals that as elsewhere the plant can be artificially explored to several other dry temperate areas of Sikkim with similar climatic conditions. The cultivation of this economic and eco-friendly plant with scientific approaches may fulfill the requirements of high altitude inhabitants. The germination of *Hippophae* seeds using perlite type soil or brown soil media, use of pretreated cuttings for mass propagation for easy and massive plantation programmes and thus, development of greenery at difficult areas, may be taken as viable package of practice for seabuckthorn cultivation in Sikkim and thus, may added to the economy of the marginalized farmers of the remote terrain. The study reveals that if the plants grow well in sandy or coarse sandy soil media or in fields with desirable soil reclamation. In a nutshell, plant creates more biomass in stressed soil and climatic condition. Further studies in details may enable the species to propagate in other localities with varied topographical conditions.

ACKNOWLEDGEMENT

The authors are thankful to DBT, Government of India for Project support, and the Sikkim State Council of Science and Technology, Gangtok.

REFERENCES

1. Andreeva, I.N., Fedorova, E.E., Il'yasova, V.B. and Tibilov, A.A. 1982. Ultrastructure of nitrogen-fixing and wintering nodules in one-year seedlings of seabuckthorn and oleaster *Hippophae rhamnoides, Elaeagnus argentea. Soviet Plant Physiol. (USA)* 29: 109-116.
2. Anonymous, 1982. *State Report on Agriculture Census, 1980-81. Vol. I.* Bureau of Economic and Statistics, Gangtok, Sikkim.
3. Auauzato, D. Megharini, R. and Lodoli, E. 1987. Studies on the germination potential of seeds and the rooting ability of cuttings of *Hippophae rhamnoides. Ministero dell' Agricoltura e delle Foreste* 1987: 411-419.
4. Basistha, B.C. 2001. Vivipary in seabuckthorn (*Hippophae salicifolia* D. Don). *Journal of Hill Research* 14 (1): 67.
5. Basistha, B.C., Adhikari, I.M. and Thapa, K.K. 2001. A case study of *Hippophae* Linn. with special references to agro-techniques in the Sikkim Himalayas. In: *Proceedings of International Workshop on Seabuckthorn*, February 18-21, 2001, New Delhi, 274p.
6. Dhakal, L. P. 2001. In: *Mountain Sciences and Ecotourism/Biodiversity.* Proceedings of the International Symposium on the Himalayan Environments, p. 64-71, 277p.
7. Donald, D.G.M. 1997. Vegetative propagation of *Pinus* using cuttings. *South Africa Hort. J.* 140: 16-23.
8. Jeppsson, N., Bartish, I.V. and Persson, H.A. 1999. DNA analysis as a tool in seabuckthorn breeding. In: *Perspective on New crops and New uses* (J. Janick, Ed.), p. 338-341, ASHS Press, 528p.
9. Nepal, V., Poudyal, D.P., Subedi, C.K., Adhikary, K., Dhakal, L.P. and Gupta, V.N. 2001. In: *Mountain Sciences and Ecotourism/Biodiversity*: Proceedings of the International Symposium on the Himalayan Environments, p.72-81, 277p.
10. Rongsen, L. 1990. Seabuckthorn resources and it's under exploited potential in the Himalayan region. In: *International Centre for Integrated Mountain Development*, MFS Series No. 12, Kathmandu, Nepal.
11. Rongsen. L. 1992. *Seabuckthorn-A Multipurpose Plant for Fragile Mountains*. ICIMOD Occasional Paper No. 20, Kathmandu, Nepal, 62p.
12. Sharma, E., Rai, L.K., Lachungpa, S.T. and Awasthi, R.P. 1995. Status of medicinal plants and their cultivation potential in Sikkim. In: *Cultivation of Medicinal Plants and Orchids in Sikkim Himalaya*, p. 43-51, G.B. Pant Institute of Himalayan Environment and Development, Dehradun, India, 139p.
13. Singh, V. 1998. Seabuckthorn-A wonder plant of dry temperate Himalayas. *Indian Horticulture* 48 (2): 6-9.
14. Singh, V., Singh, H.S. and Awasthi, C.P. 1995. Studies on distribution, taxonomy and nutritional values of seabuckthorn growing in dry temperate Himalayas. In: *Proceedings of International Workshop on Seabuckthorn*, p.52-59, Beijing.
15. Tsar'kova, T.F. 1988. The effect of physiologically active substances on rooting and growth of lignified cuttings of *Hippophae rhamnoides. Agrotekhnika Selektsiya iMekhanizatsiya v Yagodovodstve Nechernozem'ya*, p: 41-46. (In Russian).

Chapter 3

Effect of Irrigation, Cultivars and Mulch Types on Soil Nutrient Availability and Vegetative Growth of Seabuckthorn (*Hippophae rhamnoides* L.)

Carl Boivin[1], Helene Rousseau[1], Jacques-Andre Rioux[2] and Daniel Bergeron[3]

[1]Institut de recherche et de developpement en agroenvironnement (IRDA), 2700, rue Einstein, Quebec, QC, Canada, G1P 3W8
[2]Departement de phytologie, Universite Laval, Quebec, QC, G1K 7P4
[3]Ministere de l'Agriculture, des Pecheries et de l'Alimentation du Quebec (MAPAQ), 1685, boul. Wilfrid-Hamel Ouest, Quebec, QC, Canada, G1N 3Y7

SUMMARY

Soil nutrient availability, leaf nutrient status and vegetative growth of six *Hippophae rhamnoides* L. cultivars ("Askola, Hergo, Golden Rain, Mary, Sunny and Tatjana") were measured in 2004 and 2005, under combinations of two irrigation regimes and two mulch types, on Ile d'Orleans, Quebec, Canada. Irrigation was triggered by tensiometers when soil moisture tension reached either 25 kPa or 60 kPa. Black plastic and chipped tree residues were compared as mulching materials. Experimental design was a split-strip-split-plot, with irrigation in main plot, cultivar in subplot and mulch in sub-subplot. Factorial combinations were replicated three times in plots bearing ten plants each, for a total of 720 plants. Irrigation regime and cultivars had no significant effect on soil N, P, K, Mg and Ca. In year 2004, plastic mulch significantly increased N mineralization, whereas wood mulch led to significant increase in soil K content. In year 2004, foliar Mg levels were significantly greater on plastic than on wood mulch, whereas both foliar K and Ca were enhanced by wood mulch. In year 2005, wood mulch led to a significantly greater foliar K level only. In year 2004, all cultivars' increment in trunk diameter was greatest under a combination of wood mulch and soil moisture tension of 25 kPa, whereas in year 2005, this irrigation regime improved this parameter only for "Hergo and Sunny" cultivars. In year 2004, height growth of "Hergo, Sunny and Tatjana" cultivars was more important on plastic than on wood mulch, whereas "Golden Rain" grew best on wood mulch. In year 2005, all cultivars grew better on wood mulch.

Keywords: *Seabuckthorn (Hippophae rhamnoides), Irrigation, Mulch, Soil nutrient availability, Leaf nutrient status and Tensiometer.*

Seabuckthorn (Hippophae L.): *A Multipurpose Wonder Plant, Vol. 4: Emerging Trends in Research and Technologies* (Virendra Singh, *Editor-in-Chief*, 2014), p. 30–40, Daya Publishing House, New Delhi, India.

INTRODUCTION

Seabuckthorn (*Hippophae rhamnoides* L.), a plant introduced in Canada and was first used as wind-break or ornamental, is now grown commercially in orchards. In Quebec, the first such orchards were set up in the late 1990s and are mainly home to seedlings of Cultivar "Indian Summer", developed in western Canada. Only few seabuckthorn orchards currently operating in North America, an expansion of its production in Quebec requires the selection of cultivars adapted to conditions prevailing in Quebec. In addition, further knowledge must be developed with respect to managing the crop. As a result, this industry's sustainability and market competitiveness requires in a first phase, to carrying out of cultivar trials. Such trials provide an excellent means of developing our expertise in the management of the seabuckthorn crop and can lead to a greater understanding of conditions favourable to its growth.

When planted, the seabuckthorn is subject to strong competition from surrounding vegetation (Singh, 2003), such that weeds constitute the main factor limiting its growth and survival (Li, 1999). However, as this shrub is shallow-rooted (Levandovsky, 2003), therefore, mechanical weed control can be damaging. Consequently, it is recommended that the soil should not be disturbed deeper than 8 cm (Gonchar and Sadan, 1986). As no herbicide is registered in Canada for use in seabuckthorn orchards, one is expected to use mulch for weed control, a practice which has been well documented (Carter and Johnson, 1988; Skroch *et al.*, 1992; Hembry and Davies, 1994).

Water availability is another important factor in establishing an orchard. Under our climatic conditions, total annual precipitation is generally sufficient to meet a seabuckthorn crop's needs; however, its temporal distribution remains varying and unpredictable. Even if the plant's small size limits transpirational water losses, depending on climatic conditions, evaporative losses from the soil surface may become significant. As the soil volume trapped by the roots is rather limited in the first years after planting, the plant quickly suffers from drying of the topsoil. A number of cultural practices, such as mulching the soil surface (Larsson and Bath, 1996), favour an effective maintenance of soil moisture and limit water losses. As long periods of drought can endanger young seedlings, it may become necessary when water conservation practices and rainfall fail to provide additional water through irrigation.

The primary aim of this study was to determine, in support of stakeholders in the development of the production of this crop in Quebec,using the factorially combined effects of two irrigation regimes, six cultivars and two types of mulch on soil nutrient availability, mineral nutrition and vegetative growth of young seabuckthorn shrubs.

MATERIALS AND METHODS

Plant Material

The current trials involved six female seabuckthorn cultivars: Two German and four Latvian. These were chosen on the basis of their high fruit yield, potential to adapt to winter conditions in Quebec and commercial availability.

The German female cultivars "Askola and Hergo" and pollinators "Pollmix 1, 3 and 4" were developed at the Friesdorf Floriculture and Orchard Management Experiment Station, near Bonn, Germany. Two year old seedlings purchased for the study in the spring of 2003, were propagated by cuttings by the Spath'sche Baumschulen Co. (Berlin, Germany). At planting, seedlings of cultivars "Hergo and Askola" were 30 and 50 cm tall, respectively, excluding their root systems.

The originally developed cultivars "Sunny, Golden Rain, Mary, Tatjana" and the pollinator "Lord" by Dr. T. Trifimov at Moscow State University by crossing subspecies *rhamnoides*, *mongolica* and *fluvialis*, were selected by Andrejs Bruvelis and Karlis Blums after a series of backcrosses with subspecies *rhamnoides* (Bruvelis, 2003). One to five year old seedlings of these five cultivars were obtained in Newfoundland, Canada, where they had been propagated by cuttings. At planting, the seedlings were between 10 and 15 cm tall, excluding their root systems.

Irrigation Regime

The two irrigation regimes were managed on the basis of soil moisture tension, as monitored by an array of Irrometer © tensiometers. A total of 24 tensiometers (2 × 2 × 3 × 2-irrigation regime × mulch × block × Number per plot) were installed in orchard plots housing the Latvian cultivar "Sunny". Tensiometers were grouped in pairs, so as to obtain adjacent measurements at depths of 15 cm and 30 cm.

Under the first irrigation regime, good water availability was maintained throughout with irrigation being triggered when root zone soil moisture tension (15 cm depth) exceeded 25 kPa (*in situ* field capacity [(θ_{fc}) + 15 kPa]). Under the second irrigation regime, more limited water availability conditions prevailed with irrigation only being triggered when soil moisture tension exceeded 60 kPa. When readings from four of six tensiometers equalled or exceeded the threshold soil moisture tension, irrigation was applied for a period of 90 minutes allowing the soil moisture tension to return to near zero (~0 kPa). Irrigation water was applied through a drip irrigation system installed along the base of the stems. This system was installed beneath the mulch prior to the seabuckthorn planting and was spaced 20.3 cm apart, the emitters could provide irrigation at a rate of 1l h^{-1} at an operating pressure of 103 kPa.

Mulches

Two types of mulch were investigated: Forestry-grade black plastic mulch and 0.15 m thick wood chip mulch. Both mulches extended 0.45 m to either side of the base of plants. The wood chip mulch originated from branches and twigs of several deciduous tree species (*Acer* spp., *Sorbus* spp. and *Fraxinus* spp.)

Experimental Set up and Monitoring

The experimental orchard was set up near Saint-Laurent, Ile d'Orleans, Quebec (46° 52′ N latitude, 70° 54′ W longitude), in plant hardiness zone 4b. The mean annual precipitation was 1100 mm that falling between November and April largely in the form of snow. The mean monthly precipitations May to September were 118, 108, 129, 114 and 127 mm, respectively. The Monthly mean of daily maximum and minimum air temperatures are shown in Figure 3.1 (30 years, 1971-2000).

The planting density was 2,184 plants ha^{-1} and row orientation was north-south. Throughout the period of study, vegetative cover was maintained between the mulched rows and this was mown on a regular basis during the growing season. Both mulches provided adequate weed suppression throughout the period of study. No fertilizer and pesticide applications were made, as in Quebec no such products are registered for use on this crop.

In order to raise its soil pH to near neutrality in October 2002 and 2003, the sandy clay loamy site was amended with 2.2 mg ha^{-1} and 6.7 mg ha^{-1} of lime, respectively. Preliminary soil sampling was done in May 2004 to establish initial field fertility levels (Table 3.1). Nutrient analysis showed the soil to be low in P, moderately rich in K and Mg, and adequate in Ca content (CRAAQ, 2003).

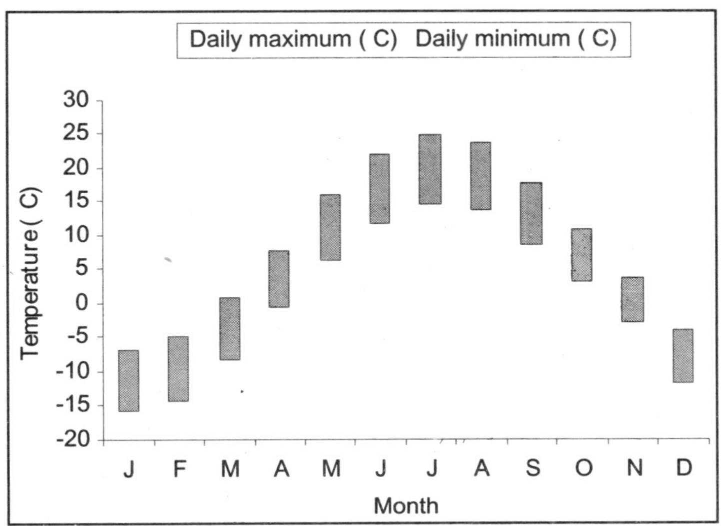

Figure 3.1: Monthly Mean of Daily Maximum and Minimum Air Temperatures.

Table 3.1: Content of Major Soil Nutrients (mg/kg) at the Saint-Laurent, Quebec Orchard, Prior to Seabuckthorn Planting in Spring Season 2004

NH_4^+	NO_3^-	P(mg/kg)	K	Mg	Ca
1.26	8.69	53	209	133	4223

No pruning was done at planting. In the spring of 2004, prior to bud-break, all female plants were pruned for shape, so as to favour a dominant apex. In the spring of 2005, a light pruning was done to eliminate smaller branches at the base of the trunk. For the plants of Latvian cultivars, which were younger than those of the German cultivars, an apex was chosen and two lateral branches were maintained. For the German cultivars, a dominant apex was selected. The plants showed no winter freeze damage, even though the German cultivars aouting was rather later.

Three factors were investigated in this study. Main plots were split into two sub-plots, according to their irrigation regime, and further into four sub-sub-plots according to the type of mulch applied (Figure 3.2). The randomization of the six cultivars was constrained by the fact that these were planted in single rows across the breadth of the main plots. Such a design was justified, given the orchard's commercial avocation. Indeed, since the fruits of different cultivars ripen at different times, harvesting is much simplified if there is only one cultivar per row. The randomization of mulch types in the sub-subplots suffered from no such constraints. The experimental design was thus, a *split-strip-split-plot* with repeated measures, the strip representing the cultivar rows within the main plots. The layout included 72 experimental units, each housing 10 seabuckthorn for a total of 720 plants. To these were added one guard row each between blocks A and B, and between blocks B and C. These guard rows were made up of 160 pollinator plants, including cultivars "Pollmix 1, 3 and 4 as well as Lord".

Plant Growth and Nutrition Variables Measured

In both years of the experiment, each shrub height defined as the tallest point on the stem, and stem diameter 10 cm above the soil surface were measured in the spring prior to budbreak and in the

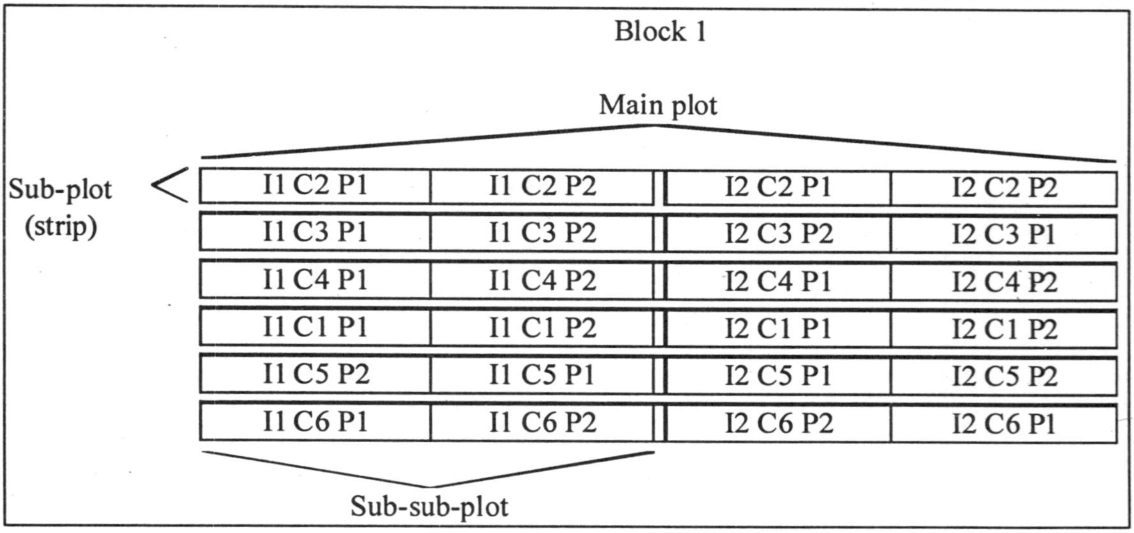

Block 1

Main plot

Sub-plot (strip)

I1 C2 P1	I1 C2 P2	I2 C2 P1	I2 C2 P2
I1 C3 P1	I1 C3 P2	I2 C3 P2	I2 C3 P1
I1 C4 P1	I1 C4 P2	I2 C4 P1	I2 C4 P2
I1 C1 P1	I1 C1 P2	I2 C1 P1	I2 C1 P2
I1 C5 P2	I1 C5 P1	I2 C5 P1	I2 C5 P2
I1 C6 P1	I1 C6 P2	I2 C6 P2	I2 C6 P1

Sub-sub-plot

Figure 3.2: Experimental Design (*Split-strip-split-plot*).

Main plots: Two irrigation regimes, one triggered at 25 kPa of soil-moisture tension (I1) and other triggered at 60 kPa (I2). Sub-plots: Six cultivars, namely "Askola (C1), Golden Rain (C2), Hergo (C3), Mary (C4), Sunny (C5) and Tatjana"(C6). Sub-sub-plots: Plastic mulch (P1) or wood chips mulch (P2).

fall after growth cessation. The increment in diameter or height was calculated as the difference between the spring and fall measurements. In both years, levels of soil minerals (N-NH_4^+, N-NO_3^-, P, K, Ca and Mg) were measured in composite samples drawn at two week intervals, from the top 15 cm of soil of each of the 72 experimental units. Six such samplings were completed in 2004 and four in 2005.

Similarly, leaf mineral (N, P, K, Ca and Mg) levels were measured in composite samples drawn from all 72 experimental units. Two foliar samplings were completed in each year: July 19 and September 4 in 2004, July 18 and August 29 in 2005. These dates correspond to a period of active plant growth and to one when growth has ended, but nutrients have yet to be translocated to the roots.

Statistical Analysis

Statistical analysis employed the PROC MIXED procedure in SAS (Littell *et al.*, 1996).

RESULTS AND DISCUSSION

Annual differences in weather conditions over 2-year period study resulted in the proportion of soil water contribution by precipitation differing significantly from year to year. In 2004, water inputs from rainfall were regular and sufficient to maintain soil moisture levels above irrigation thresholds until July 27, when the season's first 25 kPa threshold-triggered irrigation occurred. The 60 kPa threshold was never reached in 2004, so this treatment received no irrigation over the entire season. Early in 2005 and throughout the season, irrigation was necessary to maintain soil moisture levels above irrigation thresholds. For the period of June through August 2005, 60 days with no or less than 2 mm of rain were recorded.

Soil Nutrients

In 2004 season, NO_3^--N and K levels differed significantly between the two mulch types, but those of P, Mg and Ca did not. At all 6 sampling dates in 2004, significantly greater ($p \leq 0.001$) soil NO_3^--N levels were found under black plastic mulch than under wood chips mulch (Figure 3.3).

The lower soil NO_3^--N levels observed under the wood chips (*vs.* plastic) mulch suggested less effective nitrification as reported by Halvorson (2002). Such a drop in the nitrification rate could explained the significantly higher levels of soil NH_4^+-N measured under wood chips (*vs.* plastic) mulch (Figure 3.4) in late June (T2; $p \leq 0.001$), mid-July (T3; $p \leq 0.006$), mid-August (T5; $p \leq 0.005$) and late August (T6; $p \leq 0.002$).

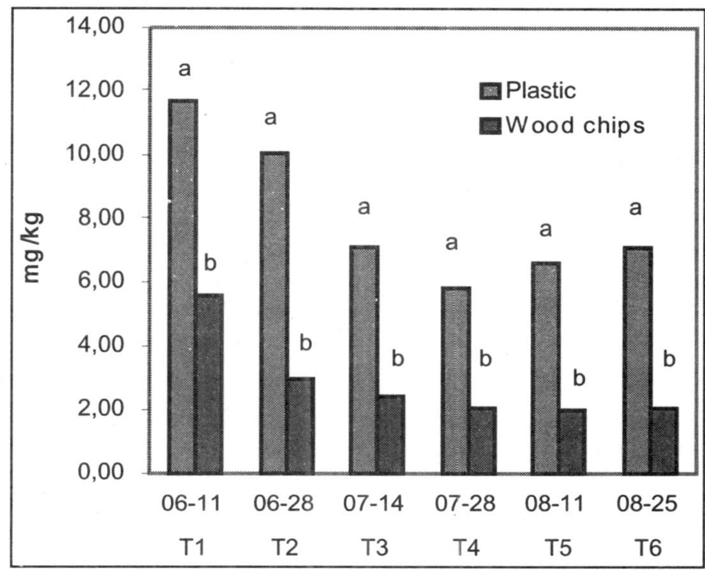

Figure 3.3: Seabuckthorn Orchard Topsoil (0-15 cm) NO_3^--N Levels (mg kg^{-1}) as a Function of the Interaction between Mulch Type and Sampling Date (P×T), for the 2004 Season.

For all 2004 sampling dates (T1-T6), soil K levels were significantly greater ($p \leq 0.001$) under wood chips (*vs.* plastic) mulch (Figure 3.5). The rise in soil K over the season can be attributed to the leaching of this element from the wood chips (Fraedrich and Ham, 1982; Merwin *et al.*, 1995).

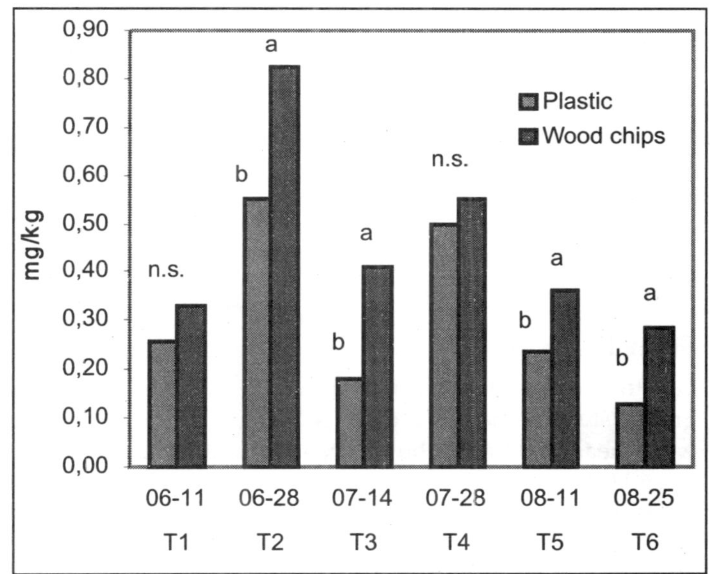

Figure 3.4: Seabuckthorn Orchard Topsoil (0-15 cm) NH_4^+-N Levels (mg kg^{-1}) as a Function of the Interaction between Mulch Type and Sampling Date (P × T), for the 2004 Season.

Mineral Nutrition

Mulch type also clearly influenced leaf mineral uptake; indeed in 2004, leaves from wood chips mulched seabuckthorn bore more K than their plastic mulched counterparts (Figure 3.6), confirming the results of Merwin *et al.* (1995). Leaf K and Ca levels were significantly ($p \leq 0.001$) enhanced by wood chips (vs. plastic) mulch, while the converse was the case for leaf Mg ($p \leq 0.04$).

In year 2005, withstanding mulch type was not significantly influencing top soil K levels. K levels in the leaves of seabuckthorn plants grown on wood chips mulch were once again greater than those

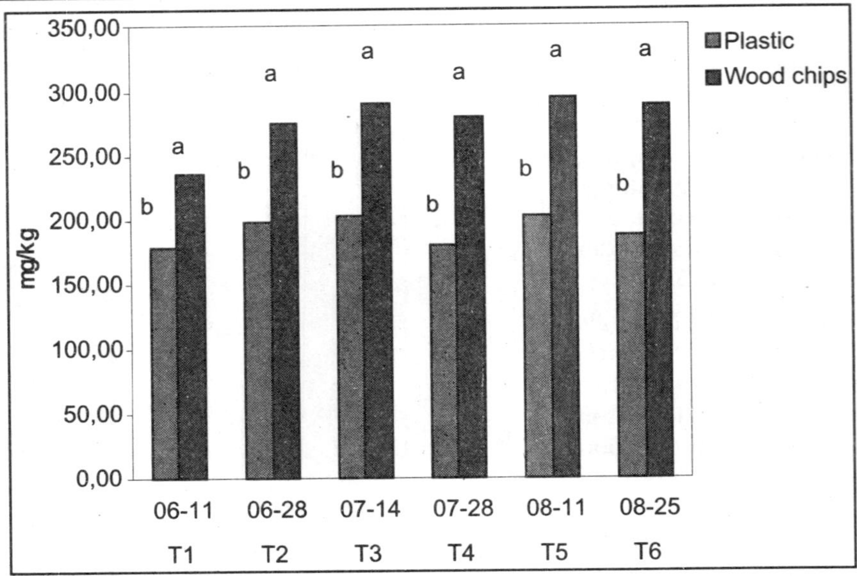

Figure 3.5: Seabuckthorn Orchard Topsoil (0-15 cm) K Levels (mg kg^{-1}) as a Function of the Interaction between Mulch Type and Sampling Date (P × T), for the 2004 Season.

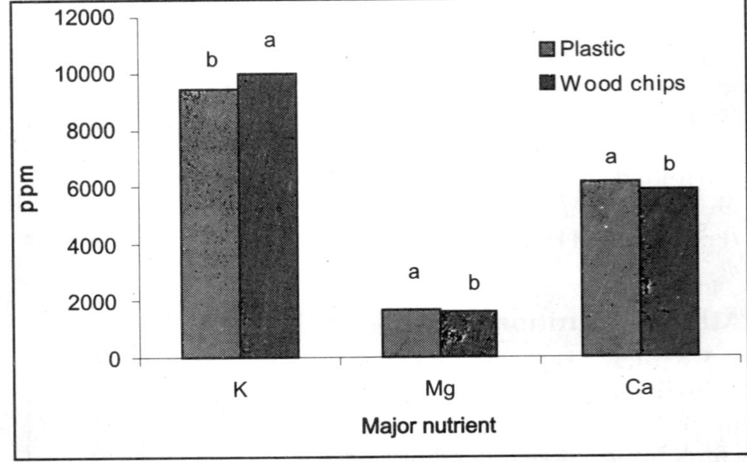

Figure 3.6: K, Mg, and Ca Levels (ppm on dry weight basis) in Seabuckthorn Leaves as a Function of Mulch Type for the 2004 Season.

grown on plastic mulch (Figure 3.7). In this regard, soil moisture tension measurements showed a quicker drying out of soil under plastic (*vs.* wood chips) mulch. Given that transport of K through the soil occurs by mass flow (Marschner, 1986), greater soil moisture content would tend to enhance plant K uptake.

Stem Diameter

In year 2004, significantly greater stem diameter increments were measured under the 25 kPa (*vs.* 60 kPa) soil moisture tension irrigation threshold, but mulch type only showed a significant effect within the 25 kPa irrigation regime, where diameter increased more with wood chips than plastic mulch ($p \leq 0.001$) (Figure 3.8). By allowing better nutrient uptake (Marschner, 1986; Patterson *et al.*, 1990; Bignami *et al.*, 2000), the elevated moisture conditions under the wood chips (*vs.* plastic) mulch may have enhanced radial growth of seabuckthorn plant stems.

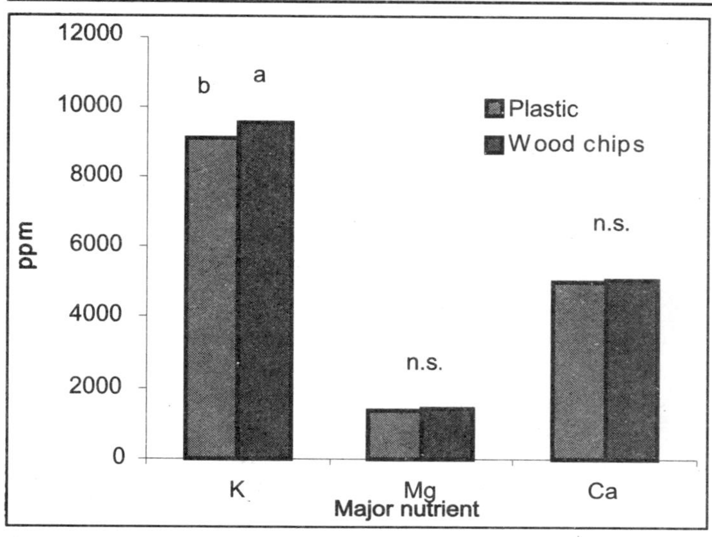

Figure 3.7: K, Mg, and Ca Levels (ppm on dry weight basis) in Seabuckthorn Leaves as a Function of Mulch Type for the Season 2005.

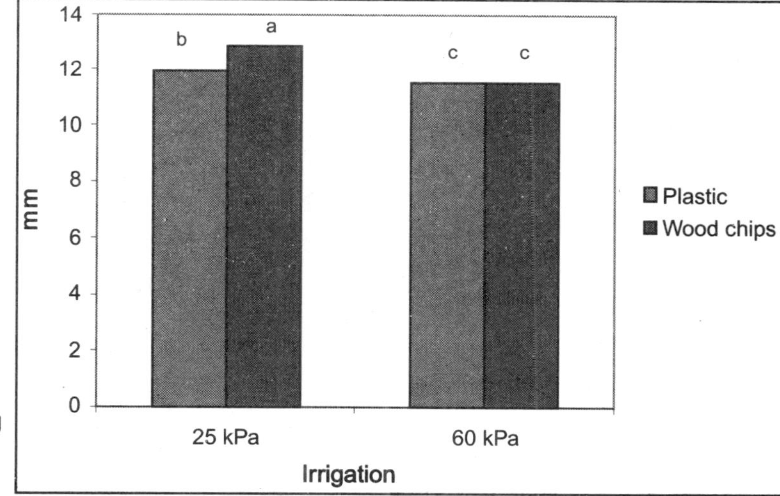

Figure 3.8: Increase in Seabuckthorn Stem Diameter (mm) as a Function of Interacting Irrigation × Mulching Treatments (I × P) for the 2004 Season.

In year 2005, cultivars were differentially affected by the irrigation regime; indeed, the increment in stem diameter of cultivars "Hergo and Sunny" was significantly greater under the 25 kPa (*vs.* 60 kPa) soil moisture tension threshold irrigation regime ($p \leq 0.04$, $p \leq 0.02$, respectively), whereas the other cultivars showed no such difference (Figure 3.9).

Plant Height

In year 2004, the increment in height of seabuckthorn plants of cultivars "Hergo, Sunny and Tatjana" was significantly greater ($p \leq 0.03$, $p \leq 0.002$, and $p \leq 0.02$, respectively) on plastic (*vs.* wood chips) mulch. These growth differences were of 6.9 cm, 8.9 cm and 6.2 cm, respectively. The growth of seabuckthorn, a symbiotic nitrogen-fixing plant, may have been enhanced by the warmer soil temperatures under the plastic mulch (Gupta and Singh, 2003). However, cultivar "Golden Rain" grew better on the wood chips (*vs.* plastic) mulch ($p \leq 0.03$) (Figure 3.10).

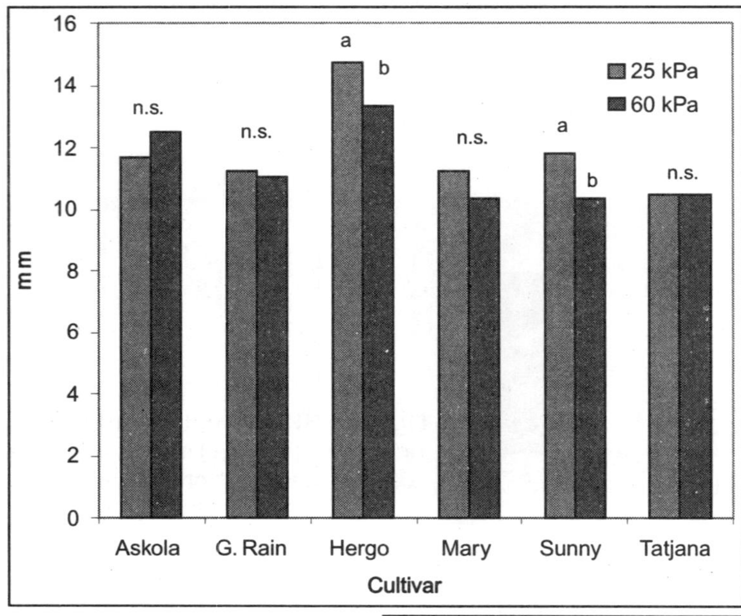

Figure 3.9: Increase in Seabuckthorn Stem Diameter (mm) 10 cm from the Soil Surface as a Function of Interacting Irrigation × Cultivar Treatments (I × C), for the Season 2005.

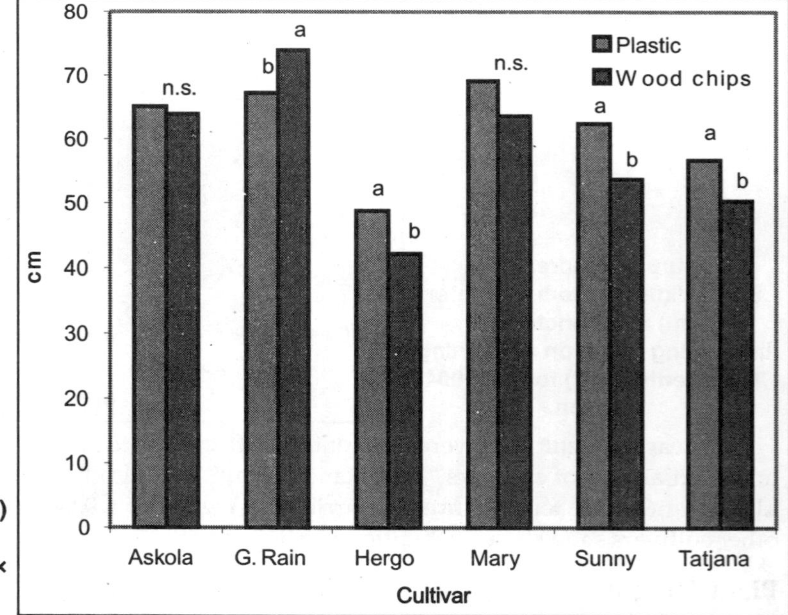

Figure 3.10: Increase in Seabuckthorn Plant Height (cm) as a Function of Interacting Cultivar × Mulch Treatments (I × C), for the Season 2004.

In the drier year (2005), the mean increments in plant height across cultivars were 57.5 and 61.9 cm on plastic and wood chips mulches, respectively (Table 3.2). The improved soil moisture conditions under the wood chips mulch, and the resulting enhancement in plant nutrient uptake, may account for the greater increment in height with this type of mulch. However, it is difficult to attribute these results to a single specific factor. Results obtained in other studies were varied and the explanations postulated by their authors remain similarly unspecific. Larsson *et al.* (1997) showed lesser black currant bush growth with wood chips (*vs.* plastic) mulch, whereas Merwin *et al.* (1995) and Foshee *et al.* (1999) showed the contrary results.

Table 3.2: Increment in Height (cm) of Seabuckthorn Plants Mulched with Wood Chips *vs.* Black Plastic for the Growing Seasons 2004 and 2005

Mulch (P)	*2004*	*2005*
Plastic	61.6	57.5
Wood chips	58.0	61.9
F value	**	***

Note: **, *** Differences between mulches in a given year, significant at 0.01 or 0.001, respectively.

CONCLUSION

The use of irrigation to supplement natural rainfall and of mulch to control weeds are the effective method in the successful establishment of seabuckthorn orchards. While such interventions clearly modify the plants' immediate environment, complex plant-soil interactions, and the present study has only addressed a limited number of the issues at hand, namely the impact of two irrigation regimes and two types of mulching material on the availability of soil nutrients, their uptake by seabuckthorn plants, and the latter's vegetative growth.

The two irrigation regimes differentially affected increments in stem diameter and height of young seabuckthorn plants, whereas the different mulch types differentially affected mineral nutrition and growth of these plants. However, both the seabuckthorn cultivar and the meteorological conditions influenced treatment effects under which it grew.

ACKNOWLEDGEMENT

The project was undertaken in collaboration with la Ferme Maurice et Philippe Vaillancourt. The authors thank Drs. Adrien N'Dayegamiye and Mohammed Lamhamedi as well as Mrs. Marie-Pierre Lamy and Mr. Marcel Giroux for their collaboration. Mrs. Michele Grenier is thanked for her help in the statistical analysis of the results. The project was funded by the Institut de recherche et de developpement en agroenvironnement (IRDA), Economic Development Canada (EDC), and l'Association des producteurs d'argousiers du Québec (APAQ).

REFERENCES

1. Bignami, C., Cammilli, C., Moretti, G. and Romoli, F. 2000. Irrigation of *Corylus avellana* L.: Effects on canopy development and production of young plants. *Acta Horiculturae* 537: 903-910.
2. Bruvelis, A. 2003. Seabuckthorn cultivation in Baltic States. In: *Proceeding of the 1st Congress of the International Seabuckthorn Association* (Morseland Thies, Ed.), p. 64-66, Berlin, Germany, 214p.
3. Carter, J. and Johnson, C. 1988. Influence of different types of mulches on eggplant production. *HortScience* 23: 143-145.
4. Centre de reference en agriculture et agroalimentaire du Quebec (CRAAQ), 2003. *Guide de reference en fertilisation* (1st Ed.) CRAAQ Ste-Foy, Quebec, 298 p.

5. Foshee, W.G., Goff, W.D., Patterson, M.G., Tilt, K.M., Dozier Jr, W.A., Tucker, L.S. and Bannon, J.S. 1999. Organic mulches affect soil and leaf nutrient levels of young pecan trees. *Journal of Arboriculture* 25 (2): 81-84.

6. Fraedrich, S.W. and Ham, D.L. 1982. Wood chip mulching around maples: Effect on tree growth and soil characteristics. *Journal of Arboriculture* 8: 85-89.

7. Gonchar, M.T. and Sadan, B.A. 1986. Choice of tree and shrub species for afforestation of sites after opencast mining of sulphur. *Lesnoe-Khozyaistvo* 9: 47-49.

8. Gupta, R.K. and Singh, V. 2003. Nitrogen fixation in seabuckthorn (*Hippophae rhamnoides* L.). In: *Seabuckthorn (Hippophae L.)-A Multipurpose Wonder Plant. Vol. I: Botany, Harvesting and Processing Technologies* (V. Singh, Ed. in Chief, 2003), p. 286-299, Indus Publishing Company, New Delhi, 518p.

9. Halvorson, A.D. 2002. Nutrient-water interactions. In: *Encyclopedia of Soil Science* (R. Lal, Ed.), p. 888-890, Marcel Dekker Inc., New York, Basel, 1476p.

10. Hembry, J.K. and Davies, J.S. 1994. Using mulches for weed control and preventing leaching of nitrogen fertiliser. *Acta Horticulturae* 371: 311-315.

11. Larsson, L. and Bath, A. 1996. Evaluation of soil temperature moderating and moisture conserving effects of various mulches during a growing season. *Acta agriculturae scandinavica, Soil and plant sciences* 46: 153-160.

12. Larsson, L., Stenberg, B. and Torstensson, L. 1997. Effects of mulching and cover cropping on soil microbial parameters in the organic growing of black currant. *Communication in Soil Science and Plant Analysis* 28: 913-925.

13. Levandovsky, V.N. 2003. Application of herbicides in seabuckthorn nurseries and plantations. In: *Seabuckthorn (Hippophae L.)-A Multipurpose Wonder Plant. Vol. I: Botany, Harvesting and Processing Technologies* (V. Singh, Ed. in Chief, 2003), p. 365-368, Indus Publishing Company, New Delhi, 518p.

14. Li, T.S.C. 1999. Seabuckthorn: New crop opportunity. In: *Perspectives on New Crops and New Uses* (J. Janick, Ed.), p. 335-337, ASHS Press, Alexandria, VA, 528p.

15. Littell, R.C., Milliken, G.A., Stroup, W.W. and Wolfinger, R.D. 1996. SAS[a] *System for Mixed Models*, Cary, NC: SAS Institute Inc, 636p.

16. Marschner, H. 1986. *Mineral Nutrition of Higher Plants*, Academic Press, London, 674p.

17. Merwin, I.A., Rosenberger, D.A., Engle, C.A., Rist, D.L. and Fargione, M. 1995. Comparing mulches, herbicides, and cultivation as orchard groundcover management systems. *HortTechnology* 5 (2): 151-158.

18. Patterson, M.G., Wehtje, G. and Goff, W.D. 1990. Effects of weed control and irrigation on the growth of young pecans. *Weed Technology* 4: 892-894.

19. Singh, V. 2003. Plantation and management technologies of seabuckthorn. In: *Seabuckthorn (Hippophae L.)-A Multipurpose Wonder Plant. Vol. I: Botany, Harvesting and Processing Technologies* (V. Singh, Ed. in Chief, 2003), p.352-359, Indus Publishing Company, New Delhi, 518p.

20. Skroch, W.A., Powell, M.A., Bilderback, T.E. and Henry, P.H. 1992. Mulches: Durability, aesthetic value, weed control, and temperature. *Journal of Environmental Horticulture* 10 (1): 43-45.

Chapter 4

Effect of Fertilizers, Mulch and Land Contours on Young Seabuckthorn Cultivars in Organic Farming

Merja Heinaaho

Faculty of Biosciences, Natural Product Research Laboratories, University of Joensuu, Post Office Box 111, Joensuu FIN-80101, Finland

SUMMARY

The effects of different organic farming methods on leaf and berry phenolic contents, growth and berry production of two Finnish seabuckthorn (*Hippophae rhamnoides* ssp. *rhamnoides*) cultivars, "Terhi" and "Tytti", were studied in a field experiment for four years in Merikarvia, Western Finland and two experiments were conducted. First, the fertilization experiment included the fully-crossed effects of four different fertilizers and two land contours on the cultivars, "Terhi" and "Tytti". The other, the mulch experiment included the fully-crossed effects of five different mulches and two land contours on the cultivar "Tytti".

During the first growing season, cultivar "Terhi" formed longer shoots than "Tytti". In the second growing season, the cultivars increased their shoot number, and "Terhi" grew thicker than "Tytti". Ridged beds increased shoot length in both growing seasons. Plants grown with plastic mulch had more shoots as compared to other mulch treatments. Bushes started to produce a small number of berries in the second growing season. The yield increased from year to year, and was highest in the last growing season, 2007. Cultivar "Terhi" produced bigger berries and larger yields than "Tytti". Ridged beds increased the yield from "Terhi".

The number of phenolic compounds in leaves varied between the cultivars. Altogether, 23 phenolic compounds were detected in the cultivar "Terhi" and 24 compounds in "Tytti". The amount of hydrolyzable tannins 1, 2, 4, 5 and isorhamnetin-3-glucoside was higher in "Terhi" than in "Tytti". On the other hand, there were more hydrolyzable tannins 7, 10, 11 and pentagalloylglucose in "Tytti" than in "Terhi". The concentrations of gallic acid, pentagalloylglucose, quercetin 3-rhamnoside, monocoumaroyl astragalin A, total hydrolyzable tannins and condensed tannins were higher on flat land than on the ridged bed. The use of plastic mulch led to a decrease in the concentration of gallic acid, hydrolyzable tannins and condensed tannins as compared to the other mulch used.

---

*Seabuckthorn (*Hippophae* L.): A Multipurpose Wonder Plant, Vol. 4: Emerging Trends in Research and Technologies* (Virendra Singh, *Editor-in-Chief*, 2014), p. 41–58, Daya Publishing House, New Delhi, India.

The concentrations of quercetin derivatives, such as 1, 3-isorhamnetin 3,7-diglucoside, quercetin 3-glucoside 7-rhamnoside, quercetin 3-glucoside, isorhamnetin 3-glucoside and flavonoid derivative 3 were higher in "Tytti" berries than in "Terhi" berries. On the other hand, the concentrations of isorhamnetin-glucosides 2 and 3 were higher in "Terhi" than in "Tytti". Growing on flat land increased concentrations of quercetin derivatives such as 2,4-isorhamnetin-3,7-diglucoside, and isorhamnetin-glucoside 1 and of condensed tannins.

The results of this study indicate that the selection of cultivation methods may strongly modify the growth, yield and phenolic composition of seabuckthorn and consequently, its economic importance for farmers.

Keywords: *Seabuckthorn, Cultivars, Fertilizers, Mulches, Land contours, Organic farming and Yield.*

INTRODUCTION

Background

Seabuckthorn (*Hippophae rhamnoides* L.) is a unique and valuable plant species, which is interesting as a crop for the farmers, consumers, health food markets and food processing industry (Harrison and Beveridge, 2002). Seabuckthorn domestication started in Siberia in the early 1930s by selection and breeding (Kalinina and Panteleyeva, 1987). Since the 1950s, the biochemical analysis of leaves, bark, berries and oil increased the amount of information available on its nutritional and medicinal value, and the potential for its multiple uses. In view of these promising results, many countries, such as Mongolia, Hungary, Romania and East Germany started seabuckthorn breeding and cultivation in the 1960s. Later, the interest in seabuckthorn research and cultivation was also spread to Canada, China, Japan and various European countries, such as Estonia and Finland (Yao, 1994; Li and Schroeder, 1996; Piir, 1996; Harrison and Beveridge, 2002; Yang and Kallio, 2002).

In Finland, seabuckthorn breeding has been carried out by the Institute of Horticulture in Piikkio since the 1960s. Finnish farmers have mainly used Russian cultivars in their cultivations. However, they have been sensitive to fungal diseases and the changing Finnish climatic conditions during winter (Prokkola, 2003). The first, totally Finnish origin cultivars were bred using X-ray on seeds of wild seabuckthorn (*Hippophae rhamnoides* ssp. *rhamnoides*) originating in the Baltic Sea region in Finland. The females "Terhi" and "Tytti" and a male "Tarmo" were released in 2000. These cultivars fulfill important breeding expectations; they adapt to Finnish climatic conditions and they are resistant to the worst fungal disease (*e.g. Stigmina* sp.) affecting the seabuckthorn in Finland. In addition, their vitamin C concentration is higher and berry size bigger as compared to previous cultivars bred in Finland (Karhu and Prokkola, 2004). Their harvesting time is long, spread over one month. In addition, females reach two meters in height, which make them easy to harvest (Karhu, 2000).These characteristics increase their reliability in berry production.

Most of the farmers cultivate seabuckthorn conventionally (non-organically). However, with consumer´s increased environmental awareness in the 1990s, interest in organically farmed seabuckthorn grew rapidly. The organic farming area increased from 0.37 ha (in 1995) to 50 ha (in 2004) in Finland. Since 2004, the cultivation area has declined slightly and was 39 ha (in 2007), making up 19 per cent of seabuckthorn cultivations in Finland (Ministry of Agriculture and Forestry, 2008). Organic seabuckthorn cultivation is also carried on Germany, Estonia and Russia (personal communication by Thomas Morsel, Toivo Univer and Yury Zubarev).

The basic idea of farming is understood differently in organic and conventional farming. The major challenge for organic farming is to take care of long-term ecological health, such as biodiversity, activity of microbial populations and soil quality, rather than to achieve short-term productivity gains (IFOAM, 2006; Srivastava *et al.*, 2007). In conventional farming, on the other hand, short-term productivity gain forms the major challenge. Organic farming excludes the use of synthetic fertilizers, pesticides and herbicides (IFOAM, 2006). However, a number of synthetic chemicals have been used to increase the growth of seabuckthorn and decrease the growth of weeds in conventional seabuckthorn plantation (Singh, 2003). Generally, the yields in organic farming have been found to be 10-40 per cent lower than in conventional farming (Mader *et al.*, 2002; Avery *et al.*, 2005). However, the lower cost of fertilizer, energy and pesticides together with the premiums paid by consumers for organic produce, help to create higher profits for organic farmers (Mader *et al.*, 2002).

Seabuckthorn (*Hippophae rhamnoides* L.) is a particularly suitable plant for organic farming because, like other members of the family *Elaeagnaceae*, it is a symbiotically nitrogen fixer (*Frankia* sp.), which enables it to grow not only in barren soil, but also to improve soil fertility (Rousi, 1971; Yang, 2001; Berry *et al.*, 2002; Khabarov, 2003). The seabuckthorn's nitrogen fixing capability has been found to be around 72-120 kg/ha (Khabarov, 2003). This capability reduces the need for nitrogen fertilization and thereby nutrient leaching close to the farm. However, various slowly soluble mineral powders that are used in organic cultivations, such as rock phosphates, potassium and limestone are also recommended for seabuckthorn (Li, 2008). Other agro-technical constructions for organic cultivation, such as airy, well-drained soil are essential for its fast developing root system (Singh, 2003). Because of this extensive root system, it is usable plant for soil conservation and erosion prevention (Singh and Gupta, 2008). In addition, seabuckthorn has very few pests in Finland and there are no chemicals registered for pest control, but organic control measures are available for insect control (Li, 2008). The major problem during organic cultivation of seabuckthorn is weed control, especially for a newly planted field, since seabuckthorn is sensitive to shade and is easily outcompeted in a few years (Rousi, 1965; Rousi, 1971; Li, 2008).

Phenolic Compounds of Seabuckthorn

The phenolic compounds of seabuckthorn can be divided into two main groups: Non-tannic and tannic phenolics. The flavonoids comprised the biggest group of non-tannic phenolics (Keskitalo *et al.*, 2001). Their subgroup, flavonols (kaempherols, quercetins, myricetins, isorhamnetins), occur widely in seabuckthorn (Figure 4.1) (Krolikowska, 1972; Tolkachev and Sheichenko, 2006; Tsybikova *et al.*, 2006).

Most of the flavonoids are in the form of water soluble glycosides, where one or more hydroxyl groups of aglycones bind to a sugar. Hydrolyzable and condensed tannins represent tannic phenolics, which are formed of flavonoids and phenolic acids. They differ from non-tannic phenolics in that their molecular weight is higher. Hydrolyzable tannins are either gallotannins or ellagitannis, depending on their phenolic acids. Condensed tannins (proanthocyanidins) are oligomers of catechins (Keskitalo *et al.*, 2001; Puupponen-Pimia *et al.*, 2005). Phenolic compounds are synthesized along with normal plant growth, and their concentration changes gradually. However, abiotic (poor main nutrients, low temperature and dryness) and biotic factors (insects, pathogens, herbivores) can cause sudden changes in the activity of metabolic pathways (glycolytic, pentose phosphate, shikimic acid, phenylpropanoid and mevalonate) (Lawanson *et al.*, 1972; Bell and Charlwood, 1980; Tan, 1980; Dixon and Paiva, 1995; Geetha *et al.*, 2008) (Figure 4.2). Synthesis of these defensive compounds is costly in terms of energy allocation (Gershenzon, 1994). Various models explain the variation in plant responses to resource availability. Three of the hypotheses, *i.e.* carbon/nutrient balance (CNB) (Bryant *et al.*, 1983), growth differentiation balance (GDB) (Herms and Mattson, 1992), and protein competition

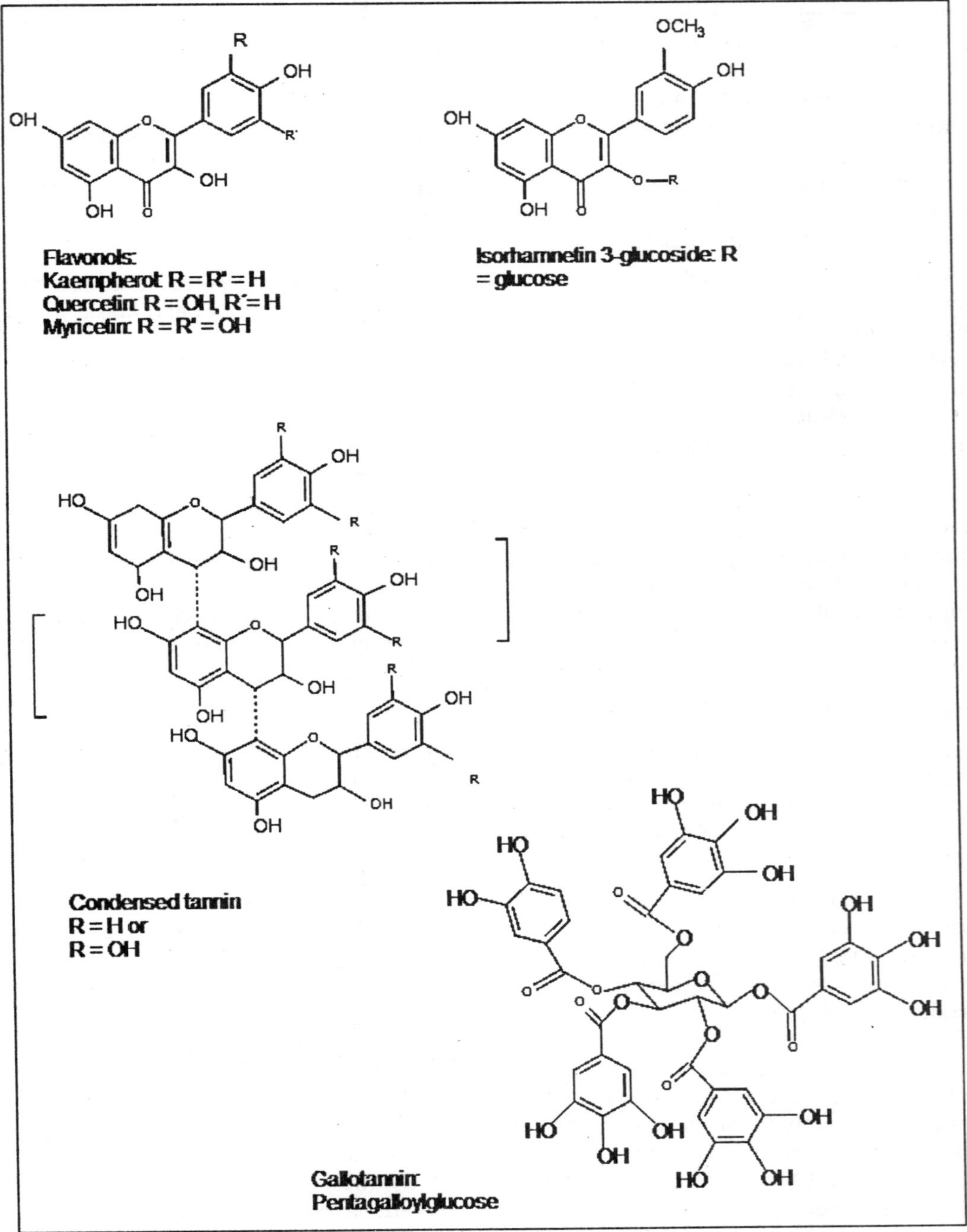

Figure 4.1: The Basic Structures of Phenolics Detected in Leaves and Berries of Finnish Seabuckthorn Cultivars.

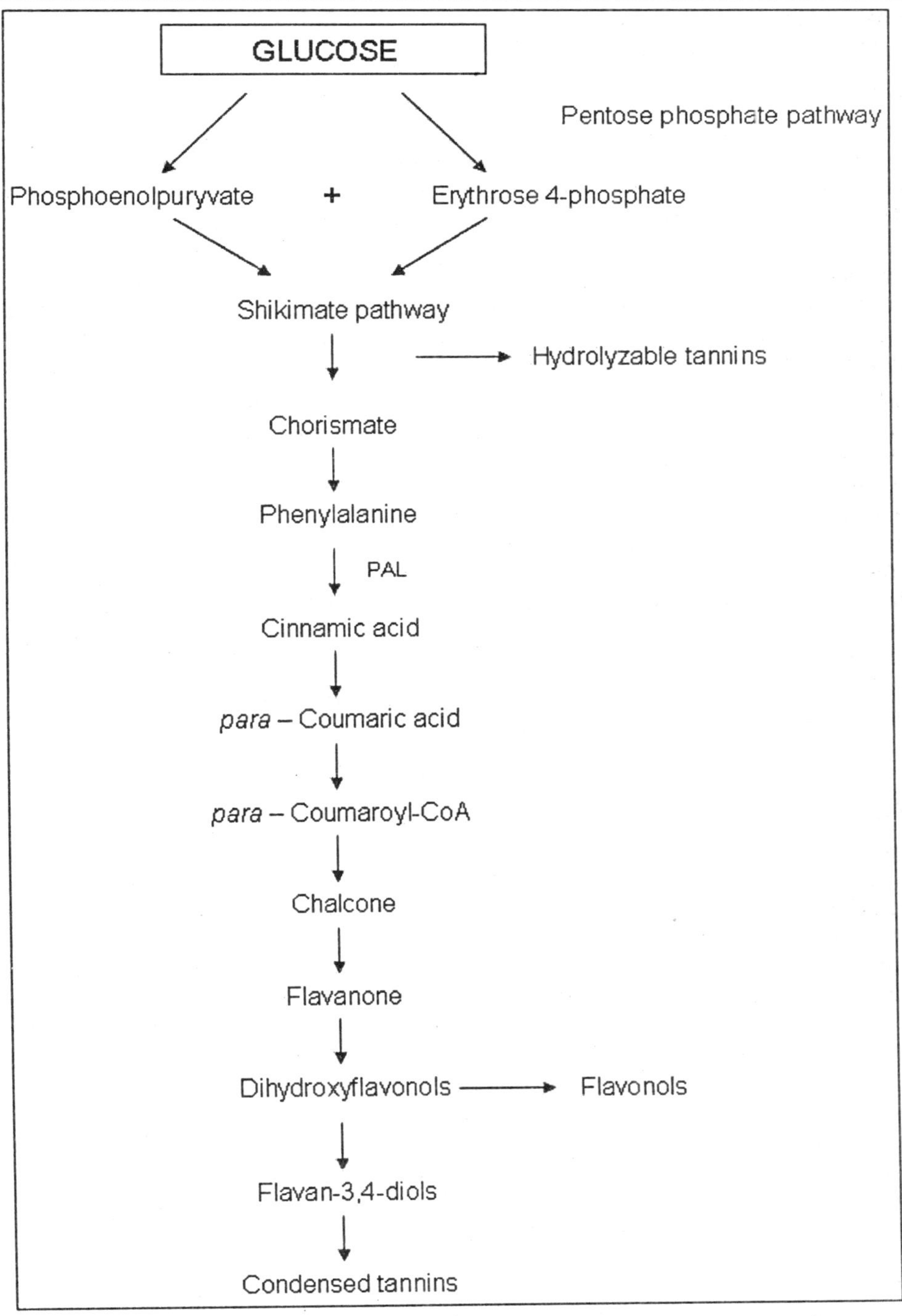

Figure 4.2: Biosynthesis of Phenolic Compounds.

(PCM) (Jones and Hartley, 1999), predict that fertilization with nitrogen should reduce the concentration of carbon-based secondary metabolites, such as phenolic compounds.

Purpose of the Study

During the past 80 years, since the domestication of seabuckthorn started, it has turned out to be one of the most favourable berry plants in Asia and Europe. Researchers worldwide have bred multiple cultivars of this plant to satisfy the basic requirements of industry. Agro-technical methods have increased the efficiency of berry production, the quality of the fruits and the processing of raw material. In this way, a solid information base has been created for the global spread of this plant (Zubarev, 2008). However, research on organic farming methods for cultivating seabuckthorn has been minor or non-existent during these years. In this study, the different organic farming methods were studied in a field experiment in order to increase knowledge of reliable cultivation methods for Finnish seabuckthorn cultivars. Now, their amount is less than 10 per cent of the cultivars grown in Finland.

The aims of this study were to test the effects of selected organic farming methods and different land contours and their combinations on the phenolic composition of seabuckthorn leaves and berries, on growth and berry yield. The effects of combinations of cultivation methods were tested because they were expected to be effective in the organic farming of seabuckthorn. The different farming methods were carried out and comprised the use of conifer chips, dry grass, plastic and straw as mulch, and apatite, bioapatite and Biolan as fertilizer, flat land and ridged beds as land contours. Also, studies showed the possible difference in response between the two Finnish cultivars to the combinations of the fertilizers and land contours. Low-nitrogen fertilizers were chosen to test their effects on this plant as this species exhibits a symbiosis with nitrogen-fixing bacteria. Mulch was chosen to test their efficacy for weed control. In addition, the mulches used are easily available and affordable for farmers. The potential effects of these variables, together with planting beds (land contours), on seabuckthorn phenolics, growth and yield have not been studied earlier.

MATERIALS AND METHODS

Study Site, Experimental Design and Plant Material

The seabuckthorn field study was established in a coastal area in Merikarvia, western Finland (61°52´N, 21°30´E) in autumn 2003. Two experiments were conducted in a flat (90 m × 90 m) field. The field had not been farmed since 25 years and it was growing as small deciduous trees (*Salix* sp., *Alnus incana*), which were removed. The top 30-40 cm of the soil was a 1:2 (v/v) mix of humus and fine sand. The readily available levels of P and K were 2.4 mg^{-1} and 152 mg^{-1}, respectively (in 2003). The soil pH was 5.5 and soil was to approximately 20 cm deep before the land contours (planting beds) were established. Ridged beds were raised or the land contour treatment using an excavator. The ridged beds were approximately 30 cm high and 100 cm wide. For the flat land, 0.5 m^2 plot was tilled to a depth of 20 cm for each sapling during planting. Mean day temperature of the growing season from May to September was 13.3 °C in 2004, 13.6 °C in 2005, 14.9 °C in 2006 and 13.8 °C in 2007. The annual precipitation in the area was over 690 mm in the years 2004-2007 (Meteorological Institute, Tuorila, 2008). Two Finnish seabuckthorn female cultivars, "Terhi" and "Tytti" and a male cultivar "Tarmo" were used in this study. The cuttings used had been grown to plants in a peat-sand-perlite substratum at the Agrifood Research Finland, Plant Production Research facility in Laukaa. They were elite plants without any actinomyces addition. However, they had 1-2 root nodules before planting. In totality, 560 1.5-year old saplings were planted in the field at the end of September in 2003 (691 plants ha^{-1}). Five saplings were arranged in each of 112 rows in such a way that the first was a male ("Tarmo") and the remaining four were females ("Terhi" or "Tytti"). The distance between the plants was 2.5 m with 5 m between the rows. The study area consisted of nine blocks so that each block received one of the two female cultivars "Terhi" (4 blocks) and "Tytti" (5 blocks) at random (Figure 4.3).

In each block, four different fertilizers [control (no fertilizer), apatite, bioapatite (Kemira GrowHow Oy, Siilinjarvi, Finland) or Biolan™ (Biolan Oy, Kauttua, Finland)] and two land contours (flat land or ridged bed) were arranged at random according to a fully-crossed design in eight rows with four female plants in each block row. In addition, each row was covered with plastic mulch. The chemical composition of commercial fertilizers and the amounts added were as follows: Apatite (0:14:0 NPK, corresponds to 1000 kg ha^{-1}), bioapatite (0:2:4 NPK, corresponds to 5000 kg ha^{-1}) and Biolan (3:3:15 NPK, corresponds to 1670 kg ha^{-1}). These fertilizers were chosen, because as an organic or stone fertilizer, they are recommended for organic farming. Their chemical composition is suitable for young seabuckthorn growth. It has been found earlier that seabuckthorn responds well to phosphorus, potassium and even nitrogen fertilizers (Li, 2008).

The blocks with "Tytti" had a further eight rows, each with four female plants, which were randomized to fully-crossed combinations of the two contours and five different mulch options (*i.e.* control without mulch, straw, dry grass, conifer chips, or plastic mulch). The observed nutrient contents (g kg^{-1}) were: Oat straw (N7.5 P1.4 K20), dry grass, which consisted of *Phleum pretense, Festuca pratensis* and *Alopecurus pratensis* (N16, P2.2, K19) and conifer chips, which consisted of pine chips (No.86, P0.1, K 0.7). The chemical composition of the mulch was determined in 2004 (Viljavuuspalvelu Oy, Mikkeli, Finland). These rows received Biolan fertilizer. The mulches mentioned above were chosen as they are very common and successful in organic cultivation. The plastic mulch used was an ultraviolet-stabilized woven black polypropene material that blocked sunlight, but allowed moisture, nutrient and gas flow (MyPex 125). In the control treatment for mulching, the grass growing in the area was simply cut and thrown away, the rest being left to decompose during the winter.

Two experiments were conducted. In the first, fertilization experiment, the effects of cultivar ("Terhi" or "Tytti"), fertilizer and land contour, and their interactions were studied. In the second, mulch experiment, the effects of mulch and land contour, and their interactions were studied using the cultivar "Tytti" alone. To keep the experimental design balanced with both cultivars, only four randomly chosen "Tytti" blocks were used in the fertilization study.

RESULTS AND DISCUSSION

Variation in Vegetative Growth, Berry Yield and Phenolic Compounds

The effect of cultivar was found significant in this experiment. "Terhi" shoots grew longer in the first growing season and its stem grew thicker in the second growing season as compared to "Tytti". "Terhi" also yielded better and produced larger berries than "Tytti". Overall, both cultivars increased their vegetative growth in the second growing season. It is typical for young seabuckthorn to develop an extensive root system during the first growing season. Later, it will increase its growth above the root layer. The growth habit changes from monopodial to sympodial branching (Li, 2008). Increased vegetative growth was decisive in the second growing season, as the first good berry yield was obtained one year later in 2006. This hypothesis could be proved by correlations between shoot length in 2005 and yield in 2006 (n=248, r=0.282, $P<0.001$), and between shoot number, in 2005 and yield, in 2006 (n=248, r=0.585, $P<0.001$). The highest yield was obtained in the fourth growing season, in 2007 ("Terhi" 3.4 ± 0.9 kg, "Tytti" 0.8 ± 0.3 kg per bush). The first commercially important yield is not usually obtained until the fifth year after planting (Karhu, 2003). This kind of pattern of berry production

Terhi								Tytti							
bia (f)	apa (f)	1ctrl (r)	biol (r)	1ctrl (f)	bia (r)	apa (r)	biol (f)	chips (f)	apa (f)	bia (f)	2ctrl (r)	chips (r)	bia (r)	dg (f)	1ctrl (r)
O	O	O	O	O	O	O	O	O	O	O	O	O	O	O	O
X	X	X	X	X	X	X	X	X	X	X	X	X	X	X	X
X	X	X	X	X	X	X	X	X	X	X	X	X	X	X	X
X	X	X	X	X	X	X	X	X	X	X	X	X	X	X	X
X	X	X	X	X	X	X	X	X	X	X	X	X	X	X	X
Tytti															
chips (r)	apa (f)	1ctrl (r)	straw (f)	bia (r)	2ctrl (f)	dg (r)	apa (r)	biol (f)	1ctrl (f)	apa (r)	straw (r)	biol (r)	straw (f)	2ctrl (f)	dg (r)
O	O	O	O	O	O	O	O	O	O	O	O	O	O	O	O
X	X	X	X	X	X	X	X	X	X	X	X	X	X	X	X
X	X	X	X	X	X	X	X	X	X	X	X	X	X	X	X
X	X	X	X	X	X	X	X	X	X	X	X	X	X	X	X
X	X	X	X	X	X	X	X	X	X	X	X	X	X	X	X
								Terhi							
straw (r)	biol (f)	biol (r)	1ctrl (f)	2ctrl (r)	bia (f)	dg (f)	chips (f)	1ctrl (r)	biol (r)	bia (r)	apa (r)	1ctrl (f)	apa (f)	bia (f)	biol (f)
O	O	O	O	O	O	O	O	O	O	O	O	O	O	O	O
X	X	X	X	X	X	X	X	X	X	X	X	X	X	X	X
X	X	X	X	X	X	X	X	X	X	X	X	X	X	X	X
X	X	X	X	X	X	X	X	X	X	X	X	X	X	X	X
X	X	X	X	X	X	X	X	X	X	X	X	X	X	X	X
Terhi								Tytti							
bia (f)	biol (f)	1ctrl (f)	apa (r)	1ctrl (r)	bia (r)	biol (r)	apa (f)	biol (r)	dg (r)	2ctrl (f)	straw (r)	chips (f)	bia (f)	dg (f)	1ctrl (r)
O	O	O	O	O	O	O	O	O	O	O	O	O	O	O	O
X	X	X	X	X	X	X	X	X	X	X	X	X	X	X	X
X	X	X	X	X	X	X	X	X	X	X	X	X	X	X	X
X	X	X	X	X	X	X	X	X	X	X	X	X	X	X	X
X	X	X	X	X	X	X	X	X	X	X	X	X	X	X	X
Terhi															
biol (r)	bia (f)	apa (f)	1ctrl (r)	1ctrl (f)	biol (f)	bia (r)	apa (r)	biol (f)	bia (r)	apa (r)	straw (f)	2ctrl (r)	1ctrl (f)	apa (f)	chips (r)
O	O	O	O	O	O	O	O	O	O	O	O	O	O	O	O
X	X	X	X	X	X	X	X	X	X	X	X	X	X	X	X
X	X	X	X	X	X	X	X	X	X	X	X	X	X	X	X
X	X	X	X	X	X	X	X	X	X	X	X	X	X	X	X
X	X	X	X	X	X	X	X	X	X	X	X	X	X	X	X
Tytti								Tytti							
straw (r)	2ctrl (r)	chips (r)	biol (r)	dg (f)	apa (r)	bia (f)	straw (f)	1ctrl (f)	biol (f)	straw (f)	bia (r)	2ctrl (r)	biol (r)	apa (f)	chips (f)
O	O	O	O	O	O	O	O	O	O	O	O	O	O	O	O
X	X	X	X	X	X	X	X	X	X	X	X	X	X	X	X
X	X	X	X	X	X	X	X	X	X	X	X	X	X	X	X
X	X	X	X	X	X	X	X	X	X	X	X	X	X	X	X
X	X	X	X	X	X	X	X	X	X	X	X	X	X	X	X
1ctrl (f)	dg (r)	apa (f)	2ctrl (f)	1ctrl (r)	biol (f)	chips (f)	bia (r)	dg (r)	2ctrl (f)	apa (r)	1ctrl (r)	chips (r)	dg (f)	bia (f)	straw (r)
O	O	O	O	O	O	O	O	O	O	O	O	O	O	O	O
X	X	X	X	X	X	X	X	X	X	X	X	X	X	X	X
X	X	X	X	X	X	X	X	X	X	X	X	X	X	X	X
X	X	X	X	X	X	X	X	X	X	X	X	X	X	X	X
X	X	X	X	X	X	X	X	X	X	X	X	X	X	X	X

Figure 4.3: Experimental Design of Seabuckthorn Study Field.

Two experiments were conducted in a (90m × 90m) field, which was split into nine blocks. In the first, the fertilization experiment, the fertilizers were apatite (apa), bioapatite (bia), Biolan (biol) and control (1ctrl). In the second, the mulch experiment, the mulches were straw, dry grass (dg), conifer chips (chips), plastic mulch (biol) and control (2ctrl). The land contours were flat land (f) and ridged surface (r). Five plants were arranged in each of 112 rows such that the first plant in each row was a male (O) and the four remaining plants were female [× ("Terhi" and "Tytti")].

for seabuckthorn grown from cuttings has also been previously reported (Bernath and Foldesi, 1992; Li, 2008). However, "Terhi" bears a good yield one year earlier than other cultivars tested, previously. Even though, "Terhi" and "Tytti" are from the same origin, there are some genetically differences between them. Cultivar "Terhi" was early maturing and can exploit the tested agricultural conditions for growth and berry production more readily than "Tytti". Fully-grown "Terhi" and "Tytti" bushes have been found to yield 4-12 kg of berries in field conditions, but no variation in the amount of yield between them has been reported (Karhu and Prokkola, 2004). It is expectable that the difference in growth and berry yield will later balance out between them.

There was not a high variation between the cultivars in the number of leaf phenolic compounds. Altogether, 23 phenolic compounds were detected in "Terhi" and 24 compounds in "Tytti". In addition, both cultivars had condensed tannins. The concentrations of individual phenolic compounds differed between the cultivars. The concentration of hydrolyzable tannins 1, 2, 4, 5 and isorhamnetin-3-glucoside was significantly higher in "Terhi" than in "Tytti". On the other hand, there were significantly more hydrolyzable tannins 7, 10, 11 and pentagalloylglucose in "Tytti" than in "Terhi". Quercetin made up of about 46 per cent of the total flavonoids in "Terhi", but only 13 per cent in "Tytti". The quercetin content of leaves has also been studied in Asian seabuckthorn (*H. salicifolia* and ssp. *yunnanensis*), and has been found to vary from 84 per cent to 66 per cent, respectively (Singh, 2006). In *H. gyantsensis*, quercetin concentration was 47 per cent (Tolkachev and Sheichenko, 2006).

The number of phenolic compounds varied between cultivars found more in berries than in leaves. In "Terhi", 15 different phenolic compounds were identified and 21 compounds in "Tytti". In addition, both cultivars had condensed tannins. The concentrations of quercetin derivatives, such as 1, 3-isorhamnetin-3,7-diglucoside, quercetin 3-glucoside 7-rhamnoside, quercetin 3-glucoside, isorhamnetin 3-glucoside and flavonoid derivative 3 were higher in "Tytti" than in "Terhi". On the other hand, the concentrations of isorhamnetin-glucosides 2 and 3 were higher in "Terhi" than in "Tytti". It has been found that there is a significant variability in the flavonoid composition of various populations (in fruits and leaves), depending on their origin, geographical distribution and ecological conditions (Novruzov, 2001; Tolkachev and Sheichenko, 2006). "Terhi" and "Tytti" have the same Finnish origin. In addition, they were grown under similar ecological field conditions. It is well known that the chemical composition of seabuckthorn berries and leaves is genetically controlled (Kawecki *et al.*, 2004; Tiitinen *et al.*, 2005; Tsybikova *et al.*, 2006). This leads to the conclusion that variation in phenolic concentration is most likely a result of differences in their genotype.

Changes in Vegetative Growth and Berry Yield Caused by Fertilization

The effect of fertilization on vegetative growth and berry yield was minor in seabuckthorn. This is probably associated with its ability to receive nitrogen from its symbiont organisms, which are susceptible to excess nitrogen in their environment. However, bioapatite and Biolan fertilizers tended to improve the growth of stem diameter as compared with the control. This effect might have been due to the advantageous nutrient content of these fertilizers. They contained potassium which has been reported to increase the growth of seabuckthorn (Gentili and Huss-Danell, 2002; Li, 2002). In addition, the positive effect of Biolan fertilizer might have been emphasized due to its nitrogen content which has been predicted to have beneficial effects, especially for the growth of newly planted seabuckthorn (Li, 2008). Later, when the growth of shoots and leaves is most active in June-July, the nitrogen-fixing ability has been found to increase (Khabarov, 2003). Bioapatite contains biotite, which when added to potassium is also a calcium fertilizer. As a slowly soluble fertilizer, biotite reduces soil acidity (Rajala, 1995). Calcium has been found to be effective in promoting the growth of seabuckthorn (Li and

Schroeder, 1996). Seabuckthorn has been observed to nodulate at pH 6.0-8.8, although the plant can survive in more acidic conditions (Li, 2008). The soil pH was 5.5 and thus, it is possible that the added bioapatite improved the growth of stem diameter instead of weak nodulation.

Apatite produced the highest yield on flat land in "Terhi" (II). Apatite contains phosphate which has been found to increase the berry size and yield (Singh, 2003). Phosphate is necessary for the symbiotic microbes of seabuckthorn, because it stimulates N_2 fixation increasing the size of nodules (Gentili and Huss-Danell, 2002). It is also important for flowering and yield (Karhu, 2000), and thus, it was regarded as being beneficial for berry production in "Terhi". However, in the fourth growing season, the control treatment increased berry size in "Terhi" as compared to the Biolan fertilizer treatment. In spite of the small berries produced, Biolan increased the yield slightly as compared to the control. This is possibly due to the advantageous content of Biolan fertilizer. The small yield in the control indicates a trade-off between berry yield and size. In general, since the berry yield was small in this young seabuckthorn, it is not possible to make a stronger generalization on the effect of fertilization on berry production.

Effect of Mulch on Vegetative Growth, Berry Yield and Phenolic Compounds

In the first growing season, plastic mulch increased the growth of shoot length and stem diameter (I). In the second growing season, it also increased the shoot number. In the fourth growing season, the plastic mulch increased the yield when compared to straw, dry grass and conifer chips. The effect of plastic mulch clearly increased from the first growing season, perhaps because it prevented weed growth more efficiently than the other mulches, as suggested by Orava *et al.* (2006). The competition between root systems increases year-on-year, especially with newly planted fields. As a weak competitor seabuckthorn benefits from plastic mulch (Kivijarvi *et al.*, 1999; Li, 2008). Other benefits of plastic mulch are probably associated with its ability to equalize the temperature and moisture conditions in the soil (Herms *et al.*, 2001; Orava *et al.*, 2006). Moreover, Miles *et al.* (2003) found that the difference between day and night temperatures was greater, if there was no mulch than with any other mulch treatments. Under conditions of ample water supply, seabuckthorn was able to invest more in branches and leaves, while drought greatly reduced the total biomass, especially in young plants (Li *et al.*, 2004; Guo *et al.*, 2007). Organic mulch, such as dry grass, releases nitrogen while decomposing, and this probably disturbs the function of symbiotic microbes and thus, growth and yield (Jaakkola, 1996; Kato *et al.*, 2007). In addition, dry grass encouraged the growth of weeds, which increased the nutrient competition and thus reduced growth. Of the organic mulches conifer chips reduced shoot growth as compared to the control. Conifer chips absorb soil moisture and slow down frost melting, which reduces aeration and soil temperature (Jaakkola, 1996; Orava *et al.*, 2006). Low soil temperature has been found to decrease nitrogenase activity and thus, nitrogen fixation (McNeil and Carpenter, 1980; Kato *et al.*, 2007), which may lead to retarded vegetative growth and yield. Straw mulch increased the yield on flat land in the second growing season. The yield was only a few berries, which discounts the result. However, there are some benefits of using straw. It has the ability to keep soil moisture stable, as described by Cook *et al.* (2006). It has been found that in Russian studies, besides increasing moisture capacity in straw, also effectively prevented weed growth (Zubarev, 2008). Straw has also been found to isolate ground temperature and reduce maximum temperatures (Jaakkola, 1996; Orava *et al.*, 2006).

Mean berry weight was higher in the second (2005) and fourth (2007) growing seasons as compared with that in the third (2006) growing season in the mulch blocks (II). The largest berries were obtained in the fourth growing season. The variation in berry size between the years can be influenced by weather conditions; especially inadequate soil moisture may cause a reduction in berry size (Li, 2008).

Precipitation from June to August was lower in year 2006 as compared to the other two growing seasons, which explains the smaller berry size in that year [280 mm (2005); 111 mm (2006); 256 mm (2007)].

Berry yield in year 2006 correlated strongly with shoot number in year 2005 ($n=150$, $r=0.144$, $P<0.001$). Flower buds developed during the previous growing season (Igoshina and Selekhov, 2003; Li, 2008), and thus the increased number of shoots predicts increased berry production one year later. However, the number and size of flowers is mostly dependent on the conditions of the vegetative period. In addition, mild winters with unstable temperatures have been found to cause bud damage from 20 per cent in the female plants to 80 per cent in the male plants (Igoshina and Selekhov, 2003).

Mulch caused significant changes in phenolic composition only in leaves but not in berries. In leaves, a total of 25 phenolic compounds were found. The concentration of total (HPLC-analyzed) phenolic compounds was significantly higher in straw, dry grass and conifer chips as compared to plastic mulch treatment on the ridged beds. The total concentration of phenolic compounds did not differ between the mulches on flat land. In addition, plants grown with straw mulch had a significantly higher total phenolic concentration as compared to plants grown with no mulch.

Of the other individual compounds in leaves, the concentrations of gallic acid and hydrolyzable tannins 1, 8, 9 and 11 were significantly higher in the control, straw, dry grass and conifer chips treatments than in the plastic mulch treatment. The concentration of hydrolyzable tannin 5 was higher in the straw, dry grass and conifer chips treatments than in the plastic mulch treatment. Straw and dry grass treatments increased the concentration of hydrolyzable tannin 4 as compared to the plastic mulch treatment. Hydrolyzable tannin 7 and pentagalloylglucose were induced only by straw when compared to plastic mulch treatment. Only isorhamnetin derivative 4 was induced by plastic mulch as compared to the other mulches used.

The plants with plastic mulch had a lower concentration of condensed tannins than those with other mulches on flat land. On the ridged beds, the plants mulched with straw or dry grass had a higher concentration of condensed tannins than those with plastic mulch or no mulch. In addition, mulching with dry grass produced more condensed tannins than mulching with conifer chips.

The concentrations of many phenolic compounds in leaves were significantly lower when the plants were mulched with plastic than with other mulch treatments. The organic mulch used in this study contained different amount of nutrients. According to chemical analyses, straw and dry grass had the highest levels of N, P, K and Ca. The concentrations of nitrogen in straw and dry grass were 7.6 $g\ kg^{-1}$ and 16 $g\ kg^{-1}$, respectively. Straw provides nitrogen at the beginning of its decomposition and it decomposes more slowly than dry grass (Jaakkola, 1996). Nitrogen concentration in the conifer chips treatment was only 0.86 $g\ kg^{-1}$. Conifer chips decompose slowly and they may utilize nitrogen during decomposition rather than release it (Jaakkola, 1996; Kivijarvi *et al.*, 1999; Lehtonen, 2004). The control plots were not free of added fertilizer, as grass growing in the plots was weeded out and left to release nutrients while decomposing (Jaakkola, 1996; Kivijarvi *et al.*, 1999).

According to earlier studies, fertilizers, especially nitrogen, reduce the concentration of carbon based secondary compounds, such as tannins (Bryant, 1987; Keinanen *et al.*, 1999; Keski-Saari and Julkunen-Tiitto, 2003) (III). These observations support the hypotheses that explained plant secondary product accumulation based on nutrient status, such as the CNB hypothesis. The results of our experiment seem to contradict this pattern on the basis of seabuckthorn leaf phenolics. However, there is evidence that high levels of nitrogen in soil have negative effects on the activity of nitrogen-fixing symbionts (Li, 1999; Gentili and Huss-Danell, 2002; Li, 2002; Kato *et al.*, 2007). It may be that the nitrogen release from organic mulch disturbed the function of the nitrogen-fixing bacteria typically found in symbiosis with seabuckthorn. Thus, the effect of organic mulch on symbiosis could explain the high concentrations of condensed tannins and some hydrolyzable tannin in the control and mulched plants. It may be that phenolic concentration in leaves changes during nutrients release from mulch. Plants with nitrogen-fixing symbionts may in general, be inadequately described by the existing hypotheses for the interaction between nutrient levels and phenolics accumulation.

Nitrogen stress triggers the gene expression of flavonoid pathway enzymes and may boost the production of condensed tannins (Margna *et al.*, 1989; Lavola and Julkunen-Tiitto, 1994; Bongue-Bartelsman and Phillips, 1995). However, some phenolic compounds, such as flavonoid glycoside (kaempherol, quercetin) do not seem to be affected to the same extent (Bryant *et al.*, 1983; Bryant, 1987; Keinanen *et al.*, 1999), and phenolic compound levels may vary and independent of nutrition levels (Bryant, 1987; Waterman and Mole, 1989). In our study, isorhamnetin derivative was independent of nutrient level when plants were grown with plastic mulch.

Variation in Vegetative Growth, Berry Yield and Phenolic Compounds depending on Land Contours

Planting in ridged beds increased shoot length and stem diameter in both growing seasons, but shoot number increased only during the second growing season (I). Cultivar "Terhi" yielded better on the ridged bed (4.6 ± 1.4 kg per bush) as compared to flat land (2.3 ± 1.3 kg per bush). In the third growing season, berry weights in ridged beds increased as compared to flat land.

There are several reasons why ridges could be a more favourable substratum for the growth of seabuckthorn than flat land. Seabuckthorn cannot tolerate flooding or standing water and therefore, slight slopes with good drainage are essential for its survival and growth (Singh, 2003; Li, 2008). The vertical roots of 1-2 year old seabuckthorn have been found to extend up to 47 cm deep (Singh, 2006). Later, the maximum number of root nodules lies at a depth of 10-60 cm below the soil surface (Gupta and Singh, 2003). Thus, the ridges permit the development of roots mostly in soil above water-logged or compact zones. The better drainage of ridges as compared to flat land helps to accelerate soil warming, especially in the springtime and allows longer period of growth (Starbuck, 2003). The increase in temperature may enhance decomposing processes inside the ridge, and this may increase the plant growth, because they receive more nutrients, as suggested by Starbuck (2003). The results suggested that the soft, warm, well-aerated soil on the ridged bed was beneficial for the fast developing root system in "Terhi", and thus, for its growth and berry production (Starbuck, 2003; Freeman *et al.*, 2007). It has been reported that in general, yields may be at least 10 per cent better on the ridged beds than on flat land (Sayne, 2001). However, our results indicated that in "Terhi", the yield was as much as 50 per cent lower on the flat land as compared to ridged beds. Due to the higher profile of ridged beds as compared to flat land, they offer better weed control than flat land, which improves photosynthesis and thus, berry development.

The concentration of total (HPLC) phenolic compounds was increased more in plants grown on flat land than that of those grown on the ridged beds. Among the individual compounds, the concentrations of gallic acid, hydrolyzable tannins 1-11, pentagalloylglucose, quercetin-3-rhamnoside, monocoumaroyl astragalin A and condensed tannins were higher in plants grown on flat land than in those grown on the ridged beds. According to carbon/nutrient balance, environmental factors, such as low temperature, dryness and nitrogen deficiency may induce the synthesis of phenolic compounds (Tan, 1980; Bryant *et al.*, 1983; Dixon and Paiva, 1995). Even though, seabuckthorn is self-sufficient as regards nitrogen, root growth and nitrogen fixing may have been disturbed by the tight soil structure and lower temperature on flat land (McNeil and Carpenter, 1980; Kato *et al.*, 2007). Therefore, it seems that the differences between the treatments in terms of the environmental conditions for nitrogen-fixing symbionts and thus, nitrogen availability most likely explain our results on flat land.

Growing on a flat land increased the concentrations of quercetin derivatives, such as 2, 4, isorhamnetin 3,7-diglucoside, and isorhamnetin-glucoside 1, and of condensed tannins. In some cases, a significant cultivar × land contour interactions indicated that the effect of land contour depended on cultivar. Cultivation on flat land increased the concentration of quercetin derivative 2, isorhamnetin-glucoside 3 and flavonoid derivative 3 in "Tytti" but not in "Terhi". In addition, there were six fertilization × land contour interactions. This indicates that the effect of fertilization on the concentrations of quercetin 3-glucoside 7-rhamnoside, quercetin 3-glucoside, isorhamnetin 3-glucoside 7-rhamnoside, isorhamnetin-3-glucoside, isorhamnetin-glucoside 3 and flavonoid derivative 3 depended on land contour. There are some differences between flat land and ridged bed as a growth substratum, which may affect the accumulation of phenolic compounds. Vegetative growth and berry yield were lower on flat land than on ridged beds (I, II). There was a negative correlation between berry yield and content of phenolic compounds ($n=30$, $r= -0.433$, $p=0.017$) (in both varieties). All these differences indicated that flat land offered a low-resource environment as compared to ridged beds and thus, partly increased the accumulation of phenolic compounds in berries (Bryant *et al.*, 1983). In addition, environmental stresses on flat land, such as low temperature and limited photosynthesis have been found to increase oxidative stress and therefore, to induce an accumulation of different phenolic compounds (Close and McArthur, 2002). Especially, some quercetin derivatives have been found to be highly active against radicals (Yokozava *et al.*, 1998). In spite of this, there was significant land contour effect on the concentration of quercetin derivative 2. The concentration was higher in all mulch treatments when the bushes were grown on flat land as compared to those on the ridged beds.

CONCLUSION

The results of this experiment suggested the ways to cultivate seabuckthorn in order to ensure that the bushes grow well and produce a good berry yield. The results also suggested the ways to increase the chemical value of the leaves and berries. The results of the growth experiment demonstrated that fertilization with Biolan and bioapatite increased only the growth of stem diameter. The cultivar "Terhi" increased its vegetative growth faster than in "Tytti". Ridged beds and plastic mulch provided the most favourable effects on biomass growth and when used together, they formed a particular effective growth substratum. When seabuckthorn bushes began to produce berries, the yield was enhanced by fertilization, land contour and mulch. At the beginning of berry production (in 2005), apatite fertilizer and straw mulch increased the yield. Overall, the yield increased every year and was highest in year 2007. Cultivar "Terhi" berries yielded better as compared to "Tytti". The ridged beds clearly increased yields as compared to those on the flat land in "Terhi".

The chemical composition of seabuckthorn leaves and berries depended on both the cultivar and the cultivation method applied. Overall, hydrolyzable and condensed tannins were the dominant phenolic compound groups in leaves and their amount was increased by growth in organic mulch on flat land. Flavonoids, such as quercetins and condensed tannins were dominant in berries. Their concentrations were increased with the cultivar "Tytti" and growing on flat land.

The growth on flat land increased the concentrations of phenolic compounds in leaves and berries. Ridged beds and plastic mulch formed a favourable growth substratum and provided an advantage in seabuckthorn berry productivity that will possibly continue for several years. At this moment, the latter combination seems to support organic farming better than the first farming method, since farmers aim at a good yield. Later, it is possible that the pharmaceutical industry will use seabuckthorn berry and leaf phenolics for medicinal purposes, and some farmers may begin to cultivate seabuckthorn for that purpose.

REFERENCES

1. Allman, S., Strass, G., Kranl, K., Frank, T., Bitsch, I., Bitsch, R. and Netzel, M. 2006. Bioactivity and bioactive compounds of seabuckthorn fruit juice. In: *Seabuckthorn (Hippophae L.)-A Multipurpose Wonder Plant. Vol. II: Biochemistry and Pharmacology* (V. Singh, Ed. in Chief, 2006), p. 402-412, Daya Publishing House, New Delhi, 600p.
2. Avery, A.A., Prakash, C.S., McHughen, A., Trewavas, A.R. and DeGregori, T.R. 2005. What kind of farming works best? *Science Letters* 307: 1409-1312.
3. Bell, E.A. and Charlwood, B.V. 1980. Secondary plant products. In: *Encyclopedia of Plant Physiology,* p. 35.85, Springer-Verlag, Berlin, Germany, 674p.
4. Bernath, J. and Foldesi, D. 1992. Seabuckthorn (*Hippophae rhamnoides* L.): A promising new medicinal and food crop. *Journal of Herbs, Spices and Medicinal Plants* 1: 27-35.
5. Berry, P.M., Sylvester-Bradley, R., Phillips, L., Hatch, D.J., Cuttle, S., Raynes, F. and Gosling, P. 2002. Is the productivity of organic farms restricted by the supply of available nitrogen? *Soil Use and Management* 15: 137-143.
6. Bongue-Bartelsman, M. and Phillips, D.A. 1995. Nitrogen stress regulates gene expression of enzymes in the flavonoid biosynthetic pathway of tomato. *Plant Physiology and Biochemistry* 33: 539-546.
7. Bryant, J.P. 1987. Felt leaf willow-snowshoe hare interactions: Plant carbon/nutrient balance and floodplain succession. *Ecology* 68: 1319-1327.
8. Bryant, J.P., Chapin, F.S. and Klein, D.R. 1983. Carbon/Nutrient balance of boreal plants in relation to vertebrate herbivory. *Oikos* 40: 357-368.
9. Close, D.C. and McArthur, C. 2002. Rethinking the role of many plant phenolics-protection from photodamage not herbivores? *Oikos* 99: 166-172.
10. Cook, H.F., Valdes, G.S.B. and Lee, H.C. 2006. Mulch effects on rainfall interception, soil physical characteristics and temperature under *Zea mays* L. *Soil and Tillage Research* 91: 227-235.
11. Dixon, R.A. and Paiva, N.L. 1995. Stress-induced phenylpropanoid metabolism. *Plant Cell* 7: 1085-1097.
12. Freeman, K.W., Girma, K., Tea, R.K., Arnall, D.B., Klatt, A. and Raun, W.R. 2007. Winter wheat (*Triticum aestivum* L.) grain yield and grain nitrogen as influenced by bed and conventional planting. *Journal of Plant Nutrition* 30: 611-622.
13. Geetha, S., Basu, M., Jayamurthy, A.S., Malhotra, A.S., Pal, K., Prasad, R., Kumar, R. and Sawhney, R.C. 2008. Protective and therapeutic potentials of seabuckthorn (*Hippophae rhamnoides* L.). In: *Seabuckthorn (Hippophae L.)-A Multipurpose Wonder Plant. Vol. III: Advances in Research and Development* (V. Singh, Ed. in Chief, 2008), p. 245-253, Daya Publishing House, New Delhi, 566p.
14. Gentili, F. and Huss-Danell, K. 2002. Phosphorus modifies the effects of nitrogen on nodulation in split-root systems of *Hippophae rhamnoides*. *New Phytologist* 153: 53-61.

15. Gershenzon, J. 1994. The cost of plant chemical defense against herbivory: A biochemical perspective. In: *Insect-Plant Interactions. Vol. 5* (E.A. Bernays, Ed.), p. 105-173, CRC Press, Boca-Raton, FL, USA, 240p.

16. Guo, W., Li, B., Zhang, X. and Wang, R. 2007. Architectural plasticity and growth responses of *Hippophae rhamnoides* and *Caragna intermedia* saplings to simulated water stress. *Journal of Arid Environments* 69: 385-399.

17. Gupta, R.K. and Singh, V. 2003. Nitrogen fixation in seabuckthorn (*Hippophae rhamnoides* L.). In: *Seabuckthorn (Hippophae L.)-A Multipurpose Wonder Plant. Vol. I: Botany, Harvesting and Processing Technologies* (V. Singh, Ed. in Chief, 2003), p. 286-299, Indus Publishing Company, New Delhi, 518p.

18. Hagerman, A.E. 1995. *Tannin Analysis*, p. 24-25, Department of chemistry, Miami University, Miami, OH, USA. *http://www.users.muohio.edu/hagermae/.*

19. Harrison, J.E. and Beveridge, T. 2002. Fruit structure of *Hippophae rhamnoides* cv. Indian summer (seabuckthorn). *Canadian Journal of Botany* 80: 399-409.

20. Herms, D., Gleanson, M., Ilesm, J., Lewis, D., Hoitink, H. and Hartman, J. 2001. Using mulches in managed landscapes, Iowa State University, Ames, IA, USA, 12p. *http//:www.extension.iastate.edu/ Publications/SULI2.pdf.*

21. Herms, D.A. and Mattson, W.J. 1992. The dilemma of plants: To grow or defend. *Quaternary Review of Biology* 67: 283-335.

22. IFOAM, 2006. *Principles of Organic Agriculture. http://www.ifoam.org/organic_facts/principles/pdfs/ Principles_Organic_Agriculture.pdf.*

23. Igoshina, V.G. and Selekhov, V.V. 2003. Studies of some factors influencing frost tolerance of seabuckthorn. In: *Seabuckthorn (Hippophae L.)-A Multipurpose Wonder Plant. Vol. I: Botany, Harvesting and Processing Technologies* (V. Singh, Ed. in Chief, 2003), p. 273-280, Indus Publishing Company, New Delhi, 518p.

24. Jaakkola, S. 1996. Eloperaiset maanpintakatteet viljelyksilla ja viheralueilla, Tyotehoseuran maataloustiedote, p.4, Karprint, Huhmari, Finland 12. (In Finnish)

25. Jones, C.G. and Hartley, S.E. 1999. A protein competition model of phenolic allocation. *Oikos* 86: 27-44.

26. Kalinina, I.P. and Panteleyeva, Y.I. 1987. Breeding of seabuckthorn in the Altai. In: *Advances in Agricultural Science*, p. 76-78, Moscow, Russia.

27. Karhu, S. 2000. Odotetut uutuudet-Testatusti kestavat tyrnit. *Puutarha ja kauppa* 4, 37: 6-7. [In Finnish]

28. Karhu, S. 2003. New seabuckthorn cultivars Terhi, Tytti and Tarmo. In: *Harvest- Horticultural Research Results 2000-2002-Suomen maataloustieteellisen seuran julkaisuja* (T.Hovi, S. Karhu, M.M. Linna and T. Suojala, Eds.), p.83-84, Datacom Finland Oy, Jokioinen, Finland, 42.

29. Karhu, S. and Prokkola, S. 2004. Tyrnilajikkeet Suomen oloihin. In: *Maataloustieteen paivat. Suomen maataloustieteellisen seuran julkaisuja* (A. Hopponen and M. Rinne, Eds.), p.4, Jokioinen, Finland, 19. (In Finnish)

30. Kato, K., Kanayama, Y., Ohkawa, W. and Kanahama, K. 2007. Nitrogen fixation in seabuckthorn (*Hippophae rhamnoides* L.) root nodules and effect of nitrate on nitrogenase activity. *Journal of the Japanese Society for Horticultural Science* 76: 185-190.

31. Kawecki, Z., Szalkiewicz, M. and Bieniek, A. 2004. The common seabuckthorn-A valuable fruit. *Journal of Fruit and Ornamental Plant Research* 12: 183-193.

32. Keinanen, M., Julkunen-Tiitto, R., Mutikainen, P., Walls, M., Ovaska, J. and Vapaavuori, E. 1999. Trade-offs in phenolic metabolism of silver birch: Effects of fertilization, defoliation, and genotype. *Ecology* 80: 1970-1986.

33. Keski-Saari, S. and Julkunen-Tiitto, R. 2003. Resource allocation in different parts of juvenile mountain birch plants: effect of nitrogen supply on sapling phenolics and growth. *Physiologia Plantarum* 118: 114-126.

34. Keskitalo, M., Hyvarinen, H. and Pihlava, J.M. 2001. Yleista fenolisista yhdisteistaa. In: *kasviperaiset biomolekyylit-fenoliset yhdisteet ja terpeenit. Kirjallisuuskatsaus* [H.Hyvarinen, (Ed.)], p. 9-13, Jyvaskylan yliopistopaino, Jyvaskyla, Finland. (in Finnish)

35. Khabarov, S.N. 2003. Elements of commercial cultivation technology of seabuckthorn. In: *Seabuckthorn (Hippophae L.)-A Multipurpose Wonder Plant. Vol. I: Botany, Harvesting and Processing Technologies* (V. Singh, Ed. in Chief, 2003), p. 347-351, Indus Publishing Company, New Delhi, 518p.

36. Kivijarvi, P., Heiskanen, L. and Dalman, P. 1999. Cultivation technique of organically grown currant-Literature review. *Maatalouden tutkimuskeskus*, Jokioinen, Finland A 19, 43 p. (In Finnish)

37. Krolikowska, M. 1972. Flavonoid glycosides from common seabuckthorn berries *Hippophae rhamnoides*. *Planta medica* 22: 418-427.

38. Lavola, A. and Julkunen-Tiitto, R. 1994. The effect of elevated carbon dioxide and fertilization on primary and secondary metabolites in birch, *Betula pendula* (Roth). *Oecologia* 99: 315-321.

39. Lawanson, A.O., Akindele, B.B., Fasalojo, P.B. and Akpe, B.L. 1972. Time-course of anthocyanin formation during deficiencies of nitrogen, phosphorus and potassium in seedlings of *Zea mays* Linn. Var. E.S.1. *Zeitschrift fur Pflanzenphysiologie* 66: 251-253.

40. Lehtonen, U. 2004. Kateviljely. In: *Luonnonmukaisesti omassa kotipuutarhassa (2nd Ed.)*, p. 25-29, WS Bookwell Oy, Porvoo, Finland. (in Finnish)

41. Li, C.Y., Ren, J., Luo, J.X. and Lu, R.S. 2004. Sex-specific physiological and growth responses to water stress in *Hippophae rhamnoides* L. populations. *Acta Physiologiae Plantarum* 26: 123-129.

42. Li, T.S.C. 1999. Seabuckthorn: New crop opportunity. In: *Perspectives on new crops and new uses* (J. Janick, Ed.), p. 335-337, ASHS Press, Alexandria, Virginia, VA, USA, 528p.

43. Li, T.S.C. 2002. Product development of seabuckthorn. In: *Trends in New Crops and New Uses* (J. Janick and A. Whipkey, Eds.), p. 393-398, ASHS Press, Alexandria, Virginia, VA, USA.

44. Li, T.S.C. 2008. Plantation and management practices of seabuckthorn. In: *Seabuckthorn (Hippophae L.)-A Multipurpose Wonder Plant. Vol. III: Advances in Research and Development* (V. Singh, Ed. in Chief, 2008), p. 27-48, Daya Publishing House, New Delhi, 566p.

45. Li, T.S.C. and Schroeder, W.R. 1996. Seabuckthorn (*Hippophae rhamnoides* L.): A Multipurpose Plant. *Hort. Technology* 6: 370-380.

46. Mader, P., Fliedbach, A., Dubois, D., Gunst, L., Fried, P. and Niggli, U. 2002. Soil fertility and biodiversity in organic farming. *Science* 296: 1694-1697.

47. Margna, U., Margna, E. and Vainjarv, T. 1989. Influence of nitrogen nutrition on the utilization of L-phenylalanine for building flavonoids in buckwheat seedling tissues. *Journal of Plant Physiology* 134: 697-702.

48. McNeil, R.E. and Carpenter, P.L. 1980. Effect of temperature of acetylene-reducing activity associated with root nodules on woody trees and shrubs. *Journal of the American Society for Horticultural Science* 105 (4): 543-546.

49. Miles, C., Garth, L., Sonde, M. and Nicholson, M. 2003. Searching for alternatives to plastic mulch. In: *Publications of WSU Vancouver Research and Extension Unit*, p. 1-7, WA, USA. *http://agsyst.wsu.edu*.

50. Ministry of Agriculture and Forestry. 2008. Marjan-ja hedelmänviljely avomaalla. Horticultural Enterprise Register. (In Finnish). *http://www.mmmtike.fi/attachments/5lLjNrWgp/5nEqLBzH3/Files/CurrentFile/070328_pure_yhteenveto.pdf*.

51. Novruzov, E.N. 2001. Flavonoids of different forms of seabuckthorn (*Hippophae rhamnoides* L.) growing in Azerbaijan. In: *Seabuckthorn-A Resource for Health and Environment in Twenty First Century*. Proceedings of International Workshop on Seabuckthorn, p.140-146, Palampur, India.

52. Orava, H., Kivijarvi, P., Lahtinen, R., Matilainen, A., Tillanen, A. and Kuopanportti, H. 2006. Hajoavan katteen kehittaminen riviviljelykasveille. *Mikkelin ammattikorkeakoulun tutkimuksia ja raportteja*. p.10-14, Interkopio Oy, Mikkeli, Finland. (In Finnish)

53. Piir, R. 1996. *Hippophae rhamnoides* L. in Estonia for fruit growing. *Journal of Agricultural Sciences* 2: 162-176.

54. Prokkola, S. 2003. Reliability of Russian cultivars in North Ostrobothnia, Finland. *Acta Horticulturae* 626: 389-395.

55. Puupponen-Pimia, R., Nohynek, L., Alakomi, H.L. and Oksman-Caldentey, K.M. 2005. Bioactive berry compounds-Novel tools against human pathogens. *Applied Microbiology and Biotechnology* 67: 8-18.

56. Rajala, J. 1995. *Luonnonmukainen maatalous*, Raamattutalo, Pieksamaki, Finland, 309 p. (in Finnish)

57. Rousi, A. 1965. Observations on the cytology and variation of European and Asiatic populations of *Hippophae rhamnoides*. *Annales Botanici Fennici* 2: 1-18.

58. Rousi, A. 1971. The genus *Hippophae* L.: A taxonomic study. *Annales Botanici Fennici* 8: 177-227.

59. Sayne, K.D. 2001. Bed planting system-An overview. CIMMYT, Mexico. *http://www.css.cornell.edu/faculty/hobbs/file per cent 20posters/posterBP1.pdf*.

60. Singh, V. 2003. Plantation and management technologies of seabuckthorn. In: *Seabuckthorn (Hippophae L.)-A Multipurpose Wonder Plant. Vol. I: Botany, Harvesting and Processing Technologies* (V. Singh, Ed. in Chief, 2003), p. 352-360, Indus Publishing Company, New Delhi, 518p.

61. Singh, V. 2006. Free radicals, diseases, anti-oxidants and anti-oxidant properties of seabuckthorn (*Hippophae rhamnoides* L.). In: *Seabuckthorn (Hippophae L.)-A Multipurpose Wonder Plant. Vol. II: Biochemistry and Pharmacology* (V. Singh, Ed. in Chief, 2006), p. 3-69, Daya Publishing House, New Delhi, 600p.

62. Singh, V. and Gupta, R.K. 2008. Micropropagation of seabuckthorn (*Hippophae rhamnoides* L.). In: *Seabuckthorn (Hippophae L.)-A Multipurpose Wonder Plant. Vol. III: Advances in Research and Development* (V. Singh, Ed. in Chief, 2008), p. 3-13, Daya Publishing House, New Delhi, 566p.

63. Srivastava, R., Roseti, D. and Sharma, A.K. 2007. The evaluation of microbial diversity in a vegetable based cropping system under organic farming practices. *Applied Soil Ecology* 36: 116-123.

64. Starbuck, C.J. 2003. *Raised-bed gardening*. MU Guide, University of Missouri-Columbia, MO, USA. *http://extension.missouri.edu/explore/agguides/ort/g06985.htm.*

65. Tan, S.C. 1980. Phenylalanine ammonia-lyase and the phenylalanine ammonium-lyase inactivating system: Effect of light, temperature and mineral deficiencies. *Australian Journal of Plant Physiology* 7: 159-167.

66. Tiitinen, K., Hakala, M.A. and Kallio, H.P. 2005. Quality components of seabuckthorn (*Hippophae rhamnoides*) varieties. *Journal of Agricultural and Food Chemistry* 53: 1692-1699.

67. Tolkachev, O.N. and Sheichenko, O.P. 2006. Flavonoids of seabuckthorn (*Hippophae rhamnoides* L.): Chemistry and Pharmacology. In: *Seabuckthorn (Hippophae L.)-A Multipurpose Wonder Plant. Vol. II: Biochemistry and Pharmacology* (V. Singh, Ed. in Chief, 2006), p. 159-167, Daya Publishing House, New Delhi, 600p.

68. Tsybikova, D.T., Rasputina, D.B. and Komissarenko, N.F. 2006. Flavonoids and other biologically active substances of seabuckthorn leaves. In: *Seabuckthorn (Hippophae L.)-A Multipurpose Wonder Plant. Vol. II: Biochemistry and Pharmacology* (V. Singh, Ed. in Chief, 2006), p. 168-176, Daya Publishing House, New Delhi, 600p.

69. Waterman, P.G. and Mole, S. 1989. Extrinsic factors influencing production of secondary metabolites in plants; In: *Insect-Plant Interactions. Vol. I.* (E.A. Bernays, Ed.), p. 107-134, CRC Press: Boca Raton, Florida, FL, USA, 164p.

70. Yang, B. 2001. *Lipophilic Components of Seabuckthorn (Hippophae rhamnoides) Seeds and Berries and Physiological Effects of Seabuckthorn Oils*, Doctoral thesis, University of Turku, Finland, 110 p.

71. Yang, B. and Kallio, H. 2002. Effects of harvesting time on triacylglycerols and glycerophospholipids of seabuckthorn (*Hippophae rhamnoides* L.) berries of different origins. *Journal of Food Composition and Analysis* 15: 143-157.

72. Yao, Y. 1994. *Genetic diversity, Evolution and Domestication in Seabuckthorn (Hippophae rhamnoides L.)*, Academic Dissertation in Plant Breeding, University of Helsinki, Finland, 30p.

73. Yokozava, T., Chen, C.P., Dong, E., Tanaka, T., Nonaka, G.I. and Nishioka, I. 1998. Study on the inhibitory effect of tannins and flavonoids against the 1,1-diphenyl-2-picrylhydrazyl radical. *Biochemistry and Pharmacology* 56: 213-222.

74. Zubarev, Y. 2008. Commercial cultivation of seabuckthorn in western Siberia, Russia. In: *Seabuckthorn (Hippophae L.)-A Multipurpose Wonder Plant. Vol. III: Advances in Research and Development* (V. Singh, Ed. in Chief, 2008), p. 49-60, Daya Publishing House, New Delhi, 566p.

Chapter 5

Seabuckthorn (*Hippophae* L.) Propagation and Plantation is a New Avenue to Farmers and Future Scope of Cultivation in Cold Desert Area of Himachal Pradesh, India

R.K. Rana[1], Ashok Singh[2], Virendra Singh[4], L.K. Sharma[3], Reena Devi[4], Pankaj Katoch[5] and Manohar Lal[6]

[1]CSK Himachal Pradesh Agricultural University, Krishi Vigyan Kendra, Bajaura (Kullu), H.P., India
[2]CSIR-Centre for High Altitude Biology, Ribling, P.O.Tandi, Lahaul and Spiti, H.P., India
[3]CSK Himachal Pradesh Agricultural University,
Krishi Vigyan Kendra Lahaul and Spiti, Kukumseri, Palampur, India
[4] Department of Biology and Environmental Sciences,
CSK Himachal Pradesh Agricultural University, Palampur, India
[5]CSK Himachal Pradesh Agricultural University, HAREC Kukumseri, Lahaul, H.P., India
[6]G.B. Pant Institute of Himalayan Environment and Development,
Himachal Unit, Mohal-Kullu, H.P., India

SUMMARY

The present study is the pioneer step towards commercial plantation of the seabuckthorn (*H. salicifolia and H. rhamnoides)* by selection, evaluation, cultivation and conservation of genotypes for socio-economic development of tribal communities of cold desert areas in India. About 1,50,000 saplings of high yielding selection of *H. salicifolia* were planted in 75 ha in Lahaul. About 150 farmers participated in plantation programme. The survival rate varied from 40 to 60 per cent during 3 years of the plantation programme. The selected genotype of *H. salicifolia*, used for plantation, was mild thorny, high yielding (5.5 kg/plant), and rich in vitamin C. The cultivation of selected superior land-races will enhance the quality and quantity of the produce in future.

Seabuckthorn (*Hippophae* L.): *A Multipurpose Wonder Plant, Vol. 4: Emerging Trends in Research and Technologies* (Virendra Singh, *Editor-in-Chief*, 2014), p. 59–68, Daya Publishing House, New Delhi, India.

In the cold desert area, we are looking forward to increase more area under plantation on marginal lands. To fulfill target, time to time training of farmers, exposure visits and interaction with different stakeholders including buyers of the produce, were conducted.

Keywords: *Superior land-races, Selection, Mass multiplication, Conservation and plantation.*

INTRODUCTION

Seabuckthorn is a golden bush found in the high altitudinal Trans Himalayas of Indian Himalayan region, has got attention for its wonderful medicinal, nutraceutical and ecological properties during last two decades. At present, different research and developmental organizations (about 35 numbers) are working to harness its potential for socio-economic development of masses in rural parts of cold desert in India. In cold desert environment, only a few plant species are able to cope with the harsh climatic conditions. Among these, 3 *Hippophae* species (*Hippophae rhamnoides* ssp. *turkestanica* L., *H. salicifolia* D. Don. and *H. tibetana* Sch.) are adapted very well. It is a cold hardy plant which contains long fibrous and tap root system having nitrogen fixation properties in the soil layers (Rongsen, 1992). By understanding the importance of species for the area and to improve ecological balance, Defence Research and Development Organization and Ministry of Environment and Forest, Government of India is trying to explore its available potential and to increase the area under plantation. It has been reported that the chemical constituents of Indian *Hippophae* species have high potential than the Chinese and Russian *Hippophae* (Arimboor *et al.*, 2006). In spite of its importance, it is not so much popularized in India. A few research work on taxonomical (Aswal and Mehrotra, 1994), ecological (Singh *et al.*, 1991; Roy *et al.*, 2001; Singh, 2007), ethnobotanical (Chauhan *et al.*, 2001; Dhyani *et al.*, 2010; Singh *et al.*, 2012), medicinal, health product and environmental potential (Chauhan *et al.*, 2001; Zeb, 2004; Singh, 2006; Dhyani *et al.*, 2007; Stobdan *et al.*, 2008) and nursery technique (Singh *et al.*, 1997) is available for the species in India. However, at ground level, comprehensive study on selection of high yielding land-races from natural populations/habitats, evaluation of promising forms, mass multiplication and conservation of superior genotypes of seabuckthorn in India are few so far (Singh, 2006; Singh *et al.*, 2010). Domestication of wild natural populations is a challenging work that we have undertaken for the first time under National Agriculture Innovative Programme in Lahaul valley of Himachal Pradesh. The present work is supposed to be a role model for the future research and planning of fragile ecosystem of the cold desert area in India. The main aim of study is to increase the area under plantation and production of seabuckthorn.

MATERIALS AND METHODS

Experimental Area

The present study was carried out in Lahaul valley of Lahaul and Spiti district of Himachal Pradesh, a part of Cold Desert Biosphere Reserve. The cold desert of Lahaul valley of district Lahaul and Spiti extends within 31° 44′ 34" N to 32° 59' 57" N latitude and 76° 46' 29" E to 78 ° 41' 34" E longitude. The climate of the area is dry temperate to alpine type with scanty rainfall (10 to 300 mm/annum) during the summer season, high snowfall (up to 450 cm/annum) in winters and annual temperature ranging from -25 °C to + 34 °C. The whole district has typical topography of Trans Himalayan Mountains with highest peak up to 6600 m amsl. In such a climate, seabuckthorn species are found growing gregariously in the wild natural habitats like along the river side, nallahs and water channel, along the roads side and on the bunds of agricultural fields. To popularize and harnessing its potential, CSK Himachal Pradesh Agricultural University, Palampur has continuously working for last nineteen years.

Selection of Promising Land-races of Sea-buckthorn

A detailed field survey in different villages/ locations (approximately 70) of Lahaul and Spiti district was undertaken during 2008-2012. For selection of superior land races of *Hippophae* species (*Hippophae rhamnoides* ssp. *turkestanica* L., *H. salicifolia* D. Don. and *H. tibetana* Sch.; locally called as *Chharma, Chasterlulu, Sarla, Chermen, Tirkug and Rwasak*), physical parameters were recorded on the spot and geographical locations of each sites were carried out by GPS system. The novel characters for selection of superior land races were large fruit size, maximum fruit weight, more than 4 kg fruit yield per plant and less number of thorns per 10 cm shoot length. The selected plants were tagged and the fruit samples were collected for analysis of the characteristics of fruit.

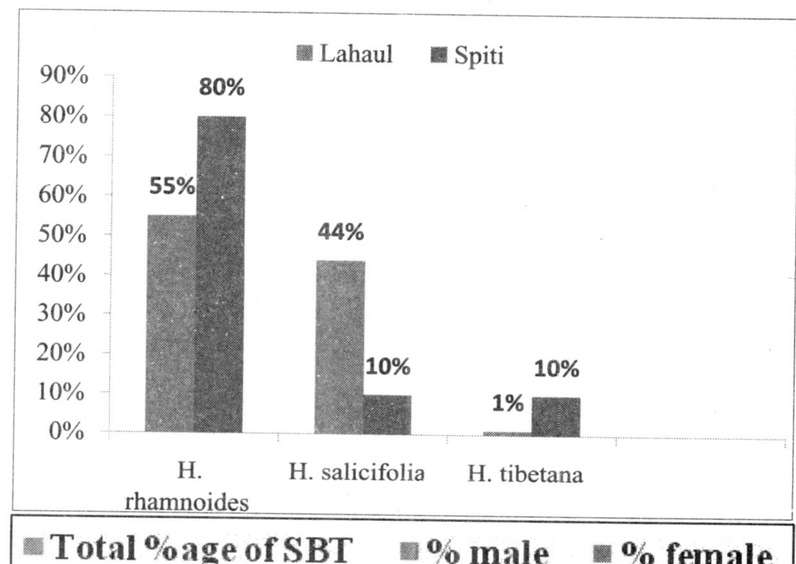

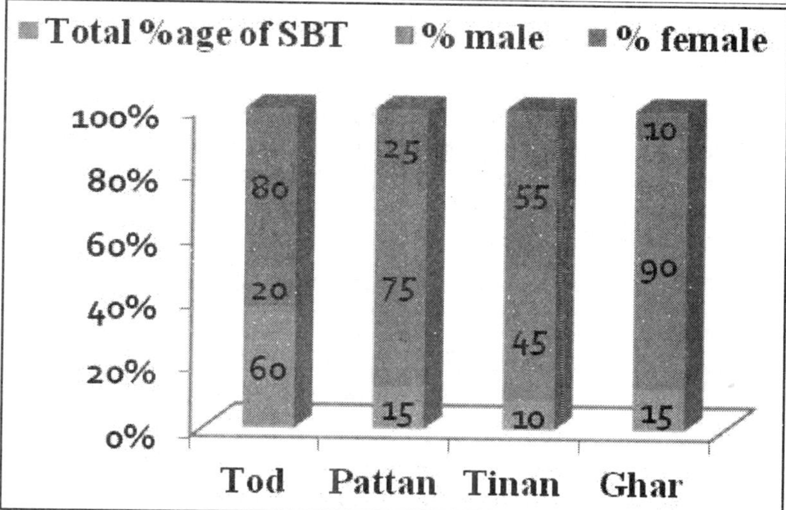

Figure 5.1: Distribution of Seabuckthorn Species in Lahaul and Spiti, H.P.

RESULTS AND DISCUSSION

Distribution pattern showed that in Lahaul and Spiti, these species, *i.e. H. rhamnoides* ssp. *turkestanica, H. salicifolia* and *H. tibetana* are found at different locations in isolation as well as in combinations. Among all the species, *H. rhamnoides* ssp. *turkestanica* showed wider range of distribution followed by *H. salicifolia* and *H. tibetana*. The valley wise distribution of species and their sex ratio is shown in Figure 5.1.

In general, variation in fruit weight of *H. salicifolia* varied from 13.75-30.56 g/100 berries; *H. tibetana* (24.58-29.23 g/100 berries) and *H. rhamnoides* ssp. *turkestanica* (11.57-25.41 g/100 berries) are

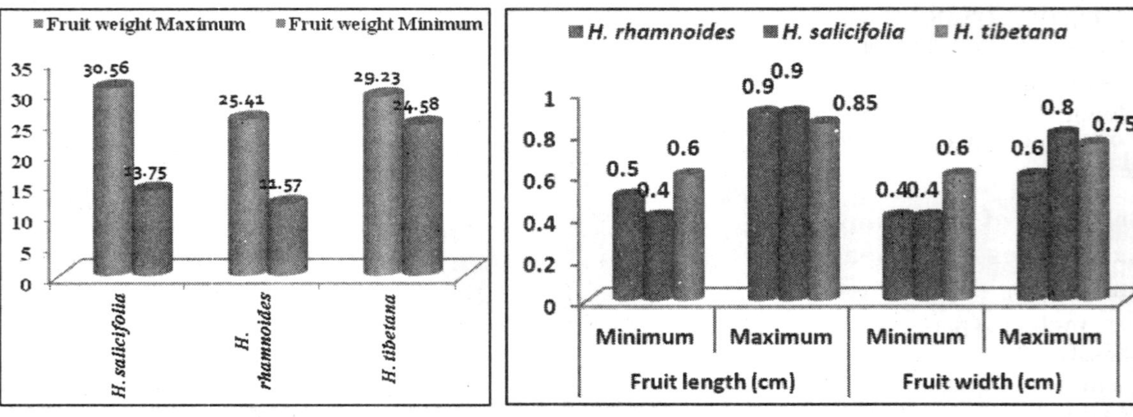

Figure 5.2: Variations in the Fruit Weight of 100 Fruits (g) (Left) and Fruit Size in Different Species of *Hippophae* (Right).

shown in Figure 5.2. Among all the species, *H. salicifolia* showed maximum fruit weight (30.56 g/100 berries) selected from Tinu village followed by *H. tibetana* (29.23 g/100 berries) from Lossar and *H. rhamnoides* ssp. *turkestanica* (25.44 g/100 berries) from Jispa. The fruit colour and shape of *H. rhamnoides* ssp. *turkestanica* varied from red, orange and yellow with ovoid, oblong, depressed shape; for *H. salicifolia* yellowish with rounded shape; however, for *H. tibetana* orange with rounded shape was recorded. *H. rhamnoides* ssp. *turkestanica* showed highest number of thorns per 10 cm shoot length followed by *H. tibetana* and *H. salicifolia*. Fruit size of *H. rhamnoides* ssp. *turkestanica*, varied from 0.5 × 0.4-0.9 × 0.6 cm; *H. salicifolia* 0.4 × 0.4-0.9 × 0.8 cm and for *H. tibetana* from 0.6 × 0.6-0.85 × 0.75 cm was recorded (Figure 5.2).

Table 5.1: Morphological Characteristics of different Land-races of Seabuckthorn Species

Location	*Altitude (m amsl)*	*Yield (kg/plant)*	*Fruit Weight (g/100 fruit)*	*Maximum Fruit Size (cm) (Length × Width)*	*Colour*	*Thorns/10 cm Shoot Length*
		Lahaul valley *H. rhamnoides* ssp. *turkestanica*				
Gemur	3020	4.2	21.42	0.7 × 0.6	Orange	6
Jispa	3050	5.0	25.41	0.9 × 0.7	Red	7
Darcha	3210	4.5	19.72	0.8 × 0.7	Red	6
Chaaling	3200	4.0	19.24	0.7 × 0.6	Yellow	6
			H. salicifolia			
Tinu	3100	5.5	29.69	0.9 × 0.8	Yellow	4
		Spiti valley *H. rhamnoides* ssp. *turkestanica*				
Schiling -1	3420	5.0	14.80	0.8 × 0.5	Orange	7
Schiling -2	3420	5.2	15.00	0.7 × 0.5	Red	7
Shego -1	3535	4.6	14.00	0.9 × 0.6	Red	9
Shego -2	3535	4.3	17.28	0.9 × 0.6	Yellow	8
Rangrik	3615	4.5	17.00	0.7 × 0.6	Orange	7

amsl: above mean sea level.

On the basis of novel morphological characters, ten land-races were selected. Out of these land races, one is of *H. salicifolia* from Tinu village and 9 of *H. rhamnoides* ssp. *turkestanica, i.e.* 4 from Lahaul valley (Darcha, Jispa, Gemur, Chaaling) and 5 from Spiti valley (Schiling-1, Schiling-2, Shego-1, Shego-2, Rangrik) were selected (Table 5.1).

Mass Multiplication

Mass multiplication was carried out through hardwood cuttings at Krishi Vigyan Kendra at Kukumseri (2730 m asl) in Lahaul. Out of the total species of *Hippophae*, two species (*H. rhamnoides* ssp. *turkestanica* and *H. salicifolia*) were selected. In the wild populations, most of the sites contained less number of superior plants with exceptional physical characteristics and hence, the cuttings for mass multiplication were taken from these sites, where fruit yield per plant was comparable up to 3 to 4 kgs. During survey, the potential sites were identified for both the species for mass multiplication. For mass multiplication, the cuttings of *H. rhamnoides* ssp. *turkestanica* were taken from Jispa, Gemur, Chaaling and Darcha in Lahaul valley; Shego, Schiling, Lara and Rangrik in Spiti valley; and the cuttings of *H. salicifolia* from Tinu villages at Lahaul valley. Since, the right of local people for seabuckthorn stands is more and hence, the help of trained local self help groups were taken for collection of cuttings. Every year, the cuttings of female and male plants in the ratio of 9:1 were collected from different identified sites either in October-November for winter storage purpose or March-April for fresh plantation without storage. The collected cuttings during October-November were tagged and stored in the pit for winter season (15th November to March end) in alternate layer with river sand. To reduce water loss by transpiration, regular watering was done before snow fall and after snow melts (Figure 5.3).

Figure 5.3: Storage Method of Hardwood Cuttings of Seabuckthorn for Winters.

The mass propagation in the nursery was done by two methods as below:

Direct Plantation of Cuttings in the Nursery Beds

Before plantation of cuttings in the nursery, preliminary operations like preparation of sunken beds of size 3 m × 1 m, removal of pebbles/stones followed by mixing of river sand (approximately 1quintal/bed), vermicompost (approximately 3-4 kg/bed) and 1 kg soil containing *Frankia* bacteria is incorporated and then gentle mixing of composition was carried out. Every day depending upon the requirement, stored cuttings were taken out for treatment with running water for 24 hours. After water treatment, the basal end is treated up to for 30 seconds either with Indole-3-butyric acid (IBA @ 500 ppm) or Naphthalene acetic acid (NAA @ 500 ppm) solutions. Simultaneously, the cuttings were planted in the beds at a distance of 6 inches. To get maximum survival rate, regular irrigation by sprinkler irrigation system for whole of the season (April to October end) till senescence of leaves in dormant season will be ensured. For obtaining maximum growth, hoeing and weeding operations were carried out for every 10-15 days interval in the rainy and summer season.

In Polybags

To increase the survival rate in the field, the cuttings were raised in polybags of size 9"x 4". Before planting of cuttings in polybags, preliminary operations like development of land, preparation of sunken beds of size 3 × 1 m for adjusting polybags, sieving of soil and sand, preparation of potting mixture in the ratio of 5 : 4 : 1 ratio (soil : river sand : FYM) operations were completed timely. To enhance the growth of rooted cuttings, a little soil containing *Frankia* was added into the mixture. Other cultural operations were similar to that of direct planting method.

Effect of Hormonal Treatments to Plant Survival Rate

The rooting success of freshly collected cuttings and the cuttings collected in winter, and stored in sand during winter, and planted in the nurseries during spring were tested by either Indole-3-butyric acid (IBA @500 ppm) or Naphthalene acetic acid (NAA @500 ppm). The cuttings collected fresh and treated with IBA followed by plantation in the month of April showed maximum success rate of 94.74 per cent in comparison to treatment by NAA with maximum success rate of 80.45 per cent. The planting of cuttings collected in the month of May resulted in low percentage of rooting, *i.e.* 82.17 per cent when treated with IBA and 70.94 per cent with NAA. However, the cuttings collected in winter and stored in the alternate layers with sand and planted in April showed comparatively less rooting percentage (79.83 per cent with IBA and 71.34 per cent with NAA treatments) than the freshly collected cuttings. Similarly, delay in planting, *i.e.* in May month recorded less rooting of cuttings (59.68 per cent by IBA and 48.43 per cent by NAA treatment).

Among the two species of *Hippophae* species (*H. salicifolia* and *H. rhamnoides* ssp. *turkestanica*), *H. salicifolia* showed higher success rate in both the conditions (polybags as well as direct filed plantation methods) in the nursery after treatment with IBA. The genotype of high yielding and nutritionally rich *H. salicifolia* was identified and morphologically characterized in 1994 (Singh, 1994) and later

Figure 5.4: Seabuckthorn Plants Raised by Direct and Polybags Methods.

biochemically characterized (Singh *et al.*, 2011). Fresh collected cuttings of *H. salicifolia* species planted in the polybags showed higher rooting percentage, *i.e.* 97.23 per cent when planted in April and 88.78 per cent when planted during May month. However, success rate of winter stored cuttings of *H. salicifolia* showed 88.73 per cent in April and 57.68 per cent in May planting. Similarly, fresh cuttings of *H. rhamnoides* ssp. *turkestanica* showed 74.34 per cent survival for April and 49.43 per cent for May month planting. Similar trend as that of *H. salicifolia* for winter stored cuttings was recorded in *H. rhamnoides* ssp. *turkestanica* (64.53 per cent in April and 48.73 per cent in May). The results for success rate of direct planting were at par with that of polybags (Figure 5.4).

Conservation of Superior Genotypes of Seabuckthorn

In order to conserve the superior genotypes, gene banks has been established at two different locations, *i.e.* one at KVK farm at Kukumseri (2730 m asl) and one in farmer's field at Tinu in Lahaul (3200 m asl). These two locations were selected by sampling technique in the Lahaul valley with varying aspects, such as altitude and habitats. Fifty plants of each selected land-races have been planted at a distance of 2 × 1 m, which will act as a mother plant for the propagation of these land races in near future. To reduce mortality, regular cultural operations like irrigation, weeding, hoeing, time to time visit by the experts are being carried out. Among these two locations, mother orchard at Kukumseri showed better survival rate (99.11 per cent) than Tinu (78.25 per cent). These mother orchards will be helpful in providing the soft and hardwood cuttings for further multiplications and for breeding purposes in future.

Systematic Plantation on Marginal Lands

The seabuckthorn in Lahaul and Spiti district is found growing in wild and there is no cultivated variety and cultivars, and therefore, there is no systematic plantation. As it is a new crop and recently been declared as horticultural crop in India. To undertake the systematic plantation in the farmer's field, farmers were first trained, after that the plants were supplied to them. In the month of March-April, one-year old rooted plants in the ratio of 9 : 1 (female : male) quality planting materials of two species of *Hippophae* were supplied to the farmers according to their demand with the help of local NGO "Lahaul Kla Sangam Avam Rojgar Srijan Munch" based at Jhalma. The plants were supplied to farmers after compilation of preliminary requirement like preparation of pits (45 cm^3 size) at the distance of 3 m x1.5 m, assurance of irrigation facilities. Simultaneously, time to time feedback were also been taken at local meetings/trainings and by visiting the plantation/demonstration sites.

Till date, more than 1,50,000 plants of seabuckthorn have been planted by about 150 farmers on marginal land and has covered approximately 75 ha area in Lahaul. Every year, the survival rate of plants supplied to farmers in the Lahaul valley was conducted and increasing trend for survival rate from 45 per cent to 60 per cent (2010-12) was recorded. It is not so good as far as the farmer's feedback is concern. The reason behind is that it is the preliminary experiment for commercial cultivation in the farmers field, which we have undertaken in the Himalayan region, with obvious less interest by the local people. Further, the plantations have been carried out on wasteland type of land, with little moisture. Efforts are needed to motivate the local farmers and give them market assurance and better irrigation facilities. The other reason is the availability of other commercially cultivated crops like *Solanum tuberosum* (potatos), *Pisum sativum* (peas), *Inula racemosa* (manu), *Saussurea costus* (kuth), *Humulus lupulus* (hops), *Malus domestica* (apple) and vegetables crops.

FUTURE APPROACH

The plantation raised by the listed farmers will get maximum benefit from the outcome of the project. As the berries and leaves collected after harvesting from these plantations, they will be purchased by the private companies like Seabuckthorn India Pvt. Ltd., Arctic-Bio-Pro., Shimla Hills Pvt. Ltd. and many more as they are the co-partners of the project. They will purchase raw materials through seabuckthorn societies and other cooperative societies of the area. To increase the area under cultivation of seabuckthorn in India, local selections of *H. salicifolia* from Tinu site and high yielding thornless varieties from Russia for mass multiplication will be given preference. Since year 2010 onwards, our focus is shifted on large scale cultivation of selected land-races of *H. salicifolia* from Tinu area and also in a process to introduce the high yielding Russian varieties. Beside this, breeding programme for hybridization between Russian female varieties and local male selections of *H. rhamnoides* are required to be initiated immediately to develop own varieties.

CONSTRAINTS

As per target of achieving maximum area under plantation and to increase the productivity, some of the main constraints that we faced till now are (i) old mind set of local inhabitants for cultivation of native selections of seabuckthorn; (ii) availability of other much profitable agricultural cash crops than seabuckthorn; (iii) most of barren lands for plantation is under forest department with less peoples' rights; (iv) difficulties in harvesting; (v) low shelf life of fruits and (vi) lack of irrigation facilities in marginal lands in many sites.

CONCLUSION

The present study was done in the Himalayan region to increase area under cultivation as well as to increase productivity of seabuckthorn stands. It has wide implications and impact for further developmental activities of the cold desert area, because of fragile ecosystem and occurrence of sparse vegetation. Since, aim of the ICAR (NAIP) project was to make a value chain from production to consumption system and this study is considered to be the backbone for the implementation of project. Many organizations are looking forward to see initial impact of this study for further programmes. The study is concluded with following points:

- Ten land-races, *i.e.* one of *H. salicifolia* and 9 of *H. rhamnoides* have been selected from Lahaul and Spiti district of Himachal Pradesh and conserved in the mother orchard at Krishi Vigyan Kendra, Kukumseri.
- Among the different land-races, *H. salicifolia* selection from Tinu village, Lahaul is the most valuable resource for breeding and cultivation purposes.
- Mass multiplication of available good stands of native populations of seabuckthorn species and till date, 1,50,000 plants have been raised and distributed among the local inhabitants of this area.
- Further, it is concluded that hardwood cuttings of seabuckthorn should be collected fresh during March-April and should be treated with IBA @ 500 ppm instead of NAA @ 500 ppm.
- Farmers needed to be educated about the potential of *H. salicifolia* selection, and Russian varieties needed to be introduced in the region.

ACKNOWLEDGEMENT

The authors are grateful to ICAR (NAIP) for providing the requisite funds for making this study possible. Help received from Dr. Jagdish Chand, Mr. Durga Dass and Mr. Ram Chand of Krishi

Vigyan Kendra Kukumseri (L&S) are duly acknowledged. All Krishi Vigyan Kendra Kukumseri staff, especially Programme Coordinator, scientists and technical staff are also acknowledged for their support and cooperation to conduct R and D work.

REFERENCES

1. Arimboor, R., Venugopalan, V.V., Sarinkumar, K., Arumughan, C. and Sawhney, R.C. 2006. Integrated processing of fresh Indian seabuckthorn (*Hippophae rhamnoides*) berries and chemical evaluation of products. *J. Sci. Food and Agri.* 86 (14): 2345-2353.
2. Aswal, B.S. and Mehrotra, B.N. 1994. *Flora of Lahaul Spiti*, Bishen Singh Mahendra Pal Singh, Dehradun, India, 761p.
3. Chauhan, A.S., Rekha, M.N., Ramteke, R.S. and Eipesen, W.E. 2001. Potential of seabuckthorn in processing of health food. In: *Seabuckthorn-A Resource for Health and Environment in Twenty First Century*. Proceedings of International Workshop on Seabuckthorn, p. 255-263, February 18-21, 2001, New Delhi, 274p.
4. Dhyani, D., Maikhuri, R.K., Misra, S. and Rao, K.S. 2010. Endorsing the declining indigenous ethnobotanical knowledge system of seabuckthorn in Central Himalaya, India. *J. Ethnopharmacol.* 127: 329-334.
5. Dhyani, D., Maikhuri, R.K., Rao, K.S., Kumar, L., Purohit, V.K., Sundriyal, M. and Saxena, K.G. 2007. Basic nutritional attributes of *Hippophae rhamnoides* (seabuckthorn) populations from Uttarakhand Himalaya, India. *Curr. Sci.* 92: 1148-1152.
6. Rongsen, L. 1992. *Seabuckthorn-A Multipurpose Plant Species for Fragile Mountains*, Occasional Paper No. 20, ICIMOD, Kathmandu, 62p.
7. Roy, P.S., Porwal, M.C. and Sharma, L. 2001. Mapping of *Hippophae rhamnoides* L. in the adjoining areas of Kaza in Lahaul and Spiti using remote sensing and GIS. *Curr. Sci.* 80: 1107-1111.
8. Singh, A. 2007. *Assessment of Plant Diversity and Conservation Status of Forest Vegetation in a Cold Desert Biosphere Reserve of Western Himalaya*, Ph.D. Thesis, Kumaun University, Nainital, India, 250p.
9. Singh, A., Butola, J.S., Samant, S.S., Sharma, P., Lal, M. and Marpa, S. 2012. Indigenous techniques of product development and economic potential of seabuckthorn: A case study of cold desert region of Himachal Pradesh, India. *Proceedings of the National Academy of Sciences, India Section B: Biological Sciences*: 82 (3): 391-398.
10. Singh, R.P., Bahar, N., Prasad, B. and Gupta, M.K. 1991. Ecological studies on *Hippophae rhamnoides* L. (seabuckthorn) in the cold desert of Himachal Pradesh. *Ann. Arid Zone* 30: 99-122.
11. Singh, R.P., Jishtu, V. and Negi, D.V. 1997. Studies on nursery techniques of *Hippophae rhamnoides* L. in Spiti valley of Himachal Pradesh. *Ann. Arid Zone* 30: 99-122.
12. Singh, V. 1994. A report on arboreal seabuckthorn (*Hippophae salicifolia*) from dry temperate Himalayas. *Journal of Tree Sciences* 3 (1): 67-68.
13. Singh, V. 2006. Research and development of seabuckthorn (*Hippophae* L.) in India. In: *Seabuckthorn (Hippophae L.)-A Multipurpose Wonder Plant. Vol. II: Biochemistry and Pharmacology* (V. Singh, Ed. in Chief, 2006), p. 542-575, Daya Publishing House, New Delhi, 600p.

14. Singh, V., Gupta, R.K., Arumughan, C., Sawhney, S.C., Rana, R.K., Singh, A., Lal, M., Devi, R. and Sharma, N. 2011. Biochemical evaluation of *Hippophae salicifolia* and *H. mongolica* as horticultural crops in dry temperate Himalayas. In: *Proceedings of National Conference on Seabuckthorn: Emerging Trends on R&D on Health Protection and Environmental Conservation"*, p.75-92, CSK HPKV, Palampur, HP.

15. Singh, V., Sharma, V.K., Sharma, M., Tyagi, S.P., Dhaliwal, Y.S., Rana, R.K., Saini, J.P., Pathania, P., Lal, M., Singh, A., Sharma, R.K., Sharma, V., Devi, R. and Kumar, R. 2010. Fifteen years of research on seabuckthorn in CSK Himachal Pradesh Agriculture University, Palampur. In: *Proceedings of National Conference on Seabuckthorn,* p.1-12, Febuary 16-18, 2010, CSK HPKV, Palampur, 82p.

16. Stobdan, T., Angchuk, D. and Singh, S.B. 2008. Seabuckthorn: An emerging storehouse for researchers in India. *Curr. Sci.* 94 (10): 1236-1237.

17. Zeb, A. 2004. Important therapeutic uses of seabuckthorn (*Hippophae*)-A review. *J. Biol. Sci.* 4: 687-693.

PART II:
GENETICS AND BREEDING

Chapter 6

Breeding Strategies of Russian Seabuckthorn (*Hippophae rhamnoides* ssp. *mongolica*) Varieties and their Global Introduction

Virendra Singh and Eury Zubarev*

Department of Biology and Environmental Sciences,
CSK Himachal Pradesh Agricultural University, Palampur – 176 062, H.P., India
*The Lisavenko Research Institute of Horticulture for Siberia, Barnaul, Altai, Russia

SUMMARY

In view of the findings of global research on seabuckthorn, the plant has emerged as an important crop for food, nutrition, health and cosmetic industries, besides being utilized in environmental conservation. Many genotypes of seabuckthorn were identified with high yield, and vitamins and anti-oxidants in countries like Russia, China, India, Germany, Mongolia, etc., as there was interest in local Seabuckthorn varieties due to their wide adaptation to local environmental conditions. But presence of thorns, small fruit size and high harvesting cost kept growers away from seabuckthorn for long in many countries, despite best efforts of research organizations of many countries. Soon it was realized that Russian seabuckthorn varieties (*Hippophae rhamnoides* ssp. *mongolica*) have advantage in characteristics of economic importance like few thorns or no thorns, large fruits (30-120 g/100), high yield (5-12 kg/plant) and high oil content (4-7 per cent) and carotenoids and easy harvesting. Russian varieties were introduced into the republics of former Soviet Union and then China decided to go ahead on large scale on introduction and cultivation of Russian seabuckthorn varieties, which proved very successful for their commercial cultivation to meet growing demands of industries. Russian varieties have also been introduced in Europe like Germany, Finland, Sweden and North America (USA and Canada). Lately, India has also started trials on introduction and evaluation of Russian varieties in Himalayan state of Himachal Pradesh. It is important to understand the breeding strategies of Russian experts and characteristics of Russian seabuckthorn varieties before their introduction in orchards.

Keywords: *Russian seabuckthorn varieties (Hippophae rhamnoides ssp. mongolica), Breeding strategies, Features, Introduction, Evaluation, Russia, Europe, North America and Asia.*

Seabuckthorn (*Hippophae* L.): *A Multipurpose Wonder Plant, Vol. 4: Emerging Trends in Research and Technologies* (Virendra Singh, *Editor-in-Chief*, 2014), p. 71–88, Daya Publishing House, New Delhi, India.

INTRODUCTION

Although, seabuckthorn has been quite popular in folk medicines in Siberia since centuries, it was the vitamin C discovery by Russians in early 1940s that led to its commercial utilization. As wild seabuckthorn is thorny and low yielding, therefore, there was a need for the improved varieties of seabuckthorn for commercial utilization. Seabuckthorn grows widely in Altai mountains, Buryatia, Tuva, central Asia, Kazakhstan, Caucasus, Baltic republics, Kaliningrad region, and south-western Ukraine, all republics of former Soviet Union. Siberia makes 57 per cent of the Russia geographical area and produces 16.4 per cent of Russia agricultural production. The region is characterized by extreme climatic conditions, low fertility of soil and humid to semi-humid conditions. In Russia, main institutes, *i.e.* the Lisavenko Research Institute of Horticulture for Siberia (LRIHS), Institute of Cytology and Genetics, Novosibirsk, Nizhny Novgorod State Agricultural Academy (NNSAA), Nizhny, Novgorod, Buryat Fruit and Berry Growing Experimental Station (BFBGES), Fruit and Berry Growing Experimental Station, Novobirsk (NFBGES) took up research are working on seabuckthorn.

Large variations in morphological and biochemical characters in natural populations of seabuckthorn offer us an opportunity for the genetic improvement of this plant. The researchers at the Lisavenko Research Institute of Horticulture in Siberia were the first in the world to develop the thornless, large fruited and high yielding varieties of seabuckthorn, which helped the country to cultivate the improved forms in over 5000 ha orchards in 15 regions from Moscow to Krasnoyarsk. The institute tested the improved varieties in Baltic republics, Byelorussia, Ukrain, Azerbaijan, Armenia and other regions. Kalinina (1988) has reviewed the breeding work on seabuckthorn in the CIS states. The present papers aimed to analyze the useful economic characteristics of Russian seabuckthorn varieties, breeding work carried out in Russian institutes, mainly at Lisavenko Institute and introduction of Russian seabuckthorn (*Hippophae rhamnoides* ssp. *mongolica*) varieties in other countries like China, Europe and North America.

HISTORY OF BREEDING OF SEABUCKTHORN

On recommendation of I.V. Michurin, the first attempts on selection of promising forms of seabuckthorn in this regard were made by M.A.Lisavenko (1897-1967) in Russia, as early as 1933. Lisavenko started the pioneering work on selection of high yielding forms in 1934 in Altai mountains of Siberia, Russia. Later on J. Gatin and E.I. Panteleeva joined the work. Plant breeders started breeding work with selection of promising forms from wild forms of seabuckthorn from Katun (Gorny Altai) and produced a number of better forms "Dar Katuni, Novost Altaya and Zolotoy Pochatok", the first 3 selections in former Soviet Union. This team did excellent work on breeding and produced a large pool of over 400 genotypes belonging to 10 ecotypes of seabuckthorn from different regions of CIS states. There was special emphasis on the selection of female forms possessing good characteristics like absence of thorns, dwarfness, compact crown, winter hardiness, large fruit (>0.3 g), strong fruit skin, solid pulp and long peduncle (5-10 mm), easily detachable from the branch, high oil content (>7 per cent), high vitamin C (>100 mg/100 g) and carotene (>6 mg/100 g). The male forms were selected for winter hardiness and their flowering synchronizes with flowers of female plants. As a first step, they collected 3,00,000 seeds from 148 selected forms from different regions and sowed them under similar conditions at the institute. After evaluation of fruits of 15,300 seedlings with free pollination, they made selections as first improved forms by the names of "Novost, Altaya, Dar Katuni, Zolotoi pochatok, Maslichnaya, Vitaminnaya and Chuyskaya" (Kalinina, 1988).

During second phase of their work since 1959, J. Gatin and E.I. Panteleeva started work on hybridization of distant eco-geographical forms of seabuckthorn. The female forms were taken from the improved elite forms selected earlier and naturalized in the conditions of Altai area. The male forms were taken from the central Asia and European parts of former Soviet Union. For next 28 years, they artificially crossbred 350 combinations and obtained 2 million hybrid seeds, raised 1,45,000 seedlings and evaluated the fruiting of over 65,000 first generation hybrid plants under the orchard conditions. These days study is in progress on second and third generation hybrids, while making selections of plus trees for a number of desirable characters. Breeders have been doing saturation crossbreeding to intensify some of the characters, which provides for a multiple pairing of the obtained hybrids with one of the primary parental forms, therefore, they have succeeded in producing dozen varieties of intense type.

Later, they started work on hybridization of distant forms of seabuckthorn from "Katun" (Altai) and "Tunkin" (Trans-Baikal), producing better forms with higher adaptability, productivity and richer in oil and carotenoids (Goncharov, 1995). In next programmes, there was emphasis on developing forms with dwarfness, early maturing, large fruits, compact crown, thornlessness, firm fruit skin, large peduncle, easy fruit separation by machines, and resistance to pests and adaptation to mechanic harvesting. In Siberia, breeders could produce 20 improved varieties of seabuckthorn, among them 16 were produced by breeders in Altai. These 20 varieties have been introduced in almost 20 regions of Russia and other CIS states. Experts have raised these improved varieties in 2000 ha in Altai, which is about third of total area under seabuckthorn in Russia. Therefore, people are raising orchards of seabuckthorn in different parts of Russia, mainly varieties from Altai province, because of better economic characteristics. After long period of research trials, the Lisavenko institute produced improved varieties of "Dar Katuni, Novost Altaya and Maslichnaya", which were registered as early as 1965. During next stage, "Oranzhevaya" was released in 1974. "Chuyskaya and Obilnaya" were released in 1979. This institute released 2 more varieties "Samorodok and Zolotistaya" in 1985. The success of breeding programme of seabuckthorn has been because of clearly defined goals, use of promising parent materials, use of effective breeding methods and selection for economic properties of seabuckthorn.

It can be seen that for the production of 21 improved varieties, the parent materials of seabuckthorn were used from Gorny Altai, Trans-Baikal and Siberian selections. Among these varieties, 3 varieties were developed through selection and 8 varieties through mutagenesis. During these breeding programmes, irrespective of parent material and methods used in the breeding, selection for main economic and biological properties were applied during the final stage.

Despite the release of improved varieties, breeders continue their work on the genetic improvement of genotypes for important characteristics like winter hardiness, early ripeness, higher productivity, richer in oil, vitamin C, carotenoids and pectin. On the one hand, further survey of wild populations for selection of better forms continues. On the other hand, work on hybridization between different better forms continues including methods of mutagenesis and biotechnology. Efforts have also been made on breeding of sweet forms of seabuckthorn. The fruits should contain sugars more than 9 per cent and acid more than 1.3 per cent. The most promising forms are "Tenga" and forms like "45-15-13" and "91-81-3" (Zubarev, 1998).

During the last 70 years, since 1933, more than 40 improved varieties have been developed at Lisavenko institute. Fruit yield varies from 7.5 to 18.0 t/ha, weight of 100 fruits varies from 62 to 120 g (maximum of 140 g), length of peduncle 3-6 mm, oil content 4-8 per cent, carotenoids 15-48 mg/100 g, sugar 5-10 per cent and acidity 1.0-1.9 per cent. The most of the varieties are thornless or mild thorny. The detail of characters of some of the Russian varieties are given below in Table 6.1 (Zubarev, 2005).

Table 6.1: Characteristics of Seabuckthorn Varieties Bred at Lisavenko Institute of Horticulture for Siberia, Barnaul, Russia

Sl.No.	Variety	Oil Per cent	Vit. C (mg/100 g)	Carotene (mg/100 g)	Sugar (per cent)	Acid (per cent)	Weight of 100 Fruits (g)	Fruit Yield* (t/ha)
1.	Novost	4.5	50	14.3	5.5	1.6	50	9.8
2.	Maslichnaya	5.7	64	10.6	4.0	1.5	37	10.2
3.	Dar Katuni	6.9	66	13.0	5.3	1.6	40	9.6
4.	ZolotoyPochatok	7.1	68	12.8	4.8	1.5	40	10.2
5.	Vitaminnaya	5.9	125	13.0	4.6	1.6	57	10.0
6.	Zivko	6.3	53	48.2	6.2	1.2	56	13.0
7.	Chuyskaya**	6.2	134	13.7	6.4	1.7	89	18.0
8.	Chulishmanka**	6.2	169	23.3	8.0	1.4	62	12.5
9.	Chechek**	7.8	157	24.7	7.8	1.3	77	15.1
10.	Tenga**	4.9	110	21.0	7.0	1.3	67	13.0
11.	Inya**	4.0	80	25.0	5.2	1.7	85	14.9
12.	Elizaveta**	4.8	80	19.0	6.2	1.3	100	12.7
13.	Altayskaya**	7.0	98	18.0	9.7	1.1	75	13
14.	Avgustina**	6.7	82	20.0	9.6	1.6	120	7.5
15.	Agurnaya**	6.2	112	12.7	8.3	1.9	110	7.5
16.	Dgemovaya**	8.0	154	29.3	7.6	1.0	75	7.5

* 1250 plants/ha, ** New varieties.

Source: Zubarev, LIHS, Barnaul.

Three methods were applied in the breeding of seabuckthorn, *i.e.* selection, hybridization and mutagenesis. After hybridization by over 100 crosses, about 50,000 hybrid seeds are obtained annually. Presently, there are 45,000 hybrid specimens of seabuckthorn at the institute.

In the breeding programme of seabuckthorn, it has been emphasized that the improved forms should adapt in the expected regions of introduction, must be superior in economic and biological characteristics to the existing forms of the that region, while meeting the local requirements, like early ripeness in the north and drought tolerance in the south and resistance to the attack of pests and diseases of introduction areas. The improved varieties released should be having wide genetic homeostatic.

CHARACTERISTICS OF SOME HIGH YIELDING SEABUCKTHORN VARIETIES AT LIHS, BARNAUL

There are many seabuckthorn varieties developed and adapted for the cultivation in Russia, features of some varieties, including new ones have been described below.

1. Avgustina

Cultivar was bred by selection of seedlings from open pollination of large-fruited forms ("Co-1 x Shcherbinka Katun" ecotype). It is a dwarf form without thorns, stem light brown, with a silvery gloss.

Fruits are large (weight of 100 fruit 110-140 g), egg-shaped, orange with a soft pulp, thin peel sweet and sour taste, and long peduncle (5.0-6.0 mm). It contains sugar-up to 9.6 per cent, acid-1.5 per cent, vitamin C-111.6 mg/100 g, carotenoids-20.0 mg per cent, and oil-6.7 per cent. It is winter hardy. Fruit yields at 5 years of age is 5.2 kg per bush (6.0 t/ha). It is early ripening, the fruit ripens in mid-August. Fruit is multipurpose; the fruits are suitable as table fruit for fresh consumption and for processing of different types of products.

2. Elizabeveta

Cultivar was bred by using the method of chemical mutagenesis. Seeds of varieties "Panteleevskaya" were treated in 1981 with a mutagen DES (0.05 per cent). Plant is bush type, oval crown with medium density and branches are dark brown. Fruits are large, weight of 100 fruits is 81.5-110.0 g, cylindrical, orange, and long peduncle (5.0-6.0 mm). Fruits have sweet and sour taste with a pleasant aroma. It contains 5.9-8.9 per cent sugar, acid 1.1-1.6 per cent, vitamin C 71.3-100.0 mg per cent, and oil 4.4-5.1 per cent. Winter hardiness is high. Yields in the 6 years of age is 14.7 and 18.0 t/ha after 8 years. Bearing of fruit starts in the 4th year after planting. Fruits are of multipurpose, it is suitable as table fruit for fresh consumption and for different types of products processing.

3. Chuskaya (Star of Altai)

It is the most widely cultivated variety of seabuckthorn in Russia and many countries. It is sweet, fruits are bright orange in colour. Fruits are large in size (weight of 100 fruits is 70-90 g, oval in shape and bright orange in colour. It contains vitamin C-up to 134 mg/100 g of fresh fruit, carotenoids -13.7 mg per cent and oil-6.2 per cent. Yields in the age of 7 year is 12.5-14.7 t/ha. It is easy to harvest through hand picking and one can picking up to 100-150 kg in 8 hours.

4. Chechek

Cultivar was bred by selection of seedlings from open pollination in 1973 of selected forms "7-66-321 (1 x Shcherbinka Chulyshman ecotype) and Katun" ecotypes. It is bush type with a compact crown, without spines. Fruits are large, weight of 100 fruits is 77.0-95.6 g, broadly oval, orange, with large rosy spots at the base of the calyx and stem. Peel is thick. Peduncle length is 4.0-5.0 mm. Taste is sweet and sour. It contains sugar-up to 7.8 per cent, acid 1.3 per cent, vitamin C-up to 157.0 mg per cent, carotenoids-24.7 mg per cent and oil-up to 7.8 per cent. It ripens in late August-early September. The average yield after 4th year of fruiting is 16.8 t/ha.

5. Chulyshmanka

Cultivar was bred by using the geographically distant hybridization by crossing varieties in 1966 Shcherbinka-1 form of environmental Chulyshman and Gorny Altai. Variety was adopted in 1990. It is a dwarf form and has a strong and compact crown. Fruits are medium size (weight of 100 fruits of fruit 62.4-67.2), oval and bright orange. Taste is sweet and sour, with the aroma. Peduncle is long (5.0-6.0 mm). It matures during August 25-September 10. It contains sugar- up to 7.96 per cent, acid-1.39 per cent, vitamin C-169.4 mg per cent, carotenoids-23.3 mg per cent and oil-up to 6.2 per cent. The average yields in the age of 7 years is 12.5 t/ha.

6. Dzhemovaya

The cultivar was bred by selection of seedlings from open pollination varieties. Fruits are oval, orange-red, with bright, large spot on the top and bottom of the fruit stalk. Weight of 100 fruits is 60.0-73.0 g. It is a late ripening variety. It contains sugar-up to 5.8 per cent, acid-1.3 per cent, vitamin C-

154.0 mg per cent, carotenoids-29.3 mg per cent and oil up to 10.2 per cent. Yields in the 6 years of age is up to 6.0 t/ha and 7-year onward the yield is 16.0 t/ha. Plants come in bearing after 4 years of planting. Fruit is used for multi-purposes.

7. Tenga

Cultivar was bred in using the geographically distant hybridization by crossing the 1968 class, "Shcherbinka-1 and Aleem" ('Katun" ecotype). Fruits are medium size (average weight of 100 fruits 58.7 g, maximum 69.0 g), oval, bright orange, with medium-sized bright spots on the top and bottom of the fruit stalk. It is an early ripening variety. It contains sugar up to 5.8-9.8 per cent, acid 1.3-1.5 per cent, vitamin C 101.6-264.1 mg per cent and carotenoids 12.5-31.7 mg per cent. It is a high yielding variety; the yield after 6 year of planting is to the tune of 26.8 t/ha. It bears fruit every year, starting in the 4th year after planting.

8. Golden Siberia

Cultivar was bred by the method of geographically distant hybridization due to interbreeding in the 1961 class "Shcherbinka-1" with selective form of the "Katun" ecotype. Plant is bush with oval crown of medium density. Weight of 100 fruit is 80 g; peduncle length is 4-5 mm. Fruits ripens in early September. It contains up to 7.2 per cent sugars, titratable acids-1.76 per cent, oil-6.4 per cent, vitamin C-165.0 mg/100 g, 5.5 mg/100 g of carotene, vitamin E-5.2 mg/100 g and vitamin B-1.01 mg/100 g. Winter hardiness is high. Yields in the 6 years of age is 14.6 t/ha and after 7 years, it yields 25.6 t/ha.

9. Panteleevskaya

Cultivar was bred by crossing in 1968, ("Shcherbinka 1 x Katun" ecotype) with selective form of the "Katun" ecotype. Fruits are large, weight 100 g fruit is 80.5-110.8, oblong-oval, red-orange, sweet and sour. Peduncle length 3.0-4.0 mm. Flesh is firm and it contains: sugar - up to 5.8 per cent, acid - 1.9 per cent, vitamin C - 87.5 mg per cent, carotenoids - up to 17.2 mg per cent, oil-5.7 per cent, vitamin E-to 10.5 mg per cent. Fruit ripen from 5th to 25th September. Yields in the 6 years of age 10.9t/ha and after 7 years, it yields 22.9 t / ha. The fruits are suitable for various kinds of processing.

10. Dewdrop

Cultivar was bred by selection of seedlings from open pollination of selected forms of "30-61-1363" ("Shcherbinka 1 x Katun"). It is a dwarf type with oval crown and sparse foliage. Fruit are large (up to 0.8 g /berry), wide-oval, orange, firm flesh with a sweet-sour taste. Peduncle length is 4.0-6.0 mm. It contains sugar 5.6-8.0 per cent, acid 1.3-2.0 per cent, vitamin C 40.2-92.0 mg per cent, carotenoids-up to 17.3 mg per cent, and oil 4.1-7.0 per cent. Fruit yield in 6-7th year after planting is 15.7 t/ha.

Besides above varieties, the most recent varieties are "Cloudaya and Padarok Sadu", which are high in carotene content, have large fruits and are high yielding.

SEABUCKTHORN BREEDING STRATEGIES AT LIHS, BARNAUL

Scientists working at the Lisavenko Institute of Horticulture for Siberia (LIHS), Barnaul adopted the strategies for the development of seabuckthorn varieties, which are high yielding with large fruits and high oil, carotenoids and adaptable in varied environmental conditions, and resistant to cold and diseases and pests. Some of the aspects of breeding strategies are as follows:

Thorns

LIHS has evaluated over 1,000 seedlings of seabuckthorn obtained from various crosses, with varying degree of thorniness and observed that after crossing thornless varieties with those with few

Figure 6.1: Some Important Russian Seabuckthorn Varieties at LIHS, Barnaul.

thorns produced a generation (up to 74 per cent of total numbers) with lesser thorns. For example, when experts crossed thornless female "Novost Altaya" with mild thorny male "Dar Katuni", they also got thornless plants, numbering 0.9-1.9 per cent. Consequently, these workers could obtain 5 thornless varieties, *i.e.* "Novost Altaya, Vitaminaya, Velikan, Prevoskhodnaya and Alei" and 13 mild thorny varieties (Kalinina, 1988).

Fruit Peduncle Length

Pedicle length and abscission affect the efficiency of mechanical harvesting. Therefore, fruits with long peduncle with easy separation are desirable for selection. Fruit skin toughness is an important trait, which affects the suitability for mechanical harvesting in seabuckthorn. Skin damage during fruit harvesting invites the growth of microbes, therefore, affecting the quality of raw materials and also causing crop losses. Forms were selected with not only larger fruit size, but also easily detachable by mechanical means. On crossing the "Chuyskaya" and "Kuddyrga-1" and other varieties with "Katun" ecotype, experts could produce promising hybrids. Improvement of length of peduncle helps in easy separation of fruits. The length of peduncle may vary from 1 mm to 10 mm. Some varieties with large peduncles (6-10 mm) are "Orangevaya, Luchezarnaya, Chiulishmanka and Ulala", which makes it easier to pick up the fruits (Panteleeva, 1998).

Fruit Weight

Large fruits are considered easy to harvest. Breeders in order to increase yield, looked for large fruit size, especially large fruit pulp size. As a result of breeding work, they could produce the fruits with weight of 100 fruits from 37 to 120 g/100 (Zubarev, 2005). The "Maslichnaya" had smallest fruits (37 g/100 g), whereas it was as high as 100 g/100 in "Elizaveta" and 110 g in "Agurnaya" and "Avgustina" has largest fruits (110-140 g/100). There are varieties with large fruits with high oil content.

Yield

Breeders also focused on the fruit yield. Fruits are produced on the branches, buds of which are borne a year earlier. The fruit yield depends on the size of fruits, their density on the branches, number and length of fruit bearing branches. The Lisavenko Institute has the distinction of producing all high yielding varieties of seabuckthorn. For example, "Novost Altaya and Dar Katuni" varieties produced 11.3 tonnes and 12.7 tonnes per ha, respectively, for over 10 years of period at 4 m × 2 m spacing, while peaking at 23.1 and 27.6 tonnes per ha respectively, without any irrigation. Effect of reduced spacing was better, as "Chuyskaya" variety produced higher yield of fruits varying from 28.7 tonnes at 4 m × 2 m to 33.0 tonnes at 4 m × 1 m and even higher (49.2 tonnes) on being irrigated. The institute has recommended the following high yielding female improved varieties for orchard raising, "Vitaminnaya, Dar Katuni, Shcherbinki-1, Chuskaya, Oranzhevaya and Obilnaya" and following male varieties, "Katun and Sayan" (Kalinina, 1988). Their Evaluation studies on introduction in introduction areas found that seabuckthorn varieties have produced an average of 12.9 tonnes of fruits per ha in 5th year after plantation under orchard conditions at the experimental farm, Barnaulskoye, Altai, 16.0 tonnes in the Chistoozerny State Farm, Novosirbirsk region and 13.1 tonnes in Magnitogorsky state farm in Chelyabink region. After 6 years of plantation, 3 varieties, "Samorodok, Chuyskaya and Oranzhevaya" produced 11.1-16.0 tonnes per ha on an average and a maximum to 21.1-26.5 tonnes per ha in Gorky region. A 3-year old variety produced 6.7-9.4 tonnes fruits per ha in Yaroslavl region (Kalinina, 1988,). Most successful in Russia and China and promising in fruit yield has been "Chuskaya" with maximum fruit yield of 18 tonnes/ha. However, "Avgustina" with largest fruit of 120 g/100, had 7.5 tonnes of fruits per ha (Zubarev, 2005).

Winter Hardiness

Seabuckthorn, which belongs to temperate zone, may or may not tolerate winter frost. Seabuckthorn forms of Altai region, where Lisavenko institute is situated do not have significant winter damage due to extreme harsh conditions (-40 °C). However, in some regions of Siberia, floral buds of male plants of seabuckthorn are often frostbitten, sharply decreasing the number and viability of pollen, causing reduction of fruit yield in female plants. After evaluation of floral buds of 10 hybrid forms of male seabuckthorn, the breeders could select "Katun and Alei male" varieties (Kalinina, 1988).

Fruit Quality

Quality of fruit is an important parameter in seabuckthorn breeding programme. Seabuckthorn fruit is known mainly by presence of high contents of vitamin C, oil and carotenoids. Russian varieties are considered very rich in oil content (4.5-8 per cent) and carotenoids, they are generally not that rich in vitamin C (50-170 mg/100 g) (Zubarev, 2005), as compared to Chinese and Indian seabuckthorn. Variations in vitamin C contents have been widely found in geographical regions (Korovina and Fefelov, 2006). A low vitamin C correlation with fruit size has been reported (Karhu and Ulvinen, 1999). In hybrids, vitamin C content has been found related to the genotype of a male population. Other compounds of interests for a breeder may be vitamin E, riboflavin, niacin, folic acid, carotenoids and unsaturated fatty acids.

Bioactive compounds may be related to the characteristics of a fruit. In seabuckthorn, pigments and aromatic compounds occurs in fruit skin and hypanthium. Therefore, in case we want to extract mainly these compounds, then fruit skin characteristics and colour (red or reddish orange), etc. may be useful parameters for selection. For oil extract from the seeds, oil content and seed size are important criteria. Seed shape has not been studied to have any influence on oil content.

Initially, breeders believed that small fruits are richer in oil than large fruits. Soon, they discovered that even large fruits are richer in oil content than small fruits, as there was a positive correlation between fruit size and oil content. Consequently, breeders could produce improved forms with high oil content of 4.7-6.9 per cent, 50-330 mg/100 g vitamin C and 2.8-7.6 mg/100 g carotene. Vitamin C rich varieties are "Oranzhevaya, Sibirskaya and Yantarnaya, Zolotistaya Sibiri". Carotene rich forms are "Luchezarnaya, Panteleevskaya, Yantarnaya and Zolotistaya Sibiri" (Kalinina, 1988). "Zivko" has 48.2 mg/100 g carotenoid content and "Dgemovaya" has about 29.3 mg/100 g carotenoids in the fruits. "Chechek" and "Dgemovaya" have 7.8 per cent and 8 per cent oil, respectively. Varieties with medium oil content, but with high fruit productivity or yield like "Chuyskaya" (6.2 per cent oil and yield 18 t/ha) have been considered useful for raising orchards, as total oil production will be higher than from oil-rich forms with lower fruit yield. However, consideration of high contents of carotenoids and tocopherols are important priority.

Pollen

The number of pollens and duration of their release, compatible with the development of female flower, directly affect the fruit yield. Therefore, it is an important parameter for selection that male plants produce optimum quantity of pollen at the time of female plants is in flowering stage with receptive stigma (Buglova, 1981). Russian breeders have developed a number of male varieties with promising characteristics "Dar Katuni, Katun, Sayan and Aley"".

Varieties for Handpicking

Presently, seabuckthorn varieties are provided to the growers, which collect fruit by handpicking. Varieties of seabuckthorn have been developed which are easy for hand-picking for fruits, *i.e.*

"Chuyskaya, Inya and Avgustina", etc., in which a person can collect 100-150 kg fruits in 8 hours by handpicking method. Some high yielding cultivars have been developed, which are suitable for hand-picking of their fruits and a person can collect as high as 200-250 kg in 8 hours.

Varieties for Mechanical Harvesting

Handpicking of seabuckthorn is very expensive and time consuming in Russia, therefore, the Lisavenko institute has developed a mechanical harvester, which can harvest 10 tonnes of fruits per 8 hours. However, most of the varieties are not suitable for mechanical harvesting; therefore, breeders have been working on varieties rich in economic and biological characters and also suitable for mechanical harvesting of fruits. The main characters for mechanical harvesting of seabuckthorn are easy and dry separation of fruits, long peduncles, compact crown, moderate growth and strong branches.

Harvesting Period

In order to expand harvesting period, breeders have produced early and late harvesting varieties of seabuckthorn. Consequently, they could extend the harvesting to 2.5-3.0 months of fruiting period, *i.e.* from second half of July to the middle of October. Further, varieties are also being developed, which can be mature and develop fruits during winter, while maintain their biochemical quality.

Pest and Diseases Resistant

For the higher yield and better acceptability in market, breeders have given special attention to the development of varieties, which are resistant to pests and diseases like *Fusarium*, which infect plants even in nursery stage.

Taste

Taste of seabuckthorn fruits is an important parameter for utilization in various products development. However, the sugar-acid index (SAI) can be used to decide the taste of seabuckthorn fruits. The sugar-acid index of seabuckthorn fruits under the Altai conditions is on the higher side. On the basis of the studies, the experts at this institute have made a table to decide the taste of the fruit on the basis of sugar-acid index (Table 6.2).

Dissert Varieties

Seabuckthorn has high demand in fresh juice, juices without addition of sugar, beverages, jams, etc., therefore, considering requirement of health food products, dissert varieties are being developed under the breeding programme. For health food production, vitamin C rich varieties are given preference. However, for cosmetic and medicines production, seabuckthorn varieties having oil with high content of oil, carotenoids and tocopherols are preferred.

Table 6.2: Sugar-Acid Index and Taste Relationship in Seabuckthorn Fruits

Sl.No.	*SAI*	*Taste*
1	<4	Sour
2	4-6	Sweet sour
3	6-8	Sour sweet
4	>8	Sweet

Source: Zubarev (2005).

BREEDING OF SEABUCKTHORN AT OTHER INSTITUTES IN RUSSIA

Nizhny Novgorod State Agricultural Academy (NNSAA), Nizhny, Novgorod

Like the Lisavenko Institute, NNSAA, earlier known as Gorky Agricultural Institute, has also been in front of research work on breeding of seabuckthorn in Russia. Work on selection and introduction was started in 1949 by I.P. Eliseev. The first cultivars "Scherbinky-1 and Scherbinky-2" were selected from Sayan and Irkut rivers region. Scherbinky-1 was used for many hybrid varieties development. Five varieties were developed by mutagenesis methods. These varieties have large fruits, few thorns and dwarf type. The institute developed hybrids from crosses between "T-50 × Katun and T-50 × Sayan". The pollinator was obtained from seeds of "Katun 24" irradiated with by gamma rays (50 cR) (Fefelov, 1998). The institute has selected hybrid cultivars and elite forms of seabuckthorn with high content of carotenoids like "Zarevo (259 ppm), Nadezhda (316 ppm), Plamennaya (224 ppm) and Ryabinka" (333 ppm). The forms with red colour and high contents were developed like "Gomelskaya, T-50, Karatal 8" and a male form "K-24 and Otradnaya" (Fefelov and Selekhov, 2009).

The Buryat Fruit and Berry Experiment Station (BFBES), Ulan Ude

The focused work on seabuckthron breeding started since 1976 at BFBES, Ulan Ude. More than 80 forms were tested from various regions like Temnik, Selenga, Irkut rivers, etc. 7 varieties like "Atsula, Ajayganga, Stepnaya, Sayana, Bajyan-gol, Bakal Ruby and Naran" have been finalized and introduced. "Zerya Dabat" was developed by mutagenesis (Myahanova, 1998). The oil content in fruits was quite high in "Sayana (7 per cent) and Naran" (9 per cent). The best results of fruit production were at spacing of 4 m × 1.5 m (Shripnimbueva and Myakhanova, 2009). The cultivars "Novost Altaya, Maslichnaya and Prevoshodnaya" were high wilt resistant (Fefelov and Selekhov, 2009).

Institute of Cytology and Genetics, Novosibirsk

The institute developed a number of varieties of seabuckthorn in collaboration with Novosibirsk Zonal Fruit and Berry Experimental Station. The main focus was to develop hybrid varieties with large fruits and few or no thorns. Some of the varieties are- "Rumyanets" with red fruit and high carotene content, maturing in mid August; "Druzhina" is an early ripening (mid August), dwarf plants and large fruits; "Kapriz", early ripening with high sugar content (up to 10 per cent) and pleasant aroma; "Podruga" with very large fruits (0.9-1.1 g); and "Zarnitsa" late ripening, orange-red colour fruit with high carotene content (28.8 mg per cent) and oil (4.5 per cent), etc. (Kreymer *et al.*, 2009).

"Triumf", a hybrid form was developed by Shchapov (2009). Fruits are large (0.7-0.8 g) with red colour, large pedicle (5-8 mm), vitamin C 93 mg/100 g, carotene 3.4 mg/100 g and oil 4.3 per cent. It is late ripening (mid September). The fruit yield is 11 kg per plant. As a results of mutagenesis, a number of seabuckthorn varieties were produced in Russia, among them "Zyrianka" being prominent, which has been introduced in Russia and other countries. The weight of 100 fruits was 61-64 g, vitamin C-110 mg/100 g, carotenoids-19.7 mg/100 g and fruit yield-13.9 kg/plant (Privalov, 1986).

Minusinskaya ESHM, Naberezhnaya

U.G. Leonova selected the first 2 varieties "Skorospelka and "Kukuruzka", were early selections from Baikal ecotype, but planted in limited area due to many thorns and small fruit size. Varieties like "Obilnaya, Prevoskhodnaya and Chyuiskaya", and pollinator "Aley" collected from LIHS, Barnaul were introduced in southern area of Krasnoyarskiy region. But, there were problems like attack by seabuckthorn fly. Many varieties with high economic characters were introduced with promising results. New varieties like "Minusa and Solnechnaya" were developed and introduced in middle Siberia. They achieved as high as 25 tonnes/ha yield (Smykova, 2009).

Michurin Zonal Fruit and Berry Growing Experimental Farm, Novosibirsk

Research work on selection and breeding was started with the purpose of development of varieties, with high economic characters and high contents of biologically active compounds. The institute developed collaboration with Institute of Cytology and Genetics, Novosibirsk. Work started in 1995 with new varieties "Podruga and Zolotoi Kaskad" and 5 other commercial varieties. The maximum vitamin C content was found in "Krasny Fakel (148 mg per cent) and Ivushka" (115 mg per cent). A high content of carotenoids (46 mg per cent) was estimated in red fruits of "Sibirsky Rurnya-nets". A high content of fruit oil (6.3 per cent) was found in "Krasny Fakel" (Karpova, 1998).

Moscow State University, Moscow

Work started on introduction of seabuckthorn at Botanical Garden of Moscow State University, Moscow in 1952. The main focus was to introduce high yielding seabuckthorn varieties able to grow in stress conditions. Out of the 32 seabuckthorn varieties, 7 have been brought under cultivation in orchards, which are "Trophimovskaya, Botanicheskaya, Botanicheskaya Aromatnaya, Botanicheskaya Lubitelskaya, Moskvichka, Ortradnaya and Perchik" (Aksenova and Dolgacheva, 1998).

INTRODUCTION OF RUSSIAN SEABUCKTHORN VARIETIES IN OTHER COUNTRIES

Due to their useful economic characters like few or no thorns, large fruits, high fruit yield and oil content, Russian seabuckthorn varieties have been introduced in over 20 countries like CIS countries, China, Mongolia, Canada, USA, India, Finland, Germany etc. Experience of introduction and evaluation of some the countries are given below.

China

As local seabuckthorn species like *H. rhamnoides* ssp. *sinensis*, the most popular one, have small fruits and many thorns, they are difficult to harvest and could not be introduced in orchards for commercial purpose on scale, as initially planned in China. Therefore, improved Russian varieties, mostly thornless have been introduced in many provinces of China at several places. They are being used as pure cultures for commercial cultivation or for cross hybridization. Generally, local names have been given to Russian varieties being cultivated. For breeding purposes, further selections are being carried out after evaluation tests as well as female Russian seabuckthorn, which have good economic characters, are being used for cross with male Chinese ssp. *sinensis*, which are widely adapted.

About 20 seabuckthorn varieties with large fruits and thornless had been imported by Institute of Soil and Water Conservation, Chinese Academy of Sciences and Ministry of Water Resources, Yanling from Novosibisk, Moscow and Ulan-Ude of Russia during 1995-2004. They have tentatively selected 9 fine seabuckthorn varieties for evaluation in the 7 experimental sites in this area (Daiqiong *et al.*, 2014). They have characters of fine growth, resistance and higher fruit yield. They have tentatively selected 9 fine Russian varieties of seabuckthorn ("Zirianka, Druzhina, Red torch, Ivshca, Batanical Garden, Buliyate, Panjielieyewa and Yalishanda") and developed 7 experimental bases in the "three Northern Areas" (Qiqihaer, Fuxin, Dalian, Huhehaote, Eerduosi, Wuqi, Ansai) and Loess plateau in China. These regions belong to moist, semi moist and semi-arid areas within dry temperate zone and temperate zone, respectively. There are natural forest and huge plantations of seabuckthorn in these regions. The early evaluation studies found the diameter of fruits of Russian varieties have been found to be 9-14 mm, length of fruit stalk is 2-4 mm, weight of per 100 fruits is 20-60 g and early fruit yield per plant is 0.9-3.1kg. These values are 1.8-2.5, 1.3-2.7, 4.6-8.8 and 3.8-6.2 times higher than local wild Chinese seabuckthorn, respectively (Table 6.3). Russian seabuckthorn has the characteristics of larger fruit, no thorns and high yield. The quality, oil contents and active substances of Russian seabuckthorn varieties are higher than local wild Chinese seabuckthorn, therefore, they are important genetic materials for introduction and selecting economic forms for improving the Chinese seabuckthorn.

Table 6.3: Early Results of Characteristics of Fruits of Introduced Russian Seabuckthorn Varieties in some Regions of China

Name of Russian Varieties	*Fruit Length (cm)*	*Fruit Breadth (cm)*	*Length of Fruit Stalk (cm)*	*Weight of 100 Fruits (g)*	*Fruit Yield (kg/plant)*
Zirianka	0.9-1.4	0.7-1.0	0.2-0.4	32.7-52	1.9-2.2
Druzhina	0.9-1.3	0.7-0.9	0.2-0.4	26-40	2.7
Podruga	0.8-1.1	0.7-0.9	0.3-0.4	30-39	2.3
Sibirian Rumianes	0.8-1.1	0.7-0.9	0.2	32-41	0.9
Gold Cascad	0.8-1.1	0.6-0.8	0.3	40	1.2-2.6
Red torch	0.8-1.1	0.6-0.9	0.2-0.3	42	1.8
Ivshca	0.9-1.3	0.7-0.9	0.4	60	2.9
Ornistaya	1.1-1.2	0.8	0.2-0.3	60	2.6
Zarnitsa	1.0-1.2	0.7-0.9	0.3	43	1.5-2.5
Botanical Garden	1.1	0.8	0.3	45	1.6
Panjielieyewa	1.4	1.0	0.3	46	1.9
Buliyate	1.0	0.8	0.3	42	1.3
Yalishandai	1.3-1.5	0.9	0.4	46-60	3.1
Non-seed SBT	1.0	0.8	0.4	23-52	0.6-2.0
Orange×Chinese SBT	1.0	0.9	0.3-0.5	31-503	2.1
Azhula×Chinese SBT	0.7-0.9	0.9	0.4-0.5	23-52	0.6-2.0
Chinese SBT(CK)	0.4-0.6	0.3-0.5	0.1-0.2	7-7.4	0.4-0.65

SBT: Seabuckthorn.

Source: Daiqiong *et al.* (2014).

However, the nutritional values of Russian seabuckthorn are lower than Chinese wild seabuckthorn. The height, crown and diameter at butt end of Chinese seabuckthorn are 1.5-2.5 m, 1.3×1.5 m, and 4-5 cm, respectively and these values are 1.2-1.9, 1.3-2 and 1.3-2.2 times of Russian seabuckthorn. The main reason was that the introduced Russian seabuckthorn has lower adaptability and resistant capability, especially lower resistance to dry conditions and high temperature during summer, and also to plant diseases and insect pests. On the other hand, Chinese seabuckthorn has strong adaptability and higher content of vitamin C. The adaptability and growth potential of hybrid varieties of the Russian seabuckthorn and Chinese seabuckthorn are promising. The plant height of 4 years hybrid "Orange × Chinese seabuckthorn" was 1.7-2.3 m, while the length of new branch was 30-48 cm, diameter of fruit was 0.7-1.0 cm and weight of 100 fruits was 23-50 g. These properties and stability of inheritance should be observed successively.

Institute of Yancheng Technology introduced 1-year old seedlings of 8 cultivars from the Institute of Cytology and Genetics Siberian Branch, Novossibirsk, Russia in 1999 (Ruan *et al.*, 2004). Plants were maintained at the Institute of Yancheng Technology's Test Garden, Yancheng, Jiangsu Province,

China. They were evaluated with cultivars from China and Mongolia. The names of Russian varieties have been changed to local names. The performance and yield of these cultivars are given in Table 6.4.

Table 6.4: Performance Evaluation of Russian Varieties and Hybrids of Russian and Chinese Seabuckthorn

Code	*Variety*	*Native*	*Growth and Yield*
XY	Xiangyang	Russia (ssp. *mongolica)*	Height about 2-3 m, resistant withered disease, no thorn, orange and columnar fruit, mean weight per fruit of 0.98 g and yield of fresh fruit over 22.5 t/ha
CS	Chengse	Russia (ssp. *mongolica)*	Height about 3 m, late maturing, few and in firm thorn, salmon pink and oblong of fruit, mean weight per fruit of 0.66 g and yield of fresh fruit of 12.0-33.0 t/ha
CY	Chuyi	Russia (ssp. *mongolica)*	Height about 2.5 m, no thorn, round fruit, mean weight per fruit of 0.90 g and yield of fresh fruit over 9.0 t/ha
ALY	Aleiyi	Russia (ssp. *mongolica)*	Male pollinator type, strong growth vigour, height over 3.8 m, cold resistance in flower, plentiful pollen and strong viability (95.4 per cent)
ZL	Zeliang	Russia (ssp. *mongolica)*	Selection from wild seabuckthorn, treated by gamma-ray for breeding, height about 2.5 m, longer fruit stalk (6-7 mm), small and infirm thorn, shallowly orange-yellow and columnar fruit, mean weight per fruit of 0.64 g and yield of fresh fruit of 11.4 t/ha
HGG	Houguang	Russia (ssp. *mongolica)*	Hybrid of "Huoju × Zeliang No. 104", no thorn, oval and orange-red fruit, mean weight per fruit of 0.60-0.70 g
NY	Nuyou	Russia (ssp. *mongolica)*	Hybrid of "118/4 × 120/2" of the cultivars, strongly resistant to cold, oval and orange-yellow fruit, no thorn, mean weight per fruit of 0.90-1.10 g and yield of fresh fruit of 12.7 t/ha
HY	Hongyun	Russia (ssp. *mongolica)*	Seedling of open-pollination of "101T", red-fruit type, strongly resistant to cold, resistant withered disease, oval and red fruit,mean weight per fruit of 0.60–0.70 g
LHYH	Liaohuerhao	China (Hybrid)	Hybrid of "Qiuyisike" from Russia × *H. rhamnoides* subsp. *sinensis* from China, height about 2.2-2.3 m, thorn in top branch and no or few thorns below, dark saffron-yellow and columnar fruit, fruit stalk of 0.3-0.4 cm, mean weight per fruit of 0.55-0.61 g

Source: Ruan *et al.* (2004).

As early as 1990, the Institute of Berries of Heilongjiang, Academy of Agricultural Sciences, Suiling, Heilongjiang, China had introduced the varieties of seabuckthorn from the Lisavenko Research Institute of Horticulture of Russia, to carry out experiments on the breeding and cultivation (Shan *et al.*, 2014). The promising Russian varieties, locally named were "Chuyi, Hunjin, Chengse and Wulangemu" to cultivate in the region.

Canada

Selections of Russian varieties (ssp. *mongolica*) were released for propagation in nurseries for commercial cultivation in 2000 in Saskatchewan, Manitoba and Alberta. The nurseries were raised at Prairie Plant Systems in Saskatoon, D'nA Gardens in Elnora, AB and Day Spring Nurseries in Teulon, MB. The orchards are in early stage of evaluation (Wang, 2007). The early results of evaluation some of the selections are as follows:

Chuyskaya

It was introduced from Russia in 2001 and fruit is harvested during mid to late August. It has few or no thorns. For pollination, "Sergei" variety is used. The weight of 100 fruits is 97.3 g, moisture content being 87.4 per cent and yield is 1.3-1.9 kg/plant.

Samarodok

The fruit yield is 0.8-1.3 kg/plant, weight 72.7 g/100 and moisture 87 per cent.

Vitaminnaya

It was also introduced in 2001 and harvested by mid to late August. It is also thornless or mild thorny. It is pollinated by "Sergei". The fruit yield is 2 kg/plant and fruit weight being 61.7 g/100 with moisture content 88.9 per cent.

Orangevaya

Introduced from Russia in 2001, harvested during mid to late August, cross pollinated by "Sergei" and has few or no thorns. The fruit weight is 58.8 g/100 and yield being 6 kg/plant.

Obilnaya

It has similar features like "Vitaminnaya", but less yield of 2.2 kg, and but larger fruit weight 86.4 per cent and moisture content 86.6 per cent.

USA

The cuttings of 5 cultivars of Russian seabuckthorn ("Dar Katugne, Tchyskaja, Orangevaja, Maslitchnaja and Tchuiskaja") were introduced from the Altai region of Russia to Trout Lake (sandy loam soil of volcanic origin, pH 5.6) and White Salmon (alluvial soil, pH 5.5), Washington in 1996 (Lobatcheva *et al.*, 2002). The saplings were transplanted to the fields at the end of April, 1997. All the plants were treated with standard agronomic practices, with additional irrigation during hot and dry summer seasons. Plant and fruit performance of each cultivar was closely monitored. The yield, oil content and composition of fruit of cultivars were evaluated after 4 years of plantation.

Performance of 3 cultivars "Dar Katugne", "Orangevaja" and "Tchuiskaja" was most successful. These cultivars achieved 115 to 125 cm height after 4 years. Most of the morphological features were similar to original ones as estimated in Siberia. Leaf and berry size, and colour and plant thorniness are presented in Table 6.5. The fruit yield (Table 6.6) was about 25 per cent less than estimated in Siberian growing conditions. However, oil percentage of the dried fruit, total carotenoid content, and flavour was similar with plants grown under the western Siberian conditions (Table 6.6).

Table 6.5: Features of 3 Seabuckthorn Cultivars Introduced from Western Siberia (Altaiskii Krai, Russia) Grown Washington State after Five Years of Growth

Cultivar	*Plant Height (cm)*	*Leaf Length (cm)*	*Leaf Colour*	*Fruit Colour*	*Fruit Weight (g/100)*	*Plant Thorniness*	*Start of Flowering*	*Start of Ripening*
Dar Katugne	115	16	silver green	Red	46	Thorny	May 27	Sept. 25
Orangevaja	125	17	dark green	Yellow orange	67	Thorny	May 29	Sept. 28
Tchuiskaja	118	17	silver green	Orange	68	Less thorny	May 22	Sept. 22

Source: Lobatcheva *et al.* (2002).

Table 6.6: Characters of 3 Russian Seabuckthorn Cultivars in USA

Cultivar	*Fruit Yield (kg/plant)*	*Fruit Size (mm²)*	*Dry Matter (per cent)*	*Oil Content (per cent of dry fruit)*	*Total Carotenoids (mg per cent)*	*Taste*
Dar Katugne	3.6	28	31	27.6	25.6	Acidic sweet
Orangevaja	5.1	26	39	38.9	35.8	Like orange
Tchuiskaja	6.52	25	33	31.3	42.8	Bitter sweet

Source: Lobatcheva *et al.* (2002).

Belarus

Belarus was a part of former Soviet Union, therefore, seabuckthorn varieties were introduced from Russia to Belarus. Institute of Fruit Growing, Samokhvalovichy, Belarus collected 19 Russian seabuckthorn varieties during 1980s. Further during 1992-2006, evaluation work of 23 Russian seabuckthorn varieties was done. Seven promising hybrids were developed by V.A. Phephelov. Recently, variety "Plamennaya" having high fruit productivity (17 t/ha), large fruit (0.8 g) and rich in compounds was released for cultivation (Shalkevich and Radkevich, 2009).

India

The first Russian seabuckthorn variety was introduced through seeds at Research Farm of CSK Himachal Pradesh Agricultural University at Kukumseri, Lahaul, a dry temperate region of Himalayan state of Himachal Pradesh (Singh and Singh, 2004; Singh *et al.*, 2011). Named with code word as "HI-1" (H. *rhamnoides* ssp. *mongolica*), a mild thorny form, presently it is being multiplied and will be provided to growers. The fruit weight is about 30 g/100, oil content of yellowish fruits is about 5 per cent. The fruit yield is about 5.4 kg/plant and productivity is 10.2 tonnes/ha. It is harvested in August month. Further two more Russian varieties have been introduced, which are under evaluation process.

CONCLUSION

Russia has a long history of breeding work on the improvement of seabuckthorn forms. Several cultivars and varieties were developed over the last 70 years. Russian seabuckthorn varieties are characterized by large fruits (30-120 g/100), high productivity (6-18 tonnes/ha), mild thorny or no thorns and high oil content (4-9 per cent) and carotenoids and easy harvesting, however, lesser in vitamin C content (50-150 mg/100 g). The Altai region of Siberia is the main area of natural distribution of many promising forms of seabuckthorn. Varieties have been developed by selection, hybridization and mutagenesis methods, mainly at 5-6 institutes. There are some important economic characters, which need to be considered before their introduction in other regions. Russian varieties have been introduced successfully in CIS states, China, India, Europe and north America. There have been some difficulties too like failures, or small size and fruit yields and re-occurrence of thorns in dry regions like Loess plateau, China. However, hybridization of Russian female plants (ssp. *mongolica*) and local male plants (ssp. *sinensis*) have been very successful in such regions. Among the various Russian varieties, the most successful varieties are "Chuyskaya", which has been introduced in Russia and other countries on large scale. Some of the Altai forms have also been susceptible to diseases and pests too on introduction in new regions. But largely, Russian seabuckthorn varieties have been very successful and becoming the basis of commercial cultivation globally.

REFERENCES

1. Aksenova, N.A. and Dolgacheva, V.S. 1998. New specimens of the seabuckthorn (*Hippophae rhamnoides* L.) in the botanical garden of Moscow State University. In: *Material of International Symposium on Seabuckthorn* (*Hippophae rhamnoides*), p.173-174, Russia, 248p.

2. Buglova, T.L. 1981. Effect of male plants of *Hippophae rhamnoides* on the yield of a plantation. *Lesn. Khoz.* 11: 39-40.

3. Daiqiong, L, Wu, Q., Zhang, J., Guo, C., Guo, Y., Chen, Y., Li, Z., Sun, X., He, D., Ruan, C. and Zhou, Z. 2014. Studies on introduction and evaluation of improved Russian seabuckthorn varieties in China. In: *Seabuckthorn-A Multipurpose Wonder Plant. Vol. IV.* (V. Singh, Ed. in Chief, 2014), New Delhi. (In press)

4. Fefelov, V.A. 1998. Summary and selection purpose of *Hippophae rhamnoides* in the Volgo-Vyatka region. In: *Material of International Symposium on Seabuckthorn (Hippophae rhamnoides),* p.149-151, Russia, 248p.

5. Fefelov, V.A. and Selekhov, V.V. 2009. Genetic and selective estimation of hybrid seabuckthorn seedlings by fruit colour. In: Proceedings of ISA conference on *"Seabuckthorn on the Way Between Science and Industry Interaction"*, p. 55-56, Russia, 199p.

6. Goncharov, P.L. 1995. Seabuckthorn in Siberia: Problems and solution. In: *Proceedings of International Workshop on Seabuckthorn*, p. 90-93, Beijing, China, 206 p.

7. Kalinina, I. 1988. Seabuckthorn. *Science in the USSR*: 80-87.

8. Karhu, S.T. and Ulvinen, S.K. 1999. Vitamin C: A variable quality factor in seabuckthorn breeding. In: *Agri-Food Quality II: Quality Management of Fruits and Vegetables: From Field to Table* (M. Hagg, R. Ahvenainen and A.M. Evers, Eds.), p. 360-367, April 22-25, 1998, Turku, Finland.

9. Karpova, E.A. 1998. Breedings of seabuckthorn for biochemical composition improvement in west Siberia. In: *Material of International Symposium on Seabuckthorn* (*Hippophae rhamnoides*), p. 166-169, Russia. Russia, 248p.

10. Korovina, M.A and Fefelov, V.A. 2006. Biochemical variations in seabuckthorn (*Hippophae rhamnoides* L.) growing in different regions of CIS states. In: *Seabuckthorn* (*Hippophae L.*)*-A Multipurpose Wonder Plant. Vol. II. Biochemistry and Pharmacology*, (V. Singh, Ed. In Chief, 2006) p. 108-132, Daya Pub. House, N. Delhi, 600p.

11. Kreymer, V.K., Shchapov, N.S. and Belih, A.M. 2009. Varieties of seabuckthorn for industrial and individual gardening. In: Proceedings of ISA conference on *"Seabuckthorn on the Way Between Science and Industry Interaction"*, p. 66-67, Russia,199p.

12. Lobatcheva, I.I., Letchamo, W., Huszar, L., Muchortov, S.A., Malkova, N.N. and Panteleeva, E.I. 2002. Evaluation of Siberian seabuckthorn cultivars in Washington State. In: *Trends in New Crops and New Uses* (J. Janick and A. Whipkey, Eds.), ASHS Press, Alexandria, VA.

13. Myahanova, N.T. 1998. The results of seabuckthorn selection in Buryatia. In: *Material of International Symposium on Seabuckthorn* (*Hippophae rhamnoides*), p. 146-147, Russia, 248p.

14. Panteleeva, E.I. 1998. Seabuckthorn breeding on lengthening of pedicle and diminishing of a breakaway (taking off effort). In: *Material of International Symposium on Seabuckthorn* (*Hippophae rhamnoides*), p. 144-145, Russia, 248p.

15. Privalov, G.F. 1986. Zryianka: A mutant variety of seabuckthorn. *Mutation Breeding Newsletter*. 28: 4-5.

16. Ruan, C., Qin, P., Zheng, J. and He, Z. 2004. Genetic relationships among some cultivars of seabuckthorn from China, Russia and Mongolia based on RAPD analysis. *Scientia Horticulturae* 101: 417-426.

17. Shalkevich, M.S. and Radkevich, D.D., 2009. Seabuckthorn introduction in Belarus. In: Proceedings of ISA conference on *"Seabuckthorn on the Way Between Science and Industry Interaction"*, p. 52-53, Russia, 199p.

18. Shan, J., Gao, Q. and Ding, J. 2014. A Comprehensive evaluation study on the adaptability of the introduced Russian and Mongolian seabuckthorn varieties. In: *Seabuckthorn (Hippophae L.)-A Multipurpose Wonder Plant. Vol. IV.* (V. Singh, Ed. in Chief, 2014), Daya Publishing House, New Delhi. (In Press)

19. Smykova, T.K. 2009. New seabuckthorn varieties for Minusinsk (Russia) Area. In: Proceedings of ISA conference on *"Seabuckthorn on the Way Between Science and Industry Interaction"*, p. 68-69, Russia, 199p.

20. Shchapov, N.S. 2009. Triumf: A seabuckthorn variety with remarkable transportability and keeping capacity of fruits. In: Proceedings of ISA conference on *"Seabuckthorn on the Way Between Science and Industry Interaction"*, p. 65, Russia, 199p.

21. Shripnimbueva, B.T. and Myakhanova, N.T. 2009. Seabuckthorn in back of Baikal. In: Proceedings of ISA conference on *"Seabuckthorn on the Way Between Science and Industry Interaction"*, p. 55-56, Russia, 199p.

22. Singh, V. and Singh, R.K. 2004. Evaluation studies on the exotic and local genotypes of seabuckthorn (*Hippophae rhamnoides* L.). *J. Non Timber Forest Products* 11 (3): 161-165.

23. Singh, V., Gupta, R.K., Arumughan, C., Sawhney, S.C., Rana, R.K., Singh, A. Lal, M., Devi, R. and Sharma, N. 2011. Biochemical evaluation of *Hippophae salicifolia* and *H. mongolica* as horticultural crops in dry temperate Himalayas. In: Proceedings of National Conference on Seabuckthorn, *"Emerging Trends on R&D on Health Protection and Environmental Conservation"*, p. 75-92, CSK HPKV, Palampur, HP, 324p.

24. Wang, H. 2007. Seabuckthorn planted at experimental station Deschambault, Quebec, Canada. In: *Proceedings of International Seabuckthorn Association Conference*, Quebec, Canada.

25. Zubarev, Y.A. 1998. Selection estimate of sweet seabuckthorn. In: *Material of International Symposium on Seabuckthorn* (*Hippophae rhamnoides*), p. 172-173, Russia, 248p.

26. Zubarev, Y.A. 2005. The main directions of seabuckthorn breeding programme in Siberia. In: *Proceedings of invited Speeches of the Second International Seabuckthorn Association Conference*, p. 15-19, August 26-29, 2005, Beijing, China, 63p.

Chapter 7

Characteristics of Russian Seabuckthorn (*Hippophae rhamnoides* subsp. *mongolica*) Varieties

Yury A. Zubarev, Alexey V. Gunin and Elena V. Oderova

The Lisavenko Research Institute of Horticulture for Siberia,
49 Zmeinogorskiy Tract, Barnaul, Russia

SUMMARY

Since 1933, 46 cultivars and varieties of seabuckthorn have been released by the Lisavenko Research Institute of Horticulture for Siberia, Russia. 57,000 hybrid saplings have been raised and 200 forms were evaluated. Generally, the weight of fruits is in range of 0.6-0.8 g, may reach to 1.5 g. The progenies of parent variety "Avgustina" have large fruits. Some thornless varieties like "Chechek", as a parent form for hybridization purpose has been developed. Varieties with high tear-off qualities like hybrids of "Druzhina × Gnom", have also been released. Most of the seabuckthorn varieties have yellow or orange berries. Few red coloured varieties with large fruits and high content of carotenoids for good quality oil are also bred. Oval or cylindrical shaped fruits are easy to harvest. 6 per cent varieties have sweet taste. Harvesting period is from end of August to September. *H. rhamnoides* subsp. *mongolica* is a valuable source for selection and breeding for commercial cultivation. Over 3500 ha land has been brought under cultivation of improved seabuckthorn varieties in Altai region of Siberia.

Keywords: *H. rhamnoides subsp. mongolica, Berry, Characteristics and Siberia.*

INTRODUCTION

Altai territory is a well known region of Siberia, Russia for seabuckthorn breeding, propagation and cultivation. Approximately 3,500 hectares of seabuckthorn plantations have been carried out in the region, *i.e.* 70-80 per cent of total plantations in Russia. Annually, more than 400 hectares of new plantations are being established in this territory. Most of the regions of this territory was planted with varieties, which were bred by the Lisavenko Research Institute of Horticulture for Siberia, the biggest

Seabuckthorn (*Hippophae* L.): *A Multipurpose Wonder Plant, Vol. 4: Emerging Trends in Research and Technologies* (Virendra Singh, *Editor-in-Chief*, 2014), p. 89–98, Daya Publishing House, New Delhi, India.

scientific center on Horticulture not only in Siberia (Altai is a part of great Siberian region), but also in all eastern part of Russia as well.

The selection work on seabuckthorn was initiated by M.A. Lisavenko in 1933. A number of experts contributed in the development of useful varieties (Kalinina, 1988). Since 1933, work on selection and breeding had been started on seabuckthorn, 46 cultivars as well as numerous varieties have been released by the Lisavenko Institute (Table 7.1). It is interesting to know that all of varieties bred at Altai belong to very common subspecies *mongolica* of species *Hippophae rhamnoides* L. This subspecies is naturally distributed in Russia (mainly in Siberian region), China (Altai mountains) and Mongolia. It is difficult to say that another species or subspecies were not included in hybridization over the years. That is why the varieties bred at the Lisavenko Institute should be considered as pure ssp. *mongolica* and their features could be referred to entire group.

Table 7.1: List of Seabuckthorn Varieties Bred at the Lisavenko Institute of Horticulture for Siberia, Barnaul

Variety	*Year*	*Variety*	*Year*
Novost Altaia *(Novelty of Altai)*	1963	Tenga (Village at Altai)	1992
Dar Katuni *(Gift of the Katun river)*	1963	Ayula (Mountain at Altai)	1992
Zootoy Pochatok *(Golden branch)*	1963	Ruet*October (unkn. Ing.)*	1992
Maslichnaya *(Full of oil)*	1964	Ulala (Village at Altai)	1994
Vitaminnaya *(Full of vitamins)*	1972	Inja (The Inja river)	1994
Orangevaya *(Orange-coloured)*	1979	Solnechnaya (Sunny)	1996
Obilnaya *(Abundant)*	1979	Gnom (male) (Dwarf)	1997
Chuyskaya *(From Chuya river)*	1979	Elizaveta (Proper name)	1997
Zolotistaya Sibiri *(Golden of Siberia)*	1979	Altayskaya (From Altai)	1997
Samorodok *(Nugget)*	1979	Ogni Eniseya (Lights of the Enisey r.)	2000
Prevoskhodnaya *(Magnificent)*	1979	Avgustina (Ripen in August)	2001
Velikan *(Giant)*	1979	Azhurnaya (Delicate)	2001
Yantarnaya *(Amber coloured)*	1979	Dzhemovaya (For jam)	2001
Obskaya *(From the Ob river)*	1981	Zhemchuzhnitsa (Pearl box)	2004
Sibirskaya *(From Siberia)*	1981	Sudarushka (Madam)	2004
Aley (male) *(The Aley river)*	1985	Rosinka (Drop of dew)	2004
Panteleevskaya *(Proper name)*	1985	Sentyabrinka (Ripen in September)	2005
Luchezarnaya *(Radiant)*	1985	Klavdia (Proper name)	2006
Lubimaya *(Favorite)*	1987	Essel (Proper name)	2006
Altaechka *(Altai inhabitant)*	1988	Etna (The Etna Volcano)	2009
Zhivko *(Proper name)*	1990	Zlata (Gold)	2009
Chulishmanka *(The Chulishman river)*	1990	Aurelia (Aurum)	2011
Chechek*Flower (Altai Ing.)*	1990	(Ognivo) (Flint)	2011

The cultivars bred at the Institute are distinguished by different features, which have been already described in many publications. The present study is an attempt to summarize the features not only of cultivars, but overall total collection. Obviously, it is difficult to highlight all characteristics, that is why we focus on most important economic characteristics of them.

Here, it is necessary to emphasize on approximately 57,000 hybrid seabuckthorn plants were grown at the Institute's research orchards and more than 200 of them were under annual comprehensive evaluation (refered as *varieties under selection*). Each year, this collection extends up to 2000-3000 hybrid plants. There was noticed a great variability and diversity in most of features even within one subspecies. Because of selection, the most valuable characteristics of varieties have been improved to the better level, and today they are a bit different as compared to that of 20-30 years ago.

CHARACTERISTICS OF SEABUCKTHORN VARIETIES

Over a period of last six decades, several useful seabuckthorn varieties mainly of ssp. *mongolica* have been developed at LIHS, Barnaul. The main focus for selection and breeding has been on the weight of fruits, fruit yield, fruit tearing off, thornlessness, high contents of fruit oil and carotenoid, etc.

Weight of Berries

Significant progress has been made on breeding of seabuckthorn during the last decade. Compared with selections of early 30th of last century, today berries' size has been increased up to 5-7 times and reached to 1.5 g. Only 11.3 per cent of all evaluated varieties have weight of berries lower than 0.6 g (wild forests have size of about 0.2-0.3 g), half of varieties have weight of berries varying from 0.6 to 0.8 g. The most promising result was 11.3 per cent of varieties under selection have berries larger than 1.0 g (Figure 7.1). In another words, it is considered that subspecies *mongolica* is a source of large berries.

Undoubtedly to obtain the prospective results, the proper initial forms with desirable features are necessary for the hybridization. For last several years, the comprehensive progeny analysis has been done to find out the sources of different selected characteristics including berries size. It has been found out that progeny of "Avgustina" variety distinguished for large berry. More than 30 per cent of plants have weight of berries 0.9 g and higher. There were no plants with the feature below 0.6 g. Variety under selection "1320-86-6" provides 11.8 per cent plants with berries weight more than 0.9 g and no samples with berries below 0.6 g (Figure 7.1).

Table 7.2: Variation in Key Characteristics of Berries of Seabuckthorn Varieties

Colour	*Per cent*	*Shape of Berry*	*Per cent*	*Taste*	*Per cent*	*Pedicle Length (mm)*	*Per cent*
Yellow	1.9	Roundish	4.3	Sore	58.1	<3	6.2
Orange-yellow	7.5	Oval	49.1	Sweet-sore	25.0	3-5	76.3
Yellow-orange	10.6	Wide-oval	20.5	Sore-sweet	10.6	>5	17.5
Orange	55.6	Cylindrical	14.3	Sweet	6.3		
Reddish-orange	10.6	Inversely egg-shaped	11.8				
Orange-red	5.0						
Red	9.4						

The comparative analysis of initial mother forms and progeny plants has shown that berries weight of hybrid plants are mostly lower as comparing to parent forms. It means that weight of berries is not in protogene. That is why male component has the same importance as female. The major problem for searching of proper male source is the absence of berries. We were looking for the male varieties from population, where female part provides lot of plants with large berries. Using this

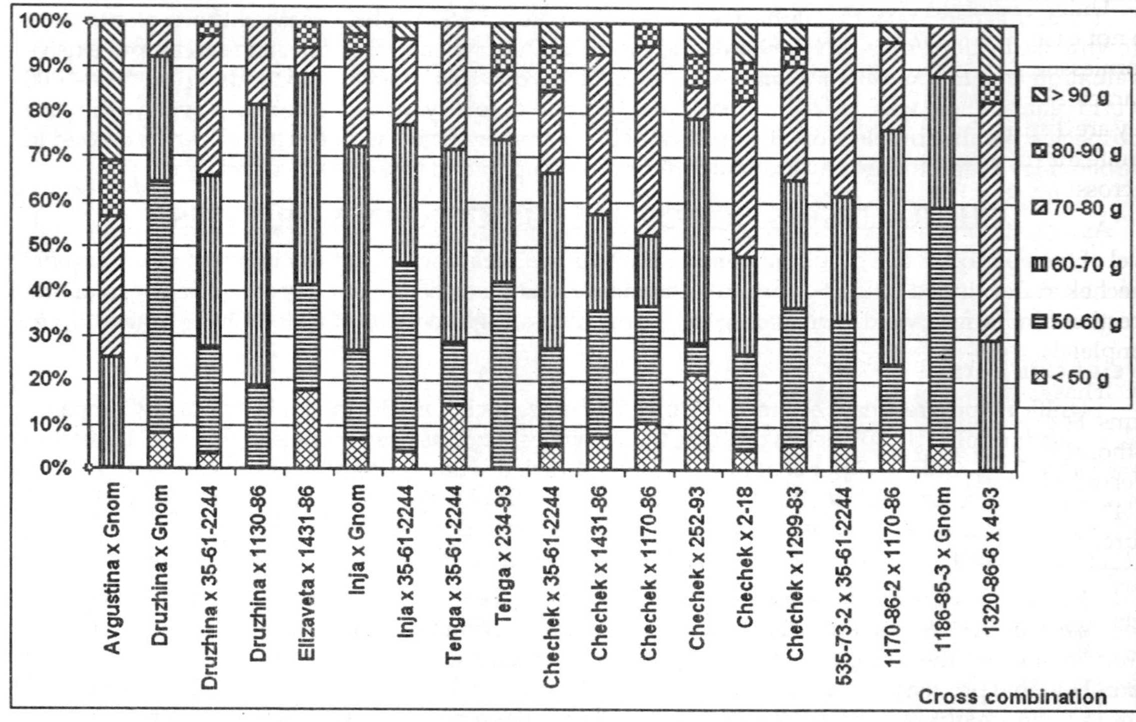

Figure 7.1: Weight of Berries (g/100 berries) of Seabuckthorn Varieties Bred, as Compared to Parents Forms.

method, it was found that male variety "35-61-2244" significantly increases the weight of berries in progeny. For instance, about 60 per cent of hybrid plants in cross combination "535-73-2×35-61-2244" have weight of berries from 0.6 to 0.8 g, though initial mother variety has only 0.55 g. It is very important notice for breeders, who would like to carry out interspecies crossings using subsp. *mongolica* as the male component.

Thornlessness

Thornness is a well known dominant character for seabuckthorn and searching of thornless forms is extremely difficult task. This feature is common to most of species and subspecies of seabuckthorn, but it significantly differs between groups. It is necessary to note that estimation of thornness is a bit tricky because there is no universal scale for such estimation. For instance, moderate level for "Altai" varieties according to our method of estimation could be considered as low level for other species at different research centers and vice-versa. For the purpose of unification, it is proposed to use scale which are as follows: 0 points-complete absence of thorns; 1 points-thorns are short and soft, and mostly appear on the top of sprouts; 2 points-thorns without branching from 2 to 4 thorns per 40 cm of sprout's length; 3 points-thorns with single branching from 2 to 4 per 40 cm of sprout's length, or thorns without branching but more that 4 per 40 cm of sprout's length; 4 points-thorns with double branching from 2 to 4 per 40 cm of sprout's length. From time to time, we recorded some samples with thornness more than 4 points, but considered them only as defensive shield on the state border.

Unfortunately, thornlessness is not a dominating character and as a result, most thornless varieties do not carry this feature to descendant. It is especially difficult task to find the samples combined thornlessness and other valuable characters. It is considered that thornness, as one of the most limiting character in selection process and discard samples with level of thornness more than 3 points, even they are distinguished for high productivity, large berry size, palatability and so on. Also, it is necessary to note that to obtain seedlings with fewer thorns as possible, we do not use male varieties with thorns in crossings.

As a result of long period selection, some combinations have been achieved which provide high level of thornless plants. They were "Avgustina × Gnom, Chechek × 1170-86, Chechek × 252-93 and Chechek × 2-18". It was deduced that thornless variety "Chechek" is a most promising initial form in direction of reducing thorn level. 13.1 per cent hybrids from cross combination "Chechek × 2-18" were completely without thorns as well as 70 per cent have only 1 point of thornness level (Figure 7.2).

It has been found that male initial (parent) forms also provide a significant influence to descendant forms. For instance, progeny of cross combination "535-73-2 × 35-61-2244" has more than 10 per cent of thornless plants and about 40 per cent having low thornness (1 point), while initial female form referred as thorny (2 points). Here, we could find the direct influence of thornless male form "35-61-2244". Therefore, from above studies, it is clear that *H. rhamnoides* subsp. *mongolica* is a promising source for absence of thorn on seabuckthorn plants.

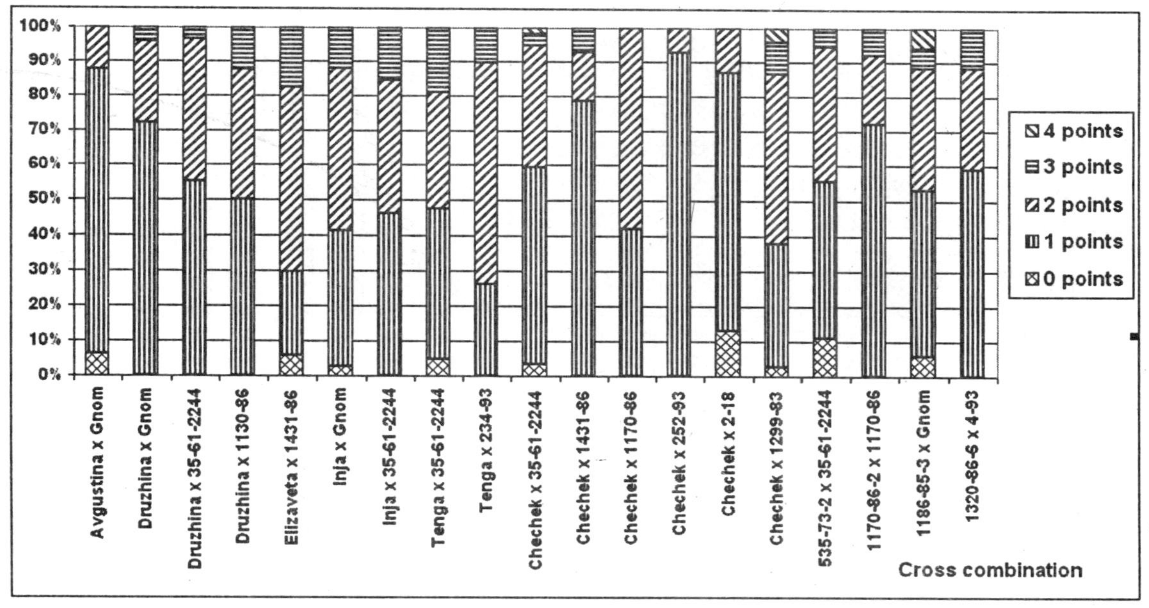

Figure 7.2: Seabuckthorn Varieties Thornness Level as Compared to Parent Forms.

Tear-off Quality

As hand-picking is a main harvesting method for seabuckthorn, a great attention has been paid on berries collection. Tear-off quality is dramatically complicated feature and depends on many issues and several of them were: Tear-off force, skin breaking point, period of ripening, type of branches, length of pedicle, density of berries inside particular branch, and hand-picking and skill level of harvester. To summarize these, we called them as wet or dry tear-off character. The weight of berries collected by ordinary methods consist of squeezed samples including free juice, or berries collected without obvious damages (sometimes even with pedicle). The characters were divided into three groups: Dry-berries collected with part of pedicles; semi-dry-berries collected without pedicles and some of them could lost a part of skin in pedicle-berry contact point, but juice do not leak out; and wet-berries collected without pedicles and significant part of them are squeezed and juice leak out from berries. While, dry tear-off of berries is more than expecting features, because it provides both efficient harvesting and berries quality. Unfortunately, most of varieties in our study of collection (more that 80 per cent) belong to third undesirable group and mainly could be used only for technical purpose but not for table. It means that subspecies *mongolica* could not be recommended as reliable source of dry tear-off character. However, several varieties have been selected as a source of dry and semi-dry tear-off character. For instance, "Druzhina (Novosibirsk station variety) × Gnom" combination provide 68 per cent dry and semidry tear-off level seedlings (Figure 7.3). Several other cross combinations where "Druzhina" used as initial form also showed high percentage of dry tear-off plants. The variety under selection "1170-86-2" which get mentioned character also convey it to descendant (56 per cent).

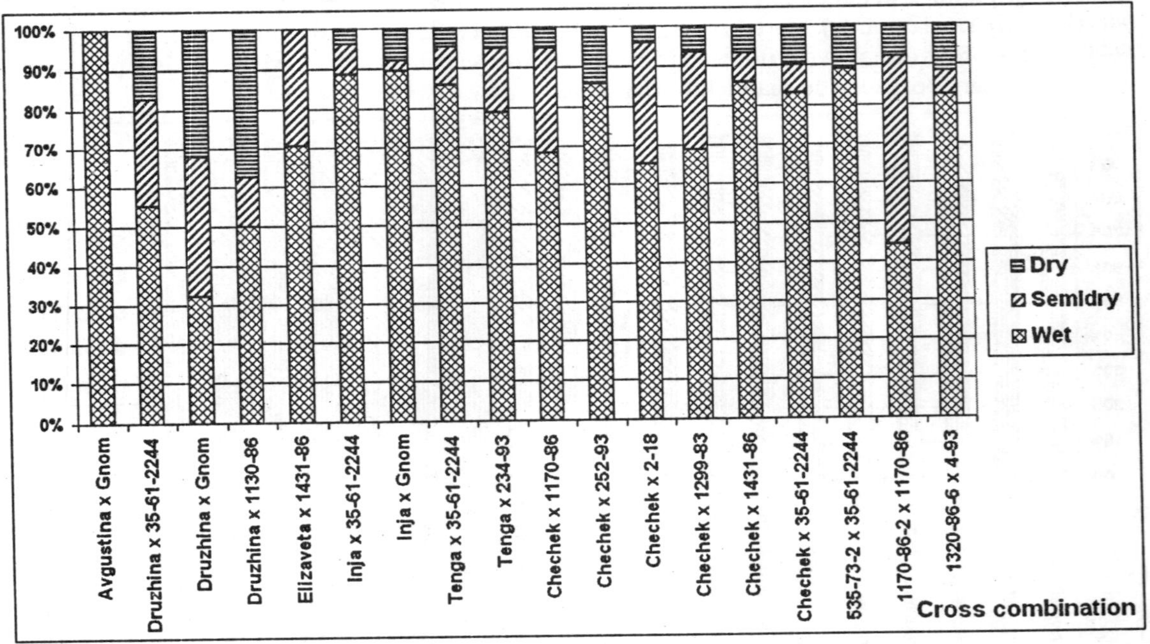

Figure 7.3: Tear-off Level of Berries of Seabuckthorn Varieties (Progeny analysis).

The relative character to tear-off quality and hand picking efficiency is the length of pedicle. In most cases, the longer pedicle the easier is hand-picking. Most of varieties in our collection have short pedicle from 3 to 5 mm (Table 7.2). Nevertheless, 17.5 per cent of them with relatively long pedicle from 5 to 8 mm are promising initial forms, which have been found for further selection.

Another relative character to tear-off quality is tear-off force, *i.e.* force you need to apply for separating the berries from the branch. It is a bit tricky feature, which depends on the way and equipment for estimation. Previously (till the end of 90th), "school dynamometer" was used as special force measured device. The only problem with that device is the active hook situated in flat level and results mostly show breaking force of pedicle, but not tear-off force as itself. It is very important for subspecies *mongolica* breaking force of pedicle mainly much higher comparing to tear-off force. More than 80 per cent of varieties in our collection characterized by wet tear-off quality, that pedicle is not breaking at all during harvesting, and stay at branches with small part of berries skin. That is why previously no correlations between breaking force of pedicle and tear-off quality was found. Nowadays, we exploited specially designed electronic device for such purposes, which is called "DINA-1" (last version "DINA-2"). Besides electronic filling, the critical difference of this device is that active hook positioned in perpendicular plain, allowing to pull particular berry and do not touch the pedicle. The data obtained by mentioned device has strong correlation with tear-off quality.

Tear-off force below 150 g was considered as a promising one, which provided efficiency at harvesting in most cases. It has been found that such cultivars as "Avgustina, Inja and Tenga" characterized by low level of tear-off force, and all of them in different range convey this feature to descendant. Most promising results have been obtained by using cultivar "Avgustina" as one of the initial forms. More than 12 per cent of plants in cross combination "Avgustina × Gnom" had tear-off force below 100 g, and the rest of them have no more than 150 g (Figure 7.4). These results open up new horizons in the selection of promising forms.

Despite of relatively high tear-off force for cultivar "Chechek", some cross combinations have shown significantly high quantity of plants with moderate tear-off force. For instance, cross combination "Chechek × 252-93" provides 93 per cent of seedlings with tear-off force from 100 to 150 g. It confirms the significant influence and importance of male component in hybridization.

Colour of Berries

The best colour for seabuckthorn berries depends on the purpose of berries utilization. In Russia, main direction of seabuckthorn utilization is the processing to oil. The value of seabuckthorn oil in Russia is estimated by carotenoids content. The more is the carotenoids, more is the valuable oil. The

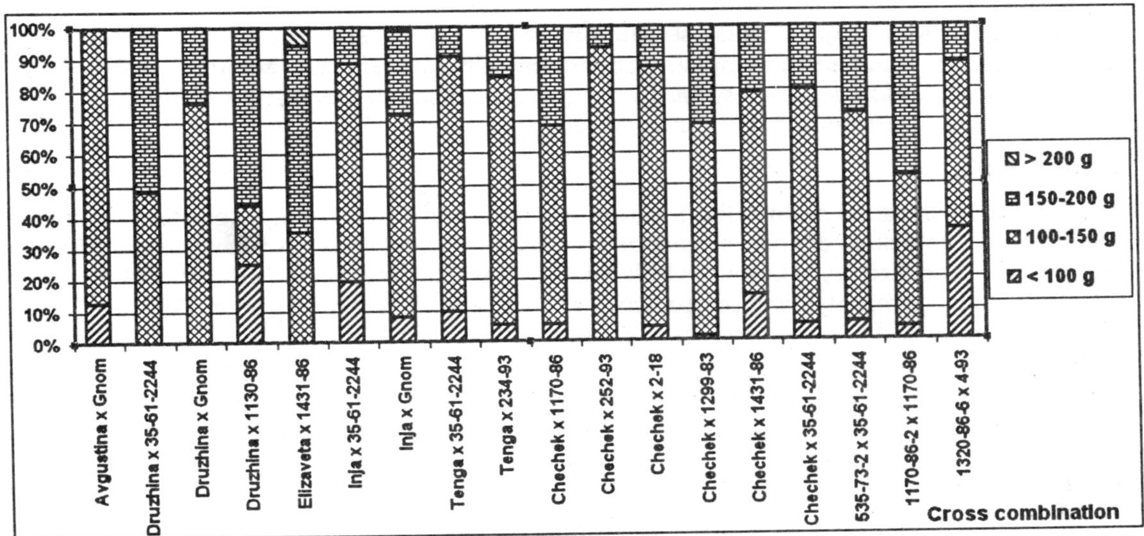

Figure 7.4: Tear-off Force of Berries on Seabuckthorn Varieties (Progeny analysis).

strong correlation between berries colour and carotenoids content had been previously found. It has been proved that red-coloured berries are richer in carotenoids content. That is why we focused our selection to red coloured samples. For *H. rhamnoides* subsp. *Mongolica,* the most common colour is orange tends to yellow, but red colour of berries is quite rare. Moreover, a strong inverse correlation between berries size and red colour had been established. In another words, it is hard to find red coloured varieties with large berries.

Table 7.2 shows the colour ranging of our total collection. Due to long time targeting selection, we obtained 9.4 per cent red coloured varieties and 15.6 per cent varieties with different level of reddish (from reddish-orange to orange-red). Some of red coloured varieties have weight of berry up to 0.8-0.9 g. At the same time, they have significantly reduced yellow coloured samples and volume of which falls to 1.9 per cent.

Shape of Berries

The shape of berries is not only an extremely valuable character for seabuckthorn, but also plays a significant role in harvesting efficiency, as well as is very important from point of view of consumer preferences. Wild seabuckthorn characterized by small and round-shaped berries, which is not quite attractive both for farmers and for ordinary customer in supermarket. It seems more interesting to have oval or cylindrically-shaped berries or even more exotic egg-shaped (or inversely egg-shaped) berries. To discuss unusual shape of berries, only *mongolica* provided such an interesting character. It has been recorded that varieties with egg-shaped berries are characterized by relatively large berries. Moreover, strong correlation has been found between shape of berries and tear-off level. It has been proved that inversely egg-shaped varieties have lower tear-off level. That is why it is considered that this feature is a key character in choosing initial forms for cross combinations. Fortunately, the initial lines have been determined with dominant gene which provides such shape to most of progeny. As a result, for this period of time more than 11 per cent of evaluated varieties (not total collection) in Institute's collection are egg-shaped (Table 7.2).

Besides, cylindrical shape of berries, which is quite rare and 14.6 per cent of varieties were distinguished for such shape. Because, selection of prevalent berries shape in our collection is oval, *i.e.* about half of total volume, but roundish shape was mostly eliminated and accounts only for 4.3 per cent. In nutshell, *H. rhamnoides* subsp. *mongolica* is a valuable source for selection on different shapes of berries including extra-ordinary cylindrical and egg-shaped.

Taste of Berries

Estimation of taste of seabuckthorn berries is extremely difficult concept because of absence of universal grade as well as common experience for estimation. It is hard to say whether taste of particular sample sweet or delicious, balanced or little bit sour. It is the same situation like that of thornness. Mainly, it is subjective matter and comparative analysis depends on expert person. Perhaps, the taste of berries of most seabuckthorn species is more or less permanent character and described as sour with narrow variations. But, for subspecies *mongolica* the situation is not the same. The taste diversity in our seabuckthorn collection varies from extremely sour and even bitter to really sweet with pleasant aroma. Previously, the scale for taste estimation according to sugar-acid index has been suggested by Zubarev (2005). However, according to further and deep investigation, we recommended to make some corrections for the scale as follows: Sour: Below 2 points; Sweet-sour: 2-4; Sour-sweet: 4-6; and Sweet: More than 6 points.

Table 7.2 presents organoleptic taste estimation that is provided by laboratory's experts, during last several years. It was shown that more than 10 per cent of varieties are sour-sweet and more than 6 per cent were sweet. *H. rhamnoides* subsp. *mongolica* is one of the most promising subspecies in the direction of taste enhancement.

Period of Ripening

It is understandable that because of rapid increase of scales of seabuckthorn plantations. It is extremely important to have varieties with different period of ripening to prolong the efficient harvesting season. For instance, seabuckthorn harvesting period in Siberia lasts from the end of August till the end of September. For about 30 days, it is necessary to collect all berries from plantation otherwise their quality is significantly reduced. Not considering winter time, the climatic conditions of region allow us to prolong harvesting time till the end of October.

It is not a surprise but most of varieties in our collection belong to middle ripening group. However, within last two decades, we selected samples significantly different in a period of maturity. As an example, many times mentioned before cultivar "Avgustina" ripen in the middle of August and recently released cultivar "Ethna" ripen in the beginning of August. Moreover, we selected varieties which ripen in the end of July, providing at the same time sufficient productivity. On another hand, we recorded several hybrids, which stay green until the first frosts in the middle of October. Hereby, it was noticed a great variability in this feature, which differ within 3-3.5 months.

All the varieties belong to 3 groups, according to time of maturing, which are as follows: a) Early ripen-From July till the last ten days of August; b) Middle ripen-From last ten days of August till the middle of September; and c) Late ripen-From the middle of September till the end of October. The progeny analysis shows that most of varieties had all groups of ripening from early to late, but main part is middle ripen seedlings. "Chechek" cultivar tends to convey late ripening character, "Avgustina"

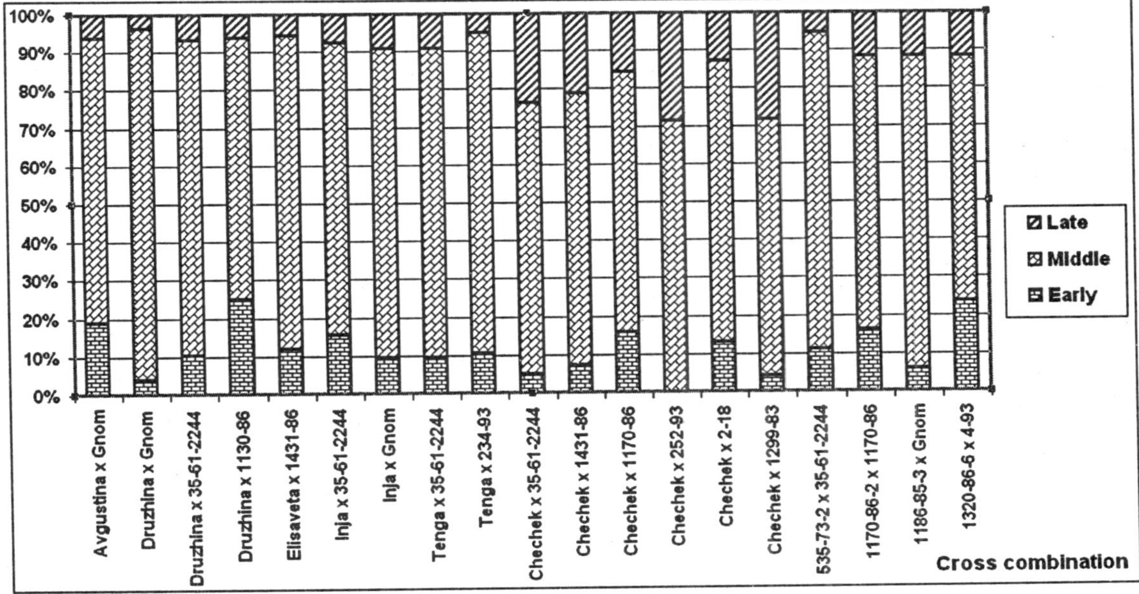

Figure 7.5: Period of Berries Maturation of Seabuckthorn Varieties (Progeny analysis).

cultivar in contrary, had yearly ripening (Figure 7.5). Male initial forms are also important and for instance, variety, such as "35-61-2244", provides more late ripen samples rather than early ripen.

CONCLUSION

The present study emphasized that from the industrial utilization point of view, *H. rhamnoides* subsp. *mongolica* is one of the most promising group for further selection, which provided much valuable characters, as not of any other known species on earth. Experts at the Lisavenko Research Institute of Horticulture for Siberia have been able to select and develop hybrids with promising economic characters like thornlessness, large berries, easy to separate the berries, acceptable taste, suitable maturation period and different coloured berries, particularly red with high carotenoid content and higher content of fruit oil.

REFERENCES

1. Kalinina, I. 1988. Seabuckthorn. *Science in the USSR*, p. 80-87.
2. Zubarev, Y. 2005. The main directions of seabuckthorn breeding program in Siberia. In: *Proceedings of the Second International Seabuckthorn Association Conference*, p.15-19, Beijing, China.

Chapter 8

Studies on Introduction and Evaluation of Improved Russian Seabuckthorn Varieties in China

Li Daiqiong[1], Wu Qinxiao[1], Zhang Jun[2], Guo Chunhua[3], Guo Yuxiao[1], Chen Yunming[1], Li Zhongyi[2], Sun Xiaochun[3], He Dengyao[4], Ruan Chengjiang[5] and Zhou Zizhi[2]

[1]Institute of Soil and Water Conservation,
Chinese Academy of Sciences and Ministry of Water Resources, Yanling, Shaanxi, China,712100
[2]Fuxin Institute of Selection and Breeding of Fine Seabuckthorn Species, Fuxin, Liaoning 123000
[3]Qiqihaer Horticulture Institute, Qiqihaer, Heilongjiang 161005
[4]Wuqi Forestry Department, Wuqi, Shaanxi 717600
[5]Dalian National Institute, Liaoning 116600

SUMMARY

The Loess plateau is the main region of distribution of the Chinese seabuckthorn subspecies. Over the years, seabuckthorn forest has been planted for the environmental conservation in a large area in this region. It is generally wild forms, having the characteristics of small fruit, more thorns, difficult to harvest and low fruit output. In order to solve the problem of deficiency of fine varieties of seabuckthorn in the area of northwest, northeast and north China (including the Loess plateau), during the experiment of 10 years, 20 fine forms of seabuckthorn had been introduced in this area from Russia. We have tentatively selected 9 fine varieties of seabuckthorn and developed 7 experiment bases in the "three north areas" and Loess plateau. They have the characters of fine growth, adaptation and higher fruit yield. Then, the development of crossbred varieties for new generation for wide adaptation was made through crossbreeding of Russian seabuckthorn (female) with Chinese seabuckthorn (male). On the basis of the introduction experiments and evaluation on its introduction results in regions of "three north areas" and Loess plateau, and other regions of China, 4 ecological regions for mass cultivation have been advanced. It is an important direction to breed and introduce the hybrids of Russian seabuckthorn with Chinese seabuckthorn in the region for the cultivation.

Keywords: *Areas of northwest, northeast and north China ("three north areas"), Loess plateau, Russian seabuckthorn varieties, Introduction, Selection and Breeding.*

Seabuckthorn (*Hippophae* L.): *A Multipurpose Wonder Plant, Vol. 4: Emerging Trends in Research and Technologies* (Virendra Singh, *Editor-in-Chief*, 2014), p. 99–106, Daya Publishing House, New Delhi, India.

INTRODUCTION

Seabuckthorn is a multipurpose plant of high altitude cold regions of China and Himalayas and Europe (Singh, 2003). Fruits and leaves are quite rich in vitamins and anti-oxidants (Kallio *et al.*, 2002), therefore, it has health protection values. Its medicinal values were discovered by Tibetan doctors, as described in Tibetan System of Medicines as early as 8th century. It has been widely harnessed for its rich nutritional values of fruits in health food, cosmetics and medicines in China, Russia and Europe. However, China has shown the ecological potential in controlling soil erosion and providing greenery in dry mountains, besides conservation of biodiversity and wild life (Rongsen, 1992).

In order to solve the problem of scarcity of improved seabuckthorn varieties in the area of northwest, northeast and north China ("three north areas", including the Loess plateau), China had undertaken the project assigned by Ministry of Water Resources for the introduction of Russian seabuckthorn varieties, and a Cooperation Research Project between China and Russia in 1997-2006. Through the experiment of 10 years, 20 varieties of seabuckthorn from Russian had been introduced in the "three north areas". We have tentatively selected 9 fine seabuckthorn varieties for evaluation in the 7 experimental sites in this area. They have characters of fine growth, resistance and higher fruit yield. It has produced good ecological, economical and social benefits. It has important significance for developing large-scale cultivation and quality seabuckthorn forests and orchards in the "three north areas" and Loess plateau area, for developing local economy.

MATERIALS AND METHODS

Experimental Area and Climatic Conditions

The 20 seabuckthorn varieties with large fruit and thornless had been introduced from Novosibisk, Moscow and Ulan-Ude of Russia during 1995-2004 (Table 8.1).

Table 8.1: Natural Local Conditions of Seed Source Areas of Russian Seabuckthorn in Russia

Variety	*Longitude and Latitude*	*Annual Precipitation (mm)*	*Altitude (m)*	*Mean Annual Temperature (°C)*	*Annual Aunshine (h)*	*Drying Index*	*Non-frost Duration (d)*
Novosibirsk	E82°002 N55°102	400-550	200-650	0.1 and 3.0	1627	0.8-0.7	90
Moscow	E37°342 N55°452	500-600	156	3.6-4.0	1538	2.0-0.5	95
Ulan-Ude	E107°02 N51°072	240	400	1.7-1.9	2472	1.3	97

The fine varieties of seabuckthorn had been introduced at 7 different experimental sites in the "three north areas" (Qiqihaer, Fuxin, Dalian, Huhehaote, Eerduosi,Wuqi, Ansai) in China. These regions belong to moist, semi moist and semi-arid areas within dry temperate zone and temperate zone, respectively. There are natural forest and huge plantations of seabuckthorn in these regions (Table 8.2).

Material Used

We have introduced 20 thornless fine Russian seabuckthorn forms with large fruits from Russia. We analyzed their ecological adaptability and varied characteristics of bio-ecology and genetics, contrastively and brought up fine seabuckthorn forms suitable to these regions. The area of experimental plot was 20-60 m^2 and plantation density of plants was 3 m×1 m. The genealogy selection was used for measuring introduction and selection.

Table 8.2: Natural Conditions of Experiment Areas in China for Introduced Russian seabuckthorn Varieties

Introduced Experiment Area	*Longitude and Latitude*	*Altitude (m)*	*Mean Annual Temperature (°C)*	*Annual Precipitation (mm)*	*Drying Index*	*Annual Sunshine Time (d)*	*Non-frost duration (d)*
Qiqihaer	E123°24′ N 47°08′	200	4.7	374.9	1.30	2740	120-125
Fuxin	E 121°48′ N42°26′	380	8.5	393.8	1.48	2718	150
Dalian	E122°01′ N39°38′	42	11.7	434.1	1.33	2763	169
Huhehaote	E111°36′ N40°52′	1130	7.9	419.0	1.29	2877	127-133
Eerduosi	E110°09′ N39°40′	1400	7.3	348.1	1.30	3121	140
Wuqi	E108°32′ N37°24′	1365	8.6	463.9	1.10	2400	130
Ansai	E109°16′ N36°51′	1431	9.5	467.6	1.14	2416	159

In order to select fine individual seabuckthorn forms as the gene resource for sexual reproduction and clone, we bred fine seabuckthorn varieties with the characteristics of fine inheritance, strong adaptability, high output and quality. Then, expanded experiment of selection was implemented.

RESULTS AND DISCUSSION

Through the experiment for 10 years, 20 improved forms of seabuckthorn had been introduced in the 3 north areas of China from Russia. The 9 fine seabuckthorn varieties and forms had been tentatively selected in this area. It included varieties of Russian seabuckthorn "Zirianka, Druzhina, Red torch, Ivshca, Batanical garden, Buliyate, Panjielieyewa, Yalishanda" and Non-seed seabuckthorn (Table 8.3). Then, hybridization for F1 generation was made through crossbreeding of Russian seabuckthorn with Chinese seabuckthorn. On the basis of introduction, selection and breeding of seabuckthorn, we popularized and cultivated them, which has produced good early results.

The experiment result showed that the introduced fine Russian seabuckthorn varieties have better economic properties. The diameter of fruits was 9-14 mm, length of fruit stalk was 2-4 mm, weight of per 100 fruits was 20-60 g and early fruit yield per plant was 0.9-3.1kg (Table 8.4). These values were 1.8-2.5, 1.3-2.7, 4.6-8.8 and 3.8-6.2 times higher than local wild Chinese seabuckthorn, respectively. Russian seabuckthorn has the characteristics of larger fruit, no thorns and high yield. The quality, oil contents and active substances of Russian seabuckthorn varieties were higher than local wild Chinese seabuckthorn and therefore, they are important genetic materials for introduction and selecting economic forms for improving the Chinese seabuckthorn.

Table 8.3: Evaluation Results of Introduced Russian Seabuckthorn Varieties in China

Name of Varieties	Plant Height (cm)	Crown (cm)	Diameter at Butt End (cm)	Annual New Top (m)	Leaf Length and Width (cm)	Adaptability
Zirianka	140.6	78.9×91.6	2.7	37.8	5.60.9	good
Druzhina	125.7	68.9×75.8	2.8	35.8	5.70.9	good
Podruga	127.8	74.0×81.7	2.6	35.6	5.51.0	middle
Sibirian Rumianes	121.9	71.1×84.1	2.5	35.4	5.50.9	middle
Gold Cascad	123.4	75.0×85.4	2.7	32.1	5.91.0	middle
Red torch	130.4	77.5×89.9	2.7	33.8	6.11.0	good
Ivshca	128.2	75.4×85.4	2.6	30.6	6.00.8	good
Ornistaya	121.9	71.1×84.1	2.5	35.4	5.51.0	middle
Zarnitsa	121.9	76.7×96.1	3.0	31.1	5.60.9	middle
Botanical Garden	204.3	137.8×146	4.3	48.2	6.30.8	good
Panjielieyewa	152.8	104.5×118.3	4.0	32.1	5.50.8	good
Buliyate	195.5	180.0×187.5	5.3	40.2	6.11.0	good
Yalishanda	166.0	125.3×146.5	3.2	38.6	6.10.8	good
Non-seed SBT	125.0	95.0×102.0	2.6	22.8	5.60.9	good
Orange×Chinese SBT	201.7	196.2×226.0	4.0	33.3	5.60.7	fine
Azhula ×Chinese SBT	204.6	171.2×194.2	5.2	33.7	5.70.6	fine
Chinese SBT(CK)	234.8	184.6×197.0	5.6	41.2	4.50.8	fine

SBT: Seabuckthorn.

But, the nutritional values of Russian seabuckthorn were lower than Chinese wild seabuckthorn. The height, crown and diameter at butt end of Chinese seabuckthorn were 1.5-2.5 m, 1.3×1.5 m, and 4-5 cm, respectively and these values were 1.2-1.9, 1.3-2 and 1.3-2.2 times of Russian seabuckthorn. The main reason was that the introduced Russian seabuckthorn has lower adaptability and resistant capability, especially lower resistance to dry conditions and high temperature during summer, and also to plant diseases and insect pests. On the other hand, Chinese seabuckthorn has strong adaptability and higher content of vitamin C. The adaptability and growth potential of hybrid varieties of the Russian seabuckthorn and Chinese seabuckthorn are promising. The plant height of 4 years hybrid Orange × Chinese seabuckthorn was 1.7-2.3 m, while the length of new branch was 30-48 cm, diameter of fruit was 0.7-1.0 cm and weight of 100 fruits was 23-50 g. These properties and stability of inheritance should be observed successively.

The experimental results have showed that the introduced varieties of Russian seabuckthorn have different performances at the 7 experiment bases, which belong to different latitude sites or ecological conditions (Tables 8.5–8.8). These different results were related to bio-ecology and inheritance of Russian seabuckthorn, and also varied environmental factors (water, heat, light and soil). We analyzed the effects of latitude, altitude, soil property, illumination and bio-ecological conditions of the introduced regions on Russian forms as compared with Russian seabuckthorn growing regions.

Table 8.4: Characteristics of Fruits of Introduced Russian Seabuckthorn Varieties

Name of Varieties	*Fruit Length (cm)*	*Fruit Breadth (cm)*	*Length of Fruit Stalk (cm)*	*Weight of 100 Fruits (g)*	*Fruit Yield (kg/plant)*
Zirianka	0.9-1.4	0.7-1.0	0.2-0.4	32.7-52	1.9-2.2
Druzhina	0.9-1.3	0.7-0.9	0.2-0.4	26-40	2.7
Podruga	0.8-1.1	0.7-0.9	0.3-0.4	30-39	2.3
Sibirian Rumianes	0.8-1.1	0.7-0.9	0.2	32-41	0.9
Gold Cascad	0.8-1.1	0.6-0.8	0.3	40	1.2-2.6
Red torch	0.8-1.1	0.6-0.9	0.2-0.3	42	1.8
Ivshca	0.9-1.3	0.7-0.9	0.4	60	2.9
Ornistaya	1.1-1.2	0.8	0.2-0.3	60	2.6
Zarnitsa	1.0-1.2	0.7-0.9	0.3	43	1.5-2.5
Botanical Garden	1.1	0.8	0.3	45	1.6
Panjielieyewa	1.4	1.0	0.3	46	1.9
Buliyate	1.0	0.8	0.3	42	1.3
Yalishandai	1.3-1.5	0.9	0.4	46-60	3.1
Non-seed SBT	1.0	0.8	0.4	23-52	0.6-2.0
OrangexChinese SBT	1.0	0.9	0.3-0.5	31-50	2.1
AzhulaxChinese SBT	0.7-0.9	0.9	0.4-0.5	23-52	0.6-2.0
Chinese SBT(CK)	0.4-0.6	0.3-0.5	0.1-0.2	7-7.4	0.4-0.65

SBT: Seabuckthorn.

Table 8.5: Performance and Fruit Characteristics of Russian Seabuckthorn Variety "Zirianka" and Chinese Seabuckthorn

Year of Growth	*Place/ Site*	*Plant Height (cm)*	*Crown (cm)*	*Diameter at Butt End (cm)*	*Thorns per 10 cm of Branch (cm)*	*Fruit Length (cm)*	*Fruit Breadth (cm)*	*Fruit Yield per Plant (kg)*	*Annual Precipitation (mm)*
3-5	Qiqihaer	150.9	119x145	4.0	No thorn	1.1	0.8	2.2	374.9
3-5	Fuxin	142.6	77.5x89.8	3.0	No thorn	1.4	0.9	1.3	393.8
Chinese SBTCK 3-5	Fuxin	151.0	138.9x151	4.3	2-4 0.6-1.5	0.6	0.4	0.2	393.8
3-4	Dalian	138.6	54.4x67.3	2.0	No thorn				434.1
3-4	Huhehaote	122.2	45.5x48.2	1.7	No thorn	1.0	0.8	1.2	419.0
2	Eerduosi	38.7	29.3x37	1.3	No thorn				348.1
3-5	Wuqi	148.1	74.0x84.1	3.5	1.0-2.0 0.2-0.8	1.3	0.9	1.1	463.9
3-4	Ansai	141.3	103.0x115.3	2.0	1.0-2.0 0.5-1.3	1.0	0.8	1.0	467.9
Chinese SBT(CK) 3-5	Ansai	250.7	130.3x151.3	4.8	2.0-4.0 0.8-2.0	0.4	0.4	0.13	467.6

SBT: Seabuckthorn.

Table 8.6: Growing Performance and Fruit Characteristics of Russian Seabuckthorn Variety "Ivshca"

Growing Year	Site	Plant Height (cm)	Crown (cm)	Diameter at Butt End (cm)	Thorns and Length of 10 cm of Branch (cm)	Fruit Length (cm)	Fruit Breadth (cm)	Fruit Yield per Plant (kg)
3-5	Qiqihaer	135.7	112.0×121.8	3.8	No thorn	1.3	0.9	2.9
3-5	Fuxin	109.5	62.7×67.2	2.0	No thorn	1.2	0.8	2.0
3-4	Dalian	153.3	53.4×72.6	2.1	No thorn			
3	Huhehaote	137.8	83.8×99.5	3.2	No thorn	1.0	0.8	1.8
3	Eerduosi	65.0	56.0×62.0	1.0	No thorn			
3-5	Wuqi	156.2	72.7×91.8	4.0	1-3 0.5-0.9	1.2	0.9	0.9
3-4	Ansai	140.0	81.3×88.8	1.9	1-3 0.6-1.0	1.0	0.8	0.7

Table 8.7: Growing Performance and Fruit Characteristics of "Zirianka" during different Growth Years in Qiqihaer

Growing Year	Year	Plant Height (cm)	Crown (cm)	Diameter at Butt End (cm)	Thorns and Length of 10 cm Branch (cm)	Fruit Length (cm)	Fruit Breadth (cm)	Fruit Yield/ Plant (kg)	Annual Precipi- tation (mm)	Mean Annual Temperature (°C)
1	1999	67.7	5.0×14.4	0.5	No thorn	–	–	–	413.5	4.6
2	2000	73.7	39.5×44.0	1.0	No thorn	–	–	–	344.7	4.3
3	2001	80.3	58.2×72.5	1.9	No thorn	–	–	–	248.9	4.4
4	2002	118.0	99.5×116.3	3.3	No thorn	1.1	0.8	2.0	369.9	4.5
5	2003	163.8	122.5×150.0	4.2	No thorn	1.1	0.8	1.1	600.2	5.2
6	2004	171.0	135.0×169.0	4.6	No thorn	1.2	0.8	2.2	272.4	4.9

Table 8.8: Growing Performance and Fruit Characteristics of "Zirianka" during different Growth Years in Wuqi

Growing Year	Year	Plant Height (cm)	Crown (cm)	Diameter at Butt End (cm)	Thorns in 10 cm of Branch (cm)	Fruit Length (cm)	Fruit Breadth (cm)	Fruit Yield/ Plant (kg)	Annual Precipi- tation (mm)	Mean Annual Temperature (°C)
2	2000	60.5	9.0×19.5	1.0	No thorn	–	–	–	398.6	8.6
3	2001	93.8	37.5×41.8	1.7	No thorn	–	–	–	538.6	8.8
4	2002	134.8	68.0×83.0	2.5	1-20.5-0.8	–	–	–	522.2	8.6
5	2003	168.7	106.9×112.5	4.2	1-20.5-0.8	1.0	0.8	0.5	590.8	8.3
6	2004	169.5	72.3×85.3	4.9	1-20.5-1.2	1.0	0.9	1.0	335.1	8.4

To introduce Russian fine seabuckthorn varieties, we selected Qiqihaer (N latitude 47°) of Heilongjiang Province of China, which is close to Novosibirsk (N latitude 55°), Moscow (N latitude

55°), Ulan-Ude (N latitude 51°) and the best introducing results were obtained. The experimental results have showed that Qiqihaer is a suitable region for cultivation of Russian seabuckthorn. In this region, introduced Russian seabuckthorn forms had the characteristics of big fruit, no thorn and high yield. In the another experiment in regions of Fuxin (N latitude 42°) and Dalian (N latitude 39°) of Liaoning Province, Huhehaote (N latitude 40°) and Eerduosi (N latitude 39°) of Inner Mongolia, the better introducing results have also been obtained. These regions are also suitable for introducing Russian fine seabuckthorn. In these regions, the characteristics of introduced Russian seabuckthorn had some morpho-physiological changes, such as appearance of thorns, lower fruit size and yield are not very good. The result of introducing experiment in Ansai (N latitude 36°) and Wuqi (N latitude 37°) of Shaanxi province showed that introducing Russia seabuckthorn directly to the semi-arid Loess hilly region had worse effect. In these regions, however, introducing and domesticating Russian varieties in transitional regions, like Qiqihaer and Fuxin is better. The rates of germination and survival of seedlings of Russian seabuckthorn from Siberian region for continuous 3 years were 40 per cent -50 per cent during 1993-1995. The young plants grew slower, had poor potential of growth and its height was 6-10 cm in the first year. Most of the young plants died from freezing injury and drought during winter, only there were few survivals. The main reason is that there is a huge difference in geographical location and environmental conditions between Russia and Ansai, Wuqi Loess hilly region of China. Therefore, the fine Russian seabuckthorn varieties introduced directly to semi-arid Loess region can't obtain better results. However, hybrids of Russian seabuckthorn with Chinese seabuckthorn have better adaptability in this region. This is an important direction for selecting and breeding of fine seabuckthorn varieties in semi-arid Loess region of China.

During the process of introduction, selection and breeding of fine seabuckthorn varieties, proper techniques of planting and managing seabuckthorn were given more attention and done well. It is necessary to establish a germplasm banks and mother orchards at various locations for doing good work of introduction, selection, breeding and reproduction of seabuckthorn. It is important to grasp the techniques of dry-resistant nursing seedlings and twig, and hard branch interposition in "three north areas". The resistant capability of young seedlings of Russian seabuckthorn varieties in nurseries to drought and high temperature on the ground surface are very low therefore, proper attention to the managerial measures of watering, overshadowing, scarifying, weeding and fertilization are needed after seeds of seabuckthorn has been sown and interposed. After plantation of saplings, when the introduction garden is established, the spacing between plants and row was 3×1 m or 3×2 m. The depth of pits for planting was 0.4-0.5 m for maintaining definite soil water. The planting of seabuckthorn is better in earlier spring and last stage of autumn. After the planting of seabuckthorn, it is necessary to grasp the techniques of foster management, especially water and fertilizer management and scientifically fruit-picking techniques. In order to improve genetic quality, it is also necessary to establish high yielding seabuckthorn garden, where it have a definite area to select fine seabuckthorn group and individuals, and do breeding work.

CONCLUSION

The introduced fine Russian seabuckthorn varieties have better economic properties. The diameter of fruits, length of fruit stalk, weight per 100 fruits and fruit yield per plant of Russian seabuckthorn are 1.8-2.5, 1.3-2.7, 4.6-8.8 and 3.8-6.2 times of local wild Chinese seabuckthorn forms, respectively. However, the adaptability to environmental conditions and the nutritional properties of Russian seabuckthorn are lower than the Chinese wild seabuckthorn. The height, crown and diameter at butt end of Chinese seabuckthorn are 1.2-1.9, 1.3-2.0 and 1.3-2.2 times of Russian seabuckthorn, respectively. Qiqihaer (N latitude 47°) is suitable region for introducing Russian seabuckthorn. Fuxin (N latitude

42°), Dalian (N latitude 39°), etc. are also suitable region for introducing Russian seabuckthorn. Ansai (N latitude 36°) and Wuqi (N latitude 37°) are the possible region for introducing Russian seabuckthorn. It was important to hybridize the Russian seabuckthorn (female) with Chinese seabuckthorn (male) in this region. It will provide a reliable scientific basis for introduction, selection and breeding of seabuckthorn, step by step in a planned way. Russian seabuckthorn has an important significance for constructing large-scale high production and quality seabuckthorn forests and garden, and improving the environment of the "three north areas" and Loess plateau area, and developing local economy.

REFERENCES

1. Kallio, H., Yang, B., Peippo, P., Tahvonen, R. and Pan, R. 2002. Triacylglycerols, glycerophospholipids, tocopherols, tocotrienols in berries and seeds of two species (ssp. *sinensis* and *mongolica*) of seabuckthorn (*Hippophae rhamnoides*). *Journal of Agricultural and Food Chemistry* 50 (10): 3004-3009.

2. Rongsen, L. 1992. *Seabuckthorn-A Multipurpose Plant for Fragile Mountains*. Occasional paper No. 20, ICIMOD, Kathmandu, Nepal, 62p.

3. Singh, V. 2003. Geographical adaptation and distribution of seabuckthorn (*Hippophae* L.) resources. In: *Seabuckthorn (Hippophae L.)-A Multipurpose Wonder Plant. Vol. I: Botany, Harvesting and Processing Technologies* (V. Singh, Ed. in Chief, 2003), p. 21-34, Indus Publishing Company, New Delhi, 518p.

Chapter 9

Role of Intraspecific Variability in Introduction of Seabuckthorn (*Hippophae rhamnoides* L.) in Belarus

I.M. Garanovich and T.V. Shpitalnaya

Central Botanical Garden of Academy of Sciences, 2A Surganov Street, Minsk 220012, Belarus

SUMMARY

In the present study, five climatypes were taken for the study of intraspecific variability in the introduction of seabuckthorn, which were Baltic, Danube, North Caucasian, South Caucasian and Siberian climatype. A long-term research on bio-morphological characteristics of five climatic types of *Hippophae rhamnoides* L. during their introduction in the climatic conditions of Belarus proved the genetical nature of their polymorphism, which determines the differences in morphogenesis at all stages of ontogenesis. The Baltic climatic type is found to be a valuable source of breeding material because of presence of comparatively small shrubs, low or medium amount of thorns and big fruits that can harvest easily. All these characteristics showed its perceptiveness for zonation and usage as basic material for selection on standing power and productivity.

Keywords: *Seabuckthorn, Five climatypes, Intraspecific variability, Introduction and Belarus.*

INTRODUCTION

Central Botanical Garden (CBG), NASB is one of the leading seabuckthorn research institutions in the Republic of Belarus. As a result, new mother plantations and conservation zones for varietal plants have been established. A large collection of plant forms of various geographical origins has been preserved, and the first industrial plantations have been organized. Using the method of remote hybridization and chemical mutagenesis, the CBG scientists have got a range of large-fruited and productive types.

The complex work on seabuckthorn introduction in Belarus has been started since 1975. As a result, a huge number of sorts and forms has been involved, a perspective to introduce this species into

Seabuckthorn (*Hippophae* L.): *A Multipurpose Wonder Plant, Vol. 4: Emerging Trends in Research and Technologies* (Virendra Singh, *Editor-in-Chief*, 2014), p. 107–139, Daya Publishing House, New Delhi, India.

cultivation as a source of raw material has been scientifically established. The result achieved was thoroughly studied and involved into selection process. Henceforth, sorts such as "Chuyskaya, Samorodok, Velikan, Obilnanaya, Zolotistaya Sibiri, Prevoschodnaya, Obskaya and Yantarnaya" had been introduced into cultivation. These sorts exceeded old ones in their fruit size, weight and fruit stalk length. The promising sorts have been developed on the basis of Baltic and western European types in the MSU botanical garden ("Aromatnaya, Botanicheskaya, Otradnaya, Podarok Sadu, Perchik and Trofimovskaya"). These sorts are the examples of positive results of introduction of large fruited and high vitamin richer, thornless and dwarf growing types of seabuckthorn.

Though, the results achieved were fruitful, but the question of stability and longevity of this crop hadn't been solved in the republic to its full extent. The edaphic factors in winter were the most unfavourable abiotic factors for this plant. Endless, reproduction of varieties and types has played its negative role, which led to the loss of the initial stability potential and decrease of forms-generating processes.

Along with the involvement of a huge number of varieties, the scientists have been carrying out research on a range of *Hippophae rhamnoides* L. types of different geographical origins. The gene pool of this species population has a great stock of hereditary variability according to many indices and therefore, there is a great necessity to continue purposeful research on the gene pool of the botanical garden for new initial materials collected from different regions.

Under the influence of climatic conditions of the botanical garden, new seabuckthorn genotypes were established. However, in contrast to the results achieved in the previous years of studying the ecological and biological varieties, the knowledge of introduced varieties is fragmentary. To preserve and increase the potential of cultivated introduced plants, population plantations of the first, second and the following generations should be established and the forms which are most important in practical terms should be included into selection process. Though, the issue is theoretically and practically important, the specific characters of the development of introduced populations in general and *Hippophae rhamnoides* L. in particular, is still poorly studied.

The fundamental nature of such research lies on studying the adaptability of introduced plants in new environmental conditions. The experimental material collected is a basis of studying the objective laws of species adaptation according to their systematic and geographical origins including some anthropogenic factors.

The studying of ontogenetic features is an important part of plant introduction theory. The age aspects of adaptogenesis provide a more detailed evaluation of species potential and help to contribute into adaptation theory. On the scientific value, there is an assessment of the speed of degradation processes in introduction and studying of species specific characters in the introduction process. The data recorded was of practical value, with increasing number of plant resources and biodiversity of woody plants, preferential involvement of ecotypes and sorts, varieties and forms useful for agriculture from natural populations and cultivation places, which were the closest to Belarusian climatic and edaphic conditions.

MATERIALS AND METHODS

EXPERIMENTAL MATERIALS

The subjects of the present study were seabuckthorn plants from F_0 to 15-17 years and from different geographical origins. The samples of the Baltic climatype were collected from Kaliningrad region (Yantarny settlement); the Danube climatype from the Danube delta; the North Caucasian from

Kabardino Balkaria; the South Caucasian from the Shaki region (North Asia); and the Siberian from the Altaic region (Gorny Altai). From all the above mentioned climatypes, F_1 seed progeny of F_0 was used. All seeds were of the same year and from plants grown in the same agro-climatological and technological conditions, and the seeds were taken from plants, aged 17 years, cultivated in the Central Botanical Garden. 1000 specimens of 27 types and 5 climatypes of *Hippophae rhamnoides* L. were used in the experiment. The seed quality identification is a very important task in applied work as well as in scientific research. A seed quality test is being executed in accordance with the National Standards.

EXPERIMENTAL METHODS

Seed Sowing Qualities

Seed sowing qualities includes germinating capacity, germination energy, good quality and vital capacity, which were identified by GOSTs (State Standards) 13056.6-75, 13056.7-93, 13056.8-97 (GOST, 1995). The seed samples were selected in accordance with GOST 13056.1-67.

Ontogenetic Changes

All ontogenetic changes are age-related and the classification of age was done by Rabotnov T.A. scale (Rabotnov, 1950).

Phenological Spectrum

Phenological spectrum of development was studied in accordance with Methods used in Botanical Gardens of the USSR (1975).

Growth Analysis

The amount of growth of different tree parts, in particular; shoots linear growth was recorded by Molchanov and Smirnov (1967) method alongwith the recommendations of Maltseva and Fomenko (1990).

Winter Hardiness

Winter hardiness was estimated according to the scale used by Dendrology department of the CBG (Lapin, 1967).

STATISTICAL ANALYSIS

All observations were conducted in triplicate and the results of the research were statistically analyzed by mathematical package MatchCad 2000 at $p \leq 0.05$.

RESULTS

INTRASPECIFIC VARIABILITY IN *HIPPOPHAE RHAMNOIDES* L. AND ITS ADAPTIVE POTENTIAL

It is commonly known that species with a vast and disjunctive habitat possesses adaptive qualities. A real expansion of every species depends on the presence and degree of new habitat adaptability and under the influence of reproductive potential of particular populations. The interrupted habitats reflect century long changes in plants existence conditions. According to this theory, species that grow in immense territories, occupy a more progressive evolutional place. They are represented by a large number of polymorphous populations, which are the result of genetic mutation and possess huge ecological variation, high level of genetic heterogeneity and ability to resist the influence of various conditions. Seabuckthorn possesses all these qualities because it is an extremely polymorphous species with a vast and disjunctive habitat. Therefore, the adaptive potential of this species is high.

Historically, this species grows along the river valleys and sea coasts that are located in different eco-geographical regions. Some genetic-geographical centers, which suit such ecological category, are singled out in habitat boundaries, known as climatypes and appear under the influence of climatic differences (Eliseev, 1982). The long-term research had single out a range of climatypes and their specific features.

The Baltic climatype is characterized by life cycle process that ends together with Siberian climatype; by high winter resistance, large fruits, high contents of vitamin C and seabuckthorn oil, and spring-autumn thaw resistance. The disadvantage is high growth and thorns. The Danube climatype is characterized by dwarfness (1.5-2.0 m) and high yielding types with a moderate or small number of thorns, large-fruited and high-oiled fruits with less colourful peel, and fruits are sour with a distinct bitter taste. The North and South Caucasian climatypes are characterized by different ecological types of seabuckthorn. The ecotypes with shorter vegetative period but higher level of frost resistance grow in highlands. The ecotypes with xenomorphic features grow in semi-arid zones and are characterized by a long growth period, a low level of winter resistance, small fruits, a low level of vitamin C and carotenoids after introduction. The Siberian climatype has a shorter vegetative period, high winter resistance, short growth and small number of thorns. It has the largest fruits, high content of oil and carotene. The Central Asian climatype has a longer vegetation period than the Siberian one but a low level of winter resistance. Xenomorphic features are well-developed. It is a richer source of carotenoids and ascorbic acid.

For several years, CBG NASB research workers have been studying the population species changeability in expeditions to different regions: Kaliningrad region, the Danube delta, Kabardino-Balkaria, southern Azerbaijan and Gorny Altai.

PHENOLOGICAL OBSERVATIONS

On the basis of phenological phase observations and average long-term data, a phenospectrum of five climatypes and *Hippophae rhamnoides* L. types has been made (Tables 9.1 and 9.2).

The active growth of vegetation of seabuckthorn in these ecological conditions begins in the third 10 days of March to the first 10 days of April (gemmation phase). The specimens from the Siberian climatype and "Maslichny" sort begin the vegetative growth 5 days earlier as compared to other experimental samples. The blossoming starts 15-20 days after the beginning of vegetative phase. Different climatypes and sorts have different durations of this period. The blossoming starts in the first 10 days of May, when the temperature rises to 16 °C, and goes simultaneously with the budding and leafiness. The period of blossoming depends on the weather and moreover, in dry warm weather it lasts from 2 to 6 days.

In the existing environmental conditions, specimens from Siberian climatype and "Maslichnyj and Aley" sorts have earlier blossoming terms. The plants of Baltic and Danube climatypes and the "Botanichesky" type are in an intermediate position. The plants of North Caucasian climatype blossom 5-8 days later than the Siberian climatype plants. Also, leaf isolation of the male plants starts 2-3 days earlier as compared to female ones. The flowers on the sprout axis do not burst simultaneously.

Table 9.1: Phenospectrum of Seasonal Development of Various Climatypes and Sorts of *Hippophae rhamnoides* L. (Average for 1989-1990)

Climatype, Sort	*Gender*	*Phenological stages*						
		Buds Swelling	*Buds Opening*	*Leafing*	*Mass Blossoming*	*Fruit Ripening*	*Leaf Fall*	*Linear Sprouts Growth*
				Year 1989				
Baltic	f	24.III	12.IV	29.V	3.V	24.VIII	28.VIII	7.V
	m	24.III	12.IV	27.IV	1.V	–	28.VIII	5.V
Danube	f	24.III	12.IV	29.IV	4.V	28.VIII	28.VIII	8.V
	m	24.III	12.IV	28.IV	2.V	–	28.VIII	5.V
Noth Caucasian	f	28. III	15.IV	1.V	6.V	24.IX	4.IX	11.V
	m	28. III	15.IV	30.IV	4.V	–	4.IX	9.V
SouthCaucasian	f	24.III	12.IV	29.V	4.V	15. IX	24.VIII	7.V
	m	24.III	12.IV	27.IV	2.V	–	24.VIII	7.V
Siberian	f	19.III	6.IV	24.IV	1.V	11.VIII	22.VIII	4.V
	m	19.III	6.IV	22.IV	29.IV	–	22.VIII	2.V
Botanicheskaya	f	24.III	12.IV	27.IV	3.V	11.VIII	28.VIII	8.V
Maslichnaya	m	19.III	6.IV	28.IV	1.V	5.VIII	24.VIII	6.V
Aléy	f	19.III	6.IV	24.IV	1.V	–	24.VIII	7.V
				Year 1990				
Baltic	f	22.III	14.IV	2.IV	5.V	28.VIII	5.X	10.V
	m	22.III	14.IV	30.IV	3.V	–	5.X	8.V
Danube	f	27.III	16.IV	28.IV	5.V	28.III	27.X	4.V
	m	27.III	16.IV	27.IV	5.V	–	27.X	4.V
Noth Caucasian	f	3.IV	23.IV	1.V	8.V	14.IX	10.X	13.V
	m	3.IV	23.IV	3.V	6.V	–	10.X	12.V
SouthCaucasian	f	4.IV	19.IV	30.IV	2.V	19.IX	28.IX	7.V
	m	4.IV	19.IV	28.IV	1.V	–	28.IX	8.V
Siberian	f	15.III	6.IV	23.IV	29.IV	10.VIII	3.X	28.V
	m	15.III	6.IV	20.IV	27.IV	–	3.X	26.V
Botanicheskaya	f	2.IV	14.IV	30.IV	5.IV	14.VIII	3.X	5.V
Maslichnaya	m	15.III	6.IV	27.IV	29.IV	10.VIII	3.X	3.V
Aley	f	15.III	6.IV	25.IV	30.IV	–	20.IX	1.V

Note: f: Female; m: Male.

One of the main features of individual seabuckthorn evolution is its sprout growth. Looking at this index, one could judge about the success of introduction of this species. The sprout linear growth in Belarus starts at the beginning of May and it is short, because the main resources are spent on photosynthetic lamina surface and initiation of buds on side sprouts. The third 10 day period of

August is the earliest date of sprout growth ending of Siberian climatype plants, while the seabuckthorn from North Caucasian climatype has the latest date of growth ending. The term of fruit ripening is the chief factor and this process is not simultaneous, that is why in each group only the beginning of this phenological stage has been registered. Some difference in cultivars and natural plants ripening term is worth mentioning.

Table 9.2: Phenospectrum of Seasonal Development of Various Climatypes of *Hippophae rhamnoides* L. (average for 1991-1993)

Climatype	*Buds Swelling*	*Buds Opening*	*Leafing*	*Mass Blossoming*	*Fruit Ripening*	*Leaf Fall*	*Lenear Sprouts Grouth*
Year 1991							
Baltic	20.IV	7.V	5.V	–	–	1.IX	15.V
Danube	15.IV	25.IV	10.V	–	–	I.IX	5.V
Noth Caucasian	25.IV	5.V	15.V	–	–	15.X	25.V
South Caucasian	20.IV	25.IV	5.V	–	–	5.IX	10.V
Siberian	5.IV	15.IV	28.IV	–	–	25.VIII	5.V
Year 1992							
Baltic	2.IV	13.IV	10.V	–	–	25.IX	10.V
Danube	13.IV	17.IV	10.V	–	–	25.IX	15.V
Noth Caucasian	13.IV	25.IV	10.V	–	–	10.X	18.V
South Caucasian	2.IV	13.IV	5.V	–	–	15.IX	5.V
Siberian	2.IV	10.IV	5.V	–	–	10.IX	5.V
Year 1993							
Baltic	15.IV	27.IV	10.V	2.V	10.IX	20.X	16.V
Danube	15.IV	27.IV	10.V	10.V	15.IX	20.X	16.V
Noth Caucasian	20.IV	1.V	12.V	–	–	1.XI	16.V
South Caucasian	5.IV	22.IV	7.V	30.IV	25.VIII	20.X	10.V
Siberian	5.IV	22.IV	7.V	27.IV	15.VIII	20.IX	10.V

The "Maslichny and Botanicheski" sorts have earlier ripening terms (in the first or second ten-day period of August). The Baltic climatype had the third ten-day period of August or the first ten-day period of September; South Caucasian climatype had the second ten-day period of September; North Caucasian climatype had the second or third ten-day period of September. It was noticed that North Caucasian species ripening term is falling behind significantly.

The leaf fall begins in the second part of August and Siberian climatype species were the first ones to shed their leaves. It was noticed that North Caucasian plants hadn't got dead sprouts and stems that shows their low winter resistance. Such late leaf fall, especially after the first autumn frosts shows that seabuckthorn ancestors were evergreen from tropical and subtropical zones. The leaf fall of all climatypes starts from the plant bottom. Many years of studies have proved that leaf fall begins before leaf colouring. According to some authors, it happens because of water regime disturbance during dry or hot weather, as mature plants lack water.

The sorts of seabuckthorn were under observation along with seabuckthorn types of the first generation. The female "Botanicheski and Maslichny" sorts and the male "Aley" sort were the experimental samples. A range of regularities in passing through certain phenological stages has been observed and a correspondence has been registered: "Maslichny" sort and the Siberian climatype budding stages were the same; "Botanicheskij" sort and Baltic climatype blossoming stages were the same; "Aley" variety and South Caucasian climatype sprout linear growth stages are found the same.

Thus, the method of the studied material separation on the basis of phenological rhythm lets us to analyze the introduction results of different climatypes of *Hippophae rhamnoides* L. in cultivation conditions. All stages of individual climatypes evolution go through the whole period in Belarus in their natural succession and are determined by all eco-climatic region conditions.

SEASONAL DYNAMICS IN SPROUT GROWTH

Plant growth is the most integral process of all physiological processes and is one of the most important indices of adaptation degree to new environmental conditions. The plant growth duration is a part of plant phenorhythm and it defines to a great extent the plant's resistance to unfavourable conditions. The apical sprout observations showed that these sprouts have a longer growth period and higher absolute and relative fluctuation amplitudes, which influence the result of sprout growth rhythm study.

The apical and side sprout increase of both male and female species was studied simultaneously (Table 9.3). The annual average increase in length is characterized by low changeability by North, South Caucasian and Siberian climatypes, "Botanicheski" female varieties, North and South Caucasian climatypes male varieties. The male and female plants of Baltic and female plants of Siberian climatype, and "Maslichny" sorts possess an average changeability degree. The Danube climatype female samples have a high changeability degree.

Every ten days, climatypes and varieties dynamics was characterized by multivertex process, *i.e.* the sprout increase of seabuckthorn has several peaks. The male and female specimens showed a different increase. The apical male plant sprouts of the Baltic climatype had two maximum impulses of growth at every ten days (at the end of May and at the end of June). The female specimens grow intensively at the beginning of June and at the beginning of August. The side sprout growth peak follows the apical peak but with a slight delay. This climatype has a longer duration till the second ten-day period of September. The Siberian climatype sprouts grow steadily and reached the peak in the middle of June (males), and at the beginning of July (females). Then, the increase steadily slows down and stops at the end of August. This steady decrease in the Siberian climatype sprout growth at the end of August is caused by active preparation to wintering in these conditions. The Danube climatype sprout growth reaches to 1.1 cm in a ten-day period with peak in May and middle of June, while the Caucasian type grows slowly. The North Caucasian climatype had its peak at the end of June for both male and female varieties, besides the side sprouts are larger than apical sprouts. The latter feature is also common for South Caucasian type. Its highest sprout increased peak is at the end of May (males), and then slows down till the first ten-day period of September. The apical sprout increase of female species was observed at the beginning of August. A general trend is that all mean values of North and South climatypes characteristics decreased.

The sprouts of "Botanicheskij" sort grow slowly and without quick growth. The "Maslichny" variety is, on the contrary, characterized by multivertex sprout growth dynamics (at the middle of May, the middle of June, and the beginning of July). The "Aley" sort had one growth peak of apical sprouts at the end of June and side sprouts in the middle of July. The sprout growth slows down and acceleration could be explained by meteorological factors fluctuation. Though, as it was mentioned above, this process depends on endogenous factors.

Table 9.3: Climatype and Variety differences of *Hippophae rhamnoides* L. with Average Annual Increase of Apical and Side Sprouts in F_0(1993)

Climatype and Sort	*Gender*	*Sprout Type*	*Average Annual Increase (cm)*	*V (per cent)*	*P (per cent)*
Baltic	female	apical	35.3 ± 1.53	13.7	4.3
		side	30.3. ± 1.31	13.6	4.3
	male	apical	35.9 ± 1.74	15.3	4.8
		side	21.1 ± 0.90	13.5	4.3
Danube	female	apical	12.5 ± 1.69	12.8	2.3
		side	10.5 ± 0.26	8.0	2.5
	male	apical	13.8 ± 0.67	15.3	4.8
		side	9.0 ± 0.12	4.3	1.4
North Caucasian	female	apical	9.2 ± 0.13	4.5	1.4
		side	14.4 ± 0.49	10.8	3.4
	male	apical	9.9 ± 0.17	5.4	1.7
		side	10.4 ± 0.24	7.2	2.3
South Caucasian	female	apical	10.8 ± 0.20	5.8	1.9
		side	10.5 ± 0.29	8.8	2.8
	male	apical	10.9 ± 0.16	4.6	1.5
		side	11.7 ± 0.53	14.3	4.5
Siberian	female	apical	31.9 ± 0.93	9.2	2.9
		side	24.3 ± 1.09	13.3	4.2
	male	apical	26.3 ± 1.32	15.9	4.9
		side	25.7 ± 1.05	12.9	4.1
Botanicheskaja	female	apical	9.2 ± 0.18	6.2	2.0
		side	12.5 ± 0.60	15.3	4.8
Maslichnaja	female	apical	17.5 ± 0.84	15.1	4.8
		side	14.9 ± 0.64	13.5	4.3
Alej	male	apical	21.3 ± 0.63	9.3	3.0
		side	23.0 ± 1.06	14.6	4.6

Thus, some features of seasonal and age annual sprouts growth of some seabuckthorn climatypes has been determined. The plants of different geographical origins which were introduced into the same conditions possess different adaptive characteristics. The study of seasonal dynamics of biometric sprout characteristics showed an increase in the first part of the vegetative period and a distinct stabilization in the second part, showing the plant growth mainly takes place in the first part of the year.

The total sprout length increased during the vegetative period exceeded the initial length of the observed sprouts significantly. The male and female specimens had a different index and it is also different for every climatype. In general, the total growth of male plants was more than that of female ones. The degree of plant resistance to low temperatures is connected with the terms of growth process ending and autumn growth of sprouts. The gradual sprout growth minimization of the Siberian climatype specimens at the end of August is an indicator of an active preparation to wintering in these conditions.

The North Caucasian climatype transplants were distinguished by significant lagging (they are the last that end their cycle, *i.e.* at the end of September) behind the other specimens in both zero and first generation. A growth domination of side sprouts over apical sprouts of both male and female specimens has been noticed, which leads to a specific crown structure (simpodial branching type).

Hippophae rhamnoides L. preserves its specific home rhythm of seasonal evolution and acclimatized in the Belarusian climate.

MORPHOLOGICAL FEATURES OF SEABUCKTHORN CLIMATYPES

Baltic Climatype

At the age of 12 years, the maximum plant height reaches to 2.5 m (Figure 9.1A). The average female sprout increase was found to be 35.3 cm, while the average male sprout increase was 35.9 cm (Table 9.3), *i.e.* the increase is practically the same. The average stem diameter of the main sprout was found 6.4 cm at the bottom. The branches had a specific greyish colour and small number of thorns with their length of 3.5 cm. The maximum leaf length was 100.0 mm, width 9.0 mm with an average size of 66.6×6.6 mm. Due to such wide and long leaves, dark-green colour and thick brown downiness, and the stem crown had a thick, compact and oval form. The Kaliningrad types of this climatype could be regarded as large-fruited, their maximum length of 11.0 mm and minimum length of 5.0 mm, with an average length of 9.1 mm.

The fruit shape can be round or oblong, but the oval form is the most common (Figure 9.1B). The fruit colour changes from yellow to orange-red with the majority of the orange colour. The average weight of 100 fruits was 30.8 g. The blossom cluster could bear up to 3.6 fruits, but the average quantity was 2.6 fruits. In the 10 cm of sprout region, 12.4 to 17.0 fruits were recorded. The fruit disposition density of Baltic climatype sprouts (23.3 pieces) was found lower than that of fruit disposition density of Siberian climatype sprouts (46.2 pieces). The fruits with prolonged (5.0 mm) and shortened (1.3 mm) fruit stalk and dry fruit tearing may be found. A range of fruits were sour but odourous and palatable.

The seed characteristics of the seabuckthorn types studied confirm that their seeds were large with an average length of 5.9 mm and diameter of 2.9 mm. The types with longer seeds (7.0 mm, 7.1 mm) as well as smaller ones (5.0 mm) were found and seeds were of egg-shaped, oval and oblong form, though the egg-shaped and oblong form seeds dominate.

The seed colour varied from light-brown bistre (25 per cent) to dark-brown umber (8 per cent) while, other forms (67 per cent) had sepia (brown) colour. The external structure studies of Baltic climatype seeds revealed a furrow between seed lobes from the abdominal side and a specific seed lobes shift towards each other. The furrow was distinct and goes through the whole seed. The seed surface was mainly smooth and glossy. The weight of 1000 seeds was ranged from 18.4-29.7 g. The types of this particular climatype had an average weight of 24.4 g, which exceeds seeds weight of North Caucasian climatype significantly.

Danube Climatype

The seabuckthorn types, introduced from Danube delta (Figure 9.2A), are characterized by a small shrub height (2.0 m, aged 11), and by oblong, sprawling and medium-thick crown. The stem diameter at the bottom was 3.3 cm and thorns length was 2.8 cm. A huge number of thorns were found

Figure 9.1: Morphological Features of Baltic Climatype Specimens (A) and Fruits (B).

Figure 9.2: Morphological Features of Danube Climatype Specimens (A) and Fruits (B).

in the 10 cm of a sprout region (3.2 thorns). The vegetative sprout increase was low for both male and female specimens (13.8 and 12.5, respectively).

The average leaf length was 57.3 mm and width was 6.6 mm. The Danube Climatype leaves differ significantly from other climatypes in colour, as they were light-green, with a greyish tint, long and with grey downiness. It was noticed that taking into account the quantity of buds located on 10 cm of a sprout, they were close to the Siberian climate.

The fruits were located densely on the sprout (43.1 on the 10 cm of a sprout) besides, the dry fruit tearing dominates. The Danube climatype forms were characterized by a long fruit stalk (3.6 mm). The most common fruit forms were oval and oblong, and the most common taste was sour, tart and bitter taste. The fruit colour was also specific, from lemon-yellow to dark-orange. The average fruit size was 9.2×8.0 mm and weight of 100 fruits was 26.5-41.5 g (Figure 9.2B).

The study of 100 seed weight had showed that the average seed weight was 20.5 g for this climatype, which is in an intermediate position. The seeds studied were small (5.2×2.6 mm), seed form was egg-shaped and oblong, and colour was dark-brown and brown. The surface structure was glossy and ribbed.

North Caucasian Climatype

The tree height (Figure 9.3A) was varied with an average height of 3.6 m (aged 14) and the crown was oval, sprawling and thin. The stem diameter was 3.0 cm at the bottom. The sprouts were thin, their length was 9.9 cm for males and 9.2 cm for females. The numbers of thorns was high, and were not longer than 1 cm. The leaves were small, light-green, narrow-leaved, their average length was 47.1 mm and width was 4.9 mm, the greyish colour of leaves and rind was distinct.

The fruit was of medium size, its length ranged from 7.3 to 9.0 mm and the diameter ranged from 5.4 to 7.00 mm (Figure 9.3B). The orange fruit colour dominates and ranges from oval to prolong. The sour and bitter taste prevails due to high fruits acidity. The fruit stalk was 2.6 mm with dry fruit tearing. The inflorescence usually had 3 fruits but the number can be increased. The 100 fruits weight was found to be 23.9 g, *i.e.* these forms were small-fruited.

The seeds were small, an egg-shaped or a flat egg-shaped form, the surface was smooth, glossy and sometimes matte. The seed colour was brown or light-brown, the weight of 1000 seeds ranged from 10.4 g to 19.7 g and the average weight of 14.1 g.

South Caucasian Climatype

The seabuckthorn specimens height of this climatype were not higher than 2.5 m (aged 13) and usually have a bushy form (Figure 9.4A). The stem diameter reaches to 3.0 cm. The crown was oblong, compact and thick. The current sprout increase was found to be 10.8 cm for males and 10.9 cm for females. The Caucasian types fall behind than Kaliningrad in sprout increase. The number of thorns was less and has the shortest thorns (0.5 cm), as they were rare. The leaves were long, wide and due to the presence of this characteristic, they were the biggest leaves among all the groups. Their leaf length was 69.8 mm, width was 6.2 mm and they were dark-green, thick, with little downiness.

The fruit shape (Figure 9.4B) was oval or oblong, besides that there were 25 per cent less fruits with an oval form. The fruit size of different forms could ranged from 6.0 mm to 11.0 mm in length and from 6.0 to 8.8 mm in diameter. The bitter-sweet palatable taste dominates, only one had a bitter taste and two had a sour taste. Only three forms out of ten have dry fruit tearing, the rest had wet tearing and no fruit stalk. The fruit stalk length was small and the average length was 2.7 mm. The fruits are

Figure 9.3: Morphological Features of North Caucasian Climatype Specimens (A) and Fruits (B).

Figure 9.4: Morphological Features of South Caucasian Climatype Specimens (A) and Fruits (B).

loosely located on the sprout, recorded 26 pieces on 10 cm of the sprout and each bunch had on an average 2.3 fruits. The weight of 100 fruits ranged from 33.7 to 52.5 g. The weight of the fruits exceeds the weight of North Caucasian climatype types.

The fruit length ranged from 4.2 to 5.3 mm and width from 2.2 to 3.6 mm. The seeds studied are not big, and the average size (4.7×2.7 mm) was only bigger than North Caucasian climatype seed size. The fruit form ranges from egg-shaped to oval, the egg-shaped form domineers. The colour ranges from dark-brown to brown with light spots. The seed surface is smooth and glossy.

Siberian Climatype

The tree height of this climatype form at the age of 13 years was 2.3 m (Figure 9.5A). The stem diameter was 4.2 cm and the crown was oval, compact and of an average thickness. The annual increase of males was 26.3 cm on an average and of females was 31.9 cm, *i.e.* a bit higher. The number of thorns was averaged and thorns amount on 10 cm of the sprout was 2.5 cm on an average, and their length is 1.7 cm. The maximum leaf length was 92.7 mm and width was 9.4 mm with an average size of 69.2×6.2 mm.

The fruit size of Siberian climatype ranged from 5.4 mm to 13.0 mm in length and from 3.1 mm to 7.2 mm in diameter (Figure 9.5B). The colour of fruits ranged from yellow to orange-yellow, but the orange colour dominates. The fruit form also ranged from oval to oblong.

The weight of 100 fruits ranged from 18.5 to 57.5 g. The fruits density on sprouts ranges from 9.0 to 106.4 pieces on 10 cm of the sprout. This index led us to estimate the productivity of seabuckthorn specimens. In this type of climatype, the fruit stalk length ranged from 1.4 to 7.4 mm with an average of 3.2 mm. The dry fruit tearing dominates which shows that the rind is consistent enough, though samples number 2-14, 2-26, 4-23 have wet fruit tearing and are without fruit stalks. The taste ranged from very sour, sour, with little bitter, odourous and sweet, palatable. The average amount of fruits on the bunch was not higher than 2.3 pieces, though single fruits with 4.2 pieces occur. These fruits grew densely on sprouts.

The variation in length of seed occurs from 4.2 to 7.7 mm in length and from 2.0 to 4.4 mm in diameter. The weight of 1000 seeds of Siberian climatype ranged from 9.6 to 29.7 g, with an average index of 22.1 g.

The seed form was also varied and were of egg-shaped, rhombic, and cariniform, but the most common forms were egg-shaped or oblong egg-shaped. Some different seed groups had been singled out according to their colour, which ranged from light brown to brown. The seeds of this colour were represented by a nearly equal number: 50 per cent to 50 per cent. The seed surface of this climatype is lustrous, smooth and glossy (sometimes glossy-matte).

Thus, the plant seabuckthorn height ranged from 2.0 m (Danube climatype) to 3.6 m (North Caucasian climatype). The shrub or the tree size of this species together with the morphological factor was of great value in selective work. The crown of Baltic and South Caucasian climatypes was the most developed one and are oval, oblong, compact and thick. The Baltic climatype sprouts of male and female species were the largest, which ranged from 35.9 cm to 35.3 cm. The North Caucasian and Danube climatypes possess the highest number of thorns. The South Caucasian and Siberian climatypes species had large leaves. There is no doubt that Baltic climatype seabuckthorn had large-fruits (9.1 × 6.2 mm) and seeds (5.9 × 2.9 mm).

During the study of the morphological features of some seabuckthorn climatypes in F_1, there was a task to find out the existence of general and individual type features of the cultivar and to observe the

Figure 9.5: Morphologicai Features of Siberian Climatype Specimens (A) and Fruits (B).

process of hereditary data transmission in next generations. The plant height at aged 2 ranged from 44.1 cm (North Caucasian climatype) to 85.1cm (Danube climatype). In this case, the plant height depended not just on their genetic features but on the weather conditions of the year in question as well. The Baltic climatype specimens had the maximum increase of annual sprouts (40.4 cm). The distinguishing feature of these types is their initial opposite leaf arrangement transformed into mixed one and then into alternate leaf arrangement. The Baltic type plants have alternate leaf arrangement from the 6th and the 11th leaf; Siberian type from the 8th and the 11th leaf. The two-year old plants showed leaf size characteristics at the end of a vegetative period. The leaf size ranged from 59.2×7.4 mm for the Baltic climatype to 46.0 × 5.8 mm for North Caucasian climatype. The South Caucasian forms had large lamina of 57.0 × 7.8 mm.

One of the important characteristics of seabuckthorn is the presence of thorns, which could be of different length and firmness. This feature is taken into account while analyzing a variety and type, because thorns are an obstacle in fruits gathering. Historically, the features, such as thorns and high downiness are formed in the process of the long natural selection and preserve themselves in a cultivar. The Danube climatype plants differ from the other ones by their large number of thorns.

During the period of study, the plants of Danube climatype were the most resistant plants in our conditions, *i.e.*100 per cent of preservation, the plants of South Caucasian climatype go second (56.3 per cent). Weight and size characteristics have been estimated as well as their descriptive characteristics.

The large-fruited forms of South Caucasian climatype have been defined on the basis the morphological cultivar description, the average fruit length was 11.1 mm and the Baltic climatype fruit length, which ranged from 8.7 to 9.5 mm. The Baltic, Siberian and Danube climatypes are characterized by single fruiting and the fruit number on a knop was not larger than 11.9 pieces. This index for Baltic climatype sowing has been changed, since 1990 from 6.4 to 12.4 pieces. The same index was found for the Siberian climatype. The weight of 100 fruits of Baltic and South Caucasian climatype was 37.0 g and 36.7 g, respectively. The orange colour and oval fruit form dominate. The fruit stalk length ranged from 3.1 mm for the Siberian to 3.7 mm for South Caucasian and Danube climatypes. The data above showed the specific character of each climatype.

The morphometrical seed characteristics (length, diameter) in F_1 showed that each climatype has seeds of the same size (Table 9.4). The average seed length of Baltic climatype was 5.8 mm (1989) and 5.4 mm (1990). The Danube climatype seeds were smaller (5.3 mm). Some outer seed morphological features, such as size and colour, ranged from oblong to obovate shape and from dark-brown to light brown colour.

The weight of 1000 seeds is one of the main indices of sowing qualities. The weight estimation has showed the following characteristics. The specimen location according to the weight of 1000 seeds is stable in zero and the first generation (the Baltic and Danube climatypes). This is an indicator of genetic constancy and reveals an advantage of the type selection in accordance with their selection value.

The fruit stalk length ranged from 2.6 to 3.9 mm, fruit diameter from 5.7 to 8.0 mm, fruit length from 7.2 to 9.2 mm and weight ranged from 23.9 to 39.4 g. The variation in fruit stalk length was much lower than the fruit diameter characteristic (Table 9.5).

The fruit form has been distinctively changed. Zero and the first generation have fruits of oval, rounded and oblong forms. However, it remained unchanged throughout the whole research period and for every climatype.

Table 9.4: Comparative Seed Characteristics of some *Hippophae rhamnoides* L. Climatypes in F_0 and F_1 (1992)

Climatype	Length (mm)	Diameter (mm)	Weight/100 Fruits (g)	Seed Weight in Fruit Weight (per cent)
1	2	3	4	5
		F_0		
Baltic	5.9±0.22 5.0-7.1	2.9±0.09 2.1-4.0	24.4 18.4-29.7	8.1 4.9-10.9
Danube	5.1±0.21 4.8-5.3	2.6±0.16 2.2-3.0	18.0 15.2-2.08	7.5 6.2-8.8
North Caucasian	4.2±0.45 3.0-5.1	2.1±0.24 1.5-2.6	14.1 10.4-19.7	6.2 4.2-10.2
South Caucasian	4.7±0.15 4.2-5.3	2.7±0.20 2.2-3.6	9.6 12.7-25.8	5.3 3.3-7.5
Siberian	5.8±0.13 4.2-7.7	3.1±0.08 2.0-4.4	22.1 9.6-31.6	8.2 4.1-13.5
		F_1		
Baltic	5.6±0.26 5.0-6.8	3.4±0.30 3.0-4.3	22.0 18.3-26.4	6.9 4.2-9.3
Danube	5.0±0.55 4.4-6.3	2.5±0.27 2.2-3.1	20.9 19.4-22.7	7.2 6.9-7.6
South Caucasian	5.5±0.29 4.8-6.1	3.1±0.19 2.7-3.5	26.2 22.3-30.3	7.4 6.3-9.0
Siberian	5.5±0.31 5.1-6.1	2.6±0.22 2.2-2.9	13.5 10.3-16.4	5.1 3.6-7.9

Note: Above the denominator-Average amounts; and below the denominator-Limit amounts.

Table 9.5: Fruits Characteristics and Yield Components of some Seabuckthorn Climatypes

Climatype	Fruits		Fruit Stalk Length (mm)	Number, Pieces		Weight/ 100 Fruits (g)
	Length (mm)	Diameter (mm)		10 cm of a Sprout	Fruit Bumber on a Bunch	
Baltic	9.1 7.5	6.2 6.5	2.8 2.8	23.3 14.8	2.6 2.6	30.8 24.9
Danube	9.2 7.2	8.0 5.9	3.6 3.5	43.1 22.1	2.7 3.6	32.6 24.2
North Caucasian	8.1 –	6.2 –	2.6 –	24.4 –	2.5 –	23.9 –
South Caucasian	8.9 9.1	7.1 7.0	2.7 3.2	26.0 19.2	2.3 2.3	39.4 31.6
Siberian	8.1 7.7	5.7 6.5	3.2 3.9	46.2 20.1	2.3 3.5	28.3 34.3

Note: Above the denominator-F_0; and below the denominator-F_1.

The quality characteristics, such as colour ranges distinctively from one climatype to another. The fruit specimens with yellow, orange and red colour of different types have been found. Within each climatype, this feature ranged insignificantly and is stable for generations. The differences are constant when various climatypes are grown in the same zone with the same conditions. It proves the genetic conditionality for this feature.

Thus, large fruit forms of South Caucasian and North Caucasian climatypes of seabuckthorn have been defined as a result of intrapopulation variety analysis of morphometric features of some types.

The seed weight evaluation had revealed the following fact: The specimen location according to the weight of 1000 seeds was stable in zero and the first generation (Baltic and Danube climatypes). This is an indicator of genetic stability and reveals the advantage of the type selection in accordance with their selection value.

The variability was caused mainly by genetic factors. Different quality conditions of the climatypes are the reason for different plant population heterogeneity. The polymorphism of these characteristics gives an ability to select plants with better characteristics for further selection work. The particular morphological features could be a key to determining some seabuckthorn climatypes.

The practical value of Baltic climatype has been found. It has comparatively short shrubs, a small or an average number of thorns, large fruits and dry tearing, thus it is a valuable selection source. The maximum practical indices are registered in different climatypes that is why hybridization of these types is promising.

CLIMATYPE POLYMORPHISM ACCORDING TO THEIR LEAF FORM

We had studied morphological characteristics changes of seabuckthorn lamina form during the change of their formation condition in generations and the connection of these changes with the genetically conditioned types. The leaf growth and development (length, width) of the male and female seabuckthorn specimens of different climatypes has been studied in zero and the first generation. The lamina form change has been defined from the short to the oblong form, in accordance with the leaf form indicator. These characteristics are one of the diagnostic indications of different quality of introductive seabuckthorn populations.

The leaf size indices (Table 9.6) have been taken to identify different qualities of climatypes. As a result, testing areas for male and female specimens of 5 climatypes in F_0 have been built (Figures 9.6–9.12).

The testing grounds were built for male and female specimens at zero generation. In F_0, Baltic, South Caucasian, Siberian climatypes and the "Aley" sort had higher leaf characteristics for males than for females. The Baltic leaf form had the most distinct differences in length, South Caucasian, Siberian climatypes and "Aley" sort had the most distinct difference in leaf width. In Danube climatype, all the parameters demonstrate a notable prevalence of leaf size of female specimens. The North Caucasian female samples had larger leaf length and width in the first order and males have larger length and width in the second order. The South Caucasian climatype leaf length of the second order was larger for females, and males had larger leaf width. The male Siberian samples had larger leaf width than the female ones, especially the leaf width of the second order. The analysis of the "Botanicheski, Maslichny and Aley" male and female variety characteristics shows that the male "Aley" variety possesses the largest leaves. The female variety which had the largest leaves is the "Maslichnaja" variety.

Table 9.6: Leaf Size of different Seabuckthorn Climatypes in F_0

Climatype Sort	*Gender*	*Sprout Type*	*Leaf Size (mm)*			
			I Order		*II Order*	
			Length	*Width*	*Length*	*Width*
1	*2*	*3*	*4*	*5*	*6*	*7*
Baltic climatype	f	apical	78.0	7.0	65.0	6.0
		side	81.0	7.0	55.0	7.0
	m	apical	98.0	7.0	70.0	6.0
		side	114.0	9.1	73.0	6.0
Danube climatype	f	apical	63.0	6.0	60.0	6.0
		side	73.0	6.2	52.0	5.0
	m	apical	50.0	4.1	50.0	4.9
		side	60.0	5.6	40.0	5.0
North Caucasian climatype	f	apical	50.0	5.3	42.0	4.0
		side	42.0	5.0	60.0	5.2
	m	apical	40.0	3.4	42.0	4.0
		side	55.0	5.8	60.0	5.2
South Caucasian climatype	f	apical	74.0	8.0	58.0	6.0
		side	90.0	8.0	65.0	6.0
	m	apical	70.0	8.0	50.0	7.2
		side	100.0	9.4	65.0	7.4
Siberian climatype	f	apical	67.0	6.0	52.0	6.0
		side	72.0	6.0	58.0	6.0
	m	apical	60.0	5.4	50.0	5.5
		side	72.0	7.8	58.0	7.0
Botanicheskaya	f	apical	70.0	4.8	78.0	6.0
		side	70.0	5.3	83.0	4.0
Maslichnaya	f	apical	90.0	6.0	92.0	7.0
		side	100.0	6.0	90.0	6.0
Aley	m	apical	90.0	6.9	80.0	7.4
		side	113.0	7.1	94.0	7.0

The dimensional indices for every climatic type have their own peculiarities. For example, the leaf length of the Baltic climatype reached to 78.00 mm for female and 98.00 mm for male plants. This is the maximum amongst all climatypes. Only some cultivated varieties have leaves of such a length. They are female plants of seabuckthorn "Maslichnaya" and male plants of "Aley" cultivated variety. The minimal lamina length is observed in North Caucasian climatic type-50.00 mm for female and 40.00 mm for male plants. These data apply to apical leaves of male and female plants of *Hippophae rhamnoides* L. of the aforementioned climatic types.

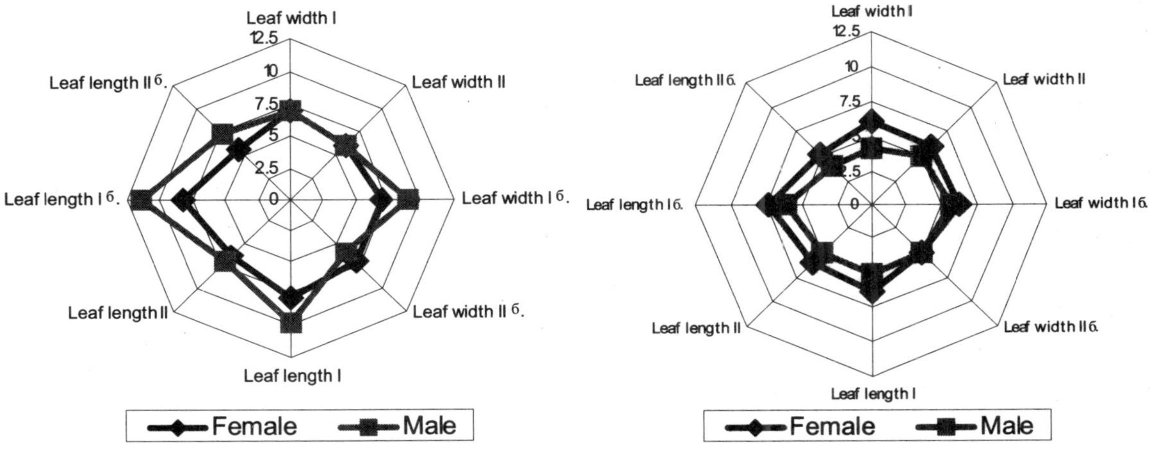

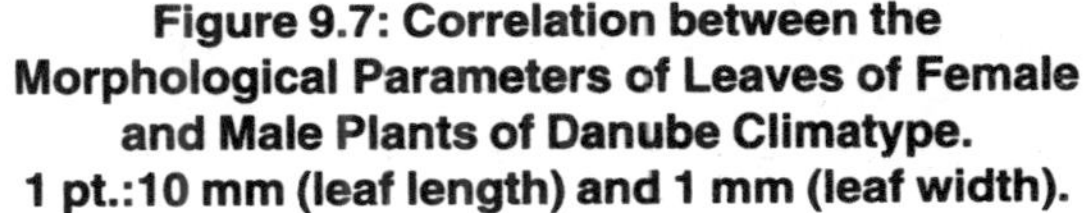

Figure 9.6: Correlation between the Morphological Parameters of Leaves of Female and Male Plants of Baltic Climatype. 1 pt.:10 mm (leaf length) and 1 mm (leaf width).

Figure 9.7: Correlation between the Morphological Parameters of Leaves of Female and Male Plants of Danube Climatype. 1 pt.:10 mm (leaf length) and 1 mm (leaf width).

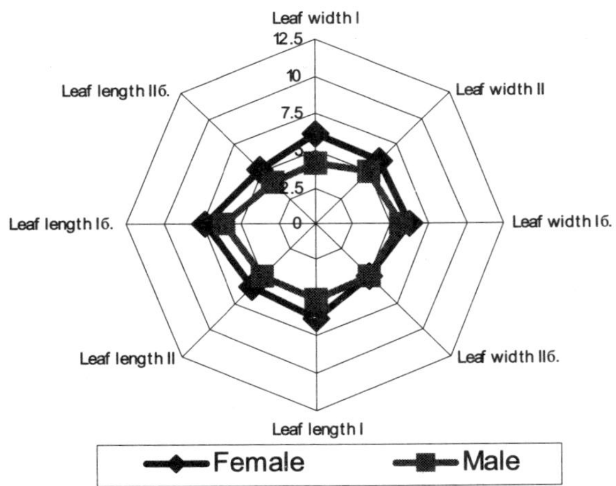

Figure 9.8: Correlation between the Morphological Parameters of Leaves of Female and Male Plants of North Caucasian Climatype. 1 pt.:10 mm (leaf length) and 1 mm (leaf width).

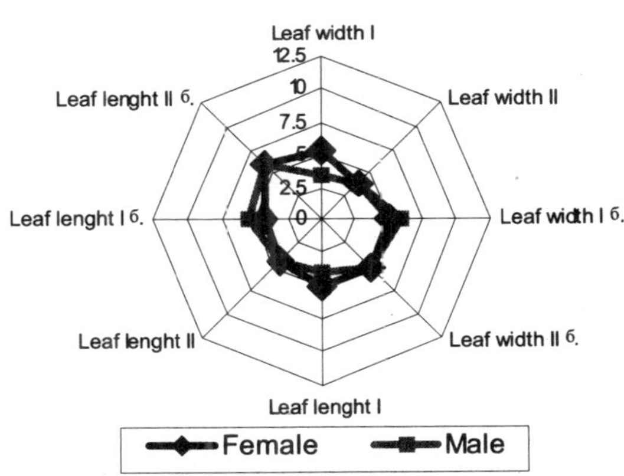

Figure 9.9: Correlation between the Morphological Parameters of Leaves of Female and Male Plants of South Caucasian Climatype. 1 pt.:10 mm (leaf length) and 1 mm (leaf width).

Variations in the duration of lamina growth lengthwise were also observed in side shoots (Table 9.7). For example, the length of side leaves of Baltic climatype reaches to 81.00 mm for female and 114.00 mm for male plants. The minimal length was observed for North Caucasian climatype 42.00 mm for female and 55.00 mm for male plants, respectively. The South Caucasian climatype had a leaf length of 90.00 mm for female and 100.00 mm for male plants. The leaves of Siberian climatype were equal in length for both male and female plants of 72.00 mm. Lamina length of "Aley" cultivar reaches

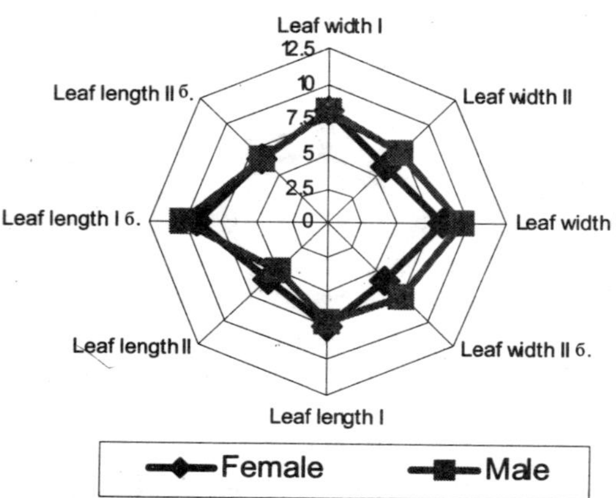

Figure 9.10: Correlation between the Morphological Parameters of Leaves of Female and Male Plants of Siberian Climatype. 1 pt.:10 mm (leaf length) and 1 mm (leaf width).

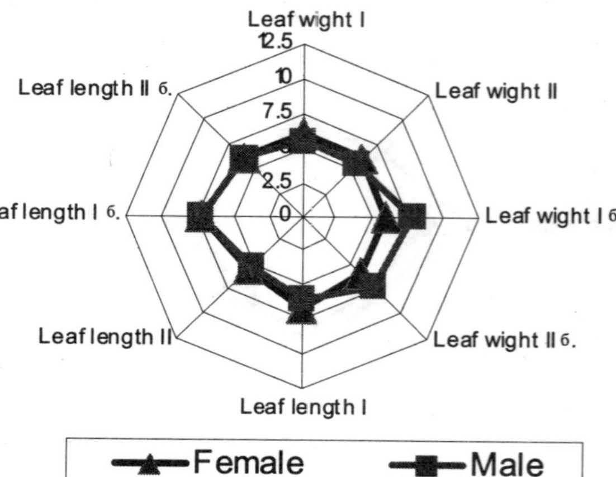

Figure 9.11: Correlation between the Morphological Parameters of Leaves of Female and Male Plants of "Aley, Maslichnaya and Botanicheskaya" Sorts. 1 pt.:10 mm (leaf length) and 1 mm (leaf width).

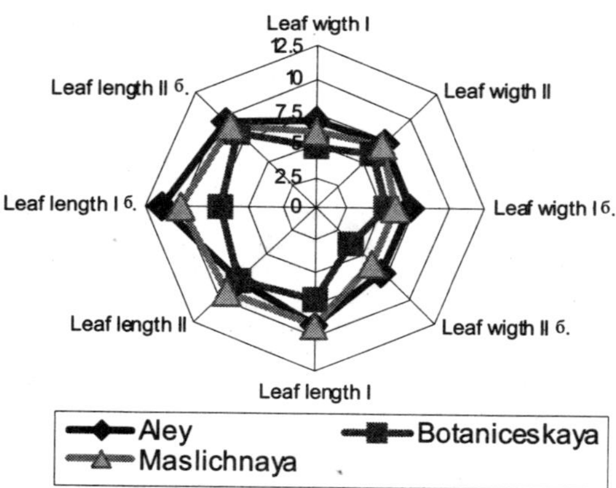

Figure 9.12: Correlation between the Morphological Parameters of Leaves of Specimens of Baltic, South Caucasian and Siberian climatic Types. 1 pt.:10 mm (leaf length) and 1 mm (leaf width).

to 113.00 mm, which considerably exceeds the length of apical leaves. It ensures that side leaves exceed apical leaves in their length and it is applied both for male and female plants. This is the same for the cultivated species as well.

The leaves of South Caucasian climatype excel in their width. The maximum was 8.0 mm for the female and male plants. Male plants of North Caucasian and Danube climatypes possess the minimal width of 3.4 mm and 4.1 mm, respectively. It has been observed that in comparison with female plants, the male ones had wider side leaves. This regularity is applied for every climatic type except for the Danube one. It is less characteristic for cultivated varieties. For example, "Maslichnaya" seabuckthorn has equal lamina width of 6.00 mm both for the apical and side leaves.

Therefore, the evaluation of the adaptation potential of various forms, climatic types and cultivated varieties of *Hippophae rhamnoides* L. in different growing conditions is given as the example for their laminas. The observed distinctions can be referred to the origin of the genetic material given, its wide polymorphism and gender dimorphism. All these factors are clearly observed. Polymorphous specimens have higher evolutionary prospective and ecological adaptivity. The usage of the leaf shape index as a diagnostic factor of the alteration of leaves added to determining outer changes, which had taken place during the evolvement of the introducent during a particular period.

Table 9.7: Leaf Size of various Climatypes of *Hippophae rhamnoides* L. in F_0

Climatype Sort	*Gender*	*Sprout Type*	*Leaf Size (mm)*			
			I Order		*II Order*	
			Length	*Width*	*Length*	*Width*
1	*2*	*3*	*4*	*5*	*6*	*7*
The Baltic climatype	f	apical	78.0	7.0	65.0	6.0
		Side	81.0	7.0	55.0	7.0
	M	Apical	98.0	7.0	70.0	6.0
		Side	114.0	9.1	73.0	6.0
The Danube climatype	M	Apical	63.0	6.0	60.0	6.0
		Side	73.0	6.2	52.0	5.0
	M	Apical	50.0	4.1	50.0	4.9
		Side	60.0	5.6	40.0	5.0
The North Caucasian climatype	M	Apical	50.0	5.3	42.0	4.0
		Side	42.0	5.0	60.0	5.2
	M	Apical	40.0	3.4	42.0	4.0
		Side	55.0	5.8	60.0	5.2
The South Caucasian climatype	f	Apical	74.0	8.0	58.0	6.0
		Side	90.0	8.0	65.0	6.0
	M	Apical	70.0	8.0	50.0	7.2
		Side	100.0	9.4	65.0	7.4
The Siberian climatype	f	Apical	67.0	6.0	52.0	6.0
		Side	72.0	6.0	58.0	6.0
	M	Apical	60.0	5.4	50.0	5.5
		Side	72.0	7.8	58.0	7.0
Botanicheskaya	f	Apical	70.0	4.8	78.0	6.0
		Side	70.0	5.3	83.0	4.0
Maslichnaya	f	Apical	90.0	6.0	92.0	7.0
		Side	100.0	6.0	90.0	6.0
Aley	M	Apical	90.0	6.9	80.0	7.4
		Side	113.0	7.1	94.0	7.0

Hippophae rhamnoides L. uses not only the newly arisen changes but also its hidden mutation reserve to adapt to various growth conditions. This is exerted in all generations starting from the zero.

BIOLOGICAL FEATURES OF SEABUCKTHORN SEEDS

Germination Studies in Laboratory

The role of endo-and exogenous factors in the adaptation of various climatic types of seabuckthorn becomes most evident during the change of generations in the conditions of introduction. New conditions influence the evolution of fruits and seeds greatly; they result in certain changes of adaptive nature. A displacement of the basic phases of morphogenesis can take place. It often occurs so that the seedlings grown from the seeds F_0 had more thriving growth and are more resistible in comparison with their stools (Nesterovich, 1955).

The samples of seeds of all the climatic types: Baltic, Danube, North Caucasian, South Caucasian, and Siberian were used. The optimal regime for seed germination is the temperature within 18-25 °C.

During the research, the following peculiarities were found. The first seedlings appear on the 3rd-4th day. This characteristic is for some forms of Baltic, South Caucasian and the Siberian climatypes. The terms of appearance of plantlets for the other climatypes are as follows: Danube on 6th day and North Caucasian on 7th day. The average time of the appearance of first radical roots differs and fluctuates within 5-7 days for Baltic and Siberian; 6-7 days for South Caucasian and Danube; and 10-11 days for North Caucasian. The distinctions observed are referred to various qualities of seeds of the five climatic types. The year variations are connected with the quality of seeds formed in different weather conditions.

Another characteristic of seeds is their germinative energy. The experimental samples can be therefore arranged as follows: The seeds of Danube climatic type possess high germinative energy up to 67 per cent (X=48.6 per cent) and several forms of South Caucasian climatype up to 56 per cent (X=45.0 per cent). The latter has small changeability throughout the years (X=35.0-56.0 per cent). The distinctive individual changeability was found out and ranged within 29.0-89.5 per cent. The seeds of Baltic climatic type occupy the intermediate position among the samples examined and have the average germinative energy of 35.9 per cent. The forms of Baltic climatype were characterized by low germinative energy from 4.0 to 11.3 per cent (X=8.5 per cent) throughout the years. This characteristic is the most varied one for the seeds of Siberian climatype ranged from 8.0 to 69.8 per cent (X=28.9 per cent).

The duration of seed germination in the laboratory conditions doesn't exceed from 30 days. The seeds of South Caucasian and Danube climatype germinate most evenly, more than a half during the first 10 days. The vegetation period for the samples of Siberian and North Caucasian climatypes is more prolonged. On an average, the main weight of the seeds vegetates within the period of 20 days.

The variations were observed in the duration of seed germination during the study of various years. The samples of North Caucasian climatype belong to group II according to the speed of germination. The forms of Danube climatype have the highest speed of germination alongside with the shortest period of germination and are mostly completed by the 13th day. The laboratory germination on an average reaches to 85.7 per cent.

The research on seed germination of the five aforementioned climatic types of *Hippophae rhamnoides* L. to determine that all of them are characterized by high vital capacity and most samples have seeds with the germination of more than 90 per cent. The germination of the samples of Baltic climatic type reached to 95.3 per cent, while Siberian type reached to 91.6 per cent. An interesting peculiarity is observed in all forms of North Caucasian climatype; irrespective of the later beginning of germination (10-11 day) and lower germination energy, the germination constituted X=88.7 per cent by 25th day. All the forms of the climatypes studied, apart from Danube and North Caucasian climatypes can be referred to as 1st class of sowing standard (more than 90 per cent). No considerable differences have been observed in the germination of the seeds. The adaptive advantages of the climatypes are more clearly seen, particularly in Baltic climatype and the germination of its seeds was found to 96.0 per cent.

We considered the seeds of self-reproduction, which were obtained in the conditions of botanical garden (F_0), to have passed a certain period of adaptation to local conditions. A range of experimental data gives us grounds to assume that each group of seeds possesses certain energy and terms of germination. For example, Danube climatype had maximum germination energy among the test samples (48.6 per cent). South Caucasian forms were characterized by high germination energy (45.0 per cent), comparatively fast completion of main germination period, and high rates of laboratory germination (93.0 per cent). Most samples were characterized by high vital capacity (germination more than 90 per cent).

The data obtained confirmed the fact that the differences between five climatic types of *Hippophae rhamnoides* L. were closely connected with bio-morphological characteristics of seeds, the variations in the starting material were greatly influenced by the conditions of reproduction.

Field Germination

In relation with the solution of a large number of practical tasks, a great importance is connected with the research of field germination of seeds. The index of ground germination gives a more clear idea of the plants ability to renew by seeds altogether with more sufficient information on the potential productivity.

Field germination of the seeds of Baltic, Danube, North Caucasian, South Caucasian, and Siberian climatic types F_0 had been studied for 3 years under the conditions of both protected and open ground. In these conditions, the seeds behaved differently. The seeds of Baltic climatype had the best field germination (40.2 per cent), while the seeds of North Caucasian climatype had the worst field germination (9.7 per cent). Earlier, during the research on germination in laboratory, these seeds also possessed the least germination energy (8.5 per cent), which is a crucial index of their quality.

The production of seedlings in a greenhouse had a lot of advantages in comparison with the production in the open ground. For example, sowing can start two weeks earlier than usual. The average day temperature in a greenhouse is 4 °C higher that in the open ground.

The average seed germination index was recorded to 43.7 per cent for Baltic climatype and 13.6 per cent for North Caucasian one. The duration of the period of seedling growth increases by 2-3 weeks. The germination of all seeds in the greenhouse conditions is spurred and the growth begins earlier.

During the research on laboratory and field germination, a complicated and not always rational dependence between the studied indices was established. For the seeds of Baltic climatic type, this dependence is direct, but the situation with Danube and North Caucasian climatypes is different, the field germination is 2.4 per cent and 9.1 per cent lower than the laboratory one, respectively.

It ensures that the seeds of the zero generation are sure to reflect the alteration and the mutation of parent plants undergo during their growth and development in new conditions. There exists certain dependence between laboratory and field germination. The rate of laboratory germination doesn't let us to predict the behaviour of the plants in field conditions. Field germination of the seeds depends on the characteristics, they acquire from the period of formation on the parent plant. The seeds with higher germination energy possess higher field germination rates. The seeds of Baltic climatic type are characterized by the maximum field germination.

INFLUENCE OF STORAGE LIFE OF SEEDS ON THEIR GERMINATION

It is important to characterize the rates of vitality loss of the seeds during their extended storage and to observe the growth of plants cultivated from the seeds which were stored at room temperature for various periods of time. The vitality of *Hippophae rhamnoides* L. seeds F_0 for long storage had been estimated. The seeds were of various climatic types (Baltic, Danube, North Caucasian, South Caucasian and Siberian). The results of the research on laboratory germination of the seeds after two, four, six, eight and ten years of storage let us specify the peculiarities of seed aging depending on various genetic, physiological and ecological factors.

The seabuckthorn seeds used in the research were of various geographic origins and therefore, possessed various morphometric parameters: Baltic climatic type (5.9x2.9 mm); Danube climatype (5.1 × 2.6 mm); North Caucasian (4.2 × 2.1 mm); South Caucasian (4.7 × 2.7 mm) and Siberian (5.8 × 3.1 mm). It is possible that the comparatively small size of the seeds in combination with peculiarities of their morphology contribute to a long storage period. The vitality of seabuckthorn seeds of different storage periods was evaluated according to their germination. Laboratory germination of F_0 seeds grown on an experimental patch of introductory seed field is different. The fact that draws attention is irregular reduction of the germination index of the climatic types analyzed (Table 9.8).

Table 9.8: Seed Germination (per cent) of Five Climatypes of *Hippophae rhamnoides* L. Depending on their Storage Life

Climatype	*Seeds Storage Life (Years)*				
	2	*4*	*6*	*8*	*10*
Baltic	98±1.07	79±5.04	51±2.88	44±3.48	32±5.35
Danube	92±3.16	85±3.85	23±3.02	0	0
North Caucasian	82±5.10	66±3.76	22±3.39	8±0.91	0
South Caucasian	88±3.39	70±4.22	63±3.76	21±1.82	6±1.58
Siberian	96±1.47	83±3.24	80±4.93	47±6.65	29±5.78

In the sixth year, the germination of North Caucasian and Danube climatypes fell to 22 per cent and 23 per cent, respectively, while the germination of the Siberian climatic type constituted 80 per cent. In the eighth year of storage, no cases of germination of Danube climatic type were registered. After ten years of storage, the seeds of North Caucasian climatype became unviable as well, and only one third of Siberian climatype seeds germinated in comparison with the initial number. The condition of the seeds of South Caucasian climatype deteriorated dramatically. In the eighth year, the germination was 21 per cent, and in the tenth year was only 6 per cent.

The dynamics of germination loss is different for seeds of Baltic climatype. It fell gradually in proportion to the prolongation of the storage period ranged from 98 per cent to 32 per cent. The seeds of Baltic climatype possessed maximum germination (32 per cent) in the tenth year of storage. Vitality loss of the seeds is connected with genetic in homogenity of the specimens and the characteristics of the variety.

BIO-MORPHOLOGICAL FEATURES OF SEABUCKTHORN SEEDLINGS

The aim of the present study was to study peculiarities of seabuckthorn seedlings for different climatic types in early stages of the morphogenesis (seedling stage, sprout stage and juvenile stage).

Firstly, seedlings appeared in III quarter of May, but North Caucasian climatype seedlings-only in I[st] quarter of June. The growth outbreak took place in the middle of the month (II quarter of June). Germination is epiterranean, two seed lobes are above the ground; they are naked, fleshy, initially twinned on the top, protecting the bud from damage. The form of seed lobes is defined by the length-width ratio. On this ground, lanceolate, narrow-ovate, ovate and wide ovate forms of seed lobes are defined. The seedlings of Baltic, Siberian and Danube climatypes had large, ovate or narrow-ovate, fleshy seed lobes. The Siberian climatype had seed lobes without leafstalks. The seed lobes of North Caucasian climatype were of lanceolate or narrow-ovate form, they were opaque and not fleshy. The North Caucasian samples had seedlings with noticeably thin hypocotyledonary part and less fleshy cotyledonary leaves. Such features are common for seabuckthorn forms that are noted for great number of thorns and small fruits.

The gradual enlarging of seed lobes took place during the growth and development of seedlings of every specimen of Baltic, Danube, North Caucasian, South Caucasian and Siberian climatic types. The seed lobes of North Caucasian climatype reached their maximum size in IInd quarter of June and all the others in IIIrd quarter of June.

The green colour of seed lobes and half-yellow hypocotyl are typical in this period. It had been found that the longevity of a seed lobe doesn't exceed more than two months. After that they have played their role in the orthogenetic development of seedlings and sprouts, they gradually die off and fall. The longevity of a seed lobe of Siberian climatype seedlings is estimated to 60 days. The change of colour towards brown begins from Ist quarter of July, and falls down from IIIrd quarter. The seedlings of other climatic types have the longevity of 50 days, but the change of colour and the fall occur at different times.

According to the dimensional parameters of seed lobes, a lot of significant differences were found. The size of cotyledonary leaves can serve as a specific morphological criterion, on the basis of which the seedlings from different ecological-geographical regions are distinguished. For example, North Caucasian climatype has minimal dimensional parameters of cotyledonary leaves in their seedlings (6.5 × 3.0 mm). The largest and fleshiest seed lobes with grown hypocotyledonary part were characteristics of Baltic (9.3 × 3.7 mm), Danube (11.2 × 4.1 mm) and Siberian (9.6 × 4.5 mm) samples.

After the completion of the sprout stage and the beginning of the juvenile stage, the seedlings grew more intensively. At the end of June, the seed lobes of each climatic type apart from the Baltic one were still green. In Ist quarter of July, the seed lobes changed their colour and became brown (Siberian, South Caucasian climatypes), or partially fell on IInd quarter of July (Baltic, Danube climatypes). The seed lobes of North Caucasian climatic type were still yellow-green in colour and fell only at the beginning of August. During the specification of the seed lobes form index, this parameter didn't fluctuate much during the period of the formation of cotyledonary leaves.

As given in Figure 9.13, the forms of Baltic and South Caucasian climatic types were slightly higher. In Istquarter of July, Siberian and Baltic climatic types had the indices of seed lobes form in the biggest contrast. During this period, the forms of Siberian climatype had the longest cotyledonary leaves. On the whole, this parameter reached its maximum in the seedlings of Siberian climatype (4.0), and minimum in the seedlings of Baltic and North Caucasian climatypes (2.5) (Figure 9.13).

The initial growth of radicle of the germinated seeds is an interesting case. In favourable conditions, the radicle starts growing and then, the hypocotyl draws out, the seed lobes unfurl and vegetate. Afterwards, the bud starts growing, giving birth to the system of the main axis. This sequence is biologically feasible because when the radicle evolves, it attaches the growing seedling to understratum and provides it with water and mineral substances. When the hypocotyl draws out, it lifts the first assimilative member of the seedling (the seed lobes). Only after that, the growth of the crown bud and

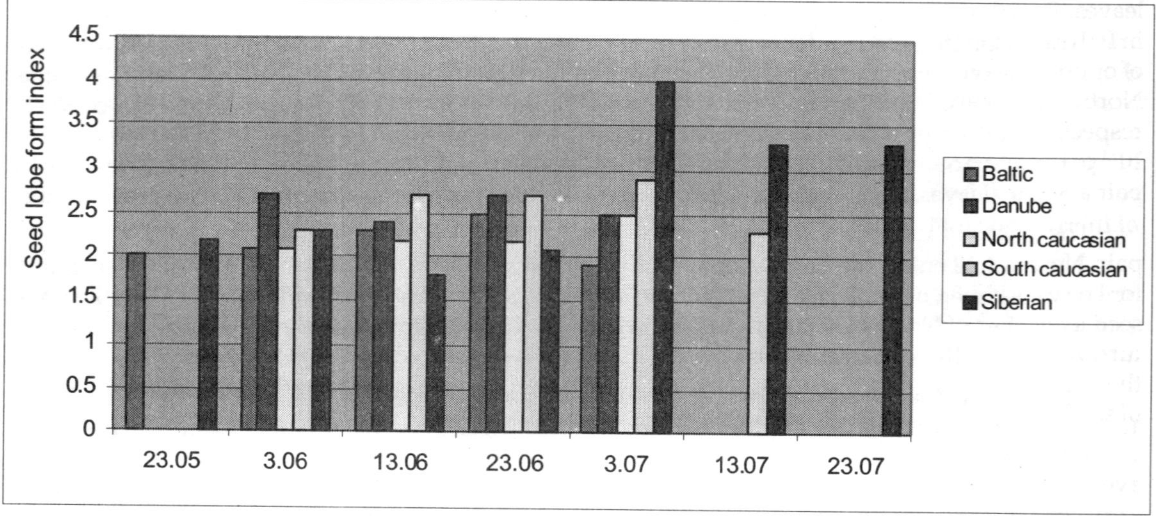

Figure 9.13: Dynamics of the Seed Lobes form Index for Yearlings of *Hippophae rhamnoides* L. of 5 Climatypes (Horizontally-Average dates; and vertically-Average seed lobes form indices).

the formation of the epiterranean sprout become possible.

In the very first stage of life cycle, the seedlings closely resemble the embryo morphologically because all the organs remain, though being slightly modified. The evolution towards the seedling stage is in close connection with the formation of the first-order branch roots and with the beginning of the autotrophic nutrition.

During the early stages of ontogenesis, the seedlings of different climatic types show their peculiarities in the development of the root system. The general features are well marked major root and its size, which fluctuates from 27.0 cm for North Caucasian climatic type to 47.0 for Danube climatic type. The average length of the central root for the seedlings of Siberian, Danube and South Caucasian climatic types is almost the same (30.0, 33.0 and 34.0 cm). It is essential that seabuckthorn develops its branch roots during the formation of the second pair of leaves.

The dynamics of root growth resembles to one of the overground portion of the plant. The roots of annual plants were smaller in length than that of their height. But by the end of vegetation, these differences were essential only for Baltic and North Caucasian samples. For South Caucasian and Siberian climatypes, they stand out less.

The roots begin and finish the period of intensive growth was quite earlier than the overground part of the plant. At the initial stage, the roots of the Baltic climatic type growing the most intensively growing, which is useful from our point of view for the fixation of a plant in the ground and its survival?

The Danubian seedlings developed the longest roots while North Caucasian seedlings were the shortest ones. This feature along with the capability of early germination and growth of the overground portion of the plant shows the lesser adaptiveness of North Caucasian seedlings. The roots of Siberian climatic type grow gradually and durably, the intensiveness of their growth is low at the beginning. By this feature, they differ from other forms (the adaptiveness to frigid climate).

The sprouting stage, as it was mentioned above, begins with the appearance of the first pair of leaves. The samples of Siberian climatype had their leaves sprouted earlier than in the other climatypes in IInd quarter of July and Danube climatype ranks second. During the period of vegetation, 4-7 pairs of opposite leaves were formed. The seedlings of Baltic climatic type had 7 pairs of leaves; Siberian and North Caucasian had 6 pairs of leaves; Danubian and South Caucasian had 5 and 4 pairs of leaves, respectively. The dimensional parameters (length and width) of the leaves were defined as Ist, IInd and IIIrd pairs. The Baltic climatype possessed the largest leaves with the average length of 38.00 mm for I pair and 37.00 for IInd pair.

The leaves of Siberian climatype had maximum width of 5.1 mm for Ist pair and 6.0 mm for IInd pair. North Caucasian seabuckthorn was found to be small-leaved, having lamina length of 25.0 mm for I pair and 21.0 mm for II pair, and width of 3.5 mm in both cases. It was noticed that by the time, the seed lobes start falling down from the end of July till the beginning of August, the lower pairs of leaves turn yellow and fade, and IIIrd pair and the successive ones fall by the middle of August. Moreover, for the seedlings of Siberian and Danube climatypes, this process begins earlier just like it was in the case of seed lobes.

The length of internodes was also found different for the seedlings of all climatic types. The average length of the internodes of I-II and II-III pairs differs insignificantly for all the experimental samples apart from Danube climatype. The average dimensions of I and II; II and III pairs differs mostly for the seedlings of Baltic climatic type. The minimal length of internodes is the characteristic of Siberian climatype.

The maximum length of leaves of annual plants was several times smaller than that of plants that entered into the reproductive period. The appearance of shoot of sylleptical type from the auxiliary bud is discovered in the case of presence of V-VI pairs of leaves. By the end of the season, the sprouts are well-formed and have thorns in their apical part.

The seedlings of all (Baltic, Danube, North Caucasian, South Caucasian and Siberian) climatic types are characterized by 3 types of phyllotaxis: Alternate, opposite and combined. At the initial stage of development, the phyllotaxis was opposite for the lower part of the sprout and then, IV-VI pairs of leaves had combined phyllotaxis. The upper part of the sprout which is formed at the end of the vegetation period has alternate phyllotaxis. At this stage, the seedlings have monopoidal branches. The Danubian and North Caucasian seedlings had high rates of lateral shoots branching (up to 100 per cent).

There are several groups of buds distinguished in seabuckthorn. They differentiate in the ways of their formation, location and size. Hypocotyls and radical buds belong to the group of adventitious buds, which appear closer to the end of summer. The cotyledonary buds are able to give sprouts by the end of August. These sprouts are attached and stand out by their vertical orientation. They were visually noticed on the seedlings of Siberian, Baltic and North Caucasian climatic types. Besides, the most grown and abundant they were on the seedlings of Siberian climatype. This type of buds raises the vitality of seedlings and gives them an opportunity to restore in the case of damage inflicted to the overground part of the plant.

The terms of the beginning of the growth and of its intensiveness differ significantly throughout the climatic types. The seedlings of most climatic types grow intensively and regularly during the late period of growth. A significant leap was observed only at the beginning of growth. It was noticed that the samples of Siberian climatic type begin to evolve early and then, stop their growth. This can be considered the reason for their death in the unfavourable conditions of spring and is the consequence of their lesser need in spring warmth.

The Baltic seedlings are adapted for ambient conditions better than other ones. They begin to grow a lot earlier than south samples and grow gradually during the whole season, not showing any slowdown like the Siberian ones. The Danubian and Caucasian specimens begin growing later, but with more intensiveness. But by end of the season, their height is smaller than of other samples. The seedlings of Siberian climatype are also comparatively small.

The research on the annual plants altogether with biennial and triennial plants described above, confirms the domination of Baltic climatype and the significant lag of Siberian and in particular, North Caucasian climatic types. The index of germinative power can serve as the degree of the adaptiveness of a seedling and of how the biology of the introducents corresponds with environmental factors. It should be mentioned that the seedlings turned out to be different in their height and can be ranged as follows: Baltic (14.2 cm), Danubian (12.4 cm), South Caucasian (10.5 cm), Siberian (9.1 cm) and North Caucasian (8.6 cm).

The analysis of the growth and development of hypocotyl deserves particular attention. At the beginning of its growth, it is practically equal to the roots in length. The Danube climatype is an exception; it has a very small hypocotyl. At first, the underground part of the hypocotyl evolves faster, especially in Baltic climatype and it serves a good protection from outer conditions for young plants. The hypocotyls of Caucasian samples grow longer when underground, but lag behind Baltic climatype is in their length. The hypocotyl of Siberian samples is underground for the longest time. But, the reason for that is another and is connected with the adaptive reaction of the organism. Afterwards, Siberian seedlings develop the overground part of the hypocotyl very quickly, and then, it starts retracting in the soil. On the other hand, above ground part of the hypocotyl of Caucasian samples evolves slowly and is retracted later than that of other ones. The Danube samples rank intermediate between the Baltic and the Caucasian ones.

The research shows that the evolvement of the hypocotyl of every sample of the studied climatypes went according to a certain scheme. The growth of overground part of the hypocotyl takes place up to a certain extent, precisely to 3.07 and for the samples of Baltic and Siberian climatypes to 23.06. Besides, its maximum height fluctuates from 5.0 to 11.0 mm. From the second part of July, the hypocotyl is retracted in the soil and by the end of summer, its height is minimal ranged from 2.0 mm for North Caucasian climatype to 5.0 mm for Siberian, Danubian and South Caucasian climatypes.

The knowledge on the biological peculiarities of seabuckthorn on the initial stages of ontogenesis, gives us the opportunity of early prediction of the introduction of several climatypes of the specimen. The Baltic climatype can be considered as the most adaptable climatype and can be judged by early and intense growth of its overground part, the maximum growth of the seedlings and the dynamics of the hypocotyl evolvement (active growth of the underground part of hypocotyl in spring and its early and intense retraction).

The present study shows the possibility of using the morphometrical parameters of the seed lobes as a diagnostic character during the introduction of seabuckthorn in different climatic types. This method of the research on the annuals of seabuckthorn can be widely applied to the seedlings of other groups of woody and frutescent plants.

ASSESSMENT OF SEABUCKTHORN STABILITY IN THE CONDITIONS OF INTRODUCTION

The analysis of seabuckthorn seeds showed their contamination with pathogenic fungi. The fungi of *Fusarium* species turned out to be the most abundant and harmful for the culture. The contaminated seeds become hollow and lose their vitality, give rise to weak seedlings that blacken and rot afterwards. In the case of severe parasitic attack, the seeds lose their germinative capacity. The biological agents included were: *F. avenaceum (Fr.) Sacc. F. solani App. et. Wr., F. sambucinum Fuck. var. minus Wr., F. culmorum (W.G. Sm.) Sacc. F. gibbosum App. et Wr. Emend. Bilai.* The following diseases observed on the surface of the seeds were: Grey villose spore bearing by *Botrytis cinerea Pers.*; dark-grey velvety taint by fungi of *Alternaria Nees* species; carnation pannose taint similar to small tussocks by *Trichothecium roseum Lk.* and greenish taint by fungi species *Penicillum*. The seeds contaminated with these fungi are mainly germinative, but their sowing and reproductive qualities differ.

The phytopathological research of annual and biennial seedlings of seabuckthorn in a dendrological introductory seedling nursery developed the symptoms of the plant contamination that ensued in colour degradation, exfoliation, cease of sprout growth, loss of turgour and the sequential fading of new twigs. The seedlings that were contaminated emerge more easily out of the ground. The root system was half-grown and had browned roots. The mycological examination of fresh parts of plants (roots, the scrapings of root crown and affected sprouts) that demonstrated the traces of diseases showed the presence of *Fusarium* fungi species: *F. oxysporum (Schlecht.) Snyd. et Hans. var. orthoceras App. et Wr. Bilai comb. Nova., F. solani (Mart.) App. et Wr., F. gibbossum App. et Wr. emend. Bilai. var. bullatum (Sherb.) Bilai comb. Nova.* fungus was developed from the rotten root crown. The biological agents caused the destruction of vascular system of plants, which had contributed to their fading and afterward death.

On the basis of the phyto-pathological analysis, it was found that the contamination of seabuckthorn seeds with the afterward attack on the vascular system of annual seedlings is caused by the fungi of *Fusarium* species. Heavy pruning and damage of the plants close inter-row cultivation, growing on heavy clay soil, shortage of light and water, and water-logged soil favours to the contamination of plants with fusariosis. The selection of stable cultivars, the growing of healthy planting material and advanced agricultural methods are needed to combat this disease. Against fusariosis of seabuckthorn, one can use the following mycocides: Fundasolum (0.2 per cent) and tecto (0.2 per cent). These preparations have an inhibitory effect on the development of pathogens. There were no strict distinctions observed in the species composition of the furiosis causative agents, apart from the fungi of *Alternaria Nees* species.

CONCLUSION

On the basis of complex research on the variations and ecological-biological characteristics of seabuckthorn, the most stable and productive forms were selected, their reproduction was accomplished, their needs for environmental factors were brought to light and recommendations on the optimization of environment to gain their maximal productivity were worked out. The methods of reproduction and the technologies of the cultivation of the planting material were improved. The initial planting material was multiplied and passed on to the growers.

During the cultivation of specimens of different climatic types on the sites with identical ecological condition, the dissimilarities are permanent and confirm the presence of genetically determined characteristics. The presence of polymorphism of the above mentioned characteristics gives us the possibility of selection of the plants with better results for the afterward selection. *Hippophae rhamnoides* L. of Baltic climatic type is a valuable source of breeding material because of its comparatively small shrubs, low or medium amount of thorns and big fruits that come off easily.

A certain complex of morphological characteristics may serve as a key to definition of different climatic types of seabuckthorn. A long-term research on biomorphological characteristics of five climatic types of *Hippophae rhamnoides* L., *i.e.* Baltic, Danube, North Caucasian, South Caucasian and Siberian, during their introduction in the climatic conditions of Belarus proved the genetical nature of their polymorphism, which determines the differences in morphogenesis at all stages of ontogenesis.

Seabuckthorn uses not only its phenotypic variation to adapt in the region of introduction, but also the potential of autogenous variation, which is exhibited in the zero and other sequential generations. Its integration and polymorphism are specified by the presence of similar as well as distinct biomorphological characteristics of single forms of different climatic types. The level of variability of these characteristics within climatic types is lower than that between different climatic types.

The seeds of seabuckthorn at the whole, reflects the genetically determined biomorphological characteristics that are typical of the climatic types and by the extent of the transformation, the level of intraspecific variability during the process of their ecological adaptation to new living conditions is defined. The stability of these characteristics, which is typical for the seeds of the zero generation, remains in the sequential generations.

The plants of all the investigated climatic types of *Hippophae rhamnoides* L. during the period of vegetation in Belarusian climatic conditions penetrate through seasonal cycle of evolvement with the natural change of phenological stages. This is the sign of correspondence in the natural environment of the region to the biological needs of introducents from different geographical points. The differences observed in the pheno-rhythmics of particular climatypes have a genetically determined nature that is in close connection with long-standing adaptation of the plants to the conditions of a particular region. This explains the fact that the specimens of Siberian climatype go through phenological stages earlier, while North Caucasian climatype later than others. The terms of most phenodates of the released sorts and the specimens of the climatypes investigated corresponds. The only difference is the late ripening of the latter. The rates of growth between male and female specimens differ clearly. The male specimens have earlier (by 2-3 days) insulation of leaves, and female ones grow more intensely on the generative stage.

On the basis of the comparative analysis of quantitative and qualitative characteristics of growth and morphogeny of annual seedlings of the investigated climatypes, their biological difference was deduced. The dominating characteristics of juvenile and pubescent plants clearly coincided. On these grounds, a new method of estimation of the successful introduction was suggested which took into account the intraspecific variation that could be extrapolated onto the seedlings of other species of woody and frutescent plants.

As a result of long-term complex research on biomorphological peculiarities of the evolvement of five climatic types of *Hippophae rhamnoides* L. in the climatic conditions of Belarus, their extensive polymorphism was found out and ensured a high level of ecological adaptability of that particular botanical species, which contributed to comparatively full realization of its adaptive potential in the region of introduction. The Baltic climatype of *Hippophae rhamnoides* L. has the highest level of correspondence with the climatic conditions of Belarus. It possesses increased winter resistance, has few thorns, its fruits come off easily. All these characteristics show its perspectiveness for zonation and usage as the basic material for the selection on standing power and productivity.

REFERENCES

1. GOST (State Standard) 13056.7-93.1995. Seeds of trees and shrubs. *Methods of Determining Vital Capacity*, Minsk, 37p. (In Belarussian)

2. Lapin, P.I. 1967. Seasonal rhythm of woody plants development and its role for introduction. *Bull. MBG of the AS of the USSR* 65: 13-18.

3. Maltseva, A.N. 1990. *Morphogenesis of Hippopae rhamnoides Introduced in Conditions of the Lower Don*, Abstract of dissertation, Doctor of Biology: 03.00.05. Moscow. 22 p. (In Russian)

4. *Methods of Phenological observation in Botanical Gardens of the USSR*. 1975. p. 27, Moscow. (In Russian)

5. Molchanov, A.A. and Smirnov, V.V. 1967. *Methods of Studying Growth of Woody Plants*, Moscow, 9 p. (In Russian)

6. Nesterovich, N.D. 1955. *Fruits of Introduced Woody Plants and Prospects of using them in BSSR*, p. 265-287, Minsk. (In Belarussian)

7. Rabotnov, T.A. 1950. Aspects of studying population composition for phytocenology purposes. *Issues of Botany* 1: 465-483.

8. Eliseev, I.P. 1982. Some genetic peculiarities of *Hippophae rhamnoides* and their role in selection. *In: Issues of Biochemistry and Physiology of Agricultural Plants-Collection of sc.works GAI*, p.75-77, Gorkiy. (In Russian)

Chapter 10

Studies on Genetic Variation in Hybrid Offsprings of Large Fruit Seabuckthorn Forms

Jinyou Shan

Institute of Berries, Heilongjiang Academy of Agricultural Sciences, Suiling, Heilongjiang 152204, China

SUMMARY

The main genetic characteristics of offsprings of eight large fruit seabuckthorn hybrids, such as "Chuyi × Mengyou 1, Chuyi × Mengyou 2, Chuyi × Mengyou 3, Chuyi × Mengyou 4, Hs-21 × Mengyou 1, Hs-22 × Alieyi, Suiji 1 ×Mengyou 7 and Hs-20 × Alieyi" selected by Heilongjiang Academy of Agricultural Science, Suiling, have been studied. The results showed genetic variation in the fruit density, thorn density, fruit weight and fruit stalk length of hybrid offsprings had a tendency of lesser affinity to parents, while plant height had a tendency of high affinity to parents. But, both fruit stalk and thorn density had the combinations of ultra-high parent about 25 per cent and 38 per cent, respectively; for the range of variation, the fruit stalk length, fruit density and thorn density had high-parent individual in various proportions, while the plant height with low-parent individual, which provided a chance for selecting good species, subspecies or varieties.

Keywords: *Seabuckthorn, Hybrid offspring, Large fruit, Genetic characteristics and Variation.*

INTRODUCTION

Seabuckthorn is a deciduous shrub or tree, which grows widely in Europe and Asia and has strong ecological adaptability with sand-fixing role, is an excellent ecological species. In addition, seabuckthorn fruit, roots, stems and leaves contain a variety of biochemical components (Yang and Kallio, 2006). It is an excellent raw material for processed food, medicine and health products. In the seabuckthorn applications, most domestic scholars studied seabuckthorn for its ecological functions, chemical composition and processing characteristics, etc. In the seabuckthorn breeding, some work has been done on the selection of seabuckthorn resources, seedling selection and introduction of new varieties, and have achieved encouraging results in Russia and China, etc. It is still in its infancy for

Seabuckthorn (*Hippophae* L.): *A Multipurpose Wonder Plant, Vol. 4: Emerging Trends in Research and Technologies* (Virendra Singh, *Editor-in-Chief*, 2014), p. 140–145, Daya Publishing House, New Delhi, India.

crossbreeding and genetic variation of seabuckthorn research. We have engaged in seabuckthorn breeding for many years, bred number of seabuckthorn varieties and good individuals. In this study, genetic variation characteristics of the progenies have been discussed, and to provide a theoretical basis for seabuckthorn breeding.

MATERIALS AND METHODS

Experimental Area and Climatic Conditions

The experiment was conducted at the Institute of Berries, Heilongjiang Academy of Agricultural Sciences, Suiling, Heilongjiang, which is located at North latitude 47°142' and East longitude 127°06' and altitude at 2027 m, having continental monsoon climate. The annual average temperature is 1.4 °C; in January, average temperature is -22.6 °C and in July, average temperature is 21.8 °C. The extreme minimum lowest temperature is -42.4 °C and the extreme minimum highest temperature is 37.3 °C. The annual sunshine hours are 2821 h and effective accumulated temperature is about 2400 °C-2600 °C. The annual precipitation is 551 mm, frost-free period is 118 d, and annual evaporation is 1242 mm. The depth of snow in winter is about 0-11 cm. The soil is viscous, leached chernozem, pH value is about 6.8-7.3 and organic matter content is 4.04-4.61 per cent.

Test Materials

From eight large fruit hybrids, plants were randomly selected from each combination. The fruit weight, fruit density, stalk length, thorns, plant height and other major characters were determined during seabuckthorn growth period, and then analysis and evaluation were done. During fruit ripening in mid-August, we picked the fruits of female plants and quickly freezed them in the refrigerator. The studied biochemical composition had the tendency to show genetic variation (Zeyuan *et al.*, 2007). The selected test materials are shown in Table 10.1.

Table 10.1: The Number of Plant Hybrids and Generations of Seabuckthorn

Hybrids Combination	*Number of Female Plants (Strains)*	*Number of Male Plants (Strains)*
ChuyixMengyou 1	19	11
ChuyixMengyou 2	16	14
ChuyixMengyou 3	20	10
ChuyixMengyou 4	20	10
Hs-21xMengyou 1	14	16
Hs-22xAlieyi	15	15
Suiji1xMengyou 7	12	18
Hs-20xAlieyi	12	18

Measurement of Morpho-physiological Parameters

We weighed 100 fruits randomly and expressed as g, and measured fruit stalk length, expressed in cm. We selected 10 fruit bearing branches from each combination randomly, measured the average number of fruits between 10 cm of the central branch. The fruit density is expressed as number/10 cm. We selected 10 representative annual branches, measured the average number of thorns between 10 cm of the central branch and expressed as number/10 cm. We measured the height of each plant and expressed in cm.

Statistical Analysis

Coefficient of variation is calculated by

$$CV\ (per\ cent) = \frac{S}{\overline{X}} \times 100$$

where,

S: Standard deviation; and

$\overline{X}$: Mean.

RESULTS

Genetic Variation in the Seabuckthorn Fruit Weight

The fruit weight of "Chuyi×Mengyou" hybrids was significantly lower than that of their parents. On the other hand, the fruit weight of "Hs-21×mengyou 1, Hs-22×Alieyi, Suiji 1×Mengyou 7 and Hs-20×Alieyi" hybrids was similar to their parents (Table 10.2). In nutshell, genetic variation in the hybrids fruit weight was lower than their parents, though some fruit weight of "Chuyi×Mengyou1" hybrids was over than its parents. Seabuckthorn is similar to the most of fruit trees and crops, and only in a few hybrids, fruit weight was higher than their parents.

Table 10.2: Variation in Fruit Weight of Seabuckthorn Hybrids

Hybrids Combination	*Mean of Parents (g)*	*Variation Range of Hybrid Offspring (g)*	*Mean of Hybrid Offspring $\overline{X}$ (g)*	*Standard Deviation S*	*Coefficient of Variation C.V. (per cent)*
ChuyixMengyou 1	74	24-45	36.22	5.68	15.68
ChuyixMengyou 2	74	28.25-46	36.11	5.71	15.81
ChuyixMengyou 3	74	21-51.5	36.80	10.01	27.20
ChuyixMengyou 4	74	23.25-58.5	42.88	9.82	22.90
Hs-21×Mengyou 1	48	27.25-55.8	41.71	8.20	19.66
Hs-22xAlieyi	67	32.5-60.5	49.69	7.67	15.44
Suiji1×Mengyou 7	39.5	26-33.5	37.83	12.18	32.20
Hs-20xAlieyi	48	24-45	37.21	6.67	17.93

Genetic Variation in the Seabuckthorn Fruit Stalk Length

The genetic variation in fruit stalk length of eight progenies is presented in Table 10.3. The average fruit stalk of hybrids "Chuyi×Mengyou 3 and Hs-22×Alieyi" was longer than their parents, while the fruit stalk of other hybrids was shorter than their parents. From the point of variation range, except "Suiji 1×Mengyou 7 and Hs-20×Alieyi", the fruit stalks of all other hybrids were longer than their parents. In totality, in 75 per cent hybrids, fruit stalk was shorter than their parents, while in 25 per cent hybrids, fruit stalk was longer than their parents (Table 10.3).

Table 10.3: Variation in the Length of Fruit Stalk of Seabuckthorn Hybrids

Hybrids Combination	*Mean of Parents (g)*	*Variation Range of Hybrid Offspring (g)*	*Mean of Hybrid Offspring $\overline{X}$ (g)*	*Standard Deviation S*	*Coefficient of Variation C.V. (per cent)*
Chuyi×Mengyou 1	0.3108	0.2436-0.386	0.2916	0.0527	18.07
Chuyi×Mengyou 2	0.3108	0.2322-0.345	0.2682	0.0363	13.53
Chuyi×Mengyou 3	0.3108	0.2424-0.4406	0.3173	0.0580	18.28
Chuyi×Mengyou 4	0.3108	0.2246-0.3200	0.2635	0.0326	12.37
Hs-21×Mengyou 1	0.3050	0.2032-0.3092	0.2474	0.0354	14.31
Hs-22×Alieyi	0.2558	0.2578-0.2943	0.2738	0.0112	4.09
Suiji1×Mengyou 7	0.3200	0.2386-0.3118	0.2648	0.0277	10.46
Hs-20×Alieyi	0.2960	0.1928-0.2534	0.2188	0.0175	8.00

Genetic Variation in the Density of Fruits of Seabuckthorn Hybrids

As evident from the Table 10.4, the average fruit density of all the hybrids was lower than their parents, but some of hybrids "Chuyi×Mengyou 1, Chuyi×Mengyou 2, Chuyi×Mengyou 3, Suiji1×Mengyou 7 and Hs-20×Alieyi" had more density than their parents. It accounts for 62 per cent of the total number of hybrid combinations that is good for selecting high yield individuals (Table 10.4).

Table 10.4: Variation in the Fruit Density of Seabuckthorn Hybrids

Hybrids Combination	*Mean of Parents (g)*	*Variation Range of Hybrid Offspring (g)*	*Mean of Hybrid Offspring $\overline{X}$ (g)*	*Standard Deviation S*	*Coefficient of Variation C.V. (per cent)*
Chuyi×Mengyou 1	30.7	15.6-36.8	21.4	8.1360	38.02
Chuyi×Mengyou 2	30.7	15.1-39.6	26.2	7.6810	29.43
Chuyi×Mengyou 3	30.7	19.7-39.1	25.5	6.0109	23.57
Chuyi×Mengyou 4	30.7	10.9-24.3	16.8	4.0058	23.84
Hs-21×mengyou 1	29.5	2.5-13.4	7.6	4.2849	56.38
Hs-22×Alieyi	29.4	5.0-27.0	15.3	6.7173	43.90
Suiji1×Mengyou 7	26.3	4.8-44.7	23.8	14.0904	59.20
Hs-20×Alieyi	29.8	12.7-38.5	26.6	8.7181	32.77

Genetic Variation in the Thorns Density of Seabuckthorn Hybrids

The genetic variation in the density of thorns among the 8 hybrid combination progenies is shown in Table 10.5. Besides "Chuyi×Mengyou 1, Chuyi×Mengyou 2 and Suiji 1×Mengyou 7", the density of thorns of other 5 hybrid combination progenies is found lesser than their parents. The genetic variations showed the tendency of low affinity, and accounted for 62 per cent of the number of hybrid combinations. There is a large variation range in the thorn density of 8 hybrid combination progenies, and there are some individuals in which thorn density was more than their parents. From the coefficient of variation point of view, there is a large coefficient of variation in the density of thorns, and wide variation in largely optional samples, which favours the choice of individual with few thorns.

Table 10.5: Variation in the Thorn Density of Seabuckthorn Hybrids

Hybrids Combination	*Mean of Parents (Thorns/10 cm)*	*Variation Range of Hybrid Offspring (Thorns/10 cm)*	*Mean of Hybrid Offspring $\overline{X}$ (Thorns/10 cm)*	*Standard Deviation S*	*Coefficient of Variation C.V. (per cent)*
ChuyixMengyou 1	6.3	3-10	7	1.5325	21.89
ChuyixMengyou 2	4.8	1-9	6	2.4018	40.03
ChuyixMengyou 3	6.9	2-8	5	1.1172	22.34
ChuyixMengyou 4	4.3	1-6	4	1.0399	26.00
Hs-21xMengyou 1	6.3	4-9	6	1.0545	17.58
Hs-22xAlieyi	5.2	3-8	5	0.8987	17.97
Suiji1xMengyou 7	5.0	4-8	6	1.2961	21.60
Hs-20xAlieyi	4.9	2-7	4	1.0400	26.00

Genetic Variation in the Plant Height of Seabuckthorn Hybrids

It is clear from Table 10.6, besides "Chuyi×Mengyou 3 and HS-21×Mengyou 1", that the plant height of the other 6 hybrid combination progenies was taller than their parents. The genetic variations showed the tendency of high affinity and accounted for 75 per cent of the number of hybrid combinations. The plant height variation range of hybrid progenies showed that 88 per cent individuals of hybrid combination progenies exceeded their parents.

Table 10.6: Variation in the Plant Height of Seabuckthorn Hybrids

Hybrids Combination	*Mean of Parents (cm)*	*Variation Range of Hybrid Progenies (cm)*	*Mean of Hybrid Progeny $\overline{X}$ (cm)*	*Standard Deviation S*	*Coefficient of Variation C.V. (per cent)*
ChuyixMengyou 1	197.90	192-251	216.39	16.47	7.61
ChuyixMengyou 2	181.90	170-240	204.70	21.50	10.50
ChuyixMengyou 3	220.40	190-250	209.63	20.16	9.62
ChuyixMengyou 4	139.90	160-235	207.60	22.87	11.02
HS-21xMengyou 1	235.00	183-300	231.06	36.40	15.75
HS-22xAlieyi	196.92	200-274	232.89	21.88	9.39
Suiji1xMengyou 7	205.56	150-260	207.89	34.73	16.71
HS-20xAlieyi	194.84	120-257	217.84	38.02	17.45

CONCLUSION

From the analysis of genetic variation of 8 big fruit hybrid combination progenies of seabuckthorn, it is clear that the genetic variation in the density of fruits and thorns, fruit weight and fruit stalk length showed the tendency of low affinity, while the genetic variation of the plant height showed the tendency of high affinity with parents. There are some combinations that have exceeded over their parents, such as fruit stalk length and density of thorns that accounts for 25 per cent and 38 per cent of the total combinations. From the variation range, there are differentially proportionate individuals with high affinity in fruit stalk length, the density of fruits and the density of thorns, contrary to this, there are some individuals with low affinity in plant height. Therefore, we have provided the opportunity of the choice of individuals with excellently comprehensive characteristics.

REFERENCES

1. Yang, B. and Kallio, H. 2006. Lipophilic components of seabuckthorn (*Hippophae rhamnoides*) seeds and berries. In: *Seabuckthorn (Hippophae L.)-A Multipurpose Wonder Plant. Vol. II: Biochemistry and Pharmacology* (V. Singh, Ed. in Chief, 2006), p. 70-97, Daya Publishing House, New Delhi, 600p.
2. Zeyuan, Y., Xingguo, L., Jinyou, S. and Hongling, Q. 2007. Study on hereditary variation trends of fruits main compositions the seabuckthorn crossed progenies. *The Global Seabuckthorn Research and Development* (2): 22-25.

Chapter 11

A Comprehensive Evaluation Study on the Adaptability of the Introduced Russian and Mongolian Seabuckthorn Varieties

Jinyou Shan, Qingyu Gao* and Jian Ding

Institute of Berries, Heilongjiang Academy of Agricultural Sciences, Suiling, Heilongjiang 152204, China
*College of Horticulture, Northeast Agricultural University, Harbin 150030, China

SUMMARY

In the present investigation, the introduced Russian and Mongolian seabuckthorn varieties were taken as the study material and the dynamic changes of conservation rate, shoot growth, plant height, crown diameter and fruit yield were evaluated. Then, the test was conducted on a comprehensive evaluation of their ecological adaptation in the Heilongjiang Province. The results showed that among the tested varieties, there was a huge difference and most of the tested varieties were suitable for the cultivation in Heilongjiang Province. The promising varieties were "Chuyi, Hunjin, Chengse and Wulangemu" to cultivate in the region. The adaptability of seabuckthorn varieties was studied in order to provide a scientific basis for the production and cultivation of suitable seabuckthorn varieties in Heilongjiang Province.

***Keywords:** Seabuckthorn, Russian varieties, Adaptability and Comprehensive evaluation.*

INTRODUCTION

The excellent seabuckthorn varieties have been bred by Russia, Mongolia and other countries, which have big fruit, thick skin, fewer thorns or without thorns and other fine features. However, by using the introduced varieties from the high latitudes of Russia and Mongolia into China, the ecological adaptation of varieties have become a key issue in production. Current production, the introduction and cultivation of big fruit seabuckthorn were still in the blind and thus, leading to failed reforestation, causing a great economic loss. Therefore, as early as 1990, the Institute of Berries of Heilongjiang

Seabuckthorn (*Hippophae* L.): *A Multipurpose Wonder Plant, Vol. 4: Emerging Trends in Research and Technologies* (Virendra Singh, *Editor-in-Chief,* 2014), p. 146–153, Daya Publishing House, New Delhi, India.

Academy of Agricultural Sciences had introduced the varieties of seabuckthorn from the Lisavenko Institute of Horticulture of Russia, to carry out experiments on the breeding and cultivation. In 1998, we collaborated with the Forestry Institute of Forestry Administration in China, to carry out the introduction of improved varieties of seabuckthorn and the research of regional test. The purpose is to ascertain the adaptability, and ecological and economic values of the main varieties of seabuckthorn, which were introduced as early as possible and to provide a scientific basis for the cultivation and promotion of big fruit seabuckthorn varieties. This study expatiate the adaptability of big fruit seabuckthorn in China's Heilongjiang Province, and to provide the basis for the development of seabuckthorn industry in Heilongjiang province.

MATERIALS AND METHODS

Experimental Materials

Testing materials are the seabuckthorn varieties which were the main cultivars introduced from Russia and Mongolia. Specific tested varieties include "Chuyi (Qiuyisike), Fengchan, Jinse, Hunjin, Juren, Katunilipin, Alieyi (male plant), Suiji No.1, Xiangyang, Chengse, Aertaixinwen, Yousheng and Wulangemu" (Lijuan and Yanqing, 2008). First of the 8 varieties were bred by cutting propagation by the Institute of Berries of Heilongjiang Academy of Agricultural Sciences, others were bred by cutting propagation by Fuxin of Liaoning. The large fruit, no thorns and superior "Chinese seabuckthorn", and male plant were chosen as the control varieties (Fanlin *et al.*, 2005; Fanlin *et al.*, 2006; Jianguo, 2006; Baoqi and Shuzhen, 2007). Tested seedlings were 2 years cutting saplings.

Experimental Design

Experimental design was randomized complete block design: Each line (of each variety) with 16 sampling as one test area with 4 replications. It was slightly different from the conventional randomized complete block design, equipped with no thorn and big fruit male plant in the normal random arrangement, every two treatments (variety), into a male plant line to ensure the normal pollination. In each replication, the order of the varieties is random but every 2 lines equipped with one line of big fruit male plant without thorns was fixed. Around the test area was set up two protection lines which had big fruit male plant without thorns. Thus, each tested variety included 64 saplings planted at line spacing of 3 m × 2 m.

Forestation, Management and Protection

Forestation method includes soil preparation by using conventional methods. The size for the planting hole was 40 cm × 40 cm × 40 cm. The saplings were planted in late April and then provided irrigation to strengthen the management and provided protection to prevent the destruction by humans and animals. Weeding was done 2 to 3 times per year, and gave attention to control the pests and diseases.

Investigation and Measurement of Growth

We investigated the survival rate in the planting year and in the end of growth 2 years of plantation. We did detail investigation about the plant survival, plant height, shoot length and crown diameter of each variety in every test area. Measurement of crown diameter includes from north to south and from east to west, and then calculated the average of crown diameter.

Adaptability Evaluation

The survival rate, plant height, crown diameter and fruit yield of the tested varieties were taken as the evaluation index, and data of 5 years was taken as the basis to evaluate the adaptabilities of various varieties.

RESULTS AND DISCUSSION

Dynamic Changes of Survival Rate

The dynamic changes of survival rate in tested varieties from 2000 to 2002 for three consecutive years is given in Figure 11.1.

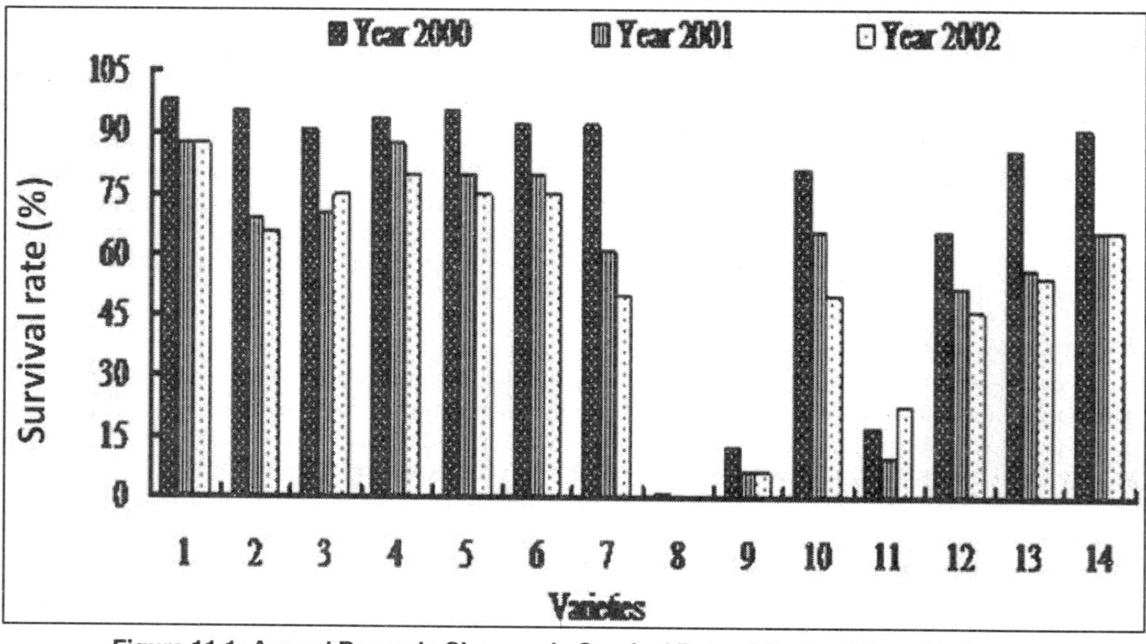

Figure 11.1: Annual Dynamic Changes in Survival Rate of Seabuckthorn Varieties.

1: Chuyi; 2: Fengchan; 3: Jinse; 4: Hunjin; 5: Juren; 6: Katunilipin; 7: Suiji No.1; 8: Xiangyang; 9: Yousheng; 10: Chengse; 11: Aertaixinwen; 12: Wulangemu; 13: Chinese seabuckthorn; 14: Alieyi.

As it can be evident from Figure 11.1, the survival rate of "Xiangyang, Yousheng and Aertaixinwen" was relatively low, especially "Xiangyang" survival rate had reached zero per cent in the 2nd year (in 2001). It shows that "Xiangyang" adaptability is worst, however, these are mainly due to poor winter hardiness rather than poor growth. At the same time, these cannot exclude the quality of seedlings of several varieties, so the accurate test results need further study. The survival rate of "Chinese seabuckthorn" was more than 80 per cent in 2000. Since 2001, the survival rate has been stabilized. But, there was the phenomenon of poor growth of new shoots in few years after planting and in later years, the phenomenon was normal. This shows that the winter hardiness of "Chinese seabuckthorn" is limited, and lower than the number of varieties introduced. All other tested varieties have relatively high survival rates. In addition, the 2002's survival rate is higher than 2001's in "Jinse and Aertaixinwen" due to the reason that these two varieties have been replanted in 2001 (Figure 11.1). The overall trend of survival rate has declined.

Dynamic Changes of Plant Height

The annual dynamic change of plant height of varieties is shown in Figure 11.2. Beside "Xiangyang" which don't adapt to local environmental conditions, is dead. In the other introduced varieties, plant height was above 140 cm in 2002, and grew strong. The overall plant height increased.

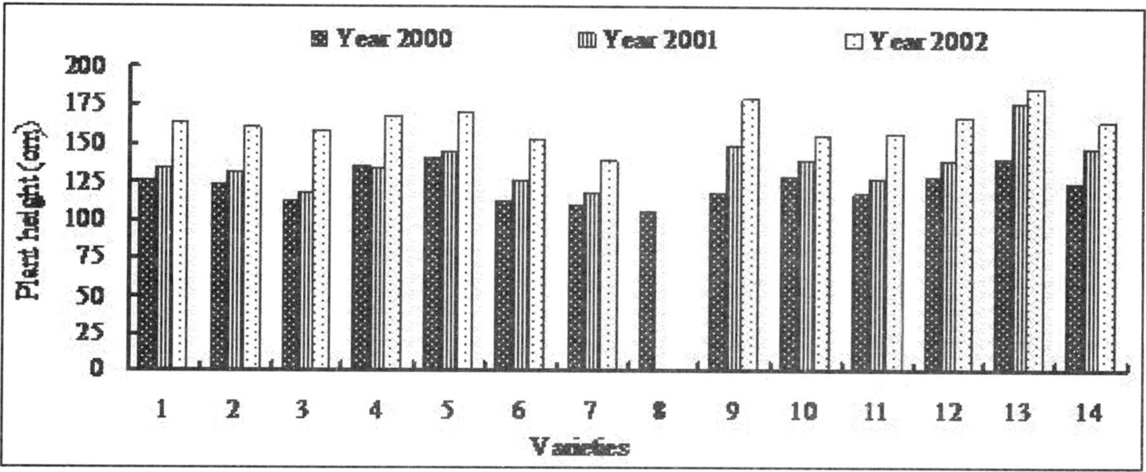

Figure 11.2: Annual Dynamic Changes in Plant Height of Seabuckthorn Varieties.

1: Chuyi; 2: Fengchan; 3: Jinse; 4: Hunjin; 5: Juren; 6: Katunilipin; 7: Suiji No.1; 8: Xiangyang; 9: Yousheng; 10: Chengse; 11: Aertaixinwen; 12: Wulangemu; 13: Chinese seabuckthorn; 14: Alieyi.

Dynamic Changes of Shoot Growth

The dynamic change of shoot growth is shown in Figure 11.3. The results showed that there is a dynamic rhythm in the change of shoot growth. The maximum shoot growth of tested varieties was

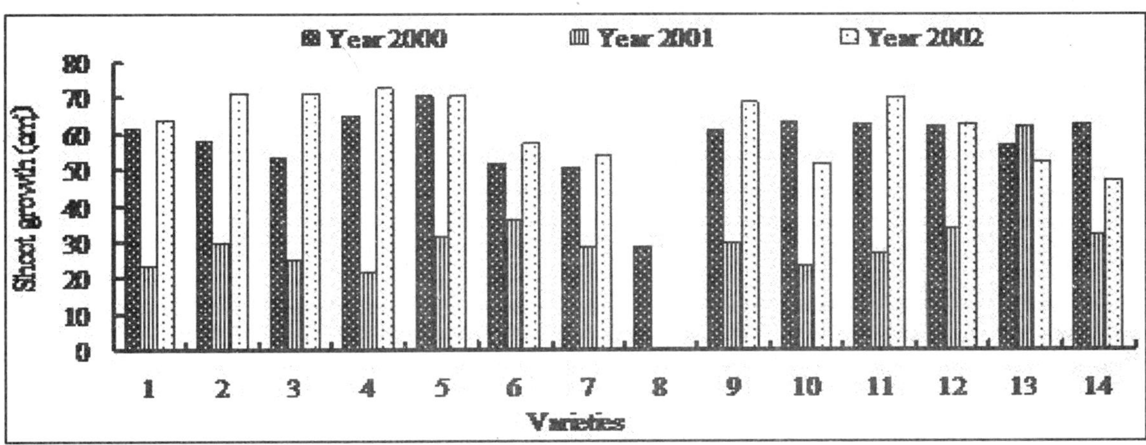

Figure 11.3: Annual Dynamic Changes in Shoot Growth of Seabuckthorn Varieties.

1: Chuyi; 2: Fengchan; 3: Jinse; 4: Hunjin; 5: Juren; 6: Katunilipin; 7: Suiji No.1; 8: Xiangyang; 9: Yousheng; 10: Chengse; 11: Aertaixinwen; 12: Wulangemu; 13: Chinese seabuckthorn; 14: Alieyi.

observed in 2000 and this year is the 2nd year after planting. In year 2001, besides the control varieties ("Chinese seabuckthorn"), the shoot growth of other varieties declined significantly. In year 2002, the shoot growth is recovered significantly again and beside this, the shoot growth in 2002 is less than the growth in 2000 in "Chengse, Chinese seabuckthorn and Alieyi", while the shoot growth in all other varieties was close to or over the growth in 2000. The reason is that trees grow more vigorously in early growth stages and a large number of new shoots were developed in 2000, while in the 2nd year, many of new shoots grew in the two-year old branches and therefore, the shoots in the 2nd year are relatively short, on the contrary, the shoots in the 1st year are relatively long.

Dynamic Changes of Crown Diameter

The dynamic change of crown diameter is shown in Figure 11.4. The figure shows that crown diameter increases with the age of tree; however, there is a huge difference among the varieties. Beside "Xiangyang" which don't adapt to local environments and is dead while the crown diameter of other varieties ranged between 130 cm to 200 cm. There is a close relationship between the size of crown diameter and the fruit yield; thus, to some extent, the trend of crown diameter reflects the trend of fruit yield.

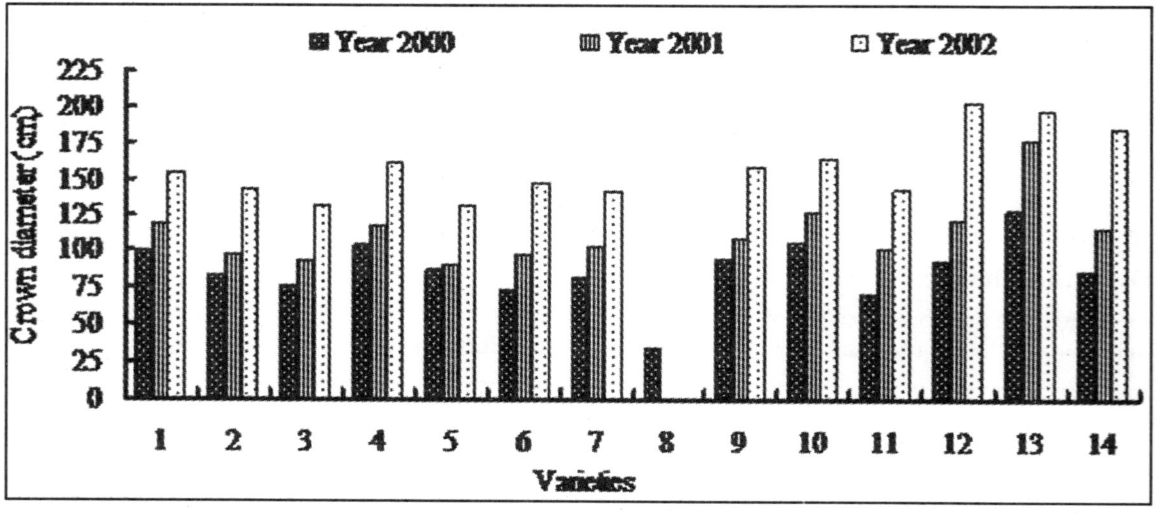

Figure 11.4: Annual Dynamic Changes in Crown Diameter of Introduced Seabuckthorn Varieties.
1: Chuyi; 2: Fengchan; 3: Jinse; 4: Hunjin; 5: Juren; 6: Katunilipin; 7: Suiji No.1; 8: Xiangyang; 9: Yousheng; 10: Chengse; 11: Aertaixinwen; 12: Wulangemu; 13: Chinese seabuckthorn; 14: Alieyi.

Adaptability Evaluation

The test site of varieties is established in Suiling County, which is located in cool temperate zone, high latitude and low altitude region. The results of comprehensive adaptability evaluation of varieties mainly reflect their winter hardiness. According to the test objective, we can choose survival rate, plant height, crown diameter, yield, etc., as well as the comprehensive evaluation index. If one can pay attention only to the adaptability of varieties, one can choose survival rate and growth as the evaluation index; or if one pays attention to the adaptability of varieties and the economic benefits, one can chose survival rate, growth, and yield for the evaluation index. In this study, we selected survival rate, plant height, crown diameter and yield for calculating the evaluation index, and took the 5th year's data as the basis to comprehensive evaluation of the adaptabilities. During the comprehensive evaluation, the following membership function was constructed (Lijuan and Yanqing, 2008).

a. Survival rate

$$\mu_1(x_1) = \text{Survival rate}/100$$

b. Plant height

$$\mu_2(x_2) = \begin{cases} \text{Plant height}/300, & x_2 < 300 \\ 1, & x_2 e \geq 300 \end{cases}$$

c. Crown diameter

$$\mu_3(x_3) = \begin{cases} \text{Crown diameter}/250, & x_3 < 250 \\ 1 & x_3 e \geq 250 \end{cases}$$

d. Comprehensive index

$$E = [\mu_1 + (\mu_2 + \mu_3)/2]/2$$

According to the membership functions, the comprehensive index of the tested varieties can be calculated and the results of comprehensive evaluation is shown in Table 11.1. By the comprehensive index (*E* value), the adaptability of the tested varieties can be broadly classified into three grades. All plants of variety "Xiangyang" were died, which reflects the worst adaptability and winter hardness is rather poor. The *E* value of "Yousheng and Aertaixinwen" was below 0.5 and this is mainly due to the survival rates of these two varieties are relatively low. However, they are not less than the other varieties in plant growth index. So, the two varieties had certain adaptability, but they will be further studied. The *E* values of other varieties was between 0.586 and 0.756, they show very strong adaptability. Although, there was the occurrence of freezing injury on the shoots of "Chinese seabuckthorn" in the 2nd year after planting, but *E* value was still larger and reached 0.733, which explains its strong adaptability.

Table 11.1: Results of Comprehensive Adaptability Evaluation of Introduced Seabuckthorn Varieties

Varieties	μ_1	μ_2	μ_3	*E Value*
Chuyi	0.875	0.580	0.692	0.756
Fengchan	0.750	0.580	0.601	0.670
Jinse	0.766	0.544	0.599	0.669
Hunjin	0.797	0.568	0.699	0.719
Juren	0.859	0.644	0.621	0.746
Katunilipin	0.891	0.540	0.653	0.744
Alieyi	0.656	0.603	0.625	0.635
Suiji No.1	0.641	0.508	0.555	0.586
Xiangyang	0	0	0	0
Yousheng	0.297	0.579	0.638	0.453
Chengse	0.609	0.567	0.690	0.619
Aertaixinwen	0.172	0.554	0.584	0.370
Wulangemu	0.526	0.605	0.737	0.598
Chinese seabuckthorn	0.563	0.897	0.908	0.733

Comparison and Evaluation of Yield

The main purpose of the introduction of big fruit seabuckthorn varieties is to consider the economic characteristics. Therefore, in addition to the adaptability of varieties investigated, the focus was on the yield and economic benefits. The comparison of yield of tested varieties between 2002 and 2005 is shown in Table 11.2. The comprehensive evaluation is needed because there are some differences in the yield of different varieties and different years. As shown in Table 11.2, varieties "Chuyi, Hunjin, Chengse and Wulangemu" yielded above 6,000 kg/hm^2; varieties "Fengchan, Jinse, Yousheng and Aertaixinwen" yielded between 4,500 kg/hm^2 and 6,000 kg/hm^2; varieties "Katunilipin and Suiji No.1" yielded below 4,500 kg/hm^2.

Table 11.2: Results of Fruit Yield of Introduced Seabuckthorn Varieties during 2002-2005 (kg/hm^2)

Varieties	*2002*	*2003*	*2004*	*2005*
Chuyi	5 550	6 349.5	4 275	10 309.5
Fengchan	4 215	3 474	5 940	6 774
Jinse	4 440	3 408	3 982.5	8 158.5
Hunjin	6 210	5 317.5	8 052	7 050
Juren	3 765	4 029	6 960	4 950
Katunilipin	4 650	1 210.5	5 283	3 465
Suiji No.1	5 100	433.5	4 279.5	4 897.5
Yousheng	9 105	1 942.5	5 452.5	3 748.5
Chengse	7 770	3 474	8 328	4 899
Aertaixinwen	7 320	2 775	5 521.5	5 964
Wulangemu	7 320	2 431.5	8 562	6 091.5

CONCLUSION

According to the study of the index variation of survival rate, plant height, crown diameter and yield in big fruit seabuckthorn, we can see that there are big differences among the varieties, which show that the adaptabilities of different varieties have significant differences. It is clear from all the test indicators of varieties "Aertaixinwen, Yousheng and Xiangyang" that their adaptabilities are relatively poor, whereas the adaptabilities of other varieties are relatively better. It indicates that most of the tested varieties are suitable for cultivation in Heilongjiang, and also shows that the climatic conditions in high-latitude Heilongjiang are similar to that of the climatic conditions of the varieties of origin. Therefore, the introduction will be more likely to be successful.

The introduced seabuckthorn varieties with big fruit, no thorn and high-yielding can be directly used for the production at high latitudes in the region. So, the varieties "Chuyi, Hunjin, Chengse, Wulangemu, Fengchan, Jinse, Juren, Aertaixinwen, Yousheng, Katunilipin, Alieyi (male), and Chuyi, Hunjin, Chengse and Wulangemu" were found promising ones for this region.

Although, the comprehensive adaptability evaluation index of "Xiangyang, Aertaixinwen and Yousheng" is lower, especially "Xiangyang" which was dead, it may be due to the poor quality of seedlings and the low survival rate. Therefore, the adaptabilities of these varieties needs further study.

REFERENCES

1. Baoqi, Z. and Shuzhen, J. 2007. The germplasm resources of *Hippophae* and utilization. *Modern Agricultural Science and Technology* 9: 198-199.
2. Fanlin, M., Baoqi, Z., Rongsen, L. and Guizhen, G. 2005. The germplasm resources of *Hippophae* and utilization. *Liaoning Forest Science and Technology* 4: 41-43.
3. Fanlin, M., Baoqi, Z., Xiangli, M., Rongsen, L., Zhiyuan, X. and Qiang L. 2006. Preliminary report on the cross breeding of Chinese seabuckthorn and Russian fine seabuckthorn. *Seabuckthorn* 4: 5-8.
4. Jianguo, Z. 2006. *Study on Adaptability and Introduction of Excellent Big Fruit seabuckthorn*, p.112-113, Science Press, Beijing, 133p.
5. Lijuan, D. and Yanqing, J. 2008. Research report on the introduction and breeding of seabuckthorn and the utilization of heterosis. *Modern Horticulture* 5: 50-51.

PART III:
MOLECULAR BIOLOGY AND BIOTECHNOLOGY

Chapter 12

Application of SSR Markers for the Assessment of Genetic Diversity in Latvian Seabuckthorn (*Hippophae rhamnoides* L.)

Gunars Lacis, Irita Kota and Dainis Rungis*

Latvia State Institute of Fruit-Growing, Graudu Str. 1, Dobele, LV 3701, Latvia
*Genetic Resource Centre, Latvian State Forestry Research Institute "Silava",
Rigas Str. 111, Salaspils, LV 2169, Latvia

SUMMARY

Nine polymorphic SSR (microsatellite) primer pairs previously described, developed in *Hippophae rhamnoides* ssp. *sinensis* were tested on 36 seabuckthorn accessions developed from crosses among *H. rhamnoides* ssp. *mongolica*, ssp. *rhamnoides* and ssp. *fluviatilis* to adapt a fast, reliable method for screening of seabuckthorn germplasm collections. Eight out of nine primer pairs gave a successful amplification and were highly polymorphic with 4 to 22 different alleles with 10.25 alleles on an average. Observed heterozygosity values ranged from 0.781 to 0.972. Five distinct seabuckthorn accession groups were detected in the analyzed collection. This study demonstrated that DNA fingerprinting with SSR markers developed in *H. rhamnoides* ssp. *sinensis* is suitable for application in other *H. rhamnoides* subspecies, as well as in crosses among *H. rhamnoides* ssp. *mongolica*, ssp. *rhamnoides* and ssp. *fluviatilis*. The eight tested microsatellite markers were found suitable for the evaluation of genetic diversity and relatedness of seabuckthorn accessions developed and utilized within the Latvian breeding programme.

Keywords: *Seabuckthorn (Hippophae rhamnoides L.), Genetic diversity, Germplasm, Microsatellites and Lativa.*

INTRODUCTION

Seabuckthorn (*Hippophae rhamnoides* L.) (2n=24) is becoming a promising crop in Latvia. The first trials of introduced seabuckthorn under Latvian horticultural conditions took place in the 1970's (Bruvelis, 2003; 2007). Initially, varieties developed from *Hippophae rhamnoides* ssp. *mongolica* were

Seabuckthorn (*Hippophae* L.)*: A Multipurpose Wonder Plant, Vol. 4: Emerging Trends in Research and Technologies* (Virendra Singh, *Editor-in-Chief*, 2014), p. 157–166, Daya Publishing House, New Delhi, India.

introduced; however, these were not successful because of low adaptation to the variable Baltic maritime climate, characterized by frequent late winter thaws. Therefore, during the next attempt to introduce seabuckthorn to Latvia in the 1980's, varieties derived from crosses among *H. rhamnoides* ssp. *mongolica*, ssp. *rhamnoides* and *ssp. fluviatilis* were utilized. This hybrid material was developed by Prof. T. Trofimov at the Moscow State University. The local breeding of seabuckthorn varieties was performed by Latvian breeders Andrejs Bruvelis and Karlis Blums, who continued to strengthen the adaptation of introduced varieties by backcrosses with *H. rhamnoides* ssp. *rhamnoides* (Bruvelis, 2003; 2007). Several well adapted cultivars have been developed, *e.g.* cultivars Mary and Tatjana. Later, several open pollinated elite seedlings selected at the M.A. Lisavenko Scientific Research Institute of Barnaul were introduced for field evaluation and further utilized in growing and breeding. This germplasm represents a geographically distant group of seabuckthorn in comparison with the previously introduced material. As a result, the seabuckthorn plant material used in Latvia both for breeding and growing is geographically highly diverse and potentially genetically variable.

Traditionally, cultivar identification and assessment of breeding material has relied on the morphological and agronomic characteristics of the plant material. In fruit crops including seabuckthorn, these are associated with several important limiting factors: Phenotypic characters are generally influenced by environment and growth stage of the plant; resulting in long and expensive evaluation during the whole vegetative growth period to obtain satisfactory morphological data for genetic diversity and relatedness evaluation. Therefore, different molecular markers are very useful to distinguish between accessions and for investigations of genetic diversity or relatedness. Markers, such as RAPD (Persson and Nybom, 1998; Jeppsson *et al.*, 1999; Bartish *et al.*, 2000; Ruan *et al.*, 2004; Sheng *et al.*, 2006; Singh *et al.*, 2006; Sun *et al.*, 2006), ISSR (Tian *et al.*, 2004a; 2004b), AFLP (Ruan and Li, 2005) and SSR (Wang *et al.*, 2008) have been applied to seabuckthorn species for genetic diversity studies.

Weber and May (1989) developed a general method for the detection of polymorphic Simple sequence repeats (SSR) or microsatellites. They have become the marker of choice in many plant species due to several advantages over other molecular markers: They are abundant in most genomes, uniformly distributed, multi-allelic and co-dominant. Therefore, the polymorphism information content is very high, which enables the discrimination of cultivars and genetic diversity analysis using a limited number of SSR markers. SSRs are PCR-based, thus requiring little DNA for the amplification. Every SSR locus is defined by a unique pair of primers; therefore information exchange between laboratories is easy and allows individuals to be uniquely genotyped (Powell *et al.*, 1996). One disadvantage of SSR markers is that the unique nucleotide sequences flanking the repeat regions that must be determined. Due to sequence conservation between species, it may be possible to transfer SSR markers developed in one species to other related species. The success of cross species by SSR marker transferability must be empirically determined and can vary from locus to locus. Problems can be caused by disappearance of the repeated sequence (loss of polymorphism), degeneration of the primer binding sites (null alleles) and loss of primer specificity (non-specific fragment amplification) (Peakall *et al.*, 1998).

The aims of the present study was to use SSR markers to (1) develop a method for germplasm analysis of varieties developed from crosses among *H. rhamnoides* ssp. *mongolica*, ssp. *rhamnoides* and ssp. *fluviatilis*; (2) estimate the level of genetic diversity of *Hippophae* L. and assess interspecific and intraspecific relationships. The SSR markers used in this study were developed in *H. rhamnoides* ssp. *sinensis* (Wang *et al.*, 2008), and which are the only SSR markers currently available for the *Hippophae* L. genus. We assessed the transferability of these markers to other *H. rhamnoides* subspecies, and their utility in differentiating varieties and breeding lines as well as assessing their genetic relationships.

MATERIALS AND METHODS

Plant Material

Thirty six accessions of seabuckthorn (*Hippophae rhamnoides* L.) grown in Latvia were used in the present investigation. These accessions include cultivars widely grown in Latvia ("Avgustinka, Botaniczeskaja Lubitelskaya, Lomonosovskaya, Luczistaya, Mary, Podarok Sadu, Prozracznaya, Sjurpriz Pribaltiki, Tatjana, Trofimovskaya"), local selections ("Edgars"), introduced breeding material ("Vir 3, Vir 4, Vir kv I, Vir kv II"), wild samples or cultivation escapees ("Skibes siev, Skibes vir") as well as open pollinated elite seedlings selected at the M.A. Lisavenko Scientific Research Institute of Barnaul (all accessions with designation 'H'), which represents Altai (Siberia, Russia) type of seabuckthorn adapted to continental climate (Table 12.1). All seabuckthorn accessions used in the investigation represents different level of inter-subspecies crosses among *H. rhamnoides* ssp. *mongolica*, ssp. *rhamnoides* and ssp. *fluviatilis*.

Table 12.1: Accessions of the Latvian Seabuckthorn Collection

Sl.No.	*Accession Name*	*Country of Origin*	*Accession Type**	*Description of Accession***
1.	Avgustinka	Russia	A	MGU, selected sample from Leningrad region o.p.
2.	Botaniczeskaya Lubitelskaya	Russia	A	MGU, selected sample from Leningrad region o.p.
3.	Edgars	Latvia	W	unknown origin cultivated male plant
4.	H 1.1.	Russia	B	Barnaul, Altai type
5.	H 1.10.	Russia	B	Barnaul, Altai type
6.	H 1.11.	Russia	B	Barnaul, Altai type
7.	H 1.12.	Russia	B	Barnaul, Altai type
8.	H 1.13.	Russia	B	Barnaul, Altai type
9.	H 1.14.	Russia	B	Barnaul, Altai type
10.	H 1.2.	Russia	B	Barnaul, Altai type
11.	H 1.3.	Russia	B	Barnaul, Altai type
12.	H 1.4.	Russia	B	Barnaul, Altai type
13.	H 1.5.	Russia	B	Barnaul, Altai type
14.	H 1.6.	Russia	B	Barnaul, Altai type
15.	H 1.7.	Russia	B	Barnaul, Altai type
16.	H 1.8.	Russia	B	Barnaul, Altai type
17.	H 1.9.	Russia	B	Barnaul, Altai type
18.	H 2.1.	Russia	B	Barnaul, Altai type
19.	H 2.2.	Russia	B	Barnaul, Altai type
20.	H 2.3.	Russia	B	Barnaul, Altai type
21.	H 3.1.	Russia	B	Barnaul, Altai type
22.	H 3.2.	Russia	B	Barnaul, Altai type
23.	Lomonosovskaya	Russia	A	MGU

Contd...

Table 12.1–*Contd...*

Sl.No.	Accession Name	Country of Origin	Accession Type*	Description of Accession**
24.	Luczistaya	Russia	A	MGU
25.	Mary	Latvia	A	Botanicheskaya Ljubitelskaya o.p.
26.	Podarok Sadu	Russia	A	MGU, selected sample from Leningrad region o.p.
27.	Prozracznaya	Russia	A	MGU
28.	Sjurpriz Pribaltiki	Russia	A	Mitzcurinsk, selected sample from Kaliningrad region o.p.
29.	Skibes siev	Latvia	W	unknown origin wild sample
30.	Skibes vir	Latvia	W	unknown origin wild male sample
31.	Tatjana	Latvia	A	Botanicheskaya Ljubitelskaya o.p.
32.	Trofimovskaya	Russia	A	MGU
33.	Vir 3	Latvia	B	MGU
34.	Vir 4	Latvia	B	MGU
35.	Vir kv I	Latvia	B	MGU
36.	Vir kv II	Latvia	B	MGU

* A: Advanced cultivar, B: Breeding line, W: Wild sample.

** Barnaul: Open pollinated elite seedling selected at the M.A. Lisavenko Scientific Research Institute of Barnaul; MGU: Variety originated by T. Trofimov at the Botanical Garden of Moscow State University; Mitzcurinsk: Variety originated by V.T. Kondratov at the I.V. Mitzcurin Scientific Research Institute, Mitzcurinsk; o.p.: Open pollinated.

Isolation of Genomic DNA and PCR Analysis

Young leaves were collected during May-June. Total DNA was isolated using a genomic DNA Purification Kit (Fermentas, Lithuania). PCR reactions were performed in a 20 µl reaction with 25 ng DNA, 2 mM each primer, 200 mM of each nucleotide, 1.5 mM $MgCl_2$ and 0.5 U *Taq* DNA polymerase (Sigma) per reaction, in the ep gradient thermal cycler (Eppendorf) for 40 cycles with denaturation at 95 °C for 30 s, annealing at 48, 50, 51°C for 30 s and extension at 72 °C for 30 s, with a final extension step of 10 min at 72 °C (Wang *et al.*, 2008). PCR products were first checked on 1 per cent agarose gels in 1X TAE buffer and visualized by staining with ethidium bromide to test for the presence of PCR products. The same PCR products were subsequently analyzed on an ABI PRISM® 3100 Genetic Analyzer (Applied Biosystems, USA) and genotyped using GeneMapper® Software v4.0 (Applied Biosystems, USA).

Data Analysis

Number of effective alleles, loci information index, heterozygosity, probability of identity and fixation index values was calculated using the computer program GENALEX 6.1 (Peakall and Smouse, 2006). Potential population structure was analyzed using PowerMarker (Liu and Muse, 2005) and phylogenetic tree was constructed based on Nei's genetic distance (Nei, 1973) and Neighbour-Joining clustering method and presented using MEGA version 4 (Tamura *et al.*, 2007).

RESULTS AND DISCUSSION

Nine SSR markers were tested on the 36 Latvian *Hippophae rhamnoides* accessions, eight amplified PCR fragments, only locus *Hr09* failed to amplify detectable PCR products (Table 12.2). In some cases, more than the expected 2 alleles were amplified from one individual, in this case, the two most distinct and intense fragments were scored. The level of polymorphism was generally high, with the number of alleles identified ranging from 4 (*Hr08*) to 22 (*Hr04*) (10.25 on an average) (Table 12.3). The number of effective alleles ranged from 2.653 (*Hr05*) to 8.945 (*Hr04*). The high number of detected alleles could be explained by the origin of the tested plant material, as combines the varieties of different and probably diverse genetic origin, including accessions of geographically distant origin (*e.g.* seabuckthorn) varieties collected or developed from the European *Hippophae rhamnoides* material as well as varieties originating from the Altai region (Siberia) and reciprocal hybrids between these seabuckthorn groups. The observed heterozygosity was high, ranged from 0.781 (*Hr07*) to 0.972 (*Hr03*). The fixation index was negative for all loci except *Hr04* (0.083) and *Hr07* (0.016). The probability of identity for these eight tested markers was 7.56E-10, which is the probability that two samples with an identical genotype are in fact unrelated individuals. Previously published genetic diversity analysis has used a similar number or even fewer SSR markers to determine genetic relationships with in genetic resource collections (Bianchi *et al.*, 2004; Karaagac *et al.*, 2010). It has been reported that in diverse sweet cherry rootstock material, only five SSR primer pairs were required to discriminate among the rootstock accessions (Bianchi *et al.*, 2004).

Table 12.2: Putative Alleles in Eight SSR loci Identified in Accessions of the Latvian Seabuckthorn Collection

Loci	*Putative Alleles*
Hr01	147,151,153,158,164,166,167,169,171,173,175,177,179,181
Hr02	152,158,160,162,164,166,168,172,174,178
Hr03	85,87,89,91,93,95,97,99,101,103
Hr04	146,152,174,182,202,204,208,210,214,216,222,224,232,234,236,238,242,244,262,272,274,294
Hr05	153,155,172,185,201,203
Hr06	65,71,77,79,85
Hr07	168,172,180,231,256,265,274,290,294,301,306
Hr08	129,133,209,225

The length of alleles (in base pairs) was widely distributed for some loci, particularly in *Hr04* (146-294 bp), *Hr07* (168-306 bp) and *Hr08* (129-225 bp). The sizes of alleles overlapped the size ranges originally reported, except for markers *Hr04* and *Hr07*, where the allele sizes identified in this study were larger than in the previous report.

The aim of this study was to investigate the utility of SSR markers developed for *H. rhamnoides* ssp. *sinensis* in investigations on different subspecies-*H. rhamnoides* ssp. *mongolica*, ssp. *rhamnoides* and ssp. *fluviatilis*, and hybrids. The high rate of successful cross species amplification of these SSR markers (8 from 9) is somewhat unexpected given the large differences in distribution and phylogeny between the two subspecies (Sun *et al.*, 2002). These markers revealed a high level of polymorphism within the investigated material and in some cases, higher polymorphism level was recorded than described in the previous study. A unique genotype was obtained for each tested accession showing the high discrimination power of the utilized SSR markers. The number of alleles identified was higher for all loci except for *Hr03* and *Hr06*; however, this could be a result of the limited number of individuals assayed in the previous study or it may reflect the diversity of the material included in this study, which consisted of accessions from a variety of breeding programmes, and included hybrid material. More striking observation was the differences in heterozygosity. The observed heterozygosities ranged from 0.1397-0.2997 (Wang *et al.*, 2008), whereas in the present study, they ranged from 0.781-0.972 (Table 12.3). The heterozygosity revealed in this study is high, which is probably due to the hybrid origin of all the accessions included in this study. These accessions were based upon inter-subspecies crosses followed by selection from naturally pollinated offspring. As seabuckthorn is an out-crossing species, this could explain the high level of heterozygosity found in this study. The hybrid background could also explain the large size range in some of the SSR fragments, with the markers amplifying loci derived from different sub-species. Another possibility is that these SSR markers are amplifying non-specific fragments or multiple loci from *H. rhamnoides* subspecies. This was noticed with the amplification of non-specific fragments at lower annealing temperature, and which was only partially rectified by increasing the stringency of the PCR conditions. The size range of alleles was large which may be an indication of the amplification of non-specific fragments or multiple loci. However, all fragments were polymorphic which suggests that they are not non-specific, and may be multiple loci detected by these primer pairs. The fixation index was negative for all loci except *Hr04* and *Hr07*, indicating an excess of heterozygotes. However, as these samples were not from natural populations, so the deviation from Hardy-Weinberg equilibrium could not be determined. In order to better characterization of these SSR markers, it would be desirable to apply them to a segregating population in order to identify amplification of multiple loci or non-specific fragments as well as to identify null alleles.

Table 12.3: Genotyping Results for the Latvian Seabuckthorn Accessions using Eight SSR loci

	Loci							
	Hr01	*Hr02*	*Hr03*	*Hr04*	*Hr05*	*Hr06*	*Hr07*	*Hr08*
Number of alleles	14	10	10	22	6	5	11	4
Number of effective alleles	7.043	5.684	5.153	8.945	2.653	3.713	4.853	2.723
Information index	2.171	1.932	1.870	2.643	1.235	1.425	1.861	1.103
Observed heterozygosity	0.889	0.861	0.972	0.815	0.861	0.889	0.781	0.833
Expected heterozygosity	0.858	0.824	0.806	0.888	0.623	0.731	0.794	0.633
Unbiased expected heterozygosity	0.870	0.836	0.817	0.905	0.632	0.741	0.807	0.642
Fixation index	–0.036	–0.045	–0.206	0.083	–0.382	–0.216	0.016	–0.317

The SSR data was used in the estimation of Latvian seabuckthorn accession genetic diversity and relatedness. Seabuckthorn accessions grown in Latvia showed rather high genetic diversity. The average pair-wise value of Nei's genetic distance (Nei, 1973) among seabuckthorn accessions was 0.3373, ranged from 0.0625-0.5357. In 68.4 per cent of cases, genetic distances among accessions had values over 0.30; only 0.48 per cent of cases, genetic distances had values under 0.01. The lowest genetic distance values were found between three pairs of accessions: "Podarok Sadu and Vir kv II" (distance 0.0625), which also form close mutual clusters (Figure 12.1). Neighbour-Joining Clustering did not identify clear accession groups, probably due to the hybrid nature of the analyzed material.

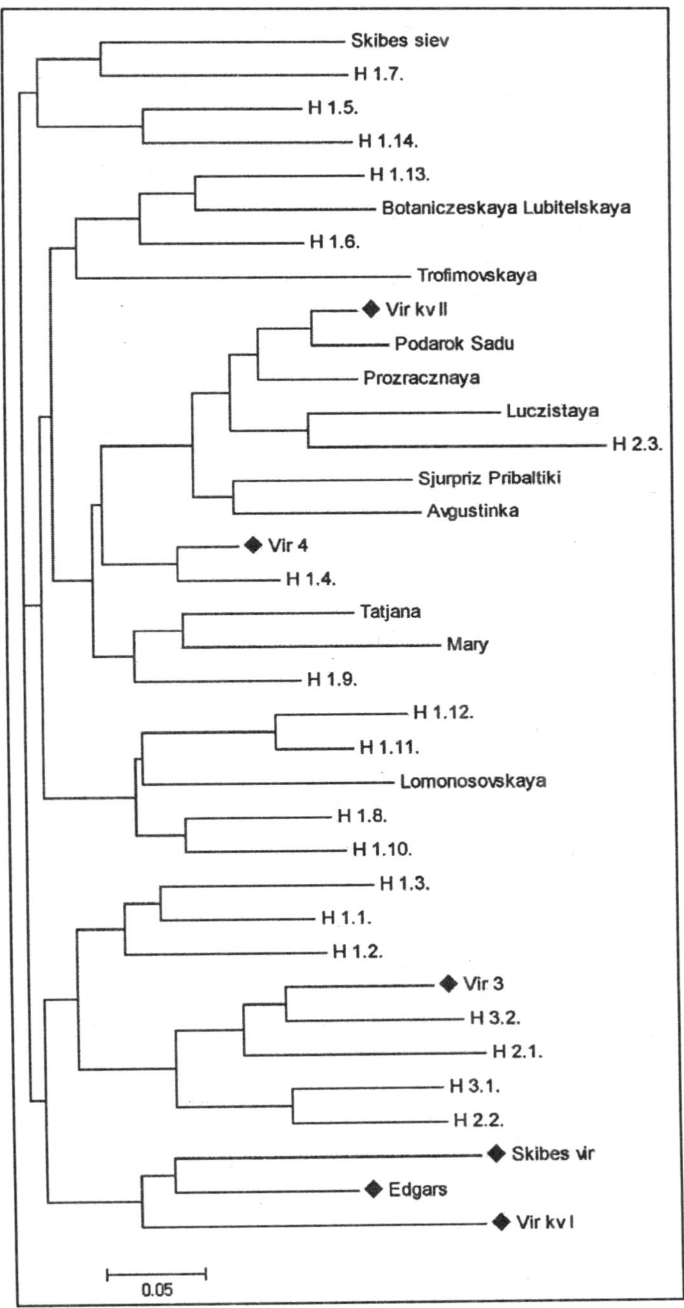

Figure 12.1: Dendrogram of Genetic Relatedness of Latvian Seabuckthorn Accessions using Eight SSR loci Based on Nei's Similarity Index and Neighbour-Joining Clustering Method. ◆: Seabuckthorn male plants.

However, some correlation between the cluster analysis and the plant material origin and known pedigree was found, which demonstrated the suitability of the utilized SSR markers for the characterization of seabuckthorn genetic resources. High values of bootstrapping (over 50 per cent) were found for varieties "H 1.7. and Skibes siev (57 per cent), H 1.11 and H 1.12 (63 per cent), and Padarok Sadu and Vir kv II" (75 per cent). A separate cluster was formed by two Latvian varieties ("Mary and Tatjana"), which have a common pedigree-both are open pollinated seedlings from cultivar "Botanicheskaya Lubitelskaya", although the mother plant was located in a different cluster. Close genetic relatedness was also identified for most of the seabuckthorn cultivars developed by T. Trofimov at the Botanical Garden of Moscow State University ("Botanicheskaya Lubitelskaya, Trofimovskaya, Vir kv II, Podarok Sadu, Prozrachnaya, Luczistaya and Vir 4") indicating a possible common origin. Exceptions were observed for varieties "Lomonosovskaya, Vir 3 and Vir kv I", which showed higher genetic distance. The male plants "Skibes vir, Edgars and Vir kv II" formed a separate cluster but in other cases, no particular grouping for male plants was observed as they were distributed over all clusters of the dendrogram. These phylogenetic clustering patterns were at least partially also confirmed by RAPD marker analysis of the same plant material, where cultivars "Mary and Tatjana" also form a separate cluster. Common grouping was also observed for the widely grown varieties ("Avgustinka, Luczistaya, Podarok Sadu, Prozrachnaya, Sjurpriz Pribaltiki"), similar to the results obtained by the SSR markers. This confirms that these SSR markers were useful in identifying genetic relationships between the tested accessions, despite of being developed on a different subspecies of seabuckthorn. The low bootstrap values of seabuckthorn accessions could be generally explained by the characteristics of the tested plant material which includes geographically distant subspecies of *H. rhamnoides* and their reciprocal hybrids with different hybridization levels. This artificial hybridization of geographically widely separated accession, would tend to obscure any genetic clustering according to geographical origin and only maintain high bootstrapping values for accession with shared pedigrees.

CONCLUSION

Microsatellite (SSR) markers developed in *H. rhamnoides* ssp. *sinensis* are suitable for the application in other *H. rhamnoides* subspecies as well as in crosses among *H. rhamnoides* ssp. *mongolica*, ssp. *rhamnoides* and ssp. *fluviatilis*. The eight tested microsatellite markers allowed a complete discrimination of the tested seabuckthorn accessions and showed suitability for evaluation of genetic diversity of Latvian seabuckthorn accessions as well as determination of genetic similarity and relatedness. These DNA markers will provide additional tools to Latvian seabuckthorn breeders for cultivar identification and assessment of breeding material.

ACKNOWLEDGEMENT

This research was supported by the Latvia State Research Program No. 9 "Innovative Technologies for the development of high-value, safe and healthy food products from genetically, physiologically and biochemically various plant and animal material".

REFERENCES

1. Bartish, I.V., Jeppsson, N., Bartish, G.I., Ronsen, L. and Nybom, H. 2000. Inter- and intraspecific genetic variation in *Hippophae* (*Elaeagnaceae*) investigated by RAPD markers. *Plant Systematics and Evolution* 225: 85-101.

2. Bianchi, V.J., Sansavini, S. and Fachinello, J.C. 2004. Microsatellite markers for identification of *Prunus* ssp. rootstocks. *Scientia Agricola* 61 (3): 303-306.

3. Bruvelis, A. 2003. Cultivation of seabuckthorn in Baltic States. In: "*Seabuckthorn-A Resource of Health a Challenge to Modern Technology*". Proceedings of 1st Congress of the International Seabuckthorn Association, p. 64-66, Berlin, Germany.

4. Bruvelis, A. 2007. Seabuckthorn (Hippophae rhamnoides L.)-Taxonomy, distribution and introduction in Baltic States. *Vegetation of Latvia* 13: 33-38.

5. Jeppsson, N., Bartish, I.V. and Persson, H.A. 1999. DNA Analysis as a tool in seabuckthorn breeding. In: *Perspectives on New Crops and New Uses* (J. Janick, Ed.), p. 338-341, ASHS Press, Alexandria, VA, 528p.

6. Karaagac, E., Yilma, S. and Vales, M.I. 2010. SSR-based DNA fingerprinting of potato clones from the Pacific north-west potato variety development programme. *Acta Horticulturae* 859: 121-128.

7. Liu, K. and Muse, S.V. 2005. Power Marker: An integrated analysis environment for genetic marker analysis. *Bioinformatics* 21 (9): 2128-2129.

8. Nei, M. 1973. Analysis of gene diversity in subdivided populations. *Proceedings of the National Academy of Sciences of the United States of America* 70: 3321-3323.

9. Peakall, R., Gilmore, S., Keys, W., Morgante, M. and Rafalski A. 1998. Cross-species amplification of soybean (*Glycine max*) simple sequence repeats (SSRs) within the genus and other legume genera: Implications for the transferability of SSRs in plants. Molecular Biology and Evolution 15 (10): 1275-1287.

10. Peakall, R. and Smouse, P.E. 2006. GENALEX 6: Genetic analysis in Excel. Population genetic software for teaching and research. *Molecular Ecology Notes* 6: 288-295.

11. Persson, H.A. and Nybom, H. 1998. Genetic sex determination and RAPD marker segregation in the dioecious species seabuckthorn (*Hippophae rhamnoides* L.). *Hereditas* 129: 45-51.

12. Powell, W., Machray, G.C. and Provan, J. 1996. Polymorphism revealed by simple sequence repeats. Trends in Plant Science 1: 215-221.

13. Ruan, C. and Li, D. 2005. AFLP fingerprinting analysis of some cultivated varieties of seabuckthorn (*Hippophae rhamnoides*). *Journal of Genetics* 84 (3): 311-316.

14. Ruan, C., Qin, P., Zheng, J. and He, Z. 2004. Genetic relationships among some cultivars of seabuckthorn from China, Russia and Mongolia based on RAPD analysis. *Scientia Horticulturae* 101: 417-426.

15. Sheng, H.M., An, L.Z., Chen, T., Xu, S.J., Liu, G.X., Zheng, X.L., Pu, L.L., Liu, Y.J. and Lian, Y.S. 2006. Analysis of the genetic diversity and relationships among and within species of *Hippophae* (*Elaeagnaceae*) based on RAPD markers. *Plant Systematics and Evolution* 260: 26-37.

16. Singh, R., Mishra, S.N., Dwivedi, S.K. and Ahmed, Z. 2006. Genetic variation in seabuckthorn (*Hippophae rhamnoides* L.) populations of cold arid Ladakh (India) using RAPD markers. *Current Science* 91 (10): 1321-1322.

17. Sun, K., Chen, X.L., Ma, R., Li, C., Wang, Q. and Ge, S. 2002. Molecular phylogenetics of *Hippophae* L. (*Elaeagnaceae*) based on the internal transcribed spacer (ITS) sequences of nrDNA. *Plant Systematics and Evolution* 235: 121-134.

18. Sun, K., Chen, W., Ma, R., Chen, X., Li, A., Ge, S., 2006. Genetic variation in *Hippophae rhamnoides* ssp. *sinensis* (*Elaeagnaceae*) revealed by RAPD markers. *Biochemical Genetics* 44 (5/6): 186-197.

19. Tamura, K., Dudley, J., Nei, M., Kumar, S., 2007. MEGA 4: Molecular Evolutionary Genetics Analysis (MEGA) software version 4.0. *Molecular Biology and Evolution* 24: 1596-1599.

20. Tian, C., Lei, Y., Shi, S., Nan, P., Chen, J. and Zhong, Y. 2004a. Genetic diversity of seabuckthorn (*Hippophae rhamnoides*) populations in northeastern and northwestern China as revealed by ISSR markers. *New Forest* 27: 229-237.

21. Tian, C., Nan, P., Shi, S., Chen, J. and Zhong, Y. 2004b. Molecular genetic variation in Chinese populations of three subspecies of *Hippophae rhamnoides.* *Biochemical Genetics* 42 (7/8): 259-267.

22. Wang, A., Zhang, Q., Wan, D., Yang, Y. and Liu, J. 2008. Nine microsatellite DNA primers for *Hippophae rhamnoides* ssp. *sinesis* (*Elaeagnaceae*). *Conservation Genetics* 9: 969-971.

23. Weber, J.L. and May, P.E. 1989. Abundant class of human DNA polymorphisms which can be typed using the polymerase chain reaction. *The American Journal of Human Genetics* 44 (3): 388-396.

Chapter 13

Application of DNA Technologies for Improvement of Seabuckthorn

Ankit Jain, Saurabh Chaudhary and Prakash C. Sharma

University School of Biotechnology, Guru Gobind Singh Indraprastha University, Dwarka Sector 16 C, New Delhi – 110 078, India

SUMMARY

Seabuckthorn, a multipurpose plant inhabiting diverse geo-climates, is known for its richness in bioactive compounds and ecological value as ecosystem restorer. Due to its immense medicinal and ecological values, efforts have been initiated to establish seabuckthorn as a main stream crop plant by improving various agronomic traits. In the present study, we have focused on the use of DNA based technologies in seabuckthorn was aimed to assess interspecific and intraspecific genetic diversity, DNA fingerprinting of available germplasm, and gene tagging. With the advent of next generation sequencing, whole transcriptome profiling has allowed identification and characterization of genes regulating various pathways involved in the production of important bioactive compounds and imparting resistance to various biotic and abiotic stresses. Application of molecular markers in conjunction with advanced functional genomics tools is certain to provide opportunities not only for effective marker assisted breeding for useful traits in seabuckthorn, but also for the introgression of genes of economic importance in other crop plants.

Keywords: *Seabuckthorn, DNA based technologies, Molecular markers and Functional genomics.*

INTRODUCTION

Seabuckthorn, a major inhabitant of Himalayas, is also found growing in geographically diverse areas ranging from seacoasts, riverbeds to cold deserts by virtue of its ability to withstand a variety of harsh environmental conditions of cold and heat stress, salinity, drought, etc. (Rongsen, 1997). Apart from being an abiotic stress tolerant plant, seabuckthorn is a potential source of many bioactive compounds also. In general, seabuckthorn contains vitamins (C, E and K), minerals, amino acids, fatty acids (omega -3, 6, 9, palmitoleic acid, palmitic acid, stearic acid) and antioxidants, carotenoids

Seabuckthorn (*Hippophae* L.): *A Multipurpose Wonder Plant, Vol. 4: Emerging Trends in Research and Technologies* (Virendra Singh, *Editor-in-Chief*, 2014), p. 167–178, Daya Publishing House, New Delhi, India.

(lycopene, β-carotenes, zeaxanthin), different flavonoids (quercetin, kaempferol, isorhamnetin, myricetin, etc.), and other secondary metabolites (Zeb, 2004; Kumar *et al.*, 2011; Fatima *et al.*, 2012). High content of such beneficial bioactive compounds makes it a medicinally useful shrub. Numerous health benefits include anti-tumor activity (Teng *et al.*, 2006), potent antioxidant activity (Gorbatsova *et al.*, 2007), protection against radiation induced DNA damage (Goel *et al.*, 2003; Shukla *et al.*, 2006), gamma rays (Goel *et al.*, 2007), and intestinal pathogens (Pimia *et al.*, 2005). Seabuckthorn berry extract has been found to be immuno-modulator (Mishra *et al.*, 2008) and anti-carcinogenic (Zeb, 2006). Besides its medicinal value, seabuckthorn also carries immense ecological importance. Seabuckthorn is well known to prevent soil erosion in open as well as flood prone areas due to its extensive root system (Trajkovski and Jeppsson, 1999). The symbiotic association of actinomycetes *Frankia* with seabuckthorn roots improves the fertility of the soil by fixing atmospheric nitrogen. Hence, seabuckthorn is also known as an ecosystem restorer. Considering its immense importance, efforts have been initiated worldwide to improve and establish seabuckthorn as a multipurpose crop. Moreover, there have been certain issues ranging from correct taxonomic classification to genome characterization that need to be addressed with priority.

Beginning with Swedish taxonomist Linnaeus, who in 1753 recorded seabuckthorn, distributed in Baltic coast and named it *Hippophae rhamnoides* Linn. and included in the family *Elaeagnaceae*, several modification have been made till date. Servettaz (1909) classified genus *Hippophae* in one species, *H. rhamnoides* with three subspecies, namely *rhamnoides, salicifolia,* and *tibetana.* Later in 1971, Finnish taxonomist Arne Rousi (1971) modified this classification by elevating these subspecies to species and further introducing seven new subspecies of *H. rhamnoides.* Liu and He (1978) added another species, *H. neurocarpa* to this group. Lian *et al.* (1998) reported one new species, *H. goniacarpa* and two subspecies *H. goniacarpa* ssp. *litangensis* and *H. neurocarpa* ssp. *stellatopilosa.* Currently, seabuckthorn is represented by 7 species and 8 subspecies (Swenson and Bartish, 2002). Taxonomist still remains inconclusive about the correct taxonomic rank of this genus, as many of the subspecies deserve an independent rank as species. Wide geographic distribution of seabuckthorn and its capability to grow in diverse environmental conditions could have an impact over the morphological characteristics, which in turn could make taxonomic classification of genus more complicated. Similarly, determination of gender of seabuckthorn seedlings is not possible until flowering stage, which takes a long period in field. Thus, plant breeders have to maintain large number of plants till flowering for gender identification. Since, seabuckthorn is a dioecious plant with males being heteromorphic, occurrence of distinguishable sex chromosomes suggests that gender could be determined by X/Y system (Shchapavn, 1979).

In recent past, efforts have been initiated to develop an appropriate classification system involving traditional morphological and present day DNA based technologies. Further development of genetic marker (s), which could lead to correct identification of gender at an early stage for screening seabuckthorn seedlings and discarding the males, would be very useful in plant breeding programmes.

Another critical area where any development could be a major breakthrough not only for improvement in seabuckthorn but for other crops also has focused on understanding the ability of seabuckthorn to sustain harsh environmental conditions. Extreme environmental stresses force plants to adopt alternate molecular mechanism manifested as various physiological and morphological changes that help in survival and adaptation. Adaptive behaviour of seabuckthorn to sustain extreme environmental conditions is well known but the molecular mechanisms underlying such complex traits remain unresolved. Similarly, web of genes involved in biosynthesis of bioactive compounds are also not known so far. In this chapter, we made an attempt to review briefly the various DNA based technologies that have been used in addressing different issues concerned with present and future objectives of seabuckthorn improvement.

DNA BASED TECHNOLOGIES

In view of current challenges, DNA based technologies could be exploited as two-way approach for improvement of seabuckthorn. On one hand, molecular marker based genome characterization may allow assessment of genetic diversity in natural and cultivated gene pool, marker assisted selection of improved characters and identification of specific regions of chromosome carrying genes contributing to traits of agronomic interest (such as resistance to certain biotic/abiotic factors, high fruit yield, high biochemical content, etc.). On other hand, functional genomics tools may assist in resolving internal blueprint (genes) of seabuckthorn to identify gene networks that are involved in controlling genetic variation for agronomically valuable traits in seabuckthorn.

MOLECULAR MARKERS

Molecular markers represent segments of DNA located at specific positions in the genome. Such markers generally have no apparent effect on the phenotype of the individual, are least affected by the environment and follow Mendelian inheritance. Molecular markers have the capability to show allelic variation, *i.e.* polymorphism of nucleotide sequence at same locus among different accessions of a species or other related species. The advent of different genetic tools, such as restriction enzymes, polymerase chain reaction, abundance of DNA sequence data coupled with automated high-throughput assays have facilitated development of several classes of molecular markers including restriction fragment length polymorphisms (RFLPs), random amplified polymorphic DNA (RAPDs), microsatellite DNA or simple sequence repeats (SSRs), amplified fragment length polymorphism (AFLP), single nucleotide polymorphisms (SNPs), and their variants. A brief idea of different marker systems and their applications in seabuckthorn research is presented in Table 13.1.

With the advent of high throughput sequencing techniques, development of molecular markers has recently become a thrust area in biological sciences. Molecular markers are being used in wide areas of research spanning genome characterization and mapping, population biology, ecology and phylogeny. An ideal marker is one which is cheap, provides an immense information, reproducible, suffers with very low error rate and easy, unambiguous to score. In the following subsections, various molecular markers have been discussed along with their utility in seabuckthorn research.

Random Amplified Polymorphic DNA (RAPD)

RAPD (Williams *et al.*, 1990) opened the door for PCR based molecular marker systems, and are still in use albeit with decreased popularity. These markers are preferred to assess the genetic diversity/ DNA fingerprinting, when no sequence information is available about organism under investigation. A single arbitrary oligonucleotide primer of approximately 10 bp length is used for PCR amplification from unspecified loci in target genotype DNA as primer anneals to its complementary sequences present throughout genome. Amplicons are generated, if annealing sites are not far away from the 3′ end of the other primer. Series of amplicons thus, generated under low stringency conditions are resolved on agarose gels. Despite, being an easy as well as cost effective method of assay, RAPD markers have certain disadvantages, including dominant nature and reproducibility among laboratories as any change in the experimental conditions may alter the results. Due to reproducibility problems, RAPD have found limited use in genome mapping research, although conversion of a single RAPD band of interest into a corresponding SCAR (sequence characterized amplified region) marker overcomes this problem, particularly in gene tagging experiments.

Table 13.1: Characteristics of different Molecular Markers and their Application in Seabuckthorn

Marker Type	Polymor-phism/ Abundance	Reproduci-bility	Mode of Inheri-tance	Develop-mental Cost	Locus Specificity	Application in Seabuckthorn
RAPD	High	Low	Dominant	Low	No	Sex determination (Persson and Nybom, 1998; Dong *et al.*, 2007; Sharma *et al.*, 2010), Diversity assessment and germplasm characterization (Sheng *et al.*, 2006; Ercisli *et al.*, 2008; Chen *et al.*, 2010a)
ISSR	High	High	Dominant	Low	No	Assessment of genetic variability and population structure (Tian *et al.*, 2004a; Tian *et al.*, 2004b; Li *et al.*, 2009); Association with DSD resistance (Ruan *et al.*, 2009)
SSR	High	High	Codominant	Medium	Yes	Diversity assessment (Yanlin, 2007; Wang *et al.*, 2008; Jain *et al.*, 2010; Jain *et al.*, 2013)
AFLP	High	High	Dominant	High	No	Diversity assessment and phylogenetic relationship (Ruan and Li, 2005; Ruan, 2006; Shah *et al.*, 2009)
SCAR	Medium	High	Codominant	Medium	Yes	Sex Determination (Korekar *et al.*, 2012)
SRAP	Medium	High	Codominant	Medium	Yes	DSD resistance (Li *et al.*, 2010)
SAMPL	Medium	High	Dominant	Medium	Yes	Diversity assessment (Raina *et al.*, 2012)
SNP	High	High	Codominant	Low	Yes	–

RAPD markers have extensively being used in seabuckthorn due to unavailability of molecular data until recent past. Subsequent to first report of identification of four RAPD markers discriminating among plants of different sexes (Persson and Nybom, 1998), there has been a continuous rise in development of these markers in seabuckthorn. Dong *et al.* (2007) identified a female specific marker in *H. goniocarpa*, while Sharma *et al.* (2010) reported one male specific marker in *H. rhamnoides*. In addition to this important use in providing sex specific markers for early detection of sex of seabuckthorn seedlings, RAPD technology has been used in the assessment of taxonomic status, genetic structure and genetic diversity that lies within and among the seabuckthorn species (Ganghal *et al.*, 2008; Kalia *et al.*, 2011; Ruan *et al.*, 2012). Ercisli *et al.* (2008) used fatty acid methyl ester (FAME) analysis in conjunction with RAPD markers for characterizing and grouping of seabuckthorn genotypes collected from Eastern Anatolia. Some of the landmark studies have inferred that a relationship exists between the altitude and genetic distance in seabuckthorn (Sheng *et al.*, 2006; Chen *et al.*, 2010b).

Amplified Fragment Length Polymorphism (AFLP)

AFLP (Vos *et al.*, 1995), are DNA fragments obtained as a result of restriction digestion by endonucleases (preferably a combination of hexa-cutter and a tetra-cutter enzyme) followed by ligation of double stranded adaptors to the overhangs of restriction fragments and selective amplification using primers complementary to adaptor and restriction site sequence, extended to a few arbitrarily chosen nucleotides extension at 3' ends into the restriction fragments. The amplicons thus obtained could be resolved on denaturing polyacrylamide gels through autoradiography or fluorescence detection. Automated capillary sequencing instruments are also available for this purpose. AFLP technology provides reproducibility of RFLP and robustness of PCR, while dominant inheritance and multi-locus amplification have some disadvantages associated with AFLP markers. In parallel to

RAPDs, AFLP markers have been used to resolve taxonomic issues involving seabuckthorn and to assess interspecific and intraspecific genetic diversity in *Hippophae*.

In the very first study (Vrieling *et al.*, 1997), AFLP protocol was tested on three different plant species, namely *Senecio jacobaea, Ehiumc vulgare and Hippophae rhamnoides*. The number of scorable bands was highest in *S. jacobaea*, followed by *E. vulgare* and *H. rhamnoides*. In subsequent studies completely focused on seabuckthorn, fifteen cultivated varieties of seabuckthorn from China, Russia and Mongolia were fingerprinted using AFLP markers (Ruan and Li, 2005). A total of 731 AFLP bands including 645 polymorphic bands were resolved using eight primer combinations. In a similar manner, nine primer combinations were used that generated 451 polymorphic bands in cultivated varieties from China, Russia and Mangolia. Cluster analysis revealed *H. rhamnoides* ssp. *sinensis* from China as the most distantly related to the 14 other varieties (Ruan, 2006). These results thus could provide a more rigorous scientific basis for the characterization of seabuckthorn varieties and improvement strategies in seabuckthorn breeding programs. On the basis of AFLP fingerprinting of 25 ecotypes of *H. rhamnoides* from northern Pakistan, Shah *et al.* (2009) suggested that these ecotypes with considerable genetic variability expressed common heritage in their phylogeny. However, a few ecotypes showed close relationship irrespective of their geographic distances and morphological attributes.

Simple Sequence Repeats

SSRs/Microsatellites (Tautz, 1989), are the tandem repeats of 1-6 bp distributed in all prokaryotic as well as eukaryotic genomes. Microsatellite markers are considered to be markers of choice as they are fast, robust, reproducible and highly polymorphic with codominant mode of inheritance. Polymerase slippage at microsatellite sites is the most common reason behind their length variability (Ellegren, 2000). However, unequal crossing over and duplications are some other molecular events that are common to these sites. Microsatellite sequences can either be isolated by screening partial genomic libraries (Tautz, 1989; Rassmann *et al.*, 1991) by colony hybridization with repeat containing probes, or by enrichment approaches that use hybridization capture procedures (Kandpal *et al.*, 1994; Kijas *et al.*, 1994). Availability of abundant sequence data in the public domain in recent times has facilitated *in silico* mining of microsatellites in large number required to undertake extensive genome mapping experiments. Analysis of DNA polymorphism is performed by amplification of target loci with locus specific primers flanking microsatellite regions. However, as stated above, due to unavailability of substantial sequence data in seabuckthorn till recent past, very limited reports are available on microsatellite marker development so far. Yanlin (2007) reported successful cross amplification of 107 grapes microsatellite markers in seabuckthorn, 46 of which were found polymorphic in *H. rhamnoides*. Wang *et al.* (2008) used enrichment procedures to develop nine polymorphic markers in *H. rhamnoides* ssp. *turkestanica*, *H. gyantsensis* and *H. neurocarpa*, resolving 3-12 alleles per locus.

In our laboratory, we have developed microsatellite markers both by enrichment methods as well as by screening the sequence data (EST Sequencing and Next generation transcriptome sequencing data). An alternative cost effective and efficient approach has been developed allowing isolation of more than 90 per cent microsatellite positive genomic sequences. These microsatellite markers are currently being validated on the seabuckthorn accessions collected from different regions of Leh (Unpublished). Earlier, our pilot project on EST sequencing in *H. rhamnoides* generated 3412 high quality EST sequences, which could be clustered into 1665 putative unigenes using CAP3 (Huang and Madan,1999). On screening these putative unigene with a perl script MISA (Thiel *et al.*, 2003), 62 SSRs were identified in 59 microsatellite positive putative unigenes. Out of 30 primer sets tested, nine and two microsatellite markers were found polymorphic in *H. rhamnoides* and *H. salicifolia* genotypes, respectively (Jain *et al.*, 2010). Further, to exploit next generation sequencing methods, whole transcriptome sequencing of seabuckthorn has been performed, which has provided us an opportunity to look at the occurrence and abundance of microsatellite in the expressed genome of seabuckthorn. Of the 88297 putative unigenes, 7.69 per cent unigenes were found microsatellite positive. Dinucleotide repeats were most abundant followed by trinucleotide repeats. Microsatellite density was higher in the coding regions in comparison to untranslated regions. AG and AAG type repeats were the most frequently represented repeats in seabuckthorn transcriptome. Additionally, Unigene Specific Microsatellite Markers (USMM) are being developed to assess the genetic diversity of seabuckthorn (Jain *et al.*, 2013) (Figure 13.1). With the availability of transcriptome sequences (Fatima *et al.*, 2012; Sharma *et al.*, 2012) in large number, development of gene based markers for seabuckthorn can be a routine task facilitating experiments aiming at assessment of genomic structure, genetic diversity and gene tagging.

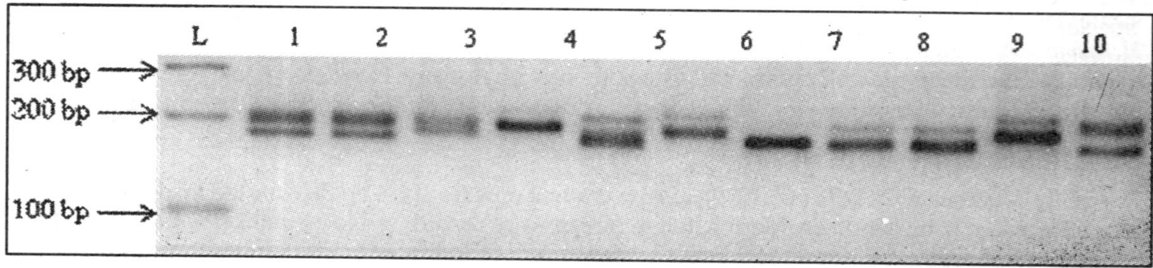

Figure 13.1: Polymorphism Displayed by Microsatellite Locus USMM5. L: Molecular size marker; 1-10: Accessions of *H. rhamnoides* Collected from Diverse Locations.

Inter Simple Sequence Repeats

Inter Simple Sequence Repeats (ISSR; Meyer *et al.*, 1993), is a dominant marker system almost similar to RAPD marker system in a manner that prior knowledge of genome sequence is not required. The method allows amplification of region(s) between two inversely oriented microsatellite repeats. Mostly, single primer is used to generate array of amplicons that could be resolved on agarose as well as polyacrylamide gels. Oligonuclotide primers for ISSR amplification are motifs of microsatellites, *e.g.* $(CT)_{10}$, $(CG)_8$. However, these primers could also be anchored, *e.g.* $(CT)_{10}AG$, $(CG)_7GT$, as anchoring of primers provide more reproducibility and high resolution of ISSR profile.

ISSR is the second most frequent marker system used in seabuckthorn after RAPDs. Tian *et al.* (2004a) screened eleven natural populations of *H. rhamnoides* from northeastern and northwestern China for genetic variability among and within populations using eight ISSR markers that revealed 207 alleles. The coecient of gene dierentiation showed that the total molecular variance mainly existed within populations. No significant correlation between genetic and geographic distances of the populations was found using ISSR markers. Working on similar lines, Tian *et al.* (2004b) analyzed 300 individuals of fifteen natural populations of seabuckthorn (*Hippophae rhamnoides* ssp. *yunnanensis*, ssp. *sinensis*, and ssp. *gyantsensis*) from China. Results suggested occurrence of different genetic structure of these subspecies. Using 15 primers, Li *et al.* (2009) amplified 384 polymorphic bands and discovered the presence of different Chinese cultivars with the same name, and identical cultivars with different names. They also compared genetic relationships of cultivars with unknown and known parentage. Ruan *et al.* (2009) screened 15 ISSR primers on 52 accessions of seabuckthorn to identify markers associated with dried-shrink resistance. Fifteen ISSR primers generated 346 bands, with an average of 23 bands per primer. Four ISSR markers were signicantly correlated with resistance to dried-shrink disease. These markers provided a major breakthrough towards identification of potential breeding varieties and lines when no linkage maps and Quantitative Trait Loci are available.

Internal Transcribed Spacer

ITS were developed by Baldwin (1992). Plant ribosomal RNA (rRNA) genes are arranged in tandem array of highly repeated sequences, each consisting of sequences for the 18S, 5.8S, and 25S ribosomal subunits. The spacer between the 18S and 5.8S genes are known as Internal Transcribed Spacer 1 (ITS 1) and the spacer between the 5.8S and 25S genes are ITS2. A third spacer is also considered sometimes, the large inter-genic spacer (IGS) or non-transcribed spacer (NTS), which are found between the 3′ end of the 25S and the 5′ beginning of the 18S genes. Sequence comparison of the ITS region has widely been used for molecular systematic studies at the species level, and even within species, because of its higher degree of variation than other genic regions of rDNA. In seabuckthorn, Sun *et al.* (2002) studied 15 taxa of the genus *Hippophae* by comparing ITS region of nrDNA. Parsimony analysis suggested that the species *H. tibetana, H. neurocarpa* and *H. salicifolia* were all distinct. Sequence divergence among subspecies of *H. rhamnoides* was found remarkably high. Sun *et al.* (2003) investigated the phylogeny of the putative hybrid species *H. goniocarpa*. ITS sequencing results strongly support the hypothesis that *H. goniocarpa* was derived through hybridization between *H. rhamnoides* subssp. *sinensis* and *H. neurocarpa*. Further, Chen *et al.* (2010a) suggested that genetic diversity of *H. rhamnoides* ssp. sinensis mainly existed within populations and with the results from sequence comparison, it could be inferred that geographic distances did not have any clear effect on their genetic differentiation which contradicts the result obtained from some of the previous studies with RAPD markers (Sheng *et al.*, 2006; Chen *et al.*, 2010b).

Other Miscellaneous Marker Systems

The conversion of AFLP and RAPD and other molecular markers to Sequence Characterized Amplified Regions (SCARs; Paran and Michelmore, 1993) makes markers more reproducible and reliable, and further improves efficiency of these markers to be utilized in applications such as marker assisted selection (MAS) and cultivar identification. Codominant in inheritance, these markers require amplification of the target locus by the primers developed with the sequence information obtained as a result of sequencing of selected RAPD/AFLP bands. Korekar *et al.* (2012) developed female specific HrX1 and HrX2 markers. Capability of these markers to determine gender of seabuckthorn in early stages could be of great use for plant breeders.

Selective amplification of microsatellite polymorphic loci (SAMPL) is a microsatellite based AFLP marker system. Selective amplification of a locus is achieved with one AFLP primer and one SAMPL primer (Witsenboer *et al.*, 1997). Like AFLPs, SAMPL makers are dominant, but quite frequently being used in the assessment of genetic diversity. However, in case of seabuckthorn, this marker system has been used in conjunction with AFLP markers to conclude that level of inter-population genetic diversity of *H. rhamnoides* ssp. *turkestanica*, is higher than intra-population diversity (Raina *et al.*, 2012).

Sequence related amplified polymorphism (SRAP) is a unique PCR-based molecular marker method that is based on the amplification of open reading frames (ORFs) producing a number of co-dominant markers per amplification (Li and Quiros, 2001). Li *et al.* (2010) developed eleven SRAP markers significantly associated with dried- shrink disease (DSD) resistance in 77 accessions of 22 seabuckthorn cultivars. These markers provide a viable option for breeding programs that select lineages with DSD resistance, especially when no other genetic information such as linkage maps and quantitative trait loci are available.

FUNCTIONAL GENOMICS

One of the thrust areas of present day plant biology research involves identification and characterization of genes involved in various molecular processes that help the plant in either better adaptation to adverse environmental condition or enhancement of economic yield. Presently, limited numbers of reports are available with the focus on identification and expression of genes in seabuckthorn. In the pioneer report on functional genomics of seabuckthorn, glycerol-3-phosphate acyltransferase (GPAT) gene was sequenced and its expression was assessed by northern blotting showing an increase in cold stressed leaves (Gupta, 2009). In our laboratory, under a project on EST sequencing in seabuckthorn, we sequenced 4700 clones selected from cDNA library using Sanger's method, which resulted in the generation of 3412 high quality EST sequences, submitted to dbEST of NCBI in 2010. Assignment of Gene Ontology (GO) terms to unigenes was done using BLAST2GO (Conesa *et al.*, 2005). In total, 914 putative unigenes could be assigned GO terms. Nearly 43 genes were found involved in important biological processes like response to biotic and abiotic stress stimulus. On the basis of GO annotation and Gene Investigator software (Zimmermann *et al.*, 2004), 16 genes related to cold stress were identified from the clustered unigenes and validated through RT-PCR (Ghangal *et al.*, 2011). Extending our study from few thousand unigenes, we exploited next generation sequencing of seabuckthorn leaves and root tissue libraries using Illumina GA platform, that yielded more than 90 million short reads comprising of approximately 8.5 billion nucleotides, which in turn could be assembled into 88,297 putative unigenes (Sharma *et al.*, 2012; Jain *et al.*,2013). Further, to get a clearer and larger picture of the genes expressed during cold stress, Deep SAGE was used wherein tag based sequencing of RNA isolated from control and stressed leaf tissues of germinated seabuckthorn revealed differential expression of approximately 20,000 genes (Sharma *et al.*, 2012). Validation of some of the important genes though real time PCR is under progress. In another report, Fatima *et al.* (2012) performed sequencing of cDNA of mature seabuckthorn berries using 454 sequencing accumulating 89,141 putative unigenes represented by 37,482 contigs and 51,659 singletons. However, this study mainly focused on identification of genes involved in fatty acid biosynthesis. Thus, vast amount of transcriptome data that has been generated from these recent studies (Ghangal *et al.*, 2011; Fatima *et al.*, 2012; Sharma *et al.*, 2012) will certainly increase our understanding of plant biology thereby enhancing our ability to exploit genomic information for future improvements in seabuckthorn. Further advances in microarray technologies as well as other sequencing based gene expression analysis methods such as RNA-seq (Wang *et al.*, 2009) are opening up new roads for simultaneous expression studies and comparative transcriptomics that will elucidate gene function (s) and enhance our understanding of the complex web of interactions between genes that regulate different physiological responses under diverse environmental conditions. All these efforts will contribute towards multifaceted improvement in seabuckthorn leading to its acceptability as a crop in near future.

REFERENCES

1. Baldwin, B.G. 1992. Phylogenetic utility of the internal transcribed spacers of nuclear ribosomal DNA in plants: An example from the *Compositae*. *Molecular Phylogenetics and Evolution* 1: 3-16.
2. Chen, L.H., Yu, Z. and Jin, H.P. 2010a. Comparison of ribosomal DNA ITS regions among *Hippophae rhamnoides* ssp. *sinensis* from different geographic areas in China. *Plant Molecular Biology Reporter* 28: 635-645.

3. Chen, W., Su, X., Zhang, H., Sun, K., Ma, R. and Chen, X. 2010b. High genetic differentiation of *Hippophae rhamnoides* ssp. *yunnanensis* (*Elaeagnaceae*), a plant endemic to the Qinghai-Tibet plateau. *Biochemical Genetics* 48: 565-576.

4. Conesa, A., Gotz, S., Garcia-Gomez, J.M., Terol, J., Talon, M. and Robles, M. 2005. Blast2 GO: A universal tool for annotation, visualization and analysis in functional genomics research. *Bioinformatics* 21: 3674-3676.

5. Dong, L.N., Sun, K., Su, X., Zhang, L., Ding, S.S. and Ma, R.J. 2007. RAPD markers related to sex locus in *Hippophae goniocarpa*. *Bulletin of Botanical Research* 27: 73-76.

6. Ellegren, H. 2000. Microsatellite mutations in the germline: Implications for evolutionary inference. *Trends in Genetics* 16: 551-558.

7. Ercisli, S., Orhan, E., Yildirim, N. and Agar, G. 2008. Comparison of seabuckthorn genotypes (*Hippophae rhamnoides* L.) based on RAPD and FAME data. *Turkish Journal* of *Agriculture and Forestry* 32: 363-368.

8. Fatima, T., Snyder, C.L., Schroeder, W.R., Cram, D., Datla, R., Wishart, D., Weselake, R.J. and Krishna, P. 2012. Fatty acid composition of developing seabuckthorn (*Hippophae rhamnoides* L.) berry and the transcriptome of the mature seed. *PLoS ONE* 7: e34099.

9. Ganghal, R., Raghuvanshi, S. and Sharma, P.C. 2008. Exploring seabuckthorn genome using molecular tools. In: *Seabuckthorn-The Wonder Plant* (S.K. Dwivedi, T. Perimalazagan, S.B. Singh and Z. Ahmad, Eds.), p.347-371, SSPH Publishers, Delhi, 389p.

10. Ghangal, R., Raghuvanshi, S. and Sharma, P.C. 2011. Expressed sequence tag based identification and expression analysis of some cold inducible elements in seabuckthorn (*Hippophae rhamnoides* L.). *Plant Physiology and Biochemistry* 51: 123-128.

11. Goel, H.C., Kumar, P., Samanta, N. and Rana, S.V.S. 2007. Induction of DNA-protein cross-links by *Hippophae rhamnoides*: Implications in radioprotection and cytotoxicity. *Molecular and Cellular Biochemistry* 245: 57-67.

12. Goel, H.C., Samanta, N., Kannan, K., Kumar, I.P. and Bala, M. 2006. Protection of spermatogenesis in mice against gamma ray induced damage by *Hippophae rhamnoides*. *Andrologia* 38: 199-207.

13. Gorbatsova, J., Lougas, T., Vokk, R. and Kaljurand, M. 2007. Comparison of the contents of various antioxidants of seabuckthorn berries using CE. *Electrophoresis* 28: 4136-4142.

14. Gupta, S.M. 2009. Isolation of cDNA fragment of glycerol-3-phosphate acyltransferase gene from seabuckthorn. *Defence Science Journal* 59: 147-151.

15. Huang, X. and Madan, A. 1999. CAP3-A DNA sequence assembly program. *Genome Research* 9: 868-877.

16. Jain, A., Chaudhary, S. and Sharma, P.C. 2013. Mining and validation of microsatellite markers from seabuckthorn (*Hippophae rhamnoides* L.) transcriptome. *Molecular Biology Reports*, (Communicated).

17. Jain, A., Ghangal, R., Grover, A., Raghuvanshi, S. and Sharma, P.C. 2010. Development of EST-based new SSR markers in seabuckthorn. *Physiology and Molecular Biology of Plants* 16: 375-378.

18. Kalia, R.K., Singh, R., Rai, M., Mishra, G.P., Singh, S.R. and Dhawan, A.K. 2011. Biotechnological interventions in seabuckthorn (*Hippophae* L.): Current status and future prospects. *Trees* 25: 559-575.

19. Kandpal, R.P., Kandpal, G. and Weissman, S.M. 1994. Construction of libraries enriched for sequence repeats and jumping clones, and hybridization selection for region-specific markers. *Proceedings of the National Academy of Sciences of the United States of America* 91: 88-92.

20. Kijas, J.M.H., Fowler, J.C.S., Garbett, C.A. and Thomas, M.R. 1994. Enrichment of microsatellites from the citrus genome using biotinylated oligonucleotide sequences bound to streptavidin-coated magnetic particles. *Biotechniques* 16: 656-662.

21. Korekar, G., Sharma, R.K., Kumar, R., Bisht, N.C., Srivastava, R.B., Ahuja, P.S. and Stobdan, T. 2012. Identification and validation of sex-linked SCAR markers in dioecious *Hippophae rhamnoides* L. (*Elaeagnaceae*). *Biotechnology Letters* 34: 973-978.

22. Kumar, R., Kumar, G.P., Chaurasia, O.P. and Singh, S.B. 2011. Phytochemical and pharmacological profile of seabuckthorn oil: A review. *Research Journal of Medicinal Plants* 5: 491-499.

23. Li, G. and Quiros, C.F. 2001. Sequence-related amplified polymorphism (SRAP), a new marker system based on a simple PCR reaction: Its application to mapping and gene tagging in *Brassica*. *Theoretical and Applied Genetics* 103: 455-461.

24. Li, H., Ruan, C.J. and Teixeira da Silva, J.A. 2009. Identification and genetic relationship based on ISSR analysis in a germplasm collection of seabuckthorn (*Hippophae* L.) from China and other countries. *Scientia horticulturae* 123: 263-271.

25. Li, H., Ruan, C.J., Teixeira da Silva, J.A. and Liu, B.Q. 2010. Associations of SRAP markers with dried-shrink disease resistance in a germplasm collection of seabuckthorn (*Hippophae* L.). *Genome* 53: 447-457.

26. Lian, Y.S., Chen, X.L. and Lian, H. 1998. Systematic classification of the genus *Hippophae* L. *Seabuckthorn Research* 1: 13-23.

27. Liu, S.W. and He, T.N. 1978. The genus *Hippophae* L. from Quing-Zang plateau. *Acta Phytotaxonomica Sinica* 16: 106-108.

28. Meyer, W., Mitchell, T.G., Freedman, E.Z. and Vilgays, R. 1993. Hybridization probes for conventional DNA fingerprinting used as single primers in the polymerase chain reaction to distinguish strains of *Cryptococcus neoformans*. *Journal of Clinical Microbiology* 31: 2274-2280.

29. Mishra, K.P., Chanda, S., Karan, D., Ganju, L. and Sawhney, R.C. 2008. Effect of seabuckthorn (*Hippophae rhamnoides*) flavone on immune system: An *in vitro* approach. *Phytotherapy Research* 22: 1490-1495.

30. Paran, I. and Michelmore, R.W. 1993. Development of reliable PCR-based markers linked to downy mildew resistance genes in lettuce. *Theoretical and Applied Genetics* 85: 985-993.

31. Persson, H. A. and Nybom, H. 1998. Genetic sex determination and RAPD marker segregation in the dioecious species seabuckthorn (*Hippophae rhamnoides* L.). *Hereditas* 129: 45-51.

32. Pimia, R.P., Nohynek, L., Schmidlin, S.H., Kahkonen, M., Heinonen, M., Riihinen, K.M. and Caldentey, K.M.O. 2005. Berry phenolics selectively inhibit the growth of intestinal pathogens. *Journal of Applied Microbiology* 98: 991-1000.

33. Raina, S.N., Jain, S., Sehgal, D., Kumar, A., Dar, T.H., Bhat, V., Vaishnavi, S., Bhargav, A., Singh, V., Rani, V., Tandon, R., Tewari, M. and Mahmoudi, A. 2012. Diversity and relationships of multipurpose seabuckthorn (*Hippophae* L.) germplasm from the Indian Himalayas as assessed by AFLP and SAMPL markers. *Genetic Resources and Crop Evolution* 59: 1033-1053.

34. Rassmann, K., Schlotterer, C. and Tautz, D. 1991. Isolation of simple sequence loci for use in polymerase chain reaction-based DNA fingerprinting. *Electrophoresis* 12: 113-118.

35. Rongsen, L. 1997. Eco-geographical distribution of seabuckthorn and prospects of International cooperation. In: *Worldwide Research and Development of Seabuckthorn*, (S. Lu, M. Li, J. Hu and S. Liu, Eds.), p.11-22, China Science and Technology Press, Beijing, China.

36. Rousi, A. 1971. The genus *Hippophae* L.-A taxonomic study. *Annales Botanici Fennici* 8: 177-227.

37. Ruan, C.J. 2006. Genetic relationships among some seabuckthorn cultivars from China, Russia and Mongolia using AFLP markers. *Journal of Horticultural Science and Biotechnology* 81: 409-414.

38. Ruan, C.J. and Li, D.Q. 2005. AFLP Fingerprinting analysis of some cultivated varieties of seabuckthorn (*Hippophae rhamnoides*). *Journal of Genetics* 83: 311-316.

39. Ruan, C.J., Li, H. and Mopper, S. 2009. Characterization and identification of ISSR markers associated with resistance to dried-shrink disease in seabuckthorn. *Molecular Breeding* 24: 255-268.

40. Ruan, C.J., Rumpunen, K. and Nybom, H. 2012. Advances in improvement of quality and resistance in a multipurpose crop: Seabuckthorn. *Critical Reviews in Biotechnology, doi: 10.3109/07388551.2012.676024.*

41. Servettaz, C. 1909. Monographie des Eleagnacees. *Beihefte zum Botanischen Centralblatt* 25: 1-40.

42. Shah, A.H., Ahmad, S.D., Khaliq, I., Batool, F., Hassan, L. and Pearce, S.R. 2009. Evaluation of phylogenetic relationship among seabuckthorn (*Hippophae rhamnoides* spp. *turkestanica*) wild ecotypes from Pakistan using amplified fragment length polymorphism (AFLP). *Pakistan Journal of Botany* 41: 2419-2426.

43. Sharma, A., Zinta, G., Rana, S. and Shirko, P. 2010. Molecular identification of sex in *Hippophae rhamnoides* L. using isozyme and RAPD markers. *Forestry Studies in China* 12: 62-66.

44. Sharma, P.C., Jain, A. and Chaudhary, S. 2012. Transcriptome analysis in seabuckthorn (*Hippophae rhamnoides* L.), a medicinally important plant. In: *Proceedings of International Conference on Environmental and Biological Sciences*, p. 21-24, Bangkok.

45. Shchapavn, S. 1979. On the karyology *of Hippophae rhamnoides* L. *Tsitologiika Genetika* 13: 45-47.

46. Sheng, H.M., An, L.Z., Chen, T., Xu, S.J., Liu, G.X., Zheng, X.L., Pu, L.L., Liu, Y.J. and Lian, Y.S. 2006. Analysis of genetic diversity and relationships among and within species of *Hippophae* (*Elaeagnaceae*) based on RAPD markers. *Plant Systematics and Evolution* 260: 25-37.

47. Shukla, S.K., Chaudhary, P., Kumar, I.P., Samanta, N., Afrin, F., Gupta, M.L., Sharma, U.K., Sinha, A.K., Sharma, Y.K. and Sharma, R.K. 2006. Protection from radiation induced mitochondrial and genomic DNA damage by an extract of *Hippophae rhamnoides*. *Environmental Mutagen Society* 47: 647-656.

48. Sun, K., Chen, X.L., Ma, R.J., Li, C., Wang, Q. and Ge, S. 2002. Molecular phylogenetics of *Hippophae* L. (*Elaeagnaceae*) based on the internal transcribed spacer (ITS) sequences of nrDNA. *Plant Systematics and Evolution* 235: 121-134.

49. Sun, K., Ma, R., Chen, X., Li, C. and Ge, S. 2003. Hybrid origin of the diploid species *Hippophae goniocarpa* evidenced by the internal transcribed spacers (ITS) of nuclear rDNA. *Belgium Journal of Botany* 136: 91-96.

50. Swenson, U. and Bartish, I.V. 2002. Taxonomic synopsis of *Hippophae* (*Elaeagnaceae*). *Nordian Journal of Botany* 22: 369-374.

51. Tautz, D. 1989. Hypervariability of simple sequences as a general source for polymorphic DNA markers. *Nucleic Acids Research* 17: 6463-6471.

52. Teng, B.S., Lu, Y.H., Wang, Z.T., Tao, X.Y. and Wei, D.Z. 2006. *In vitro* anti-tumor activity of isorhamnetin isolated from *Hippophae rhamnoides* L. against BEL-7402 cells. *Pharmacological Research* 54: 186-194.

53. Thiel, T., Michalek, W., Varshney, R.K. and Graner, A. 2003. Exploiting EST databases for the development of cDNA derived microsatellite markers in barley (*Hordeum vulgare* L.). *Theoretical and Applied Genetics* 106: 411-422.

54. Tian, C., Lei, Y., Shi, S., Nan, P., Chen, J. and Zhong, Y. 2004a. Genetic diversity of seabuckthorn (*Hippophae rhamnoides*) populations in northeastern and northwestern China as revealed by ISSR markers. *New Forests* 27: 229-237.

55. Tian, C., Nan, P., Shi, S., Chen, J. and Zhong, Y. 2004b. Molecular genetic variation in Chinese populations of three subspecies of *Hippophae rhamnoides*. *Biochemical Genetics* 42: 259-267.

56. Trajkovski, V. and Jeppsson, N. 1999. Domestication of seabuckthorn. *Botanica Lithuanica Suppl*. 2: 37-46.

57. Vos, P., Hogers, R., Bleeker, M., Reijans, M., Lee, T.V.D., Hornes, M., Frijters, A., Pot, J., Peleman, J., Kuiper, M. and Zabeau, M. 1995. AFLP-A new technique for DNA fingerprinting. *Nucleic Acids Research* 23: 4407-4414.

58. Vrieling, K., Peters, J. and Sandbrink H. 1997. Amplified fragment length polymorphisms (AFLPs) detected with non-radioactive digoxigenine labelled primers in three plant species. *Plant Molecular Biology Reporter* 15: 255-262.

59. Wang, A., Zhang, Q., Wan, D. and Liu, J. 2008. Nine microsatellite DNA primers for *Hippophae rhamnoides* ssp. *sinensis* (*Elaeagnaceae*). *Conservation Genetics* 9: 969-971.

60. Wang, Z., Gerstein, M. and Snyder, M. 2009. RNA-Seq: A revolutionary tool for transcriptomics. *Nature Reviews Genetics* 10: 57-63.

61. Williams, J.G.K., Kubelik, A.R., Livak, K.J., Rafalski, J.A. and Tingey, S.V. 1990. DNA polymorphisms amplified by arbitrary primers are useful as genetic markers. *Nucleic Acids Research* 18: 6531-6535.

62. Witsenboer, H., Michelmore, R.W. and Vogel, J. 1997. Identification, genetic localization, and allelic diversity of selectively amplified microsatellite polymorphic loci in lettuce and wild relatives (*Lactuca* spp.). *Genome* 40: 923-936.

63. Yanlin, S. 2007. Exploitation of SSR marker of *Hippophae rhamnoides*. *Journal of Anhui Agricultural Sciences* 35: 45.

64. Zeb, A. 2004. Important therapeutic uses of seabuckthorn (*Hippophae*): A review. *Journal of Biological Science* 4: 687-693.

65. Zeb, A. 2006. Anticarcinogenic potential of lipids from *Hippophae*: Evidence from recent literature. *Asian Pacific Journal of Cancer and Prevention* 7: 32-35.

66. Zimmermann, P., Hirsch-Hoffmann, M., Hennig, L. and Gruissem, W. 2004. GENEVESTIGATOR. *Arabidopsis* microarray database and analysis toolbox. *Plant Physiology*: 2621-2632.

Chapter 14

Characterization and Micro-propagation of Some Wild Seabuckthorn Genotypes Selected from the Danube Delta, Romania

Liviu Adrian Vescan, Cristian Radu Sisea and Doru Pamfil

University of Agricultural Sciences and Veterinary Medicine Cluj-Napoca
Calea Manastur 3-5, 400372, Cluj-Napoca, Romania

SUMMARY

Studies on the selection of wild Romanian seabuckthorn were carried out for breeding purposes and for direct introduction in culture of the best genotypes. Thirty individuals with certain unique traits were selected and the most promising were kept under observation over a period of three years. Morphological and biometric characterization was done for features such as plant height, number of thorns, fruit colour and size, number of fruits per bud and productivity. Biochemical compounds of the fruits were analyzed (*e.g.* soluble solids, vitamin C, carotenoids, minerals, oil content and its composition of fatty acids and tocopherols). Some genotypes of great interest were introduced in a nursery and were also cultivated *in vitro* after complex micro-propagation protocols were developed. Axillary shoot propagation and leaf regeneration were successfully achieved. Acclimatization stage proved to be the most difficult one; however, the promising results were obtained regarding direct *ex vitro* rooting and acclimatization. Micro-propagation of seabuckthorn proved to be less efficient as compared to the conventional propagation methods and not suitable to be used on a large scale.

Keywords: *Seabuckthorn, Morphological and biochemical characterization, Wild genotypes, Micro-propagation, Danube Delta and Romania.*

INTRODUCTION

Seabuckthorn is a thorny, deciduous, dioecious and anemophilous berry-producing shrub of high economical value. The genus *Hippophae* is a very diverse one, comprising six species, of which

Seabuckthorn (*Hippophae* L.): *A Multipurpose Wonder Plant, Vol. 4: Emerging Trends in Research and Technologies* (Virendra Singh, *Editor-in-Chief*, 2014), p. 179–201, Daya Publishing House, New Delhi, India.

the most common is *H. rhamnoides* (Lian *et al.*, 2003). Within this species, large populations of ssp. *carpatica* can usually be found in Romania, from the Black Sea shore in the Danube Delta to Sub-Carpathian region (Rati and Rati, 2003).

Because of its large spectrum of uses, from ecological and ornamental to nutritional and therapeutic, seabuckthorn is a multi-purpose species of great importance. Raw materials and fruit-, leaves-, and bark-derived products, generate an entire industry based on this species (Dwivedi and Singh, 2003). Seabuckthorn berries can be undoubtedly ranked among the most valuable soft fruits due to their high content of nutritive and active substances, such as minerals, vitamins, fatty acids and proteins (Rongsen, 2006; Yang and Kallio, 2006a). The large quantities of vitamin E and carotenoids found in seabuckthorn oil, along with vitamin C in berries, have strong antioxidant properties (Gao, 2006). Seabuckthorn oil has several pharmacological properties: diminishing inflammations, relieving pain, reducing the toxic effect of traditional drugs, improving the cardiovascular conditions, protecting against cancer and diabetes (Rongsen, 1992; Brad *et al.*, 2002; Singh, 2006; Yang and Kallio, 2006b).

In order to obtain true to type seabuckthorn planting material, different vegetative propagation methods can be used. Conventional propagation techniques using softwood and hardwood cuttings are successfully used for all subspecies, including *carpatica* (Rati and Rati, 2003; Singh and Gupta, 2003; Lebeda, 2008). The use of micro-propagation is recommended when high amounts of planting material are required starting from only one elite individual, as in our case. However, many problems, such as hyperhydricity, low proliferation and acclimatization rates, were encountered in seabuckthorn micro-propagation (Montpetit and Lalonde, 1988; Lummerding, 2001), only few positive results being recorded (Singh and Gupta, 2008; Sriskandarajah and Lundquist, 2009). No complex study was previously developed regarding the micro-propagation of the Romanian seabuckthorn. To our knowledge, no seabuckthorn plantation was ever established using micro-propagated planting material.

Even if many varieties of Romanian seabuckthorn are already homologated, most of them cannot compete with German or Russian cultivars in terms of productivity and biochemical properties. This, along with the market demand for seabuckthorn berries and their main products (the seed and pulp oil) encouraged us to select new high yielding genotypes with large fruits, few thorns and high nutritional value for introducing them directly in culture or in a breeding process.

A high variability was found in the Danube Delta between Romanian seabuckthorn populations and even between individuals of the same population, reason why the selection process was focused in this region. Genotypes with unique characteristics were selected based on the several morphological and biometric criteria (*e.g.* productivity, number of thorns, fruit colour and size). Several experiments were performed in order to determine the biochemical composition (*e.g.* vitamins, acidity, minerals, oil content and its composition). A nursery comprising the genotypes of great interest was created. They were further established *in vitro* and complex micro-propagation protocols were developed.

MATERIALS AND METHODS

EXPERIMENTAL MATERIALS

Biological material was selected from three different populations from Danube Delta region, located in Sulina (Black Sea shore), Cardon and Sfistofca (further referred to as M, C and S, followed by F for females and M for males genotypes; Figure 14.1). Basic selection was made with regards to productivity, fruit dimension, shape and colour, number of fruits per bud, number of thorns, height of the plants, pedicle length etc. Thirty individuals were selected, mainly from the Sfistofca population.

EXPERIMENTAL METHODS

Analysis of Biochemical Characters

Soluble Solids

Soluble solids (°Brix) of the fresh berry juice were determined using a portable refractometer, while the dry weight was obtained by drying the fruits at 60 °C.

Total Titratable Acidity

Total titratable acidity was performed by titration with NaOH (Marca, 1990).

Vitamin C Content

Vitamin C content in fruits was determined by titration with potassium iodate (Stanila *et al.*, 2001).

Mineral Composition in Dried Fruits

K, Ca, Mg, Zn and Fe were determined directly using an AA-6300 Shimadzu atomic absorption spectrophotometer. Phosphorous was determined using the method proposed by Marghitas and Balutiu (1996).

Total Carotenoids

Total carotenoids in fresh fruits were extracted with methanol:ethyl acetate:petroleum ether solution (1:1:1, v/v/v) according to Parlog *et al.* (2009). The total amount of carotenoids was determined using a Jasco V 530 spectophotometer.

Seed and Pulp Oil

Seed and pulp oil was extracted using chloroform:methanol solution (2:1, v/v).

Fatty Acids

The fatty acids were determined with gas chromatography.

Tocopherols

Tocopherols were determined by HPLC (Vescan *et al.*, 2010).

Micro-propagation

***In vitro* Culture Establishment**

More than 10 selected genotypes were established *in vitro*. Cuttings were prelevated directly from *in situ* plants or from the newly established nursery. Small cuttings containing 1-2 axillary dormant buds were surface sterilized for 20 minutes using 2 per cent sodium hypochlorite solution and then rinsed three times with distilled water. Meristems with leaf primordium were aseptically removed and inoculated in test tubes on Woody plant medium (WPM) media supplemented with 8 g/l plant agar, 30 g/l sucrose and several types of plant growth regulators (PGRs). The pH of the media was adjusted to 5.8 prior to autoclaving at 121°C for 20 minutes. Cultures were maintained in the growth chamber at 24 ± 1°C under 16 h photoperiod with a light intensity of 2000 lx provided by cool white fluorescent tubes.

Multiplication and Adventitious Shoot Regeneration

Several experiments regarding the influence of culture media, gelling agents and PGRs were performed to assess the proliferation capacity of different seabuckthorn genotypes. The experiments were focused on the axillary shoot multiplication in order to secure true to type plantlets. Proliferation rate, shoot length and hyperhydricity ratio were evaluated. The plantlets were sub-cultured for at least three times every 8 weeks. Due to many variants tested, high amounts of data were generated, reason why only representative examples will be presented in this paper.

Well developed leaves obtained *in vitro* were transferred onto regeneration media comprising several cytokinins, alone or in combination with auxins. The regeneration percentage was calculated after six weeks and the shoots with a length of at least 2 mm were counted. Randomly selected leaves with regenerated shoots were transferred to multiplication/elongation media for the next 8 weeks. Proliferation rate and shoot length were evaluated.

Rooting and Acclimatization

Shoots with a length of 15-20 mm were transferred to rooting media comprising several variants with or without auxins. The rooting capacity, number of roots and the vigour of the shoots were evaluated. Rooted and unrooted plantlets were transferred *ex vitro* for acclimatization in water, Jiffy-7 pots or in different combinations of sand, horticultural perlite and peat.

STATISTICAL ANALYSIS

Data were analyzed by non-parametric one-way or two-way ANOVA followed by specific comparison tests for mean separation. The software used was Prism 5 for the Windows.

RESULTS

MORPHOLOGICAL CHARACTERIZATION

A high variability was found for all the measured characteristics, especially within the Sfistofca population from which 20 individuals with certain unique characteristics were selected.

Plant Height

The female plants from Danube population were characterized by average height and vigour (Table 14.1), while some male individuals reached higher values than those reported in the literature, having vigour comparable to tree species (up to 7 m, Figure 14.1).

Table 14.1: Morphological Characteristics of Female Seabuckthorn Selections

n=23	*Maximum Height (m, ±SD)*	*No. of Thorns/ 30 cm*	*Leaf Dimension (l/l-cm)*	*Pedicle Length (mm)*	*Fruits per Bud*	*Fruit Length (mm)*	*Fruit Diameter (mm)*	*Weight of 100 Berries (g)*	*Weight of 1000 Seeds (g)*
Mean	2.41±0.71	7.8±3.2	6 × 0.7	3.1±0.8	4.7±1.2	10.7±1.2	8.4±0.7	46.3±11.4	22.4±3.8
Range (average)	1-4	0-15	4.5 × 0.4-8 × 0.9	2-5	3-7	8.8-13.2	6.8-9.7	23.1-73.1	14.5-29.9

Root Suckering

An average root suckering capacity of 3.2±1 was registered; 5 points were given to plants with no suckers and 1 point to individuals with more than 6. Individuals with higher suckering capacity were found at Black Sea shore, while lower values were recorded within the Sfistofca population.

Figure 14.1: Location of the Seabuckthorn Populations (A), Seabuckthorn Male Individual with Height above Average (B), Different Size, Shape and Colour of Berries (C and D) and Productive Genotype (SF7) (E).

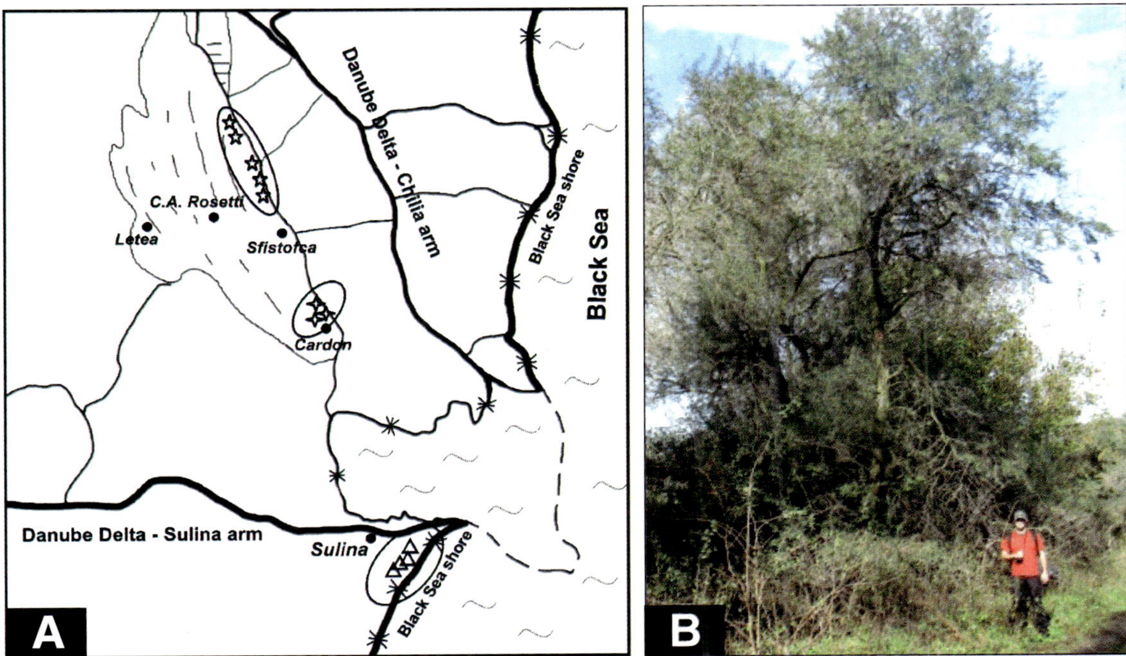

Contd...

Figure 14.1–*Contd...*

Number of Thorns

As shown in Table 14.1, an average of around 8 thorns can be found on every 30 cm of branch. The majority of thorns were very aggressive, strong and with multiple branches. Only a single female individual was found without any thorns. Clones derived from this are tested in other climate to see if the characteristic is maintained. Sexually propagated plants derived from the thornless genotype were also obtained and kept under surveillance.

Leaf Characteristics

The majority of plants had typical leaf aspect and rust leaf spots on abaxial side, as previously described for ssp. *carpatica* (Rati and Rati, 2003), with an average leaf surface of about 3.04 cm^2. Leaf dimensions are given in Table 14.1.

Fruit Productivity

Since the productivity was not the sole parameter desired for further breeding process, this characteristic greatly varied from hundred of grams up to more than 10 kg per plant. The genotypes tested for direct introduction in commercial plantations produced from 6.5 kg of fresh berries per plant (small vigour individuals) up to 12 kg/plant for higher vigour ones. A highly productive cultivar (*i.e.* SF7) is presented in Figure 14.1. In a plantation with a density of 2200 plants/ha, SF6 genotype has a fruit bearing potential of over 25 t/ha.

Fruit Colour and Shape

A high diversity in fruit colour and shape was observed for individuals from all the three Danube Delta populations (Figure 14.1). The predominant fruit colour is orange (3±1, where 1 point is given to light yellow colour and 5 to light red) and the basic fruit shape is oval. Even if no red fruits were found, rare individuals with dark orange/pale red fruit colour were encountered. Fruit shape varied from round to elongated oval or egg-shaped.

Fruit Size and Weight

Most of the selections had bigger and heavier fruits compared to wild genotypes found in other regions of Romania (Transylvanian selection-0.25 g/berry). Out of 24 selected female individuals, 9 had fruits weighting more than 0.5 g/berry. The largest individual fruit reached almost 1.4 g (SF14); however, only one genotype produced berries that averaged more than 0.7 g. The fruit weight and size varied a lot during the three picking years (2009-2011) (Table 14.1 and Figure 14.1).

Pedicle Length and Number of Fruits per Bud

The pedicle length proved to have direct influence over the ease of picking the berries and was negatively correlated to the number of fruits per bud (Table 14.1). Although an average of no more than 7 fruits per bud was recorded, rare clusters of up to 10 fruits per bud were seen in certain genotypes (*e.g.* CF5, SF4).

Weight of 1000 Seeds

Seabuckthorn commonly found in Romania, especially in the Transylvanian region, are characterized by very small seeds, of about 10 g/1000 seeds. On the other hand, genotypes with extremely large seeds (almost three times heavier) were found in the Danube Delta (Table 14.1). Their colouration differed from light brown to almost black, and also the shape from egg-shaped to elongated. The weight of the seeds was positively correlated to weight of the fruits (r=0.8***).

BIOCHEMICAL CHARACTERIZATION

Dry Weight and Soluble Solids

In the fresh fruits, lower total oil content was found, ranging from 2.2 to 3.3 per cent. Large differences were recorded between samples with regards to dry weight and soluble solids (Table 14.2, Figure 14.2). Between the two components, a significant positive correlation was found (r=0.77). The soluble solids content in fruit was also significantly influenced by the picking year (2010***>2009, p<0.0001, Two-way ANOVA, three genotypes tested).

Table 14.2: Biochemical Composition of the Seabuckthorn Fruits

	Dry Weight (per cent)	*Soluble Solids (°Brix)*	*pH*	*TTA (g Malic Acid/ 100 fruits)*	*Vitamin C (mg/100 g fruits)*	*Total carotenoids (mg/100 g fruits)*	*Oil Content (per cent in fruit)**
Mean (±SD)	21.3±2.1	10.9±1.4	3.06±0.08	2.93±0.59	83.3±68.1	10.93±4,42	6.4±1.5
Range	18.1-24.4	8.6-12.8	2.85-3.23	2.2-4.4	12.1-170.4	4.13-19.68	4.25-8.56
Samples (n)	19	15	13	15	6	14	5

* Using different extraction protocols (LAREX official standards).

Total Titratable Acidity (TTA) and pH

TTA had an average of 2.9 g malic acid/100 g of fresh fruit with large discrepancies between genotypes (Table 14.2), being also highly influenced by the picking year (Figure 14.2). The pH values of the seabuckthorn juice samples were around 3 for all the selections (Table 14.2).

Vitamin C Content

The content of vitamin C in the fresh fruits collected in September ranged from 12 to 170 mg per cent (Table 14.2 and Figure 14.2). The vitamin C content decreased very fast when fruits were kept at 4 °C. The best method of preserving high values of vitamin C in seabuckthorn fruits was by fast freezing them after picking.

Mineral Content

Seabuckthorn fruits proved to have high amounts of minerals, especially potassium (2200-18000 ppm). The mineral content in fruits can be seen in Figure 14.1. The mineral composition is similar to that of other conventional consumed fruits except for phosphorous, which has lower values (Table 14.3).

Total Carotenoids

Total carotenoids in fruits ranged from 4.1 to 19.7 mg/100 g of fresh fruit, with an average of 10.9 (52 per cent on dry weight basis). Carotenoid content was positively correlated with berry colour (r=0.81***) and the pH of the juice (r=0.73*) (Table 14.4).

Lipophilic Components

Oil Content

The solvent extraction yielded high amounts of oil from both seabuckthorn fresh pulp and seeds. The seeds produced an average of 10.7 per cent oil, the pulp 6.2 per cent and the whole fruit 6.4 per cent (Vescan *et al.*, 2010, Table 14.2).

Figure 14.2: Soluble Solids in Fresh Fruits (A), Mineral Composition of Dried Fruits (B), TTA Annual Differences (C, means with the same letter are not statistically different), and Vitamin C Content in Fresh Fruits of Seabuckthorn (D).

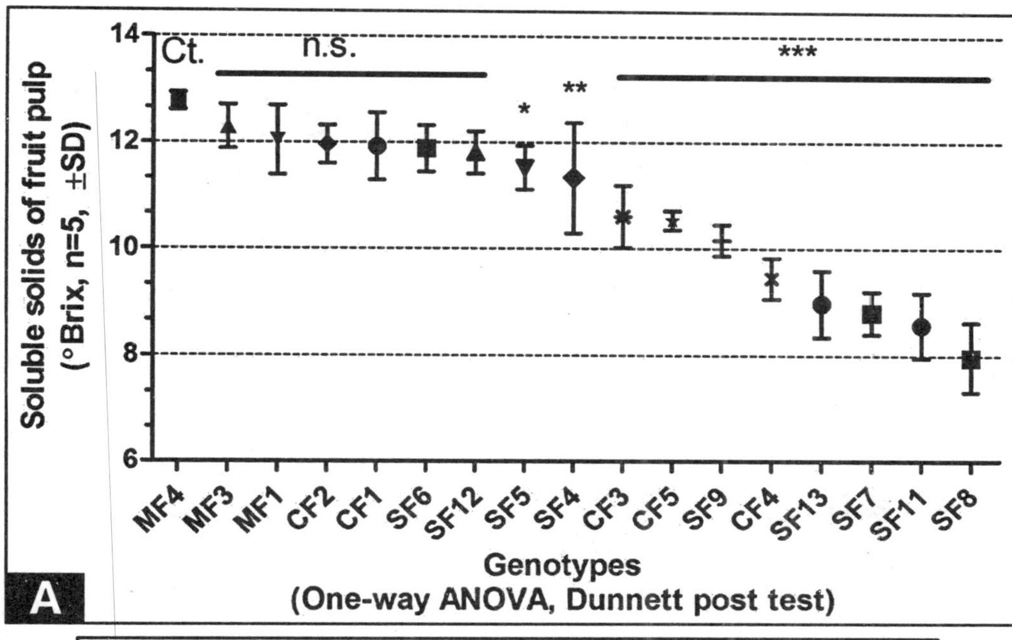

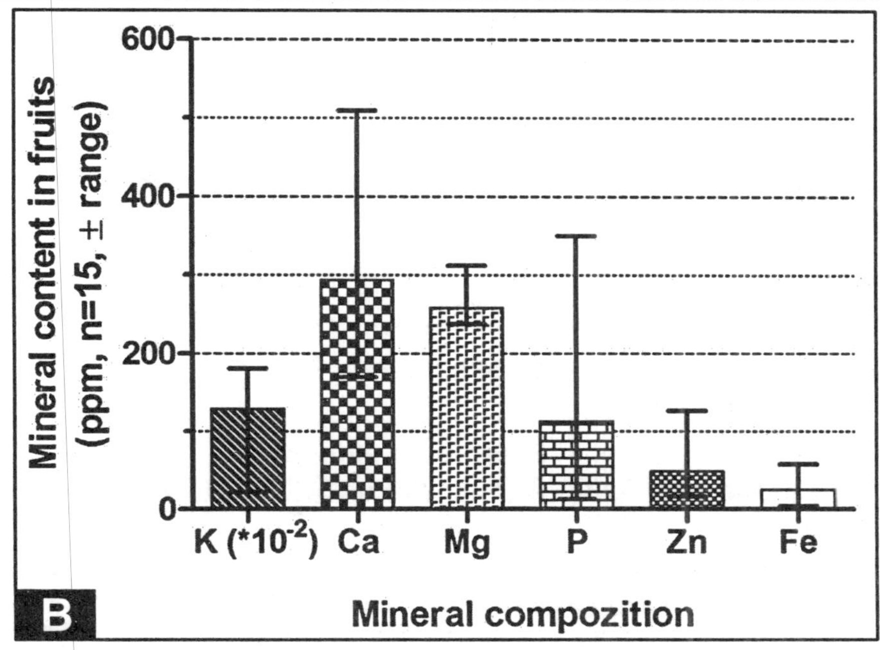

Contd...

Figure 14.2–Contd...

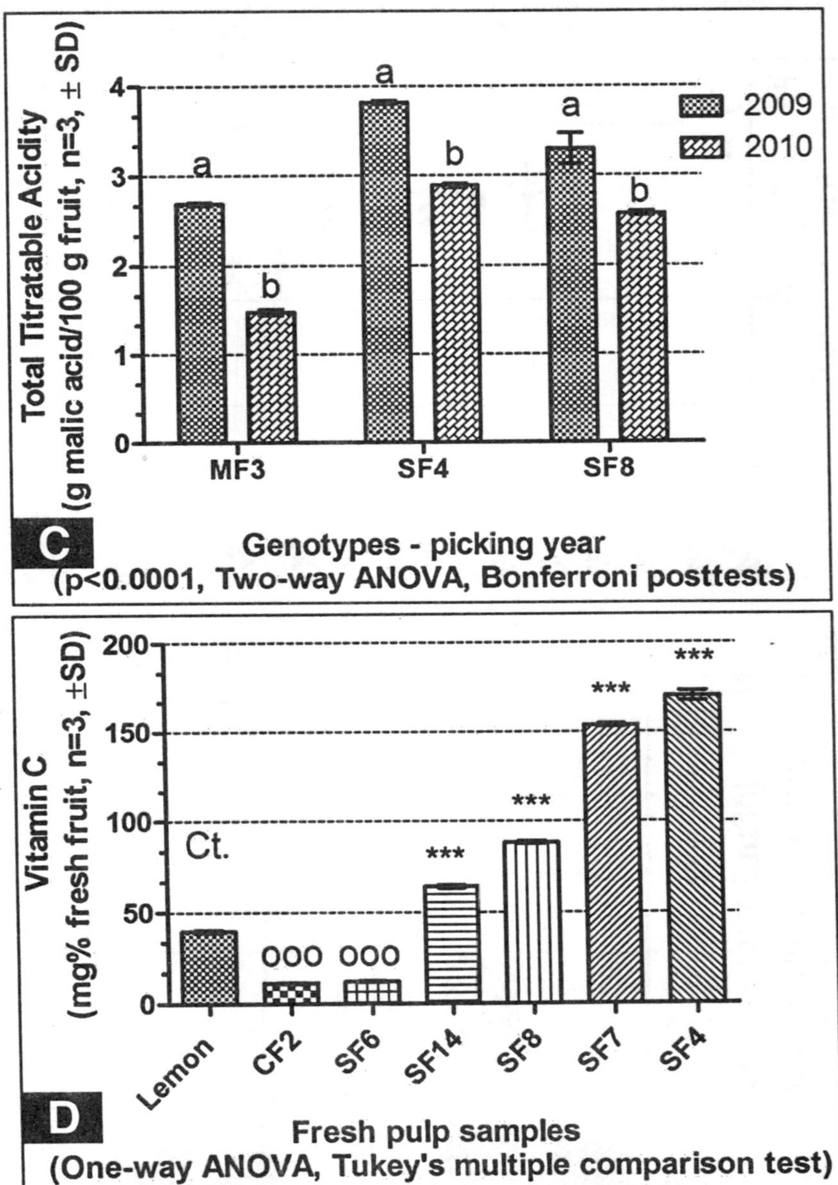

Fatty Acids Composition (FA)

In the seed oil, the polyunsaturated FA were dominant (67.1 per cent), while in the pulp oil, the saturated (37.6 per cent) and monounsaturated FA (57.9 per cent) (Figure 14.3). Linoleic (40.1 per cent), linolenic (26.3 per cent) and oleic acids (17.9) were dominant in the seed oil while in the pulp oil, the palmitic (36.55 per cent), oleic (30.9 per cent) and palmitoleic acids (20.45 per cent) (Vescan *et al.*, 2010).

Table 14.3: Mineral Content of Seabuckthorn Fruits Compared to Other Fruits

Elements (mg per cent Fresh Fruit)	Dry Weight (per cent)	K (mg per cent)	Ca (mg per cent)	Mg (mg per cent)	P (mg per cent)	Zn (mg per cent)	Fe (mg per cent)
Seabuckthorn	**21.3**	**271.6**	**6.2**	**5.4**	**2.4**	**0.39**	**0.20**
Apricots*	13.65	259	13	10	23	0.2	0.39
Bananas*	25.09	358	5	27	22	0.15	0.26
Apples*	14.44	107	6	5	11	0.04	0.12
Red currant*	16.05	275	33	13	44	0.23	1.00
Kiwi*	16.93	312	34	17	34	0.14	0.31
Grapes* (*Vitis vinifera*)	19.46	191	10	7	20	0.07	0.36

* Data from USDA databases (United States Department of Agriculture)-Agricultural Research Service.

Source: http://www.nal.usda.gov/fnic/foodcomp/search, 2011.

Table 14.4: Total Carotenoids in Seabuckthorn Fruits of different Subspecies

Subspecies (H. rhamnoides)	Total Carotenoids (mg per cent fresh fruit)	References
caucasica	0.32-16.7	Novruzov (2006)
turkestanica	0.8-28.8	
carpatica (Danube Delta)	0.32-3.87	
carpatica	10.93 (4.13-19.68)	Present study

Tocopherols

In the seed oil, both α and β+λ tocopherols were found in similar ratio with a content of up to 360 mg per cent oil, while in the pulp oil α-tocopherol was dominant (87.8 per cent). Total tocopherols counted up to 221 mg per cent in the pulp oil (Vescan *et al.*, 2010, Figure 14.3).

MICRO-PROPAGATION

Culture Establishment

The sterilization protocol generated high amounts of viable plantlets and low rates of infection (0-20 per cent), when meristems were used for the *in vitro* culture establishment. This protocol was found to be less suitable for buds and inefficient for cuttings. Best establishment media for most cultivars was ¾WPM (salts and vitamins) supplemented with BAP 0.15 mg/l (WG media, Figure 14.4). On an average, no statistical differences were noticed between culture media (Figure 14.4), so various BAP concentrations should be tested for new genotypes. The hyperhydricity was generally low on WG media, except for certain genotypes (*e.g.* SF8-in some repetitions up to 100 per cent). WPM supplemented with BAP and IBA 0.5:0.1 mg/l generated, in some cases, proliferation and a high number of adventitious basal buds. Some genotypes proved to be more recalcitrant to BAP and were successfully established *in vitro* by using zeatin. No viable shoots were regenerated from meristems using MS media.

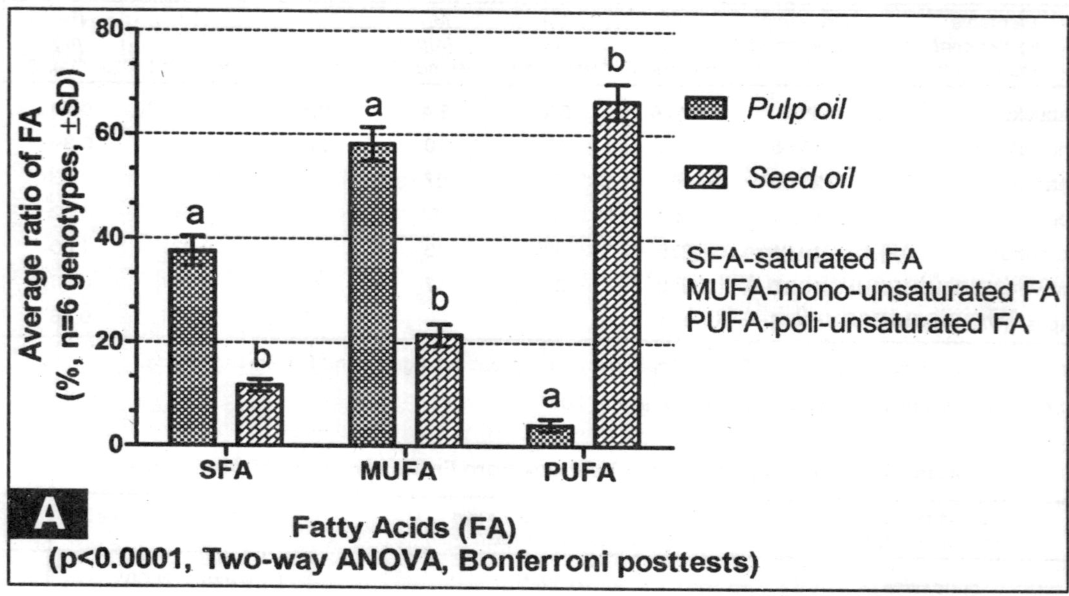

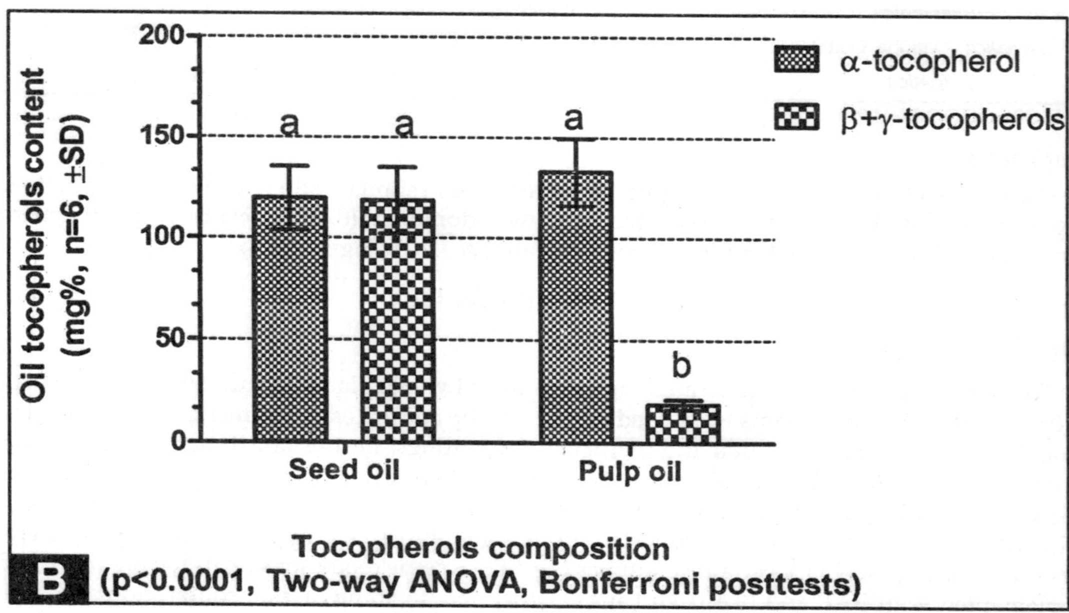

Figure 14.3: Seabuckthorn Seed and Pulp Oil Composition in Fatty Acids (A) and Tocopherols (B). Means with the same letter are not statistically different.

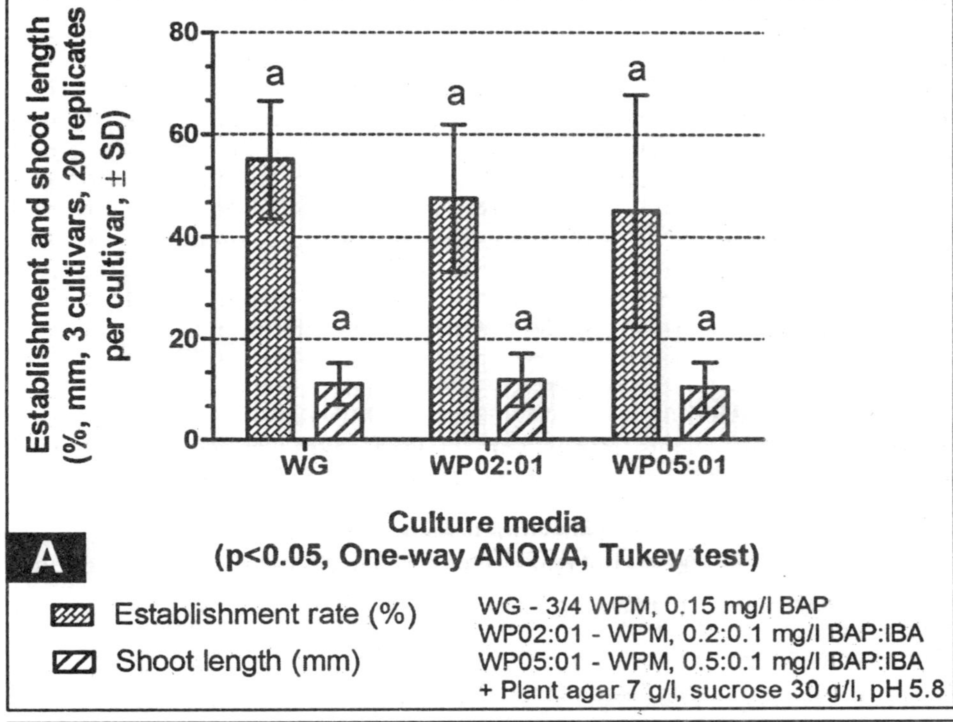

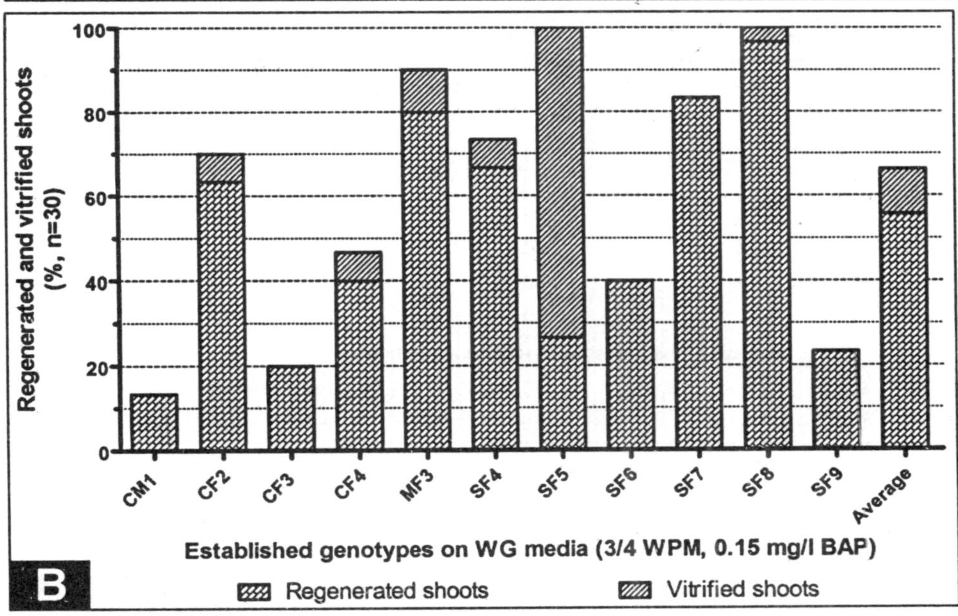

Figure 14.4: Culture Establishment Rates and Vigour on Several Media (A), and Regeneration Rates and Hyperhydricity on WG Media (B).

Multiplication Stage

Gelrite alone or in combination with plant agar generated high percentage of vitrification (Figure 14.5). Plant agar alone proved to be the most suitable gelling agent; however, even at 8 per cent doses some of the genotypes manifested strong hyperhydricity. For most of the cultivars, WPM supplemented with BAP and IBA (0.5:0.1 mg/l) generated the highest proliferation rate with an average of almost 3 shoots per initial explant (maximum 8 axillary shoots >2mm). The maximum axillary proliferation rate (26 shoots per initial explant) was obtained for SF6 genotype on WPM media supplemented with 0.5 mg/l BAP alone. Increased BAP concentrations (*e.g.* 1 mg/l) promoted high adventitious proliferation rate in some cases, but due to somaclonal risks, lower doses were preferred for further experiments. A very high phenotypic diversity was observed between different individuals with regards to vigour and leaves (Figure 14.5). None of the culture media promoted proliferation for some of the genotypes.

In vitro Leaf Regeneration

Regeneration of viable shoots from *in vitro* obtained leaves was achieved when using WPM supplemented with various PGRs concentrations and combinations (Table 14.5 and Figure 14.6). No positive results were obtained for direct regeneration when using MS or WPM media supplemented with various concentrations of TDZ (Thidiazuron) and Kinetin, alone or in combination with auxins (data not presented). Best results were obtained when WPM media was supplemented with BAP and AIB at a ratio of 0.5:0.1 mg/l, variant which also generated the longest shoots. Shoots obtained *via* direct regeneration from *in vitro* leaves were successfully proliferated and elongated (Figure 14.6) on the same regeneration media or in variants with lower BAP concentration (0.2 mg/l).

Table 14.5: Effect of different PGRs on Shoot Regeneration from *in vitro* Obtained Leaves

PGR (mg/l)				*No. of Explants (n)*	*Shoot Regeneration Rate (per cent)*	*Type of Regeneration**	*Callus Formation (per cent)*	*No. of Shoots per Explant (± range)*	*Shoot Length (mm, ± range)*
BAP	*IBA*	*IAA*	*NAA*						
0.5	–	–	–	10	80	0	0	3.4 (2-6)	4.9 (2-25)
0.5	–	–	0.1	10	40	0	100	1.75 (1-3)	4.6 (2-20)
0.5	–	0.1	–	20	70	0	0	3.4 (1-9)	4.2 (2-20)
0.5	0.1	–	–	40	90	0	0	3.3 (1-9)	5.0 (2-50)
1	0.1	–	–	10	80	0+1	100	4.4 (2-7)	4.8 (2-42)
Sub-cultivation of the leaves with regenerated shoots									
0.2	0.1	–	–	10		–	–	5[a] (1-10)	11.5[a] (2-45)
0.5	0.1	–	–	10		–	–	4[a] (1-8)	16.6[b] (2-44)

* 0: direct regeneration; 1: indirect regeneration (from callus).Values within each column followed by the same letter are not significantly different at the 5 per cent level by Bonferroni post-tests, Two-way ANOVA.

BAP: 6-Benzylaminopurine; IAA: Indole-3-acetic acid; IBA: Indole-3-butyric acid; NAA: α-Naphthaleneacetic acid; PGR: Plant growth regulator.

Rooting and Acclimatization

Spontaneous rooting occurred in almost all genotypes during the proliferation stage in about 20 per cent of the explants, even in variants with high amounts of BAP and no auxins. Roots were noticed

Figure 14.5: The Influence of Gelling Agents on Hyperhydricity (A), Proliferation and Shoot Length on different Proliferation Media (B), and Proliferation Stage Showing different Proliferation Rates and Phenotypic Differences (C, D, E and F).

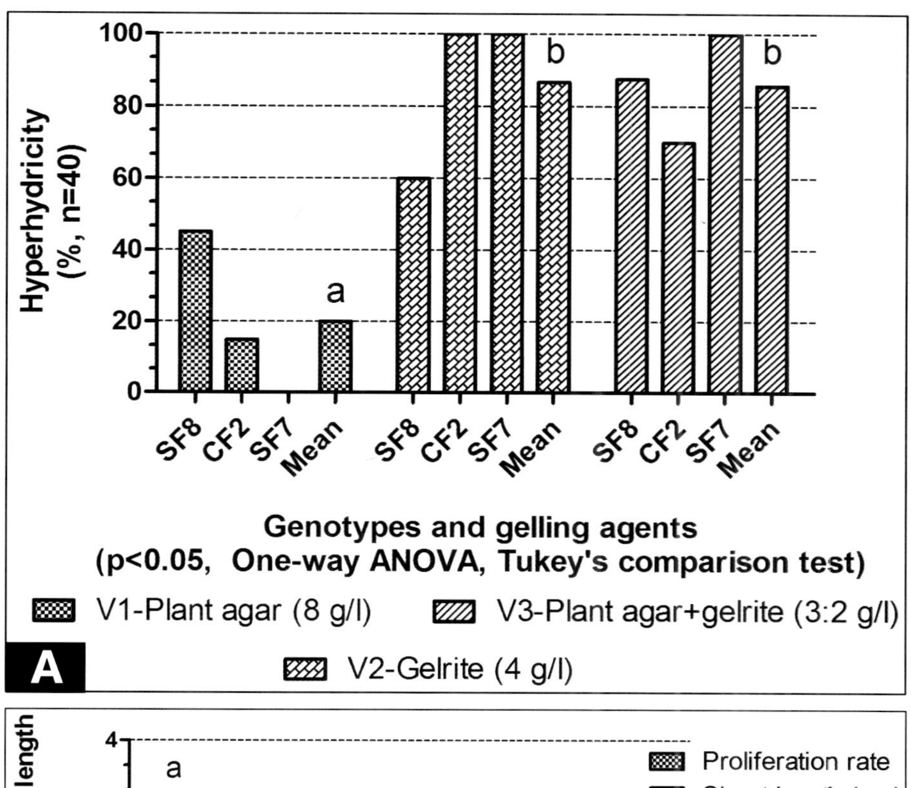

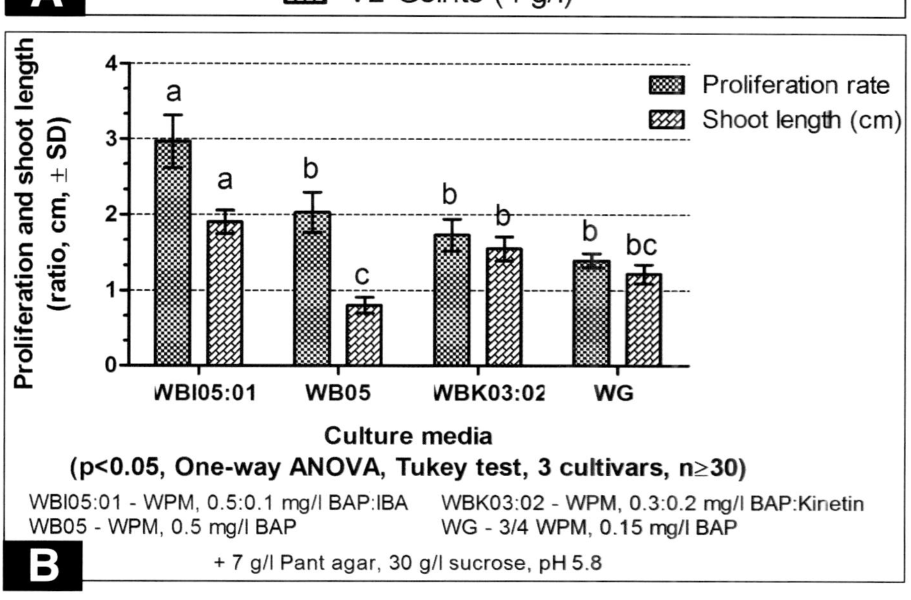

Contd...

Figure 14.5–*Contd...*

even in the shoot regeneration stage from leaves, at the bottom of the shoot clusters (Figure 14.6). Roots obtained in this stage, in general, were more vigourous as compared to those induced in the rooting stage (Figure 14.7). No more than 50 per cent of the plantlets rooted within the rooting stage and both the shoots and roots were not vigourous enough to be acclimatized. The roots were in general poorly

Figure 14.6: Direct Shoot Regeneration from *in vitro* Leaves (A and B) and Shoot Elongation (C and D).

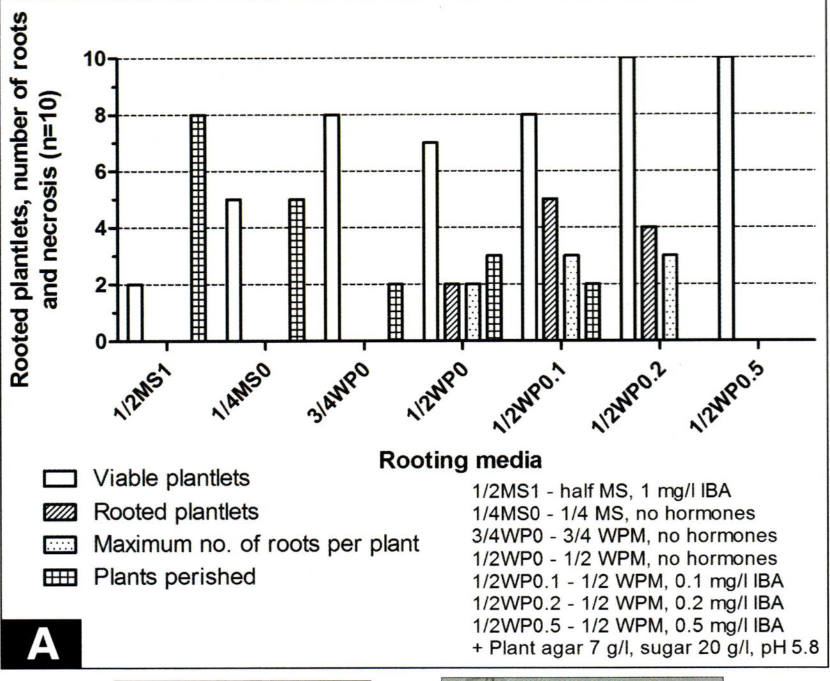

Figure 14.7: ***In vitro*** **Rooting Stage (A), Rooted Plantlet (B) and Acclimatized Plant in Jiffy-7 Pellet (C).**

attached to the shoots and in most of the cases, these were lost when washed to remove the plant agar. *Ex vitro* rooting in water occurred with high frequency when vigourous shoots with lignified base were used. Acclimatization in peat, perlite and sand occurred at extreme low rates. Best results were achieved when strong shoots were kept for three weeks in water and then moved in Jiffy-7 pellets under controlled environment for another 3 weeks.

DISCUSSION

Characteristics of the Romanian Seabuckthorn Selections

Romanian seabuckthorn found in the Danube Delta region is characterized by a great variability for both phenotypic and biochemical traits. Some of our selections have very big fruits, larger as compared to most Romanian and German cultivars (Albrecht, 2003; Rati and Rati, 2003). However, they are smaller than some wild Russian and Finnish genotypes or cultivated varieties (Korovina and Fefelov, 2003; Zubarev, 2008; Kaupinnen, 2010). Individual fruits from one variety reached more than 1.3 g *in situ*, reason why it has great potential for further breeding purposes. Even if more thornless varieties are reported in Russia and China (Huang, 2003; Singh, 2003), only one thornless female was found in our selection process. If almost 7 per cent of the Russian seabuckthorn cultivars found in the Altai region have red fruits (Korovina and Fefelov, 2003; Zubarev and Shelkovskaya, 2010), only individuals with dark orange fruits were found in our region. It seems that this characteristic may be enhanced only by crossing our subspecies with others. Some of the selections proved to be suitable for direct introduction in culture due to their very high productivity, similar to that of some of the best cultivars in Europe (Hornig *et al.*, 2010).

Our most promising selections prove to have a similar or even higher biochemical value as compared to Romanian varieties already approved for cultivation (Rati and Rati, 2003). Seabuckthorn berries were found to have a mineral content similar to other traditionally consumed fruits (Table 14.3). Even if seabuckthorn fruits are often associated with high content in vitamin C, concentrations of more than 150 mg per cent are not commonly found in our selections. The results are still comparable to those obtained for other European varieties (Zubarev and Shelkovskaya, 2010). Even if no individuals with red fruits were found, a high content of carotenoids was determined, resembling to that obtained for other subspecies (Novruzov, 2006).

A high content of seed and pulp oil was found, similar in composition to the oil extracted from the berries of other subspecies (Vescan *et al.*, 2010). Both fatty acids and tocopherols profiles can be easily used to determine possible frauds or the provenience of the oil (seed or pulp oil).

The best three selections based on morphological and biochemical criteria were already multiplied and are in the evaluation process of being licensed as new cultivars. All the genotypes showing potential for breeding purposes were propagated by seeds and another selection process within the resulted plants is being conducted.

Micro-propagation

Very low infection rates and high shoot viability can be obtained using axillary meristems tip culture along with a simple sterilization procedure. As compared to other disinfection protocols previously performed on Romanian seabuckthorn (Rati and Rati, 2003; Vantu, 2008), where high infection rates were reported, ours can be recommended as being fast and reliable for all ssp. *carpatica* genotypes. If culture establishment using entire buds is desired, more complex sterilization protocols (Singh and Gupta, 2008), are to be used for better results.

Good establishment rates were obtained on several culture media for the majority of the tested genotypes. Problems such as high phenolics accumulation in the culture media during culture establishment or proliferation (Singh and Gupta, 2008), did not affect our cultures. However, a major problem for some of the cultivars was the hyperhydricity level, which could not be eliminated even at very high concentrations of plant agar. Combinations of plant agar and Gelrite, as recommended by Sriskandarajah and Lundquist (2009) promoted, in our case, strong hyperhydricity even when less sensitive genotypes were used. High vitrification and mortality rates were obtained on MS, making this culture media not suitable for the micro-propagation of the Romanian seabuckthorn.

None of the variants tested generated high axillary proliferation rates. The majority of Romanian seabuckthorn genotypes proved to be very reticent with regards to axillary proliferation. The results were still comparable to those specified in the literature for other subspecies (Montpetit and Lalonde, 1988; Lummerding, 2001; Gupta and Singh, 2003). The averages did not exceed 3 shoots per initial explant. Better proliferation rates were obtained by Singh and Gupta (2008), when using higher concentrations of PGRs; however, in our case, the same combination (BAP plus IBA at 1:0.5 mg/l) produced high amounts of callus and very short adventitious shoots. High adventitious proliferation rates were also achieved when using zeatin (2 mg/l), but the results were very inconstant.

We successfully induced a high direct shoot regeneration rate from leaves obtained *in vitro* when using WPM supplemented with BAP with or without auxins. Zeatin supplemented with IAA (2:0.1 mg/l) also induced direct and indirect regeneration, but the plantlets were very small and difficult to elongate. Combinations of Kinetin and IAA, as recommended by Liu *et al.* (2007), did not induce regeneration or somatic embryos. Our protocol produced lower number of regenerated shoots as compared to protocols developed for other subspecies (Sriskandarajah and Lundquist, 2009), but it can be used as a viable alternative to axillary propagation.

Spontaneously rooted plantlets within multiplication stage developed stronger rooting system as compared to those rooted with auxins or in the absence of PGRs. When higher doses of auxins were used as suggested for other subspecies (Singh and Gupta, 2008; Sriskandarajah and Lundquist, 2009), the growth of the shoots was inhibited resulting in plantlets unsuitable for acclimatization. In our study, acclimatization of both rooted and unrooted plantlets was the most difficult micro-propagation stage. A good success rate was accomplished only when using very vigourous and semi lignified shoots, which are rather difficult to obtain.

Regardless of the plant hormones combinations, Romanian seabuckthorn proved to be very reticent to micro-propagation as compared to other berry fruit species, such as raspberries, blackberries or blueberries (Vescan *et al.*, 2012a; 2012b).

CONCLUSION

Romanian seabuckthorn is characterized by a high morphological and biochemical variability. Some of the Danube Delta selections had similar or even superior characteristics to Romanian or international varieties. Seabuckthorn fruits have high contents of oils, minerals, unsaturated fatty acids and vitamins, etc., which recommend this species as a complex food source. Most of the selections proved to have valuable traits useful for further breeding purposes, such as large fruits, no thorns and high fruit oil content. However, selection alone is not enough for characteristics like fruit colour, reason why crossings with other subspecies are to be considered. Due to excellent productivity and biochemical characteristics of the fruits, three elites were multiplied and are now in the evaluation trials for being approved as new cultivars (Golden Abundent, Carmen and Colosal).

Micro-propagation of Romanian seabuckthorn presents a high potential as a research tool, but due to low multiplication and acclimatization rates, is not suitable for commercial purposes.

ACKNOWLEDGEMENT

This work has benefitted from financial support through the 2008-2011 POSDRU/6/1.5/S/20 project, 'Doctoral Scholarships for 50 Ph.D. surdents and 2010 POSDRU/89/1.5/S52432 project, 'Organizing the National Interest Postdoctoral School of Applied Biotechnologies with Impact on Romanian Bio-economy', projects co-financed by the European Social Fund through the Sectoral Operational Programme Human Resources Development 2007-2013.

REFERENCES

1. Albrecht, H.J. 2003. Research on seabuckthorn (*Hippophae rhamnoides* L.) in Germany. In: *Seabuckthorn (Hippophae L.)-A Multipurpose Wonder Plant. Vol. I: Botany, Harvesting and Processing Technologies* (V. Singh, Ed. in Chief, 2003), p. 178-186, Indus Publishing Company, New Delhi, 518p.

2. Brad, I., Brad, I.L. and Radu, F. 2002. *Romanian Seabuckthorn-A Pharmacy within One Plant*, p. 112-129, Ed. Tehnica, Bucuresti, 180p.

3. Dwivedi, S.K. and Singh, B. 2003. Research on seabuckthorn in Ladack. In: *Seabuckthorn (Hippophae L.)-A Multipurpose Wonder Plant. Vol. I: Botany, Harvesting and Processing Technologies* (V. Singh, Ed. in Chief, 2003), p. 499-503, Indus Publishing Company, New Delhi, 518p.

4. Gao, X. 2006. Anti-oxidants and anti-oxidative activity of seabuckthorn (*Hippophae* L.). In: *Seabuckthorn (Hippophae L.)-A Multipurpose Wonder Plant. Vol. II: Biochemistry and Pharmacology* (V. Singh, Ed. in Chief, 2006), p. 390-401, Daya Publishing House, New Delhi, 600p.

5. Gupta, R.K. and Singh, V. 2003. Studies on micro-propagation in seabuckthorn (*Hippophae rhamnoides* L). In: *Seabuckthorn (Hippophae L.)-A Multipurpose Wonder Plant. Vol. I: Botany, Harvesting and Processing Technologies* (V. Singh, Ed. in Chief, 2003), p. 338-344, Indus Publishing Company, New Delhi, 518p.

6. Hornig, R., Hohne, F. and Jalakas, M. 2010. Results of a German-Estonian seabuckthorn cultivar trial. In: *EuroWorks Conference*, Conference Abstracts, p.17, Potsdam, Germany, 35p.

7. *http://www.nal.usda.gov/fnic/foodcomp/search*, 2011.

8. Huang, Q. 2003. Breeding of seabuckthorn (*Hippophae rhamnoides* L.) in China. In: *Seabuckthorn (Hippophae L.)-A Multipurpose Wonder Plant. Vol. I: Botany, Harvesting and Processing Technologies* (V. Singh, Ed. in Chief, 2003), p. 187-193, Indus Publishing Company, New Delhi, 518p.

9. Kaupinnen, S. 2010. Seabuckthorn cultivation in Finland, now and the future progress. In: *EuroWorks Conference*, Conference Abstracts, p.18, Potsdam, Germany, 35p.

10. Korovina, M.A. and Fefelov, V.A. 2003. Morphological variations in the fruits of seabuckthorn in different regions of CIS States. In: *Seabuckthorn (Hippophae L.)-A Multipurpose Wonder Plant. Vol. I: Botany, Harvesting and Processing Technologies* (V. Singh, Ed. in Chief, 2003), p. 72-83, Indus Publishing Company, New Delhi, 518p.

11. Lebeda, A.F. 2008. Propagation of seabuckthorn (*Hippophae rhamnoides* L.) in Ukrain. In: *Seabuckthorn (Hippophae L.)-A Multipurpose Wonder Plant. Vol. III: Advances in Research and Development* (V. Singh, Ed. in Chief, 2008), p. 22-26, Daya Publishing House, New Delhi, 566p.

12. Lian, Y., Chen, X.L. and Lian, H. 2003. Taxonomy of seabuckthorn (*Hippophae* L.). In: *Seabuckthorn (Hippophae L.)-A Multipurpose Wonder Plant. Vol. I: Botany, Harvesting and Processing Technologies* (V. Singh, Ed. in Chief, 2003), p. 35-46, Indus Publishing Company, New Delhi, 518p.

13. Liu, C.Q., Xia, X.L., Yin, V.L., Zhou, J.H. and Tang, H.R. 2007. Direct somatic embryogenesis from leaves, cotyledons and hypocotyls of *Hippophae rhamnoides* L. *Biologia Plantarum* 51 (4): 635-640.

14. Lummerding, P. 2001. Micro-propagation protocol development for seabuckthorn (*Hippophae rhamnoides* L.) selections for commercial orchard production. *Agri-Food Innovation Fund Project #19980162, Final Report, Canada,* 39p.

15. Marca, G. 1990. *Technology of Horticultural Products-Practical Guide for University*, p. 84-94, Tipo Agraria, Cluj-Napoca, 131p. (In Romanian)

16. Marghitas, M. and Balutiu, C. 1996. *Agrochemistry-Practical Guide for University*, p. 112-115, Tipo Agronomia, Cluj-Napoca, 316p. (In Romanian)

17. Montpetit, D. and Lalonde, M.1988. *In vitro* propagation of the actinorhizal *Hippophae rhamnoides* L. *Plant Cell, Tissue and Organ Culture* 15: 189-199.

18. Novruzov, E.N. 2006. Carotenoids and sterines of seabuckthorn (*Hippophae rhamnoides* L.). In: *Seabuckthorn (Hippophae L.)-A Multipurpose Wonder Plant. Vol. II: Biochemistry and Pharmacology* (V. Singh, Ed. in Chief, 2006), p.177-196, Daya Publishing House, New Delhi, 600p.

19. Parlog, R., Vodnar, D.C., Dulf, F.V., Leopold, L. and Socaciu, C. 2009. HPLC-PDA and UV-VIS spectrometry analysis used to fingerprint seabuckthorn (*Hippophae rhamnoides* L.) berries comparatively with leaves and seeds extracts. *Bulletin UASVM Agriculture* 66 (2): 409-414.

20. Rati, I.V. and Rati, L. 2003. *Romanian Seabuckthorn in Commercial Exploitation.* Anca, Urziceni, 90p. (In Romanian)

21. Rongsen, L. 1992. *Seabuckthorn: A Multipurpose Plant Species for Fragile Mountains*, p. 17-24, ICIMOD, Kathmandu, Nepal, 62p.

22. Rongsen, L. 2006. Biochemical characteristics of seabuckthorn (*Hippophae* L.). In: *Seabuckthorn (Hippophae L.)-A Multipurpose Wonder Plant. Vol. II: Biochemistry and Pharmacology* (V. Singh, Ed. in Chief, 2006), p. 98-107, Daya Publishing House, New Delhi, 600p.

23. Singh, V. 2003. Geographical adaptation and distribution of seabuckthorn (*Hippophae* L.) resources. In: *Seabuckthorn (Hippophae L.)-A Multipurpose Wonder Plant. Vol. I: Botany, Harvesting and Processing Technologies* (V. Singh, Ed. in Chief, 2003), p. 21-34, Indus Publishing Company, New Delhi, 518p.

24. Singh, V. 2006. Seabuckthorn (*Hippophae* L.) in traditional medicines. In: *Seabuckthorn (Hippophae L.)-A Multipurpose Wonder Plant. Vol. II: Biochemistry and Pharmacology* (V. Singh, Ed. in Chief, 2006), p. 505-521, Daya Publishing House, New Delhi, 600p.

25. Singh, V. and Gupta, R.K. 2003. Propagation of seabuckthorn (*Hippophae rhamnoides* L.). In: *Seabuckthorn (Hippophae L.)-A Multipurpose Wonder Plant. Vol. I: Botany, Harvesting and Processing Technologies* (V. Singh, Ed. in Chief, 2003), p. 315-333, Indus Publishing Company, New Delhi, 518p.

26. Singh, V. and Gupta, R.K. 2008. Micro-propagation of seabuckthorn (*Hippophae rhamnoides* L.). In: *Seabuckthorn (Hippophae L.)-A Multipurpose Wonder Plant. Vol. III: Advances in Research and Development* (V. Singh, Ed. in Chief, 2008), p. 3-13, Daya Publishing House, New Delhi, 566p.

27. Sriskandarajah, S. and Lundquist, P.O. 2009. High frequency shoot organogenesis and somatic embryogenesis in juvenile and adult tissues of seabuckthorn (*Hippophae rhamnoides* L.). *Plant Cell, Tissue and Organ Culture* 99: 259-268.

28. Stanila, A., Socaciu, C., Irinca, D.D. and Bobis, O. 2001. *Chemistry and Biochemistry-Practical Guide for University*. p. 152-153, Ed. Poliam, 168p. (In Romanian)

29. Vantu, S. 2008. Clonal multiplication of some varieties of *Hippophae rhamnoides* ssp. *carpatica*. In: *Biotehnologii vegetale pentru secolul XXI, Lucrarile celui de al XVI-lea Simpozion National de Culturi de Tesuturi si Celule Vegetale, Bucuresti 2007*, p. 69-75, *Ed. Risoprint, Cluj-Napoca.* (In Romanian)

30. Vescan, A., Pamfil, D., Bele, C., Matea, C. and Sisea, C.R. 2010. Some lipophilic components of five elite genotypes of Romanian seabuckthorn (*Hippophae rhamnoides* subs. *carpatica*). *Not. Bot. Hort. Agrobot. Cluj* 38 (2): 114-122.

31. Vescan, L.A., Clapa, D., Fira, A. and Pamfil, D. 2012a. Micro-propagation of cold resistant blackberry cultivar 'Gazda'. *Bulletin of UASMV Cluj-Napoca, Animal Science and Biotechnologies* 69 (1-2): 282-290.

32. Vescan, L.A., Pamfil, D., Clapa, D., Fira, A., Sisea, C.R., Pop, I.F., Berindean, I.V., Ciuzan, O. and Pop, R. 2012b. Efficient micro-propagation protocol for highbush blueberry (*Vaccinium corymbosum* L.) cv. 'Elliot'. *Romanian Biotechnological Letters* 17 (1): 6893-6902.

33. Yang, B. and Kallio, H. 2006a. Lipophilic components of seabuckthorn (*Hippophae rhamnoides* L.) seeds and berries. In: *Seabuckthorn (Hippophae L.)-A Multipurpose Wonder Plant.Vol. II: Biochemistry and Pharmacology* (V. Singh, Ed. in Chief, 2006), p.70-97, Daya Publishing House, New Delhi, 600p.

34. Yang, B. and Kallio, H. 2006b. Physiological effects of seabuckthorn (*Hippophae rhamnoides*) fruit, pulp and seed oils. In: *Seabuckthorn (Hippophae L.)-A Multipurpose Wonder Plant. Vol. II: Biochemistry and Pharmacology* (V. Singh, Ed. in Chief, 2006), p. 363-389, Daya Publishing House, New Delhi, 600p.

35. Zubarev, A.Y. and Shelkovskaya, N.K. 2010. Properties and promising ways of seabuckthorn utilization at Altai. In: *EuroWorks Conference*, Conference Abstracts, p.35, Potsdam, 35p.

36. Zubarev, Y.A. 2008. Commercial cultivation of seabuckthorn in Western Siberia, Russia. In: *Seabuckthorn (Hippophae L.)-A Multipurpose Wonder Plant. Vol. III: Advances in Research and Development* (V. Singh, Ed. in Chief, 2008), p. 49-60, Daya Publishing House, New Delhi, 566p.

Chapter 15

Proteomic Characterization of the Cold Traits in Seabuckthorn (*Hippophae rhamnoides* ssp. *turkestanica*)

Ravi Gupta and Renu Deswal

Molecular Plant Physiology and Proteomics Laboratory, Department of Botany, University of Delhi, Delhi, India

SUMMARY

Cold stress is a major abiotic stress that affects the crop productivity and plant distribution in a negative manner. Cold tolerant plants accumulate certain compounds during the process of cold acclimation. These compounds include sugars, sugar alcohols, amines and proteins. These compounds acts as osmoprotectants and prevent the cell from dehydration (osmotic stress) that is caused due to lowering of water potential of the cell after formation of ice in apoplast. Besides above changes occurring at intracellular level, cold/freeze tolerant plants secrete antifreeze proteins in their apoplast to arrest the growth of ice crystals. Antifreeze proteins bind to ice crystals and prevent their growth, a phenomenon known as ice recrystallization inhibition. As seabuckthorn is a cold hardy shrub that can tolerate freezing temperatures upto -40 °C, initiated the search for the antifreeze proteins. Seabuckthorn seeds were germinated in the laboratory conditions and total as well apoplastic proteins were isolated from control and cold acclimated seedlings. Identification of cold responsive proteins by MALDI-TOF and n-LC-MS/MS showed degradation of Rubisco in crude and presence of some putative antifreeze proteins in apoplastic extracts. Antifreeze activity was analysed using nanoliter osmometer with a phase contrast microscope and sucrose sandwich splat assay. The apoplastic extract showed cold induced antifreeze activity as observed by formation of hexagonal ice crystals and ice re-crystallization inhibition. The antifreeze proteins were enriched by ice adsorption chromatography. These results show that seabuckthorn secretes antifreeze proteins in its apoplast as a part of its freezing survival.

***Keywords**: Seabuckthorn, Cold stress, Apoplastome, Antifreeze proteins and Thermal hysteresis.*

Seabuckthorn (*Hippophae* L.): *A Multipurpose Wonder Plant, Vol. 4: Emerging Trends in Research and Technologies* (Virendra Singh, *Editor-in-Chief*, 2014), p. 202–211, Daya Publishing House, New Delhi, India.

ABBREVIATIONS

AFP: Antifreeze proteins; TH: Thermal hysteresis; CA: Cold acclimation; APF: Apoplastic fluid; IRI: Ice recrystallization inhibition; IAC: Ice adsorption chromatography.

INTRODUCTION

Seabuckthorn (*Hippophae* sp.) is a hardy, deciduous and nitrogen fixing shrub, belonging to the family *Elaeagnaceae*. It is serving as a measure of biodiversity conservation, soil conservation, medicines, food, fodder and fuel wood (Qinxiao and Hongyan, 2003). Seabuckthorn berry is a very rich source of vitamins and possesses a number of unique medicinal properties, which have a great potential to provide health-foods and a variety of medicines (Singh, 2006). There are 6 species, *i.e. H. rhamnoides, H. salicifolia, H. tibetana, H. goniocarpa, H. gyantsensis,* and *H. neurocarpa* and 12 subspecies of seabuckthorn (Singh, 2003). The distribution of seabuckthorn ranges from Himalayan regions including India, Nepal, Bhutan, Pakistan and Afghanistan, to China, Mongolia, Russia, Kazakhstan, Hungary, Romania, Switzerland, Germany, France and Britain, and northwards to Finland, Sweden and Norway (Zeb, 2004). More than 90 percent or about 1.5 million hectares of the world's seabuckthorn resources is found in China. In India, the plant is generally found in the higher Himalayan regions of Himachal Pradesh, Jammu and Kashmir, Northeast states and Uttarakhand. In Uttarakhand, it grows abundantly in the wild, mostly on river banks and sun-facing steep slopes between altitudes of 2000 and 3600 m asl. It is widely distributed in the cold desert of Ladakh region, covering an area of 11,500 ha in Leh district.

Biochemical characterization of seabuckthorn seed, berry, and fruit juice has been done for their nutritional and medicinal value including presence of ice nucleators in the fruit juice (Lundheim and Wahlberg, 1998; Abid *et al.*, 2007; Zeb and Malook, 2009). However, molecular biology studies of seabuckthorn are limited to genotyping of different species (Ercisli *et al.*, 2008).

Seabuckthorn has an extraordinary capacity to grow under adverse conditions like freezing temperatures, salinity, drought and UV radiations. Therefore, it could be a good source to understand the mechanisms of abiotic stress tolerance in plants. Proteomics and genomics study of seabuckthorn for their abiotic stress tolerance is an untouched area till date. Major bottlenecks for continuing research in this area induce the difficulty in germination of seabuckthorn seeds under laboratory conditions. It takes several months and need special presoaking treatments (Airi *et al.*, 2009). Moreover, the protein extraction procedures are also not available. Apoplastome analysis of seabuckthorn seedlings could give clues about its stress tolerance mechanisms as this is the region of the cell which is in direct vicinity of the atmosphere. As *H. rhamnoides* is the most abundant species in India, the work was initiated in it. In the present study, we describe a rapid seed germination procedure for seabuckthorn. Protein extraction procedures were optimized and proteins extracted from shoots and roots were also compared. A comparison was also made from proteins extracted from berry and leaves collected from wild. To monitor the effect of cold stress on apoplastome, apoplastic proteins were extracted from control, cold and -15 °C treated seedlings and resolved on SDS-PAGE. Polypeptides showing increased abundances after cold stress were identified by n-LC-MS/MS.

MATERIALS AND METHODS

Germination of Seabuckthorn Seeds

As germination of seabuckthorn seeds under laboratory conditions take several months and presoaking treatment, a simple and rapid procedure was optimized for germinating the seabuckthorn seeds. Seabuckthorn seeds were soaked in closed vials for five days and then plated in wet germination paper rolls. These rolls were transferred to B.O.D at 24±2 °C under 2000 *lux* light, after one day incubation in dark. Germination papers were changed twice a week as these were contaminated with the phenolics secreted by the seeds. After 20 days, seedlings were removed from the germination papers and the seedlings were used for the experimentation (Gupta and Deswal, 2012). For cold stress treatment, seedlings were exposed to 4 °C for one day and five days. For freezing stress treatment, cold stress treatment of 1 d and 5 d were followed by -5 °C treatment and one set of seedlings was given direct -5 °C.

Total and Apoplastic Protein Extraction

To monitor the effect of cold stress on seabuckthorn proteome, proteins were isolated from control, cold and freeze treated seedlings in Tris based buffer that contains 25 mM Tris (pH 7.0), 5 mM phenylmethylsulfonilfluoride (PMSF), 20 per cent Glycerol, 0.1 per cent β-mercaptoethanol and 0.5 µl of cocktail (protease inhibitor). Apoplastic proteins from control, cold and freeze treated seedlings was extracted using a vacuum infiltration method. Seabuckthorn seedlings were divided into shoots and roots, and the shoots were cut separately in segments of 1-2 cm. These segments were then repeatedly washed using deionized water to remove cytoplasmic proteins from the cut end. These segments were then kept in vacuum (260 mm Hg) with 20 mM ascorbic acid containing 20 mM of calcium chloride. After 30 min incubation in vacuum, these segments were placed in a syringe barrel. This syringe barrel was then kept inside a falcon tube and centrifuged at 4000 g (Beckman Coulter). After centrifugation, apoplastic proteins were collected from the falcon tube. Apoplastic proteins were then concentrated by acetone precipitation and quantified using Bradford's method (Bradford, 1976). The extracted proteins were resolved on SDS-PAGE (Laemmli, 1970).

Protein Identification by MALDI-ToF/ToF and n-LC-MS/MS

Protein identification was carried out at The Centre for Genomic Applications (TCGA), New Delhi, India, using Ultraflex MALDI-ToF/ToF (Bruker Daltonics, Germany) and Agilent 1100 series 2D Nano LC–MS/MS. The cold induced polypeptides were excised from the gel, destained and identified. The peak lists obtained were submitted to MASCOT (ver 2.1) search engine (*http:/ www.matrixsciences.com*) and searched against the NCBI nr database. The search parameters were as follows: Mass values-monoisotopic; protein mass- unrestricted; fixed modifications-carbamidomethylation; variable modification-methionine oxidation; peptide mass tolerance- ± 100 ppm (for PMF) and ±1.2 Da (for MS/MS); fragment mass tolerance-±0.6 Da (for MS/MS); peptide charge state-1+ (for PMF); maximum trypsin missed cleavage-1; and instrument type-ESI-TRAP (MS/ MS). Only significant hits, as identified by the MASCOT probability analysis ($p<0.05$) were accepted.

Antifreeze Activity Analysis

For detection of antifreeze activity, both IRI and ice crystal morphology were analyzed using sucrose sandwich splat assay and nanolitre osmometer at a protein concentration of 0.2 mg/ml (Gupta and Deswal, 2012). For IRI, proteins were solublized in 30 per cent sucrose in 20 mM ammonium bicarbonate (pH 8) and were sandwiched between two round coverslips. These coverslips were snap frozen at -80 °C and then transferred to a glass viewing chamber maintained at -6 °C. The ice crystals were allowed to anneal for 2 hrs and then photographed using Nikon eclipse 80i microscope with a DS-Fi1 camera (Nikon) and NIS-elements F-package (ver 3.0) software. For the quantification of the IRI activity, the diameters of the ice crystals in each image were measured using Image J software. TH and the shapes of ice crystals were analyzed using nanolitre osmometer. In brief, nanolitre volumes of the samples were loaded into the wells of the sample holder disc and kept on a freezing stage mounted on the stage of a phase-contrast microscope. Temperature was controlled by a nanoliter osmometer (Otago Osmometers). The samples were flash frozen at -20 °C to form a population of small ice crystals and then thawed until only a single ice crystal remained in the well. Temperature was gradually decreased, and the morphology of the ice crystals was photographed with the 20× objective.

RESULTS

Seabuckthorn seeds, grown under laboratory conditions were best at 20 days, therefore 20 days old seedlings were used for the experimentation. Silver stained SDS-PAGE gel of total protein extract isolated from control seedlings showed total of 26 polypeptides (Figure 15.1A, lane 2). No significant changes in the protein profile of control and 4 °C treated seedlings were observed (Figure 15.1A, lane 3). However, after freezing stress of -15 °C treatment, downregulation of two polypeptides of 80kDa and 42 kDa was observed. These proteins were identified by MALDI-ToF/MS-MS and were found to be heat shock protein and actin, respectively. Interestingly, RuBisCO degradation was observed at -15 °C suggesting inhibition of photosynthesis during subzero temperatures (Figure 15.1A, lane 4). As not much targets were identified using total protein extract, analysis of sub-proteome was done. For deep proteome, analysis of apoplastic proteins were carried out as this is the region of the cell which is more responsive to biotic/abiotic stress at proteome level.

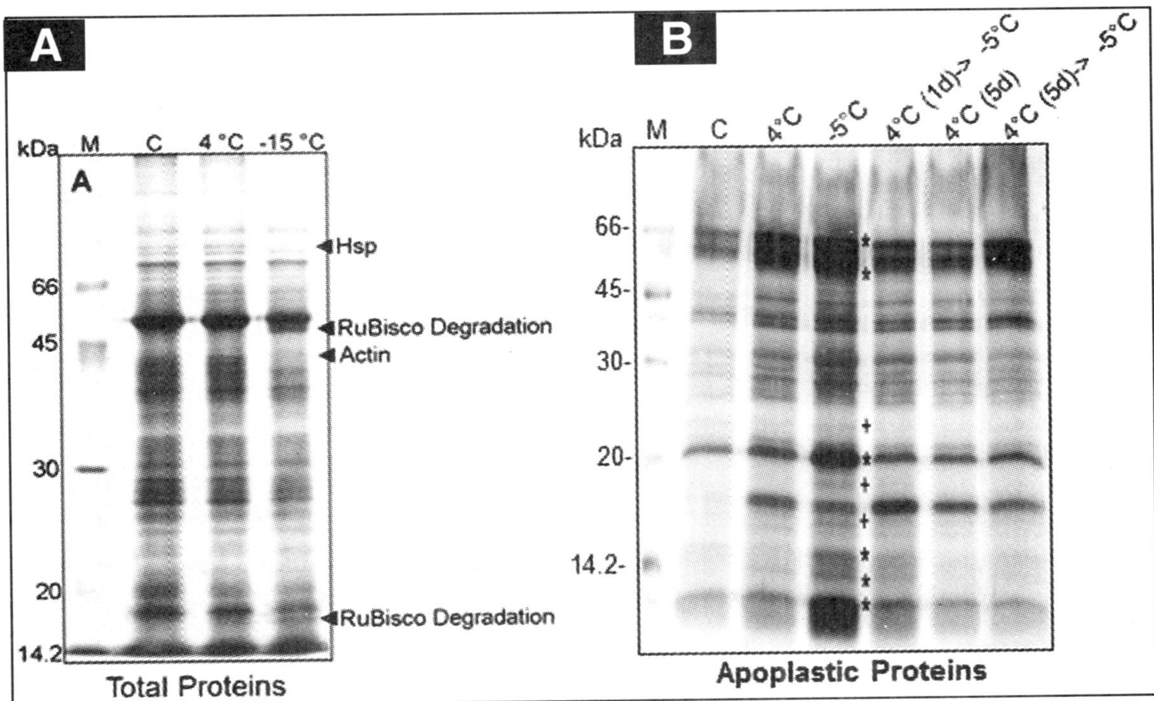

Figure 15.1: SDS-PAGE Profiles of Total (A) as well as Apoplastic Proteins (B) Isolated from Control and Cold Treated Seedlings.

After freezing stress treatment, downregulation of heat shock protein and actin was observed. In addition, degradation of RuBisCo was observed after freezing treatment. Proteins which showed accumulation after cold/freezing stress in apoplast are marked with asterisk (*).

Apoplastic proteins isolated were resolved on SDS-PAGE. Although, there was a generalized increment in the protein content, five major polypeptides of 66kDA, 27kDa, 25kDa, 15kDa and 10kDa were observed in cold/freeze treated seedlings (marked with asterisk) that showed higher abundances after cold treatment (Figure 15.1B, lane 3 and 4). These polypeptides were eluted from the gel and subjected to identification using n-LC-MS/MS. A total of 14 proteins were identified which included lipid transfer protein and chitinase as putative AFPs (Table 15.1).

Table 15.1: Identification of Cold/Freezing Stress Modulated Apoplastic Proteins by MALDI-ToF/ToF and n-LC-MS/MS

Name	*Accession No.*	*MW (Da)*	*MOWSE Score*	*Matched Peptide No.*	*Matched Peptide (ion score)*
*Lipid transfer protein	Q6IWH2	12481	375	23	CLVGAANSFPTLI (102)
Germination specific lipid transfer protein1	T07861	12419	313	22	GCCTGVTNLNNMI (55)
*Non-specific lipid transfer protein isoform II	Q9S9G0	9865	217	17	CLVGAANAFPTLI (62)
Napin BG9 precursor	PS0426	18189	79	2	IYQTATHLPR (65)
Thioredoxin PEC-2	T14379	13749	78	3	VDVDELATVAK (61)
Napin -3	2SS3	14482	76	6	IYQTATHLPK (51)
Allergen Sin a I	S65447	14627	66	4	VCNIPQVSVCPFI (54)
Glutaredoxin- likeprotein	Q9FNE2	11920	50	1	
Thioglucosidase	S26149	63266	93	21	GYAIGTDAPGR (56)
Cytosolic triose phosphateisomerase	T47683	27380	229	15	VAYALAQGLK (42)
*Chitinase	S25311	296384	105	5	AINGMECNGGNSGI (62)
Triose phosphate isomerase	A32187	27244	51	4	ALLNESNEFVGDK (51)
Ribulose bisphosphate carboxylase	RKRPS	20529	87	9	
Myrosinase-binding protein 2	T08145	67256	73	–	

* Proteins are putative AFPs.

The identified proteins were classified in 5 categories on the basis of their functional roles. Most of the identified proteins (29 per cent) were associated with defense suggesting significance of apoplastome analysis (Figure 15.2).

As putative antifreeze proteins were identified in apoplastome, antifreeze activity was analyzed to confirm their presence in seabuckthorn (Table 15.2). Antifreeze proteins (AFPs) are a class of polypeptides which bind to the ice crystals and modify their growth. In the presence of AFPs, ice crystals are hexagonal while in absence of AFPs, ice crystals are circular disc shaped. Therefore, by observing the shapes of ice crystals, presence or absence of AFPs can be detected. Also, antifreeze proteins depress the freezing point of the solution below its melting point, known as thermal hysteresis. Ice crystal grows as circular disc in apoplastic proteins isolated from control seedlings indicating absence of antifreeze activity. Interestingly, antifreeze activity accumulates in seabuckthorn apoplast after cold/freezing stress treatment as observed by formation of hexagonal ice crystal (Figure 15.3). When antifreeze activity was analyzed in 4 °C (1 d) treated seedlings, no subsequent enrichment in the antifreeze activity was observed. However, antifreeze activity accumulated when seedlings were

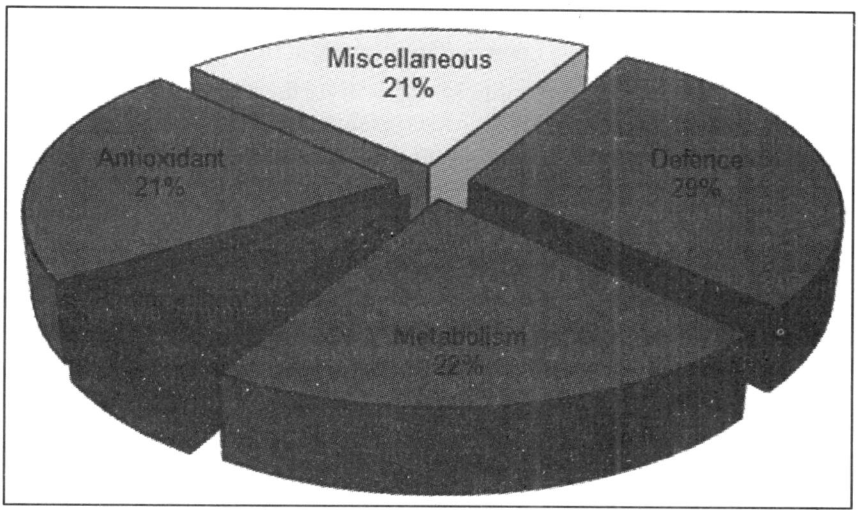

Figure 15.2: Functional Categories of the Identified Proteins. Majority of the identified proteins (29 per cent) belonged to the defense showing significance of apoplastome analysis.

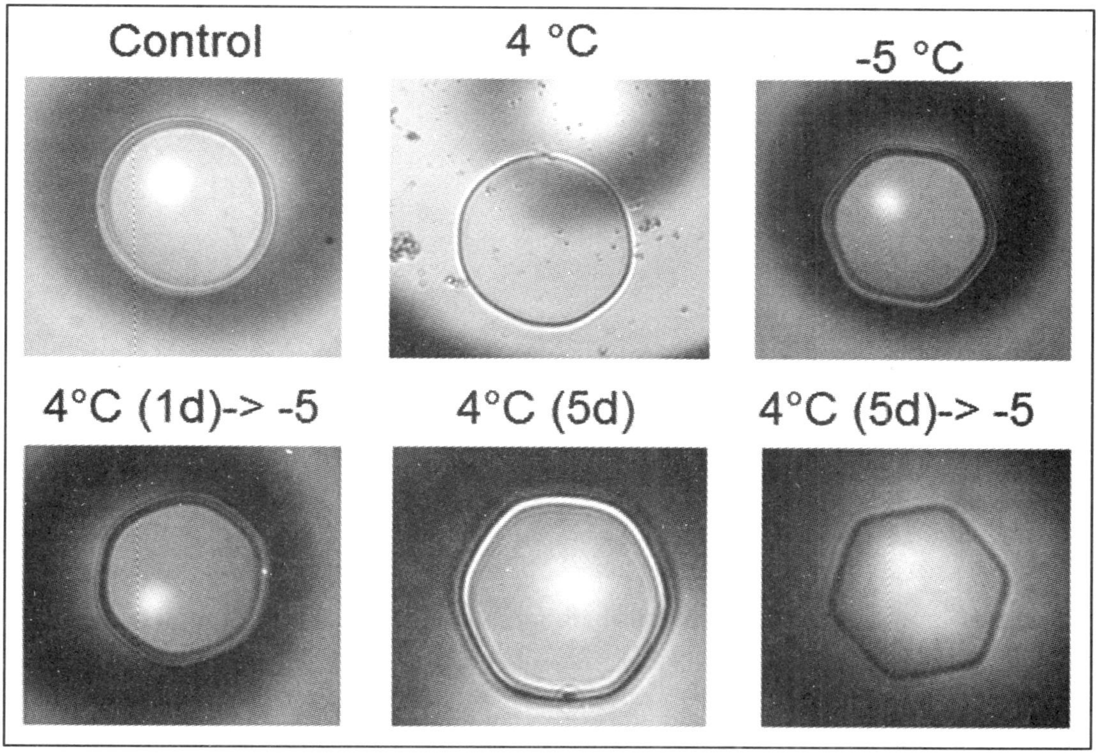

Control

4 °C

-5 °C

4°C (1d)-> -5

4°C (5d)

4°C (5d)-> -5

Figure 15.3: Ice Crystal Morphologies of Control and Cold Induced Secretome as Observed by Phase Contrast Microscopy and Nanoliter Osmometer.

subjected to direct -5 °C. A high level of antifreeze activity was observed when the seedlings were exposed to 4 °C (5 d) followed by -5 °C as observed by formation of proper hexagonal ice crystals.

The presence of antifreeze activity in seabuckthorn was further confirmed by ice recrystallization inhibition (IRI) assays (Figure 15.4). Ice forms due to crystallization of water molecules during subzero temperatures and it grows due to ice recrystallization, *i.e* larger molecules of ice grow at the expense of the smaller ones. These larger molecules of ice impose physical stress on the plasma membrane and may rupture it leading to death of the cell. AFPs inhibit this ice re-crystallization and keep a check on the size of ice crystals. Apoplastic proteins, isolated from 4 °C (1 d) and direct -5 °C treated seedlings showed 1.2 and 1.5 fold ice recrystallization inhibition activities, respectively (Figure 15.4). Similar to results obtained from ice crystal morphology, IRI activity also showed an increase up to 2.1 fold in 4 °C (5 d) followed by -5 °C treatment. These results clearly confirm presence of antifreeze proteins in seabuckthorn.

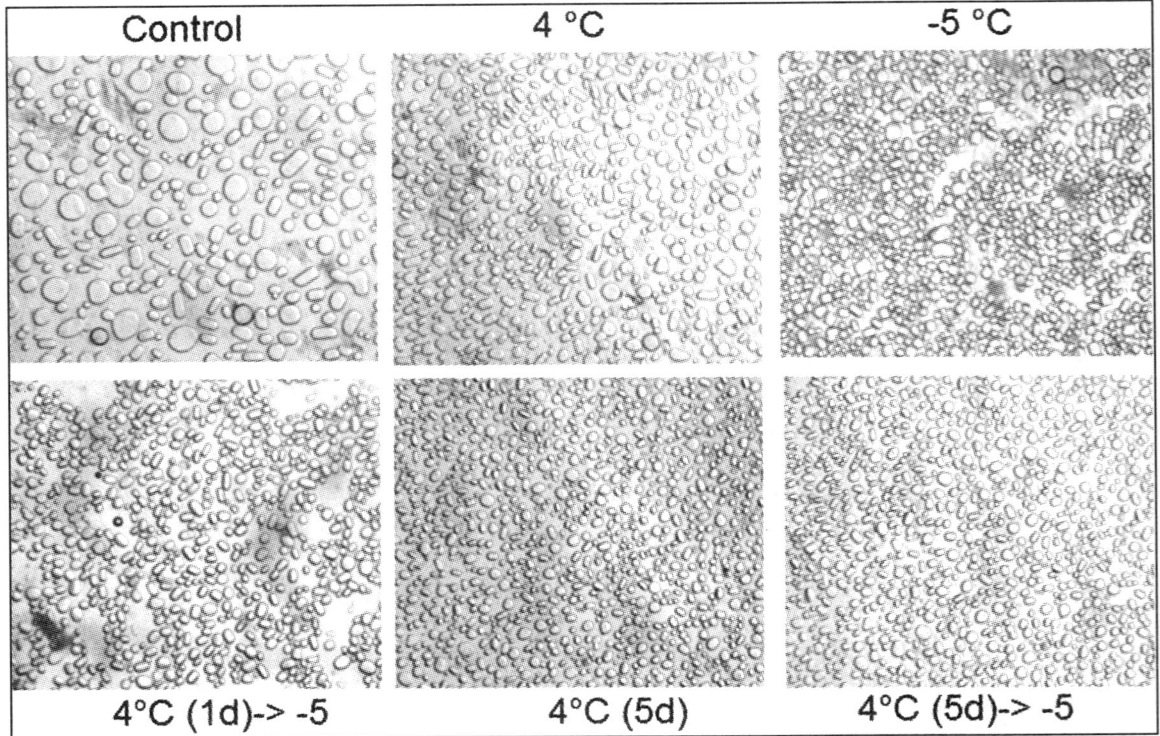

Figure 15.4: Ice Recrystallization Inhibition Observed by Sucrose Sandwich Splat Assay.

Ice crystals are larger in control while these are smaller in cold induced apoplastome showing presence of antifreeze proteins.

DISCUSSION

Overwintering plants secrete proteins in their apoplast to combat with the cold stress. These proteins are mainly involved in the signaling and defense. In addition to these, some cold tolerant plants secretes antifreeze proteins which binds to the ice crystal and inhibit their growth. Seabuckthorn

Table 15.2: A Catalogue of AFPs and their Properties in different Parts of Seabuckthorn

Name of AFPs	*Properties*				*Distribution*		
	Molecular Weight (kDa)	*TH*	*IRI*	*Ice Crystal Morphology*	*Apoplast*	*Leaf*	*Berry*
PGIP	41	0.19 °C	9 fold	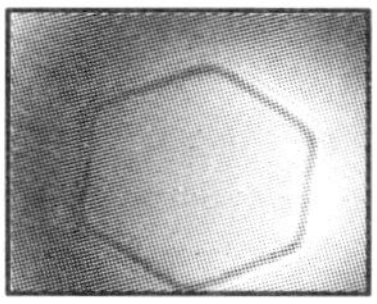	+	+	+
Chitinase	39	0.05 °C	3 fold	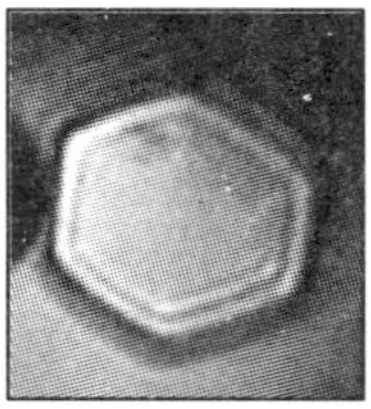	+	–	–
Chitinase	32.8	0.05 °C	2.5 fold	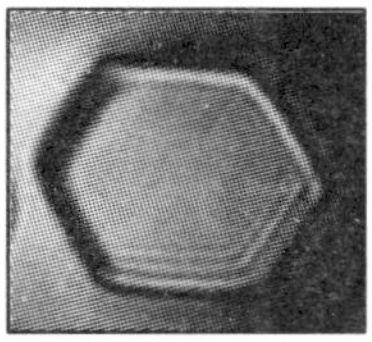	+	–	–
Desaturase	14	N.D	N.D	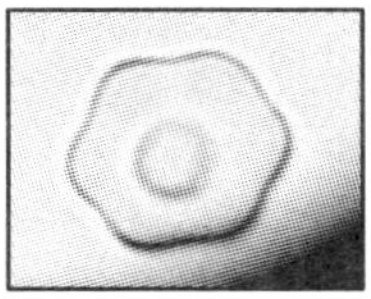	+	+	–

Plus (+) sign indicates the presence, while minus (–) sign indicates absence of AFP in the tissue analysed.

is a cold tolerant plant but whether it contains antifreeze activity or not is still unknown. To order to investigate the presence of antifreeze proteins in seabuckthorn, total as well as apoplastic proteins were isolated and resolved on SDS-PAGE. Rubisco being the most abundant plant protein, showed degradation after freezing stress. Degradation of Rubisco during cold stress has been earlier shown in rice (Yan *et al.*, 2006), spring wheat (Rinalducci *et al.*, 2011) and *Thellungiella* (Gao *et al.*, 2009). However, in apoplastic proteins Rubisco was identified as upregulated target which could be a contamination. Rubisco is absent in apoplast, however, due to membrane damage during cold/freezing stress, it is possible that Rubisco is leached out from the chloroplast to cytoplasm. The downregulation of actin was also previously reported in cold stress (Zhang *et al.*, 2012). Identification of cold induced apoplastic proteins showed presence of chitinase and lipid transfer protein. Both these proteins exhibit antifreeze activity during cold stress in winter rye (Hon *et al.*, 1995). Analysis of antifreeze activity in seabuckthorn showed that seabuckthorn exhibited cold induced antifreeze activity. Antifreeze activity has been shown in many overwintering plants including monocots like winter rye (Hon *et al.*, 1994), *Lolium perenne* (Pudney *et al.*, 2003), *Deschampsia antarctica* (John *et al.*, 2009) and some dicots like *Solonum dulcamara* (Duman, 1994), carrot (Worral *et al.*, 1998; Smallwood *et al.*, 1999), *Arachis hypogea* (Dave and Mitra, 1998).

CONCLUSION

These results of present study indicate that cold stress decreases photosynthesis in seabuckthorn as observed by the degradation of Rubisco. Besides this, the analysis of cold induced apoplastic proteins showed presence of PR-proteins suggesting their pivotal roles in cold stress tolerance. Cold/freeze treated seabuckthorn seedlings showed accumulation of antifreeze activity in the apoplast indicating that seabuckthorn secretes antifreeze proteins in its apoplast as a part of its freezing stress survival. A detailed analysis of antifreeze proteins is required to know the complete catalogue of these proteins in seabuckthorn. The most potent AFP would be exploited for its biotechnological applications.

REFERENCES

1. Abid, H., Hussain, A. and Ali, S. 2007. Physicochemical characteristics and fatty acid composition of seabuckthorn (*Hippophae rhamnoides* L.) oil. *Journal of Chemical Society of Pakistan* 29: 256-259.
2. Airi, S., Bhatt, I.D., Bhatt, A., Rawal, R.S. and Dhar, U. 2009. Variation in seed germination of *Hippophae salicifolia* with different pre-soaking treatments. *Journal of Forest Research* 20: 27-30.
3. Bradford, M. 1976. A rapid and sensitive method for the quantitation of microgram quantities of protein utilizing the principle of protein dye binding. *Analytical Biochemistry* 72: 248-254.
4. Dave, R.S. and Mitra, R.K. 1998. A low temperature induced apoplastic protein isolated from *Arachis hypogaea*. *Phytochemistry* 49: 2207-2213.
5. Duman, J.G. 1994. Purification and characterization of a thermal hysteresis protein from a plant, bittersweet nightshade *Solanum dulcamara*. *Biochimica et Biophysica Acta* 1206: 129-135.
6. Ercisli, S., Orhan, E., Yidirim, N. and Agar, G. 2008. Comparison of seabuckthorn genotypes (*Hippophae rhamnoides* L.) based on RAPD and FAME data. *Turkish Journal of Agriculture and Forestry* 32: 363-368.
7. Gao, F., Zhou, Y., Zhu, W., Li, X., Fan, L. and Zhang, G. 2009. Proteomic analysis of cold stress-responsive proteins in *Thellungiella* rosette leaves. *Planta* 230: 1033-1046.
8. Gupta, R. and Deswal, R. 2012. Low temperature stress modulated secretome analysis and purification of antifreeze protein from *Hippophae rhamnoides*, a Himalayan wonder plant. *Journal of Proteome Research* 11: 2684-2696.
9. Hon, W.C., Griffith, M., Chong, P. and Yang, D.S.C. 1994. Extraction and isolation of antifreeze proteins from winter rye (*Secale cereale* L.) leaves. *Plant Physiology* 104: 971-980.
10. Hon, W.C., Griffith, M., Mlynarz, A., Kwok, Y.C. and Yang, D.S.C. 1995. Antifreeze proteins in winter rye are similar to pathogenesis related proteins. *Plant Physiology* 109: 878-889.
11. John, U.P., Polotnianka, R.M., Sivakumaran, K.A., Chew, O., Mackin, L., Kuiper, M.J., Talbot, J.P., Nugent, G.D., Mautord, J., Schrauf, G.E. and Spangenberg, G.C. 2009. Ice recrystallization inhibition proteins (IRIPs) and freeze tolerance in the cryophilic Antarctic hair grass *Deschampsia antarctica* E. Desv. *Plant Cell Environment* 32: 336-348.

12. Laemmli, U.K. 1970. Cleavage of structural proteins during the assembly of the head of bacteriophage T4. *Nature* 227: 680-685.

13. Lundheim, R. and Wahlberg, K. 1998. Ice nucleation in fruit juice from different varieties of seabuckthorn *Hippophae rhamnoides* L. *Euphytica* 102: 117-124.

14. Pudney, P.D.A., Buckley, S.L., Sidebottom, C.M., Twigg, S.N., Sevilla, M.P., Holt, C.B., Roper, D., Telford, J.H., McArthur, A.J. and Lillford, P.J. 2003. The physio-chemical characterization of a boiling stable antifreeze protein from a perennial grass (*Lolium perenne*). *Archives of Biochemistry and Biophysics* 410: 238-245.

15. Qinxiao, W. and Hongyan, Z. 2003. Seabuckthorn for the control of soil erosion in mountain lands. *In: Seabuckthorn (Hippophae L.)-A Multipurpose Wonder Plant. Vol. I: Botany, Harvesting and Processing Technologies* (V. Singh, Ed. in Chief, 2003), p.401-408, Indus Publishing Company, New Delhi, 518p.

16. Rinalducci, S., Egidi, M.G., Karimzadeh, G., Jazii, F.R. and Zolla, L. 2011. Proteomic analysis of a spring wheat cultivar in response to prolonged cold stress. *Electrophoresis* 32: 1807-1818.

17. Singh, V. 2003. Geographical adaption and distribution of seabuckthorn (*Hippophae* L.) resources. *In: Seabuckthorn (Hippophae L.)-A Multipurpose Wonder Plant. Vol. I: Botany, Harvesting and Processing Technologies* (V. Singh, Ed. in Chief, 2003), p.21-34, Indus Publishing Company, New Delhi, 518p.

18. Singh, V. 2006. Free radicals, disease, antioxidants and antioxidant properties of seabuckthorn. *In: Seabuckthorn (Hippophae L.)-A Multipurpose Wonder Plant. Vol. II: Biochemistry and Pharmacology* (V. Singh, Ed. in Chief, 2006), p.3-69, Indus Publishing Company, New Delhi, 600p.

19. Smallwood, M., Warrall, D., Byass, L., Elias, L., Ashford, D., Doucet, C.J., Holt, C., Telford, J., Lillford, P. and Bowles, D.J. 1999. Isolation and characterization of a novel antifreeze protein from carrot (*Daucus carota*). *Biochemical Journal* 340: 385-391.

20. Worrall, D., Elias, L., Ashford, D., Smallwood, M., Sidebottom, C., Lilford, P., Telford, J., Holt, C. and Bowles, D. 1998. A carrot leucine-rich repeat protein that inhibits ice recrystallization. *Science* 282: 115-117.

21. Yan, S.P., Zhang, Q.Y., Tang, Z.C., Su, W.A. and Sun, W.N. 2006. Comparative proteomic analysis provides new insights into chilling stress responses in rice. *Molecular and Cellular Proteomics* 5: 484-496.

22. Zeb, A. 2004. Chemical and nutritional constituents of seabuckthorn juice. *Pakistan Journal of Nutrition* 3: 99-106.

23. Zeb, A. and Malook, I. 2009. Biochemical characterization of seabuckthorn (*Hippophae rhamnoides* L. spp. *turkestanica*) seed. *African Journal of Biotechnology* 8: 1625-1629.

24. Zheng, M., Wang, Y., Liu, K., Shu, H. and Zhou, Z. 2012. Protein expression changes during cotton fiber elongation in response to low temperature stress. *Journal of Plant Physiology* 169: 399-409.

Chapter 16

Assessment of Different Methods of DNA Extraction from Leaves of Seabuckthorn (*Hippophae rhamnoides* L.)

Simon-Gruita Alexandra, Constantin Nicoleta, Pavlusenco Camelia, Pojoga Daniela Maria and Duta Cornescu Georgiana

University of Bucharest, Faculty of Biology, Department of Genetics, Intr. Portocalelor 1-3, Sector 6, Bucharest, Romania

SUMMARY

The extraction of high quality DNA represents the first major difficulty to overcome in the molecular genetic studies on seabuckthorn. Four different DNA extraction methods were evaluated: Two variants of the classical SDS and CTAB protocols, a Commercial kit (innuPREP Plant DNA) and a complex method for plants with large amounts of secondary metabolites. The DNA isolates were also tested in two different PCR-based methodologies (RAPD and ISSR), currently used for the assessment of genetic variation between and within plant populations. The results showed that the method which allowed recovering the highest amount of DNA was not the best method in terms of purity or availability for molecular studies. The template DNA obtained by the complex method, which was supplemented with RNase treatment, proved to be very suitable for amplification reactions using RAPD and ISSR primers.

Keywords: *Seabuckthorn, DNA extraction methods, Leaves, RAPD, ISSR and Romania.*

INTRODUCTION

The seabuckthorn is one of the new crops in the world (Li *et al.*, 2009), because of nutritional and medicinal value of its berry and leaf, as they are rich in phytonutrients and bioactive substances (Li and Wardle, 2002). Also, the nitrogen-fixing nodules of the complex root system give this species the status of a pioneer plant in the conservation and reforestation of eroded areas (Asad *et al.*, 2007).

Seabuckthorn (*Hippophae* L.): *A Multipurpose Wonder Plant, Vol. 4: Emerging Trends in Research and Technologies* (Virendra Singh, *Editor-in-Chief*, 2014), p. 212–224, Daya Publishing House, New Delhi, India.

Over the last decades, a lot of studies on the morphology, biochemistry, utilization and cultivation of *Hippophae rhamnoides* L. populations have been done (Kallio *et al.*, 2002), but the studies at the DNA level are still at the beginning stage (Yao and Tigerstedt, 1993; Bartish *et al.*, 1999; Tian *et al.*, 2004; Sun *et al.*, 2006; Ruan *et al.*, 2007; Chen *et al.*, 2008). The natural variations in colour, size, shape and biochemical content of seabuckthorn fruits, leaves, thorns, etc. and the adaptability to extreme environments recorded within and between populations obviously results from the differences of the genetic constitution (Singh *et al.*, 2006). Thus, the identification, improvement and conservation of the economically important traits of seabuckthorn need extensive molecular studies. For example, in order to evaluate the genetic diversity of spontaneous and cultivated populations, several different DNA markers, such as RAPD, AFLP, ISSR and SSR have been used (Bartish *et al.*, 2001; Ruan *et al.*, 2004; Ruan *et al.*, 2007), but the results are not sufficient to implement coherent and appropriate breeding programmes.

For most genetic and genomic analysis, the extraction of high quality and large quantity of DNA represents the first major difficulty to overcome (Xin and Chen, 2012). In plant species, the presence of tannins, polyphenols and polysaccharides restrains the DNA isolation by their co-extraction with the nucleic acid, which requires the use of complex, labourious, time-consuming and expensive methods (Gupta *et al.*, 2011). This is also valid for seabuckthorn, which contains large amounts of vitamins C, vitamin E and carotenoids, folic acid, fatty acids (including omega 3, 6, 7 and 9), organic acids, free amino acids, carbohydrates, flavonoids (quercitin, myricetin, kaempferol, etc.), phenols, terpenes, tannins, mineral elements, etc. (Zeb, 2004). All these cell compounds with therapeutic importance may interfere with DNA extraction, affecting the subsequent molecular analysis.

In Romania, *Hippophae rhamnoides* L. spreads from the Southern Sub-Carpathian hills to the Eastern Sub-Carpathian hills, along the main rivers and their tributaries, in Transylvanian Plateau and also in Danube Delta and at the Black Sea shore (Ciocarlan, 2009). In these areas, *Hippophaë rhamnoides* subsp. *carpatica* was most often reported in literature, but subsp. *rhamnoides* and *fluviatilis* were also encountered.

The growing interest at the International level on this multipurpose species also took effect in Romania. Earlier studies were performed, primarily on the morphological and biochemical particularities (Uluitu *et al.*, 1995; Brad *et al.*, 2002; Rati and Rati, 2003). These phenotypic traits, unlike the genetic characteristics are subtle and misleading, being under the influence of the environmental changes (Ercisli *et al.*, 2008); for an accurate assessment of variability in seabuckthorn populations, the studies must be done using molecular techniques.

In one of the previous study, Nedelciu *et al.* (2011) evaluated four DNA extraction protocols in order to obtain appropriate DNA quality for the assessment of genetic diversity of seabuckthorn populations from Romania. Although, those first results were promising and we expanded that study, both by using another four DNA extraction protocols and by testing the DNA isolates in two different PCR-based methodologies.

MATERIALS AND METHODS

PLANT MATERIAL

Leaf tissue was collected during August-September 2010-2011 from eight cultivars of *Hippophae rhamnoides* subsp. *carpatica* Rousi (Table 16.1). After collection, the leaves were transferred on ice to the laboratory and then stored at -20 °C.

DNA EXTRACTION

Pre-extraction processing of the starting material was determined by grinding frozen leaf tissue to a fine powder using a mortar and pestle. For each tested protocol and each sample, 100 mg of powder was used as started material.

For DNA isolation, the following methods were tested: i) Two classical extraction methods which were slightly modified: the SDS protocol (modified from Dellaporta *et al.*, 1983) and the CTAB protocol (modified from Rogers and Bendich, 1994); ii) One commercial kit (innuPREP Plant DNA-*Analytik Jena*) and iii) A complex protocol designed for plants with high levels of secondary metabolites (modified from Dehestani and Kazemi Tabar, 2007).

Table 16.1: The List of the Seabuckthorn Cultivars

Sample Number	*Commercial Name*	*Year of Sampling*
1	Serbanesti	2010
2	Victoria	2010
3	Ovidiu	2010
4	Auras	2011
5	Silvia	2010
6	Diana	2010
7	Serpenta	2010
8	Tiberiu	2010

SDS Method

Chemicals and Solutions

(a) Lysis buffer: 100 mM Tris, 50 mM EDTA, 500 mM NaCl

(b) TE buffer: 10 mM Tris (pH 8.0), 1mM EDTA (pH 8.0)

(c) 20 per cent SDS

(d) 5 M potassium acetate (pH 5.3)

(e) RNase A solution (150 μg/ml)

(f) Chloroform:isoamyl alcohol (24:1)

(g) 5 M NaCl

(h) Ice cold ethanol (100 per cent) and isopropanol

Procedure

300 μl of preheated (65 °C) lysis buffer and 20 μl of 20 per cent SDS was added to the samples and incubated for 10 min at 65 °C. Then, 100 μl 5M potassium acetate was added and the probes were incubated on ice for 20 min and centrifuged at 12000 rpm for 5 min at 4 °C. The DNA from the supernatant was precipitated with 2 volume of ice cold ethanol, by incubating the tubes on ice for 30 min. The precipitated DNA was rehydrated in 100 μl 0.1X TE and the RNA removal was made by incubating the probes with RNase A (150 μg/ml final concentration) for 30 min at 37 °C. Deproteinization step was realized by adding chloroform:isoamyl alcohol (24:1) v/v. After a centrifugation step at 11000 rpm for 5 min at 4 °C, the supernatant was transferred into a clean tube and 0.5 volume of 5M NaCl and 2 volume of ice cold isopropanol was added followed by incubation on ice for 30 min and centrifugation at 14000 rpm for 5 min at 4 °C. Precipitated DNA was air dried and resuspended in 100 μl buffer TE and stored at 4 °C.

CTAB Method

Chemicals and Solutions

(a) 2X CTAB buffer: 2 per cent (v/v) CTAB, 20 mM EDTA (pH 8.0), 100 mM TRIS-HCl, 1.4 M NaCl, 1 per cent PVP

(b) 10 per cent CTAB solution: 10 per cent CTAB (w/v), 0.7 M NaCl
(c) Chloroform:isoamyl alcohol (24:1)
(d) Proteinase K solution (100 µg/ml)
(e) RNase A stock solution (1mg/ml)
(f) Ice cold ethanol (100 per cent) and isopropanol
(g) TE buffer: 10 mM Tris, 1 mM EDTA

Procedure

The probes transferred to Eppendorf tubes were mixed with 600 µl of preheated 2X CTAB buffer, inverted several times and then incubated for 30 min at 65 °C on water bath. Chloroform: isoamyl alcohol (24:1) v/v was added, mixed by inversion and centrifuged for 5 minutes at 11000 rpm and 4°C. For deproteinization, proteinase K was added and incubated for 1 h at 37 °C. 1/10 volume of preheated CTAB 10 per cent was added and the probes were mixed by inversion. The chloroform:isoamyl alcohol (24:1) extraction was repeated using the same mixture and the probes were centrifuged for 5 minutes at 11000 rpm and 4°C. RNA removal was made by incubating the probes with RNase A (150 µg/ml final conc.) for 30 min at 37 °C. DNA precipitation was made with an equal volume of cold isopropanol mixed well and incubated for 30 min on ice. The probes were centrifuged at 14000 rpm for 10 min at 4 °C. In order to eliminate the CTAB traces, the pellet was resuspended in 100 µl of TE buffer at 65 °C on water bath until the pellet dissolves. For DNA reprecipitation, 2 volumes of ice cold ethanol was added, mixed gently and incubated 15 min on ice and then centrifuged at 14000 rpm for 5 min at 4 °C. The pellet was air dried and resuspended in 100 µl buffer TE and stored at 4 °C.

Analytic Jena

To the homogenized material, 400 µl lysis solution and 25 µl proteinase K were added, then the mixture was vortexed and incubated at 50 °C for 45 min. The samples were moved to a prefilter column and centrifuged at 10000 rpm for 1 min. Then, 200 µl of binding solution SBS was added to the lysed samples by shaking it for several times. The samples were applied to a spin filter column and centrifuged at 10000 rpm for 2 min. After attaching the DNA to the column, one wash was given with 500 µl of washing solution HS and a second one with 750 µl washing solution MS were made, the solutions were eliminated by centrifugation at 10000 rpm for 1 min. In order to eliminate all the traces of ethanol from washing solution, a final centrifugation of 14000 rpm for 2 min was made. The DNA from the column was eluted in 200 µl of elution buffer and stored at 4 °C.

Complex Protocol

Chemicals and Solutions

(a) Extraction buffer: 120 mM Tris-HCl (pH 8.0), 80 mM EDTA (pH 8.0), 0.5 per cent Triton X-100 (v/v), 0.5 per cent β-mercaptoethanol (v/v)
(b) Lysis buffer: 120 mM Tris-HCl (pH 8.0), 10 mM EDTA (pH 8.0), 2 per cent (w/v) sarkosyl, 0.8 M NaCl, 2 per cent PVP and 0.2 per cent β-mercaptoethanol (v/v)
(c) Chloroform:isoamyl alcohol (24:1)
(d) CTAB solution: 0.7M NaCl, 1 per cent CTAB
(e) RNase A solution (150 µg/ml)
(f) Ice cold ethanol (100 per cent) and isopropanol

(g) 3 M sodium acetate (pH 5.2)

(h) High salt TE buffer: 10 mM Tris-HCl, (pH 8.0), 1 mM EDTA, 1 M NaCl

(i) TE buffer: 10 mM Tris, 1 mM EDTA

Procedure

Starting from the homogenized material, 700 µl of extraction buffer was added. The mixture was incubated at 45 °C in water bath for 20 min, vortexing the tube vigorously at every 5 min. The tubes were centrifuged at 8000 rpm for 5 min and 600 µl of lysis buffer was added to the pellet and after resuspension, the probes were incubated at 65 °C in water bath for 30 min with occasional inversion. 500 µl of chloroform:isoamyl alcohol (24:1) was added and several inversions of tubes were made, then the probes were centrifuged at 12000 rpm for 10 min. After, the upper phase was transferred to a clean tube and 1/10 volume preheated CTAB solution was added mixing the probes by inversion. Deproteinisation step continued by putting v/v of chloroform:isoamyl alcohol (24:1), mixed by inversion and centrifuged at 12000 rpm for 12 min. RNA removal was done by incubating the probes with RNase A for 30 min at 37 °C. DNA precipitation was made with 600 µl of cold isopropanol and 1/100 volume (considering supernatant and isopropanol) of 3 M sodium acetate, mixed well and incubated for 5 min at room temperature. The precipitated DNA was centrifuged at 10000 rpm for 8 min. The pellet was resuspended in 400 µl of high salt TE buffer and after DNA rehydration were added 40 µl of 3 M sodium acetate and 880 µl ice cold ethanol and incubated on ice for 20 min. The final centrifugation step consists in 14000 rpm for 10 min. The pellet was air dried and resuspended in 100 µl buffer TE and stored at 4 °C.

AGAROSE GEL ELECTROPHORESIS AND SPECTROPHOTOMETRY

The purity and integrity of isolated DNA was verified by running the samples on 0.8 per cent agarose gel in 1X TBE Buffer (pH 8.0), according to standard procedures (Vassu *et al.*, 2001). DNA concentrations were determined by reading the absorbance at 260 nm with NanoVue spectrophotometer. Also, the level of contamination with proteins and carbohydrates was estimated from A_{260}/A_{280} and A_{260}/A_{230} ratios, respectively.

PCR ANALYSIS

To verify the availability of the extracted DNA for downstream analysis, two PCR-based assays were done: RAPD (random amplified polymorphic DNA) and ISSR (inter simple sequence repeat) analysis, respectively.

PCR-RAPD and PCR-ISSR Analysis

Two decameric random primers and two ISSR primers, previously selected on their reproducibility and ability to discriminate between populations were used (Table 16.2). The analysis was carried out in a final volume of 12 µl, containing 6 µl Go Taq Green master mix (2x), 2 µl Primer (10 µM), 1 µl diluted DNA (DNA final concentration was between 15 and 20 ng/µl) and 3 µl sterile water. The programs used for amplifications were: For PCR–RAPD, 5 min/95 °C initial denaturation, followed by 45 cycles of 1 min at 95 °C, 1 min at 36 °C and 2 min at 72 °C. For PCR-ISSR, 4 min at 94 °C initial denaturation, followed by 40 cycles of 1 min at 94 °C, 2 min at 53 °C and 2 min at 72 °C. For both techniques, a final extension step of 10 min at 72 °C was included. To visualize and record the banding pattern, the agarose gel electrophoresis analysis was made on 2 per cent agarose in 1X TBE buffer (pH 8.0) at 80V for at least 150 min.

Table 16.2: RAPD and ISSR Primer Sequences

OPD 18(Bio Basic Inc.)	5'- GAGAGCCAAC - 3'
OPC 11(Integrated DNA Technologies)	5'- AAAGCTGCGG- 3'
ISSR-1(Integrated DNA Technologies)	5'- AGA GAG AGA GAG AGA GCT - 3'
ISSR-2(Integrated DNA Technologies)	5'- GAG AGA GAG AGA GAG AC- 3'

RESULTS

All the extraction methods used in this experiment (except the extraction with the commercial kit) were adapted using versions of some basic protocols, currently applied in plant molecular genetics. For simplification, according to the main characteristics of the extraction, the methods was named as follows: SDS method, CTAB method, Complex method and AJ (Analytic Jena) kit.

DNA Yield

The DNA concentration was determined on the basis of the absorbance at 260 nm (OD_{260}) with 1 OD equaling 50 µg/ml DNA (Sambrook *et al.*, 1989). For all methods and all samples, a wide range of variation was recorded (Table 16.3).

The SDS method provided the highest DNA average concentration (432.7 µg/ml DNA), as compared with the other extraction methods, although significant differences were recorded between the samples. For the other methods, the average DNA concentration was lower and varied between 67.5 µg/ml (AJ kit) to 314.2 µg/ml (CTAB method). Thus, the commercial kit provided the lowest DNA concentrations, while the CTAB-based methods allowed in obtaining intermediate values.

Table 16.3: The Concentrations of the DNA Extracted from *H. rhamnoides* subsp. *carpatica* Using Different Protocols

Sample Number	*SDS Method*		*CTAB Method*		*AJ kit*		*Complex method*	
	A_{260}	*Concentration (µg/ml)*	A_{260}	*Concentration (µg/ml)*	A_{260}	*Concentration (µg/ml)*	A_{260}	*Concentration (µg/ml)*
1	3.9	108.0	11.8	504.0	2.4	117.5	4.9	114.5
2	19.4	509.0	4.0	218.0	0.9	35.0	1.0	23.5
3	17.0	413.5	10.9	323.0	2.5	80.0	4.2	169.0
4	103.4	510.5	8.5	207.0	2.5	71.0	12.3	164.0
5	31.3	875.0	4.1	159.0	1.9	77.0	0.7	30.5
6	0.9	222.5	26.7	410.5	4.9	110.5	23.1	168.0
7	5.2	355.0	7.4	227.5	0.1	24.5	2.3	100.0
8	17.4	468.0	12.4	464.5	0.7	24.5	0.8	24.0

However, the DNA concentration computed on the basis of UV spectrophotometry has two drawbacks, which affect the results, which are i) the contribution to the total absorbance of the isolated nucleotides and single-stranded nucleic acids and ii) the interference with contaminants, such as proteins and phenols (Turci *et al.*, 2010). This might be the explanation for the differences between samples concentration when the same extraction method was used. Another explanation might be the plant material freshness. The leaves harvested in 2011 and processed in a short time, produced lower

DNA amounts than those stored at -20 °C for one year, may be because the plant material was kept frozen for a long period of time before extraction. This preservation method could reduce the presence of polyphenolic compounds (Ghaffari *et al.*, 2011).

In addition, the verification of the extracted DNA in agarose gel showed significant differences as compared to the DNA amount determined spectrophotometrically (Figure 16.1). For example, sample 5 isolated by modified SDS method showed a very small and pale band in the electrophoresis gel, whereas the concentration based on UV absorbance was 875 µg/ml, the highest concentration was observed in this study (Figure 16.1A).

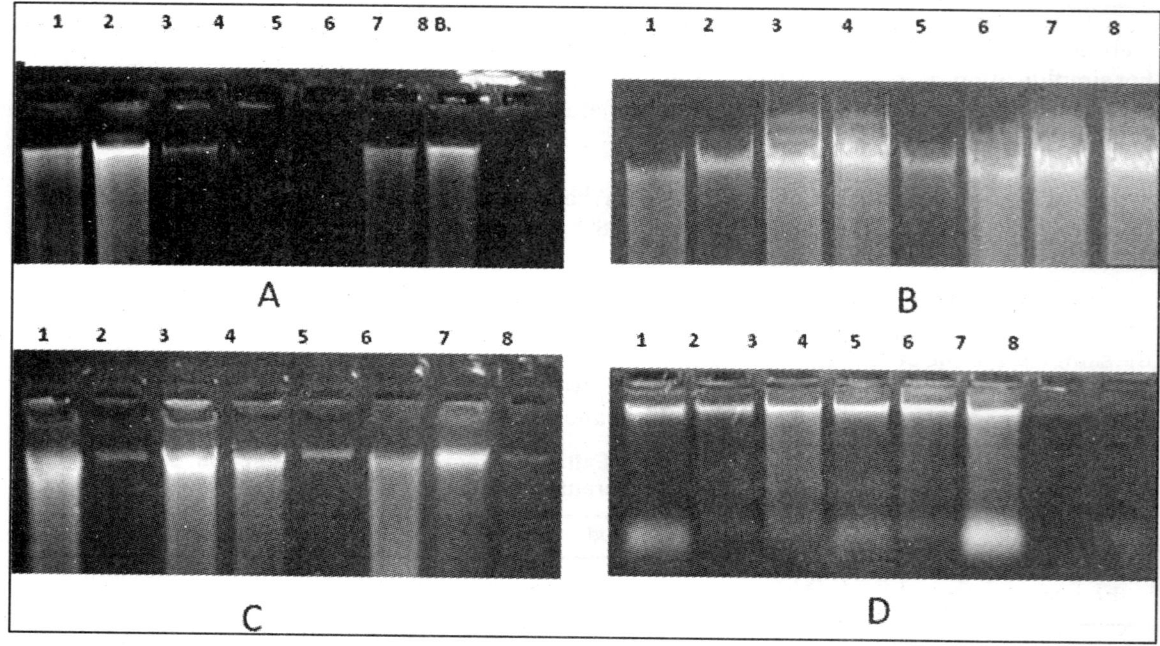

Figure 16.1: Gel Electrophoresis of the Total Genomic DNA from Eight Cultivars of *Hippophae rhamnoides* Extracted by (A) SDS Method, (B) CTAB Method, (C) Complex Method, and (D) AJ Kit.

In many situations, the analysis of agarose gel revealed a variable degree of DNA fragmentation which is common for protocols that use different chemical reagents and many successive steps.

DNA Purity

The purity of the extracted DNA was determined by the A_{260}/A_{280} and A_{260}/A_{230} ratios varied from one method to another (Table 16.4). As stated in literature (Sambrook *et al.*, 1989; Vassu *et al.*, 2001), the optimum A_{260}/A_{280} ratio is between 1.8-2.0; a ratio smaller than 1.7 indicates a protein contamination and a ratio higher than 2.0 corresponds to a contamination with RNA.

The best purity was observed for the Analytic Jena kit, followed by the complex protocol. It is necessary to underline that for the last extraction method, the values of the A_{260}/A_{280} were similar for all the samples, varying between 1.56-1.77, indicating low levels of contamination by protein and polyphenols. The variation was larger for the commercial kit, ranging between 1.31 and 2.01. The highest degree of protein and polyphenol contamination was observed for SDS method, which could be responsible for the concentration values computed after absorbance reading. The modified CTAB protocol produced good results only for three out of the eight samples (samples 1, 6 and 8, respectively), the A_{260}/A_{280} showing intermediate values as compared with the other methods (Table 16.4).

Table 16.4: The A_{260}/A_{280} and A_{260}/A_{230} Ratios of the DNA Extracted from *H. rhamnoides* subsp. *carpatica* using different Protocols

Sample Number	*SDS Method*		*CTAB Method*		*AJ kit*		*Complex method*	
	A_{260}/A_{280}	A_{260}/A_{230}	A_{260}/A_{280}	A_{260}/A_{230}	A_{260}/A_{280}	A_{260}/A_{230}	A_{260}/A_{280}	A_{260}/A_{230}
1	1.47	0.64	1.94	1.72	2.01	1.43	1.71	1.51
2	1.23	0.63	1.33	0.81	1.67	2.51	1.56	0.80
3	1.10	0.60	1.45	0.82	1.67	1.07	1.77	1.88
4	1.36	0.75	1.45	1.03	1.75	2.49	1.60	0.99
5	1.14	0.53	1.61	1.17	1.81	2.80	1.59	0.88
6	1.19	0.58	1.88	1.47	1.93	1.32	1.73	1.06
7	1.22	0.75	1.44	0.90	1.31	0.05	1.70	1.82
8	1.15	0.57	1.69	1.09	1.50	0.05	1.56	0.91

RNase treatment was not initially included in the commercial kit and the complex method, but our preliminary results have imposed this purification step. Unlike, the relative reduced protein contamination, polysaccharide contamination, as revealed by A_{260}/A_{230} ratio, is significant for all methods and samples, except samples 2, 3, and 5 extracted with the commercial kit. The threshold value for this ratio is 2 and values below 2 corresponding to the presence of carbohydrates (Vassu *et al.*, 2001).

From the analysis of A_{260}/A_{280} and A_{260}/A_{230} ratios, it became evident that the method which allows to recover a high amount of DNA is not necessarily the best method in terms of purity. Thus, all the extracts obtained with the SDS method are heavily contaminated with proteins, polyphenols and carbohydrates, which are normal secondary metabolites present in seabuckthorn leaves.

DNA Amplification

The quality of the DNA extracted from plant tissues is essential for the molecular applications. The DNA obtained from the above four presented methods, was used as template in two different PCR-based techniques: RAPD and ISSR. Both techniques are frequently used for the assessment of the genetic diversity within and between *H. rhamnoides* populations (Singh *et al.*, 2006; Li *et al.*, 2009).

RAPD primers (OPC11, OPD18) produced constant and reproducible band pattern when DNA extracted by the complex method was used (Figure 16.2). The reproducibility of RAPD fragments was lower in case of CTAB extraction method and AJ kit, whereas for SDS method, the amplification failed.

The ISSR method allowed to obtain clear and reproducible markers for three of DNA isolation methods: CTAB, complex protocol and commercial kit. The agarose gel electrophoresis for two ISSR primers is shown in Figures 16.3 (A-B).

The differences regarding the reproducibility between the two PCR assays results from the particularities of the RAPD method which is known to be highly influenced by the experimental conditions (Ghaffari *et al.*, 2011), as compared with ISSR.

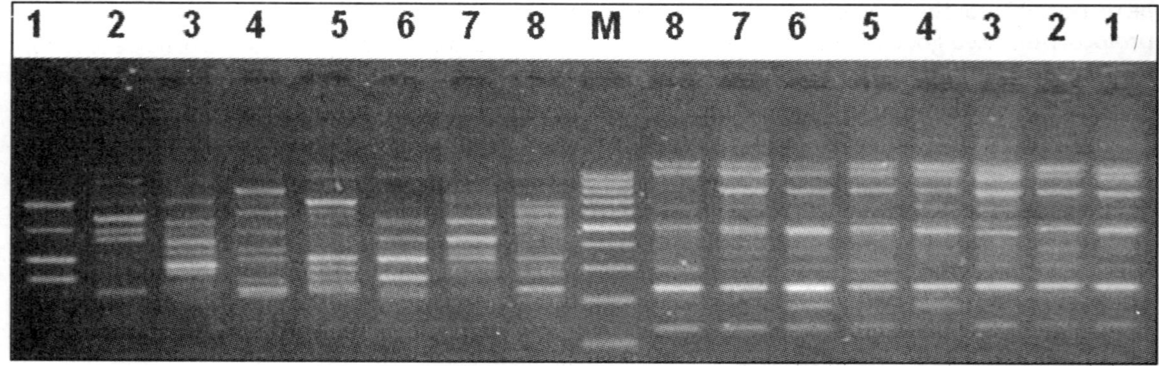

Figure 16.2: RAPD Band Pattern for DNA Extracted with the Complex Method using Primer OPC11 (Left) and Primer OPD 18 (Right). Lines: 1-8 and 8-1 seabuckthorn cultivars; M: Gene Ruler 100 DNA Ladder (Fermentas).

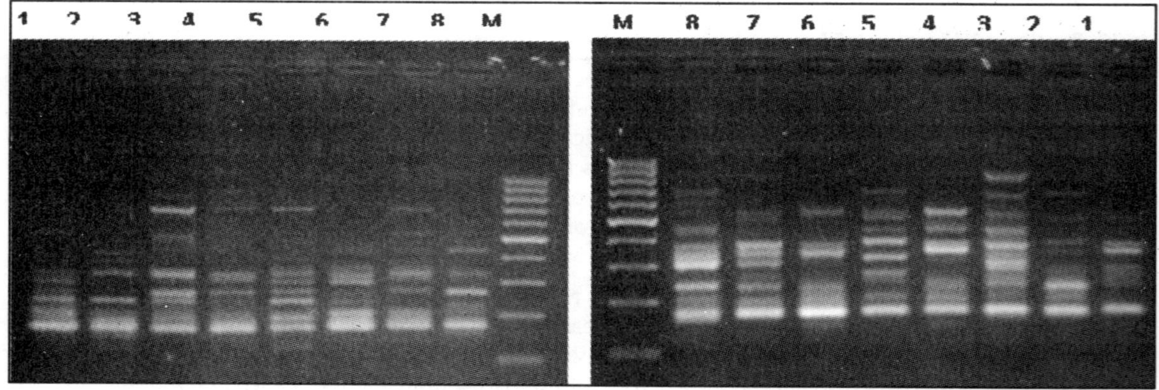

Figure 16.3: ISSR Band Pattern for DNA Extracted with the Complex Method using Primer ISSR-1a (A) and primer ISSR 2 (B). Lines: 1-8 and 8-1 seabuckthorn cultivars, M: Gene Ruler 100 DNA Ladder (Fermentas).

DISCUSSION

The four DNA extraction methods evaluated in this study have different efficiencies in terms of quantity and quality. The data analysis aimed not only the quality of isolated DNA, but also the method complexity and the relative costs. It must be mentioned that some of the extraction steps are common for two or more extraction methods, so only the relevant differences will be discussed.

The SDS method, a modified version of basic Dellaporta *et al.* (1983) protocol, is traditionally used for extraction of DNA from animal tissues. The main disadvantage of this method is the co-purification of PCR inhibitors, especially polysaccharides (Brook, 1998; Gupta *et al.*, 2011), which represents an important compound of the vegetal tissues. In a previous work, Nedelciu *et al.* (2011) established that SDS protocol represented the best DNA extraction method for seabuckthorn leaf tissue, when an enzymatic digestion of polysaccharides prior isolation using *CelluACE™XG System* kit is performed. Taken into account that *CelluACE™XG System* is no longer manufactured, it was necessary to search for an alternative DNA isolation method. In this study, the SDS protocol was tested again in its original form without *CelluACE™XG System*, but the results were unsatisfactory, especially in terms of quality, the DNA suspensions being strong coloured and viscous due to the presence of polysaccharide and polyphenol. This was a reconfirmation of Rogers's conclusions that for plant species, CTAB-based methods are more appropriate (Lodhi *et al.*, 1994).

The CTAB method and AJ kit are very similar in terms of DNA quality, the main differences was found in the concentration and reproducibility of PCR amplification. The reduced DNA concentration and the low reproducibility make AJ kit inappropriate for seabuckthorn leaves, although the time required for sample processing (30 minutes/sample) is very short. In addition, the AJ kit does not include RNase, but this treatment was necessary for obtaining clear gel band.

In nutshell, as other authors also shown (Xin and Chen, 2012), the commercial kits extract DNA with relatively good quality, but with low yield. Another aspect is the prohibitive cost, especially for small laboratories.

The modified CTAB-based method tested in this study, uses polyvinyl pyrrolidone (PVP) which has the capability to bind polyphenols. A study made on woody and herbaceous plants showed that for the first category (including seabuckthorn shrubs), the presence of PVP in the extraction buffer is essential (Sassa, 2007). The quality of DNA extracted by CTAB method was high enough for ISSR amplification, but not enough for RAPD fingerprinting, the RAPD method being more sensitive regarding the quality of DNA template than ISSR. Thus, this variant of CTAB method will be chosen depending on further DNA processing.

The best results in terms of availability for molecular studies were obtained with the complex protocol, which is a modified version of Dehestani and Kazemi Tabar (2007) extraction method. This method involves four important steps, each playing an important role in DNA purification. Thus, in the first stage, the extraction buffer consisting of Triton-X 100, β-mercaptoethanol and EDTA, removes a great portion of contaminants. It is known that the non-ionic detergent Triton is appropriate for obtaining high quantities of purified DNA, because it is able to remove most of the secondary metabolites (Gupta *et al.*, 2011). On the other hand, β-mercaptoethanol in the extraction buffer increases the amount of polysaccharides, polyphenols and pigments removed. Then, the lysis buffer prevents the oxidation of the secondary metabolites and by polyvinylpyrrolidone (Sassa, 2007) and β-mercaptoethanol acts again on polyphenols which are abundant in woody species. For better polyssaccharide decontamination, this method uses CTAB/NaCl solution and then TE washing was given and for deproteinization, chloroform:isoamyl alcohol was added. The original protocol does not include RNase treatment, but our results showed that this step is absolutely necessary.

The last method used for DNA extraction is complex, but it is not expensive because common reagents were used. Although, it allows to proceed a relatively larger number of samples in a relatively shorter period of time (8 hour for our 8 samples), similarly, for example, with other CTAB based methods.

CONCLUSION

Molecular studies on seabuckthorn are constantly expanding, both in our country and globally. The search for the most appropriate method of DNA isolation becomes a necessity, because its quality depends on the accuracy of the results obtained for further processing. It is obvious that there are several protocols that provide unfragmented and uncontaminated DNA, but each laboratory adopts a DNA extraction method according to its material possibilities and not at last, depending on the purpose of the research.

This study found that the complex method improved by the addition of RNase treatment, is the most effective for the assessment of the genetic diversity and relationships of relatedness between seabuckthorn populations. The main argument for this choice was that the DNA extract could be used with very good results in two different molecular techniques: RAPD and ISSR in which the amplification reaction being not influenced by the presence of inhibitors, possibly co-isolated with DNA.

ACKNOWLEDGEMENT

We are grateful to Professor Dr. Ioan Viorel Rati, University of Bacau, for providing the seabuckthorn samples.

REFERENCES

1. Asad, H.S., Dilnawaz, A., Mubasher, S., Shazia, A., Ishtiaque, K. and Farhat, B. 2007. Biochemical and nutritional evaluations of seabuckthorn (*Hippophae rhamnoides* spp. *turkerstanica*) from different locations of Pakistan. *Pak. J. Bot.* 39 (6): 2059-2965.
2. Bartish, G.I., Jeppsson, N., Bartish, I.V. and Nybom, H. 2001. Assessment of genetic diversity using RAPD analysis in a germplasm collection of seabuckthorn. *Agric. Food Sci. Finland* 9: 279-288.
3. Bartish, I.V., Jeppsson, N. and Nybom, H. 1999. Population genetic structure in the dioecious pioneer plant species *Hippophae rhamnoides* investigated by random amplified polymorphic DNA (RAPD) markers. *Mol. Ecol.* 8: 791-802.
4. Brad, I., Brad, I.L. and Radu, F. 2002. *Catina alba-o farmacie intr-o planta*. Ed. Tehnica, Bucuresti, 180p.
5. Brook, G.M. 1998. Total DNA isolation. In: *Molecular Genetic Analysis of Population- A Practical Approach* (A.R. Hoelzel, Ed.), p. 36-37, IRL PRESS, Oxford University Press, 445p.
6. Chen, G., Wang, Y., Zhao, C., Korpelainen, H. and Li, C. 2008. Genetic diversity of *Hippophae rhamnoides* populations at varying altitudes in the Wolong natural reserve of China as revealed by ISSR. *Silvae Genet.* 57: 29-36.
7. Ciocarlan, V. 2009. *Flora ilustratã a Romaniei. Pteridophyta et Spermatophyta*. Ed. Ceres, 419p.
8. Dehestani, A. and Kazemi Tabar, S.K. 2007. A rapid efficient method for DNA isolation from plants with high levels of secondary metabolites. *Asian Journal of Plant Sciences* 6 (6): 977-981.
9. Dellaporta, S.L., Wood, J. and Hicks, J.B. 1983. A plant DNA mini-preparation: Version II. *Plant Mol. Biol. Rep.* 1: 19-21.
10. Ercisli, S., Orhan, E., Yildirim, N. and Agar, G. 2008. Comparison of seabuckthorn genotypes (*Hippophae rhamnoides* L.) based on RAPD and Fame Data. *Turk. J. Agric. For.* 32: 363-368.
11. Ghaffari, S., Hanaoui, N. and Ferchichi, A. 2011. Rapid, high quality DNA isolation from Tunisian grapevine (*Vitis vinifera* L.) cultivars and optimization of the RAPD marker technique. *Rom. Biotechnol. Lett.* 16 (1): 5881-5890.
12. Gupta, A.K., Harish, Rai, M.K., Phulwaria, M. and Singh, S. N. 2011. Isolation of genomic DNA suitable for community analysis from mature trees adapted to arid environment. *Gene* 487: 156-159.

13. Kallio, H., Yang, B. and Peippo, P. 2002. Effects of different origins and harvesting time on vitamin C, tocopherols and tocotrienols in seabuckthorn (*Hippophae rhamnoides*) berries. *J. Agric. Food Chem.* 50: 6136-6142.

14. Li, H., Ruan, C.J. and Teixeira da Silva, J.A. 2009. Identification and genetic relationship based on ISSR analysis in a germplasm collection of seabuckthorn (*Hippophae* L.) from China and other countries. *Scientia Horticulture* 122: 263-271.

15. Li, T.S.C. and Wardle, D. 2002. Effect of harvest period on the protein content in seabuckthorn leaves. *Canadian Journal of Plant Science* : 409-410.

16. Lodhi, M.A., Ye, G.N., Weeden, N.F. and Reisch, B.I. 1994. A simple and efficient method for DNA extraction from grapevine cultivars, *Vitis* species and *Ampelopsis*. *Plant Mol. Biol. Rep.* 12 (1): 6-13.

17. Nedelciu, E., Simon-Gruita, A., Constantin, N., Duta-Cornescu, G., Popescu, M., Anastasiu, P., Rati, V., Rati, L. and Stoian, V. 2011. Genomic DNA isolation from *Hippophae rhamnoides* subsp. *carpatica* Rousi for RAPD fingerprinting. *Rom. Biotehnol. Lett.* 16 (2): 6026-6033.

18. Rati, V. and Rati, L. 2003. *Catina alba in exploatatii agricole*. Ed. ANCA, Bucuresti, 90p.

19. Rogers, S.O. and Bendich, A.J. 1994. Extraction of total cellular DNA from plants, algae and fungi. In: *Plant Molecular Biology Manual* (S.B. Gelvin and R.A. Schilperoort, Eds.), p.1-8, Kluwer Academic Publishers, Dordrecht, 152p.

20. Ruan, C.J., Qin, P., Zheng, J.W. and He, Z.X. 2004. Genetic relationships among some cultivars of seabuckthorn from China, Russia and Mongolia based on RAPD analysis. *Sci. Horti.* 101: 417-426.

21. Ruan, C.J., Teixeira da Silva, J.A., Jin, H., Li, H. and Li, D.Q. 2007. Research and biotechnology in seabuckthorn (*Hippophae* ssp.). *Med. Aromat. Plant Sci. Biotechnol.* 1: 47-60.

22. Sambrook, J., Fritsch, E.F. and Maniatis, T. 1989. *Molecular Cloning: A Laboratory Manual* (2nd Ed.), Cold Springer Harbor Laboratory, Cold Spring Harbor, New York.

23. Sassa, H. 2007. A technique to isolate DNA from woody and herbaceous plants by using a silica-based plasmid extraction column. *Anal. Biochem.* 363: 166-167.

24. Singh, R., Mishra, S.N., Dwivedi, S.K. and Ahmed, Z. 2006. Genetic variation in seabuckthorn (*Hippophae rhamnoides* L.) populations of cold arid Ladakh (India) using RAPD markers. *Current Science* 91 (10): 1321-1322.

25. Sun, K., Chen, W., Ma, R.J., Chen, X.L., Li, A.G. and Ge, S. 2006. Genetic variation in *Hippophae rhamnoides* ssp. *sinensis* (*Elaeagnaceae*) revealed by RAPD markers. *Biochem. Genet.* 44: 186-197.

26. Tian, C.J., Lei, Y.D., Shi, S.H., Nan, P., Chen, J.K. and Zhong, Y. 2004. Genetic diversity of seabuckthorn (*Hippophae rhamnoides*) populations in northeastern and north-western China as revealed by ISSR markers. *New Forest* 27: 229-237.

27. Turci, M., Sardaro, M.L.S., Vicoli, G., Maestri, E., Marmiroli, M. and Marmiroli, N. 2010. Evaluation of DNA extraction procedures for traceability of various tomato products. *Food control* 21: 143-149.

28. Uluitu, M., Brad, I., Chis, R., Brad, I.L., Rati, I.V., Rati, L. and Acatrinei, F. 1995. Variatia continutului in serotonina in fructe, frunze si extract hidrocloric de tulpini la 11 populatii de catina (*Hippophae rhamnoides*) selectionate la societatea comerciala Fructex S.A. Bacau. *Lucr. Simpoz. Inst. Agr. Timisoara* 1: 25-30.

29. Vassu, T., Stoica, I., Csutak, O. and Musat, F. 2001. *Genetica microorganismelor si inginerie genetica microbiana-Note de curs si tehnici de laborator*, p.76-82, Ed. Petrion, Bucuresti.

30. Xin, Z. and Chen, J. 2012. A high throughput DNA extraction method with high yield and quality. *Plant Methods* 8: 26.

31. Yao, Y. and Tigerstedt, P.M.A. 1993. Isozyme studies of genetic diversity and evolution in *Hippophae*. *Genet. Resour. Crop. Evol.* 40: 153-164.

32. Zeb, A. 2004. Chemical and nutritional constituents of seabuckthorn juice. *Pak. J. Nutr.* 3: 99-106.

PART IV:
PHYSIOLOGY

Chapter 17

Physiological Aspects of Seed Germination of *Hippophae salicifolia* D. Don. from Lahaul, Himachal Pradesh: Responses to Storage and Certain Chemical and Hormonal Treatments

Shashi Sharma and Suresh Katoch

Department of Biosciences, Himachal Pradesh University, Shimla – 171 005, India

SUMMARY

The seed germination behaviour of *Hippophae salicifolia* from Lahaul, Himachal Pradesh has been monitored with an emphasis on the influence of ambient storage and certain chemical and hormonal effectors. The freshly harvested seeds exhibited high viability that was maintained up to 2 years of storage. At harvest, the seed germination was poor (57 per cent) that improved with the progression of storage period suggesting the existence of non-deep physiological dormancy that was overcome by after-ripening during storage. Various seed pre-treatments namely, acid (H_2SO_4) scarification, sodium nitroprusside (SNP: a donor of nitric oxide, NO), KNO_3, GA_3 and IAA effectively improved the germination to varying extents in freshly harvested as well as stored seeds. SNP was an exception, that it was altogether ineffective in case of the stored seeds although it enhanced the germination in freshly harvested seeds. GA_3 was, by far, the most effective treatment in enhancing the seed germination and germination speed as well as reducing the mean germination time. The sensitivity of seeds towards various treatments was generally greater in 2-year stored seeds than in the freshly harvested ones. In order to get insight into the biochemical basis of storage-and effector-dependent changes in seed germination, imbibition rates, α-amylase activity and phenolic contents in the seeds were monitored. The freshly harvested seeds exhibited lower imbibition as compared to the stored seeds and imbibition could be enhanced due to acid scarification but not other treatments. The enhancement of α-amylase activity in the seeds during storage was paralleled by a concomitant decline in phenolic levels. Furthermore, all the treatments also enhanced the amylase activity and reduced the phenolic contents. Data have implications for the seed based multiplication of *H. salicifolia*.

Keywords: *Hippophae salicifolia, Seed germination, α-amylase, Phenol, SNP and storage.*

Seabuckthorn (*Hippophae* L.): *A Multipurpose Wonder Plant, Vol. 4: Emerging Trends in Research and Technologies* (Virendra Singh, *Editor-in-Chief*, 2014), p. 227–241, Daya Publishing House, New Delhi, India.

INTRODUCTION

Hippophae salicifolia (seabuckthorn) is an important multipurpose tree of western Himalayan region, particularly the cold desert areas of Himachal Pradesh, India. It is distributed between 2000-3300 m asl in cold deserts of Himachal Pradesh (Singh, 2003). Seabuckthorn berries possess high contents of vitamin C (Jeppsson and Gao, 2000), various mineral elements (Zhang *et al.*, 1989) and many important phytochemicals, such as flavonoids, carotenoids, fatty acids, vitamin K and sterols (Korovina and Fefelov, 2006). It has potent anti-oxidant, anti-cancerous and anti-cardiovascular properties. Seabuckthorn fruit extract has been reported to protect against arsenic induced tissue oxidative injury in Swiss albino mice (Gupta and Flora, 2006). *Hippophae* oil possesses UV blocking and emollient properties, and is used in anti-aging creams (Goel *et al.*, 2002), and medicines for balancing immune system (Geetha *et al.*, 2002; Singh and Moril, 2006). Being a multipurpose plant, it is extensively used by tribal people in the areas of its distribution where it is in great demand. Most of the demand is fulfilled by extraction from the wild. Its conservation and large scale cultivation needs urgent attention. The cultivation of seabuckthorn will ease the pressure on wild populations. *H. salicifolia* is generally propagated by root suckers. The natural regeneration of species through seeds is very scanty due to the poor seed germination (Sankhyan *et al.*, 2004). Seeds of *Hippophae* species possess internal dormancy which could be broken by various seed pre-treatments (Gupta *et al.*, 2011; Korekar *et al.*, 2013). Relatively more information is available on seed germination/dormancy of *H. rhamnoides* than in *H. salicifolia*, particularly from the cold desert region in Himachal Pradesh (Singh *et al.*, 2005). In the present study, seed viability and germination behaviour of *H. salicifolia* from the Pattan valley, Lahaul (Himachal Pradesh, India) have been examined with a particular reference to the changes induced by a 2-year ambient storage and diverse chemical and hormonal treatments. The metabolic basis of the changes in terms of α-amylase activity and phenolic levels has also been determined. The outcome of study could be expected to be of significance, particularly for seed-based propagation of *Hippophae salicifolia*.

MATERIALS AND METHODS

Seed Source

The mature seeds of *H. salicifolia* were collected from Pattan valley in Lahaul and Spiti district (H.P.) in the month of September-October. They were air dried for a fortnight at room temperature and stored in airtight plastic jars under ambient conditions for the subsequent studies.

Viability Test

Freshly harvested and 2-year stored seeds were tested for their viability (qualitative and quantitative) by using 2, 3, 5 triphenyltetrazolium chloride reduction test. The seeds were surface sterilized with 0.1 per cent aqueous solution of mercuric chloride for 3 min. Thereafter, they were washed and kept submerged in distilled water for imbibition for 24 h at 25 ± 2°C. The imbibed seeds were cut $1/3^{rd}$ at the micropylar end to facilitate the diffusion of TTC solution and incubated with 0.1 per cent TTC solution (aqueous) for 24 h in dark. The red/pink stained embryos were considered as viable. These assays were done in triplicate with 5 seeds (containing embryos) in each replicate. The results are presented as percent viability. For quantitative viability measurement, the stained seeds were homogenized with 5 ml methanol and centrifuged at 10,000 rpm for 10 min. Thereafter, the absorbance of supernatant was recorded at 485 nm. The quantitative viability is expressed as A_{485}/(5 seeds).

Water Imbibition

Water Imbibition takes place in terms of increase in seed weight. The uniform seeds in size and weight were soaked in distilled water/different effectors for 24 h at 25 ± 2° C. After 24 h, seeds were weighed after blotting the surface water. An increase in seed weight indicates the amount of water imbibition. It also indicates the hardness/softness and permeability/impermeability of seed coat.

Germination Assays

The uniform seeds in shape and size were surface sterilized with 0.1 per cent $HgCl_2$ for 3-4 min., washed thoroughly under running tap water and were soaked in distilled water/different treatments for 24 h at 25 ± 2°C. Thereafter, the seeds were transferred to Petriplates lined with three layers of filter paper moistened with distilled water and incubated for germination in the seed germinator/incubator at 25 ± 2°C under continuous white light (PAR: 40 µmol $m^{-2} s^{-1}$). The germination was observed daily until the final germination count. Emergence of 2.5 mm radicle was considered as seed germination (ISTA, 1966). Seed germination results are presented in terms of percent germination. The experiments were performed in triplicate of 15 seeds each. Mean germination time (MGT) (Hartmann *et al.*, 1989) and germination speed (GS) (Parmar and Kumar, 2006)) were calculated with the formulae as follows:

$$MGT = \Sigma nd/N$$

where,

n: Number of seeds that germinated after each incubation period in days (d);

N: Total number of seeds emerged at the end of germination test.

Germination speed (GS) = Germination percentage/day of complete germination.

CHEMICAL AND HORMONAL TREATMENTS FOR DORMANCY REMOVAL/ GERMINATION IMPROVEMENT

The seeds were subjected to different chemical and hormonal treatments to improve the germination percentage or to remove dormancy. The following treatments were performed.

Sodium nNitroprusside (SNP) and Potassium Nitrate (KNO_3)

The surface sterilized seeds were incubated with aqueous solutions of SNP (0.1, 1.0 and 10 mM) and KNO_3 (0.1, 1.0 and 10 mM) for 24 h at 25 ± 2°C. Thereafter, they were transferred to Petri plates lined with three layers of filter paper well moistened with distilled water and kept for germination at 25 ± 2°C.

Acid Scarification

Uniform seeds were treated with concentrated H_2SO_4 for 3 min. Thereafter, they were washed thoroughly with water to remove the acid completely and soaked in distilled water for 24 h at 25 ± 2 °C. The seeds were then transferred to moist substratum for germination at 25 ± 2 °C in a seed germinator.

Gibberellic Acid (GA_3) and Indole-3-acetic Acid (IAA)

The surface sterilized seeds were kept submerged in aqueous solutions of GA_3 (0.1 and 1.0 mM) and IAA (0.001, 0.01 and 0.1 mM) for 24 h. Thereafter, the seeds were placed in Petri plates lined with three layers of moistened filter paper and incubated in seed germinator at 25 ± 2 °C.

α-amylase Activity

α-amylase activity was determined by using the method described by Filner and Varner (1967). Seeds were homogenized with Tris-HCl buffer (pH 7.2) under low temperature conditions and centrifuged at 10,000 rpm for 10 min at 4 °C. The supernatant was taken as crude enzyme. 1 ml of enzyme was added to 1 ml of substrate (0.15 per cent starch, 0.2 mM $CaCl_2$) and incubated for 10 min at 25 °C. The reaction was quenched by addition of 3 ml of IKI reagent (0.6 per cent iodine in 6 per cent KI; 1 ml of this diluted to 50 ml with 0.05N HCl) and absorbance was recorded at 620 nm. In control, IKI was added prior to the addition of enzyme. The activity was expressed as µg starch degraded min^{-1} mg^{-1} protein.

Protein Contents

Protein contents in the seeds were estimated in accordance with the method given by Lowry *et al.* (1951).

Phenolic Contents

Phenolic contents were determined by using the method given by Goldstein and Swain (1963). Seeds were homogenized in 5 ml of 0.3N HCl in methanol and centrifuged at 10,000 rpm for 10 min. Pellet was discarded and supernatant was evaporated in a boiling water bath. Thereafter, the residue remained was dissolved in distilled water (5 ml). To this, 0.5 ml of 1N Folin-Ciocalteu reagent was added and shaken vigorously. After 3 min, 1 ml of 35 per cent Na_2CO_3 solution was added, shaken and allowed to stand for 1 h at room temperature (25 ± 2 °C) and absorbance was recorded at 630 nm. Amount of phenol contents of seeds were determined with the help of calibration curve prepared with gallic acid.

RESULTS

Viability Status

Quantitative and qualitative changes in the seed viability were monitored in freshly harvested and stored (2-year) seeds of *H. salicifolia* by using 2,3,5-triphenyltetrazolium chloride (TTC) reduction test. The freshly harvested seeds exhibited 100 per cent viability; no loss in viability was observed with storage for 2 years. However, the ability of seeds (parts containing embryo) for TTC reduction differed in freshly harvested and 2-year stored seeds. A_{485}/5 embryos were observed to be 0.607 and 0.804 in freshly harvested and stored seeds, respectively (Figure 17.1).

Seed Germination/Dormancy Status of Freshly Harvested Seeds

Freshly harvested seeds of *H. salicifolia* exhibited low germination (57 per cent). Seeds started germinating after 4 d (8 per cent) of subjecting to germination conditions. Germination increased gradually upto 57 per cent after 20 d which remained unchanged until 30 d. Among all the pre-treatments applied (H_2SO_4, SNP, KNO_3, GA_3 and IAA) for dormancy removal/germination improvement, GA_3 was found to be most effective. The effect of GA_3 was concentration dependent; the higher concentration was found to be more promotory. Thus, the GA_3 treatment led to 71 and 80 per cent germination at 0.1 and 1 mM, respectively, as compared to 57 per cent germination in control (Figure 17.2A). The germination speed was 4.4 and 4.7 seeds/day, respectively at these GA_3 concentrations whereas the germination speed in control was 3.6 seeds/day (Table 17.1). Other treatments tested namely, H_2SO_4 (3 min), SNP (1 mM), KNO_3 (1 mM), and IAA (0.001 mM) were also effective in improving the germination in terms of germination percentage and speed (Table 17.1, Figure 17.2B). In contrast,

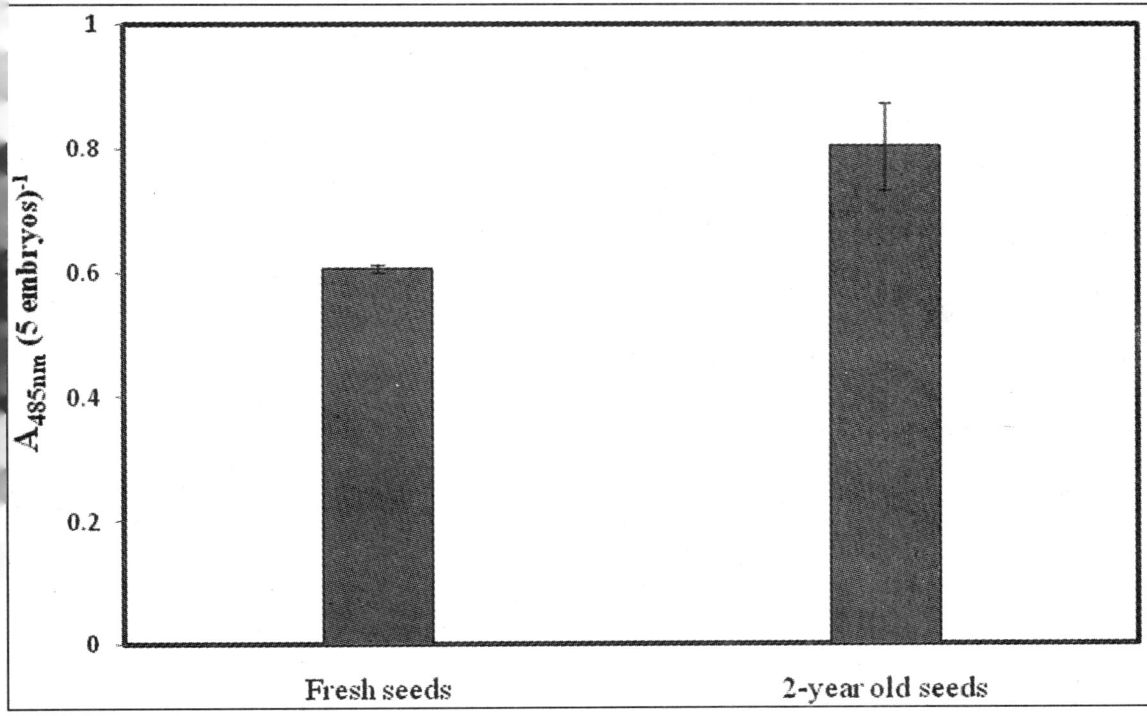

Figure 17.1: Dehydrogenase Activity in Freshly Harvested and 2-year Stored Seeds of *Hippophae salicifolia* Measured by 2,3,5-triphenyltetrazolium Chloride (TTC) Reduction Assay.

Table 17.1: Final Germination (per cent), Mean Germination Time and Germination Speed of Freshly Harvested and 2-year Stored Seeds of *Hippophae salicifolia*

Treatments	*Germination (per cent)*		*Mean Germination Time (MGT) (days)*		*Germination Speed (Seeds/day)*	
	Freshly Harvested	*2-Year Stored*	*Freshly Harvested*	*2-Year Stored*	*Freshly Harvested*	*2-Year Stored*
Control	57.7	84.4	9.3	3.5	3.6	12.0
SNP (1 mM)	68.8	82.2	11.7	3.3	4.1	13.7
KNO_3 (1 mM)	66.6	88.8	10.2	3.7	3.9	12.7
KNO_3 (10 mM)	44.4	86.6	10.4	3.7	2.6	12.4
GA_3 (0.1 mM)	71.1	77.7	10.0	3.4	4.4	13.0
GA_3 (1 mM)	80.0	100	9.6	3.2	4.7	20.0
IAA (0.001 mM)	64.4	91.1	9.2	3.3	3.8	15.2
IAA (0.01 mM)	51.1	84.4	9.6	3.6	3.0	16.8
IAA (0.1 mM)	51.1	97.7	9.4	3.5	3.0	14.0
H_2SO_4 (3 min)	64.4	86.6	10.4	2.4	3.8	17.5

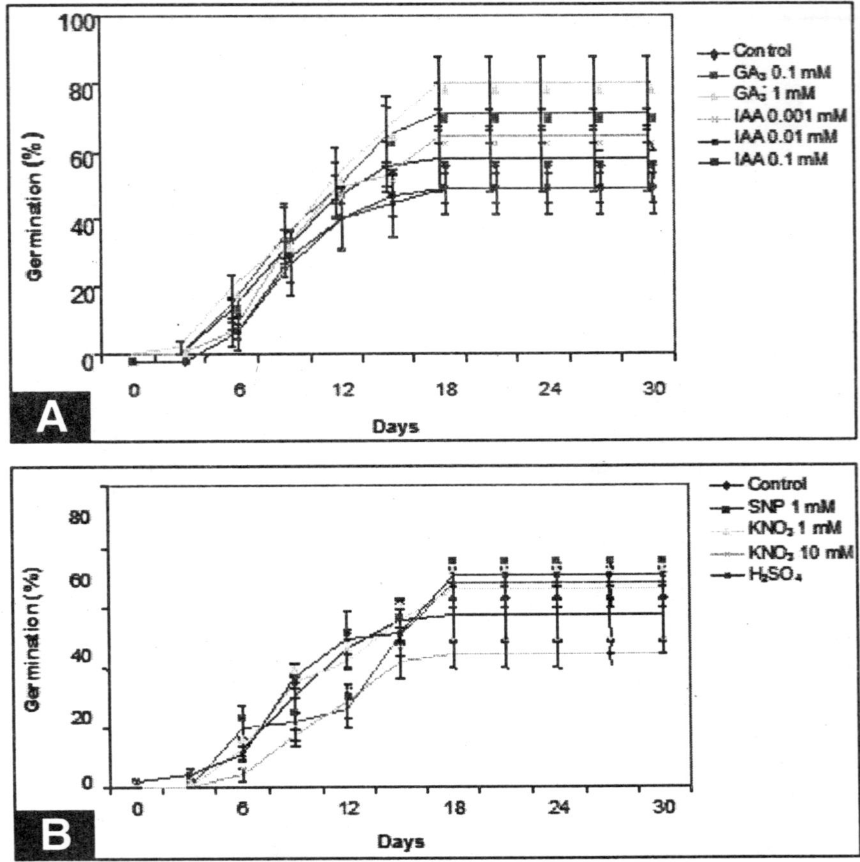

Figure 17.2: Time-Course of Seed Germination in *Hippophae salicifolia* as Affected by different Hormonal (A) and Chemical (B) Treatments in Freshly Harvested Seeds.

KNO_3 (10 mM) and IAA (0.01 and 0.1 mM) were found to be inhibitory. The order of effectiveness of different treatments in freshly harvested seeds was as follows: GA_3 (1 mM > 0.1 mM) > SNP (1 mM) > KNO_3 (1 mM) > IAA (0.001 mM) ≥ H_2SO_4 > Control.

Storage Dependent Changes in Seed Germination/Dormancy

Two-year stored seeds exhibited substantially improved germination (84 per cent) as compared to the freshly harvested ones both in terms of earliness and final count. Thus, seeds started germinating on 2 d (11 per cent) and maximum germination of 84 per cent was achieved after 7 d. Sensitivity to various effectors applied was different than that of freshly harvested seeds. All the treatments tested were found to be effective in improving germination percentage and speed. Like in freshly harvested seeds, GA_3 was found to be most effective in stored seeds also. 100 per cent germination with higher germination speed (20 seeds/day) was observed as compared to 84 per cent germination and lower germination speed (12 seeds/day) in control seeds. IAA was also found to be effective resulting in 91 and 97 per cent germination at 0.001 and 0.1 mM concentrations, respectively, accompanied by higher germination speed as compared to the control (Table 17.1, Figure 17.3A).

No substantial change in germination performance was observed with KNO_3 treatment. H_2SO_4 treatment caused early germination and marginally increased the germination. SNP was not effective in improving the seed germination in stored seeds (Figure 17.3B). Minor fluctuations in mean germination time (MGT) were observed with different treatments tested in stored as well as freshly harvested seeds. The order of effectiveness of different treatments in stored seeds was: GA_3 (1 mM) > IAA (0.1 mM > IAA 0.001 mM) > KNO_3(1 mM) > H_2SO_4 > Control.

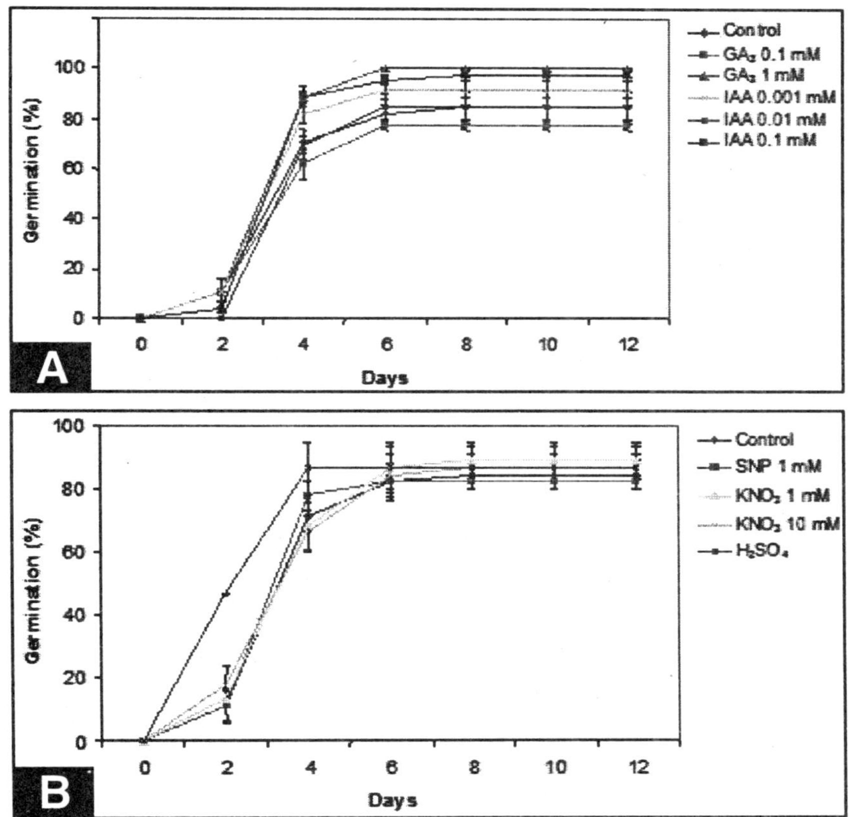

Figure 17.3: Time-Course of Seed Germination in *Hippophae salicifolia* as Affected by different Hormonal (A) and Chemical (B) Treatments in 2-Year Stored Seeds.

Storage Dependent Biochemical Changes in Seeds

In order to understand the physiological and biochemical basis of storage and diverse effector dependent changes in seed germination of *H. salicifolia*, imbibition rates, α–amylase activity and phenolic contents in treated/stored seeds were measured.

Imbibition

Imbibition is the first step for germination of seeds. To monitor the effects of various pre-treatments and storage dependent changes on water uptake/imbibition, freshly harvested and stored (2-year) seeds of H. salicifolia were treated (soaked) with different chemical and hormonal effectors for 24 h at 25 ± 2 °C. Seeds soaked in distilled water (DW) served as control. After 24 h, the imbibition was recorded by examining the per cent increase in weight of seeds. The stored seeds (2-year) showed higher imbibition than the freshly harvested ones. Among all the treatments (H_2SO_4, SNP, KNO_3, GA

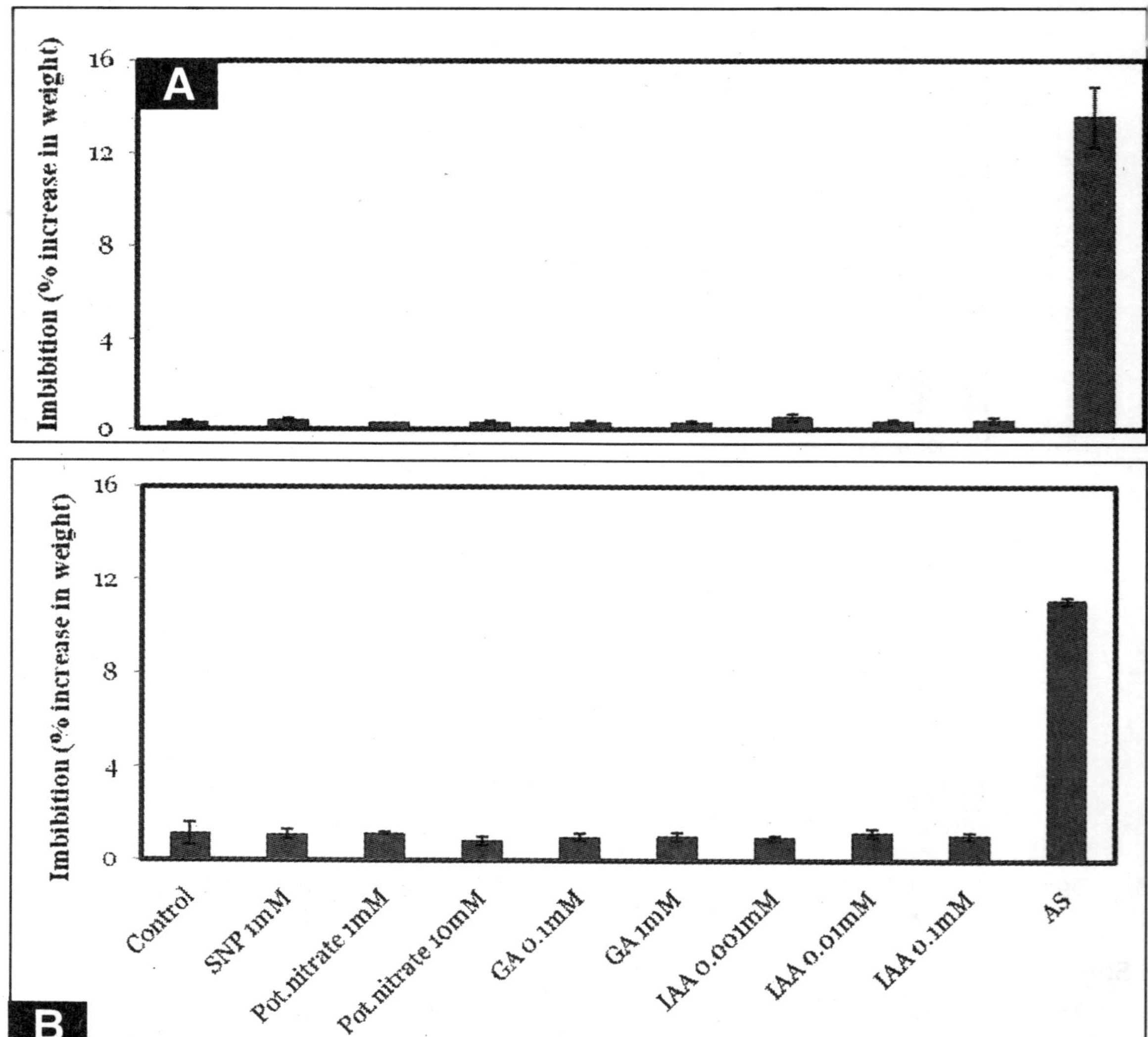

Figure 17.4: Water Imbibition in Freshly Harvested (A) and 2-Year Stored (B) Seeds of *Hippophae salicifolia* as Affected by different Treatments.
AS: Acid scarification; Pot. nitrate: Potassium nitrate; GA: Gibberellic acid (GA_3); IAA: Indole-3-acetic acid; and SNP: Sodium nitroprusside.

and IAA) applied, only H_2SO_4 treatment was found to be effective in enhancing the imbibition; the amount of imbibition was higher in freshly harvested seeds as compared to stored seeds. With acid scarification (AS), 45 and 10 times increase in seed weight was observed as compared to the control in freshly harvested and stored (2-year) seeds, respectively (Figure 17.4).

α-amylase Activity

α-amylase activity enhanced with the progression of storage period. The stored seeds (2-year) exhibited 3 times higher α-amylase activity as compared to the freshly harvested seeds. α-amylase activity was enhanced by all the treatments tested (H_2SO_4, SNP, KNO_3, GA_3 and IAA). The magnitude

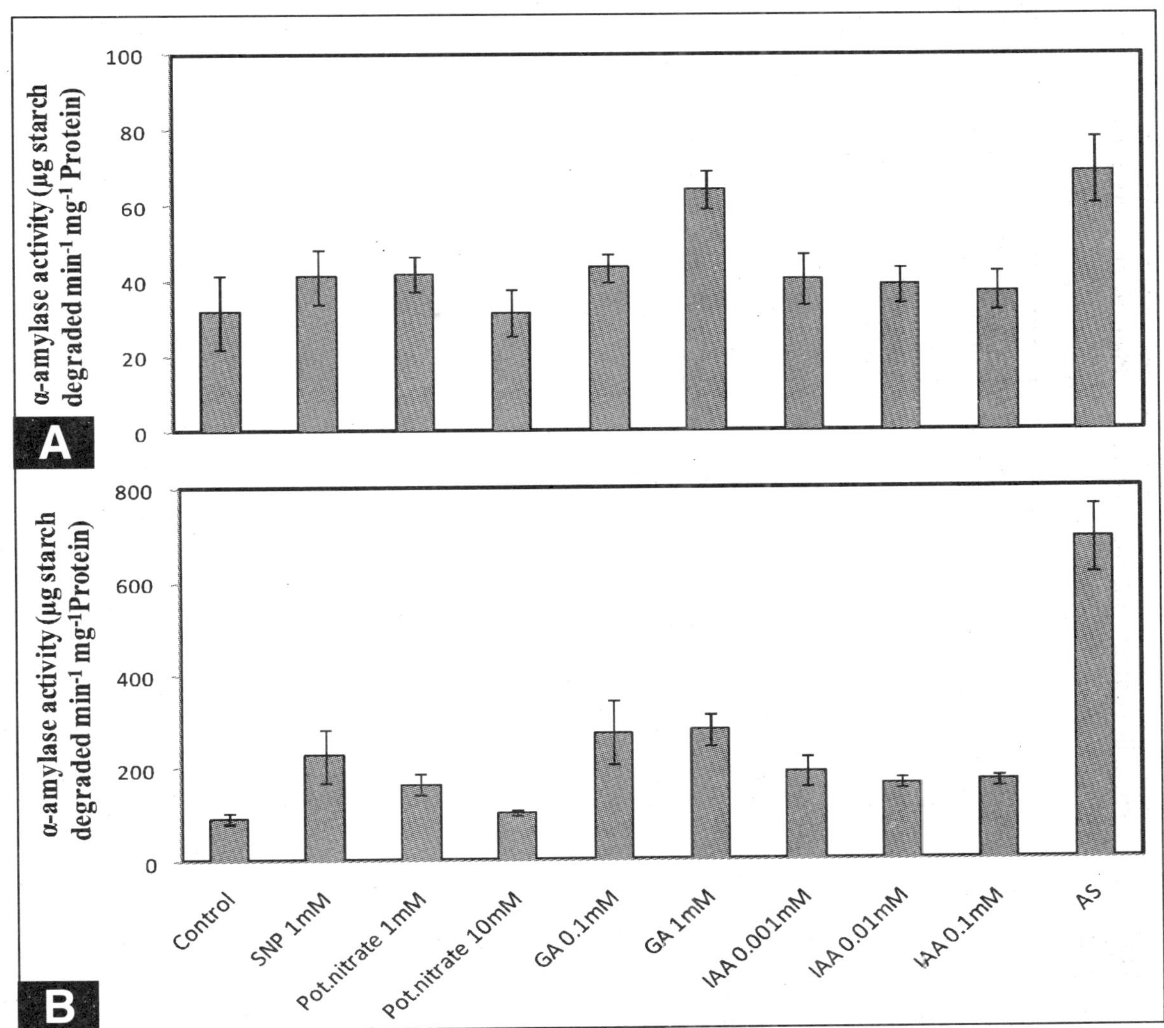

Figure 17.5: Effect of different Treatments on α-amylase Activity in Seeds of *Hippophae salicifolia* (Freshly harvested: A and 2-year stored: B). AS: acid scarification; Pot. nitrate: Potassium nitrate; GA: Gibberellic acid (GA_3); IAA: Indole-3-acetic acid; and SNP: Sodium nitroprusside.

of enhancement in α-amylase activity varied in stored and freshly harvested seeds. Acid scarification for 3 min proved to be most effective causing 2.19- and 7.44-fold increase in freshly harvested and stored seeds, respectively, as compared to the control. Marginal increase in α-amylase activity was observed in SNP, KNO_3 and IAA treated freshly harvested seeds whereas in stored seeds, SNP (1 mM) and IAA (0.001 mM) were also found to be effective that caused 2.4- and 2-fold increase in α-amylase activity. GA_3 (1 mM) enhanced the activity by 3-and 2-fold in stored and freshly harvested seeds, respectively, as compared to control (Figure 17.5).

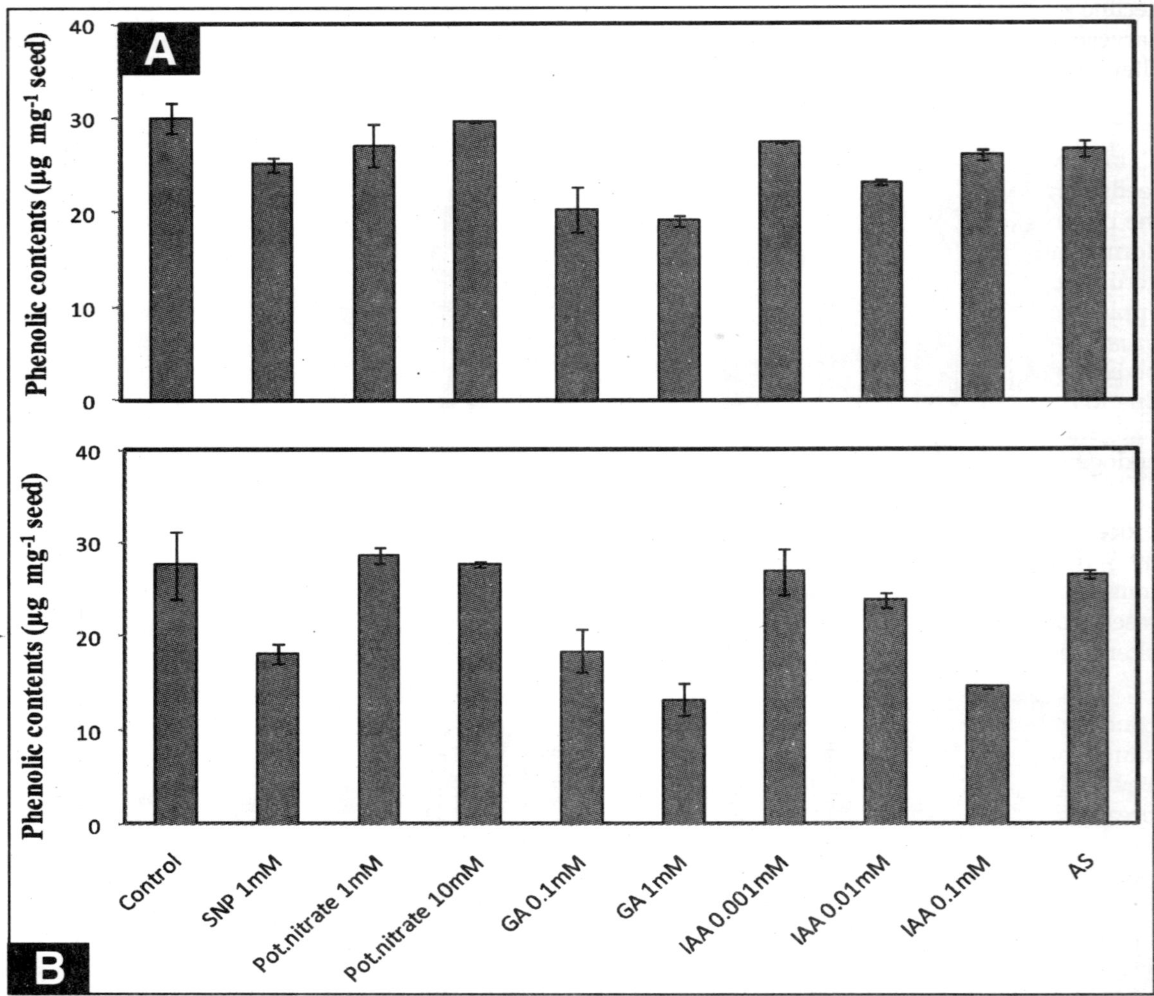

Figure 17.6: Total Phenolic Contents in Freshly Harvested (A) and 2-year stored (B) Seeds of *Hippophae salicifolia* as Affected by different Treatments. AS: Acid scarification; Pot. nitrate: Potassium nitrate; GA: Gibberellic acid (GA_3); IAA: Indole-3-acetic acid; and SNP: Sodium nitroprusside.

Total phenolic contents

Phenols categorized as germination inhibitors, were estimated in freshly harvested and stored seeds of *H. salicifolia* in order to examine their role, if any, in seed germination/dormancy. Total phenolic contents were slightly higher in freshly harvested seeds as compared to stored (2-year) seeds. Phenols were measured in 24 h imbibed seeds in distilled water (control) and various effectors. In control, 30 and 27 µg phenol/mg seed was found in freshly harvested and stored seeds, respectively. Phenolic contents were found to be reduced with GA_3, IAA and SNP treatments, however, the magnitude varied in freshly harvested and stored seeds. For example, 37, 52; 15, 47; 33, 35; and 17, 35 per cent decline was monitored in GA_3 (1 mM), IAA (0.1 mM), GA_3 (0.1 mM) and SNP (1 mM) treated freshly harvested and stored seeds, respectively, as compared to control. H_2SO_4 and KNO_3 treatment did not alter much of the total phenolic contents of the seeds (Figure 17.6).

DISCUSSION

Hippophae salicifolia is an important and multipurpose medicinal plant. The freshly harvested seeds of *H. salicifolia* exhibited low germination (57 per cent) after 20 d. Germination enhanced with the progression of storage period. Among various chemical and hormonal treatments tested for dormancy removal/germination improvement, GA_3 application was most effectively enhanced the germination (80 per cent in 20 d) followed by SNP, KNO_3, IAA and H_2SO_4 treatments. Exogenous application of GA_3 has been shown to break dormancy and improve germination in seeds of *H. rhamnoides* (Airi *et al.*, 2009) and many other plant species (Nadeem *et al.*, 2000). Higher germination percentage by GA_3 application indicates that the poor germination in freshly harvested seeds may be due to immature embryos, as GA_3 is known to be synthesized during after-ripening in embryos. Exogenous GA_3 application might replace the requirement of after-ripening by increasing the endogenous level of this hormone (Bewley and Black, 1982).

SNP is an NO (nitric oxide) donor; NO is a signaling molecule regulating a variety of plant processes (Kopyra and Gwozdz, 2003; Neill *et al.*, 2003). Marginal increase in germination percentage by SNP may be due to enhanced synthesis and secretion of α-amylase through its GA_3 signal transduction or independent of GA_3 (Zhang *et al.*, 2003). SNP promoted imbibition and caused storage reserve degeneration finally leading to enhanced germination (Zhang *et al.*, 2001). It also stimulates alternate respiratory pathway (Bewley and Black, 1982; Baskin and Baskin, 1998).

KNO_3 applied at lower concentration (1 mM) was found to be slight promotory whereas higher concentration (10 mM) inhibited the germination percentage. KNO_3 and other nitrates are known to stimulate germination by enhancing the alternate pathway of respiration (Roberts and Smith, 1977; Bewley and Black, 1982; Baskin and Baskin, 1998). KNO_3 may also replace the requirement for dormancy breaking agents such as light and temperature regimes (Evenari, 1965; Stokes, 1965). KNO_3 also acts as an oxidizing agent; it promotes the breakdown of the germination inhibiting substances such as polyphenols and short chain fatty acids (Roberts, 1969). Stimulatory effect of KNO_3 on seed germination has also been reported in *H. salicifolia* (Gupta *et al.*, 2011) and other plant species *e.g. Heracleum candicans* (Joshi and Dhar, 2003). Higher concentration of KNO_3 may become toxic causing inhibition in germination. KNO_3 can either stimulate or inhibit seed germination depending on the plant species. An adverse effect of KNO_3 has also been reported in *Terminalia sericea* (Amri, 2010).

IAA (0.001 mM) was also found to be effective in improving germination in *H. salicifolia* seeds. GA_3 and IAA level gradually increased during germination (Atici *et al.*, 2007). In some plant species *e.g.*, pea cotyledons, auxin induced α-amylase activity (Hirasawa, 1989). Acid scarification (3 min) enhanced the germination and reduced mean germination time. Olmez (2011) found H_2SO_4 to be effective in improving seed germination in *H. rhamnoides* seeds. It is likely that acid degraded the seed coat thereby reactivating the physiological and biochemical activities needed for seed germination.

In order to get an idea about the biochemical and physiological bases underlying the enhanced germination by various effectors, the effects of these effectors were studied on imbibition, α-amylase activity and phenolic contents. In GA_3 treated seeds, no effect on imbibiton, two fold increase in α-amylase activity and 37 per cent decline in phenolic contents was observed. With SNP (1 mM) and IAA (0.001 mM) application, marginal increase in α- amylase activity, 17 and 15 per cent decline in phenolic contents, respectively, with no change in imbibition as compared to control were observed. However, H_2SO_4 treatment enhanced the imbibition, α-amylase activity (2 fold) and declined the phenol content (12 per cent) as compared to control in freshly harvested seeds. The result clearly suggests that the poor germination in freshly harvested seeds was mainly due to the presence of high phenolic contents as the maximum decline (37 per cent) of phenolic contents was directly correlated with highest germination percentage (80 per cent) by GA_3 treatment. Only enhanced α-amylase activity was not sufficient to achieve enhanced germination as in case of H_2SO_4 treatment. Besides, increase in α-amylase activity along with decline in phenolic contents was responsible for enhanced germination as in case of GA_3 treatment. This suggested that a balance between growth promoter and growth inhibitor determines the removal of dormancy and germination improvement (Bewley and Black, 1982).

Stored seeds (2-year) of *H. salicifolia* exhibited higher germination percentage (84 per cent in 7 d) as compared to freshly harvested seeds (57 per cent in 20 d) along with enhanced α-amylase activity, higher imbibition and low phenolic contents as compared to freshly harvested seeds. Higher germination in stored seeds as compared to freshly harvested seeds may be due to 3 fold increase in α-amylase activity, higher imbibition (3.8 times) and low phenolic contents in stored seeds as compared to freshly harvested seeds. Sensitivity of stored seeds to various effectors was high as it was evident by 100 per cent germination with 3-fold increase in α-amylase activity and 52 per cent decline in phenolic contents in stored seeds, whereas 80 per cent germination, 2-fold increase in α-amylase activity and 37 per cent decline in phenolic contents was observed by GA_3 treatment in freshly harvested seeds as compared to control suggesting the importance of promoter and inhibitor ratio. These findings indicate that the freshly harvested seeds of *H. salicifolia* possessed non-deep physiological dormancy. Freshly harvested seeds may have physiologically immature embryos thus having lower α-amylase activity. Higher α-amylase activity in stored seeds may be attributed to the fact that during storage the embryos became fully developed and mature and were able to produce higher GA_3 leading to higher α-amylase activity and hence, higher germination. Increase in α-amylase activity by GA_3 has been shown by Baskin and Baskin (1998). Higher seed germination in stored seeds seems to be related to the changes in physiological and biochemical state of seeds during storage. The changes in the ratio of growth promotors to inhibitors during storage may play an important role in germination.

No change in the viability status was observed during storage. Freshly harvested as well as stored seeds of *H. salicifolia* exhibited 100 per cent viability. High viability status even after 2-year storage under normal conditions suggests the suitability of seeds for regeneration/cultivation of this species and possesses excellent storability.

CONCLUSION

Freshly harvested seeds of *H. salicifolia* possessed non-deep physiological dormancy due to the presence of immature embryo which became fully developed and mature during progression of storage period (2-year). Higher germination in stored seeds might be due to higher α-amylase activity and low phenolic contents as compared to freshly harvested seeds. No change in viability during 2-year storage suggests that seeds can be stored for a longer period at room temperature for the cultivation of this species. The present findings are of significance for the conservation and cultivation of *H. salicifolia.*

REFERENCES

1. Airi, S., Bhatt, I.D., Bhatt, A., Rawal, R.S. and Dhar, U. 2009.Variations in seed germination of *Hippophae salicifolia* with different pre-soaking treatments. *J. For. Res.* 20 (1): 27-30.
2. Amri, E. 2010. Germination of *Terminalia sericea* Buch ex DC seeds: Effect of temperature regime, photoperiod, gibberellic acid and potassium nitrate. *Am. Eurasian J. Agric. Environ. Sci.* 8: 722-727.
3. Atici, O., Agar, G. and Battal, P. 2007. Influence of long term storage on plant growth substances levels, germination and seedling growth in legume seeds stored for 37 years. *Ind. J. Plant Physiol.* 12:1-5.
4. Baskin, C.C. and Baskin, J.M. 1998. *Seed: Ecology, Biography and Evolution of Dormancy and Germination*. Academic Press. New York.
5. Bewley, J.D. and Black, M. 1982. *Physiology and Biochemistry of Seeds.* Vol. II. Springer-Verlag, Berlin, Heidelberg, New York, 375p.
6. Evenari, M. 1965. Light and seed dormancy. In: *Encyl. Plant Physiol. Vol.15: Differentation and Development* (W. Ruhland, Ed.), p. 804-807, Springer-Verlag, Berlin, 1362p.
7. Filner, P. and Varner, J.E. 1967. A test for *de novo* synthesis of enzyme density labeling with H_2O^{18} of barley α-amylase induced by gibberellic acid. *Biochem.* 58: 1520-1526.
8. Geetha, S., Sai Ram, M., Singh, V., Iiavazhangan, G. and Sawhney, R.C. 2002. Antioxidant and immunomodulatory properties of seabuckthorn (*H. rhamnoides*): An *in-vitro* study. *J. Ethnopharmaco.* 79: 373-378.
9. Goel, H.C., Prasad, J., Singh, S., Sagar, R.K., Kumar, I.P. and Sinha, A.K. 2002. Radio protection by herbal preparation of *Hippophae rhamnoides*, RH-3, against whole body lethal irradiation in mice. *Phytomedicine* 9: 15-25.
10. Goldstein, J.L.and Swain, T. 1963. Changes in tannins in ripening fruits. *Phytochemistry* 2: 371-383.
11. Gupta, S.M., Pandey, P., Grower, A. and Ahmed, Z. 2011. Breaking seed dormancy in *Hippophae salicifolia*, a high value medicinal plant. *Physiol. Mol. Biol. Plants* 17 (4): 403-408.
12. Gupta, R. and Flora, S.J.S. 2006. Protective effects of fruit extracts of *Hippophae rhamnoides* L. against arsenic toxicity in Swiss albino mice. *Human Exp. Toxico.* 25 (6): 285-295.
13. Hartmann, H.T., Kester, D.E. and Davis, E.T. 1989. *Plant Propagation: Principles and Practices.* Prentice Hall, USA, 600p.
14. Hirasawa, E. 1989. Auxin induced α-amylase activity in pea cotyledons. *Plant Physiol.* 91: 484-486.
15. Internaional Seed Testing Association (ISTA). 1966. *Proc. Inter. Seed Testing Assoc.* 31: 1-152.
16. Jeppsson, N. and Gao, X.Q. 2000. Changes in the contents of kaempherol, quercetin and L-ascorbic acid in seabuckthorn berries during maturation. *Agril. Food Sci.* 9:17-22.
17. Joshi, M. and Dhar, V. 2003. Effect of various pre-sowing treatments on seed germination of *Heracleum candicans* Wall., a high value medicinal plant. *Seed Sci. Technol.* 31: 737-743.

18. Kopyra, M. and Gwozdz, E.A. 2003. Nitric oxide stimulates seed germination and counteracts the inhibitory effect of heavy metals and salinity on root growth of *Lupinus lutens*. *Plant Physiol. Biochem.* 41: 1011-1017.

19. Korekar, G., Dwivedi,.K., Singh, H., Srivastava, R.B. and Stobdan, T. 2013. Germination of *Hippophae rhamnoides* L. seed after 10 years of storage at ambient condition in cold arid trans-Himalayan Ladakh region. *Curr. Sci.* 104 (1): 110-114.

20. Korovina, M.A. and Fefelov, V.A. 2006. Biochemical variations in seabuckthorn (*Hippophae rhamnoides* L.) growing in different regions of CIS states; Seabuckthorn, *Hippophae* L. a multipurpose wonder plant. *Biochem. Pharmaco.* 2: 108-132.

21. Lowry, H. Rosebrough, N.J., Farr, A.L. and Rahnall, R.J. 1951. Protein measurement with the folin-phenol reagent. *J. Bio. Chem.* 193: 265-275.

22. Nadeem, M., Palni, L.M.S., Purohit, A.N., Pandey, H. and Nandi, S.K. 2000. Propagation and conservation of *Podophyllum hexandrum* Royle-An important medicinal herb. *Biol. Conserv.* 92: 121-129.

23. Neill, S.J., Desikan, R. and Hancock, J.T. 2003. Nitric acid signaling in plants. *New Phytol.* 159: 11-35.

24. Olmez, Z. 2011. Effect of cold stratification and H_2SO_4 on seed germination of seabuckthorn (*Hippophae rhamnoides* L.). *Afr. J. Biotechnol.* 10: 4586-4590.

25. Parmar, B.S. and Kumar, J. 2006. Formulation for improved seed storage and quality. *Seed Res.* 34 (2): 215-217.

26. Roberts, E. and Smith, R.D. 1977. Dormancy and the pentose phosphate pathway. In: *The Physiology and Biochemistry of Seed Dormancy and Germination* (A. Khan, Ed), p. 385-411, North Holland public. Co., Amsterdam, 447p.

27. Roberts, E.H. 1969. Seed dormancy and oxidation processes. In: *Dormancy and Survival*, SEB Symposia, p. 45-79, Academic Press, New York.

28. Sankhyan, H.P., Sehgal, R.N. and Bhrol, N.P. 2004. Effect of growth regulators and salt solutions on seed germination and seedling vigour of seabuckthorn (*Hippophae tibetana* Schecht). *Seed Res.* 32 (1): 55-57.

29. Singh, V. 2003. Geographical adaptation and distribution of seabuckthorn (*Hippophae* L.) resources. In: Seabuckthorn *(Hippophae L.)-A Multipurpose Wonder Plant. Vol. I: Botany, Harvesting and Processing Technologies* (V. Singh, Ed. in Chief, 2003), p. 21-34, Indus Publishing Company, New Delhi, 518p.

30. Singh, V. and Morsel, J. T. 2006. Development and commercialization of seabuckthorn: A German experience. In: *Seabuckthorn (Hippophae L.)-A Multipurpose Wonder Plant. Vol. 2: Biochemistry and Pharmacology* (V. Singh, Ed. in Chief, 2006), p. 576-584. Daya Publishing House, New Delhi, 600p.

31. Singh, V., Nayyar, H., Gupta, R.K. and Uppal, R. 2005. Germination behaviour of different biotypes of seabuckthorn (*Hippophae* L.). *NonTimber Forest Products* 12 (4): 209-213.

32. Stokes, P. 1965. Temperature and seed dormancy. In: *Encyl. Plant Physiol. Vol.15/2*, p. 746-803, Springer-Verlag, New York/Heidelberg/Berlin.

33. Zhang, H., Shen, W.B. and Xu, L.L. 2003. Effects of nitric oxide on germination of wheat seeds and its reactive oxygen species metabolisms under osmotic stress. *Acta Bot. Sin.* 45: 901-905.

34. Zhang, W., Yan, J., Duo, J., Ren, B. and Guo, J. 1989. Preliminary study of biochemical constituents of berries of seabuckthorn growing in Shanxi province and their changing trend. In: *Proceedings of the First International Symposium on Seabuckthorn* (*H.rhamnoides* L.), p. 96-105, October 19-23, Xian, China.

35. Zhang, Z.G., Rui, Q. and Xu, L.L. 2001. Relationship between endopeptidases and H-$_2O_2$ during wheat leaves ageing. *Acta Bot.Sin.* 43: 127-131.

Chapter 18

Research Progress on Symbiotic Nitrogen Fixation in *Frankia*-Seabuckthorn Root Nodules

Yoshinori Kanayama and Kazuhisa Kato

Graduate School of Agricultural Science, Tohoku University, Sendai 981-8555, Japan
Faculty of Agriculture, Iwate University, Morioka 020-8550, Japan

SUMMARY

The symbiosis between seabuckthorn and *Frankia* likely guarantees the normal plant growth in infertile soil and growth may be accelerated further by the application of nitrogen fertilizer. Actinorhizal plants are symbiotic plants infected with *Frankia* of woody trees and shrubs, which is essential for the formation of root nodules that fixes nitrogen. Nitrate inhibition of nitrogenase activity may hinder high yield cultivation of seabuckthorn. Nitrogenase activity in seabuckthorn root nodules is low in November when leaf yellowing and defoliation begin, and it is not detected in January when trees have no leaves. The highest nitrogenase activity occurs in May when leaves open and current shoots grow. The high nitrogenase activity in seabuckthorn root nodules between May and September may be related to the appropriate temperature and active photosynthesis in leaves. To utilize symbiotic nitrogen fixation efficiently, seabuckthorn plants should be inoculated with *Frankia* and mycorrhizal fungi. Nitrogen fertilizer should not be supplied initially because nodulation is inhibited by nitrates. Although, nitrogenase activity in seabuckthorn root nodules is only slightly inhibited by short-term nitrate treatment, soil nitrates should be at low concentrations so that they do not persist in the soil and inhibit nitrogen fixation over the long term.

Keywords: *Seabuckthorn, Root nodules, Nitrogen fixation and Symbiosis.*

INTRODUCTION

Seabuckthorn (*Hippophae rhamnoides* L.) belongs to the family *Elaeagnaceae*, is a small fruit tree that is widely distributed throughout Europe and Asia. The genus *Hippophae* is comprised of 6 species and 12 subspecies (Li *et al.*, 2009). Seabuckthorn is a deciduous and dioecious plant with thorny branches. Seabuckthorn plants are drought-and cold-tolerant (Li *et al.*, 2005) and are also actinorhizal

Seabuckthorn (*Hippophae* L.): *A Multipurpose Wonder Plant, Vol. 4: Emerging Trends in Research and Technologies* (Virendra Singh, *Editor-in-Chief*, 2014), p. 242–247, Daya Publishing House, New Delhi, India.

plants that are infected with an actinomycetic endophyte *Frankia* of woody trees and shrubs, which is essential for the formation of root nodules that fix nitrogen (Gentili and Huss-Danell, 2002; Kato *et al.*, 2007; Kanayama *et al.*, 2009). Thus, seabuckthorn grows in nutritionally poor environments where it is difficult to grow other crops. Gupta and Singh (2003) have already reported the nodule morphology and nitrogen fixation capacity of seabuckthorn. Therefore, we focus on research progress on nitrogen fixation in the root nodules of seabuckthorn in this chapter.

NITROGEN FIXATION IN *FRANKIA*-SEABUCKTHORN ROOT NODULE

Actinorhizal plants are symbiotic plants infected with *Frankia*, which are classified into 8 families and 25 genera comprising more than 220 species including seabuckthorn (Wall, 2000). Plants of the above mentioned families can be used as a pioneer plant because they can utilize nitrogen gas by symbiotic nitrogen fixation. Mature root nodules with many lobes in a mature seabuckthorn tree and immature root nodules with a few lobes in a seabuckthorn seedling are shown in Kato *et al.* (2007). The number of lobes increases during growth. These are standard morphological characteristics of *Frankia*-actinorhizal root nodules (Bond, 1976; Wall, 2000). The diameter of the lobe is approximately 1 mm and contains a central vascular bundle and infected cells in the cortex (Kato *et al.*, 2007). Phylogenetic analysis shows that *Frankia* 16S rDNA sequences cluster into four host plant specificity groups (Clawson *et al.*, 1998), and that *Frankia* of seabuckthorn belongs to the *Elaeagnus* group (Kato *et al.*, 2007) suggesting host-symbiont cospeciation.

Legume crops have a symbiotic relationship with *Rhizobium* and they can utilize nitrogen gas by symbiotic nitrogen fixation. In legume crops, such as soybean, the percentage of nitrogen fixed in the root nodules is roughly 50 per cent of the total amount of nitrogen accumulated by the plant (Yoneyama *et al.*, 1986). On the other hand, in actinorhizal plants, such as *Ceanothus velutinus*, the percentage of nitrogen fixed in the root nodules is more than 80 per cent of the total amount of nitrogen accumulated by the plant (Busse, 2000). Furthermore, Wall (2000) reported that the annual amount of nitrogen (240-350 kg $ha^{-1}year^{-1}$) fixed in *Frankia*-actinorhizal root nodules is similar to that in *Rhizobium*-legume root nodules. The level of nitrogenase activity in seabuckthorn root nodules is similar to that in legume root nodules, indicating a high potential for *Frankia*-seabuckthorn symbiosis (Kato *et al.*, 2007). In the symbiotic relationship between seabuckthorn and *Frankia*, seedlings inoculated with a homogenate of root nodules begin to develop root nodules in about two weeks (Bond *et al.*, 1956). An apparent difference in growth is observed between the plants, with and without root nodules, one month after inoculation (Kato *et al.*, 2007). This suggests that nodulation and expression of nitrogenase activity occur within one month of inoculation. Seabuckthorn seedlings inoculated with a homogenate of root nodules grow normally under nitrogen-free conditions. The fresh weight of the root nodules is positively correlated with plant growth (Bond *et al.*, 1956; Bond and Gardner, 1957; Kato *et al.*, 2007).

NITRATE INHIBITION AND SEASONAL CHANGES IN NITROGEN FIXATION

The positive correlation between the fresh weight of root nodules and plant growth suggests that plant growth is enhanced by nodulation. The symbiosis between seabuckthorn and *Frankia* likely guarantees the normal plant growth in infertile soil. Additionally, growth may be accelerated further by the application of nitrogen fertilizer. In this case, the application of the nitrogen fertilizer may be necessary for higher yields. However, nitrates generally inhibit nodulation, nodule growth, and nitrogenase in legume root nodules (Streeter, 1988). Nitrates also inhibit nodulation and nitrogen fixation in actinorhizal plants (Huss-Danell, 1997), including seabuckthorn (Table 18.1; Gentili and Huss-Danell, 2002). Nitrate inhibition of nitrogenase activity may hinder high-yield cultivation of seabuckthorn.

Table 18.1: Relative Level of N_2 Fixation in Seabuckthorn Root Nodules

Plant Material	Nitrate Concentration (mM)	Treatment (Day)	Treatment (Month)	Relative Level of N_2 Fixation per Nodule Fresh Weight	Relative Level of N_2 Fixation per Plant	Reference
1 year old plants	0	–	2.5	–	100	Bond and Mackintosh (1975)
	0.71	–	2.5	–	66	
	1.79	–	2.5	–	35	
Rooted cuttings	0	–	1.0	100	–	Kato *et al.* (2007)
	5	–	1.0	22	–	
	10	–	1.0	21	–	
	0	10	–	100	–	
	5	10	–	104	–	
	10	10	–	56	–	
	0	5	–	100	–	
	10	5	–	161	–	
	30	5	–	111	–	

Nitrogenase activity in seabuckthorn root nodules is inhibited by low concentration nitrate treatment for periods of more than one month (Table 18.1). However, it is unclear whether nitrogenase activity in actinorhizal root nodules is inhibited by nitrate treatment in the short-term period. In the root nodules of the legumes soybean and cowpea, nitrogenase activity decreases to approximately 50 per cent with the addition of 10 and 20 mM of nitrates over short-term periods of 2 and 3 days, respectively (Kanayama *et al.*, 1990; Kanayama and Yamamoto, 1991). Seabuckthorn root nodules are apparently nitrate-tolerant as compared with legume root nodules because nitrogenase activity is inhibited neither by 30 mM nitrates in the short term of 5 days nor by 5 mM nitrates until 10 days (Table 18.1). In legume root nodules, nitrogenase is protected from oxygen by an oxygen diffusion barrier in the nodule cortex (Minchin, 1997). In addition, the symbiotic hemoglobin leghemoglobin enhances the oxygen supply to bacteroids in the infected region. The oxygen diffusion barrier and leghemoglobin may be related to the short-term inhibition by nitrates involved in nitrogenase activity in legume root nodules (Kanayama and Yamamoto, 1990; Vessey and Waterer, 1992), and we also suggest that nitrate reductase and nitric oxide are involved in the inhibition in lotus root nodules (Kato *et al.*, 2003; Kato *et al.*, 2010). On the other hand, in actinorhizal root nodules, the envelope surrounding *Frankia* vesicles in infected cells is considered to retard the diffusion of oxygen into the nitrogenase containing vesicle (Huss-Danell, 1997). Based on the colour of the infected region of the root nodules, symbiotic hemoglobin is not observed in seabuckthorn root nodules, although both symbiotic and non-symbiotic hemoglobin from *Casuarina glauca* have been identified (Guldner *et al.*, 2004). The non-symbiotic hemoglobin gene from *Myrica gale* is highly expressed in the root nodules (Heckmann *et al.*, 2006). Thus, the different responses to nitrates in the nitrogenase activity between legumes and seabuckthorn are presumably, due to the differences in the mechanism of oxygen diffusion regarding the barrier and hemoglobin.

Nitrogenase activity in seabuckthorn root nodules is low in November when leaf yellowing and defoliation begin, and it is not detected in January when trees have no leaves (Kato *et al.*, 2007). The highest nitrogenase activity occurs in May when leaves open and current shoots grow. The activity gradually decreases from July to September when the growth of the current shoots ends. In the root nodules of *Alnus incana* (*Betulaceae*), nitrogenase activity starts increasing when the leaves open in spring, and it decreases with leaf fall (Huss-Danell *et al.*, 1992). Nitrogenase activity in legume root nodules depends on temperature and current photosynthate levels (Ryle *et al.*, 1989; Guy *et al.*, 1997). The high nitrogenase activity in seabuckthorn root nodules between May and September may be related to the appropriate temperature and active photosynthesis in leaves. Plants were inoculated with a homogenate of seabuckthorn root nodules.

EFFICIENT USE OF SYMBIOTIC NITROGEN FIXATION

It is presumed that seabuckthorn cultivation increases the density of symbiotic *Frankia* in the field and also enhances nodulation as observed in legume crops. An adequate supply of phosphate is an essential requirement for nodulation of seabuckthorn (Gentili and Huss-Danell, 2002). Mycorrhizal association improves phosphate uptake in seabuckthorn, and its association with *Glomus fasciculatus* can support plant growth and nitrogenase activity (Gardner *et al.*, 1984). In fact, the growth of the Tibetan seabuckthorn seedling and nitrogen fixation are also supported by dual inoculation with *Frankia* and mycorrhizal fungi (Tian *et al.*, 2002).

To utilize symbiotic nitrogen fixation efficiently, seabuckthorn plants should be inoculated with *Frankia* and mycorrhizal fungi. Nitrogen fertilizer should not be supplied initially because nodulation is inhibited by nitrates. Although, nitrogenase activity in seabuckthorn root nodules is only slightly inhibited by short-term nitrate treatment, soil nitrates should be at low concentrations so that they do not persist in the soil and inhibit nitrogen fixation over the long term.

CONCLUSION

The seabuckthorn-*Frankia* symbiosis likely guarantees the normal plant growth in infertile soil and may be accelerated further by the application of nitrogen fertilizer. It is important to determine the appropriate timing of fertilizer application based on seasonal changes in nitrogenase activity.

REFERENCES

1. Bond, G. 1976. The results of the IBP survey of root-nodule formation in non-leguminous angiosperms. In: *Symbiotic Nitrogen Fixation in Plants* (P.S. Nutman, Ed.), p. 443-474, Cambridge University Press, Cambridge, 567p.
2. Bond, G. and Gardner, I.C. 1957. Nitrogen fixation in non-legume root nodule plants. *Nature* 179: 680-681.
3. Bond, G. and Mackintosh, A.H. 1975. Effect of nitrate-nitrogen on the nodule symbioses of *Coriaria* and *Hippophae*. In: *Proc. Royal Soc. London Series B; Biol. Sci*. 190: 199-209.
4. Bond, G., MacConnell, J.T. and McCallum, A.H. 1956. The nitrogen-nutrition of *Hippophae rhamnoides* L. *Annals Bot*. 20: 501-512.
5. Busse, M.D. 2000. Suitability and use of the ^{15}N-isotope dilution method to estimate nitrogen fixation by actinorhizal shrubs. *Forest Ecol. Manag*. 136: 85-95.
6. Clawson, M.L., Caru, M. and Benson, D.R. 1998. Diversity of *Frankia* strains in root nodules of plants from the families *Elaeagnaceae* and *Rhamnaceae*. *Appl. Environ. Microbiol*. 64: 3539-3543.

7. Gardner, I.C., Clelland, D.M. and Scott, A. 1984. Mycorrhizal improvement in non-leguminous nitrogen fixing associations with particular reference to *Hippophae rhamnoicles* L. *Plant Soil* 78: 189-199.

8. Gentili, F. and Huss-Danell, K. 2002. Phosphorus modifies the effect of nitrogen on nodulation in split-root systems of *Hippophae rhamnoides*. *New Phytol*. 153: 53-61.

9. Guldner, E., Godelle, B. and Galtier, N. 2004. Molecular adaptation in plant hemoglobin, a duplicated gene involved in plant-bacteria symbiosis. *J. Mol. Evol*. 59: 416-425.

10. Gupta, R.K. and Singh, V. 2003. Nitrogen fixation in seabuckthorn (*Hippophae rhamnoides* L.). In: *Seabuckthorn (Hippophae L.)-A Multipurpose Wonder Plant. Vol. I: Botany, Harvesting and Processing Technologies* (V. Singh, Ed. in Chief, 2003), p. 286-299, Indus Publishing Company, New Delhi, 518p.

11. Guy, S., Berger, M. and Planchon, C. 1997. Response temperature in dinitrogen fixing soybeans. *Plant Sci*. 123: 67-75.

12. Heckmann, A.B., Hebelstrup, K.H., Larsen, K., Micaelo, N.M. and Jensen, E.O. 2006. A single hemoglobin gene in *Myrica gale* retains both symbiotic and non-symbiotic specificity. *Plant Mol. Biol*. 61: 769-779.

13. Huss-Danell, K. 1997. Actinorhizal symbioses and their N_2 fixation. *New Phytol*. 136: 375-405.

14. Huss-Danell, K., Lundquist, P.O. and Ohlsson, H. 1992. Nitrogen fixation in a young *Alnus incana* stand, based on seasonal and diurnal variation in whole plant nitrogenase activity. *Can. J. Bot*. 70: 1537-1544.

15. Kanayama, Y., Ohkawa, W., Chiba, E., Sato, K., Ofosu-Anim, J. and Kanahama, K. 2009. Nutritional components and nitrogen fixation in seabuckthorn (*Hippophae rhamnoides* L.). *Acta Hort*. 806: 309-314.

16. Kanayama, Y., Watanabe, I. and Yamamoto, Y. 1990. Inhibition of nitrogen fixation in soybean plants supplied with nitrate I. Nitrite accumulation and formation of nitrosylleghemoglobin in nodules. *Plant Cell Physiol*. 31: 341-346.

17. Kanayama, Y. and Yamamoto, Y. 1990. Inhibition of nitrogen fixation in soybean plants supplied with nitrate. II. Accumulation and properties of nitrosylleghemoglobin in nodules. *Plant Cell Physiol*. 31: 207-214.

18. Kanayama, Y. and Yamamoto, Y. 1991. Formation of nitrosylleghemoglobin in nodules of nitrate-treated cowpea and pea plants. *Plant Cell Physiol*. 32: 19-24.

19. Kato, K., Kanahama, K. and Kanayama, Y. 2010. Involvement of nitric oxide in the inhibition of nitrogenase activity by nitrate in lotus root nodules. *J. Plant Physiol*. 167: 238-241.

20. Kato, K., Kanayama, Y., Ohkawa, W. and Kanahama, K. 2007. Nitrogen fixation in seabuckthorn (*Hippophae rhamnoides* L.) root nodules and effect of nitrate on nitrogenase activity. *J. Japan Soc. Hort. Sci*. 76: 185-190.

21. Kato, K., Okamura, Y., Kanahama, K. and Kanayama, Y. 2003. Nitrate-independent expression of plant nitrate reductase in *Lotus japonicus* root nodules. *J. Exp. Bot*. 54: 1685-1690.

22. Li, H., Ruan, C.J. and Teixeira da Silva, J.A. 2009. Identification and genetic relationship based on ISSR analysis in a germplasm collection of seabuckthorn (*Hippophae* L.) from China and other countries. *Sci. Hort*. 123: 263-271.

23. Li, C., Yang, Y., Junttila, O. and Palva, E. T. 2005. Sexual differences in cold acclimation and freezing tolerance development in seabuckthorn (*Hippophae rhamnoides* L.) ecotypes. *Plant Sci.* 168: 1365-1370.

24. Minchin, F.R. 1997. Regulation of oxygen diffusion in legume nodules. *Soil Biol. Biochem.* 29: 881-888.

25. Ryle, G.J., Powell, C.E., Timbrell, M.K. and Gordon, A.J. 1989. Effect of temperature on nitrogenase activity in white clover. *J. Exp. Bot.* 40: 733-739.

26. Streeter, J. 1988. Inhibition of legume nodule formation and N_2 fixation by nitrate. *CRC Critical Rev. Plant Sci.* 7: 1-23.

27. Tian, C.J., He, X.Y., Zhong, Y. and Chen, J.K. 2002. Effects of VA *mycorrhizae* and *Frankia* dual inoculation on growth and nitrogen fixation of *Hippophae tibetana*. *For. Ecol. Manage.* 170: 307-312.

28. Vessey, J.K. and Waterer, J. 1992. In search of the mechanism of nitrate inhibition of nitrogenase activity in legume nodules. Recent developments. *Physiol. Plant.* 84: 171-176.

29. Wall, L.G. 2000. The actinorhizal symbiosis. *J. Plant Growth Reg.* 19: 167-182.

30. Yoneyama, T., Nakano, H., Kuwahara, M., Takahashi, T., Kambayashi, I. and Ishizuka, J. 1986. Natural ^{15}N abundance of field grown soybean grains harvested in various in various locations in Japan and estimate of the fractional contribution of nitrogen fixation. *Soil Sci. Plant Nutr.* 32: 443-450.

PART V: CHEMISTRY AND BIOCHEMISTRY

Chapter 19

Influence of Latitude and Altitude on the Composition of Wild Chinese Seabuckthorn (*Hippophae rhamnoides* ssp. *sinensis*)

Jie Zheng, Heikki Kallio, Kaisa Linderborg and Baoru Yang

Department of Biochemistry and Food Chemistry,
University of Turku, FI-20014 Turku, Finland

SUMMARY

The latitude and altitude affected the composition of sugars, sugar alcohols, fruit acids and ascorbic acid in wild seabuckthorn (*H. rhamnoides* ssp. *sinensis*). Among the samples collected from six growth areas in China, the berries from Inner Mongolia had the highest contents of sugars and L-quebrachitol and the lowest contents of malic acid and ascorbic acid ($p < 0.05$). In contrast, the berries from Sichuan had the lowest contents of sugars and L-quebrachitol and the highest contents of malic acid and ascorbic acid ($p < 0.05$). The changes in the berry composition were oppositely associated with the variation in latitudes and altitudes. The values of fructose, glucose, L-quebrachitol, total sugar and sugar/acid ratio all correlated positively with the latitude but negatively with the altitude ($p < 0.01$). In contrast, the contents of malic acid, citric acid, ascorbic acid and total acid correlated negatively with the latitude but positively with the altitude ($p < 0.01$).

Keywords: *Chinese seabuckthorn (Hippophae rhamnoides ssp. sinensis), Latitude, Altitude and Composition.*

INTRODUCTION

Seabuckthorn has been used as an ingredient in traditional Chinese medicine since the ancient times. The berry and berry fractions have beneficial effects on the skin, mucosa, eyes, cardiovascular system and sugar metabolism. *Hippophae rhamnoides* ssp. *sinensis* is an important natural resource in China. In the present study, wild berries of *H. rhamnoides* ssp. *sinensis* were collected from nine natural growth sites in China in order to investigate the effects of the latitudes and altitudes of the growth locations on the composition of the sugars, sugar alcohols, fruit acids, and ascorbic acid in the berries.

Seabuckthorn (*Hippophae* L.): *A Multipurpose Wonder Plant, Vol. 4: Emerging Trends in Research and Technologies* (Virendra Singh, *Editor-in-Chief,* 2014), p. 251–257, Daya Publishing House, New Delhi, India.

MATERIALS AND METHODS

Wild berries of *H. rhamnoides* ssp. *sinensis* were collected from nine growth locations in six provinces in China from 2006 to 2008 (Table 19.1). The berries were harvested in duplicate or quadruplicate and were loosely frozen immediately after picking and kept at -20 °C before being analyzed. Gas chromatographic analysis of sugars, sugar alcohols, fruit acids and ascorbic acid of each sample were performed in quadruplicate as described earlier (Zheng *et al.*, 2009a). The column temperature of gas chromatography was programmed as follows: 2 min at 150 °C, rise to 210 °C at a rate of 6 °C/min, and to the final temperature of 275 °C at a rate of 40 °C/min, and hold at 275 °C for 10 min. The ethyl β-D-glucopyranoside (henceforth ethyl glucose) was quantified as glucose and methyl-*myo*-inositol as L-quebrachitol. The total sugar content was defined as the sum of sugars, sugar derivatives, and sugar alcohols. The total acid content was defined as the sum of malic acid, citric acid, quinic acid and ascorbic acid.

Table 19.1: Locations and Spatial Parameters of Sample Collection in China

Growth Site	*Area*	*Longitude*	*Latitude*	*Altitude (m)*
Suiling Town, Suiling County	Heilongjiang	127°06'E	47°14'N	210
West Channel, Liu Jianfang, Dage Town, Fengning County	Hebei	116°34'E	41°17'N	818
La'erguan, Dongxia Village, Huangyuan County, Xi'ning	Qinghai	101°23'E	36°45'N	3115
Jiucheng Palace, Hantai Town, Dongsheng District, Ordos	Inner Mongolia	109°48'E	39°47'N	1480
Beizhe Channel, Li Jiahui, San Daochuan Forestry Centre, Guan Dishan, Luliang	Shanxi	113°52'E	37°05'N	1512
Paoma Weir, Xihua Town, Qian'nian Forestry Centre, Guan Dishan, Lüliang		113°52'E	37°05'N	2182
Natural Reserve, Wolong, Wenchuan County	Sichuan	106°54'E	31°01'N	2000, 2500,3000

RESULTS AND DISCUSSION

Compositional Analysis

Fructose (0.01-7.17 g/100 ml juice) and glucose (0.05-7.85 g/100 ml) were the two major sugars and L-quebrachitol (1L-2-*O*-methyl-*chiro*-inositol, 0.21-1.09 g/100 ml) was the major sugar alcohol in the seabuckthorn berries (Figure 19.1). Malic acid (1.55-8.84 g/100 ml) and quinic acid (0.07-2.94 g/100 ml) were the two major acids in the berry juice. The content of ascorbic acid varied from 0.25-1.66 g/100 ml juice. Sucrose (≤ 0.31 g/100 ml), methyl-*myo*-inositol (≤ 0.16 g/100 ml), *myo*-inositol (≤ 0.07 g/100 ml), ethyl glucose (≤ 0.04 g/100 ml) and citric acid (≤ 0.13 g/100 ml) were the minor components in the berries.

Comparison of Berries from Different Growth Areas

As shown in Figure 19.1, the composition of seabuckthorn berries from different growth areas varied widely. The berries from Inner Mongolia had the highest contents of fructose (6.49 g/100 ml juice), glucose (6.89 g/100 ml), L-quebrachitol (0.99 g/100 ml) and total sugar (14.53 g/100 ml), and the lowest contents of malic acid (1.84 g/100ml), ascorbic acid (0.32 g/100 ml) and total acid (4.04 g/100 ml), and hence the highest sugar/acid ratio (3.73). In contrast, the berries from Sichuan contained the lowest levels of fructose (0.08 g/100 ml), glucose (0.13 g/100 ml), L-quebrachitol (0.29 g/100 ml), quinic acid (0.29 g/100 ml), total sugars (0.59 g/100 ml) and sugar/acid ratio (0.07), and the highest levels of malic acid (6.42 g/100 ml) and ascorbic acid (1.21 g/100 ml). The total acid content (7.98 g/

Figure 19.1: Contents of Sugars and Sugar Alcohols (A), Fruit Acids and Ascorbic Acid (B) and Quality Parameters (C) in Seabuckthorn Berries from different Growth Areas in China. Results of each component marked with different letters are significantly different ($p < 0.05$). *Source*: Zheng *et al.*, 2009b.

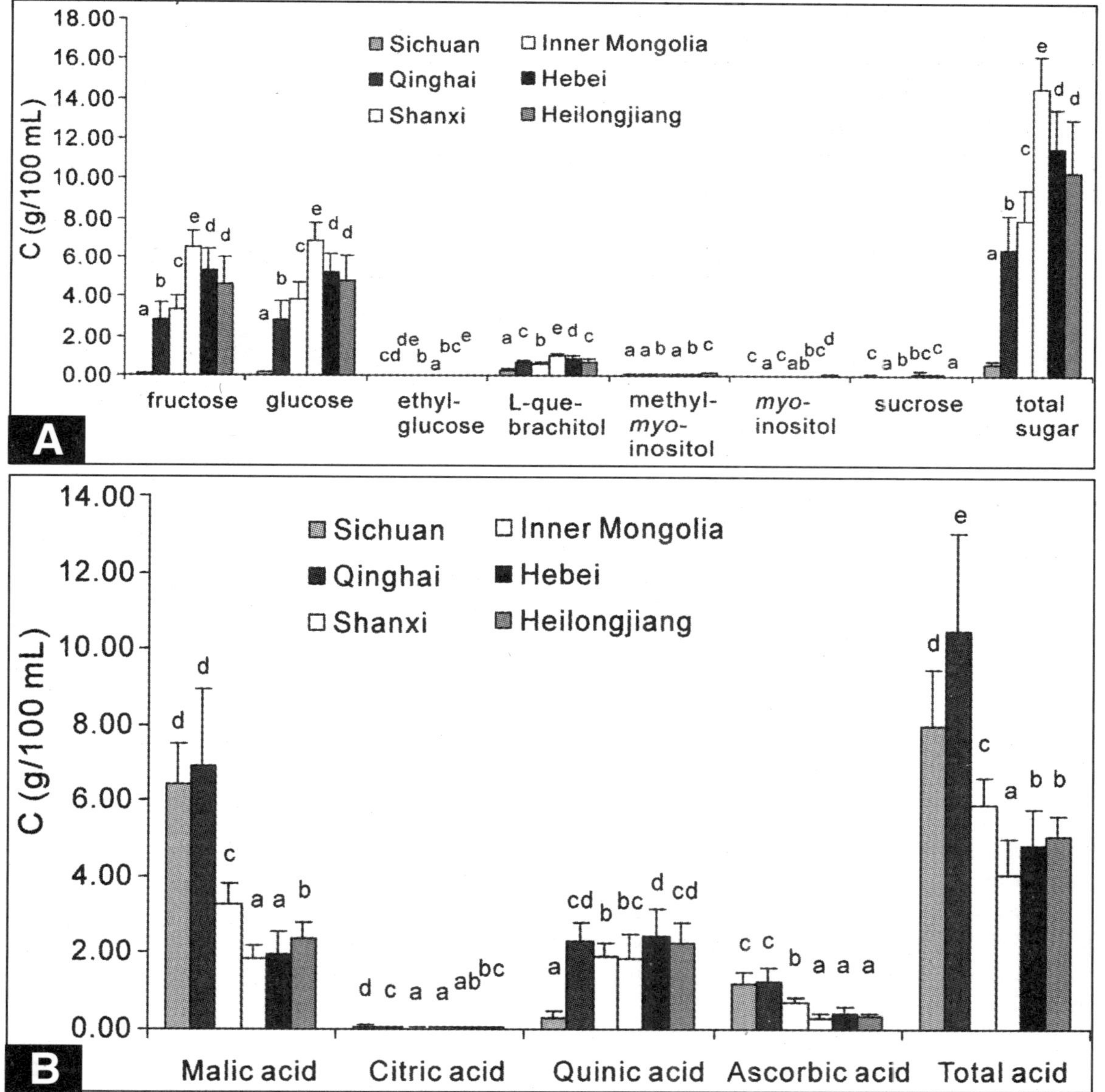

Contd...

Figure 19.1–*Contd...*

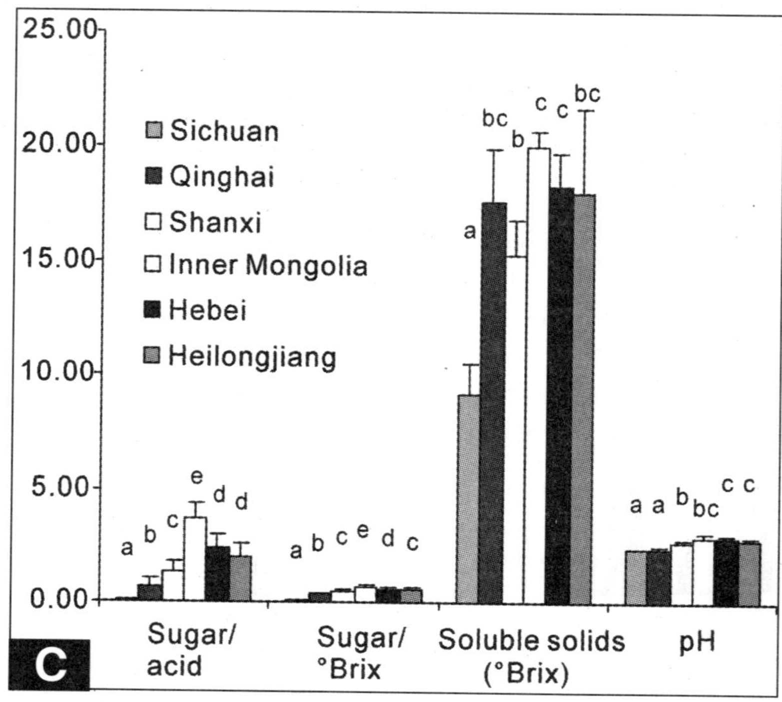

100 ml) in berries from Sichuan was significantly higher than all the other samples, except for those from Qinghai (10.51 g/100 ml) (Zheng *et al.*, 2009b). The sugar/acid ratio and the total sugar content are important factors determining the sweetness and the fruity taste of berries and fruits (Tiitinen *et al.*, 2005). Therefore, the berries from Inner Mongolia were supposed to be the sweetest and the most favoured and the berries from Sichuan were expected to be the sourest among the berry samples studied.

Spatial Effects on Berry Composition

Clearly decreasing trends in the contents of major sugars (fructose and glucose) and increasing trend in content of major acid (malic acid) were found (Figure 19.2A) as the altitude increased and as the latitude decreased. As a result, the total sugar content decreased and the total acid content increased as altitude increased but as latitude decreased (Figure 19.2B). The berries collected from high latitude and low altitude were supposed to have better taste with higher sugar/acid ratio than those from low latitude and high altitude (Figure 19.2E). Moreover, decreasing trend of methyl-*myo*-inositol content (Figure 19.2C) and increasing trend of ascorbic acid content (Figure 19.2D) were detected as the altitude increased and as the latitude decreased.

Bivariate correlation analysis showed the values of fructose, glucose, L-quebrachitol, methyl-*myo*-inositol, quinic acid, total sugars, sugar/acid ratio, sugar/°Brix ratio, soluble solids and pH all correlated positively with the latitude but negatively with the altitude ($p<0.01$). In contrast, the contents of malic acid, citric acid, ascorbic acids, total acid and juice yield correlated negatively with the latitude but positively with the altitude ($p<0.01$) (Zheng *et al.*, 2011). It seems that latitude and altitude

Figure 19.2: Correlations between Spatial Parameters and Contents of Major Sugars (Fructose and glucose) and Major Acid (Malic Acid) (A), Total Sugar and Total Acid (B), Methyl-*myo*-inositol (C), Ascorbic Acid (D), and Sugar/Acid Ratio (E). The linear regression curve was drawn in accordance to the variation of altitude.

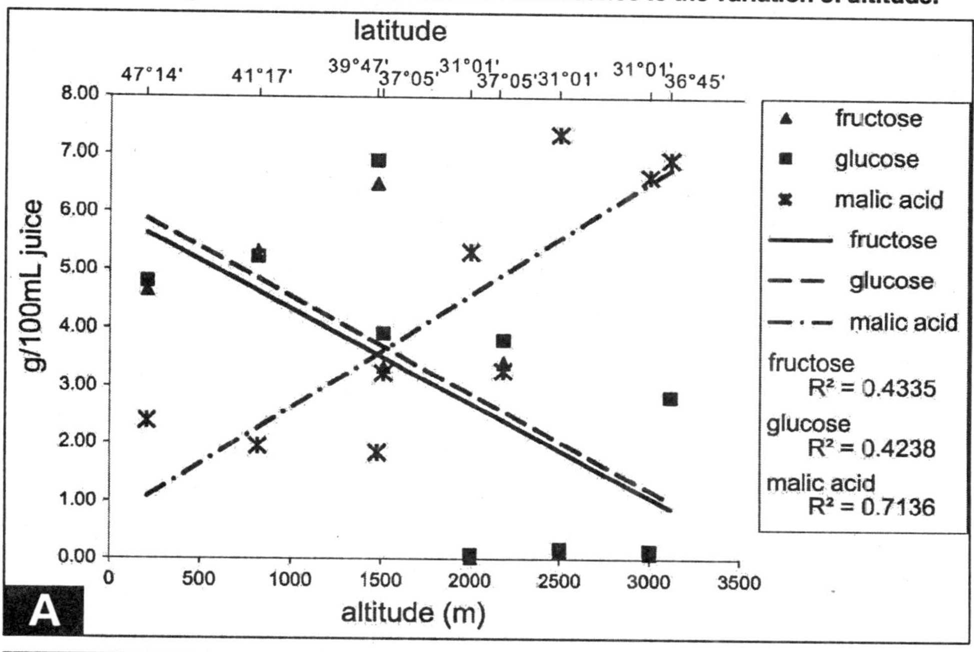

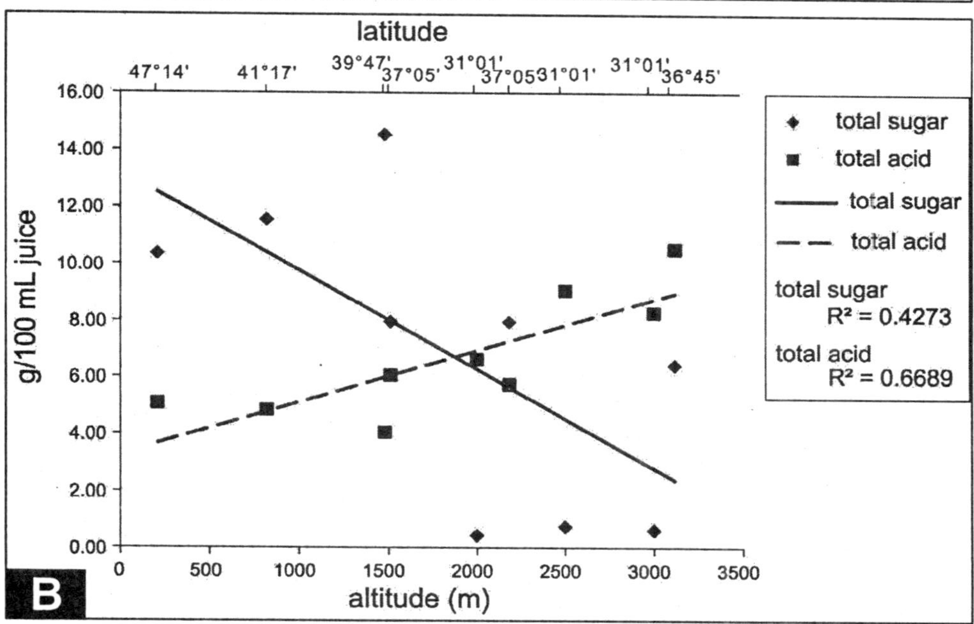

Contd...

Figure 19.2–*Contd...*

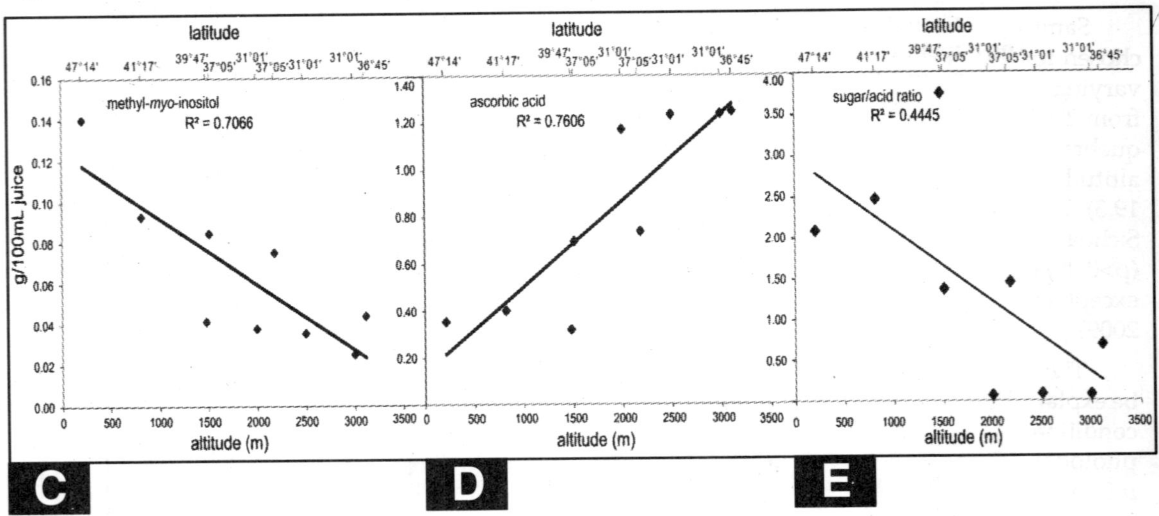

had opposite effects on the composition of the berries. This is in accordance with the results by Lian *et al.* (2000). It is important to notice that the concurrent changes in the altitude and the latitude made it difficult to clearly distinguish the effect of the altitude from that of the latitude on the composition of berries.

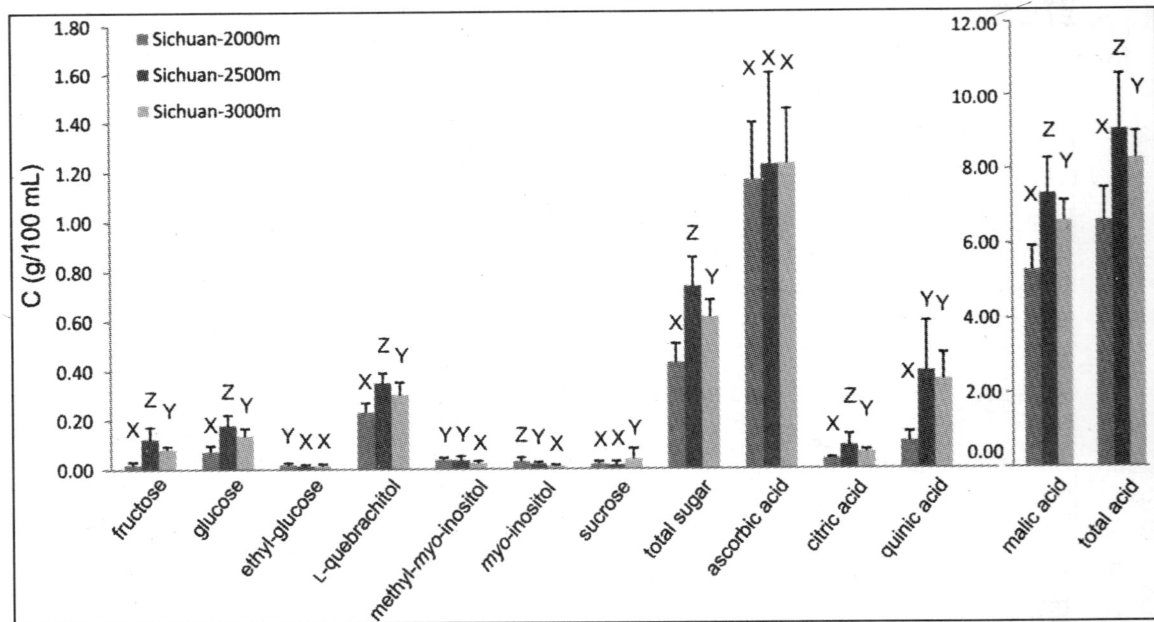

Figure 19.3: Composition of Seabuckthorn Berries Grown at different Altitudes in Sichuan Area. Results of each component marked with different letters are significantly different ($p < 0.05$).

Altitudinal Effects on Berry Composition

Samples collected from three altitudes in Sichuan area and two altitudes in Shanxi area were chosen for investigation of altitudinal effects on the berry composition without interference of the varying longitudes and latitudes. In the samples collected from Sichuan, when the altitude increased from 2000 m to 2500 m, the contents of major sugars (fructose and glucose), major sugar alcohol (L-quebrachitol), and major acid (malic acid) clearly increased ($p<0.05$). Yet, a further increase in the altitude from 2500 m to 3000 m led to a slight decrease in the contents of these compounds (Figure 19.3). The content of ascorbic acid did not vary significantly in berries grown at different altitudes in Sichuan ($p>0.05$). Concerning the seabuckthorn berries grown in Shanxi, there's no significant difference ($p>0.05$) in most of the compositional parameters between the samples collected from two altitudes except for a slightly higher content of quinic acid at the altitude of 1512 m than at 2182 m (Zheng *et al.*, 2009).

The influences of latitude and altitude on the biosynthesis of metabolites in plants and fruits may be explained by complex combinations of environmental factors, such as soil conditions and climatic conditions as well as the interaction between the genetic and the environmental factors. Light intensity, photoperiod, and temperature respond markedly to variation in latitude and altitude, and strongly influence the metabolism and concentration of metabolites in plants and fruits (Buchanan *et al.*, 2000). Yet, water supply and air humidity are also important factors affecting the composition of sugars and acids in berries and fruits (Zheng *et al.*, 2009a; Zheng *et al.*, 2009b). This study provides useful guidelines for berry breeding and cultivation as well as industrial utilization of seabuckthorn berries.

ACKNOWLEDGEMENT

We thank Mr. Tai Yuanlin from the Seabuckthorn Office, Ministry of Water Resources, China, Mr. Li Yunhai and Mrs. Xu Tao from Conseco Seabuckthorn Corporation, China, and Prof. Lu Rongsen from Chengdu Institute of Biology, Chinese Academy of Sciences, China for providing the samples.

REFERENCES

1. Buchanan, B.B., Gruissem, W. and Jones, R.L. 2000. *Biochemistry and Molecular Biology of Plants*, p. 630-675, Rockville MD: American Society of Plant Physiologists.
2. Lian, Y.S., Lu, S.G., Xue, S.K. and Chen, X.L. 2000. *Biology and Chemistry of the Genus Hippophae*, p. 107-110, Lanzhou: Gansu Science and Technology Press.
3. Tiitinen, K.M., Hakala, M.A. and Kallio, H.P. 2005. Quality components of seabuckthorn (*Hippophae rhamnoides*) varieties. *Journal of Agricultural and Food Chemistry* 53 (5): 1692-1699.
4. Zheng, J., Kallio, H. and Yang, B. 2009a. Effects of latitude and weather conditions on sugars, fruit acids and ascorbic acid in currant (*Ribes* sp.) cultivars. *Journal of the Science of Food and Agriculture* 89 (12): 2011-2023.
5. Zheng, J., Yang, B., Tuomasjukka, S., Ou, S. and Kallio, H. 2009b. Effects of latitude and weather conditions on contents of sugars, fruit acids, and ascorbic acid in black currant (*Ribes nigrum* L.) juice. *Journal of Agricultural and Food Chemistry* 57 (7): 2977-2987.
6. Zheng, J., Kallio, H., Linderborg, K. andYang, B. 2011. Sugars, sugar alcohols, fruit acids, and ascorbic acid in wild Chinese seabuckthorn (*Hippophae rhamnoides* ssp. *sinensis*) with special reference to influence of latitude and altitude. *Food Research International* 44: 2018-2026.

Chapter 20

A Comparative Study on Composition of Seabuckthorn Oil and other Edible Oils

Lu Rongsen

Chinese Academy of Sciences, Chengdu Institute of Biology, Chengdu, China

SUMMARY

With rich nutritional and bioactive substances, seabuckthorn oil is considered as one of the most valuable oils. It is necessary to determine why seabuckthorn fruit oil is unique and better than other nutritional oils. A study on the comparison of composition of seabuckthorn oil and other oils obtained from olive, walnut, wheat embryo and corn was carried out. The results of the analysis showed that seabuckthorn oil and other edible oils are rich in unsaturated fatty acids. The contents of unsaturated fatty acids in oils ranged from 70.0 per cent to 91.6 per cent, with seabuckthorn exhibited lower content of unsaturated fatty acids than others. The contents of vitamin E in seabuckthorn pulp oil and seed oil were are 206.6 and 213.0 mg/100 g, carotenoids 78.0 and 373.2 mg/100 g and β-carotene 2.5 and 82.2 mg/100 g, respectively. The values of oil-soluble vitamins in seabuckthorn oil are much higher than that of all other selected edible oils. The results of analysis suggested that seabuckthorn pulp and seed oils are a useful source of valuable nutrients (omega fatty acids, carotenoids and vitamin E, etc.), which can be a factor in oil price formulation.

Keywords: *Seabuckthorn oil, Olive, Walnut, Wheat embryo, Corn and Nutritional values.*

INTRODUCTION

Many studies have proved that seabuckthorn oil (pulp oil and seed oil) is very rich in nutritional and bioactive substances (Li *et al.*, 1997; Rongsen, 2003), which is considered as the most valuable part in seabuckthorn berry. But, the price of seabuckthorn oil is also considered much higher than other edible oils. In order to get a reasonable price mechanism of seabuckthorn oil, it is necessary to make a comparative study on the seabuckthorn oil and other edible oils. This study will report experimental results on the characteristics of seabuckthorn oil, which could be the basis to influence the price formulation.

Seabuckthorn (Hippophae L.): A Multipurpose Wonder Plant, Vol. 4: Emerging Trends in Research and Technologies (Virendra Singh, *Editor-in-Chief*, 2014), p. 258–261, Daya Publishing House, New Delhi, India.

MATERIALS AND METHODS

Experimental Materials

Seabuckthorn pulp oil and seed oil (50 g each) were estimated in seabuckthorn berries which were collected from Qingsui, Gansu, China. Olive oil (100 g) was extracted from olive fruit from Xichang, Sichuan, China. Walnut oil (40 g) was extracted from walnut collected from Yangbi, Yunnan, China. Wheat embryo (40 g) oil was extracted from wheat collected from Shuangliu, Sichuan, China. Corn oil (40 g) was extracted from corn collected from Chengdu, Sichuan, China.

Methodology

The biochemical analysis of the berries was made in the laboratory of Chengdu Institute of Biology, Chinese Academy of Sciences, which included the composition of fatty acids, carotenoids, β-carotene and vitamin E. The oils of seeds and pulp were extracted by petroleum ether, as described by Keyong (2004). The fatty acids were determined by GB/T 17377-2008 using Shimadsu GC-14B (Japan), carotenoids by HB/QS 001-94 using Shimadsu UV 2550 (Japan), β-carotene determined by HB/QS 001-83 using HPLC, Waters 510 (USA) and vitamin E by GB/T 5009-82-2003 using HPLC, Waters 510 (USA).

RESULTS AND DISCUSSION

Unsaturated Fatty Acids

The results of chemical analysis on fatty acid components of seabuckthorn (*H. rhamnoides* ssp. *sinensis*) oil and other 5 edible oils carried out in 2009 are shown in Table 20.1. It can be seen that all these oils are characterized by higher proportion of unsaturated fatty acids, whose content ranged from 70.0 to 91.6 per cent of the total oil. The unsaturated fatty acid components of seabuckthorn seed oil and pulp oil are 81.3 per cent and 70.0 per cent, respectively, which conform to the results of many reports (Li *et al.*, 1997; Lu, 2003). The unsaturated fatty acid components of olive oil, walnut oil, wheat embryo oil and corn oil are 84.5 per cent, 91.6 per cent, 82.0 per cent and 85.3 per cent, respectively. The present data accord with the results of studies conducted by Lu *et al.*, 1997; Zhou *et al.*, 1997; Wan *et al.*, 2001; You, 2004; Yao, 2005; Lu, 2005; Hui *et al.*, 2006; Mali, 2007. The highest content of unsaturated fatty acids was found in walnut oil with 91.6 per cent. The lowest contents of unsaturated fatty acids

Table 20.1: Fatty Acid Composition in Seabuckthorn Oil and Five other Edible Oils

Fatty Acids (Per cent of total oil)	*Seabuckthorn Seed Oil*	*Seabuckthorn Pulp Oil*	*Olive Oil*	*Walnut Oil*	*Wheat Embryo Oil*	*Corn Oil*
Palmitic acid	14.2	29.7	14.1	6.0	16.2	12.8
Palmitoleic acid	8.5	32.8	1.2	0.1	0.3	–
Stearic acid	1.7	1.2	1.4	2.2	0.7	1.9
Oleic acid	27.8	32.8	72.0	24.5	17.4	30.4
Linoleic acid	29.1	2.8	9.9	59.5	57.9	52.4
Linolenic acid	18.6	1.3	1.4	7.5	5.1	2.5
Arachidonic acid	–	0.3	–	0.2	1.3	–
Saturated acid	15.9	30.9	15.5	8.2	16.9	14.7
Unsaturated acid	81.3	70.0	84.5	91.6	82.0	85.3

were in seabuckthorn seed oil (81.3) and pulp oil (70.0 per cent), although they are quite rich in unsaturated fatty acids. It can be obviously seen that seabuckthorn seed oil and pulp oil are characterized by lesser contents of unsaturated fatty acids as compared to other oils.

However, the proportion of palmitoleic acid, an important unsaturated fatty acid, is maximum in seabuckthorn pulp oil (32.8 per cent) followed by seabuckthorn seed oil (8.5 per cent), as compared to negligible in others (0-1.2 per cent). Linoleic acid content was significantly higher in other oils than in seabuckthorn oil, with exception of olive oil. But seed oil of seabuckthorn has higher linolenic acid (18.6 per cent) than other oils.

Oil Soluble Vitamins

The results of chemical analysis of seabuckthorn (*H. rhamnoides* ssp. *sinensis*) oil and other 5 edible oils for oil-soluble vitamins are shown in Table 20.2. From this study, it can be seen that the contents of oil-soluble vitamins in all these oils are quite different. The contents of carotenoids are ranked in following order: Seabuckthorn pulp oil (373.2 mg/100 g) > seabuckthorn seed oil (78.2 mg/100 g) > corn oil (50.9 mg/100 g) > olive oil (11.9 mg/100 g) > wheat embryo oil (0.4 mg/100 g) > walnut oil (0.36 mg/100 g). The contents of β-carotene ranked in following order: seabuckthorn pulp oil (82.2 mg/100 g) > seabuckthorn seed oil (2.5 mg/100 g) > corn oil (0.29 mg/100 g) > walnut oil (0.17 mg/100 g) > wheat embryo oil (nil) and olive oil (nil). The contents of vitamin E are ranked in following order: seabuckthorn pulp oil (213.0 mg/100 g) > seabuckthorn seed oil (206.6 mg/100 g) > wheat embryo oil (144.5 mg/100 g) > corn oil (34.0 mg/100 g) > walnut oil (21.3 mg/100 g) > olive oil (6.0 mg/100 g), therefore, seabuckthorn oil is a better quality oil than other oils.

Table 20.2: Fat Soluble Vitamins in Seabuckthorn Oil and Five other Edible Oils

Vitamins (mg/100 g)	*Seabuckthorn Seed Oil*	*Seabuckthorn Pulp Oil*	*Olive Oil*	*Walnut Oil*	*Wheat Embryo Oil*	*Corn Oil*
Carotenoids	78.2	373.2	11.9	0.36	0.4	50.9
β-carotene	2.5	82.2	nil	0.17	nil	0.29
Vitamin E	206.6	213.0	6.0	21.3	144.5	34.0

Seabuckthorn pulp oil contains the higher contents of carotenoids, β-carotene and vitamin E than other tested oils. The second highest content of these oil-soluble vitamins was found in seabuckthorn seed oil. In the other words, seabuckthorn pulp oil and seed oil contains much higher oil-soluble vitamins than any other test edible oils.

CONCLUSION

The results of chemical analysis of all the tested oils suggested that although the fatty acid components of seabuckthorn pulp oil and seed oil are characterized by high contents of unsaturated fatty acids but those values are not higher than other tested edible oils. However, seabuckthorn pulp and seed oil are characterized by higher oil-soluble vitamins, which is the most valuable part of seabuckthorn. This character of seabuckthorn oil (carotenoids and vitamin E) could be a basis for deciding the price of seabuckthorn fruit oil.

REFERENCES

1. Hui, R. *et al.*, 2006. Preparation of corn oil and its determination of fatty acids. *Food Science* 27 (11): 418-420.

2. Li, Y. *et al.*, 1997. *An Introduction to Seabuckthorn Oil*, p. 26-59, North-west University Press. (In Chinese)

3. Lu, M. *et al.*, 1997. The nutritional value and application of wheat embryo oil. *Journal of Jilin Grain College* 14 (1): 1-3. (In Chinese)

4. Mali 2007. A comparison study on camellia oil and olive oil in nutritional value. *Agricultural Products Resources* 10 (1): 42-44.

5. Rongsen, L. 2003. *Hippophae* and its general chemical compositions. In: *Proceedings of 1st Congress of the International Seabuckthorn Association*, p.20-35, September 14-18, 2003, Germany.

6. Wan, B. *et al.*, 2001. A study on the specific property and nutritional value of walnut oil. *Science and Technology of Grain and oils in West China* 26 (5): 18-20. (In Chinese)

7. Yao, Y. 2005. A comparison study on peanut oil and olive oil in nutritional value. *China Oils* 30 (4): 66-68. (In Chinese)

8. You, X. 2004. Nutritional function of corn oil and its development perspective. *Science and Technology of Grain and Oils* 12 (2): 21-22. (In Chinese)

9. Zhu, K. 2004. *Food Detection Technology*, Beijing Science Press (In Chinese).

10. Zhou, Z. *et al.*, 1997. Extracting wheat embryo oil by supercritical fluid extraction. *Sichuan Food Science and Technology* 14 (2): 20-23. (In Chinese)

11. Zhu, Z. *et al.*, 2005. A study on extraction of walnut oil by supercritical carbon dioxide fluid. *China Oil* 30 (7): 65-68. (In Chinese)

Chapter 21

Isolation, Characterization and Chemical Fingerprinting of Bioactives from Indian Seabuckthorn (*Hippophae* L.) Species

Chinchu Bose, N. Pandurangan, Virendra Singh* and A. Banerji

Amrita School of Biotechnology, Amrita VishwaVidyapeetham,
Distt. Kollam – 690 525, Kerala, India
*Department of Biology and Environmental Sciences, CSK HPKV,
Palampur – 176 062, Himachal Pradesh, India

SUMMARY

Though, biochemical and nutraceutical properties of Indian seabuckthorn mainly *Hippophae rhamnoides* have been reported, systematic investigation on its phytochemistry is lacking. In the present study, isolation and characterization of bioactives from different Indian seabuckthorn species have been carried out. Vast variations were observed between the constituents from different species/subspecies/cultivars, thereby making it necessary to isolate and characterize them. With the availability of phytochemical profiles of seabuckthorn species, it is possible to relate the reported biological activities of the extracts with chemically identified compounds, such as sterols (anti-cholesteromic), ursolic (anti-tumour, reduces MMP-9) and oleanolic acids, quercetin (radioprotection, anti-tumour, cardioprotective), isorhamnetin (anti-obesity, anti-diabetic), kaempferol (kinase inhibitor), tannins (anti-viral), glycosides of isorhamnetin, quercetin and kaempferol (antioxidants). It may be noted that seabuckthorn contains gallo-and ellagitannins as well as free gallic and ellagic acids as major components. These constituents show high antioxygenic, radioprotective and anti-diabetic activities. Many of the bioactivities of seabuckthorn can be attributed to these components. A Thin Layer Chromatography based protocol for chemical fingerprinting of chemical constituents has been developed. As compared to RAPD fingerprinting, it is inexpensive, simple and fast, does not need any elaborate laboratory facilities. The species of seabuckthorn, such as *Hippophae rhamnoides* ssp. *turkestanica*, *H. rhamnoides* ssp. *mongolica*, *H. salicifolia* from Himachal Pradesh and Sikkim were studied using this protocol. Even small (1 g) samples like those of *H. rhamnoides* procured from Siberia, Russia, *H. sinensis* from Datong, China and *H. rhamnoides* from Germany were fingerprinted. Quantitative analysis of the individual components was done by HPLC (DAD). Molecular characterization of labile compounds, *i.e.* tannoids or glycosides was carried out using LCMS/MS. Techniques,

Seabuckthorn (*Hippophae* L.): *A Multipurpose Wonder Plant, Vol. 4: Emerging Trends in Research and Technologies* (Virendra Singh, *Editor-in-Chief*, 2014), p. 262–274, Daya Publishing House, New Delhi, India.

such as IR, UV and high-resolution NMR spectroscopy provided non-destructive methods for advanced analytical work. Antioxygenic activities of seabuckthorn extracts, isolated compounds and analogs have been carried out. The chemical syntheses of "seabuckthorn flavones" and analogs have also been developed.

Keywords: *Seabuckthorn flavones, Bioactives, Ursolic acid, Tannins, Chemical fingerprinting, Antioxidant activity, Quercetin and Isorhamnetin.*

ABBREVIATIONS

HSSK: *Hippophae salicifolia* from Sikkim; HSHP: *H. salicifolia* from Himachal Pradesh; HRL: *H. rhamnoides* ssp. *turkestanica* from Ladakh; HRMHP: *H. rhamnoides* ssp. *mongolica* from Himachal Pradesh; PE: Petroleum ether.

INTRODUCTION

Seabuckthorn (*Elaeagnaceae, Hippophae* sp.) in preventing soil erosion, water conservation and nitrogen fixation is well known. Massive plantation of seabuckthorn has helped in the reclamation of environmentally degraded lands in countries like China and Russia. In addition, seabuckthorn is a valuable biosource for an array of biologically active products (Suryakumar and Gupta, 2011). Well developed industries based on seabuckthorn products have emerged in countries like China, Russia, Germany and Finland. In India, though there are substantial natural resources of seabuckthorn, it has primarily remained an underutilized biosource. Seabuckthorn population in India is represented by mainly three species *viz., Hippophae rhamnoides* ssp. *turkestanica* (HR), *H. salicifolia* (HS) *and H. tibetana* (HT).

Recently, most of the Indian studies on seabuckthorn were mainly confined to *Hippophae rhamnoides* ssp. *turkestanica* of Ladakh and Himachal Himalayas (Singh, 2003), the most abundant species. It occurs at high altitudes (2500-3800 m) in difficult to access terrains. The presence of thorns on the plant makes its commercial harvesting very difficult. Bioprospection for seabuckthorn showed that *H. salicifolia* occurs in significant quantities in Himachal Pradesh, Uttaranchal and Sikkim at comparatively lower altitudes which are more accessible. Also, HS has lesser thorns, making its commercial harvesting easier. Compared to HR, much less is known on the agro-technology, nutraceuticals and bioactive principles of *H. salicifolia* or *H. tibetana*. In order to expand the biosource base for seabuckthorn, studies on the nutraceutical and bioactive principles from HS were undertaken. The comparative study on HR (from Ladakh and Himachal Pradesh) and HS (from Himachal Pradesh and Sikkim) has been carried out. It will ameliorate the commercial potentialities of different seabuckthorn samples. Also, the studies could lead to hitherto unreported properties of seabuckthorn. In India, very significant pharmacological and biochemical studies have been carried out on HR (Geetha *et al.*, 2002). Most of the studies were carried out on non-standardized undefined extracts. By and large, the activities have not been related to chemical constituents. Detailed phytochemical studies on Indian seabuckthorn are lacking. Most of the studies on seabuckthorn were directed on the commercially important parts, such as fruit, seed and pulp oils (Singh *et al.*, 2011). Fruits of seabuckthorn are seasonal and have a very short shelf life. The most serious problem for processing of seabuckthorn fruit lies on its fast bio-deterioration and therefore, appropriate quick processing facilities should be available for commercial exploitation.

Our attention was drawn towards the phytochemical investigation on leaves of seabuckthorn since they are available for most of the time of the year. Literature survey showed that not much investigation have been reported on the leaves. Therefore, research on leaves of seabuckthorn was initiated by us as early as in 2003. Though, seabuckthorn is systematically cultivated in countries like China and Russia but in India most of the seabuckthorn grows wild and show enormous genetic and chemo variation. Recently, the cultivation of seabuckthorn has been initiated in Himachal Pradesh under technical guidance of CSK Himachal Pradesh Agricultural University, Palampur and in Ladakh by Defence Institute of High Altitude Research. In countries, where seabuckthorn based industries are flourishing, serious attempts have been made to develop elite varieties with desirable traits and phytochemicals. Though, convenient methods for the estimation of biochemicals and nutraceuticals are available, it is not true for identification and quantization of high-value bioactive products (phytochemicals). Therefore, it is important to develop methods for the isolation and characterization of bioactive compounds.

MATERIALS AND METHODS

Sample Collection

Leaves of *Hippophae rhamnoides* ssp. *turkestanica* were collected from Himachal Pradesh (HRHP) and Ladakh (HRL); *H. salicifolia* from Himachal Pradesh (HSHP) and North Sikkim (Lachen, Lachung) (HSSK); *H. rhamnoides* ssp. *mongolica* (HRMHP) is the Russian variety which was introduced successfully in Himachal Pradesh. The leaves were shade-dried, powdered and extracted sequentially in Soxhlet extractors with petroleum ether, chloroform, methyl ethyl ketone and methanol.

Analytical Methods

Thin Layer Chromatography (TLC) was performed using silica gel plates (DC Kieselgel, CCM Gel de silice 60, F_{254}, Merck). HPLC Analyses were performed on a Shimadzu-SPD-M20A HPLC, equipped with DAD (Diode Array detector) detector using Phenomenex Luna 5µ C_{18}(2) 100A, size 250 × 4.60 nm. All the compounds were detected at 254 nm at room temperature with an eluent flow rate of 1.2 ml/min and an injection volume of 10µl. The mobile phase consisted of acetonitrile (A) and 0.03 per cent of formic acid in water. ^{1}H NMR spectra were recorded on a Bruker (Wissembourg, France) AM 400 spectrometer (400 MHz, for ^{1}H NMR) using TMS used as a standard. UV spectra were recorded on a Shimadzu UV spectrophotometer Model UV-1800 in methanol and IR spectra were recorded on a Shimadzu IR Affinity-1 in KBr or Film.

Free Radical Scavenging Activities

For evaluation of free radical scavenging activities of test compounds (20-200 µg) in methanol (100 µl) were added to a solution of DPPH (0.025 per cent ; 3.9 ml) in methanol. The mixture was left for 5 min and absorbance of the resulting solution was measured at 540 nm with a colorimeter. The results are expressed as IC_{50} (µg) and are presented in Table 21.1.

Extraction and Fractionation of Seabuckthorn

Extraction (Soxhlet) of seabuckthorn with PE gave small amount of residue which contained carotenoids and sterols. Biologically important metabolites, such as ursolic and oleanolic acids were extracted with chloroform (Ashutosh *et al.*, 2007). TLC analysis of the extract primarily showed one major spot. On a careful/high resolution/fine tuning HPLC analysis, it was found to be a mixture of two compounds which were characterized as ursolic and oleanolic acids. The individual compounds were separated by silica gel column chromatography. For polar constituents, depending on the class of compounds, different procedures were adopted. For extraction of flavonoids, their glycosides, tannins and conjugates in the leaves of *Hippophae*, extraction (3 × 24 hrs) was done by shaking with 70 per cent aqueous acetone/methanol in a rotary shaker at room temperature. By fine tuning the polarities of the extraction solvents, it was possible to bring about a broad separation of tannins from flavonoid glycosides. After 70 per cent acetone, the leaves were extracted with water to remove the polyglycosylated flavonoids, carbohydrates and more polar constituents. However, the extractives were in small amounts suggesting that aqueous acetone had extracted most of them.

Table 21.1: Antioxidant Activity of Seabuckthorn Extracts and Isolated Compounds

Sl.No.	*Compounds*	*EC$_{50}$ µg*
1.	Gallic acid	3.8
2.	Ellagic acid	4.8
3.	Gossypetin	5.5
4.	Quercetagetin	10.5
5.	Trolox	12.5
6.	P.C	12.8
7.	Quercetin	14.4
8.	Kaempferol	14.6
9.	Hibifolin	16.0
10.	Isorhamnetin	26.2
11.	Seabuckthorn Flavone	18.5
12.	Rutin	15.5
13.	Seabuckthorn leaf 70 per cent acetone extract	40.4
14.	kaempferide	97.0
15.	Tamarixetin	>100
14.	Seabuckthorn Seed oil	>100
15.	Seabuckthorn Pulp oil	>100

The native tannins could be preferentially precipitated from the 70 per cent aqueous acetone extract by continous addition with BSA solution or by adding just sufficient amount of methanolic lead acetate (10 per cent). The precipitate was separated by centrifugation. The lead complex was taken in methanol and the tannins were regenerated by the addition of aqueous sulphuric acid (8 per cent) to decompose the Pb-complex. HPLC showed the presence of ellagic acid derivative. Acid hydrolysis generated ellagic acid and glucose. Thus, it is concluded that ellagic acid is present as glucoside. Lead acetate in the supernatant was removed by precipitating with dilute sulphuric acid till the mixture became acidic (pH 5). The colour reactions of the supernatant indicated the presence of mainly phenolics and flavonoids (glycosides). TLC and HPLC analysis showed that acid hydrolysate mainly contained isorhamnetin, glucose with small amounts of quercetin and kaempferol, thereby confirming the presence of corresponding glycosides.

Procedure for Fingerprinting

Dried powdered leaves (1 g) were extracted with petroleum ether by heating on a water bath or sonication for 30 min. The leaves were filtered and fresh volume of solvent was added and extraction process was repeated three times. The solvent was removed from the combined extracts by rotary evaporation below 40 °C. The residual leaves were extracted (3 times) sequentially with chloroform and 70 per cent aq. acetone as mentioned above. The extracts from 70 per cent aq. acetone was subjected to acid hydrolysis and extracted with ethyl acetate. For non-polar extracts, TLC was performed with solvent mixture containing petroleum ether:ethyl acetate (7:3) polar fractions were analyzed with solvent mixture containing toluene:ethyl acetate:formic acid (5:5:1) and ethyl acetate:methanol:formic acid:water (95: 2: 4: 6).

RESULTS AND DISCUSSION

Leaves of *H. rhamnoides* ssp. *mongolica* (HRMHP, Russian variety grown in HP) and HRHP showed similar chemical profiles. HRHP was richer in non-polar extractives (PE extract, 4.25 per cent), when compared to *H. salicifolia* from Himachal Pradesh (HSHP) (PE extract, 2 per cent). The chloroform extract of HRMHP contains β-carotene, sitosterol, lutein and ursolic acid. *H. salicifolia* (from Sikkim) HSSK was found to be many times richer in carotenoids (Figure 21.1). Since, ursolic and oleanolic acids are structurally very similar and their separation is quite difficult. A quantitative method for simultaneous separation and quantitation of ursolic and oleanolic acids mixture has been developed using column chromatography and HPLC-DAD. Out of the leaf samples tested, HRL contains more of the terepenoids, ursolic acid (120 mg/100 g) and oleanolic acid (170 mg/100 g) whereas in comparison, HSSK contains very less amounts of urosolic acid (20 mg/100 g) and oleanolic acid (5 mg/100 g) (Figure 21.2) (Chinchu *et al.*, 2010a; 2010b). The presence of substantial quantity of ursolic acid in HRL is noteworthy, since many bioactivities including anti-tumour activity have been attributed to it (Ashutosh *et al.*, 2007; Prasad *et al.*, 2011).

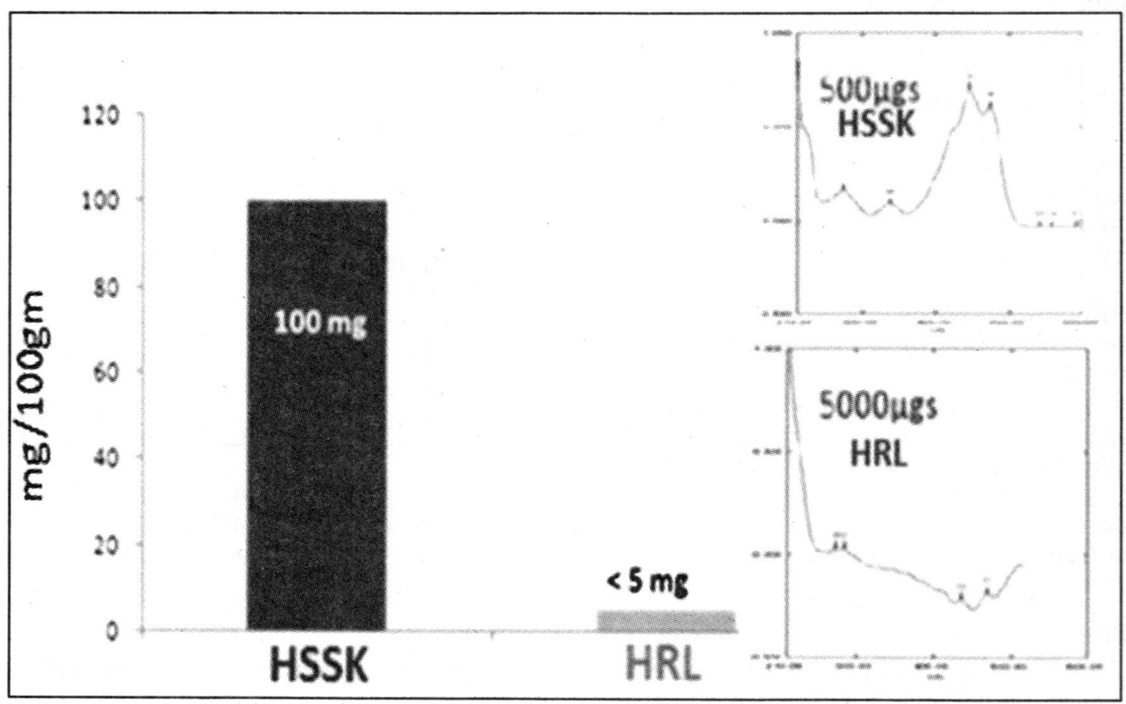

Figure 21.1: β-carotene Content of HSSK and HRL.

In the present study, vast differences in the flavonoid contents extracted from different seabuckthorn samples were noticed. The flavonoid content of HRMHP was similar to that of native HRHP though, they are different in composition. After acid hydrolysis of the extracts, both contained ellagic acid, isorhamnetin and quercetin in substantial quantities and kaempferol in minor amount. HRL contains more of isorhamnetin (150 mg/100 g), when compared to HSSK species (Figure 21.3). The discovery of isorhamnetin in seabuckthorn has elicited interest because of its novel bioactivities, such as anti-tumour, adipogenesis inhibitor (Teng *et al.*, 2006; Sanchez, *et al.*, 2007; Jongsung *et al.*, 2009; Kim *et al.*, 2011). Our studies have shown that gallic and ellagic acids had anti-diabetic and radioprotective activities (Vishnuprasad *et al.*, 2010). The presence of free gallic and ellagic acids were observed in the extracts of HSSK.

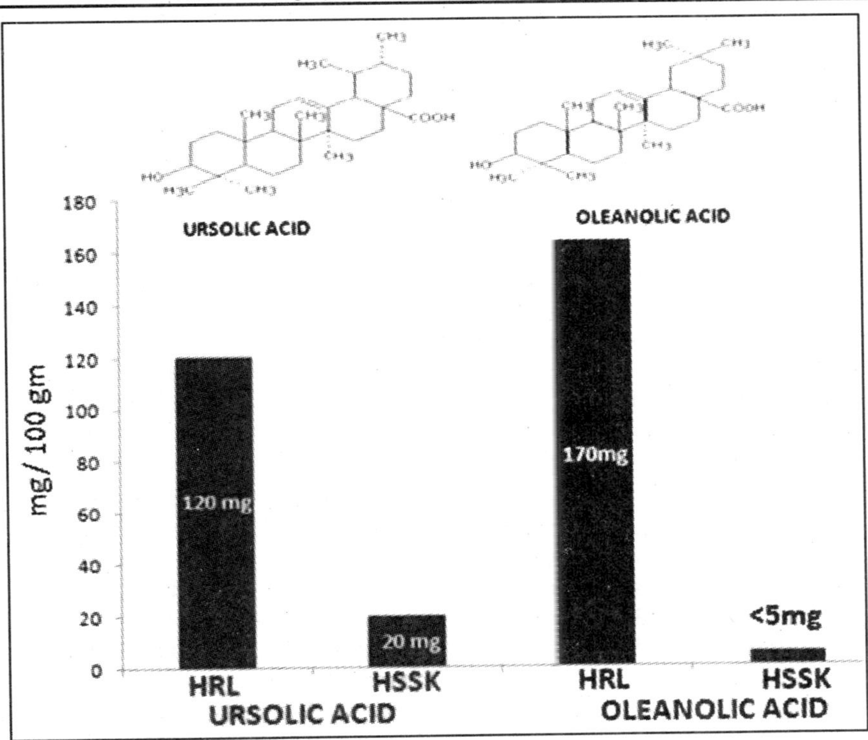

Figure 21.2: Ursolic and Oleanolic Acid Content of HRL and HSSK.

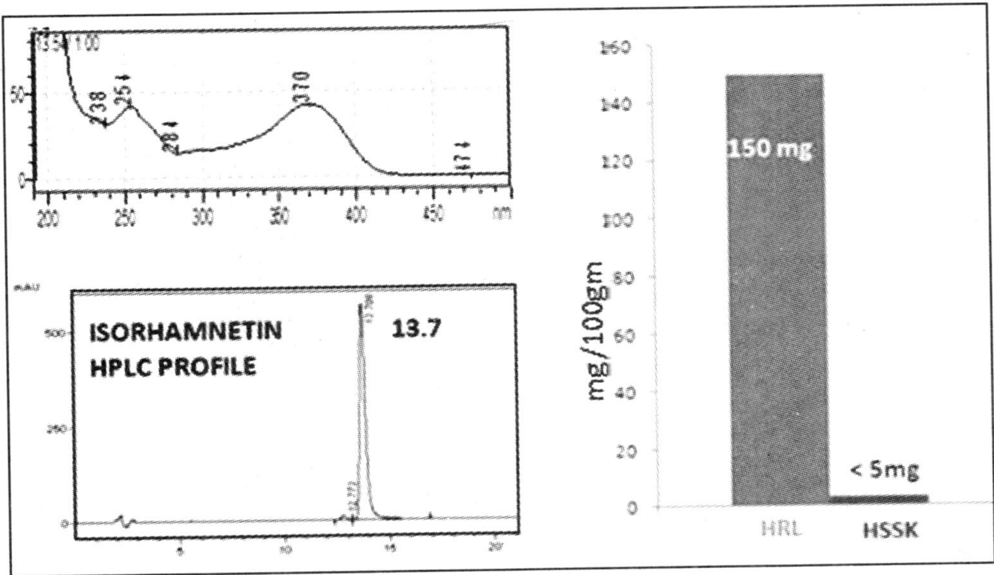

Figure 21.3: Isorhamnetin Content of HRL and HSSK.

For the separation of native tannins and flavonoid glycosides, seabuckthorn leaves were extracted exclusively with chloroform (to remove the non-polar constituents) prior to extraction with methyl ethyl ketone and 70 per cent aqueous acetone. The individual extracts were subjected to gel-permeation chromatography on Sephadex LH-20. The fractions were analyzed by TLC, HPLC and chemical reactions. HPLC-DAD of the fractions showed them to be intricate mixtures of many constituent. The presence of rutin, ellagic and gallic acids were inferred from the colour reactions, TLC and HPLC. The presence of free gallic and ellagic acids could be inferred but no evidence for free flavonoids was obtained. Therefore, flavonoids must be present in bound forms. UV spectral data and colour reactions suggest the presence of flavonoid glycosides mixtures. In order to isolate the free flavonoids, the polar extracts were subjected to acid hydrolysis. The hydrolysates were extracted with ethyl acetate followed by methyl ethyl ketone. Flavonoids were present in ethyl acetate extract while gallic and ellagic acids were present in the ethyl methyl ketone extract. The derivatization of the flavonoids (methylation and acetylation) was carried out to confirm the identities by UV, NMR and MS. Very substantial amounts of gallic and ellagic acids were also isolated. Glucose, galactose and rhamnose were identified by TLC and paper chromatography as the main carbohydrates from the hydrolysates.

The fractions obtained from LH-20 gel permeation chromatography were analyzed by LC-MS/MS and very valuable structural information could be obtained. Fraction 1 showed clusters of unresolved signals in LC. Acid hydrolysis of the fraction resulted in the formation of mainly ellagic acid with small amount of isorhamnetin. Fraction 1, therefore, contains ellagi tannin. UV spectra of Fraction 2 show flavonoid type absorption, acid hydrolysis indicated that it primarily contains isorhamnetin with smaller quantities of quercetin, ellagic and gallic acid (Figure 21.4). Thus, it may be inferred that fraction 2 mainly contains isorhamnetin glycoside. UV spectra of Fraction 3 suggested the presence of gallic acid and isorhamnetin with lesser amounts of quercetin and ellagic acid. Fraction 4 and 5 mainly contain ellagic acid and closely related compounds. Results of LC-MS/MS of flavonoid components are summarized in the Table 21.2.

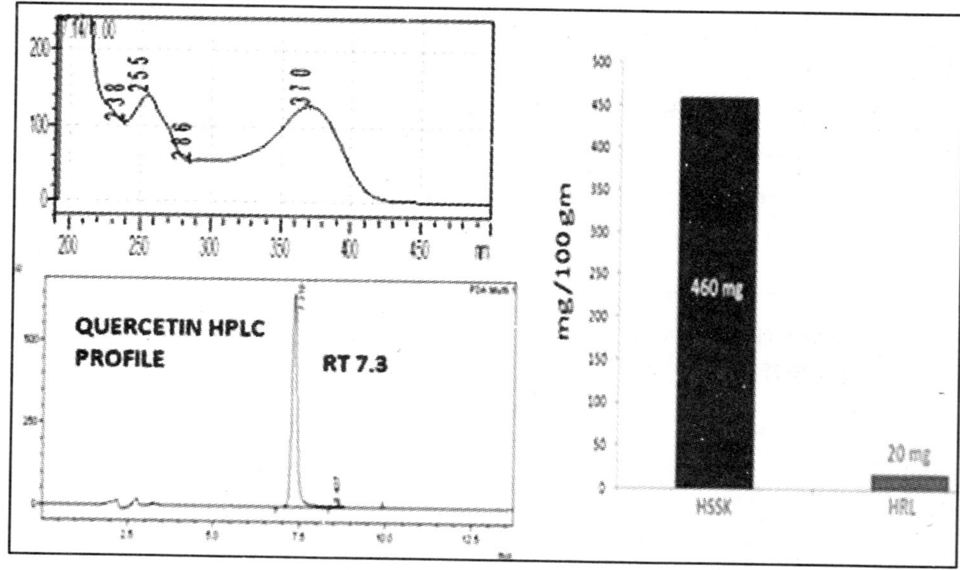

Figure 21.4: Quercetin Content of HRL and HSSK.

Table 21.2: LC-MS/MS Characterized Glycosides

Sl.No.	*Compounds*	*Molecular Weight/Fragments*
1	Kaempferol diglucoside	MW=611, fragment 611-162-162=287.
2	Isorhamnetin rhamnodiglucoside	MW=787, fragment 787-162-162-146=317
3	Kaempferol rhamnoglucoside	MW=595, fragment 595-162-146=287.
4	Isorhamnetin glucorhamnoside	MW=625, fragment 625-162-146=317
5	Kaempferol glucorhamnoside	MW=595, fragment 595-162-146=287.
6	Isorhamnetin rhamnoglucoside	MW=625, fragment 625-146-162=317
7	Quercetin diglucoside	MW= 627, fragment 627-162-162=303.
8	Isorhamnetin rhamnoglucoside	MW= 641, fragment 641-162-146=317
9	Quercetin diglucoside	MW= 627, fragment 627-162-162=303.
10	Quercetin diglucorhamnoside	MW=773, fragment 773-162-146-162=303.
11	Rutin	MW = 611. Fragment 611-465-162=303.

Quality Control and Fingerprinting

The quality of seabuckthorn products is ascertained by their contents of bioactive principles. Wide variations in the levels of bioactive constituents have been observed in different species/strains/cultivars of seabuckthorn (Korovina and Fefelov, 2006). For commercial exploitation of seabuckthorn,

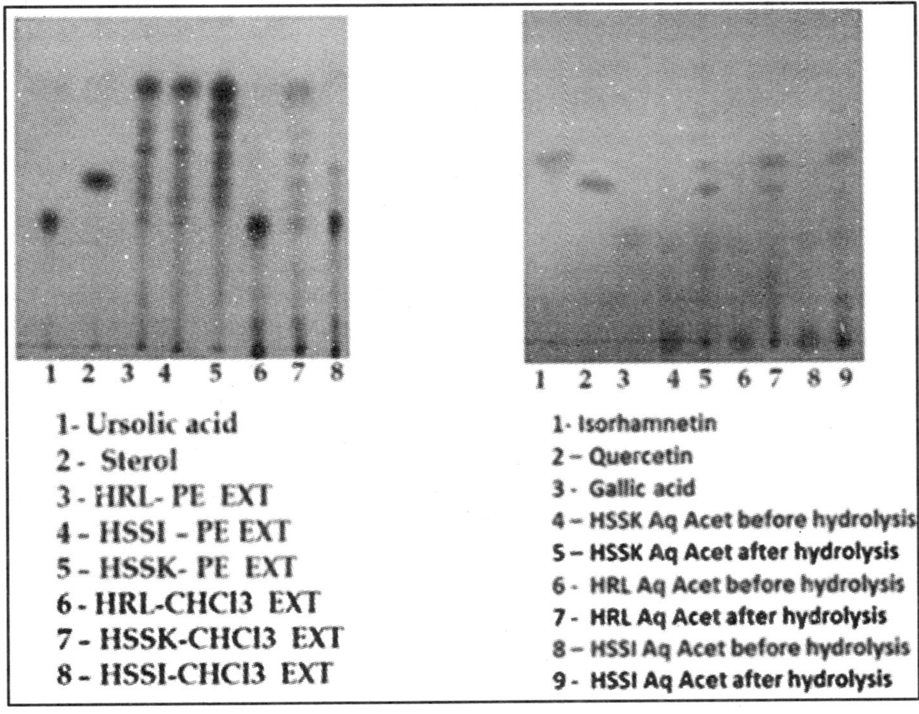

Figure 21.5: TLC Fingerprinting of HRL, HSSI and HSS.

it is essential to develop quick, affordable and efficient protocols for grading them on the basis of their bioactive constituents. Convenient TLC based methods have now been developed which are useful for fingerprinting and standardizing the seabuckthorn extracts using a variety of solvent mixtures. The protocol has been applied for a comparative study of extracts of *Hippophae rhamnoides* from Ladakh HRL, *H. salicifolia* from Sikkim HSSK *and H. sinensis* HSSI from China, Datong. Qualitative as well as quantitative differences in their constituents were observed (Figure 21.5).

The comparative thin-layer chromatography of PE extracts of different samples showed almost similar profile. However, the extract from the leaves of HSSK was several folds richer in carotenes (Figure 21.1). The carotene contents were determined by quantitative UV spectroscopy. The main constituents of chloroform extract were identified as a mixture of ursolic and oleanolic acids. Out of the three samples, HRL contains higher amount of ursolic acid. The colour reactions of aqueous methanolic extracts indicated the presence of tannins, ellagic acid, phenolics and flavonoids (mainly as glycosides). TLC of aqueous acetone extract showed some interesting spots which are under investigation but no evidence for free flavonoids were obtained. A mild acid treatment of the extracts resulted in the formation of free acids, carbohydrates and flavones (Figure 21.6).

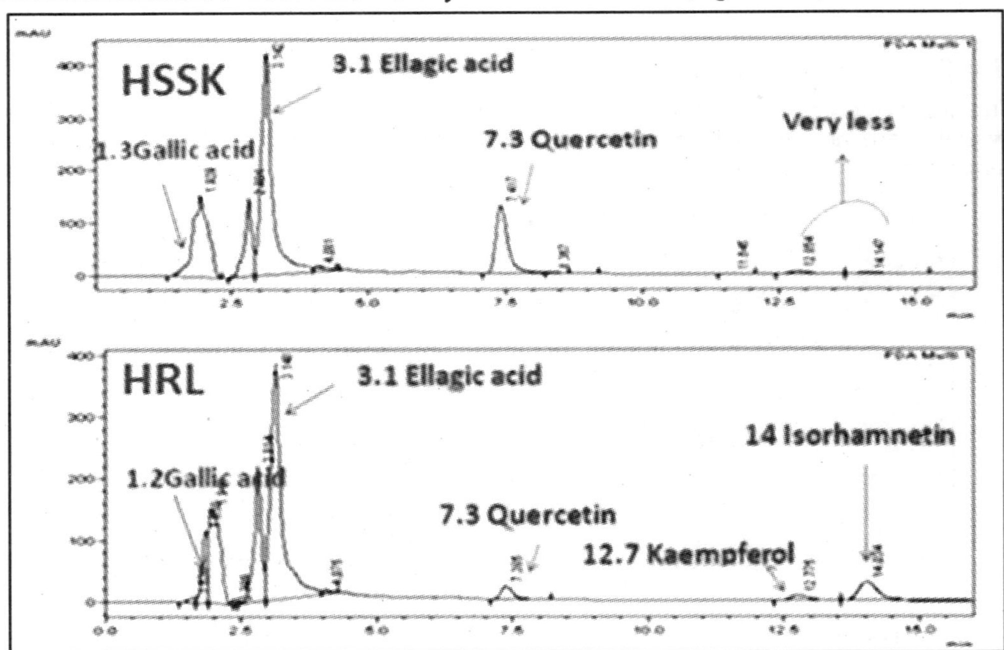

Figure 21.6: Seabuckthorn Leaves Polar Fraction Hydrolysis HPLC Comparison of HSSK and HRL.

The colour reactions and TLC analysis showed that HRL leaves contain isorhamnetin as the major flavonol along with quercetin and kaempferol. The sugar part was analyzed as glucose or galactose or rhamnose by circular paper chromatography and TLC. In contrast, HSSK contains very small quantity of isorhamnetin, whereas quercetin was the major flavonoid. 70 per cent aqueous acetone extract of HSSI showed more or less similar profile as for HRL. The judicious use of HPLC-DAD resulted in the characterization of individual compounds. Gallic and ellagic acids were also isolated in large amounts and characterized.

FLAVONOIDS OF COMMERCIAL "SEABUCKTHORN FLAVONE"

Many of the bioactivities of seabuckthorn, such as radioprotection, anti-tumour activity, cardioprotective and immunomodulatory activities have been attributed to "flavonoids" ("seabuckthorn flavones" or "total flavones", etc.). The flavone contents of the commercial samples were determined by Folin-ciocalteu assay or more recently by aluminum complex reaction (Papoti *et al.*, 2011). However, both the methods have limitations. Therefore, it is necessary to isolate and characterize the individual flavonoids. A commercial product, so called *"seabuckthorn flavone"* is extensively used as a source of flavonoids for the study of bioactivities (Gupta *et al.*, 2006; Suryakumar and Gupta, 2011), although its composition is not clearly defined. Not much attention has been paid towards the isolation and molecular characterization of the individual constituents. This attracted our attention to analyze "seabuckthorn flavone" for flavonoids and to synthesize the flavonols (flavone-3-ols), so that sufficient quantities are available for biological studies. Our analysis of commercial "seabuckthorn flavone" samples of different origins revealed that they mainly contained a mixture of isorhamnetin, quercetin and small amount of kaempferol in highly variable proportions (15-30 per cent) with more than 50 per cent intractable materials. No evidence was found for presence of glycosides or tannins in "seabuckthorn flavone" samples. In view of the above studies, the bioactivities attributed to "seabuckthorn flavone" should be rationalized and relate to the actual bioactivity in terms of its constituents. Very recently, it has been reported that quercetin provides radiation protection to red blood cell (RBC) against gamma-radiation (Das *et al.*, 2013). Thus, the radioprotection provided by the extracts of seabuckthorn could be due to quercetin and related compounds.

ANTIOXIDANT ACTIVITY OF SEABUCKTHORN EXTRACTS AND THEIR ISOLATED COMPOUNDS

Antioxygenic activities of flavonoids of seabuckthorn extracts have been attributed to many bioactivities, such as radioprotection, cancer prevention, anti-aging and prevention of cardiovascular diseases. Therefore, it was of our interest to carry out a study of antioxygenic activities of seabuckthorn extracts, "seabuckthorn flavone" and other common flavonoids. For this, the free radical scavenging activities were determined by quenching of DPPH. EC_{50} values were calculated as μg required for inhibition of 50 per cent of activity (Table 21.1).

Trolox was used as a reference antioxidant. Amongst the seabuckthorn extracts, 70 per cent aqueous acetone/methanol extract of seabuckthorn leaves was found to be the most active; the seed and pulp oils showed only low activities. A sample enriched in phenolics, was isolated from 70 per cent aqueous acetone extract. It was found to have higher antioxygenic activity (EC_{50}, 12.8) as compared to "seabuckthorn flavones" (EC_{50}, 18.5). Quercetin and kaempferol showed much better activity as compared to isorhamnetin which forms the major constituent of "seabuckthorn flavone". Isorhamnetin, tamarixetin and kaempferide are less active than the parent compounds *viz.*, quercetin and kaempferol. Gossypetin and quercetagetin showed high activities. The blocking of a hydroxyl group by glycosylation decreases the bioactivities considerably as can be seen for rutin and gossypin (gossypetin 8-glucoside). However, hibifolin (8-Glucuronide of gossypetin) showed good activity. Thus, it may be inferred that radical scavenging activities are dependent on number and location of hydroxyl groups. Presence of *o*-dihydroxy group acts as metal chelating systems and 2, 3, double bond (flavones, flavonols) also contribute to antioxygenic activity. The comparison of antioxygenic activities of flavones with corresponding flavonols (flavone-3-ols) shows that hydroxyls at 3-position augment the anti-oxygenic activities.

Amongst isolated pure compounds, gallic acid showed the highest antioxygenic activity followed by ellagic acid as compared to seabuckthorn flavones. Both these may be artifact of hydrolysable tannins present in seabuckthorn leaves. Also, we have found them to have anti-diabetic and radio-protective activity. Since, tannoids including gallic and ellagic acids form major constituents and due importance should be given to these compounds for their possible role in imparting bioactivities of seabuckthorn extracts. Since, many flavone-3-ols show better activities when compared to flavones from seabuckthorn, it should be feasible to develop improved bioactive formulations as compared to seabuckthorn products. Alternatively, they may be useful for fortifying the formulations with the required flavonoids/gallic/ellagic acids.

SYNTHESIS OF SEABUCKTHORN FLAVONOLS

The insights derived from the studies on bioactivities of seabuckthorn extracts made it imperative to locate the active molecule (s). The procurement of seabuckthorn in sufficient quantities from the natural sources for biological studies often poses logistic problems. A practical synthesis of polyhydroxy-and regio-specifically methylated flavone-3-ols, which are components of commercial "seabuckthorn flavone" has been achieved by a simple, eco-friendly process (Pandurangan, *et al.*, 2011) (Figure 21.7).

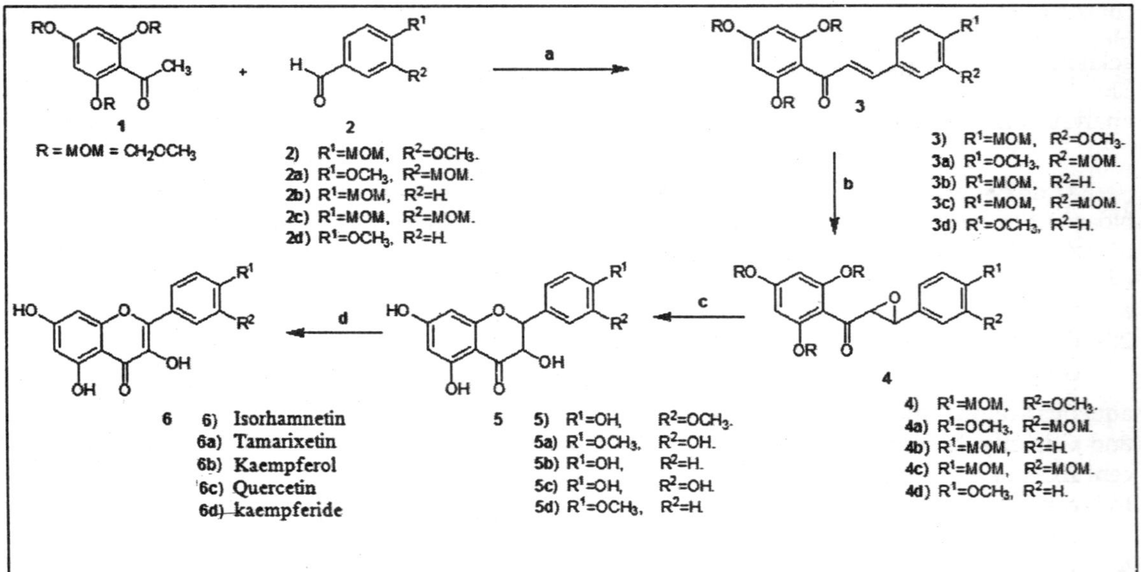

Scheme 1. Synthesis of flavone-3-ols : reagents and conditions: a) KOH, DMF, 0°C: b) 30% H_2O_2, NaOH, Methanol: c) HCl, Methanol, Reflux: d) $K_2S_2O_5$ soln, 100°C.

Figure 21.7: Pathway for Synthesis of Seabuckthorn Flavonols.

CONCLUSION

Biochemical and nutraceutical properties of Indian seabuckthorn have already been reported. Bioactivities of extracts have also been studied. However, systematic studies on phytochemistry are lacking. Presently, the detailed phytochemical investigation was undertaken and several phytochemicals, such as sterols, ursolic, oleanolic acids, quercetin, isorhamnetin, kaempferol, gallic acid, ellagic acid, their tannins and gluco-, rhamno-, rutinoside of the flavonoids have been isolated and characterized. It is now possible to relate some of the reported bioactivities of the extracts with chemically identified compounds. Various species of seabuckthorn, such as *Hippophae rhamnoides* ssp. *turkestanica* from Ladakh (J&K), Himachal Pradesh; *H. rhamnoides* ssp. *mongolica* and *H. salicifolia* from Himachal Pradesh and *H. salicifolia* from Sikkim were studied for their chemical profiles. Significant differences in the chemoprofiles were observed between the samples; even within the same species collected from different geographical locations. Therefore, it is necessary to know the phytochemical constituents before embarking on any commercial venture on bioactivity related activities. Since, it is neither feasible nor desirable to undertake detailed phytochemical analysis of each and every sample, a quick protocol which does not require elaborate laboratory facilities and expertise for chemical fingerprinting of constituents has been developed. Even very small samples like those of *H. rhamnoides* collected from Siberia and Russia, *H. sinensis* collected from Datong, China and *H. rhamnoides* from Germany were fingerprinted. Quantitative analyses were carried out using HPLC-DAD. It may be noted that seabuckthorn contains gallo-and ellagitannins as major components. These tannins show very high antioxygenic activities, radioprotective and anti-diabetic activities.

ACKNOWLEDGEMENT

Financial support provided from Department of Biotechnology, Government of India is gratefully acknowledged. We are thankful to the Dean, School of Biotechnology for his keen interest in this work. Thanks are due to Dr. Walter Shrenk, of Amrita-Agilent Analytical Research Centre, for mass spectral analysis and helpful discussions.

REFERENCES

1. Ashutosh, K.P., Manisha, B., Asha, S.N., Kwang, S.A., Arup, C., Humam, K., Sushovan, G., Gautam, S. and Bharat, B.A. 2007. Ursolic acid inhibits STAT3 activation pathway leading to suppression of proliferation and chemo sensitization of human multiple myeloma cells. *Molecular Cancer Research* 5: 943-955.

2. Chinchu, B., Pandurangan, N. and Banerji, A. 2010a. Isolation and characterization of bioactive constituents of Himalayan (Ladakh) seabuckthorn. In: *Proceedings of National Conference on Seabuckthorn*, p. 321-324, CSK Himachal Pradesh Agricultural University, Palampur, 324p.

3. Chinchu, B., Pandurangan, N., Pradhan, S., Radha Kumari, B., Basistha, B.C. and Banerji, A. 2010b. Phytochemical studies in *Hippophae salicifolia* from Sikkim Himalayas. In: *Abstract Book of National Conference on Seabuckthorn*, p. 20, Defence Institute of Physiology and Allied Sciences (DIPAS), Delhi.

4. Das, D., Chakraborty, A., Sinha, M., Manna, K., Mukherjee, A., Chakraborty, A., Bhattacharjee, S. and Dey, S. 2013. Modulatory role of quercetin against gamma irradiation mediated biochemical and morphological alteration of red blood cells. *International Radiation Biology*. (In Press)

5. Suryakumar, S. and Gupta, A. 2011. Medicinal and therapeutic potential of seabuckthorn (*Hippophae rhamnoides* L.). *Journal of Ethnopharmacology* 138: 268-278.

6. Geetha, S., Sai Ram, M., Singh, V., Ilavazhagan, G. and Sawhney, R.C. 2002. Antioxidant and immunomodulatory properties of seabuckthorn (*Hippophae rhamnoides*): An *in vitro* study. *Journal of Ethnopharmacology* 79: 373-378.

7. Gupta, A., Kumar, R., Pal, K., Singh, V., Banerjee, P.K. and Sawhney. R.C. 2006. Influence of seabuckthorn (*Hippophae rhamnoides* L.) flavone on dermal wound healing in rats. *Molecular and Cellular Biochemistry* 290: 193-198.

8. Jongsung, L., Eunsun, J., Jienny, L., Saebom, K., Sungran, H., Youngsoo, K.,Yongwoo, K., Sang, Y.B., Yeong-Shik, K. and Deokhoon, P. 2009. Isorhamnetin represses adipogenesis in 3T3-L1 Cells. *Adipocyte Biology* 17: 226-232.

9. Kim, J.E., Lee, D.E., Lee, K.W., Son, J.E., Seo, S.K., Li, J., Jung, S.K., Heo,Y.S., Mottamal, M., Bode, A.M., Dong, Z. and Lee, H. J. 2011. Isorhamnetin suppresses skin cancer through direct inhibition of MEK1 and PI3-K. *Cancer Prevention Research (Phila)* 4: 582-591.

10. Korovina, M.A. and Fefelov, V.A. 2006. Biochemical variations in seabuckthorn (*Hippophae rhamnoides L.*) growing in different regions of CIS states. In: *Seabuckthorn (Hippophae L.)-A Multipurpose Wonder Plant. Vol. II: Biochemistry and Pharmacology* (V. Singh, Ed. in Chief, 2006), p.108-132, Daya Publishing House, New Delhi, 600p.

11. Matsuda, H., Kogami, Y., Nakumura, S., Sugiyama, T., Ueno, T. and Yoshikawa, M. 2011. Structural requirements of flavonoids for adipogenesis of 3T3-L1 cells. *Bioorganic and Medicinal Chemistry* 19: 2835-2841.

12. Pandurangan, N., Chinchu, B. and Banerji, A. 2011. Synthesis and antioxygenic activities of seabuckthorn flavone-3-ols and analogs. *Bioorganic and Medicinal Chemistry Letters* 21: 5328-5330.

13. Papoti, V.T., Xystours, S., Papagianni, G. and Tsimidou, M.Z. 2011. "Total flavonoid" content assessment *via* Aluminium (Al III) complexation reactions. What we really measure? *Italian Journal of Food Science* 23: 252-260.

14. Prasad, S., Yadav, V.R., Kannappan, R. and Aggarwal B.B. 2011. Ursolic acid, a pentacyclin triterpene, potentiates TRAIL-induced apoptosis through p53-independent up-regulation of death receptors: Evidence for the role of reactive oxygen species and JNK. *Journal of Biological Chemistry* 286: 5546-5557.

15. Sanchez, M., Lodi, F., Vera, R., Villar, I.C., Cogolludo, A., Jimenez, R., Moreno, L., Romero, M., Tamargo, J., Perez-Vizcaino, F. and Duarte J. 2007. Quercetin and isorhamnetin prevent endothelial dysfunction, superoxide production and overexpression of p47 phox induced by angiotensin II in rat aorta. *Journal of Nutrition* 137: 910-915.

16. Singh, V. 2003. Geographical adaptation and distribution of seabuckthorn (*Hippophae* L.) resources. In: *Seabuckthorn (Hippophae L.)-A Multipurpose Wonder Plant. Vol. I. Botany, Harvesting and Processing Technologies* (V.Singh, Ed. in Chief, 2003), p. 21-34, Indus Publishing Company, New Delhi, 518p.

17. Singh, V., Gupta, R.K., Arumughan, C., Sawhney, S.C., Rana, R.K., Singh, A., Lal, M., Devi, R. and Sharma, N. 2011. Biochemical evaluation of *Hippopahe salicifolia* and *H. mongolica* as horticultural crops in dry temperate Himalayas. In: *Seabuckthorn: Emerging Trends on R&D on Health Protection and Environmental Conservation.* Proceedings of National Conference on Seabuckthorn, p. 75-92, CSK HPKV, Palampur, 324p.

18. Teng, B.S., Lu, Y.H., Wang, Z.T., Tao, X.Y. and Wei, D.Z. 2006. *In vitro* anti-tumor activity of isorhamnetin isolated from *Hippophae rhamnoides* L. against BEL-7402 cells. *Pharmacological Research* 54: 186-194.

19. Vishnuprasad, C.N., Anjana, T., Banerji, A. and Anil kumar, G. 2010. Gallic acid induces GLUT4 translocation and glucose uptake activity in 3T3-L1 cells. *Febs Letters* 584: 531-536.

Chapter 22

L-Querbrachitol in Seabuckthorn (*Hippophae rhamnoides* L.) Berries of Different Origins and Harvesting Dates

Baoru Yang and Heikki Kallio

Department of Biochemistry and Food Chemistry,
University of Turku, FI-20014 Turku, Finland

SUMMARY

Seabuckthorn (*Hippophae rhamnoides*) berries are known to have a wide range of beneficial effects on human health. These widely proven physiological effects are partially ascribed to the commonly known components such as fatty acids, vitamins, flavonoids, and plant sterols of the berries. Much work remains to be done to identify new bioactive compounds of the berries and to investigate their effecting mechanisms. A sugar alcohol (-)-2-O-methyl-L-*chiro*-inositol [(-)-L-quebrachitol] has been reported in seabuckthorn berries recently. The presence and the content of this compound may be an indicator of some important aspects of plant physiology and may have significant impact on the health effects of sea buckthorn berries.

The aim of the present study was to compare the content of L-quebrachitol in seabuckthorn berries of different origins and to investigate the effects of harvesting date on the content of the compound in the berries. Wild and cultivated berries of ssp. *sinensis*, ssp. *rhamnoides* and ssp. *mongolica* were harvested in China, Finland and Russia in four consecutive years. The sugar fraction was isolated with a solid phase extraction method from freshly pressed berry juice. Trimethylsilyl (TMS) derivatives of the sugars and sugar alcohols were prepared and analysed quantitatively by gas chromatography (GC-FID). The content of L-quebrachitol varied widely from 0.1 per cent to 1.6 per cent in the juice depending on the different origins and harvesting dates of the berries. The levels in the wild Chinese berries (average content in berry juice, 0.8 per cent) were significantly higher than those found in wild and cultivated berries from Finland (average content in berry juice, 0.3 per cent) and Russia (average content in berry juice, 0.2 per cent) ($P < 0.01$). During the harvesting period from late August to late November, the content of L-quebrachitol showed an increasing trend in the berries of wild ssp. *sinensis* from China and the cultivars of ssp. *rhamnoides* from Finland. Considerable annual variations were

Seabuckthorn (Hippophae L.): A Multipurpose Wonder Plant, Vol. 4: Emerging Trends in Research and Technologies (Virendra Singh, *Editor-in-Chief*, 2014), p. 275–280, Daya Publishing House, New Delhi, India.

recognized in the berries of the same origins. The accumulation of this compound in the berries and especially the high content found in Chinese berries may have been resulted from the activation of biochemical pathways in the plant in adaptation to limited supply of water in the growth environment. Based on the widely shown health effects of inositols and methylinositols, L-quebrachitol is likely to be an important bioactive compound in seabuckthorn berries. Further studies are essential in order to understand the physiological effects of this compound as well as the interactions and synergies between this compound and other components in the berries.

Keywords: *Seabuckthorn, L-quebrachitol, Different origins and Harvesting times.*

INTRODUCTION

Seabuckthorn (*Hippophaë rhamnoides*) berries are known to have a wide range of beneficial effects on human health. The health benefits are partially due to the physiological effects of the commonly known components such as fatty acids, vitamins, flavonoids, and plant sterols of the berries. Much work remains to be done to identify new bioactive compounds of the berries and to investigate their effecting mechanisms. A sugar alcohol (-)-2-*O*-methyl-L-*chiro*-inositol [L-quebrachitol] has been identified in seabuckthorn berries recently (Kallio *et al.*, 2009). The presence and the content of this compound may be an indicator of some important aspects of plant physiology and may have significant impact on both the sensory properties and the health effects of seabuckthorn berries. The aim of the present study was to compare the content of L-quebrachitol in seabuckthorn berries of different origins and to investigate the effects of harvesting date on the content of the compound in the berries.

MATERIALS AND METHODS

Berries were collected from Finnish cultivars S3003, S3006, 74006003, 74006005 (*H. rhamnoides* ssp. *rhamnoides*) and Raisa (a cross between ssp. *rhamnoides* and ssp. *caucasica*) grown at the Research Station at Satakunta (Kokemäki, Finland), Agrifood Research Finland in four consecutive years from1996 to1999. In the same years, wild berries of *H. rhamnoides* ssp. *sinensis* were collected from eighteen natural growth sites in central and northwestern China. In addition, wild berries of *H. rhamnoides* ssp. *rhamnoides* were picked from the Baltic coast in south-western Finland in 1999. Berries of seven commercial cultivars of *H. rhamnoides* ssp. *mongolica* were picked from Novosibirsk, Russia (Ruet, Luchezarnaya, Dar Katuni, Vitaminaya, and Maslichnaya, in 1997) and from Riihimäki (Oranzevaya and Tsuiskaya, in 1999), Finland. In order to follow the changes during the harvesting period, berries were collected from wild bushes of ssp. *sinensis* and cultivated bushes of ssp. *rhamnoides* from China and Finland, respectively, at different harvesting dates from late August to late November. Berries were loosely frozen immediately after picking and kept at -20 °C until being analyzed within one year after the collection. Berries of each sample were pulled from a 5 kg lot following a sample partitioning procedure.

Frozen berries were melted, juice pressed and filtered through a cheese cloth. After addition of sorbitol as the internal standard, the sugar fraction was isolated from berry juice by a solid phase extraction process on an ion exchange Isolute SAX column (International Sorbent Technology, Hengoed, UK) as previously described (Yang, 2009). After silylation, TMS-derivatives of the sugar fraction were analyzed by gas chromatography combined with a flame ionization detection system (Yang, 2009). L-quebrachitol in the samples was quantified as sorbitol without application of any correction factor.

Statistical analysis was carried out using the statistical software SPSS for Windows version 11.0. Independent samples' T-Test and Mann-Whitney Test were used for the comparison of the three subspecies. Differences reaching a minimal confidence level of 95 per cent were considered as being statistically significant.

RESULTS AND DISCUSSION

The content of L-quebrachitol varied widely from 0.3 to 1.6 g/100 ml juice in the Chinese berries of *H. rhamnoides* ssp. *sinensis*, depending on the origins and harvesting years (Table 22.1). The levels in the Finnish (ssp. *rhamnoides*) and Russian (ssp. *mongolica*) berries fell within a much narrower range, typically 0.2-0.4 g/100 ml juice.

Table 22.1: L-Quebrachitol in Wild Berries of *H. rhamnoides* ssp. *sinensis* from different Natural Growth Sites in China

Subspecies	*Origin (Growth Sites)*	*Harvesting Date*	*L-Quebrachitol (g/100 mL juice)*
sinensis	Wenshui, Shanxi	Oct. 23, 96	0.5
		Oct. 4, 97	0.6
		Sep. 15, 99	0.3
		Oct. 15, 99	0.4
		Nov. 15, 99	0.6
sinensis	Fuxian, Shanxi	Oct. 23, 96	0.6
		Oct. 5, 97	1.4
sinensis	Wangtao, Shanxi	Oct. 29, 96	0.6
		Oct. 27, 97	0.9
sinensis	Kelan, Shanxi	Oct. 30, 96	0.7
		Oct. 29, 97	0.7
sinensis	Youyu, Shanxi	Nov. 6, 96	0.8
		Nov. 7, 97	0.9
sinensis	Xixian, Shanxi	Oct. 12, 96	0.8
		Oct. 17, 97	1.0
		Sep. 15, 99	0.9
		Oct. 15, 99	0.8
		Nov. 15, 99	0.8
sinensis	Heshun, Shanxi	Nov. 20, 96	0.7
		Nov. 5, 97	0.8
sinensis	Xunyi, Shaanxi	Nov. 11, 96	0.9
		Nov. 3, 97	1.1
sinensis	Wuzai, Shanxi	Nov. 15, 96	0.5
		Nov. 26, 97	0.5
sinensis	Youyu, Shanxi	Nov. 29, 96	0.9
		Nov. 23, 97	1.3
sinensis	Wutai,Shanxi	Nov. 27, 96	0.9
		Nov. 21, 97	0.5

Contd...

Table 22.1–*Contd...*

Subspecies	*Origin (Growth Sites)*	*Harvesting Date*	*L-quebrachitol (g/100 mL juice)*
sinensis	Yongshou Shaanxi	Nov. 28, 96	1.0
		Nov. 21, 97	1.3
sinensis	Ningwu, Shanxi	Oct. 27, 97	0.5
sinensis	Datong, Qinghai	Nov. 7, 97	0.7
sinensis	Longde, Ningxia	Nov. 20, 97	1.6
sinensis	Dingxi, Gansu	Nov. 9, 97	1.1
sinensis	Zhongyang, Shanxi	Oct. 15, 98	0.7
		Oct. 15, 99	0.5
sinensis	Lishi, Shanxi	Oct. 15, 98	0.7
		Oct. 15, 99	0.5

Overall, the levels in the wild Chinese berries (average content, 0.8 g/100 ml berry juice) were significantly higher than those found in wild and cultivated berries from Finland (average content, 0.3 g/100 ml juice) and Russia (average content, 0.2 g/100 ml juice) ($P < 0.01$, Figure 22.1). Considerable annual variations were observed in the Chinese berries of the same origins indicating significant influence of weather conditions, such as rainfall and growth temperature on the level of this compound in the berries.

Methylinositols are known to be present in some plant species. The presence of methylinositols and other sugar alcohols has been shown to be related to an insufficient water supply to the plants (Pharr *et al.*, 1995; Pattanagul and Madore, 1999). Seabuckthorn is a hardy plant with special tolerance

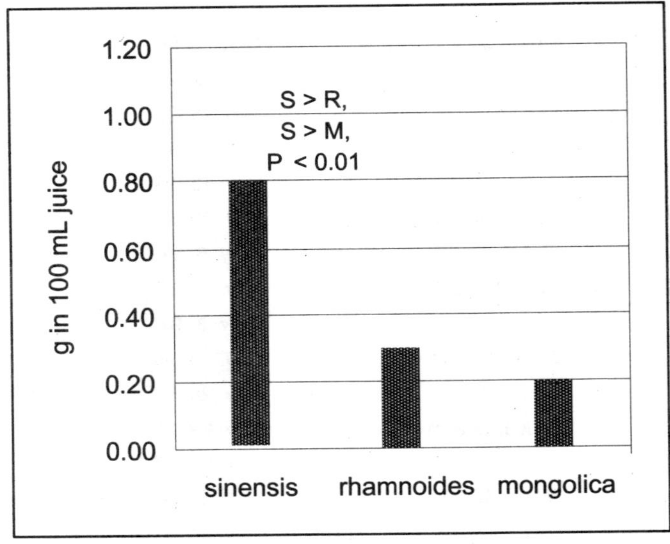

Figure 22.1: Content of L-Quebrachitol in Berry Juice of Three Subspecies of *Hippophaë rhamnoides*. S, *ssp. sinensis*; R, *ssp. rhamnoides*; M, *ssp. mongolica*.

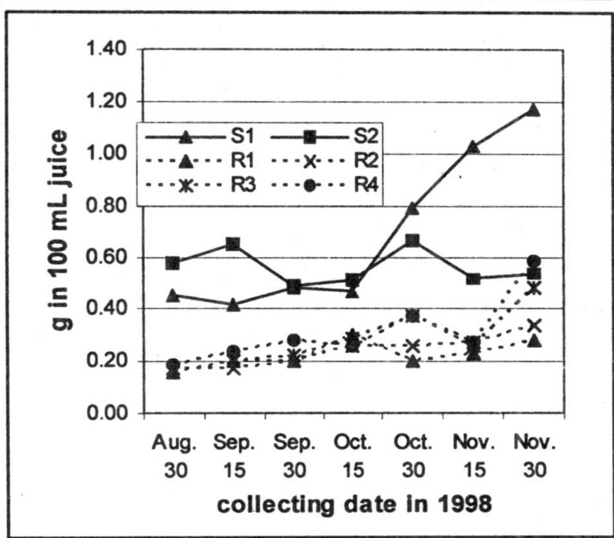

Figure 22.2: Changes in the Content of L-Quebrachitol in Berry Juice during Harvesting Period. S1, wild berries of ssp. *sinensis* from Wenshui, Shanxi, China; S2, wild berries of ssp. *sinensis* from Xixian, Shanxi, China. R1, S3003; R2, S3006; R3, 74006003; R4, 74006005.

of drought and cold. The Chinese wild seabuckthorn grows typically in dry mountainous areas. The relatively high content of L-quebrachitol in Chinese seabuckthorn may have resulted from a special mechanism developed by the plant to fight against the long-lasting drought in the growth areas. During the harvesting period from late August to late November, the content of L-quebrachitol showed an increasing trend in the berries of both wild ssp. *sinensis* from China and of the cultivars of ssp. *rhamnoides* from Finland (Figure 22.2). L-quebrachitol is known to have cryoprotective and membrane stabilizing properties (Orthen and Popp, 2000; Hincha and Hagemann, 2004). The increase in the level of this compound may play a role in helping the plant to cope with the temperature drop during the period followed by the study.

In vitro and animal studies have demonstrated detoxifying, radical scavenging and gastroprotective effects of L-Quebrachitol (Nobre Júnior *et al.*, 2006; de Olinda *et al.*, 2008). The importance of L-quebrachitol as a biologically active component as well as its interaction with other components such as vitamin C, tocopherols and tocotrienols in seabuckthorn berries and berry products shall be investigated in studies in the future.

REFERENCES

1. Kallio, H., Lassila, M., Järvenpää, E., Haraldsson, G., Jonsdottir, S. and Yang, B. 2009. Chiro-inositol and methyl-chiro-inositol in seabuckthorn berries. *J. Chromatog. B* 877: 1426-1432.

2. Yang, B. 2009. Sugars, acids, ethyl β-D-glucopyranose and a methyl inositol in seabuckthorn (*Hippophaë rhamnoides*) berries. *Food Chem.* 112: 89-97.

3. Pattanagul, W. and Madore, M. A. 1999. Water Deficit Effects on Raffinose Family Oligosaccharide Metabolism in Coleus. *Plant Physiol.* 121: 987–993.

4. Pharr, D. M., Stoop, J. M. H., Williamson, J. D., Studer Feusi, M. E., Massel, M. O. and Conkling, M. A. 1995. The dual role of mannitol as osmoprotectant and photoassimilate in celery. *Hortscience.* 30: 1182–1188.

5. Orthen, B. and Popp, M. 2000. Cyclitols as cryoprotectants for spinach and chickpea thylakoids. *Environ. Exp. Toxicol*. 44: 125-132.

6. Hincha, D.K., Hagemann, M. 2004.Stabilization of model membrane during drying by compatible solutes involved in stress tolerance of plants and microorganisms. *Biochem. J.* 383: 277-283.

7. Nobre Júnior, H.V., Cunha G.M., Moraes, M.O., Luciana, M.F., Oliveira, R.A., Maia, F.D., Nogueira, M.A. Lemos, T.L. and Rao, V.S. 2006. Quebrachitol (2-*O*-methyl-L-inositol) attenuates 6-hydroxydopamine-induced cytotoxicity in rat fetal mesencephalic cell cultures. *Food Chem. Toxicol.* 44: 1544-1551.

8. de Olinda, T.M., Lemos, T.L.G., Machado, L.L., Rao, V.S. and Santos F.A. 2008. Quebrachitol-induced gastroprotection against acute gastric lesions: Role of prostaglandins, nitric oxide and K^{+}_{ATP} channels. *Phytomedicine* 15: 327-333.

Chapter 23

Studies on Correlation of Berry Quality Parameters of Seabuckthorn in Relation to Increasing Altitude in China

Lu Rongsen

Chengdu Institute of Biology, Chinese Academy of Sciences, Chengdu, China

SUMMARY

Natural forests of seabuckthorn (*Hippophae rhamnoides* ssp. *sinensis*) are broadly distributed in many areas in China. Many ecological factors influence the quality of seabuckthorn berries. Among those factors, altitude plays an important role. A study on the correlation between seabuckthorn berry quality and altitude of its growing locations was carried out. The berry samples were collected from low altitudes in the east plain areas to high altitudes in the west mountain areas. A special study was conducted in wild seabuckthorn stands in the Wolong valley, Sichuan, at altitudes from 1900 m to 3100 m. Among quality factors, the content of soluble sugars, organic acids and vitamin C have a direct correlation with the growing altitude. At higher altitudes of 1900 to 3100 m, there are greater amounts of organic acids and vitamin C in the berries. The results suggest that establishing seabuckthorn plantations in higher altitude areas can produce better quality berries.

Keywords: *Seabuckthorn, Altitude, Berries, Correlation, Quality and China.*

INTRODUCTION

Chinese seabuckthorn (*Hippophae rhamnoides* ssp. *sinensis*) is very adaptable to harsh ecological conditions and has large populations. It is the most important species of *Hippophae* and is widely distributed in a large area of north and north-west China. It is estimated that the total area of Chinese seabuckthorn is approximately 2.13 million hectares including wild and cultivated forests which accounts for 90 per cent of total seabuckthorn resources in the world (Er, 2003).

Chinese seabuckthorn is the main resource for the seabuckthorn industry in China. It is important to know which seabuckthorn stands produce better quality berries and how to select areas to establish

Seabuckthorn (Hippophae L.): A Multipurpose Wonder Plant, Vol. 4: Emerging Trends in Research and Technologies (Virendra Singh, *Editor-in-Chief*, 2014), p. 281–285, Daya Publishing House, New Delhi, India.

new plantations that could produce the best quality of berries. Many ecological factors influence the quality of seabuckthorn berries, including the altitude at which they are grown. Some studies have shown that there are correlations between berry quality and increasing altitudes (Yang *et al.*, 1988; Yang, 1989; Grazynova *et al.*, 1991). This study will report experimental results that detail how the altitude of the growing location influences seabuckthorn berry quality.

MATERIALS AND METHODS

EXPERIMENTAL MATERIALS

Comparison of Seabuckthorn Growing Locations

Berry samples were collected from 9 locations from the north-east to the south-west of China according to the altitudes of their growing areas. Ripe berry samples were collected from low to high altitude in 2005. In order to get representative samples, berries (2 kg samples) were collected from 20 individual seabuckthorn trees from each location.

Comparison of Seabuckthorn Growing Altitudes

Ripe berry samples were collected from 7 locations at altitudes ranging from 1900 m to 3100 m in the Wolong valley, West Sichuan, China during 2004-2006. In order to get representative samples, berry samples (2 kg) were collected from 20 individual seabuckthorn trees at each altitude.

METHODOLOGY

The analysis of chemical components of the berries was made in the laboratory of Chengdu Institute of Biology, Chinese Academy of Sciences. The chemical analysis was done for soluble sugars, organic acids, vitamin C, seed oil and pulp oil contents. The soluble sugar content was determined by Abe Refract Meter; organic acids by titration; vitamin C by 2.6-dichloroindophenol sodium; oil contents was extracted from dried seeds and oil contents of pulp was extracted from fresh berries; oil of seeds and pulp was extracted by petroleum ether (Zhu, 2004). Each analysis of chemical components was done in triplicates.

RESULTS AND DISCUSSION

The results of chemical analysis of Chinese seabuckthorn berries (*H. rhamnoides* ssp. *sinensis*) growing in locations at different altitudes in 2005 is shown in Table 23.1. It can be seen that there is no trend in the weight per 100 berries, the content of soluble sugars, seed oil and pulp oil along with the increase of altitude, but the contents of organic acids and the vitamin C generally increased with an increase in altitude.

The correlation analysis between altitudes and berry quality of Chinese seabuckthorn (*H. rhamnoides* ssp. *sinensis*) grown in locations at different altitudes in 2005 is shown in Table 23.2. There are direct correlations between the altitudes and the contents of organic acids and vitamin C, the correlation coefficient being 0.734 ($p=0.05$) and 0.877 ($p=0.01$), respectively. There is also a significant correlation between the content of organic acids and content of vitamin C, the correlation coefficient being 0.932 ($p=0.01$).

In order to get a more precise picture about the relationship between the growing altitudes and berry quality, a river valley with a large difference in vertical elevation was selected for an experiment. The experiment was repeated for 3 years (2004-2006). The results of chemical analysis of seabuckthorn berries (*H. rhamnoides* ssp. *sinensis*) growing in Wolong valley, Sichuan at different altitudes in 2004-

Table 23.1: Chemical Analysis of Chinese Seabuckthorn Berries Growing in Locations at different Altitudes in 2005

Growing Location	*Altitude (m)*	*Weight per 100 Berries (g)*	*Soluble Sugars (per cent)*	*Organic Acids (per cent)*	*Vitamin C (mg/100 g)*	*Seed Oil Content (per cent)*	*Pulp Oil Content (per cent)*
Harbin, Heilongjiang	150	16.80	10.0	2.36	345.5	6.25	0.62
Suilin, Heilongjiang	300	19.20	11.5	2.96	403.0	6.20	0.93
Fuxin, Liaoning	450	18.77	10.5	3.82	460.5	6.78	1.06
Erdos, Inner Mongolia	1200	16.66	13.0	3.08	479.0	8.87	1.05
Fengning, Hebei	1400	15.08	11.5	2.78	581.0	8.14	1.08
Huanglong, Shaanxi	1500	15.52	11.5	4.07	915.0	6.65	0.87
Jiaokou, Shanxi	1600	19.64	11.0	2.73	593.5	6.27	0.64
Wuqi, Shaaxi	1700	14.38	13.0	5.39	1008.0	7.45	1.01
Qingshui, Gansu	1800	16.15	12.0	5.07	1082.0	8.44	1.65
Maoxian, Sichuan	1900	19.10	10.0	5.57	1163.0	7.46	1.03
Wolong, Sichuan	2400	15.18	11.0	6.28	1243.0	6.84	1.02

Table 23.2: Correlation Analysis between Altitudes and Berry Quality of Chinese Seabuckthorn Grown in Locations at different Altitudes in 2005

Altitude (m)	*Weight per 100 Berries (g)*	*Soluble Sugars (per cent)*	*Organic Acids (per cent)*	*Vitamin C (mg/100 g)*	*Seed Oil Content (per cent)*	*Pulp Oil Content (per cent)*
Altitude (m)	r 0.400	0.239	0.734*	0.877**	0.367	0.358
	p 0.222	0.079	0.01	0.000	0.266	0.28
Weight per 100 berries		–0.51	–0.327	–0.382	–0.409	–0.282
		0.109	0.326	0.247	0.212	0.4
Soluble Sugars			0.074	0.075	0.586	0.376
			0.829	0.826	0.579	0.255
Organic acids				0.932**	0.189	0.487
				0	0.579	0.129
Vitamin C					0.226	0.454
					0.505	0.161
Seed oil content						0.705*
						0.015
Pulp oil content						

* Correlation is significant at the 0.05 level (2-tailed); **Correlation is significant at the 0.01 level (2-tailed); r: Represents for correlation coefficient; p: represents for significant level.

2006 are shown in Table 23.3. The weight per 100 berries showed varied trend with an increase in altitude, but the content of soluble sugars, organic acids and the vitamin C generally increased along with an increase in altitude.

Table 23.3: Chemical Analysis of Chinese Seabuckthorn Berries in Wolong Valley, Sichuan at different Altitudes in 2004-2006

Altitude (m)	*Harvest Year*	*Weight per 100 Berries (g)*		*Soluble Sugars (per cent)*		*Organic Acids (per cent)*		*Vitamin C (mg/100 g)*	
		Data	*Average*	*Data*	*Average*	*Data*	*Average*	*Data*	*Average*
1900	2004	14.22	7.5	5.55	688.8				
	2005	14.00	(14.46)	6.0	(6.83)	4.56	(5.10)	748.0	(755.1)
	2006	15.15	7.0	5.19	828.6				
2100	2004	14.92	7.0	4.95	795.0				
	2005	14.98	(15.04)	7.5	(7.33)	5.39	(5.16)	973.0	(913.0)
	2006	15.23	7.5	5.15	971.0				
2300	2004	15.88	8.0	4.90	1005.0				
	2005	17.21	(6.55)	7.0	(7.83)	5.62	(5.21)	1163.0	(1059.3)
	2006	16.55	8.5	5.10	1010.0				
2500	2004	15.72	9.0	6.30	1100.0				
	2005	16.04	(16.25)	10.0	(10.00)	6.58	(6.75)	1151.0	(1089.2)
	2006	17.00	11.0	7.36	1016.7				
2700	2004	15.55	10.5	6.70	1164.5				
	2005	15.23	(15.70)	11.5	(11.00)	7.64	(7.28)	1243.0	(1206.0)
	2006	16.30	11.0	7.51	1210.5				
2900	2004	15.20	10.5	7.25	1235.0				
	2005	14.09	(15.13)	10.0	(10.17)	7.17	(7.16)	1381.0	(1304.7)
	2006	16.10	10.0	7.05	1298.0				
3100	2004	15.42	10.5	6.39	1372.5				
	2005	16.50	(15.61)	9.0	(9.83)	6.34	(6.18)	1162.0	(1305.6)
	2006	14.90	10.0	5.81	1381.0				

Table 23.4: Correlation Analysis between Altitude and Berry Quality of Chinese Seabuckthorn in Wolong Valley, Sichuan, China in 2004-2006

	Weight per 100 Berries (g)	*Soluble Sugars (per cent)*	*Organic Acids (per cent)*	*Vitamin C (mg/100 g)*
Altitude (m)	r 0.296	0.845*	0.742	0.975**
	P 0.520	0.017	0.056	0.000
Weight per 100 berries		0.366	0.205	0.410
		0.420	0.659	0.361
Soluble sugars			0.962**	0.861*
			0.001	0.013
Organic acids				0.767*
				0.044

* Correlation is significant at the 0.05 level (2-tailed); **Correlation is significant at the 0.01 level (2-tailed); r: Represents for correlation coefficient; P: Represents for significant level.

The correlation analysis between altitudes and berry quality of seabuckthorn (*H. rhamnoides* ssp. *sinensis*) growing in Wolong valley, Sichuan at different altitudes in 2004-2006 is shown in Table 23.4. There is a direct correlation between the content of soluble sugars and the content of organic acids, the correlation coefficient being 0.962 (p=0.01). There is also a direct correlation between the content of soluble sugars and the content of vitamin C, the correlation coefficient being 0.861 (p=0.05). There is also a significant correlation between the content of organic acids and the content of vitamin C, the correlation coefficient being 0.767 (p=0.05).

CONCLUSION

The results show that the berry quality attributes in this study (weight per 100 berries, soluble substances, organic acids, vitamin C, seed oil and pulp oil contents) seemed dependant on the altitudes of its growing location. Among studied quality attributes, the content of soluble sugars, organic acids and vitamin C have direct correlation with the increasing altitudes. At the elevations of 1900-3100 m, higher the altitude, higher is the content of soluble sugars, organic acids and vitamin C. The results suggest that establishing seabuckthorn plantations in higher altitude areas can produce better quality berries.

REFERENCES

1. Er, J. 2003. The important role of seabuckthorn development in improving ecological environment of West China. *The Global Seabuckthorn Research and Development* 1 (2): 1-2. (In Chinese)
2. Grazynova, E.M., *et al.*, 1991. The biochemical compositions of seabuckthorn berries in the high altitude area of West Pamir, Tajikistan. In: *The Proceedings of New Progress on Biochemistry and Pharmacology in Seabuckthorn*, p. 103-107, Novosibirsk, Science Press, Russia.
3. Yang, H. 1989. A study on dynamic changes of some bioactive components in seabuckthorn berries in Qinghai-Tibetan plateau, China. In: *Proceedings of the International Symposium on Seabuckthorn (H. rhamnoides L.)*, p. 125-128, Xian, China.
4. Yang, H., Wang, S. and Su, X. 1988. A study on dynamic changes of ascorbic acid in seabuckthorn berries in Qinghai, China. *Hippophae* 1 (4): 24-26. (In Chinese)
5. Zhu, K. 2004. *Food Detection Technology*, Beijing Science Press. (In Chinese)

Chapter 24

Studies on the Total Flavonoid Contents of Seabuckthorn (*Hippophae* L.) Berries and Leaves

Lu Rongsen

Chengdu Institute of Biology, Chinese Academy of Sciences, Chengdu, China

SUMMARY

The total flavonoid content (TFC) of *Hippophae* is found to occur in high amounts, thereby indicating a potential in development of medicines. In China, most of the medicines and health products are made mainly from *H. rhamnoides* ssp. *sinensis*. In the present study, the TFC in the berries of species and subspecies of *Hippophae* varied from 0.16 to 2.67 per cent. Due to weather and others factors, significant differences were observed over the years. The TFC in berries of cultivars and hybrids of *Hippophae* ranged from 0.13 to 1.45 per cent. In comparison with the wild species and subspecies, the TFC in the berries of cultivars and hybrids was found to be lower. The TFC in the leaves of species and subspecies of *Hippophae* varied from 0.25 to 2.67 per cent. Similarly affected by weather and others factors, obvious differences were during the study period. The TFC in leaves of cultivars and hybrids of *Hippophae* varied from 0.60 to 2.20 per cent. The leaves of seabuckthorn cultivars and hybrids showed no significant differences in their TFC as compared with the wild forms. Thus, in the seabuckthorn plant, the leaves were found to be a better source of TFC than the berries.

Keywords: *Seabuckthorn, Flavonoids, Berry, Leaves, Bioactivity, Species and subspecies.*

INTRODUCTION

The medicinal value of seabuckthorn was discovered more than 1000 years ago, but its utilization for making medicines and health products by using modern science and technology is only 30 years old. The total flavonoids of *Hippophae* have been proved to have high bio-activities and are being used in making medicines and health products (Binyi *et al.*, 1980; Jialiang *et al.*, 1982; Peiyun and Xiaohong, 1985; Lang *et al.*, 1986; Binwen and Rong, 1993; Chunshe, 1995; Tianfu, 1997; Zhao and Wu, 1997; Jinmin *et al.*, 1998). The resources of TFC are mainly come from the berries of *H. rhamnoides* ssp. *sinensis*. However, in recent years, it has been proved that TFC in the leaves of seabuckthorn possess same bio-

Seabuckthorn (Hippophae L.): A Multipurpose Wonder Plant, Vol. 4: Emerging Trends in Research and Technologies (Virendra Singh, *Editor-in-Chief*, 2014), p. 286–296, Daya Publishing House, New Delhi, India.

activities and can be used in medicines and health products (Yun *et al.*, 2002; Zhao *et al.*, 2004; Xiaoping *et al.*, 2010).The purpose of this study is to analyze the TFC in the berries and leaves of 6 species and 10 subspecies of *Hippophae*, so as to identify the better resources of TFC.

MATERIALS AND METHODS

Experimental Materials

Seabuckthorn is widely distributed in Eurasia continent and grows in extremely different ecological environments. In order to get real biological characters and useful chemical components, all fresh samples of berries and leaves of seabuckthorn were collected from their original places. Flavonoids are secondary metabolites of plants and their composition is greatly affected by ecological environment and weather. Thus, in order to study the variability in the TFC of species and subspecies of *Hippophae*, the samples were collected from distinct locations over a time period of few years.

EXPERIMENTAL METHODS

The analysis of samples was conducted in Chengdu Institute of Biology and Chengdu Test and Analysis Center, Chinese Academy of Sciences by using the standard methods.

Sample Collection and Treatment

The fresh mature berries of *Hippophae* were collected and placed in freezer for storage. At the same time, the fresh leaves too were collected, air dried and stored in the refrigerator.

Preparation of Test Samples

The frozen berries were dried and the seeds ticked out to stamp the pulp and skin, the oil was extracted using petroleum ether. The defatted sample and dried leaves were ground to a fine powder for further analysis.

Determining Test Samples

One gram of the dried sample was placed in 15 ml of 70 per cent alcohol at a temperature of 80-90 °C. for extraction that was repeated eight times. To 10 ml of the extract, added 5 ml of concentrated hydrochloric acid and 20 ml of methanol was added. This solution was put in reflowing for 1.5 hours under boiling water, cooled and placed in 50 ml capacity bottle at constant volume. Waters' HPLC determination by: Chromatogram condition using methanol:water:phosphoric acid::55:45:0.5; flowing speed 0.5 ml/min; column temperature 30 °C; sample volume 10 µl; determined at wavelength 360 nm.

RESULTS AND DISCUSSION

TFC of *Hippophae* Berries

Results of 1999

From Table 24.1, it can be seen that TFC in the berries of species and subspecies of *Hippophae* ranged from 0.16-2.67 per cent, indicating a wide variation. Among them, the higher TFC are found in *H. rhamnoides* ssp. *sinensis*, *H. rhamnoides* ssp. *yunnanensis*, *H. gyantsensis*, *H. goniocarpa* ssp. *goniocarpa*, *H. goniocarpa* ssp. *litangensis*, *H. neurocarpa* ssp. *neurocarpa*, *H. neurocarpa* ssp. *stallatopilosa*, *H. salicifolia and H. tibetana*. Their TFC values are more than 1 per cent and the highest is observed in *H. neurocarpa* ssp. *neurocarpa* (2.67 per cent). The lower TFC (≥1 per cent) are found in *H. rhamnoides* ssp. *rhamnoides*,

H. rhamnoides ssp. *mongolica and H. rhamnoides* ssp. *turkestanica* and the lowestcontent is found in *H. rhamnoides* ssp. *mongolica* (0.16 per cent).

Table 24.1: Total Flavonoid Content of Berries in Species and Subspecies of *Hippophae* in 1999

Species/Subspecies	*Location and Altitude (m asl)*	*Four Aglycones Content (mg/g)*				*ΣAgly-cones (mg/g)*	**TFC (per cent)*
		1#	*2*	*3*	*4*		
H.rham. ssp.*sinensis*	Maoxian, Sichuan (1850)	–	4.060	1.265	2.97	8.295	2.07
H.rham. ssp.*sinensis* (Red)	Maoxian, Sichuan (1850)	0.087	1.033	0.290	2.445	3.853	0.96
H.rham. ssp.*sinensis*	Miyaluo, Sichuan (2500)	0.132	1.807	0.247	2.934	5.120	1.27
H.rham.ssp.sinensis	Danba, Sichuan (1800)	0.091	0.772	0.412	1.697	2.972	0.74
H.rham. ssp.*sinensis*	Qianning, Sichuan (3200)	–	3.082	1.962	2.904	7.948	1.98
H.rham. ssp.*yunnanensis*	Zhongdian, Yunnan (3700)	0.157	1.257	0.117	3.337	4.8681	1.21
H.rham. ssp.*mongolica*	Habahe, Xinjiang (600)	0.017	0.032	0.011	0.585	0.645	0.16
H.rham. ssp.*rhamnoides*	Turku, Finland (150)	0.028	0.605	0.112	3.332	4.077	1.02
H.rham. ssp.*rhamnoides*	Balsgard, Sweden (110)	–	0.284	0.107	2.378	2.769	0.69
H.rham. ssp.*turkestanica*	Girgit, Pakistan (2800)	0.003	0.190	0.083	1.208	1.483	0.37
H.rham. ssp.*turkestanica*	Gilgit, Pakistan (1800)	–	1.635	0.945	1.620	4.200	1.05
H.rham. ssp.*turkestanica*	Yili, Xinjiang (700)	0.022	0.725	0.104	1.260	2.661	0.66
H.rham. ssp.*turkestanica*	Khashi, Xinjiang (700)	0.013	0.263	0.052	0.665	0.993	0.25
H.gyantsensis	Lhasa,Tibet (3700)	0.029	2.633	1.041	2.112	5.865	1.46
H.gyantsensis	Linzhi, Tibet (2900)	0.102	2.041	1.237	1.825	5.206	1.30
H.goniocarpa ssp. *goniocarpa*	Hongyuan, Sichuan (3700)	0.151	0.755	0.521	1.247	2.676	0.67
H.goniocarpa ssp. *goniocarpa*	Hongyuan, Sichuan (3700)	–	4.172	0.692	3.846	8.710	2.16
H.goniocarpassp.litangensis	Litang, Sichuan (3500)	0.063	2.844	0.6	3.565	7.072	1.76
H.neurocarpa ssp.*neurocarpa*	Hongyuan,Sichuan (3700)	–	4.565	0.990	5.180	10.735	2.67
H.neurocarpassp. stallatopilosa	Litang, Sichuan (3500)	–	2.896	2.506	3.538	6.940	1.72
H.neurocarpassp. stallatopilosa	Daocheng,Sichuan (3200)	0.271	2.100	0.595	4.410	7.376	1.83
H.salicifolia	Maoxian,Sichuan (1850)	–	5.145	0.545	–	5.690	1.42
H.salicifolia	Maoxian,Sichuan (1850)	–	5.780	0.535	–	6.315	1.57
H.tibetana	Hongyuan, Sichuan (3700)	–	2.760	0.900	2.495	6.155	1.53

#: 1: Myricetin; 2: Quercetin; 3: Kaempferol; 4: Isorhamcetin.

* Total Flavonoid content (TFC) of *Hippophae* is calculated by the formula: Σ (aglycones) × 2.49 × 100, 2.49 is average conversion factor (15).

Results of 2002

The data in Table 24.2 was recorded in 2002 and showed that the TFC in the berries of species and subspecies of *Hippophae* varied from 0.16 to 2.30 per cent, which is lower than that recorded in the year 1999 (0.16-2.67 per cent). The higher contents of TFC is observed in *H. rhamnoides* ssp. *sinensis, H. rhamnoides* ssp. *yunnanensis, H. gyantsensis, H. goniocarpa* ssp. *litangensis, H. neurocarpa* ssp. *neurocarpa,*

H. neurocarpa ssp. *stallatopilosa, H. salicifolia and H. tibetana,* which is more than 0.75 per cent and the highest is seen in *H. salicifolia* (2.30 per cent). The lower TFC is found in *H. rhamnoides* ssp. *rhamnoides, H. rhamnoides* ssp. *mongolica and H. rhamnoides* ssp. *turkestanica, which* is less than 0.71 per cent and the lowest one is found in *H. rhamnoides* ssp. *mongolica* (0.16 per cent).

Table 24.2: Total Flavonoid Content of Berries in Species and Subspecies of *Hippophae* in 2002

Species/Subspecies	*Location and Altitude (m asl)*	*Four Aglycones Content (mg/g)*				*ΣAglycones (mg/g)*	*TFC (per cent)*
		1#	*2*	*3*	*4*		
H.rham. ssp.*sinensis*	Fenghing, Hebei (400)	0.1390	0.8980	3.006	0.3770	4.420	1.10
H.rham. ssp.*sinensis* (Red)	Maoxian, Sichuan (1850)	–	0.5368	1.8950	0.1264	2.558	0.64
H.rham. ssp.*sinensis*	Ruoega, Sichuan (3500)	0.0140	0.6640	2.5695	0.0790	3.327	0.83
H.rham. ssp. *sinensis*	Maoxian, Sichuan (1850)	0.0301	0.8419	4.3003	0.2254	5.398	1.34
H.rham. ssp.*yunnanensis*	Zhongdian,Yunnan (3600)	0.2180	0.795	1.7880	0.2100	3.011	0.75
H.rham. ssp.*mongolica*	Burjin, Xinjiang (600)	–	0.0129	0.6483	–	0.661	0.16
H.rham. ssp.*rhamnoides*	Barsgard, Sweden (100)	0.3170	0.761	1.4670	0.3090	2.854	0.71
H.rham. ssp.*fluviatilis*	Paris, France (120)	–	0.4297	8.2739	–	8.7018	2.16
H.rham. ssp.*turkestanica*	Denkou, Inner Mongolia (800)	0.2380	0.700	1.6380	0.2390	2.851	0.70
H.gyantsensis	Linzhi, Tibet (2900)	0.1020	2.041	1.2370	1.825	5.206	1.30
H.goniocarpa ssp. *goniocarpa*	Hongyuan, Sichuan (3700)	0.1399	0.7109	1.3400	0.3732	2.564	0.64
*H.goniocarpa*ssp. *litangensis*	Litang, Sichuan (3500	0.2660	0.604	2.5000	0.3120	3.682	0.92
H.neurocarpa ssp. *neurocarpa*	Shiqu, Sichuan (4200)	–	1.0811	2.4231	2.4557	3.960	0.99
H.neurocarpa ssp. *stallatopilosa*	Litang, Sichuan (3800)	0.0570	2.101	2.5760	2.3830	7.117	1.77
H. salicifolia	Cuona, Tibet (3000)	0.2340	3.046	0.7940	0.5820	4.656	1.17
H. salicifolia	Cuona, Tibet (3000)	0.1760	4.165	0.3360	0.3310	5.008	1.25
H. salicifolia	Milin, Tibet (2800 m)	0.2007	2.3718	6.1403	0.5269	9.2400	2.30
H. tibetana	Ruorgai, Sichuan (3500)	0.0060	0.4780	1.7240	0.9370	3.1450	0.78

#: 1: Myricetin; 2: Quercetin; 3: Kaempferol; 4: Isorhamcetin.

Results of 2004

The data in Table 24.3 depicts the TFC in seabuckthorn cultivars and hybrids in the range of 0.13-1.45 per cent. Among them, the higher TFC (>0.50 per cent) are found in cultivars "Sun, Azura, Hybrid Yellow, German Hergo and German Askola", whereas other cultivars and hybrids have lower TFC (<0.50 per cent). The highest content of total flavonoids in the berries is observed in "Sun" (1.45 per cent), and lowest in "Hybrid Yellow" (0.13 per cent) which is late maturing.

TFC of *Hippophae* Leaves

Results of 1999

From Table 24.4, it can be seen that the TFC in the leaves of species and subspecies of *Hippophae* are in the range of 0.25 to 2.67 per cent, showing a wide variation. Among them, the higher contents of total flavonoids (>1.2 per cent) are found in *H. rhamnoides* ssp. *sinensis, H. rhamnoides* ssp. *yunnanensis,*

Table 24.3: Total Flavonoid Content of Berries in some Varieties and Hybrids of Seabuckthorn in 2004

Species/Subspecies (Cultivars/Hybrids)	Location and Altitude (m)	Four Aglycones Content (mg/g)				ΣAglycones (mg/g)	TFC (per cent)
		1#	2	3	4		
H.rham. ssp.*sinensis*	Harbin, Heilongjiang (150)	0.0003	0.1726	1.4901	0.0736	1.737	0.43
H.rham. ssp.*sinensis*	Fuxin, Liaoning (470)	–	0.4345	1.2881	–	1.723	0.43
Sun	Qiqihar, Heilongjiang (150 m)	–	1.951	0.779	3.101	5.831	1.45
Azora	Qiqihar, Heilongjiang (150)	–	2.056	1.145	1.532	4.733	1.18
North Europe Big Berry	Fuxin, Liaoning (470)	–	0.1430	1.6021	–	1.745	0.44
Liaofu No.1	Fuxin,Liaoning (470)	–	0.2945	1.2326	–	1.527	0.38
Hybrid Yellow	Fuxin, Liaoning (470)	0.0249	0.6151	1.3891	–	2.029	0.51
Suiji HS-7	Suilin, Heilongjiang (350)	–	0.3767	1.4075	–	1.784	0.44
Suiji HS-2	Suilin, Heilongjiang (350)	–	0.5129	1.1440	–	1.657	0.41
German Hergo	Potsdam,Germany (150)	–	0.2268	4.3369	–	4.564	1.14
German Askola	Potsdam, Germany (150)	–	0.1601	3.8827	–	4.047	1.01
Hybrid Yellow (late maturing)	Fuxin, Liaoning (470)	–	0.2018	0.4326	–	0.5030	0.13
Zhuang Yuan Yellow	Fuxin, Liaoning (470)	4.419	4.519	1.7461	–	1.9480	0.49
Hybrid Orange	Fuxin, Liaoning (470)	–	0.1988	0.6666	–	0.8650	0.22
Hybrid Orange-red	Fuxin, Liaoning (470)	–	0.3460	0.6709	0.0356	1.052	0.26

#: 1: Myricetin; 2: Quercetin; 3: Kaempferol; 4: Isorhamcetin.

Table 24.4: Total Flavonoid Content of Leaves in Species and Subspecies of *Hippophae* in 1999

Species/Subspecies	Location and Altitude (m)	Four Aglycones Content (mg/g)				ΣAglycones (mg/g)	TFC (per cent)
		1#	2	3	4		
H.rham. ssp.*sinensis*	Maoxian, Sichuan (1850)	–	2.560	1.265	2.970	8.700	2.17
H.rham. ssp.*sinensis*	Danba, Sicuan (1800)	–	3.935	0.730	2.300	6.965	1.73
H.rham. ssp.*sinensis*	Qianning, Sichuan (3200)	–	3.082	1.962	2.904	7.948	1.98
H.rham. ssp.*yunnanensis*	Zhongdian, Yunnan (3600)	–	3.785	0.417	0.597	4.799	1.20
H.rham. ssp.*mongolica*	Habahe, Xinjiang (600)	–	1.874	0.433	2.277	4.584	1.14
H.rham. ssp.*rhamnoides*	Turku, Finland (150)	–	0.982	0.594	1.962	3.538	0.88
H.rham. ssp.*rhamnoides*	Barsgard, Sweden (110)	–	1.011	0.187	2.400	3.598	0.90
H.rham. ssp.*turkestanica*	Gilgit, Pakistan (1800)	–	1.635	0.945	1.620	4.200	1.05
H.rham. ssp.*turkestanica*	Yili, Xinjiang (700)	0.022	0.725	0.104	1.260	2.661	0.66
H.rham. ssp.*turkestanica*	Khash, Xinjiang (700)	0.013	0.263	0.052	0.665	0.993	0.25
H.gyantsensis	Lhasa, Tibet (3700)	0.029	2.633	1.041	2.112	5.865	1.46
H.gyantsensis	Linzhi, Tibet (2900)	0.102	2.041	1.237	1.825	5.206	1.30

Contd...

Table 24.4–Contd...

Species/Subspecies	*Location and Altitude (m)*	*Four Aglycones Content (mg/g)*				*ΣAgly-cones (mg/g)*	*TFC (per cent)*
		1#	*2*	*3*	*4*		
H.goniocarpa ssp. *goniocarpa*	Hongyuan, Sichuan (3700)	–	4.172	0.692	3.846	8.710	2.16
*H.goniocarpa*ssp.*litangensis*	Litang, Sichuan (3500)	0.063	2.844	0.6	3.565	7.072	1.76
H.neurocarpa ssp.*neurocarpa*	Hongyuan, Sichuan (3700)	–	4.565	0.990	5.180	10.735	2.67
H.neurocarpa ssp. *stallatopilosa*	Litang, Sichuan (3500)	–	2.896	0.506	3.538	6.940	1.73
H. neurocarpa ssp. *stallatopilosa*	Daocheng, Sichuan (3200)	0.271	2.100	0.595	4.410	7.376	1.84
H. salicifolia	Maoxian, Sichuan (1850)	–	5.145	0.545	–	5.690	1.42
H. salicifolia	Maoxian, Sichuan (1850)	–	5.780	0.535	–	6.315	1.57
H. salicifolia	Maoxian, Sichuan (1850)	–	5.330	0.545	–	5.875	1.46
H. tibetana	Hongyuan, Sichuan (3700)	–	2.760	0.900	2.495	9.155	2.28

#: 1: Myricetin; 2: Quercetin; 3: Kaempferol; 4: Isorhamcetin.

Table 24.5: Total Flavonoid Content of Leaves in Species and Subspecies of *Hippophae*, 2004

Species/Subspecies	*Location and Altitude (m)*	*Four Aglycones Content (mg/g)*				*ΣAgly-cones (mg/g)*	*TFC (per cent)*
		1#	*2*	*3*	*4*		
H.rham. ssp.*sinensis*	Fuxin, Liaoning (470)	0.0548	1.0800	2.4990	0.9607	4.594	1.14
H.rham. ssp.*sinensis*	Ruorgai, Sichuan (3500)	0.0212	1.9921	4.0686	0.7065	6.788	1.69
H.rham. ssp.*sinensis*	Maoxian, Sichuan (1850)	–	1.4970	3.7492	0.9866	6.233	1.55
H.rham. ssp.*rhamnoides*	Potsdam, Germany (100)	0.202	0.758	2.263	0.207	3.430	0.85
H.rham. ssp.*turkestanica*	Dengkou, Inner Mongolia (800)	0.179	0.922	2.196	0.279	3.576	0.89
H.rham. ssp.*fluviatilis*	Paris, France (120)	–	1.4276	4.1816	0.4069	6.016	1.49
H.gyantsensis	Lhasa, Tibet (3700)	0.458	5.418	5.049	2.351	13.276	3.31
H.neurocarpa ssp.*neurocarpa*	Shiqu, Sichuan (4200)	–	2.7098	6.0632	0.5596	9.332	2.32
H.neurocarpa ssp.*stallatopilosa*	Litang, Sichuan (3800)	0.057	2.101	2.576	2.383	7.117	1.77
H.tibetana	Ruorgai, Sichuan (3500)	0.2151	1.8405	8.1158	0.7978	10.969	2.73
H.rham. ssp.*sinensis* (Male)	Maoxian, Sichuan (1850)	–	1.3058	2.0923	0.7608	4.159	1.03
H.rham. ssp.*sinensis* (Female)	Maoxian, Sichuan (1850)	0.0107	1.4883	1.7948	0.5410	3.835	0.96
H.rham. ssp.*sinensis* (Male)	Dongsheng, Inner Mongolia (1000)	–	0.8715	1.8845	0.4240	3.180	0.79
H.rham. ssp.*sinensis* (Female)	Dongsheng, Inner Mongolia (1000)	–	1.1963	2.6557	0.5620	4.414	1.10
H.rham. ssp.*sinensis* (Male)	Fuxin, Liaoning (470)	0.0134	1.1334	1.8738	0.6081	3.629	0.90
H.rham. ssp.*sinensis* (Female)	Fuxin, Liaoning (470)	0.0309	0.5509	1.3208	1.3263	3.229	0.80

Contd...

Table 24.5–Contd...

Species/Subspecies	Location and Altitude (m)	Four Aglycones Content (mg/g)				ΣAgly-cones (mg/g)	TFC (per cent)
		1#	2	3	4		
H.goniocarpa ssp.*goniocarpa* (Male)	Hongyuan, Sichuan (3700)	0.0023	1.5398	2.5779	0.8262	4.946	1.23
H.goniocarpa ssp.*goniocarpa* (Female)	Hongyuan, Sichuan (3700)	–	2.7002	2.7124	0.7072	6.120	1.52
H.gyantsensis (Male)	Mozhugongka, Tibet (3900)	0.2521	4.2722	4.5152	0.8944	9.935	2.47
H.gyantsensis (Female)	Mozhugongka, Tibet (3900)	0.1911	3.4671	4.0650	0.7735	8.497	2.12
H.neurocarpa ssp.*neurocarpa* (Male)	Hongyuan, Sichuan (3700)	–	3.0122	5.3947	0.5126	8.919	2.22
H.neurocarpa ssp.*neurocarpa* (Female)	Hongyuan, Sichuan (3700)	–	4.1896	7.7360	0.5170	12.443	3.10
H.salicifolia (Male)	Maoxian, Sichuan (1850)	–	2.6430	0.0255	0.3022	2.971	0.74
H.salicifolia (Female)	Maoxian, Sichuan (1850)	–	3.4044	–	0.4068	3.811	0.95
H.salicifolia (Male)	Cuona, Tibet (2800)	0.176	3.034	0.404	0.351	3.965	0.99
H.salicifolia (Female)	Cuona, Tibet (2800)	0.166	3.300	0.315	0.326	4.107	1.02
Hergo (Male)	Potsdam, Germany (110)	–	0.4146	2.4609	0.2752	3.151	0.79
Hergo (Female)	Potsdam, Germany (110)	–	1.1700	4.4236	0.3528	5.946	1.48

#: 1: Myricetin; 2: Quercetin; 3: Kaempferol; 4: Isorhamcetin.

H. gyantsensis, H. goniocarpa ssp. *goniocarpa, H. goniocarpa* ssp. *litangensis, H. neurocarpa* ssp. *neurocarpa, H. neurocarpa* ssp. *stallatopilosa, H. salicifolia and H. tibetana*. The highest was observed in *H. neurocarpa* ssp. *neurocarpa* (2.67 per cent).The lower TFC (<1 per cent) are observed in *H. rhamnoides* ssp. *rhamnoides* and *H. rhamnoides* ssp. *turkestanica* and the lowest is 0.25 per cent in *H. rhamnoides* ssp. *turkestanica*, in Khash, Xinjiang.

Results of 2004

The data in Table 24.5 shows the TFC in the leaves of species and subspecies of *Hippophae* and some cultivars are in the range of 0.74-3.31 per cent, showing a great variation. Among them, the higher TFC (>1.0 per cent) are found in *H. rhamnoides* ssp. *sinensis, H. rhamnoides* ssp. *yunnanensis, H. gyantsensis, H. neurocarpa* ssp. *neurocarpa, H. neurocarpa* ssp. *stallatopilosa, H. salicifolia* (female) and *H. tibetana* and the highest being 3.31 per cent in *H. gyantsensis*. The lower TFC (<1.0 per cent) is found in *H. rhamnoides* ssp. *rhamnoides, H.rhamnoides* ssp. *turkestanica, H.rhamnoides* ssp. *sinensis* and the lowest is 0.74 per cent in *H. salicifolia* (male) from Maoxian, Sichuan.

Male and Female Leaves

Seabuckthorn is a dioecious plant, where more than half of the plants are male, thereby creating a curiosity to study the gender differences of the TFC. The data in Table 24.5 depicts the results of this study. The TFC of leaves in most of the female plants is higher than that in the male plants, as also reported by others (Jiaoshe *et al.*, 2000), but in some cases it is opposite, as seen in female plants in *H. rhamnoides* ssp. *sinensis* in some areas. Thus, the leaves of male plants have a greater value, since they form the larger part of the forest.

Results of 2003

The data in Table 24.6 was determined in 2003 and shows that the contents of TFC in the leaves of cultivars and hybrids are in the range of 0.6-2.20 per cent which, in general, is lower than that in wild species and subspecies. Among them, the higher TFC values (>1.19 per cent) are found in "North Europe Big Berry, Liaofu No.1, Suiji HS-7Yellow, Suiji HS-7 Green, Suilin No.2 and Askola" and the highest is 2.20 per cent in "Askola". The other cultivars and hybrids have lower contents of TFC (<0.85 per cent) and the lowest one is 0.60 per cent in "Hybrid Orange-2".

Table 24.6: Total Flavonoid Content of Leaves in some Varieties and Hybrids of Seabuckthorn in 2003

Species/Subspecies	*Location and Altitude (m asl)*	*Four Aglycones Content (mg/g)*				*ΣAglycones (mg/g)*	*TFC (per cent)*
		1#	*2*	*3*	*4*		
Hybrid Red	Fuxin, Liaoning (470)	0.0339	1.0582	1.9511	0.3523	3.396	0.85
Hydrid Orange-1	Fuxin, Liaoning (470)	–	0.7162	1.8373	0.6766	3.230	0.80
Hybrid Orange-2	Fuxin, Liaoning (470)	0.0075	0.7611	1.2906	0.3571	2.416	0.60
North Europe Big Berry	Fuxin, Liaoning (470)	–	1.1437	2.6435	0.9945	4.762	1.19
Liaofu-No.1	Fuxin, Liaoning (470)	–	1.1429	2.8986	0.4909	4.872	1.21
Hybrid Yellow	Fuxin, Liaoning (470)	0.0080	0.6735	1.3841	0.5486	2.614	0.65
Hybrid Big yellow	Fuxin, Liaoning (470)	0.0065	1.0007	2.4259	0.4254	3.858	0.96
Suiji S-7Yellow	Suilin, Heilongjiang (350)	–	1.0988	3.5858	0.5904	5.275	1.31
Suiji HS-7 Green	Suilin, Heilongjiang (350)	0.0222	1.9424	3.3403	0.6945	5.959	1.48
Suilin No.2	Suilin, Heilongjiang (350)	0.0283	2.1430	3.5287	0.7755	6.476	1.61
Askola	Potsdam, Germany (110)	0.0012	1.4946	6.7640	0.5638	8.824	2.20

#: 1: Myricetin; 2: Quercetin; 3: Kaempferol; 4: Isorhamcetin.

Seasonal Variations

The data of TFC in above tables have shown that the leaves of *Hippophae* contain considerable flavonoids, but it is needed to know when the TFC contents are high and what season is suitable for harvest? The samples were collected from 3 species and subspecies: *H. rhamnoides* ssp. *sinensis, H. rhamnoides* ssp. *yunnanensis and H. salicifolia* raised in the same place in Maoxian, Sichuan. The leaf samples were collected during vegetative growing stage (April-November).

The data of TFC in the leaves of *H. salicifolia* ranged from 2.03-2.45 per cent in 8 months, there were no significant differences, although the contents in April and November were higher than in other months (Table 24.7). The data on TFC showed in *H. rhamnoides* ssp. *sinensis,* it varied in the range of 1.59-3.73 per cent in 8 months and there were some differences among the months. The contents in April and May were higher than in other months, after that the TFC reduced gradually, and the lowest was in November. While, the TFC in *H. rhamnoides* ssp. *yunnanensis* varied from 1.62 to 3.735 per cent in 8 months. The contents in April and May were higher than in other months, after that the contents reduced gradually, and the lowest content was found in August. But in November, its content increased. These results were similar to those reported in literature (Guixian *et al.*, 1997).

Table 24.7: Monthly Variation in the Total Flavonoid Contents of Leaves of *Hippophae* during the Vegetative Growing Stage in 2003

Species/Subspecies	*Location and Altitude (m asl)*	*Four Aglycones Content (mg/g)*				*ΣAglycones (mg/g)*	*TFC (per cent)*
		*1**	*2*	*3*	*4*		
H.salicifolia (Apr.)	Maoxian, Sichuan (1850)	0.021	6.652	–	1.855	8.528	2.12
(May.)	"	0.009	6.938	–	1.400	8.347	2.08
(Jun.)	"	0.009	6.970	–	1.153	8.132	2.03
(Jul.)	"	0.020	8.252	–	1.120	9.392	2.34
(Aug.)	"	0.007	7.414	–	0.856	8.277	2.06
(Sept.)	"	0.010	7.837	–	0.698	8.545	2.13
(Oct.)	"	0.005	8.668	–	0.742	9.415	2.34
(Nov.)	"	–	8.985	–	0.854	9.839	2.45
H.rham. ssp.*sinensis* (Apr.)	"	–	3.091	3.745	6.590	13.426	3.34
(May.)	"	0.067	3.771	4.237	6.909	14.984	3.73
(Jun.)	"	0.087	2.737	2.945	5.276	11.045	2.75
(Jul.)	"	0.104	2.897	2.967	3.404	9.372	2.33
(Aug.)	"	0.086	2.826	2.889	2.621	8.422	2.10
(Sept.)	"	0.058	2.132	2.388	2.417	6.994	1.74
(Oct.)	"	0.068	2.522	2.825	1.956	7.371	1.84
(Nov.)	"	0.029	2.360	2.636	1.353	6.378	1.59
H.rham. ssp.*yunnanensis* (Apr.)	"	0.341	7.849	3.134	2.131	13.455	3.35
(May.)	"	0.056	7.447	1.771	1.673	10.947	2.73
(Jun.)	"	0.012	5.519	1.417	1.243	8.191	2.04
(Jul.)	"	0.018	4.875	1.391	0.805	7.089	1.76
(Aug.)	"	0.001	4.680	1.148	0.681	6.509	1.62
(Sept.)	"	0.023	4.815	1.500	0.748	7.086	1.76
(Oct.)	"	0.026	5.039	1.615	0.534	7.214	1.80
(Nov.)	"	0.034	5.739	1.606	0.801	8.180	2.03

CONCLUSION

The TFC in the berries of species and subspecies of *Hippophae* varied in 0.16-2.67 per cent. Due to weather and others factors, there were significant differences in different years. The TFC of berries in cultivars and hybrids of *Hippophae* varied in the range of 0.13-1.45 per cent. In comparison with the wild species and subspecies, the TFC of cultivars and hybrids were lower than wild forms of *Hippophae*. It is found that the TFC of those species and subspecies that growing in the areas with high altitudes was higher than those growing in the areas with lower altitudes. *H. rhamnoides* ssp. *sinensis*, *H. rhamnoides* ssp. *yunnanensis*, *H. gyantsensis*, *H. goniocarpa* ssp. *goniocarpa*, *H. goniocarpa* ssp. *litangensis*, *H. neurocarpa* ssp. *neurocarpa*, *H. neurocarpa* ssp. *stallatopilosa*, *H. salicifolia and H. tibetana* usually grow in high altitude areas. *H. rhamnoides* ssp. *rhamnoides*, *H. rhamnoides* ssp. *mongolica and H. rhamnoides* ssp. *turkestanica* usually grow in the lower altitude areas.

The reason why the TFC contents of cultivars and hybrids were lower is because of their genetic relationship. The cultivars and hybrids that were studied were selected or bred from *H. rhamnoides* ssp. *rhamnoides* and *H. rhamnoides* ssp. *mongolica*. Although, the hybrids have *H. rhamnoides* ssp. *sinensis* as male parent, the characters of *H. rhamnoides* ssp. *mongolica* seems more dominant. The TFC in the leaves of species and subspecies of *Hippophae* varied from 0.25 to 2.67 per cent. Affected by weather and others factors, obvious differences over the years were observed as in the berries. The TFC of leaves in cultivars and hybrids of *Hippophae* varied from 0.60 to 2.20 per cent. Compared with the wild species and subspecies, cultivars and hybrids have no significant differences in their TFC of leaves.

Among the species and subspecies of *Hippophae*, it seems that the leaf TFC of those growing in the areas with high altitudes was higher than those growing in the areas with lower altitudes. The differences of TFC in leaves caused by altitudes, is similar to that in berries.

The TFC of leaves in female plants in most of the species and cultivars is higher than in male plants. But in some areas, the TFC of leaves in male plants of *H. rhamnoides* ssp. *sinensis* are higher than that in female plants. However, the leaves of male plants could be used for extracting TFC due to its larger proportion in the natural forests. The TFC in leaves of *H. rhamnoides* ssp. *sinensis* and *H. rhamnoides* ssp. *yunnanensis* in the months of April, May and June, was higher than that in other months. Since, the TFC of mature leaves is comparatively higher, the leaves that in late growing season can be harvested.

Not only in berries but also in leaves, the species and subspecies of *Hippophae* and their cultivars and hybrids contain considerable TFC, which can be used for making medicines and health products. Among these, *H. rhamnoides* ssp. *sinensis*, *H. rhamnoides* ssp. *turkestanica*, *H. rhamnoides* ssp. *mongolica*, *H. gyantsensis and H. salicifolia* have high potential for use in industries, because their natural forests occupied large areas in Eurasia. For those, such as *H. goniocarpa* ssp. *goniocarpa*, *H. goniocarpa* ssp. *litangensis*, *H. neurocarpa* ssp. *neurocarpa and H. tibetana*, although they contain high TFC, they cannot be used since their natural forests are scarce, however, they need to be protected because they are precious resources for environmental conservation and future research.

REFERENCES

1. Binwen, W. and Rong, Z.1993. Effect on function of cardiography and hemodynamics of normal human beings by total flavonoids. *Acta Academiae Medicinae Xian* 14 (3): 130-140.
2. Binyi, L. *et al.*, 1980. Curative effect on hyperlipemia and coronary heart disease by seabuckthorn flavonoids. *Acta Academiae Medicinae Sichuan* 11 (10): 11-14.
3. Chunshe, Y. 1995. Clinic study on reducing blood fat and antioxidation by seabuckthorn dry emulsion. *Hippophae* 8 (1): 33-35.
4. Guixian, F., Ruizhi, F. and Peigen, X. 1997. Determining flavonoid contents in the leaves of *Hippophae* species in different harvesting seasons. *China Journal of Chinese Materia Medica* (3): 18-21.
5. Jialiang, W., Maoshun, Z. and Zhiqi, X. 1982. Curative effect on ischemic heart disease by seabuckthorn flavonoids. *Acta Academiae Medicinae Sichuan* 13 (12): 6-11.
6. Jiaoshe, L., Baoxin, G., Yun, Y. and Junmin, F. 2000. Determining contents of total flavonoids in male and female plants of seabuckthorn in different seasons. *Hippophae* 13 (1): 39-40.
7. Jinmin, G., Anlin, Z., Yunsheng, L. and Wang L. 1998. Research progress on seabuckthorn flavonoids. *Hippophae* 11 (2): 34-40.

8. Lang, H., Changwu, L. and Weiheng, F. 1986. Pharmacology study on curing coronary heart disease by seabuckthorn flavonoids. *West China Pharmacy Research* 1 (3): 163-167.

9. Peiyun, Z. and Xiaohong, Z. 1985. Curative effect on hyperlipemia, ischemic heart disease and hypertension by xintonping (seabuckthorn flavonoids). *Shanxi Medical Research* (2): 54-57.

10. Tianfu, N. 1997. Curative effect on hyperlipemia by oral seabuckthorn flavonoids. *Hippophae* 10 (1): 36-38.

11. Xiaoping, T., Zhongping, W., Honghao, W. and Shengzhuo, H. 2010. Summary on progress of seabuckthorn leaf research. *The Global Seabuckthorn Research and Development* 8 (3): 17-20.

12. Yun, Y., Shumin, Q. and Yaosheng, H. 2002. Effect on regulating blood fat by seabuckthorn extraction. *Chinese Traditional Drug* 33 (9): 824-825.

13. Zhao, C., Li, Y. and Zhang, H. 2004. Animal experiment on relieving constipation with laxatives by Baitong seabuckthorn tea. *The Global Seabuckthorn Research and Development* 2 (1): 48-50.

14. Zhao, Y. and Wu, F. 1997. The flavonoids in seabuckthorn and their medicinal value. *Hippophae* 10 (1): 39-41.

Chapter 25

Biochemical Characteristics of Altai Seabuckthorn Varieties

Inessa Ershova

The Lisavenko Research Institute of Horticulture for Siberia
49 Zmeinogorskiy Tract, 656045, Barnaul, Russia

SUMMARY

Altai varieties of seabuckthorn (*Hippophae rhamnoides* ssp. *mongolica*) are known for the high level of nutritious and biological value. In the present investigation, comparative analysis of biochemical composition of fruits of 19 seabuckthorn varieties procured from the Lisavenko Research Institute of Horticulture for Siberia was carried out in search of promising varieties for breeding purpose. The amount of dry soluble substances varies between 8.9 per cent ("Chechek") to 13.4 per cent ("Sudarushka, Tenga"), the sugar contents is 4.7 ("Chechek") to 8.0 per cent ("Sudarushka"), acids is 1.0 ("Altaiskaya") to 2.0 ("Ulala"), vitamin C is 55.2 ("Essel") to 121.9 mg/100 g ("Altaiskaya"). The total amount of carotenoids ranged from 13.2 ("Chuyskaya") to 21.9 mg/100 g ("Sudarushka") and the content of oils varied from 4.4 ("Avgustina") to 7.2 mg/100 g (Sibirskaya) in the fresh fruits. The average sugar-acidic index (SAI) ranged from 3.2 ("Inya") to 8.9 ("Altaiskaya"). In seabuckthorn varieties with high contents of carotenoids and oil, etc. are "Altaiskaya, Sudarushka, Tenga and Chulishmanka".

Keywords: *Seabuckthorn, Biochemical composition, Dry soluble substances, Total sugars, Acids, Vitamin C, Oil and carotenoids.*

INTRODUCTION

Due to the presence of high contents of various vitamins in seabuckthorn fruits, it is considered to be a unique poly-vitaminic plant. Nowadays, the food and medicinal prophylactic value of this crop has removed all doubts. As it has high range of biochemical variation (Eliseev *et al.*, 1985), potential possibilities of this plant are not yet fully exhausted. Therefore, it is necessary to carry out investigations on biochemical composition of seabuckthorn fruits, which in turn open the possibilities for the

Seabuckthorn (*Hippophae* L.): *A Multipurpose Wonder Plant, Vol. 4: Emerging Trends in Research and Technologies* (Virendra Singh, *Editor-in-Chief*, 2014), p. 297–307, Daya Publishing House, New Delhi, India.

development of seabuckthorn assortments. Present socio-economic importance of the horticultural fruits and aggravation of ecological situation in the world increased the demands for more varieties of fruit and berry crops. The solution to the problem of improving the quality and biological value of the product is connected with the development and introduction of new breeding varieties with high contents of nutritional and biologically active substances in the fruits.

From the comparative studies on the estimation of biochemical composition of fruits the nutritionally valuable and vitamin rich varieties and hybrids can be selected. The selected varieties are one of the reserves of increasing the effectiveness of the breeding process, production of horticultural fruits on the whole.

The long-term biochemical investigations of seabuckthorn at the Lisavenko Research Institute of Horticulture for Siberia (LRIHS) gave a possibility to select the valuable varieties, which are rich in ascorbic acid, total carotenoids and oil content, with high content of dry soluble substances (DSS), total sugars, balanced acidity, etc.

MATERIALS AND METHODS

The investigations of biochemical composition of seabuckthorn fruits were carried out in biochemical laboratory of the LRIHS in 2000-2009. The investigation included 19 seabuckthorn varieties developed at RIHS: "Chuiskaya (control variety, C), Avgustina, Azhurnaya, Altaiskaya, Jamovaya, Elizaveta, Zhemchuzhnitsa, Inya, Panteleevskaya, Rosinka, Ruet, Sentyabrinka, Sibirskaya, Sudarushka, Tenga, Ulala, Chechek, Chulismanka and Essel". The biochemical composition of fruits was studied at the period of optimum ripening for the following indices: Total content of DSS, sugars, titrated acids, vitamin C, carotenoids and oil. Total content of DSS was determined by refractometer method; total content of sugars were estimated by method of Bertran and direct titration of water extracts; titrated acids (total acidity) was determined by titration of extracts with decinormal solution of sodium hydroxide; ascorbic acid by indophenol titration with potentiometer indication; oil in fruits was estimated by Soxhlet method and total carotenoids was determined by the method of Ermakov *et al.* (1987); Sedov *et al.* (1995) and Sedov *et al.* (1999).

RESULTS AND DISCUSSION

Having carried out long-term study, the biochemical composition of the fruits of various seabuckthorn varieties on characterization was made as follows. The amount of DSS varies between 8.9 per cent ("Chechek") to 13.4 per cent ("Sudarushka, Tenga"); the content of total sugars is 4.7 per cent ("Chechek") to 8.0 per cent ("Sudarushka"); titrated acids-1.0 ("Altaiskaya, Essel") to 2.0 ("Ulala"); vitamin C is 55.2 ("Essel") to 121.9 mg/100 g ("Altaiskaya"). The total amount of carotenoids ranged from 13.2 ("Chuyskaya") to 21.9 mg/100 g ("Sudarushka") and the content of oils from about 4.4 ("Avgustina") to 7.2 per cent ("Sibirskaya") in the raw fruits. The average sugar-acidic index (SAI) ranged from 3.2 ("Inya") to 8.9 ("Altaiskaya") and is shown in Table 25.1.

Dry Soluble Substances (DSS)

Dry soluble substances is a very important index, which determines flavoring qualities of seabuckthorn fruits. The average content of DSS in fruits was 11.4 per cent, varying from 8.9 to 13.4 per cent. The content of DSS in almost all varieties of seabuckthorn except "Chechek" (8.9 per cent) was higher than control variety, "Chuiskaya" (10.2 per cent). The level of DSS was high in varieties : "Sudarushka and Tenga" (13.4 per cent on an average). Maximum DSS was 14.7 per cent ("Tenga"). Both of these varieties were characterized by the stability of this character (V-7.6 and 5.7 per cent,

Table 25.1: Biochemical Composition of Fruits of Altai Seabuckthorn Varieties (Average indices for 2000-2009)

Variety	*Dry Soluble Substances (DSS) (Per cent)*	*Total Sugars (Per cent)*	*Titrated Acids (Per cent)*	*Sugar Acidic Index (SAI)*	*Ascorbic Acid (ÀA) (mg/100 g)*	*Total Carotenoids (mg/100 g)*	*Oil (per cent)*
Chuiskaya (c)	10.2	5.9	1.6	3.9	77.6	13.2	5.2
Avgustina	11.2	6.3	1.6	4.1	85.1	14.4	4.4
Azhurnaya	10.8	5.9	1.7	3.6	97.2	16.5	4.8
Altaiskaya	11.7	7.5	1.0	8.9	121.9	18.3	5.5
Jamovaya	11.8	6.4	1.4	4.7	71.4	27.5	4.7
Elizaveta	11.1	6.5	1.2	5.6	100.5	15.5	4.8
Zhemchuzhnitsa	11.4	6.4	1.2	6.0	64.0	18.3	4.9
Inya	11.1	5.8	1.8	3.2	83.6	17.7	5.5
Panteleevskaya	10.9	6.3	1.6	4.0	89.1	20.4	4.7
Rosinka	10.5	6.0	1.7	3.7	66.9	13.8	5.0
Ruet	12.2	6.8	1.7	4.1	75.7	–	4.9
Sentyabrinka	11.2	6.5	1.1	5.9	84.1	16.7	6.6
Sibirskaya	11.4	6.2	1.3	5.1	91.7	–	7.2
Sudarushka	13.4	8.0	1.3	6.0	67.7	29.1	5.9
Tenga	13.4	6.9	1.3	5.6	120.5	28.2	6.0
Ulala	10.4	5.8	2.0	3.3	92.8	–	5.0
Chechek	8.9	4.7	1.4	3.3	100.8	20.4	5.7
Chulismanka	12.4	6.7	1.5	4.7	100.3	21.0	5.5
Essel	12.0	7.2	1.0	7.3	55.2	16.6	4.8

respectively). It is well known that in breeding and practical use, the most important varieties are those which are able to keep their varietal peculiarities despite the change in environment and are more plastic, more adapted to the seasonal environment than other varieties. They can synthesize and accumulate a significant amount of nutrients under different conditions of vegetative period. Such varieties usually have a wider area of extension and are very economically effective. The limits of spreading of the character in different years and coefficients of variation can be classified as indexes of stability of a variety in relevance to the changing meteorological conditions of the environment. The variability in the variation column is considered to be insignificant, if variation coefficient is not higher than 10 per cent ; average, if it varies between 10 and 20 per cent ; above average, if it is 20-30 per cent ; high, 30-40 per cent and very high, if it is more than 41 per cent (Ravkin, 1987). In this case, "Sudarushka and Tenga" are the most valuable varieties because of higher amount of DSS in their fruits combines with the high stability of this character (Table 25.2).

The accumulation of DSS was higher in the fruits of seabuckthorn varieties "Chulismanka, Ruet, Jamovaya, Essel and Altaiskaya". These varieties belong to the category with the low variability of this index. Varieties "Ruet and Essel" reacted least to the change of meteorological conditions intensively.

Table 25.2: Dry Soluble Substances in the Fruits of Altai Seabuckthorn Varieties

Variety	Investigation Years	Dry Soluble Substances (DSS) (per cent)		Variation Coefficient (V) (per cent)
		Average	Variation in Years	
Chuiskaya (c)	10	10.2	8.0-11.2	9.1
Avgustina	6	11.2	9.8-13.3	12.5
Azhurnaya	5	10.8	7.2-14.2	23.5
Altaiskaya	4	11.7	10.9-12.4	7.4
Jamovaya	6	11.8	9.8-13.4	10.9
Elizaveta	7	11.1	9.1-13.2	13.3
Zhemchuzhnitsa	3	11.4	9.9-12.5	11.7
Inya	7	11.1	9.1-12.3	10.1
Panteleevskaya	4	10.9	10.0-11.6	6.2
Rosinka	6	10.5	10.3-11.0	2.5
Ruet	3	12.2	12.0-12.5	2.4
Sentyabrinka	6	11.2	9.3-12.2	9.4
Sibirskaya	4	11.4	10.5-13.0	9.8
Sudarushka	3	13.4	12.2-14.1	7.6
Tenga	6	13.4	12.4-14.7	5.7
Ulala	4	10.4	9.5-11.9	8.6
Chechek	5	8.9	8.1-9.8	9.3
Chulismanka	5	12.4	11.6-13.4	6.1
Essel	5	12.0	11.6-12.9	4.5

In some years, high level of DSS was noticed in the fruits of varieties "Avgustina, Elizaveta and Sibirskaya". The average variability of this index is estimated for these varieties. Variety "Chechek" has a low level of DSS content among all the varieties studied, which remains stable over the years. This index is unstable for the variety "Azhurnaya".

Total Sugars

In breeding of seabuckthorn, an increase of total sugar content and decrease of acidity level are important for the quality and dessert taste of fruits. Sugars are the largest part of DSS and varieties with a high content of DSS in fruits, usually have a higher level of accumulated sugars. Total sugar content in the fruits of seabuckthorn varieties studied, varied from 4.7 per cent to 8.0 per cent with an average of 6.4 per cent (Table 25.3).

According to long-term studies, variety "Sudarushka" has the highest amount of total sugars in fruits (8.0 per cent). This variety outstands for the high stability of this character (V-2.9 per cent). In relation to this, varieties "Essel and Altaiskaya" are especially valuable. In certain years of studying, the maximum amount of sugars was noticed in the fruits of the variety "Essel" (9.1 per cent) with an average of 7.2 per cent. It was found that there was a high level of sugar content in the fruits of variety "Altaiskaya", which is 7.5 per cent on an average, reaching its maximum point at 8.7 per cent. The variability of the character in these two varieties was average, which is close to low. The berries of

Table 25.3: Total Sugars in the Fruits of Altai Seabuckthorn Varieties

Variety	*Investigation Years*	*Total Sugars (per cent)*		*Variation Coefficient (V) (per cent)*
		Average	*Variation in Years*	
Chuiskaya (c)	10	5.9	5.3-7.1	9.0
Avgustina	6	6.3	4.7-7.4	16.3
Azhurnaya	5	5.9	4.0-6.6	18.6
Altaiskaya	4	7.5	6.9-8.7	11.1
Jamovaya	6	6.4	5.2-7.2	10.9
Elizaveta	7	6.5	4.7-7.4	19.9
Zhemchuzhnitsa	3	6.4	5.7-7.0	10.3
Inya	7	5.8	4.2-6.9	15.3
Panteleevskaya	4	6.3	5.0-7.5	16.3
Rosinka	6	6.0	5.8-6.7	6.1
Ruet	3	6.8	6.7-7.0	2.5
Sentyabrinka	6	6.5	5.6-8.3	15.2
Sibirskaya	4	6.2	4.7-7.3	18.2
Sudarushka	3	8.0	7.9-8.3	2.9
Tenga	6	6.9	5.6-8.2	16.6
Ulala	4	5.8	5.3-6.7	11.6
Chechek	5	4.7	3.8-5.1	11.9
Chulismanka	5	6.7	5.7-7.5	10.4
Essel	5	7.2	6.5-9.1	15.1

variety "Tenga" contained about 7 per cent of total sugars on an average; nevertheless, opportunities of this variety are greater. In favorable years, the content of sugars in its fruits reached at the level of 8.2 per cent. Varieties "Ruet and Chulismanka" can be classified as the ones with a significant and stable content of sugars (more than 7 per cent).

Variety "Chechek" is characterized by a low level of sugar accumulation in its fruits, is the lowest among all the varieties studied. Almost all varieties are characterized by an average variability of this character and variety "Elizaveta" was found to be the most unstable one (V-19.9 per cent).

Organic (Titrated) Acids

The content of titrated acids is an important indicator of the quality of berries. In combination with sugars, they determine the taste of fruits. The average content of titrated acids in all varieties studied is found to be 1.4 per cent and the limits of diversity of the character ranged from 1.0 to 2.0 per cent (Table 25.4).

According to our studies, differences in the content of organic acids between the varieties, on an average, can be double. Moreover, the acidity of the fruits very much depends on the climatic conditions during the vegetative period. The variability of this character was high in most of the cases. In relation to this, only variety "Sudarushka" was found stable (V-5.5 per cent) and its fruits were noticeable for their low acidity in all the years of studying.

Table 25.4: Organic Acids in the Fruits of Altai Seabuckthorn Varieties

Variety	*Investigation Years*	*Organic Acids (per cent)*		*Variation Coefficient (V) (per cent)*
		Average	*Variation in Years*	
Chuiskaya (c)	10	1.6	1.2-2.6	24.6
Avgustina	6	1.6	1.2-1.9	15.8
Azhurnaya	5	1.7	1.4-2.0	16.4
Altaiskaya	4	1.0	0.5-1.3	37.9
Jamovaya	6	1.4	1.2-1.5	10.6
Elizaveta	7	1.2	0.9-1.3	12.3
Zhemchuzhnitsa	3	1.2	0.8-1.5	29.5
Inya	7	1.8	1.2-2.6	26.3
Panteleevskaya	4	1.6	1.4-1.8	10.8
Rosinka	6	1.7	1.4-2.3	20.5
Ruet	3	1.7	1.4-2.1	21.2
Sentyabrinka	6	1.1	0.9-1.4	15.7
Sibirskaya	4	1.3	1.0-1.4	14.0
Sudarushka	3	1.3	1.3-1.4	5.5
Tenga	6	1.3	1.0-1.6	18.2
Ulala	4	2.0	1.5-2.7	27.4
Chechek	5	1.4	1.0-1.7	21.4
Chulismanka	5	1.5	1.1-2.0	24.5
Essel	5	1.0	0.8-1.3	21.2

The acidity of the fruits was found lower than control level, which was 1.6 per cent on an average in 50 per cent of all the varieties studied. The fruits of varieties "Altaiskaya" and "Essel" have the least content of acids and their amount was 1.0 per cent on an average. In certain years of studying, the acidity of fruits of variety "Altaiskaya" was as low as 0.5 per cent as compared to variety "Essel" (0.8 per cent). Despite the high variability of this index for these varieties (V-37.9 and 21.2 per cent, correspondingly), the level of content of organic acids in their fruits was not higher than 1.3 per cent. Variety "Sentyabrinka" recorded low acidity, which is 1.1 per cent on an average. The minimum amount of acids in its fruits was 0.9 per cent. The variety is characterized by the low stability of this character. Varieties "Elizaveta and Zhemchuzhnitsa" are characterized by a quite low content of organic acids (1.2 per cent on an average), the minimum amount observed is 0.9 and 0.8 per cent, respectively.

The greatest amount of acids was noticed in the fruits of variety "Ulala" (2.0 per cent on an average and maximum of 2.7 per cent, the accumulation of acids is quite high (up to 2.6 per cent) in the fruits of variety "Inya". It is well-known that high content of free acids makes the taste of fruits considerably bad.

Sugar-Acidic Index (SAI)

It is well known that the taste of fruits depends on many biochemical compounds including the content of sugars and acids. The proportion of their content is represented by SAI, which characterizes the level of fruit sweetness. In LRIHS, purposeful breeding for sweet fruits is conducted and the promising varieties are characterized by the high SAI that have a harmonious taste of berries, which can be consumed fresh.

Studies carried out on 19 varieties of *Hippophae* revealed that the SAI were varied between 3.2 and 8.9 (Table 25.1). The average value of SAI was 4.9 as compared to control (3.9). Varieties "Azhurnaya, Inya, Rosinka, Ulala and Chechek" had low values of SAI as compared to the majority of varieties and the control, because of the high content of acids and the quite low content of sugars in their fruits. Variety "Inya" is found to have the lowest SAI of 3.2.

Varieties "Altaiskaya and Essel" is characterized by the high content of SAI (8.9) which is the maximum followed by "Essel" (7.3). "Zhemchuzhnitsa and Sudarushka" (6.0), thus better dessert qualities. In relation to this, SAI of varieties "Sentyabrinka (5.9), Elizaveta (5.6) and Tenga" (5.6) are also found outstanding and the maximum SAI values in these varieties reached 7.0 and higher.

Ascorbic Acid (AA)

One of the advantages of seabuckthorn is a high content of vitamins and other biologically active substances in its fruits, particularly ascorbic acid (also known as vitamin C). The level of AA accumulation is a very important indication of biochemical composition of seabuckthorn. Its increasing level is one of the main objectives of seabuckthorn breeding in our institute. The average content of AA

Table 25.5: Vitamin C Content in the Fruits of Altai Seabuckthorn Varieties

Variety	*Investigation Years*	*Vitamin C (mg/100 g)*		*Variation Coefficient (V) (per cent)*
		Average	*Variation in Years*	
Chuiskaya (c)	10	77.6	61.6-109.4	17.6
Avgustina	4	85.1	65.0-100.6	18.4
Azhurnaya	3	97.2	76.0-112.3	19.5
Altaiskaya	3	121.9	101.5-155.4	24.0
Jamovaya	6	71.4	58.8-88.7	14.2
Elizaveta	6	100.5	93.8-108.7	6.8
Zhemchuzhnitsa	3	64.0	60.2-67.5	5.7
Inya	6	83.6	74.0-92.0	9.7
Panteleevskaya	4	89.1	80.0-100.8	9.9
Rosinka	5	66.9	52.3-84.4	23.4
Ruet	3	75.7	72.2-78.8	4.4
Sentyabrinka	6	84.1	73.4-94.6	9.3
Sibirskaya	4	91.7	72.4-115.2	20.5
Sudarushka	3	67.7	60.9-74.8	10.3
Tenga	5	120.5	90.8-149.6	20.7
Ulala	4	92.8	90.1-93.8	2.0
Chechek	5	100.8	95.9-105.6	4.7
Chulismanka	4	100.3	84.0-111.9	12.9
Essel	4	55.2	46.8-63.5	16.3

in the fruits of all the varieties studied was 86.3 mg/100 g, ranging from 55.2-121.9 mg/100 g (Table 25.5). According to average long-term (perennial) study, the variation in accumulation of AA within the same variety, over the years was very high.

The highest content of vitamin C was estimated in the fruits of seabuckthorn "Altaiskaya" (121.9 mg/100 g) among all the varieties studied. In certain years, the amount of AA in the fruits of this variety reached 155.4 mg/100 g. This fact presents a special interest and allows us in selecting it as a source of vitamin C. Despite the highest accumulation of AA in the fruits of this variety and considerable dependence on climatic conditions (V-24.0 per cent), the content of it was always higher than 100 mg/100 g. Variety "Tenga" was noticeable for relatively high content of AA in its fruits (120.5 mg/100 g on an average), and is able to accumulate up to 150 mg/100 g of vitamin C with an average variability of the character (V-20.7 per cent).

The AA content in the fruits of most of the other varieties, on an average, was also higher than control (77.6 mg/100 g). The content of AA was quite high in the fruits of the following varieties: "Elizaveta" having 100.5 mg/100 g on an average, with maximum index of 108.7 mg/100g; "Chechek"-100.8 and 105.6 mg/100 g, "Chulishmanka"-100.3 and 111.9 mg/100 g, respectively. It is quite valuable that the first two varieties reacted to the changes in climatic factors to the least degree (V-6.8 per cent and 4.7 per cent, correspondingly). Varieties "Azhurnaya and Sibirskaya" are able to accumulate more than 100 mg/100 g of vitamin C in their fruits.

Inspite of its many good qualities, variety "Essel" is characterized by the low content of vitamin C in fruits, which is 55.2 mg/100 g on an average. This is the minimum level of AA among all the varieties studied. Varieties "Zhemchuzhnitsa (64.0 per cent) and Sudarushka (67.7 per cent)" are noticeable for the low content of vitamin C as well.

Total Carotenoids and Oil Content

Seabuckthorn is a unique plant because of its ability to synthesize a considerable amount of carotenoids and oil in its fruits, which have a wide spectrum of biological activity, along with vitamin C. This is why the index of content of substances which belong to this group is very important while estimating the biochemical composition of fruits of this crop. In LRIHS, breeding for the high content of carotenoids and oil in fruits is one of the priorities of working with seabuckthorn.

Total carotenoids in the fruits of all the varieties studied, varied from 13.2 to 29.1 mg/100 g (19.2 mg/100 g on an average) (Table 25.6). All the varieties studied had higher content of carotenoids than the control variety. The highest content of carotenoids was found in the fruits of variety "Sudarushka" (29.1 mg/100 g), which correlates well with the high stability of this character (V-6.9 per cent). The maximum level of accumulation of carotenoids over all the years of study was 37.8 mg/100 g, in the fruits of variety "Tenga" (average perennial index is 28.2 mg/100 g). Nevertheless, high content of carotenoids in the fruits of this variety is unstable (V-33.4 per cent). However, variety "Tenga" as well as variety "Sudarushka", presents special interest as a source of high content of carotenoids. Variety "Jamovaya", which is able to accumulate 27.5 mg/100 g of carotenoids on an average in its fruits, also belongs to this group. It is necessary to mention the high annual stability of this character for this variety (V-6.1 per cent).

Varieties "Chulismanka, Chechek and Panteleevskaya" belong to a perspective group with a high content of carotenoids in fruits (20 mg/100 g and more). This character is stable for the first two varieties (V-9.5 per cent and 11.3 per cent, respectively), while its variability was higher than an average (V-21.4 per cent) for variety "Panteleevskaya". The amount of carotenoids in the fruits of other varieties has been close to an average level during all the years of studying. Variety "Chuiskaya" has been characterized by the lowest level of accumulation of carotenoids (13.2 mg/100 g).

Table 25.6: Total Carotenoids in the Fruits of Altai Seabuckthorn Varieties

Variety	*Investigation Years*	*Total Carotenoids (mg/100 g)*		*Variation Coefficient (V) (per cent)*
		Average	*Variation in Years*	
Chuiskaya (c)	7	13.2	11.4-15.7	10.2
Avgustina	4	14.4	12.2-18.5	19.3
Azhurnaya	3	16.5	11.7-20.9	27.9
Altaiskaya	3	18.3	16.7-21.2	13.7
Jamovaya	3	27.5	26.0-29.3	6.1
Elizaveta	5	15.5	13.5-19.4	15.1
Zhemchuzhnitsa	3	18.3	17.4-19.7	6.7
Inya	4	17.7	15.3-19.8	13.6
Panteleevskaya	4	20.4	16.5-25.4	21.4
Rosinka	3	13.8	11.9-17.3	22.2
Ruet	–	–	–	–
Sentyabrinka	6	16.7	15.5-17.9	8.8
Sibirskaya	–	–	–	–
Sudarushka	4	29.1	26.2-30.7	6.9
Tenga	5	28.2	19.2-37.8	33.4
Ulala	–	–	–	–
Chechek	3	20.4	18.8-23.0	11.3
Chulismanka	3	21.0	19.6-23.3	9.5
Essel	4	16.6	14.2-19.6	13.6

Studies revealed that in these seabuckthorn varieties an average perennial limit of variability of the oil content was 4.4-7.2 per cent in raw fruits, with an average of 5.3 per cent (Table 25.7).

In relation to that, varieties "Sibirskaya and Sentyabrinka" show high oil content in their fruits of 7.2 per cent and 6.6 per cent, respectively. The maximum level of oil content in the fruits of variety "Sibirskaya" combines with the high stability of this character. Although, the highest content of oil during the years of studying was estimated in the fruits of variety "Sentyabrinka" (8.6 per cent), variability of this character was higher than average (V-21.4 per cent). The synthesis of oil in the fruits of variety "Sibirskaya" reached to the level of 7.8 per cent over the years. This character makes it the most promising variety among all the studied. The following varieties have an ability to accumulate higher amounts of oil in fruits: "Tenga (6.0 per cent), Sudarushka (5.9 per cent) and Chechek" (5.7 per cent). Variety "Tenga" reacted to change of weather conditions to the least degree; variability of this character in the years of studying with two other varieties "Ruet and Zhemchuzhnitsa" was also insignificant.

Varieties "Chechek, Chulismanka, Altaiskaya, Inya, Rosinka and Ulala" are able to accumulate more than 5 per cent of oil in their fruits. In certain years, the fruits of varieties "Jamovaya and

Chulismanka" contained 7 per cent of oil and more; nevertheless, this character was quite unstable (V-28.8 per cent for "Jamovaya and V-20.4 per cent for Chulismanka").Variety "Avgustina" had the lowest level of oil accumulation with an average of 4.4 per cent.

Table 25.7: Oil Content in the Fruits of Altai Seabuckthorn Varieties

Variety	*Investigation Years*	*Oil Content in the Raw Fruits (per cent)*		*Variation Coefficient (V) (per cent)*
		Average	*Variation in Years*	
Chuiskaya (c)	8	5.2	4.4-6.4	12.8
Avgustina	4	4.4	3.9-5.4	16.3
Azhurnaya	4	4.8	4.1-5.3	12.2
Altaiskaya	4	5.5	4.8-5.9	8.7
Jamovaya	5	4.7	3.7-7.1	28.8
Elizaveta	5	4.8	3.9-5.8	16.9
Zhemchuzhnitsa	3	4.9	4.6-5.3	7.2
Inya	6	5.5	4.5-6.1	12.6
Panteleevskaya	4	4.7	3.8-5.7	16.7
Rosinka	5	5.0	4.2-5.8	15.1
Ruet	3	4.9	4.6-5.2	6.3
Sentyabrinka	4	6.6	5.4-8.6	21.4
Sibirskaya	3	7.2	6.6-7.8	8.3
Sudarushka	3	5.9	5.5-6.6	9.9
Tenga	6	6.0	5.5-6.4	6.2
Ulala	4	5.0	4.7-5.6	8.5
Chechek	5	5.7	5.2-6.4	8.5
Chulismanka	5	5.5	4.5-7.3	20.4
Essel	5	4.8	4.1-6.0	13.2

CONCLUSION

Studies on the biochemical composition of the fruits of the Altai varieties of seabuckthorn showed a considerable diversity, from which superior varieties having high content of nutritious and biologically active substances could be selected. Such varieties are noticeable for their high and stable biochemical characters. The following varieties are graded in the descending order (except organic acids which are in the ascending order) as follows:

- DSS content: Varieties "Sudarushka (13.4), Tenga (13.4), Chulismanka (12.4), Ruet (12.2), Essel (12.0), Jamovaya (11.8) and Altaiskaya (11.7)"
- Total sugar content: Varieties "Sudarushka (8.0), Altaiskaya (7.5), Essel (7.2), Tenga (6.9), Ruet (6.8) and Chulismanka (6.7)"
- Organic acids: Varieties "Altaiskaya (1.0), Essel (1.0), Sentyabrinka (1.1), Elizaveta (1.2), Zhemchuzhnitsa (1.2) and Sudarushka (1.3)"

- Sugar-acid index (SAI): Varieties "Altaiskaya (8.9), Essel (7.3), Zhemchuzhnitsa (6.0), Sudarushka (6.0), Sentyabrinka (5.9), Elizaveta (5.6) and Tenga (5.6)"
- Vitamin C: Varieties "Altaiskaya (121.9), Tenga (120.5), Chechek (100.8), Elizaveta (100.5), Chulismanka (100.3), Azhurnaya (97.2) and Sibirskaya (91.7)"
- Total carotenoids: Varieties "Sudarushka (29.1), Tenga (28.2), Jamovaya (27.5), Chulismanka (21.0), Chechek and Panteleevskaya (20.4)"
- Oil content: Varieties "Sibirskaya (7.2), Sentyabrinka (6.6), Tenga (6.0), Sudarushka (5.9), Chechek (5.7), Chulismanka (5.5), Inya (5.5), Altaiskaya (5.5), Rosinka (5.0) and Ulala (5.0)"

Varieties "Altaiskaya, Sudarushka, Tenga and Chulismanka" were found to be the most useful ones, since they are the sources of high contents of various nutrients. All varieties mentioned above have high potential abilities to be used in production, processing and breeding for the better biochemical composition of fruits.

REFERENCES

1. Eliseev, I., Kreknin, N., Mishulina, I. and Kreknina, L. 1985. Some physical and biochemical indices of seabuckthorn fruits of Gorkovskaya and Altaiskaya breeding. In: *Physiology and Biochemistry of Cultured Plants* (V. Tsvetkova, Ed.), p. 78-91, Publishing House of the Gorky Agriculture Institute, Gorky, 115p.
2. Ermakov, A., Arasimovich, V., Yarosh, N., Peruanski, U., Lukovnikov, G. and Ikonnikov, M. 1987. In: *Methods of Biochemical Investigation of Plants* (A. Ermakov, Ed.). Agroprompublishing Leningrad Branch, Leningrad, 430p.
3. Ravkin, A. 1987. Breeding on the main economic characters. In: *Black Currant: Initial Material, Breeding and Varieties* (A. Ravkin, 1987), p. 131-184, Publishing House of the Moscow University, Moscow, 216p.
4. Sedov, E., Sedova, Z. and Streltsina, S. 1995. Breeding on productivity and quality of a product. In: *Programme and Methods of Breeding Fruit, Berry and Nut Crops* (E. Sedov, Ed.), p. 48-57, Publishing House of the All-Russian Research Institute of Fruit Crops Breeding, Orel, 502p.
5. Sedova, Z., Leonchenko, V. and Astahov, A. 1999. Estimate of varieties according to chemical composition of fruits. In: *Programme and Methods of Fruit, Berry and Nut Crops Variety-Studying* (E. Sedov, Ed.), p. 160-167, Publishing House of the All-Russian Research Institute of Fruit Crops Breeding, Orel, 608p.

PART VI:
PHARMACOLOGY

Chapter 26

Seabuckthorn Based Dermaceuticals and Cosmeceuticals

Anuradha Pol, Pratik Patel, Lalit Kagliwal*, Rekha Singhal* and Vandana Patravale

Department of Pharmaceutical Science and Technology, Institute of Chemical Technology, Matunga, Mumbai – 400 019, India

*Food Engineering and Technology Department, Institute of Chemical Technology, Matunga, Mumbai – 400 019, India

SUMMARY

The current scenario in scientific research demonstrates a shift in universal trend from synthetic to herbal medicine, which can be claimed as 'Return to Nature'. The World Health Organization (WHO) has also emphasized on development and utilization of herbal products and cosmeceuticals for the benefit of world population as they ensure safety, cost effectiveness and show minimum side effects. One such herb, commonly known as seabuckthorn (*Hippophae* L.), is an ancient spiny tree, which is a potent rich source of bioactives and has lately gained attention worldwide mainly for its nutritional, cosmeceutical and medicinal properties. Seabuckthorn is a relatively obscure medicinal plant, which has recently been discovered to have previously unappreciated health benets. Its extracts are fortified with vitamins and numerous antioxidant compounds, which are useful for the preparation of a wide range of dietary supplements, dermaceuticals and cosmeceuticals. Additional phytoconstituents in all plant parts (leaves, fruits, branches and bark) are exceptionally useful for treating skin conditions. Moreover, the presence of a variety of medicinal components suggests that a wide range of skin disorders may be alleviated by development of new medications based on extracts from the plant. Some attractive seabuckthorn health products, topicals and cosmetics have become available in the marketplace. Unfortunately, despite research efforts from many groups indicating promising and considerable medicinal and market potential, the therapeutic properties of seabuckthorn are not yet well identified and explored extensively by scientic personnel. This has hindered the commercial development and popularization

Seabuckthorn (*Hippophae* L.): *A Multipurpose Wonder Plant, Vol. 4: Emerging Trends in Research and Technologies* (Virendra Singh, *Editor-in-Chief*, 2014), p. 311–351, Daya Publishing House, New Delhi, India.

of manufactured products. Keeping this in view, the present review devoted to phytochemical and therapeutic aspects of seabuckthorn, and its dermaceutical and cosmeceutical applications are extremely useful.

Keywords: *Seabuckthorn, Free radicals, Antioxidants, Flavonoids, Vitamin C, Fatty acids, Carotenes, Phytosterols, Dermaceuticals, Cosmetics and Skin aging.*

INTRODUCTION

Worldwide herbal medicines are utilized since centuries for safety, efficacy, cultural acceptability and fewer side effects. Herbs and their products have been explored with varying efficacy to cure and prevent diseases throughout history. Written literature on medicinal plants date back atleast 5000 years to the Sumerians and ancient records have even suggested earlier use of medicinal plants. Due to side effects of synthetic products, herbal formulations are gaining popularity in the world market. The demand for herbal based medicines, health products, pharmaceuticals, food supplements, cosmeceuticals and dermaceuticals has increased tremendously (Kalia, 2005). Amongst these herbals, *Hippophae rhamnoides* L. commonly known as seabuckthorn (SBT), an ancient crop with modern virtues is included. SBT is a member of the *Elaeagnaceae* family and a deciduous spiny nitrogen fixing shrub or small tree native to Europe, Russia, China and Indian Himalayas (Singh, 2003). A number of studies have been carried out on natural populations of seabuckthorn in Himalayas (Singh, 1994, Singh and Dogra, 1996; Singh and Singh, 2004; Raina *et al.*, 2011).

This plant has been domesticated worldwide and is an inimitable and valuable shrub which has lately gained worldwide acknowledgment, mainly for its medicinal and nutritional attributes (Li, 2003; Arimboor *et al.*, 2006).

SBT has been proven to be natural protective boon to human health, since all parts (leaves, fruit, bark and branches) of the plant are considered to be a rich source of large number of bioactive substances like vitamins (A, C, E, K, riboavin, folic acid), carotenoids (α, β, δ-carotene, lycopene), phytosterols (β-stiosterol, ergosterol, stigmasterol, lansterol, amyrins), organic acids (malic acid, oxalic acid), polyunsaturated fatty acids, flavonoids (myrecitin, kaempherol, isorhamnetin) and some essential amino acid (Beveridge *et al.*, 1999; Guliyev *et al.*, 2004; Yang and Kallio, 2006).

These bioactives are of interest not only from the phytochemical point of view, but also because many of them possess biological and therapeutic activities including antioxidative, cardiovascular, anticancer, wound healing, anti-inammatory, antiradiation, treatment of gastrointestinal ulcers, hepatoprotectant, platelet aggregation, and immunomodulatory, anti-atherogenic, anti-stress, and tissue repair (Zeb, 2004; Suryakumar and Gupta, 2011).

Traditionally, because of these arrays of pharmacological actions, *H. rhamnoides* L. extracts is often used as medicines in China, Europe and Russia. The future undoubtedly holds great promise for SBT bioactives. This ancient plant with its powerful and healing synergies has much to contribute. The present review is mainly oriented towards the ancient and modern dermal and cosmetic applications of SBT in relevance to scientic studies pertaining to its medicinal as well as pharmacological activities.

HISTORY AND TRADITIONAL USES OF SEABUCKTHORN FOR TREATING SKIN AILMENTS

In Greece, SBT leaves and twigs were used to feed animals and as a result there was weight gain and skin shining, chiefly seen in horses. Thus, the word *Hippophae* has emerged from a Latin word

'*Hippo*' meaning 'horse' and '*Phaos*' which means 'shine'. SBT has a prosperous history of being used for treating several medical conditions. Almost all parts of SBT shrub have been used for the treatment of diseases in traditional medicine in Europe and Asia. *Hippophae rhamnoides* L. is grown widely in northern and southwestern China, and has long been used for alleviating pain since ancient time.

Many of its medicinal properties have been recorded in ancient literatures, such as Sibu Yidian during the Tang Dynasty and Jing Zhu Ben Cao during the Qing Dynasty. It was utilized for its pharmacological actions in Tibet as early as 900 A.D. The references to therapeutic usage of SBT which were found in the primordial Tibetan medicinal texts, including "The rGyud Bzi" (The Four Books of Pharmacopoeia) dated to the times of Tang Dynasty (618-907A.D.) (Rousi, 1971; Bernath and Foldesi, 1992).

For centuries, the people of central and southeastern Asia have used SBT as a traditional medicine to prevent various ailments. The SBT berries were used as a source of herbal medicines, health food and natural skin care in Europe and Asia. Leaves were used in gastrointestinal and dermatological disorders, and have been applied topically as a constricted form in rheumatoid arthritis in the Middle Asia. Flowers of *H. rhamnoides* L. are exploited as skin softener in Tajikistan (Tsybikova *et al.*, 1983; Barnaulov and Belenovskaya, 1988). Application of seabuckthorn in traditional medicines have been reviewd by Singh (2006).

After assessment of the ancient literature, Li and Guo (1989) pointed out that SBT is a mild drug with the characteristic effects, such as diminishing inammation, counteracting toxicity and abscesses, and treating tumors in Tibetian medicine. The oil extracts obtained from fruits are used in inammatory processes and are applied externally in hemorrhage (Guliyev *et al.*, 2004).

Juice, syrup, and oil of the fruits have been used as pain killer to heal wounds, and in ulcers (Guliyev *et al.*, 2004). Oil from fruits and seeds is being utilized in the therapy of eczema, lupus erythematosus and chronic wounds that are difcult to heal, inammatory diseases, erosion of the cervix uteri, chronic dermatoses, and in the treatment of burns and frozen parts of the body (Guliyev *et al.*, 2004). Also, its berry oil is reported to treat skin disease (Yang *et al.*, 2000). SBT oils are further utilized in the treatment of keratitis, trachoma, injuries or burns of eyelid, conjunctivitis and in ophthalmology (Guliyev *et al.*, 2004).

In Indian Himalayas, the Amchi System of medicine, a part of Indian System of Medicine has been using the plant parts, *i.e.* leaves, berries, pulp oil, seed oil, etc., in different herbal formulations (Lal *et al.*, 2001). The oil of SBT has general nourishing, revitalizing and restorative actions. It can be used for acne, dermatitis, irritated, dry, itching skin, sore skin, skin ulcers, burns, scalds, cuts and tissue regeneration. SBT oil effectively combats wrinkles, dryness and other symptoms of malnourished or prematurely aging skin, and is utilized in anti-aging skin creams and lotions (Xing *et al.*, 2002).

PHYTOCHEMISTRY AND PHYTOCONSTITUENTS (COMPOUNDS OF SEABUCKTHORN ASSOCIATED WITH POTENTIAL SKIN HEALTH BENEFITS)

SBT bioactives and extracts have materialized to be of remarkable importance in treatment of skin disorders in humans and cattle. The plant as a whole has been investigated by many researchers and reported to possess many bioactives markedly carotenoids, tocopherols, tocotrienols, essential polyunsaturated fatty acids, lipids, amyrins, flavonoids, tannins, phenols, amino acids, organic acids, 5-hydroxytyptamine, minerals and other bioactive components (Chen *et al.*, 1990; Yang and Kallio, 2002a; Zheng *et al.*, 2009).

PHYTOCHEMISTRY OF PLANT PARTS

SBT is a multinutritional herb which constitutes various parts and these individually contribute to an array of bioactives. Hence, a phytochemical insight on various parts of SBT plant, such as fruit, leaves, bark is imperative.

Fruit

In addition to seabuckthorn's capacity for natural resource conservation (prevents soil erosion and acts as a nitrogen fixer and nutritive feed for animals), there is a recent interest in SBT due to the fact that the berries are considered amongst the most nutritious and vitamin rich fruits found and are therefore, termed as 'super fruit'. SBT berries have a unique composition, combining a cocktail of components usually found separately. Literature cites that the fruit of SBT is a store house of vitamins, minerals and important bioactive substances *viz.*, sterols, terpenoids, phenolics and fatty acids, etc. SBT berries normally constitute of three parts, pulp (68 per cent), seed (23 per cent), and peel (7.75 per cent) (Zadernowski *et al.*, 1997; Yang and Kallio, 2001; Oomah, 2003).

SBT berries are nourishing, but very acidic and astringent, distasteful to eat raw, unless 'bletted' (frosted to reduce the astringency) and/or mixed as a juice with sweeter materials, such as apple or grape juice. They also include mineral elements like Ca, P, Fe and especially K, which are the most plentiful among all the other herbal plants (Bal *et al.*, 2011).

In addition, fruits also include high levels of vitamins like vitamin C (695 mg/100 g, which is comparatively more than lemons and oranges), tocopherols (1-10 mg/100 g, which is mainly higher than wheat embryo, safflower, maize, and soybean) and carotenoids (3-15 mg/100 g) (Gao *et al.*, 2000; Zeb, 2004). The content of vitamin C, vitamin E and carotenoids reads like the label on a pack of multi-vitamin pills. SBT, in fact, one of the richest sources of β-carotene, in addition to containing significant levels of other important carotenoids, such as lutein and zeaxanthin (both very important for the eyes), and lycopene. The prime flavonoid classes in the SBT berries are flavonols and proanthocyanidins (condensed tannins), while smaller amounts of flavanols and phenolic acids are also present (Rosch *et al.*, 2003). The compounds, which were identified include gallic acid, protocatechuic acid, catechin, epicatechin, proanthocyanidins, querectin, and isorhamnetin. The flavonoids found in the largest quantity are isorhamnetin followed by isorhamnetin-3-O-β-D-glucoside, rutin, ustzagalin, quercetin, myricetin and kaempferol. Besides proanthocyanidins, further quantitatively important phenolic compounds are isorhamnetin-3-O-glycosides, phenolic compounds, such as quercetin-3-O-glycosides (representative flavonol), catechins, and hydroxybenzoic acids with a catechol structure exhibited good antioxidants (AO) capacities, but their concentration in SBT juice was found to be small (Larmo, 2011; Wang *et al.*, 2011).

All these components are classified as major natural antioxidants (OA) and result in one of the highest AO activities among the medicinal plants. These AOs form a vital part of the body's defense system. The vitamin C and proanthocyanidins comprises major hydrophilic AOs, while vitamin E, carotenoids and flavonoids constitute major hydrophobic AOs.

SBT berries also have higher percentage of sugars, mainly glucose and fructose that diverge widely in berry juice from 0.6 to 24.2 g/100 ml. Additionally, berries also contain certain other vitamins, such as folic acid, B_1, B_2 and K (Bekker and Glushenkova, 2001; Suryakumar and Gupta, 2011). Besides sugars, organic acids such as malic and quinic acids as well as oxalic citric and tartaric acids are also present. The peel of the berries contains 5-hydroxytryptamine, which is rare among the plants (Bal *et al.*, 2011; Kumar *et al.*, 2011). The chemical composition of SBT berries vary considerably due to their origin, the climate, the fruit size and maturity, and the method of processing (Leskinen *et al.*, 2010). Regarding the inimitable aroma of *Hippophae* berries, it is not comparable to any other ordinary fruit owing to their volatile compounds, *i.e.* ethyl dodecenoate, ethyl octanoate, decanol, ethyl decanoate and ethyl dodecanoate (Kumar *et al.*, 2011; Suryakumar and Gupta, 2011). The SBT fruit juice is nutritious and has the advantage to remain liquid even in sub-zero temperatures, because it has a freezing point of -22 °C (Bal *et al.*, 2011). Moreover, the SBT fruits are rich in unsaturated fatty acids (oleic acid, linoleic acid, linolenic acid) with an average of 86.3 per cent (Bal *et al.*, 2011) of the fruit fat. In addition to the plethora of AOs and other bioactives, the berries additionally contain phytosterols like β-sitosterol, ergosterol and amyrins and rare fatty acids (Leskinen *et al.*, 2010; Bal *et al.*, 2011). But, these fatty acids differ from the seed and pulp extracts.

SBT berries contain oil as the most important component. Generally, the oil from the pulp/peel fraction is pooled due to the complexity involved with separation. From SBT, different oils, such as pulp oil, berry oil, pomace oil and seed oil can be extracted (Zeb, 2006). The full-grown seeds contain 8-20 per cent oil, the dried fruit pulp (flesh and peel) has about 20-25 per cent oil, whereas the berries residue left after juice extraction or pomace has about 15-20 per cent oil content (Kumar *et al.*, 2011). The oil concentration is affected by the morphological features, *i.e.* size and colour of the berries as well as the harvesting time (Yang and Kallio, 2002b).

These oils are rich in vitamins E, K, carotenoids (lycopene, β-carotene), tocopherols (α-tocopherol is the most abundant, especially in seed oil), tocotrienols (more concentrated in pulp oil) and sterols (β-sitosterol, cholesterol, campesterol, stigmasterol). β-sitosterol is the major phytosterol present in pulp/peel and seed fractions (Cenkowski *et al.*, 2006; Zeb, 2006; Bal *et al.*, 2011; Kumar *et al.*, 2011).

However, the commercial value of SBT oil is commonly based on the levels of carotenoids and tocols. Tocol content, represented by tocopherols (α -T, β-T, γ-T, and δ-T) and tocotrienols (α-T3, β-T3, γ-T3, and δ-T3), is higher in pulp/peel oil than in juice or seed oil (George and Cenkowski, 2007). In addition, the pulp oil contains high levels of carotenoids (Beveridge *et al.*, 1999).

Seed oil contains unsaturated fatty acids, while it is the only oil that naturally provides a ratio 1:1 of linolenic acid (n-3) to linoleic acid (n-6) (Cenkowski *et al.*, 2006; Kumar *et al.*, 2011). Due to the abundance of the pulp and peels in a whole berry, the composition of the oil from the whole berry resembles to that of the soft part oil. The foremost fatty acids in the soft parts oil are palmitoleic (27-33 per cent), palmitic (27-28 per cent), oleic (17 per cent), linoleic (9-13 per cent), vaccenic (8-9 per cent) and α-linolenic (3-7 per cent) acids (Yang and Kallio, 2001). Within the major commercial subspecies, the main fatty acids quite frequently found in the seed oils are linoleic (39-41 per cent), α-linolenic (27-31 per cent), oleic (17-19 per cent), palmitic (7-9 per cent), stearic (3 per cent) and vaccenic acids (2-3 per cent) (Yang and Kallio, 2001; Kallio *et al.*, 2002; George and Cenkowski, 2007).

The pulp and seed oils in addition, differ in their physico-chemical and biochemical features. Although, SBT oil is a secondary product, it is a special oil used in medicine, as a nutraceutical supplement, and in cosmeceuticals too (Beveridge *et al.*, 1999).

Leaves

SBT leaves are also an asset towards SBT's medicinal, nutritional and cosmeceutical values. SBT leaves chiefly contain a range of bioactives, which mainly include flavonoids, carotenoids, free and esterified sterols, triterpenols, and isoprenols. The leaves are an evenly rich source of prime AOs including β-carotene, vitamin E, flavonoids, catechins, folic acid and significant values of calcium, magnesium and potassium.

The leaves have significant content of phenolics based phytoconstituents. These substances in the leaves are signified by flavonols; leucoanthocyanidins, (-) epicatechin, (+) gallocatechin, (-) epigallocatechin, ellagic acid, ferulic acid, and gallic acid (Upadhyay *et al.*, 2009; Suryakumar and Gupta, 2011). The tannins, hippophaenins A and B have been extracted and isolated from the leaves of SBT (Bal *et al.*, 2011; Kumar *et al.*, 2011).

The tannin fraction isolated from leaves showed hydrolysable gallo- and monomeric ellagi-tannins (strictinin, isostrictinin, casuarinin, casuarictin) as the prime compounds. Bioactive phenolic constituents, such as quercetin-3-O-galactoside, quercetin-3-O-glucoside, kaempferol and isorhamnetin, have been quantied in aqueous and hydroalcoholic extracts of SBT leaves by RP-HPLC (Upadhyay *et al.*, 2010).

The dried SBT leaves consist of 3888 mg/100 g of flavonoids (Yuzhen and Fuheng, 1997). Guan *et al.* (2005) observed that fresh leaves of SBT are rich in total carotenoids (26.3 mg/100 g) and total chlorophyll (98.8 mg/100 g), which indicates the quality for green vegetables; whereas dried leaves still contained large quantities of bioactive compounds comparable to commonly consumed vegetables (Guan *et al.*, 2005). The leaves also contain considerable amounts of proteins (20.7 per cent), amino acids (0.73 per cent lysine, 0.13 per cent methionine and cystine), minerals (Ca, Mg and K), folic acid, catechins, esterified sterols, triterpenols and isoprenols (Guan *et al.*, 2005; Biswas *et al.*, 2010; Suryakumar and Gupta, 2011).

Bark and Branches

Even the branches and bark of the SBT plant have nutritional value. They contain tannins, tocopherols, phytosterols, and other bioactive substances including hippophan (5-hydroxytryptamine). Extracts obtained from SBT branches and barks have been traditionally used internally to treat colitis and enterocolitis, and externally in topical compressions for dermatological disorders, burns and rheumatoid arthritis. In addition, studies suggested that extract of SBT bark may inhibit tumour growth (Suryakumar and Gupta, 2011).

PHYTOCONSTITUENTS: ROLE IN SKIN HEALTH AND NOURISHMENT

SBT parts constitute many phytonutrients which help in replenishing and fortifying the skin by oral as well as topical supplementation. These aspects have been discussed below in details.

Vitamin E: Tocopherols and Tocotrienols

Vitamin E is also known as tocopherols (having a saturated phytol side chain attached to the chromanol structure) and tocotrienols (having an unsaturated side chain) (Traber, 2006). But α-tocopherol, the most active form in humans and which also acts as a powerful AO, is majorly found in SBT berries.

However, recent studies have focused on the role of tocotrienols in the body. It has been proven that both tocopherols and tocotrienols help to avert illness and can be utilized in curing various skin diseases. Vitamin E is naturally present in the skin, and has been used in clinical dermatology as a therapy for a variety of skin ailments including eczema, atopic dermatitis, superficial burns, cutaneous ulscers, and other inflammatory skin diseases (Panin *et al.*, 2004).

Vitamin E plays irreplaceable AO functions, especially in cell membranes and lipoproteins. It is important to keep skin cell membranes undamaged, otherwise the skin cells and their components are destroyed by environmental reactive oxygen species (ROS), deteriorating health and increasing the need for repair and premature renewal. So as a topical treatment, vitamin E helps to moisturize the skin, optimize the skin health, helps to restore and maintain the skin barrier function, locking moisture into the skin and preventing dehydration (Panin *et al.*, 2004).

All forms of vitamin E protect the skin from free radical damage, which is implicated in many diseases and natural aging process. Vitamin E is recognized as an AO that can help prevent the formation of lipidperoxides and UV-induced free radicals. As a lipid soluble AO, it incorporates itself into the vital cell membranes and protect the integrity of these structures from lipid peroxidation. It also protects other fat-soluble vitamins and other skin elements from oxidative damage to provide generally enhanced and prolonged skin protection. Vitamin E lasts longer in the skin than other primarily important AOs, vitamin C and ubiquinone (Coenzyme Q_{10}). Vitamin E is also known to improve the appearance of age spots, elasticity and skin barrier function, and help to smoothen fine lines and wrinkles (Thiele *et al.*, 2005; Thiele *et al.*, 2007).

Free radical or oxidative damage induced by UV exposure is a major factor in pathogenesis of skin cancers and leading to skin aging. Multiple studies have documented the protective action of vitamin E against UV-induced skin damage. Vitamin E can also help to reduce the appearance of some of the wrinkling and reddening of the skin associated with the effects of UV-B exposure.

The benefits of vitamin E for healthy skin care also include its ability to regulate vitamin A in the body, which itself is important for healthy skin. Vitamin E products help to reduce the appearance of stretch marks, prevent the appearance of age spots, and maintain the skin's oil balance during the cleansing process (Thiele *et al.*, 2005; Thiele *et al.*, 2007).

Carotenoids

Various colours of ripe berries of SBT, ranging from yellow to orange, bright red, are related to occurrence of carotenoids. β-carotene, the most dominant carotenoid in SBT, is associated with reduced risk of breast, stomach, oesophageal and pancreatic cancers (Andersson *et al.*, 2009).

The group of carotenoids available in human skin includes α-, γ-, β-carotene, lycopene, lutein, zeaxanthin and their isomers (Stahl and Sies, 2005). They are potent quenchers of singlet oxygen and other free radicals (FR) in biological systems and helps in collagen synthesis and epithelialization (Krinsky and Yeum, 2003; Terao *et al.*, 2011). Recent investigations have shown that the carotenoids could serve as marker substances for the entire antioxidative network of human skin (Haag *et al.*, 2011; Lademann *et al.*, 2011). This is due to the fact that the AOs form protective chains in the human tissue, acting synergistically to protect each other against the destructive action of the free radical and in particular (ROS) (Darvin *et al.*, 2011; Haag *et al.*, 2011).

Vitamin C

SBT berry is among the richest food sources of vitamin C, and in the active form it is called L-ascorbic acid, which is a water soluble AO. The physiological functions of vitamin C derive from its strong AO activity and its action as an essential cofactor for several enzymes make vitamin C a vital molecule for skin health. Vitamin C is a normal skin constituent which is found at high levels in both the dermis and epidermis (Shindo *et al.*, 1994; Rhie *et al.*, 2001). Ascorbic acid acts as an AO by scavenging and quenching free radicals and by regenerating vitamin E from its radical form (Chan 1993; Byer, 1994).

Aging causes a decline in vitamin C content in both the epidermis and dermis (Rhie *et al.*, 2001). Excessive exposures to UV light or pollutants (*e.g.* cigarette smoke and ozone) may also lower vitamin C content, primarily in the epidermis (Shindo *et al.*, 1993; Thiele *et al.*, 1997; Podda *et al.*, 1998).

Following are the health functions of vitamin C in healthy skin.

Photoprotection

Vitamin C limits the damage induced by ultraviolet (UV) light exposure because it reacts with or suppresses the superoxide anion, the hydroxyl radical, and singlet oxygen and effectively interferes with UV-induced generation of reactive oxygen species (Cabelli and Bielski, 1983; Chou and Khan, 1983; Scarpa *et al.*, 1983). By replenishing vitamin E, vitamin C indirectly inhibits lipid peroxidation (Wefers and Sies, 1988). Vitamin C based transport proteins are increased in keratinocytes in response to UV light, suggesting an increased need for vitamin C uptake for adequate protection (Kang *et al.*, 2007; Steiling *et al.*, 2007).

Topical vitamin C, as a photoprotectant, has been studied *in vitro* and *in vivo*, demonstrating its effects in preventing sun damage by reducing sunburn cells and decreasing erythema, when exposed to both UVA and UVB irradiation (Werninghaus, 1995). The addition of topical vitamin C to either a UVA or UVB sunscreen was shown to improve sun protection, when compared with sunscreen alone (Darr *et al.*, 1996).

Furthermore, adding topical vitamin C to 'after-sun' products has been shown to scavenge UV-induced reactive oxygen species (Cabelli and Bielski, 1983). In cultured keratinocytes, the addition of vitamin C reduces UV-related DNA damage and lipid peroxidation, limits the release of pro-inflammatory cytokines, and protects against apoptosis (Stewart *et al.*, 1996; Tebbe *et al.*, 1997).

Topical applications of combinations of vitamin C and vitamin E are more effective in preventing photodamage than either vitamin alone. In particular, this combination of AO vitamins decreases the immunosuppressive effects of UV exposure, increase minimal erythemal dose (MED), and decreased cell damage (Darr *et al.*, 1996; Lin *et al.*, 2003; Lin *et al.*, 2005).

Photodamage/Wrinkling

Photoaging and intrinsic aging damage skin proteins and such oxidative damage can lead to changes in skin structure. In addition to its AO functions, vitamin C regulates the synthesis of the structural protein collagen. The role of vitamin C in the hydroxylation of collagen molecules is well characterized (Peterkofsky, 1991). Hydroxylation of collagen is necessary for its extracellular stability and support of the epidermis. Specifically, vitamin C has been shown to stabilize collagen mRNA and thus, increasing collagen protein synthesis for repair of the damaged skin (Geesin *et al.*, 1988). Vitamin C also increases the proliferation rate of fibroblasts, a capacity that decreases with age (Phillips *et al.*, 1994).

The use of vitamin C (3-10 per cent) in topical applications for atleast 12 weeks has been shown to decrease wrinkling, reduce protein fiber damage, decrease apparent roughness of skin, and increase production of collagen (Traikovich, 1999; Nusgens *et al.*, 2001; Fitzpatrick and Rostan, 2002; Humbert *et al.*, 2003; Raschke *et al.*, 2004). Topically, vitamin C has also been shown to revert some of the age-related structural changes in the interface between the dermis and the epidermis (Sauermann *et al.*, 2004).

Wound Healing

Vitamin C is critical in wound healing, acting as a cofactor for several enzymes, including lysyl and prolyl hydroxylase which stabilize collagen. Vitamin C levels decrease rapidly at a wound site (Kim *et al.*, 1994; Shukla *et al.*, 1997). Although, inflammatory responses often increase free radicals at the site of injury and the presence of vitamin C may limit free radical damage, free radicals may play a complex role in the healing response that is not yet understood (Shukla *et al.*, 1997; Sorensen *et al.*, 2010).

However, the increased demand for dermal collagen synthesis may increase utilization of vitamin C. Vitamin C may have additional roles in wound healing, for example, by promoting keratinocyte differentiation, stimulating the formation of the epidermal barrier, and re-establishing the stratum corneum (Ponec *et al.*, 1997; Boyce *et al.*, 2002; Savini *et al.*, 2002; Duarte *et al.*, 2009).

Dry Skin

Higher intakes of dietary vitamin C have been correlated with a decreased risk of dry skin, suggesting that ascorbic acid may have effects on the trans-epidermal water loss (TEWL). In cell culture models, addition of vitamin C promotes the synthesis of barrier lipids which would establish a functioning stratum corneum with low water permeability (Ponec *et al.*, 1997; Pasonen *et al.*, 2001; Cosgrove *et al.*, 2007). One study using topical vitamin C in human subjects correlated its use with decreased skin roughness (Humbert *et al.*, 2003).

Other Effects

Topical vitamin C may be useful against acne to reduce inflammatory lesions (Klock *et al.*, 2005; Woolery-Lloyd *et al.*, 2010). It may also have mild skin lightening effects that are caused by reductions in melanin production and melanin oxidation (Kameyama *et al.*, 1996).

Flavonoids, Phenolic Acids and Lignans

SBT is an excellent source of flavonoids. Although, there are many types of flavonoids in SBT, the four main ones are isorhamnetin, quercitin, myricetin and kaempferol. *In vitro* studies suggest antioxidative, anti-inflammatory, antimicrobial and anti-proliferative effects of flavonoids found in SBT berries (Puupponen-Pinia *et al.*, 2001; Rosch *et al.*, 2003; Garcia *et al.*, 2007; Hamalainen *et al.*, 2007). Epidemiological data indicates that flavonoids may have beneficial effects on the risk and development of cardiovascular diseases, whereas the evidence concerning cancer is less consistent.

Phenolic compounds may also enhance the AO activity of non-phenolic AOs by regenerating the oxidized forms of these compounds. For example, phenolic compounds have been reported to regenerate ascorbic acid from its oxidized form (Cossins *et al.*, 1998). The synergistic AO interactions between flavonoids and α-tocopherol *in vitro* have been demonstrated by Pedriella and Skibsted (2002). (+)-Catechin, (-)- epicatechin and quercetin are found to enhance the induction period for oxidation of α-tocopherol and results in longer and increased inhibition of oxidation. The regeneration of α-tocopherol from its oxidized form through the hydrogen atom transfer between metal chelating effects of phenolics (due to their catechol, galloyl, or 1,3 positioned hydroxyl and carbonyl moieties) also imparts secondary AO effects by inactivating metals (Gordon, 1990; Aruoma, 2002). Phenolic acids and flavonoids have been shown to complex with iron and copper ions to provide secondary AO effects (Rice-Evans, 1996; Brown *et al.*, 1998). Human epidermal keratinocytes express specific members of the organic anion transporting polypeptides family, transporters responsible for the uptake of a variety of xenobiotics, drugs and large amphipathic molecules (Schiffer *et al.*, 2003). It is postulated that flavonoids enter epidermal cells *via* this route. Once inside the cell, flavonoids can bind catalytic ATP-binding sites on a diverse range of proteins, thus influencing a wide range of cellular processes (Williams *et al.*, 2004; Arct and Pytkowska, 2008). For example, genistein and quercetin are known to inhibit tyrosine kinase and PI3 kinase, respectively. Specific interactions between certain flavonoids and cellular proteins also occur. Soy isoflavones, for example, bind to estrogen receptor-beta (ER-β), the ER isoform expressed in skin cells and the cardiovascular system (Pelletier and Ren, 2004; Jackson *et al.*, 2011).

Studies performed in cell culture, animals and humans demonstrate that treatment with certain flavonoids can minimize adverse skin reactions caused by UVR. Onion extract, rich in the flavonoids quercetin and kaempferol, has been used to reduce scar formation, particularly keloid scars. Cho *et al.* (2010) demonstrated that onion extract and quercetin induce matrix metalloproteinase-1 expression in cultured human skin fibroblasts and hairless mouse skin. UVR induces three MMPs: MMP-1 (collagenase), MMP-3, (stomelysin), and MMP-9 (gelatinase) that cleave and degrade skin collagen and contribute to photoaging (Fisher *et al.*, 2002). In the case of wound healing, a balance between MMP-1 and tissue inhibitor matrix metalloproteinase-1 enzymatic activity affects the amount of extracellular matrix (including collagen) formed at the wound site. Thus, quercetin may influence extracellular matrix deposition during wound healing in order to reduce hypertrophic scarring.

Isorhamnetin inhibited epidermal growth factor (EGF)-induced neoplastic cell transformation. It also suppressed anchorage dependent and independent growth of A431 human epithelial carcinoma cells. Isorhamnetin attenuated EGF-induced COX-2 expression in JB6 and A431 cells. In an *in vivo* mouse xenograft using A431 cells, isorhamnetin reduced tumor growth and COX-2 expression (Kim *et al.*, 2011).

There is evidence that flavonoids physically block UV penetration, influence DNA repair, reduce oxidative damage, attenuate the inflammatory response, preserve immune function and induce cytoprotective pathways.

Triterpenes and Sterols

SBT contains about 26 kinds of triterpenes and sterols, including ursolic acid, oleanolic acid, β-sitosterol, digitalinumverum (digitalin), vitellin, choleterin, 24-ethyl-7 beta lathosterol; lupeol, unsaturated aldehydes, alpha-amyrin; beta-amyrin, etc. Phytosterols are plant sterols, with structures similar to cholesterol, which on consumption are capable of lowering the plasma cholesterol in humans.

The original biological activity of phytosterols is in its anti-inflammatory properties due to which it is used in anti-aging cosmetics. In clinical studies, phytosterols and polyphenols improved sun-damaged skin. Shea butter and avocado oil, the richest sources of phytosterols, extend skin protection and rejuvenation benefits and thus, are used in skin care products for improving skin health. Even hair care products utilize a mixture of phytosterols, such as avocado or similar compounds, primarily for their hair softening and conditioning properties. Till date, the skin and hair benefits of phytosterols are not extensively known, and in fact, are now developing (Dweck, 2006).

In a study conducted in Germany, various topical preparations were tested on skin and the most anti-aging benefits were shown by formulation containing phytosterols and other natural fats. The study further found that phytosterols not only stopped the sun-induced slow down of collagen production, but also helped in new collagen production (Grether-Beck *et al.*, 2008). A second clinical study estimated effect of apple seed phytosterols on cutaneous reconstruction. Incubation of skin equivalents with apple seed phytosterols had significant consequences, namely i) differential regulation of a set of genes associated with keratinocyte proliferation and differentiation; ii) stimulation of hyaluronic acid synthesis; and iii) increase of epidermal thickness. Furthermore, *in vivo* studies revealed that apple seed phytosterols improved skin elasticity while decreasing skin roughness. The study concluded that phytosterols displayed distinct biological effects and significantly improved the structure and function of mature skin (Doering *et al.*, 2005).

A third clinical study examined *in vivo* spectrophotometric assessment of skin barrier recovery after topical application of soybean phytosterols. The study results clearly showed that phytosterols exerted positive results on skin repair. In fact, three days after tape stripping, the sites treated with a formulation containing phytosterols showed an appreciable recovery of barrier function as compared to those treated with a vehicle control (Puglia and Bonina, 2008).

Adding oils that are high in phytosterols to cosmetic emulsions would surely help to recover damaged skin from environment-induced injuries, such as pollution and sun damage. Phytosterol-based skin care products deliver moisture to the skin thereby softening it, strengthens skin barrier, boost elasticity, and impart anti-inflammatory benefits.

Fatty Acids

Fatty acids are classified as saturated fatty acid (SFA), monounsaturated fatty acid (MUPA), and polyunsaturated fatty acid (PUFA). SBT berries are an enriched with such fatty acids, especially PUFAs but the fatty acids composition is different in SBT oils, especially in seed oil and soft parts of the fruit. The unsaturated fatty acids make about 85 per cent of total berry oil (Yang and Kallio, 2001).

The human body absolutely requires the polyunsaturated essential fatty acids (EFAs), α-linoleic acid and α-linolenic acid. Both fatty acids repair the cell membrane after oxidation due to attack of free radicals. Fatty acids are essential components of natural lipids, which determine the physiological structure and function of the human skin (Trommer *et al.*, 2001). They are present in the epidermis, especially in the stratum corneum, the outermost layer and cell membranes (Hansen and Jensen, 1985). Many effects of fatty acids can be linked to changes in membrane lipid composition affecting cell signaling mechanisms originating from membranes (Jump, 2004). Skin aging may influence epidermal lipids and free fatty acid composition, and their physiological functions may be involved in aging process (Boelsma *et al.*, 2011).

Mechanism of action of EFAs is:

There are two classes of EFAs: Omega-6 (n-6) and omega-3 (n-3) fatty acids. Linoleic Acid (LA) is the parent compound of the n-6 PUFAs, while α-linolenic acid (ALA) is the parent compound of n-3 PUFAs. From these two parent compounds, the body synthesizes longer chain derivatives that also have important functions in healthy skin.

- Both n-6 and n-3 PUFAs give rise to potent signaling molecules called eicosanoids, which influence the inflammatory response in skin.
- The n-6 PUFAs have a particular role in structural integrity and barrier function of the skin.
- Both topical application and oral supplementation are effective in delivering EFAs to the skin and systemic circulation (Horrobin, 2000; Ziboh *et al.*, 2000).

As structural components of cell membranes, EFAs ensure fluidity and stability, and influence the permeability of substances into and out of the cell. In addition, specific EFAs in the skin provide moisture and insulating barrier by forming a matrix to maintain the epidermal permeability barrier (Ziboh and Chapkin, 1987). The omega-6 EFAs appear to be more critical than the omega-3 fatty acids for normal skin structure and function (Kwochka, 1993). EFA supplementation appeared to prevent formation of pressure ulcers, and increased hydration and elasticity of the skin in a recent double-blind clinical trial (Declair, 1997). Inflammation and skin damage in general, including atopic dermatitis, rosacea, burns, and aging have responded to treatment with EFAs. Recent patents include the use of conjugated linoleic acid as an effective treatment for skin damage, inflammation and aging (Perricone, 2001). Because of these vital features, they are ideal for moisturization, and also serve as protective agents, which may be included in both oral and topical cosmeceutical and personal care products.

The epidermis is dependent on the continual formation or supplementation of EFAs, such as γ-linolenic acid and dihomogammalinolenic acid (Ziboh and Chapkin, 1987; Miller and Ziboh, 1990; Ziboh *et al.*, 2000; Sies and Stahl, 2004; Pilkington *et al.*, 2011). PUFAs including linoleic acid and α-linolenic acid have also been found to exhibit skin-whitening activity, specifically lightening ultraviolet induced hyperpigmentation of the skin. The mechanism of action is thought to be through inhibition of melaninogenesis by unsaturated fatty acids in active melanocytes (Ando *et al.*, 1998).

Other Fatty Acids

Oils rich in oleic acid are very useful for softening skin, regenerating skin cells, moisturizing, and behaving as an anti-inflammatory. Palmitoleic acid constitutes our skin's fatty acid profile and is a building block to prevent burns, wounds and skin scratches as well as the most active anti-microbial in our sebum. It can be used on our skin to treat injured skin and irritated mucous membranes. It fortifies the skin lipidic barrier and thus, develops the skin moisturization. It has emollient and regenerative properties and is a tolerable skin conditioner (Ako and Gray, 1995). The high content of palmitoleic, oleic and omega-9 fatty acids makes macadamia nut oil (another resource of omega 7), an effectual AO due to its ability to skin's cell membranes from free radical damage in similar manner as human sebum. Recent reports suggested that palmitoleic acid to be an effective antimicrobial on the skin's surface (Wille and Kydonieus, 2003).

Phenols and Organic Acids

The organic acids in SBT are mainly malic acid, citric acid, succinic acid and oxalic acid. Besides this, it also contains beta-coumarin and over 14 kinds of phenolic acids. Thus, liquid beverages which use SBT as main constituent tastes lightly sour. Phenols and phenolic acids are clinically proven to be effective for treating ulcers, wound healing, arthritis and pain (Yang and Kallio, 2001).

Polar Lipids and Phospholipids

SBT fruit membranes are fortified in carotenolipoprotein complexes with 61 per cent phospholipids and 39 per cent galactolipids as structural components. SBT seed oil and whole fruit oil consist of 10-15 per cent phospholipids with lecithin as the major constituent (Yang *et al.*, 2000). Total eight phospholipids have been detected in SBT seed oil, which totally constitute 57 per cent of polar lipid content of SBT fruit includes phosphatidylethanolamine, digalactosyldiacylglycerol, monogalactosyldiacylglycerol and phophatidylglycerol is found in small quantity (2.5 per cent) (Goncharova and Glushenkova, 1995).

These polar lipids play a crucial role in structure and physiology in cell membranes, and may offer some significant applications as emulsifiers and nutrients in cosmetic preparations (Pintea *et al.*, 2001). Phospholipids reconstruct the skin's barrier layer and act as a reservoir for prolonged delivery of natural active nutrients and moisture to the inner layer (dermis) of the skin. Natural phospholipids serve as fantastic humectants, as they are hygroscopic (attract water from the surrounding air) and hold water where an increased level of hydration is required without being occlusive (forming a film to prevent water loss, and preventing normal cellular function). A recent study proved the value of topically applied phospholipids in skin care (Anthony *et al.*, 2003).

SEABUCKTHORN APPLICATIONS IN TOPICAL AND DERMAL THERAPY

Traditional use of SBT oil and plant extracts in promoting revival of skin injuries and support the healing of skin diseases well agrees with the data of current preclinical and clinical studies.

HEALING EFFECT ON ACUTE AND CHRONIC WOUNDS

Skin, the largest organ in the human body, plays a crucial role in protecting and homeostasis. When this barrier is disrupted by physical, chemical, thermal, microbial or immunological damage, these functions are no longer adequately performed. Thus, wound is illustrated as a disruption of structural and physiological continuity of a tissue. Open wound is caused when skin is torn, cut or punctured, and is termed as a closed wound when there is a forced trauma which causes contusion. Burn wounds are caused by fire, heat, radiation, chemicals, electricity or sunlight (Shuid *et al.*, 2005; Jalalpure *et al.*, 2008).

Appropriate healing of wounds is an intricate progression and is very crucial for the restitution of disrupted anatomical continuity and disturbed functional status of the skin. Wound healing includes constant interactions between cell-cell and cell-matrix that lets the process proceed in three overlapping phases *viz.*, inflammatory phase (0-3 days), cellular proliferation or proliferative phase (3-12 days) and remodeling phase (3-6 months) (Glynn, 1981; Martin, 1996). Healing requires the collaborative efforts of different tissues and cellular bodies. It mainly involves aggregation of platelets, clotting of blood, fibrin development, anti-inflammatory response to injury, modification in the ground substances, angiogenesis and re-epithelialization. Healing process is not complete until the disrupted surfaces are firmly knit by collagen (Martin, 1997). The fundamental principle behind optimal wound healing is i) to reduce tissue damage and provide sufficient tissue perfusion and oxygenation; ii) enough nutrition to tissue; and iii) damp the wound healing environment to reinstate the anatomical continuity and function of the wounded skin. Inflammation occurs as an acute response that results in a synchronized influx of neutrophils at the site of wound which then produce free radicals through their characteristic "respiratory burst" activity. It is agreed that free radicals are deleterious to wound healing process, as they show damaging effects on cells and tissues (Baboir, 1978; Aliyeva *et al.*, 2004). Free radicals are also produced by wound related non-phagocytic cells by NAD(P)H oxidase based mechanism (Griendling, 2000). Thus, the wound site has rich concentration of both oxygen and nitrogen centered reactive species along with their derivatives. The consequences are oxidative stress leading to lipid peroxidation, breakage of DNA, and enzyme inactivation including free-radical scavenger enzymes. Evident role of oxidants in the pathogenesis of many diseases advocated that AOs may be therapeutically beneficial in these conditions. Topical application of compounds with free-radical scavenging properties in patients has shown to improve wound healing processes significantly and protect tissues from oxidative damage (Thiem and Goslinska, 2004).

Complications associated with wounds may include infection, cellulitis, deformity, overgrowth of scar tissue (keloid formation), gangrene that may require bleeding (wound hemorrhage), overwhelming systemic infection (sepsis), and tetanus, which is a fatal infection of the nervous system.

Medications for wounds mainly consist of antimicrobial ointments, nitrofurazone, antibiotics, such as oxacillin and mezlocillin. Leukopenia has been reported when silver sulfadiazine was applied for prolonged period and could be secondary to medullar toxicity of silver sulfadiazine (Fakhry *et al.*, 1995; Chaby *et al.*, 2005). Bacterial resistance to silver nitrate has been described (Bishara *et al.*, 2007). The same is observed in case of antibiotics.

WOUND HEALING MECHANISM OF LEAF EXTRACT OF SEABUCKTHORN

The defensive and remedial effects of SBT against wounds, burns, scalds, ulcers and mucosal injuries have been extensively investigated using different animal models and by clinical trials (Suryakumar and Gupta, 2011). In recent years, SBT leaf extract has been scientifically explored and shown to augment acute and chronic dermal wound healing (burns and diabetic) in rats. The SBT treated animals exhibited faster reduction in wound area in comparison with control and silver sulfadiazine (standard care) treated animals. The topical application of SBT increased neovascularization, collagen synthesis and stabilization at wound site, as substantiated by up-regulated expression of vascular endothelial growth factor, collagen type-III, matrix metalloproteinases and improved content of hydroxyproline and hexosamine (Upadhyay *et al.*, 2011). SBT treatment also increased the endogenous enzymatic and nonenzymatic AOs and lessened the lipid peroxide levels in wound granulation tissue (Gupta *et al.*, 2005; 2006; 2008; Upadhyay *et al.*, 2009). Further, it has been reported that SBT leaf has no cytotoxicity, heavy metal contamination, and adverse effect after oral administration (Saggu *et al.*, 2007; Upadhyay *et al.*, 2009). The positive healing effect of topical application of SBT flavone (isolated from fruit pulp) (1.0 per cent, w/v) has been observed on dermal wounds in experimental rats (Gupta and Flora, 2006). Wound healing effect of topical application of leaf extract has been attributed to the existence of flavanoids. Flavanoids plays substantial role in angiogenesis at wound site; extracellular matrix deposition in cells signifies its mode of action on inflammatory phase of wound healing. It also acts on remodeling phase on burn wound healing (Saggu *et al.*, 2007).

WOUND HEALING MECHANISM OF SEABUCKTHORN OIL FROM FRUIT AND SEED

SBT oil, extracted from fruits and seeds when mixed together, contribute to high levels of favourable unsaturated fatty acids (omega-3, 6, 7), natural AOs, vitamins (E, K), carotenoids, and phytosterols (Li and Schroeder, 1996; Beveridge *et al.*, 1999). These phytochemicals make it ideal for medicinal and cosmetic industries for giving synergistic power to protect cell membrane and enhance cell regeneration. Palmitoleic acid, a component of skin is a valuable topical agent in treating burns and healing wounds.

Gupta *et al.* (2006) have studied the effect of SBT berry containing flavonoids topically on dermal wounds in rats. A dose of 20 µl of the flavonoid rich base was applied topically to the wounded area for 7-16 days after the creation of the wounds. The SBT flavonoids were found to promote the wound healing as indicated by improved wound contraction, faster epithelialization and increased levels of hydroxyproline and hexosamine (indicating collagen production and stabilization) in the wound tissue. Also, the AO status in the wound tissue was increased by SBT flavonoids and the levels of lipid peroxides were lowered. Vitamin C was among the AOs, the levels of which were greater in the SBT group as compared to that of control which may be responsible for wound healing action or the flavonoids may have aided its preservation. Vitamin C is essential component for collagen synthesis (Gupta *et al.*, 2006).

Mironov *et al.* (1989) evaluated hexane-extracted SBT berry oil on healing process of wound of rabbit skin as compared to sunflower oil. Wounds completely healed after 14 days treatment with SBT oil, while sunflower oil healed after 21 days (Mironov *et al.*, 1989). The wound healing abilities of SBT ointment in management of aseptic and cutaneous wounds was studied and compared against liquid paraffin and 5 per cent povidone-iodine in dogs. The SBT and 5 per cent povidone-iodine ointment treated wounds also showed early and sustained greater wound contraction. The histological observations of the healing tissue biopsies, however, confirmed superiority of SBT ointment response in wound healing over 5 per cent povidone-iodine.

The efficacy of SBT oil in healing of incisional cutaneous wounds was evaluated and compared with 5 per cent povidone-iodine ointment and liquid paraffin in calves also. Clinically, the SBT oil treated wounds remained relatively drier throughout the period of study and showed lesser inflammation. The increase in tensile strength of healing tissue was also found to be greater in SBT oil treated wounds at every stage of study. The efficacy of SBT oil was also evaluated in healing of infected cutaneous wounds in calves, and the gross wound healing response was found to be comparable in SBT and 5 per cent povidone-iodine ointment treated wounds (Mahajan *et al.*, 2002; Kumar *et al.*, 2003). A pre-clinical study performed on excision-type wounds in male calves concluded wound healing to be faster with SBT seed oil than under liquid paraffin and 5 per cent povidone-iodine ointment (Rana, 2006).

In a clinical experiment based on study, the effects of SBT oil on the healing process of perforation of tympanic membrane carried out by Fan *et al.* (1991), 47 out of 56 patients healed and the average healing time was 13 days. The healing effects of SBT oil were possibly related to increase in rate of regeneration of epithelial tissue of the tympanic membrane (Fan and Xu, 1989). Saturated non-branched alcohols with even numbers of carbon atoms (C22-C28) have been suggested to be the foremost bioactive components possessing tissue regenerative effects in the oils (Mironov *et al.*, 1983).

HEALING EFFECT ON BURNS, SCALDS, IRRADIATION DERMATITIS AND RADIOPROTECTIVENESS

Burns are one of the most regular and distressing forms of trauma. Burns are most commonly caused by exposure to flames, hot objects, hot liquids, chemicals, radiation or a combination of these. A burn is differentiated by a hypermetabolic state on thermal exposure which compromises the immune system and damages the skin by membrane destabilization, protein coagulation, associated with energy depletion, and hypoxia at the cellular level. This is in turn leads to extensive tissue necrosis and invasion of infectious agents. Healing impairment in burn injury is characterized by increased free radical mediated damage, delayed granulation tissue formation, reduced angiogenesis, and decreased collagen reorganization.

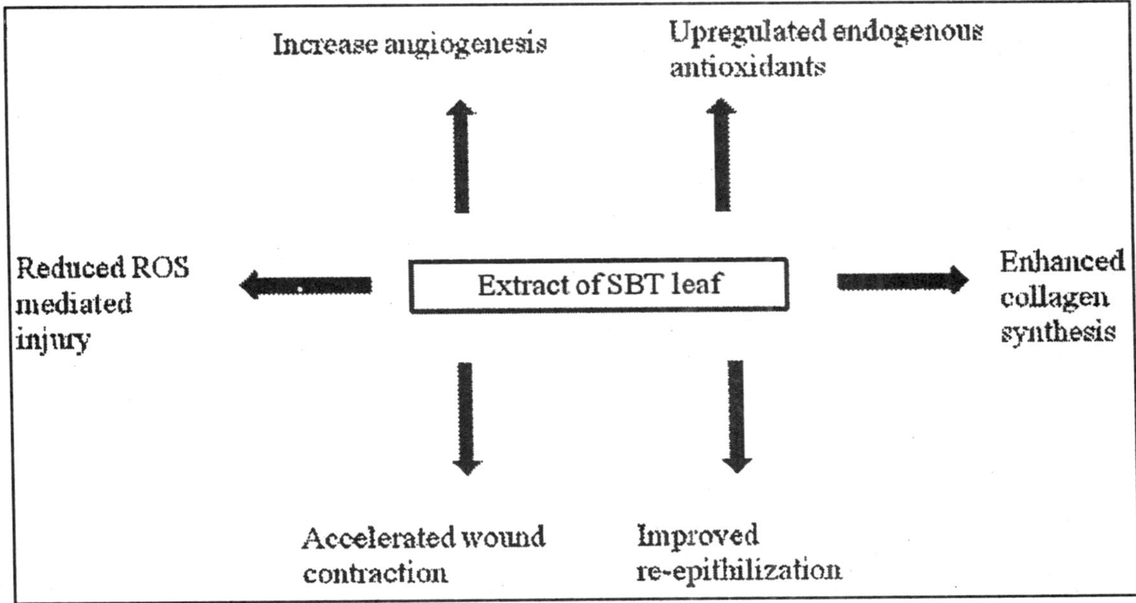

Figure 26.1: The Schematic Diagram Showing the Possible Effect of the Seabuckthorn Leaf Extract in Promoting Wound Healing Activity.

Treatment with synthetic drugs (*e.g.* antibacterial agents and disinfectants) give rise to problems, such as allergy, drug resistance, etc., forcing scientists to seek alternatives (Church *et al.*, 2006). Scalds are rooted by contact with wet heat, such as boiling fluids or steam. Electrical burns are less common, but have the potential to be more serious as the depth of the burn is usually greater than is apparent, and heart irregularities may occur.

Burns are classified as follows: Superficial seen as reddening (like sunburn), and characterized by damage to outer layer of skin only; partial thickness characterized by blistering and damage to outer layer of skin; and full thickness characterized by white or blackened areas and damage to all layers of skin, plus underlying structures and tissue.

Vlasov (1970) were among the first group of scientists who reported medicinal values of SBT oil against wound healing effects on superficial and cutaneous burns of skin as well as burn wound healing ability. Recent studies have clearly shown SBT seed oil to be effective in therapy of burns of different etiology. SBT seed oil has nearly 90 per cent unsaturated fatty acids, including ALA (ω -3) (18:3, n-3), LA (ω -6) (18:2, n-6), oleic acid (18:1, n-9), and palmitoleic acid (16:1, n-7). The wound healing activity of SBT seed oil is thought to be attributable to its high content of vitamin E, carotenoids, phytosterols, and PUFAs (ω-3 and ω -6). These phytoconstituents may be acting in an individual or in a synergistic manner to provide fortification and nutrition to the wound tissue. As reiterated earlier, palmitoleic acid nourishes the skin on topical application for treating burns and wound healing. SBT seed extracts inhibit growth of various bacteria, such as *Bacillus cereus*, *B. coagulans*, *B. subtilis*, *Listeria monocytogenes*, and *Yersinia enterocolitica*. The antibacterial effects of SBT seed implicate its potential for combating burn based wound infection (Chauhan *et al.*, 2007).

A systematic preclinical study (Upadhyay *et al.*, 2009) demonstrated the safety and wound healing efficacy of SBT seed oil extracted by supercritical carbon dioxide extraction co-administration by both route, *i.e.* topical (200 µl) and oral (2.5 ml/kg) on experimental burn wounds in rats. Treatment promoted significant re-epithelialization and wound closure, and improved formation of granulation tissue and collagen biosynthesis in burn wounds, and up-regulated expression of the vascular endothelial growth factor indicating the induction of angiogenesis by the oil. SBT seed oil has been observed to possess mitogenic potential and is involved in fibroblasts and keratinocytes proliferation at the wound site. Seed oil treatment showed an increase in endogenous AO and decrease in free radical production in burn wounds.

In a clinical study, topical treatment of SBT seed oil was used to treat 32 patients having burned or scalded skin (12 cases of first-degree burns, 18 cases of superficial second-degree burns, and 2 cases of deep second-degree burns). All the patients were cured after 7 days of treatment with no visible signs of scarring. It was suggested that SBT seed oil has functions that improve immunity, remove blood stasis, and promote blood circulation (Zhao, 1994). Labedeva *et al.* (1992) reported on the epithelization process of NaOH produced burns on mice skin. Histological and histochemical studies showed SBT oil to accelerate fibrinogenesis and collagen formation in the wounded area, suggesting that phytosterols present in SBT oil could be the responsible for tissue-regeneration.

Second degree irradiation dermatitis (with inflammation, wet peeling, secretion and partial ulcer) caused by X-ray irradiation on legs of NIH female mice were also treated with topically application of SBT seed oil, pulp oil and combined seed/pulp oil. After two weeks of treatment, ulcer areas were healed, and skin colour became normal. In contrast, irradiation dermatitis in the negative control group worsened during the two weeks (Zhang *et al.*, 1988). In another clinical study carried out on 13 patients with vulvitis and perineal inflammation, and eight patients with irradiation dermatitis in the armpit and clavicle area were treated with topically applied (three to four times a day) SBT oil, in which 85 per cent patients showed improved wound healing after treatment. The results suggested affirmative action of SBT oils on the acute dermatitis caused by irradiation therapy for different cancers (Zhang *et al.*, 1988).

The entire extract of the fresh berries of SBT has been reported to provide protection to whole mice, various tissue, cells and cell organelles against fatal irradiation. More than 50 per cent protection was observed at a dose range of 15-40 mg/kg given intra-peritoneally. It was concluded that pre-irradiation treatment of animals with SBT extracts protected the functional integrity of mitochondria from radiation induced oxidative stress (Goel *et al.*, 2005; Chawla *et al.*, 2007). SBT extracts act as an AO preventing cellular and mitochondrial free radical generation that could contribute to its ability to inhibit radiation induced apoptosis and cytotoxicity (Agrawala and Adhikari, 2009). The cellular survival, proliferation enhancement, immunostimulation and ultimately the whole body survival of mice could be connected to the free radical scavenging action SBT extracts, thus showing its efficacy in radioprotection. The radioprotective effectiveness of phytoconstituent of SBT may be attributed to preserving chromatin structure, induction of hypoxia, protects hydrogen atom donation, free radical scavenging (Shukla *et al.*, 2006), and blocking of cell cycle at G_2-M phase by interfering with topoisomerase-I activity and mitochondrial and genomic DNA (Goel *et al.*, 2003). However, intensive clinical studies must be conducted to investigate the radioprotective activity of SBT extracts.

ANTI-INFLAMMATORY, ANALGESIC AND ANTI-ARTHRITIC EFFECTS

Oxidation reactions are involved in inflammation. Hence AOs are important in protecting cells of the immune system and modulating inflammation. A few studies especially aimed at investigation of the topical effects of SBT berry fractions on inflammation and pain. Fayman (1991) and Sabynich *et al.* (1994) studied the effect of application of SBT oil on postoperative tonsillitis, chronic cervicitis, carrageenin-induced oedema, UV-induced erythema and exudative peritonitis. SBT oil was found to effectively arrest the inflammatory reaction and help in progression of recovery. Application of SBT oil induces an increase in plasma linoleic, α-linolenic acid, neutral lipids, eicosapentaenoic acids, palmitoleic acid and to small extent in pentadecanoic acid in both plasma phospholipids and neutral lipids that contribute to healing process in atopic dermatitis.

In an *in vivo* animal study conducted by Kallio *et al.* (2001), the antiinammatory and analgesic eect of SBT seed and pulp oils were used to treat skin disorders. It was found that oils reduced acetic acid induced body writhing as well as dimethylbenzene induced ear swelling in mice. However, this treatment had no eect on egg white induced metatarsal inammation in rats (Kallio *et al.*, 2001). Ganju *et al.* (2005) studied anti-inflammatory and immunomodulatory remedial effects of SBT leaf extract in adjuvant induced arthritis (AIA) rat model. SBT extract was administered intraperitoneally to treat the inflammation. Administration of SBT extract on the same day or 5 days prior to inflammatory insult into the joint appreciably reduced the inflammation as compared to the untreated animals in a dose dependent manner. These observations suggested a significant anti-inflammatory activity of the SBT leaf extract and potential in the treatment of arthritis.

Tissue regenerative, anti-inammatory and anti-microbial eects of the oils are among the mechanisms related to the eects observed. Both sterols and long chain alcohols have been reported to be the active components responsible for these eects (Zhang *et al.*, 1988; Kallio *et al.*, 2001).

ANTI-CANCER EFFECT

AOs usually avert free radical reactions which is one of the main causes of cancer induction of the respective organ. *Hippophae* contains different AOs, which can prevent such free radical reactions and oxidative damages of mitochondrial system (Goel *et al.*, 2005). Some literature is available on the activity SBT fruit to decrease carcinogen induced stomach and skin tumorigenesis. This may be due to upregulation of phase II and AO enzymes as well as DNA-binding activity of IRF-1, a known antioncogenic transcription factor causing growth suppression and apoptosis induction for its anti-cancer eect. It can be anticipated that the anti-mutagenic activity of SBT might be *via* antioxidative mechanism (Padmavathi *et al.*, 2005; Arora *et al.*, 2007). SBT juice too can block the endogenous configuration of N-nitroso compounds more eectively than ascorbic acid and thereby prevent tumor production (Li and Liu, 1991; Ren *et al.*, 1992; Zhang and Qiu, 1995).

OTHER USES

SBT oil (from the pulp and seeds) is at present used clinically in hospitals in Russia and China. These bioactive oils have also been obtained from the young branches and leaves, and have been incorporated into an ointment for treating a wide variety of skin damage including burns, bedsores, eczema, and radiation injury. For example, in 1986, many of the Chernobyl nuclear disaster victims were treated with SBT extracts (Li and Schroeder, 1996).

In ophthalmology, berry extracts have been used for keratosis, trachoma, eyelid injuries and conjunctivitis. The SBT industry has been thriving in Russia since the 1940's when scientists there began investigating the biologically active substances found in the berries, leaves and bark. The first Russian factory for SBT product development was located in Bisk. These products were utilized in the diet of Russian cosmonauts and as a cream for protection from cosmic radiation (Delabays and Slacanin, 1995; Xu *et al.*, 2001).

SBT cream made essentially from SBT oil has been reported to have therapeutic eects on chloasma, melanosis, keratoderma, xeroderma, recurrent dermatitis and other skin conditions when tested on 350 people in China (He *et al.*, 1989). On the contrary, very few reports specify the role of berry extracts in these skin orders by topical route. Hence, additional studies on SBT extracts and oil for treating dermatitis and eczema would provide an insight to SBT's role in topical therapy.

APPLICATIONS IN COMECEUTICALS

Cosmetics and skin care products developed from SBT are valued for their rejuvenating, restorative and anti-aging action. Almost the entire plant is suitable for consumption and topical application since fruit, seed and leaves are rich and vibrant source of phytonutrients, AOs and vitamins. SBT oils extracted from seeds and fruit and also leaf extracts are popular in cosmetic preparations, such as facial cream, hand cream, lip balm, deodorant and oil for body massage and hair care.

ANTI-AGING AND PHOTOAGING EFFECTS

Skin aging is a degenerative process wherein the results due to chronological/intrinsic aging are combined with effects created by environmental factors (*e.g.* sun, heat, pollution and smoking). Among the extrinsic factors, solar ultraviolet (UV) irradiation is the most significant and further aging that occurs from too much exposure to UV light is referred to as photoaging. With a decreasing ozone layer and transplant immunosuppression, there emerges an even greater demand for effective photoprotection (Madey and Pinnell, 2001).

Extensive research on the pathogenesis of photoaging has demonstrated that photodamage is mediated in a part by free radicals. ROS are generated when human skin is exposed to UV light. ROS appear to mediate their deleterious effects on skin aging by damaging DNA, cell membranes, and proteins, including collagen. Exposure to ultraviolet B (UVB) also causes direct DNA damage thereby promoting carcinogenesis (Greenway and Pratt, 2000; Farris, 2005).

Recently, research focus is on the use of natural AOs for obliterating the free radicals mediated damaging effects of UV radiation. Therefore, formulations appropriately fortified by natural AOs could serve to be useful complements to sunscreens in halting the aging process induced by UV exposure (Greenway and Pratt, 2000; Madey and Pinnell, 2001). They potentially offer fortification against damage to the tissues caused by the detrimental effects of environmental and other agents. To maintain healthy and youthful skin, it should be fortified with enough AOs to adequately protect itself which can be delivered to our skin through diet, supplements and topical products.

For our overall health, we need a healthy diet and supplements. However, for true health of our skin, internal nourishment is not enough and topical AOs are essential. The rate of delivery of these AOs to the skin is higher by topical route in comparison with the oral supplementation. Topical AOs also help in that protecting skin from UV damage, reduce inflammation, and protect against certain types of cancer as demonstrated in countless studies (Pinnell, 2003).

The best-known natural AOs are vitamin C, vitamin E, carotenes and flavonoids. Scientific studies have shown that topical vitamin C, decrease the number of abnormal 'sun burn' cells and reduce UV damage to DNA. While, vitamin E protects membranes from lipid (fatty acids) peroxidation and thus, being the first to absorb the free radicals, flavonoids and carotenes exhibit superior free radical scavenging activity (Greenway and Pratt, 2000).

SBT berries and leaves are a rich source of AOs. The AO activity of SBT berries has been reported. Various AO values of SBT berries reported by Velioglu *et al.* (1998) were antioxidant value was 0.036, antioxidant activity was 93.6 per cent, oxidation rate ratio (ORR) was 0.064, antioxidant activity coefûcient (AAC) 827.6 and total phenolics 1112 mg/100 g of seabuckthorn berries (*H. rhamnoides* L. cv. "Indian Summer") among the 28 selected plant products. It was also deduced that the SBT berries had the highest AO activity of 93.6 per cent among the other medicinal plants.

AOs possess the unique ability to neutralize free radicals, which are highly reactive molecules that result from oxidative processes and may contribute to DNA damage, carcinogenesis, inflammation, and photoaging of the skin (Nichols and Katiyar, 2010). Acute response of human skin to UVB irradiation includes erythema, edema, and pigment darkening followed by delayed tanning, thickening of the epidermis and dermis, and synthesis of vitamin D; chronic UVB effects are photoaging, immunosuppression, and photocarcinogenesis (Gil and Kim, 2000; Kullavanijaya *et al.*, 2005). Biochemical reactions that accelerate the progression of skin aging have their roots in inflammatory processes, as inflammation generates micro scars that mature into blemishes or wrinkles. Natural anti-inflammatory agents are, therefore, included in anti-aging formulations, and serve to soothe, heal and protect skin tone and integrity (Nichols and Katiyar, 2010).

SBT extracts (leaf and oils) are proven efficacious as anti-inflammatory agents, thus have the potential of anti-aging actives. Regenarative and anti-inflammatory action is due to competitive inhibition of series of leukotriens from an arachidonic acid. The most recent research showed topical use of SBT in burns to provoke a stimulation of a proliferation of fibroblasts, synthesis of collagen and an expression of a specific matrix metalloproteinase, and an angiogenic activity (Majewska and Gendaszewska-Darmach, 2011). It was suggested that this effect was in a correlation to a specific high content of unsaturated omega-3 and omega-6 fatty acids, carotenoids and tocopherols in oil (Upadhyay *et al.*, 2009).

Lipids also play an important role in maintaining the barrier functions of the skin. In general, unsaturated fatty acids reinforce the skin's barrier function, prevent moisture loss through the epidermis, provide structural integrity to the skin damaged by external influences, and show anti-inflammatory action. They also help to soften and smoothen the skin by inhibiting the formation of corneous cells. Lipid compounds that provide an occlusive effect to prevent water loss, repair lipid layers and restore barrier functions are therefore an integral part of anti-aging formulations. Given that lipids are essential to skin health, it is obvious why SBT flourishes in nourishing the body's largest organ. Its specialized fatty acid composition (it is a powerful source of omega-3, -6 and -9 fatty acids, as well as a rich source of the elusive omega-7 essential fatty acid (which is vital to collagen production), along with the high presence of AOs, replenishes essential oils and nutrients that promote skin regeneration and speed up wound healing (Yang *et al.*, 2009).

SBT's emollient properties contribute to its use in phytopharmaceuticals. SBT seed oil absorbs in the UV-B range (290-320 nm) and thus, may be used is UV skin protectant products. The seed oil contains 50-85 mg of carotenoid oil and 61-113 mg of vitamin E (approximately 50 per cent alpha, 40 per cent beta and 10 per cent gamma-tocopherols) for every 100 g of seed (Beveridge *et al.*, 1999). The *in vitro* studies conducted by Kim *et al.* (2012), on SBT alcoholic seed extracts (SBSE) showed increase in the synthesis of procollagen, decrease the expression of MMP-1, and inhibit the production of IL-1β, IL-6, and COX-2 in UVB-irradiated human fibroblasts. These findings suggest that the SBSE extract may be a potential therapeutic agent for preventing and treating skin photoaging (Kim *et al.*, 2012).

An *in vivo* study investigated the anti-aging effects of both oral supplementation and topical skin application of SBT seed oil, determining skin hydration status, elasticity, luminosity, surface roughness, and cutaneous thickness as markers for visible improvement. Both groups with oral and topical administration played significant improvement in the above measured areas, making it the first study to explore beneficial outcomes of using SBT oil orally and topically to reduce the signs of skin aging (Yang *et al.*, 2009).

Another study exploited the topical w/o emulsion of 1 per cent hydro-alcoholic extract of *H. rhamnoides* (formulation) and placebo control (base). The skin mechanical parameters determined with a cutometer, and in addition, the AO activity of the formulation was evaluated by 1, 1-diphenil-2-picrylhydrazyl (DPPH) method. A significant correlation between the active formulation and the improvement of the skin mechanical parameters was observed. The active formulation was found to be superior than the placebo control. The topical AO emulsion of *H. rhamnoides* significantly improved skin biomechanical parameters after 7 weeks of treatment. The data obtained suggested that *H. rhamnoides* could be an alternative pharmacological tool for treating age-related loss of skin elasticity in humans. These improvements might be due to the presence of AOs, such as β-carotene and tocopherols. Furthermore, increase in skin hydration improves viscous resistance against deformation and vitamin C has the advantage of stimulating dermal fibroblasts for the synthesis of collagen (Khan *et al.*, 2012). As collagen level is increased, hydration level also improves.

SKIN HYDRATION

Healthy skin is differentiated by its level of hydration and elasticity. Hydration is an important aspect in skin health. Skin hydration mainly by topical route is improved and maintained by emolliency and occlusiveness both of which are exhibited by SBT oils. SBT oil also contains lutein, lycopene and zeaxanthin. These components increase skin hydration and maintain elasticity to give a healthy glowing skin (Huang *et al.*, 2009).

An *in vivo* study was carried out in human volunteers wherein concentrated SBT fruit extract was entrapped in the inner aqueous phase w/o emulsion. Both the placebo base and formulation were applied to the cheeks of 21 healthy human volunteers for a period of 8 weeks. Stratum corneum (SC) water content and transepidermal water loss (TEWL) were measured every week to observe any effect produced by these topical creams. The vehicle showed insignificant effects, whereas the formulation showed statistically noteworthy reduction in TEWL. SC water content was significantly increased by the formulation. The newly formulated cream of *H. rhamnoides* fruit extract showed marked increase in moisture content of the skin showing that the formulation had skin moisturizing effects. The formulation was observed to decrease TEWL significantly which showed the formulation to have anti-wrinkle affects. Since, both the aforementioned parameters are involved in aging, this formulation can be used as anti-aging product (Khan *et al.*, 2011).

SKIN REJUVENATION

SBT's wealth of phytosterols, AOs and essential fatty acids, especially palmitoleic acid which is a naturally occurring component of skin enhances its ability to restore and maintain healthy skin. It promotes blood circulation and increases tissue regeneration, all of which benefit conditions, such as skin aging, acne and rosacea.

Hair Care

SBT oil is a natural way of encouraging hair growth. Extremely rich in nutrients and vitamins, the oil nourishes the scalp and is believed to stimulate hair growth, as well as heal dry and damaged hair. The high concentration of vitamin A and E, which are essential for beautiful healthy hair, has led to the development of many formulations of shampoos and conditioners with SBT oil as a featured ingredient.

Chinese scientists, in 2009, discovered that a human demodex parasite that lives in skin and hair follicles, causes problems ranging from rosacea, hair loss and premature greyness, to enlarged pores and acne (Elston, 2010). The parasites were found to feed on the nutrients from the skin. The SBT based oil kills the parasites that are believed to be the cause of some major hair loss, and in conjunction with its restorative properties, leading to better hair growth. SBT oil shampoos, soaps and conditioners are used similarly as any other skin care products for lathering and rinsing as usual. Hair masks are left on the scalp for a few minutes and then rinsed out (Aylesworth and Vance, 1982; Forton *et al.*, 2005).

Anti-acne Effects

Acne vulgaris is an inflammatory skin condition that affects the pores of the head, neck, and shoulders, causing plugs and infections that appear as blackheads, whiteheads, pimples and cysts. With a multitude of benefits of SBT oils to skin, it was only a matter of time before SBT was investigated as a treatment for persistent skin conditions, such as acne and rosacea. Chinese researchers have studied the role of a parasite called human demodex folliculorum mite, or the demodicid that lives in these skin conditions. This mite lives in pores and hair follicles usually on the cheek, chin, forehead, and nose as well as in the roots of your eyelashes (Aylesworth and Vance, 1982; Crawford *et al.*, 2004; Pelle *et al.*, 2004; Forton *et al.*, 2005). A new treatment available is SBT oil (*Hippophae rhamnoides*). Its activity is targeted against the mite to reduce the inflammation under the skin and thereby, provide relief from the mechanisms that cause the rosacea complex of symptoms. SBT contains several potent natural anti-inflammatory agents (including quercetin and salycins), as well as several groups of micronutrients (including omegas 3, 6, 7 and 9) that work together to balance and regulate the body's hormonal and immune systems. Topical SBT applications soften, thin and strengthen the skin and reduce scarring, swelling and inflammation, allowing the skin to finally break the inflammatory cycle and heal itself. SBT is also a mild analgesic which discourages scratching and picking, and is an anti-microbial (Guliyev *et al.*, 2004; Larmo, 2011).

OTHER COSMECEUTICAL USES

It is also used in cosmetics as a natural plasticizer and emulsier, in skin regenerating compositions, and as natural UV lters (He *et al.*, 1989). SBT oil can also be used as cleanser and a cosmetic agent for the lips (Zou, 1997; Gercikovs and Zoludeva, 1998).

SAFETY AND TOXICITY STUDIES

In vivo toxicity studies were carried out using SBT based formulations and extracts.

FRUIT EXTRACTS BASED TOXICITY STUDIES

Chawla *et al.* (2007) reported the highest acceptable intraperitoneal single dose of 200 mg/kg body weight/day in mice for flavonoid rich ethanol extract of SBT berries. Doses above 200 mg/kg induced mortality. A slightly higher dose of 212 mg/kg caused the death of 50 per cent of the animals within 72 h. The extract was prepared from *H. rhamnoides* berries collected from the western Himalayas.

Goel *et al.* (2002) also investigated the acute toxicity of alcoholic SBT berry extract RH-3, which was administered intraperitonially to mice as single doses of different sizes. The study results state that single doses of up to 40 mg/kg body weight were tolerated and the only apparent adverse effect was drowsyness for 3-5 min after intake. Doses of 45 mg/kg and above caused increased mortality in a dose dependent manner.

SEABUCKTHORN SEED EXTRACTS BASED TOXICITY STUDIES

Local application of SBT seed oil has been investigated for toxicity (acute toxicity, irritating effects on skin and mucosa and allergic effect on skin) in the animals. During the acute dermal toxicity (time period 7 days), the seed oil was applied topically on the dorsa of rabbits to healthy and wounded skin, 4 times daily (at three-hour intervals, two doses were used, 2.5 ml/kg and 10 ml/kg). None of the animals died, and no significant difference was found between the seed oil treated and the control solution treated groups. The result inferred that topical application of seed oil did not have any acute toxic effect on the healthy or wounded skin surface of rabbits. During the skin irritation test, the seed oil applied topically on normal and wounded skin of guinea pigs for 7 days (1 ml per day, once) did not show any irritation symptom (erythyma, inflammation). SBT seed oil was applied to the eyes (0.1 ml) and vaginal mucosa (1 ml) of rabbits in mucosa irritation test once a day for 7 days. No irritation of mucosa was observed in the oil-treated group. Repetitive topically applied seed oil on the skin surface of guinea pigs did not induce any allergic reaction (Che *et al.*, 1999).

SBT seed oil administered intramuscularly was assessed in New Zealand white rabbits for sub-chronic toxicity. SBT oil was administered intramuscularly once in a week up to seven weeks, at doses of 0.5, 1 and 1.5 ml/kg body weight. No mortality was seen in any of the treatment groups during the course of study. Various physiological, hematological, biochemical as well as histopathological parameters were examined and found to be unchanged indicating that intramuscular SBT seed oil is non-toxic even at higher dose level in rabbits and in animal muscle tissues also after IM application (Ali *et al.*, 2011).

In many of the health effect studies, the composition of the extract or other product used is not reported. In some cases, even the method of extraction or the solvents used are described in a cursory manner only. Therefore, it is not easy to compare the results of the safety evaluations between studies.

Table 26.1: *In vivo* and *In vitro* Topical Applications of SBT Extracts

Effects	*In vivo (Animal/human) and In vitro Studies*	*SBT Extracts/ Actives*	*Dose of Study*	*Observations*	*References*
Wound healing effect	Rat model	Flavonoids (isolated from fruit pulp)	1 per cent flavonoids in 20 μl PG base/d topically or 7-16 d after wound indcution	Improvement in wound contraction, faster epithelialization, increase in hydroxyproline and hexosamine, decrease in oxidative stress (GSH↑, catalase↑, vitamin C↑, lipid peroxides↓).	Gupta *et al.* (2006)
	Rabbit model	Hexane-extracted	– SBT berry oil	Wounds healed completely and faster after 14 days treatment as compared to other oils. Tissue-regenerative, anti-inflammatory and antimicrobial features may be responsible for such wound healing potentials.	Mironov *et al.* (1989)
	Effect on aseptic and cutaneous wounds in dogs and calves	SBT seed oil	–	Clinically, the SBT oil treated wounds remained relatively drier throughout the period of study and showed lesser inflammation. The increase in tensile strength of healing tissue was also found to be greater in SBT oil treated wounds.	Mahajan *et al.* (2002); Kumar *et al.* (2003)
	Clinical experiment on healing process of perforation of tympanic membrane	SBT oil	–	SBT oil's high viscosity, rejoined the valvulae of the perforated edge, assist the exudation and hematopedesis of the tissue fluid of the wound, the cellular infiltration, the metabolic process of the epithelial layers of the tympanic membrane and the full layer hyperplastic reunion of the tympanic membrane. Thus, the augmented rate of regeneration of epithelial tissue of the tympanic membrane.	Fan and Xu (1989); Fan *et al.* (1991)
Healing effect on burns and scalds	Rat model	Seed oil (CO_2 extract)	2.5 ml/kg bw/d orally and 200 μl/d topically or 7 d after burns based wound induction	Increased wound contraction hydroxyproline (collagen), hexosamine, DNA and total protein. Decrease in oxidative stress (GSH↑, ROS↓). No adverse effects and signs of sub-acute toxicity.	Upadhyay *et al.* (2009)

Contd...

Table 26.1–*Contd...*

Effects	*In vivo (Animal/human) and In vitro Studies*	*SBT Extracts/ Actives*	*Dose of Study*	*Observations*	*References*
	Clinical study in patients having burned or scalded skin	SBT seed oil	–	Wounds cured after 7 days with no visible signs of scarring which may be due to action of SBT seed oil to improve immunity, remove blood stasis, and promote blood circulation.	Zhao (1994)
Protective effect against irradiation dermatitis and radioprotec-tiveness	Mice model	SBT seed oil, pulp oil and combined seed/pulp oil	–	After two weeks of treatment, second degree irradiation dermatitis caused by X-ray irradiation on legs of mice cured and ulcer areas also healed.	Zhang *et al.* (1988)
	Mice model	Extract of the whole fresh berries	15-40 mg/kg intra-peritoneally	SBT extracts protected the functional integrity of mitochondria from radiation induced oxidative stress which may be due to preserving chromatin structure, induction of hypoxia, protects hydrogen atom donation, free radical scavenging, and blocking of cell cycle at G_2-M phase by interfering with topoisomerase-I activity and mitochondrial and genomic DNA.	Goel *et al.* (2005); Chawla *et al.* (2007)
Anti-inflammatory, analgesic and anti-arthritic effects	Effect of application of SBT oil on oedema, UV-induced erythema, exudative peritonitis and tonsilitis	SBT oil	–	SBT oil was found to effectively alleviate the inflammatory reaction and reduce the progression of such reactions.	Fayman (1991); Sabynich *et al.* (1994)
	Effects on inflam-mation and immuno-modulation in adjuvant induced arthritis rat model	SBT leaf extract	intraperitoneally	Significant anti-inflammatory activity against arthritis.	Ganju *et al.* (2005)
Anticancer effect	Skin cancer studies in mice model	SBT fruit juice		Effect may be due to upregulation of phase II and AO enzymes as well as DNA-binding activity of IRF-1 and mainly due to anti-oxidative mechanism.	Padmavathi *et al.* (2005)

Contd...

Table 26.1–*Contd...*

Effects	*In vivo (Animal/human) and In vitro Studies*	*SBT Extracts/ Actives*	*Dose of Study*	*Observations*	*References*
Antiaging and photoaging effects	UVB-irradiated human fibroblasts	SBT alcoholic seed extracts (SBSE)		Synthesis of procollagen↑, expression of MMP-1↓, and inhibit the production of IL-1β, IL-6, and COX-2 and suggesting anti-photoaging of SBSE.	Kim *et al.* (2012)
	Clinical study	Hydroalcoholic extract of *H. rhamnoides*	1 per cent in w/o emulsion	Improvement of the skin mechanical parameters (elasticity maintained) and skin hydration after 7 weeks of treatment. It may be attributed to presence of AOs, such as β-carotene and tocopherols.	Khan *et al.* (2012)
	Clinical study	SBT fruit extract	w/o emulsion	Reduction in TEWL and SC water content was significantly increased, thus showing skin moisturizing effects.	Khan *et al.* (2011)

SBT: Seabuckthorn; bw: Body weight; GSH: Reduced glutatione; mo: Month; ROS: Reactive oxygen species; wk: Week; MMP: Matrix metalloproteinase; IL: Interleukin; COX: Cyclo-oxegenase enzyme; TEWL: Trans-epidermal water loss; and SC: Stratum corneum.

COMMERCIALISED PRODUCTS

In spite of SBT's highly acidic nature and exotic avour, SBT berries have an excellent prospect for producing various processed food products, neutraceuticals and cosmeceuticals. Products on the market from SBT range from oil, juice, and food additives to candies, jellies, cosmetics, and shampoos (Schroeder and Yao, 1995). Juice, pulp oil, seed oil, cream and pigments are the main commercial products from SBT berries (Kumar, 2003). The literature describing the formation of these products is both scattered and limited.

DERMACEUTICALS

In comparison with cosmetics, the SBT's dermaceuticals market is not explored and minimal dermal products are available. Depending on different processing methods, the nutritional properties of seed and pulp oils vary (Dharmananda, 2004; St. George and Cenkowski, 2007). The SBT based oils are being utilized traditionally for various skin disorders, but in recent decades, the oils fortified with high yields of bioactives are used for topical therapy of burns, wound healing and eczema. Dermaceuticals mainly based on SBT extracts and actives are topical conventional formulations, such as gels, creams and ointments. The skin healing ointment from Thera Wise Bio Active Skin Healing/ Post Cesarean, is an anti-itch and antibacterial ointment which can be used for anything from eczema and skin irritations to burns and wounds, including surgical wounds post-cesarean contains vitamins A, C, E, and 100 per cent natural ingredients, alongwith SBT oil.

The Naqi Cica cream and gel is specially designed for daily care of burned skin. These formulations contain SBT and *Camelina sativa* seed oil (Omega-5 and -7) and are used for skin care as it has moisturizing, natural skin regrowth and repair properties. This Naqi Cica Cream was developed in conjunction with the Belgian Burns Centre and the University of Louvain. Another hypoallergenic cream, birch-rokitnikowy (Birchen and SBT) of betulin is preservatives and aromatic substances free. It is especially recommended for sensitive and allergy prone skin. The cream is to be used for topical applications for sunburns, cracked heels and feet, calluses, burns, abrasions, bedsores, insect sting, psoriasis, thermal and chemical burns, allergies, varicose wounds, frostbite, and acne.

In Russia, Dina^{+} company has marketed SBT w/Mumijo (shilajit) Forte natural cream which is claimed to be an effective remedy for skin damages (heals cuts, wounds, cracks, burns and domestic injuries, insects bites, swelling reduction, rash and peeling of the skin). Justnatural products LLC in USA have also developed Natural Psoriasis Lotion which is a natural blend of herbal extracts alongwith SBT oil. Additionally, Salcura Natural Skin Care Therapy a market leading brand for natural treatments of eczema, psoriasis, dry and itchy skin from UK, has a list of products which are mainly based on SBT oil and extracts. There are shampoo for psoriasis - Scalp psoriasis shampoo, psoriasis shower gel, psoriasis body wash, Salcura oruls spray, Salcura sinose nasal spray, eczema shower gel, eczema body wash, eczema spray-Salcura dermaspray intensive, and Salcura anti-acne spray.

SBT based creams and gels are also marketed for treatment of arthritis by various companies, such as Taylor made organics (USA), a veterinary product and OTC product, the alzean cream from Agrovetzashita Scientific (Russia).

COSMECEUTICALS

SBT berries and leaf extracts known for its anti-aging, anti-microbiological and anti-inflammatory properties are appearing in skin, nail and hair care products. Sibu Beauty offers a range of sixteen products based on the ingredient including facial cleaners, creams, serums and dietary supplements.

In Weleda cosmetics, SBT is offered in hand cream, body wash, body lotion and body oil and Fresh has a Seaberry restorative body cream. Aubrey organics also presents SBT and cucumber with Ester-C® in moisturizing cream, facial cleansing cream, moisturizing mask, nourishing hand and body lotion, facial toner, and AO serum. Seabuckwonders utilizes specifically USDA Certified Organic SBT oil from Tibetan region in their range of products from oils, creams, lotions to serums. Lavera cosmetics specialized in range of organic and gluten-free SBT products, such as shower gel, hand cream, neutral face cream, body scrub, body lotion, shampoo, lip balm, and roll on deodorant. Pai skin care offers fragonia and SBT oils based instant hand therapy cream, and SBT nail and cuticle balm. For children, Balmers Monkey balm with organic SBT oil helps to relieve dry skin and eczema and Kadria Skincare offers baby balm with cocoa butter and SBT. SBT Sea Berry natural health and beauty carries a hair care line available as a treatment oil, hair cream and hair mask. In Canada, Mont Echo Naturels specializes in 100 per cent Canadian grown SBT and formulate seven personal care products, such as creme cleanser, soothing mist and lip treatment based on their SBT seed and fruit oils. They also have a retail shop, Mont Echo Interpretive Centre and Boutique, which offers handpicked SBT tea and SBT sorbet.

Indian companies, such as Tvam Naturals and Fabindia have marketed various products based on SBT extracts. Tvam Naturals has launched aloe vera hydro gel with SBT extract which is a moisturizer gel for all types of skin and can also be used as a shaving gel and for hair styling. Fabindia, on the other hand, offers SBT seed oil and kernel based soap and lotion.

Eventhough, SBT oils have been proven to have mild sunscreen effect, very few products are available, such as Cosmetic Plant's Water-resistant sunscreen emulsion with SBT oil BIO_2 YOU's, Organic SBT light cream (with natural sunscreens) and Badger's natural sunscreen convenient lip balm stick.

Seabuckthorn International Inc., a Canadian company has also launched SBT leaf extracts based cosmetics which are mainly named SBT tea, such as creams and scrub bars for feet, elbow and face.

CONCLUSION AND FUTURE PROSPECTS

Besides health food processing industries all over the world, currently SBT has acquired great importance and recognition as one of the most sought after plant in the pharmaceutical and cosmetic based industries. The amount of scientific and pharmacological experimental data confirming the important biological activities and phytoconstituents from SBT is enormous and continues to increase rapidly. The valuable bioactives and nutritionally important constituents in SBT contribute to the scientic knowledge of importance and it is being clarified that SBT should be used as alternative nutritional and cosmeceutical source in the commercial market. The positive topical effects of SBT from its different extracts (leaves, fruits, bark, branches and seeds) on inammation, wound healing, skin aging, skin rejuvenation, acne, burn and scalds treatment, eczema, irradiation dermtitis are obvious. Due to the strong AO activities, SBT and its various formulations ensure the human and animal body's balance through the action of its various pharmacological effective compounds. Substantiation of these applications originated in traditional knowledge mainly from Russia and China and recent scientic investigations. However, there are a limited number of studies describing the bioactivities of SBT in relation with their phytochemical compositions. Additionally, most of these investigations have been carried out in animal models or in *in vitro* cell lines with clinical observations being limited. Therefore, an in depth investigation on SBT bioactives in humans with relation to dermaceutical and cosmeceutical applications need to be conducted. Since, the plants and plant products are subjected to broad variation in their phytochemical proles due to variety, geoclimatic conditions, maturity, post-harvest processing, storage and stability, it is particularly also important to conduct detailed studies on the composition and physiological signicant SBT extracts and their standardization with respect to the formulations. Further, methodical studies are essential in order to guide a pathway for estimating the efcacy of the bioactive molecules accountable for the biological activities and understand the effecting mechanisms, so that cost effective, potential herbal formulations can be manufactured commercially. There is no doubt that the future holds great promise for SBT bio-actives and the plant and thus, has much to contribute to this planet and its inhabitants.

REFERENCES

1. Agrawala, P.K. and Adhikari, J.S. 2009. Modulation of radiation-induced cytotoxicity in U 87 cells by RH-3 (a preparation of *Hippophae rhamnoides*). *Indian Journal of Medical Research* 130: 542-549.
2. Ako, H.O. and Gray, D. 1995. Healthful new oil from macadamia nuts. *Nutrition* 11: 286-288.
3. Ali, R., Ali, R., Kumar, N.D., Abhinav, J., Tyagi, A., Mittal, G., Sultana, S. and Bhatnagar, A. 2011. Safety and efficacy study of intramuscularly administered seabuckthorn (*Hippophae rhamnoides* L.) oil as depot formulation. *International Journal of Drug Development and Research* 3 (3): 356-365.
4. Aliyeva, E., Umur, S., Zafer, E. and Acigoz, G. 2004. The effect of polylactide membranes on the levels of reactive oxygen species in periodontal flaps during wound healing. *Biomaterials* 25: 4633-4637.
5. Andersson, S.C., Olsson, M.E., Johansson, E. and Rumpunen, K. 2009. Carotenoids in seabuckthorn (*Hippophae rhamnoides* L.) berries during ripening and use of pheophytin as a maturity marker. *Journal of Agricultural and Food Chemistry* 57: 250-258.
6. Ando, H., Ryu, A., Hashimoto, A., Oka, M. and Ichihashi, M. 1998. Linoleic acid linolenic acid lightens ultraviolet-induced hyperpigmentation of the skin. *Archives of Dermatological Research* 290: 375-338.
7. Anthony, J., O'Lenick, J., Siltech, L.L.C. and Smith, D. 2003. Surface-active phospholipids for personal care. *Cosmetics and Toiletries*: 43-47.
8. Arct, J. and Pytkowska, K. 2008. Flavonoids as components of biologically active cosmeceuticals. *Clinical Dermatology* 26 (4): 347-357.
9. Arimboor, R., Venugopalan, W., Arumughan, C. and Sawhney, R. 2006. Integrated processing of fresh Indian seabuckthorn (*Hippophae rhamnoides*) berries and chemical evaluation of products. *Journal of Science and Food Agriculture* 86: 2345-2353.
10. Arora, B.S., Nagpal, A., Singh, B. and Ahuja, P. S. 2007. Evaluation of *in vitro* antimutagenic activity of seabuckthorn (*Hippophae rhamnoides* Linn.) in Ames assay. *Journal of Chinese Clinical Medicine* 2 (8): 1-10.
11. Aruoma, O.I. 2002. Methodological considerations for characterizing potential antioxidant actions of bioactive components in plant foods. *Mutation Research* 523/524: 9-20.
12. Aylesworth, R. and Vance, J.C. 1982. Demodex folliculorum and demodex brevis in cutaneous biopsies. *Journal of the American Academy of Dermatology* 7 (5): 583-589.
13. Baboir, B. 1978. Oxygen dependent microbial killing by phagocytes. *New England Journal of Medicine* 298: 629-668.

14. Bal, L.M., Meda, V. S., Naik, N. and Satya, S. 2011. Seabuckthorn berries: A potential source of valuable nutrients for nutraceuticals and cosmoceuticals. *Food Research International* 44: 1718-1727.

15. Barnaulov, O.D. and Belenovskaya, L.M. 1988. In: *Medicinal Plants of USSR* (L.I. Medvedyeva, Eds.), Nauka, Leningrad, 356p.

16. Bekker, N.P. and Glushenkova, A.I. 2001. Components of certain species of the *Elaeagnaceae* family. *Chemistry of Natural Compounds* 37: 97-116.

17. Bernath, J. and Foldesi, D. 1992. Seabuckthorn (*Hippophae rhamnoides* L.): A promising new medicinal and food crop. *Journal of Herbs Spices Medicinal Plants* 1: 27-35.

18. Beveridge, T., Li, T.S.C. and Oomah, B.D. 1999. Seabuckthorn products: Manufacture and composition. *Journal of Agricultural and Food Chemistry* 47: 3480-3488.

19. Bishara, S., Atiyeh, M.C., Hayek, S.N. and Dibo, S.A. 2007. Effect of silver on burn wound infection control and healing. *Burns* 33: 39-148.

20. Biswas, A., Bharti, V.K., Acharya, S., Pawar, D.D. and Singh, S.B. 2010. Seabuckthorn: New feed opportunity for poultry in cold arid Ladakh region of India. *World's Poultry Science Journal* 66 (4): 707-714.

21. Boelsma, E., Hendriks, H.F. and Roza, L. 2011. Nutritional skin care: Health effects of micronutrients and fatty acids. *American Journal of Clinical Nutrition* 73: 853-864.

22. Boyce, S.T, Supp, A.P, Swope, V.B. and Warden, G.D. 2002. Vitamin C regulates keratinocyte viability, epidermal barrier, and basement membrane *in vitro*, and reduces wound contraction after grafting of cultured skin substitutes. *Journal of Investigative Dermatology* 118: 565-572.

23. Brown, J.E., Khodr, H., Hider, H. and Rice-Evans, C.A. 1998. Structural dependence of flavonoid interactions with Cu^{2+} ions: Implications for their antioxidant properties *Journal of Biochemistry* 330: 1173-1178.

24. Byer, R.E. 1994. The role of ascorbate in antioxidant protection of biomembranes: Interaction with vitamin E and coenzyme Q. *Archieves of Bioengineering Biomembrane* 26: 349-358.

25. Cabelli, D.E. and Bielski, B.H.J. 1983. Kinetics and mechanism for the oxidation of ascorbic acid/ascorbate by H_2O_2 radicals: A pulse radiolysis and stopped flow photolysis study. *Journal of Physical Chemistry* 87: 1805-1812.

26. Cenkowski, S., Yakimishen, R., Przybylski, R. and Muir, W.E. 2006. Quality of extracted seabuckthorn (*Hippophae rhamnoides* L.) seed and pulp oil. *Canadian Biosystems Engineering* 48: 3.9-3.16.

27. Chaby, G., Viseux, V., Poulain, J.F., De, C.B., Denoeux, J.P. and Lok, C. 2005.Topical silver sulfadiazine-induced acute renal failure. *Annales Dermatology Venereol.* 132: 891-893.

28. Chan, A.C. 1993. Partners in defense, vitamin E, and vitamin C. *Canadian Journal of Physiology and Pharmacology* 71: 725-731.

29. Chauhan, A.S. Negi, P.S. and Ramteke, R.S. 2007. Antioxidant and anti-bacterial activities of aqueous extract of seabuckthorn (*Hippophae rhamnoides*) seeds. *Fitoterapia* 78: 590-592.

30. Chawla, R., Arora, R., Singh, S., Sagar, R.K., Sharma, R.K., Kumar, R., Sharma, A., Gupta, M.L., Singh, S., Prasad, J., Khan, H.A., Swaroop, A., Sinha, A.K., Gupta, A.K., Tripathi, R.P. and Ahuja,

P.S. 2007. Radioprotective and antioxidant activity of fractionated extracts of berries of *Hippophae rhamnoides*. *Journal of Medicinal Food* 10: 101-109.

31. Che, X.P., Guo, F., Guan, X.H. and Tang, C.H. 1999. A preliminary report on the tox icity of topically used seabuckthorn seed oil on animals. *Hippophae* 12 (3): 36-39.

32. Chen, Y., Jiang, Z., Qin, W., Ni, M., Li, X. and He, Y. 1990. Chemical composition and characteristics of seabuckthorn fruit and its oil. *Chem. Ind. Forest Prod.* 10 (3): 163-175.

33. Cho, J.W., Cho, S.Y., Lee, S.R. and Lee, K.S. 2010. Onion extract and quercetin induce matrix metalloproteinase-1 *in vitro* and *in vivo*. *International Journal of Molecular Medicine* 25 (3): 347-352.

34. Chou, P.T. and Khan, A.U. 1983. L-ascorbic acid quenching of singlet delta molecular oxygen in aqueous media: Generalized antioxidant property of vitamin C. *Biochemical and Biophysical Research Communications* 115: 932-937.

35. Church, D., Elsayed, S., Reid, O., Winston, B. and Lindsay, R. 2006. Burn wound infections. *Clinical Microbiology Review* 19 (2): 403-434.

36. Cosgrove, M.C., Franco, O.H., Granger, S.P., Murray, P.G. and Mayes, A.E. 2007. Dietary nutrient intakes and skin-aging appearance among middle-aged American women. *American Journal of Clinical Nutrition* 86: 1225-1231.

37. Cossins, E., Lee, R. and Packer, L. 1998. ESR studies of vitamin C regeneration, order of reactivity of natural source phytochemical preparations. *Biochemistry and Molecular Biology International* 45: 583-597.

38. Crawford, G.H., Pelle, M.T. and James, W.D. 2004. Rosacea: I. Etiology, pathogenesis, and subtype classification. *Journal of the American Academy of Dermatology* 51 (3): 327-341.

39. Darr, D., Dunston, S., Faust, H. and Pinnell, S. 1996. Effectiveness of antioxidants (vitamin C and E) with and without sunscreens as topical photoprotectants. *Acta Dermato-Venereologica* 76: 264-268.

40. Darvin, M.E., Sterry, W., Lademann, J. and Vergou, T. 2011. The role of carotenoids in human skin. *Molecules* (16): 10491-10506.

41. Declair, V. 1997. The usefulness of topical application of essential fatty acids (EFA) to prevent pressure ulcers. *Ostomy Wound Manage.* 43 (5): 48-52.

42. Delabays, N. and Slacanin, I. 1995. Domestication and selection of new plant species of interest to the cosmetics industry. *Horticulture* 27: 143-147.

43. Dharmananda, S. 2004. *Seabuckthorn*, Institute of Traditional Medicine, Portland, Oregon. *http://www.itmonline.org/arts/seabuckthorn.htm*

44. Doering, T., Holtkotter, O., Schlotmann, K., Jassoy, C., Petersohn, D., Wadle, A. and Waldmann-Laue, M. 2005. Cutaneous restructuration by apple seed phytosterols: From DNA chip analysis to morphological alterations. *International Journal of Cosmetic Science* 27 (2): 142-142.

45. Duarte, T.L., Cooke, M.S. and Jones, G.D. 2009. Gene expression profiling reveals new protective roles for vitamin C in human skin cells. *Free Radical Biology and Medicine* 46: 78-87.

46. Dweck, C. 2006. Isoflavones, phytohormones and phytosterols. *Anthony Journal of Applied Cosmetology* 24: 17-33.

47. Elston, D.M. 2010. Demodex mites: Facts and controversies. *Clinics in Dermatology* 28 (5): 502-504.

48. Fakhry, S.M., Alexander, J. Smith, D. Meyer, A.A. and Petters On, H.D. 1995. Regional and institutional variation in burn care. *Journal of Burn Care and Rehabilitation* 16: 86-90.

49. Fan, Y L. and Xu, M. 1989. A preliminary research on the protective and curative effects of seabuckthorn oil against deafness. In: *Proceedings of International Symposium on Seabuckthorn (Hippophae rhamnoides L.)*, p. 300-302, Xian, China.

50. Fan, Y.L., Zhang, X.T. and Li, B.S. 1991. Clinical observation of the effects of seabuckthorn oil on the healing process of perforation of tympanic membrane. *Hippophae* 4 (2): 298-302.

51. Farris, P.K. 2005. Topical vitamin C: A useful agent for treating photoaging and other dermatologic conditions. *Dermatological Surgery* 31: 814-818.

52. Fayman, B.A. 1991. Treatment of operative wounds in ear, nose, throat with seabuckthorn oil. *Seabuckthorn* 4 (4): 7.

53. Fisher, G.J., Kang, S. and Varani, J. 2002. Mechanisms of photoaging and chronological skin aging. *Archives of Dermatology* 138 (11): 1462-1470.

54. Fitzpatrick, R.E. and Rostan, E.F. 2002. Double-blind, half-face study comparing topical vitamin C and vehicle for rejuvenation of photodamage. *Dermatological Surgery* 28: 231-236.

55. Forton, F., Germaux, M.A., Brasseur, T., De Liever, A., Laporte, M., Mathys, C., Sass, U., Stene, J.J., Thibaut, S., Tytgat, M. and Seys, B. 2005. Demodicosis and rosacea: Epidemiology and significance in daily dermatologic practice. *Journal of the American Academy of Dermatology* 52 (1): 74-87.

56. Ganju, L., Padwad, Y., Singh, R., Karan, D., Chanda, S., Chopra, M.K., Bhatnagar, P., Kashyap, R. and Sawhney, R.C. 2005. Anti-inflammatory activity of seabuckthorn (*Hippophae rhamnoides*) leaves. *International Immunopharmacology* 5 (12): 1675-1684.

57. Gao, X., Ohlande, M., Jeppsson, N., Bjork, L. and Trajkovski, V. 2000. Changes in antioxidant effects and their relationship to phytonutrients in fruits of seabuckthorn (*Hippophae rhamnoides* L.) during maturation. *Journal of Agricultural and Food Chemistry* 48: 1485-1490.

58. Garcia-Mediavilla, V., Crespo, I., Collado, P.S., Esteller, A., Sanchez-Campos, S., Tunon, M.J. and Gonzalez-Gallego, J. 2007. The anti-inflammatory flavones quercetin and kaempferol cause inhibition of inducible nitric oxide synthase, cyclooxygenase-2 and reactive C-protein, and down-regulation of the nuclear factor kappa B pathway in chang liver cells. *European Journal of Pharmacology* 557: 221-229.

59. Geesin, J.C., Darr, D., Kaufman, R., Murad, S. and Pinnell, S.R. 1988. Ascorbic acid specifically increases type I and type III procollagen messenger RNA levels in human skin fibroblast. *Journal of Investigative Dermatology* 90: 420-424.

60. George, S.T. and Cenkowski, S. 2007. Influence of harvest time on the quality of oil-based compounds in seabuckthorn (*Hippophae rhamnoides* L. ssp. *sinensis*) seed and fruit. *Journal of Agricultural and Food Chemistry* 55: 8054-8061.

61. Gercikovs, L. and Zoludeva, I. 1998. Cosmetic agents for the lips. *Patent in Latvia (LV12065).*

62. Gil, E.M. and Kim, T.H. 2000. UV-induced immune suppression and sunscreen. *Photodermatology, Photoimmunology and Photomedicine* 16: 101-110.

63. Glynn, L. 1981. The pathology of scar tissue formation. In: *Handbook of Inflammation, Tissue Repair and Regeneration* (L. Glynn, Ed.), p. 285-307, Elsevier/North Holland Biomedical Press, Amsterdam, 620p.

64. Goel, H.C. Gupta, D., Gupta, S., Garg, A.P. and Bala, M. 2005. Protection of mitochondria system by *Hippophae rhamnoides* L. against radiation-induced oxidative damage in mice. *Journal of Pharmacy and Pharmacology* 57: 135-143.

65. Goel, H.C., Kumar, I.P., Samanta, N. and Rana, S.V.S. 2003. Induction of DNA-protein cross-links by *Hippophae rhamnoides*: Implications in radioprotections and cytotoxicity. *Molecular and Cellular Biochemistry* 245 (1-2): 57-67.

66. Goel, H.C., Prasad, J., Singh, S., Sagar, R.K., Kumar, I.P. and Sinha, A.K. 2002 Radioprotection by a herbal preparation of *Hippophae rhamnoides*, RH-3, against whole body lethal irradiation in mice. *Phytomedicine* 9: 15-25.

67. Goncharova, N. P. and Glushenkova, A. I. 1995. Polar lipids of *Hippophae rhamnoides* leaves. *Chemistry of Natural Compounds* 31: 562-564.

68. Gordon, M.H. 1990. The mechanism of antioxidant action *in vitro*. In: *Food Antioxidants* (B.J.F. Hudson, Ed.), p. 1-18, Elsevier Applied Science, London, 317p.

69. Greenway, H.T. and Pratt, S.G. 2000. Vitamins and micronutrients in aging and photoaging skin. In: *Vegetables, Fruits, and Herbs in Health Promotion* (R.R. Watson, Ed.) p. 109-116, CRC Press, Boca Raton, FL, 339p.

70. Grether-Beck, S., Muhlberg, K., Brenden, H. and Krutmann, J. 2008. Topical application of vitamins, phytosterols and ceramides. Protection against increased expression of interstitial collagenase and reduced collagen-I expression after single exposure to UVA irradiation. *Hautarzt* 59 (7): 557-562.

71. Griendling, K. 2000. NAD(P)H Oxidase: Role in cardiovascular biology and diseases. *Circulation Research* 86: 494-501.

72. Guan, T.T.Y., Cenkowski, S. and Hydamaka, A. 2005. Effect of drying on the nutraceutical quality of seabuckthorn (*Hippophae rhamnoides* L. ssp. *sinensis*) leaves. *Journal of Food Science* 70: E514-E518.

73. Guliyev, V., Gulb, M. and Yildirim, A. 2004. *Hippophae rhamnoides* L.: Chromatographic methods to determine chemical composition, use in traditional medicine and pharmacological effects. *Journal of Chromatography B* 812: 291-307.

74. Gupta, A., Kumar, R., Pal, K., Banerjee, P.K. and Sawhney, R.C. 2005. A preclinical study of the effects of seabuckthorn (*Hippophae rhamnoides* L.) leaf extract on cutaneous wound healing in albino rats. *International Journal of Lower Extremity Wounds* 4: 88-92.

75. Gupta, A., Kumar, R., Pal, K., Singh, V., Banerjee, P.K. and Sawhney, R.C. 2006. Influence of seabuckthorn (*Hippophae rhamnoides* L.) flavone on dermal wound healing in rats. *Molecular and Cellular Biochemistry* 290: 193-198.

76. Gupta, A., Upadhyay, N.K., Sawhney, R.C. and Kumar, R. 2008. A poly-herbal formulation accelerates normal and impaired diabetic wound healing. *Wound Repair and Regeneration* 16: 784-790.

77. Gupta, R. and Flora, S.J. 2006. Protective effects of fruit extracts of *Hippophae rhamnoides* L. against arsenic toxicity in Swiss albino mice. *Human Experimental Toxicology* 25: 285-295.

78. Haag, S.F., Taskoparan, B., Darvin, M.E., Groth, N., Lademann, J., Sterry, W. and Meinke, M.C. 2011. Determination of the anti-oxidative capacity of the skin *in vivo* using resonance Raman and electron paramagnetic resonance spectroscopy. *Experimental Dermatology* 20: 483-487.

79. Hamalainen, M., Nieminen, R., Vuorela, P., Heinonen, M. and Moilanen, E. 2007. Anti-inflammatory effects of flavonoids: Genistein, kaempferol, quercetin, and daidzein inhibit STAT-1 and NF-κB activations, whereas flavone, isorhamnetin, naringenin, and pelargonidin inhibit only NF-κB activation along with their inhibitory effect on iNOS expression and NO production in activated macrophages. *Mediators of Inflammation*: 45673-45683.

80. Hansen, H.S. and Jensen, B. 1985. Essential function of linoleic acid esterified in acylglucosylceramide and acylceramide in maintaining the epidermal water permeability barrier. Evidence from feeding studies with oleate, linoleate, arachidonate, columbinate and alpha-linolenate. *Biochimica et Biophysica Acta* 834: 357-363.

81. He, X., Fang, Z., Liang, Z. and Jiang, L. 1989. Devlopment and utilization of seabuckthorn fruit cosmetic. In: *Proceedings First International Symposium on Seabuckthorn*, p. 320-321, October 19-23, 1989, Xian, China.

82. Horrobin, D.F. 2000. Essential fatty acid metabolism and its modification in atopic eczema. *American Journal of Clinical Nutrition* 71 (1): 367S-372S.

83. Huang, Z.R., Lin, Y.K. and Fang, J.Y. 2009. Biological and pharmacological activities of squalene and related compounds: Potential uses in cosmetic dermatology. *Molecules* 14 (1): 540-554.

84. Humbert, P.G., Haftek, M. and Creidi, P. 2003. Topical ascorbic acid on photoaged skin. Clinical, topographical and ultrastructural evaluation: Double-blind study *vs.* placebo. *Experimental Dermatology* 12: 237-244.

85. Jackson, R.L., Greiwe, J.S. and Schwen, R.J. 2011. Ageing skin: Oestrogen receptor beta againsts offer an approach to change the outcome. *Experimental Dermatology* 20 (11): 879-882.

86. Jalalpure, S., Agrawal, N., Patil, M., Chimkode, R. and Tripathi, A. 2008. Antimicrobial and wound healing activities of leaves of *Alternanthera sessilis* Linn. *International Journal of Green Pharmacy* 2: 141-144.

87. Jump, D.B. 2004. Fatty acid regulation of gene transcription. *Critical Reviews in Clinical Laboratory Sciences* 41: 41-78.

88. Kalia, A.N. 2005. *Text Book of Industrial Pharmacognosy*, CBS Publishers and Distributors, 285p.

89. Kallio, H., Yang, B., Peippo, P., Tahvonen, R. and Pan, R. 2002. Triacylglycerols, glycerophospholipids, tocopherols, and tocotrienols in berries and seeds of two subspecies (ssp. *sinensis* and *mongolica*) of seabuckthorn (*Hippophae rhamnoides*). *Journal of Agricultural and Food Chemistry* 50: 3004-3009.

90. Kallio, H., Yang, B., Wang, B.W., Wang, H.Y., Wang, J.X. and Song, J.E. 2001. Animal experiments on the anti-inflammatory and analgesic effects of seabuckthorn (*Hippophae rhamnoides* L.) oils. In: *Biologically Active Phytochemicals in Food: Analysis, Metabolism, Bioavailability and Function*. Proceedings of the Eurofoodchem XI Meeting, p. 69-73, Norwich, UK.

91. Kameyama, K., Sakai, C. and Kondoh, S. 1996. Inhibitory effect of magnesium L-ascorbyl-2-phosphate (VC-PMG) on melanogenesis *in vitro* and *in vivo*. *Journal of the American Academy of Dermatology* 34: 29-33.

92. Kang, J.S., Kim, H.N. and Jung, J. 2007. Regulation of UVB-induced IL-8 and MCP-1 production in skin keratinocytes by increasing vitamin C uptake *via* the redistribution of SVCT-1 from the cytosol to the membrane. *Journal of Investigative Dermatology* 127: 698-706.

93. Khan, B., Akhtar, A., Mahmood, N., Haji, T.M., Khan, S., Zaman, S.U., Rasul, A., Iqbal, M., Ali, A., Khan, S. and Qayum, M. 2011. *In-vivo* study of stratum corneum water content and transepideramal water loss using a newly formulated topical cream of *Hippophae rhamnoides* fruit extract. *African Journal of Pharmacy and Pharmacology* 5 (8): 1092-1095.

94. Khan, B.A., Akhtar, N., Braga and Valdir, A. 2012. Anti-Aging effects of *Hippophae rhamnoides* emulsion on human skin. *Tropical Journal of Pharmaceutical Research* 11 (6): 955-962.

95. Kim, H., Cho, H., Seo, Y.K., Kim, S., Yoon, M.Y., Kang, H., Park, C.S. and Park, J.K. 2012. Inhibitory effects of seabuckthorn (*Hippophae rhamnoides* L.) seed on UVB-induced Photoaging in human dermal fibroblasts. *Biotechnology and Bioprocess Engineering* 17 (3): 465-474.

96. Kim, J.E., Lee, D.E., Lee, K.W., Son, J.E., Seo, S.K., Li, J., Jung, S.K., Heo, Y.S., Mottamal, M., Bode, A.M., Dong, Z. and Lee, H.J. 2011. Isorhamnetin suppresses skin cancer through direct inhibition of MEK1 and PI3-K. *Cancer Prev. Res. (Phila)* 4 (4): 582-591.

97. Kim, M., Otsuka, M., Yu, R., Kurata, T. and Arakawa, N. 1994. The distribution of ascorbic acid and dehydroascorbic acid during tissue regeneration in wounded dorsal skin of guinea pigs. *International Journal for Vitamin and Nutrition Research* 64: 56-59.

98. Klock, J., Ikeno, H., Ohmori, K., Nishikawa, T., Vollhardt, J. and Schehlmann, V. 2005. Sodium ascorbyl phosphate shows *in vitro* and *in vivo* efficacy in the prevention and treatment of acne vulgaris. *International Journal of Cosmetic Science* 27: 171-176.

99. Krinsky, N.I. and Yeum, K.J. 2003. Carotenoid-radical interactions. *Biochemical and Biophysical Research Communications* (305): 754-760.

100. Kullavanijaya, P. Henry, W. and Lim, H.W. 2005. Photoprotection. *Journal of the American Academy of Dermatology* 52: 959-961.

101. Kumar, A., Varshney, A.C., Tyagi S.P. and Sharma D.N. 2003. Treatment of cutaneous burns with seabuckthorn oil in bovines: A histological study. In: *27th Annual Congress of Indian Society for Veterinary Surgery and National Symposium, Pantnagar*, p. 20-22, Uttaranchal, India.

102. Kumar, R., Kumar, G.P., Chaurasia, O.P. and Singh, S.B. 2011. Phytochemical and pharmacological profile of Seabuckthorn oil: A review. *Journal of Medicinal Plants Research* 5: 491-499.

103. Kumar, V. 2003. Seabuckthorn: A potential bioresource in Himalayas. *Invention Intelligence*: 159-167.

104. Kwochka, K.W. 1993. The structure and function of epidermal lipids. *Veterinary Dermatology* 4 (4): 151-159.

105. Lademann, J., Meinke, M.C., Sterry, W. and Darvin, M.E. 2011. Carotenoids in human skin. *Experimental Dermatology* (20): 377-382.

106. Lal, B., Ahuja, P.S. and Gupta, A.K. 2001. Application of seabuckthorn in Amchi System of Medicine. In: *Proceedings of International Workshop on Seabuckthorn*, p. 239-242, New Delhi, 274.

107. Larmo, P. 2011. *The Health Effects of Seabuckthorn Berries and Oil*, PhD Dissertation, Turku University, Finland.

108. Lebedeva, L., Akmolova, N., Haydarov, K. and Ismailova M. 1992. Comparision of the effects of seabuckthorn oil, β-sitosterol oleate, β-sitosterol palmitate on the healing process of burns. In: *Advances in the Biochemical and Pharmacological Research on Seabuckthorn*, p.147-148, Shaanxi Province, Wugong, Shannxi Province, Wugong Publishing House, China.

109. Leskinen, H.M., Suomela, J.P., Yang, B. and Kallio, H.P. 2010. Regioisomer compositions of vaccenic and oleic acid containing triacylglycerols in seabuckthorn (*Hippophae rhamnoides*) pulp oils: Influence of origin and weather conditions. *Journal of Agricultural and Food Chemistry* 58: 537-545.

110. Li, F. and Guo, T. 1989. Application of *Hippophae rhamnoides* L. traditional Tibetan medicine. In: *Proceedings of the International Symposium on Seabuckthorn*, p. 1-3, Xian, China.

111. Li, T.S.C. 2003. Taxonomy, natural distribution and botany. In: *Seabuckthorn (Hippophae rhamnoides L.): Production and Utilization* (T.S.C. Li and T. Beveridge, Eds.), p. 7-11, NRC Research Press, Ottawa, 133p.

112. Li, T.S.C. and Schroeder, W.R. 1996. Seabuckthorn (*Hippophae rhamnoides* L.): A multipurpose plant. *Horticulture Technologies* 6: 370-380.

113. Li, Y. and Liu, H. 1991. Prevention of tumour production in rats fed aminopyrine plus nitrite by seabuckthorn juice. *IARC Scientic Publications* 105: 568-570.

114. Lin, F.H., Lin, J.Y. and Gupta, R.D. 2005. Ferulic acid stabilizes a solution of vitamins C and E and doubles its photoprotection of skin. *Journal of Investigative Dermatology* 125: 826-832.

115. Lin, J.Y., Selim, M.A. and Shea, C.R. 2003. UV photoprotection by combination topical antioxidants vitamin C and vitamin E. *Journal of the American Academy of Dermatology* 48: 866-874.

116. Madey, D.L. and Pinnell, S.R. 2001. Using topical antioxidants to protect against photodamage and improve skin care. *Operative Techniques in Oculoplastic, Orbital and Reconstructive Surgery* 4 (2): 62-64.

117. Mahajan, A., Sharma, S.K., Tyagi, S.P. and Varshney, A.C. 2002. Efficacy of seabuckthorn (*Hippophae rhamnoides*) in the healing of infected cutaneous wounds in calves. In: *XXVI Annual Congress of "Indian Society for Veterinary Surgery"*, Mumbai, India.

118. Majewska, I. and Gendaszewska-Darmach, E. 2011. Proangiogenic activity of plant extracts in accelerating wound healing: A new face of old phytomedicines. *Acta Biochimica Polonica* 58: 449-460.

119. Martin, A. 1996. The use of antioxidants in healing. *Dermatological Surgery* 22: 156-160.

120. Martin, P. 1997. Wound healing aiming for perfect skin degeneration. *Science* 276: 75-81.

121. Miller, C.C. and Ziboh, V.A. 1990. Induction of epidermal hyperproliferation by topical n-3 polyunsaturated fatty acids on guinea pig skin linked to decreased levels of 13-hydroxyoctadecadienoic acid (13-hode). *Journal of Investigative Dermatology* 94: 353-358.

122. Mironov, V.A., Guseva-Donskaya, T.N., Amirov, N.S.H. and Nikulin, A.A. 1989. New technology and pharmacology of seabuckthorn oil production. In: *Proceedings of International Symposium on Seabuckthorn (Hippophae rhamnoides L.)*, p. 348-349, Xian, China.

123. Mironov, V.A., Vasilev, G.S., Matrosov, V., Filippova, T.M., Zamurenko, V.A., Mishchenko, V.V., Maironovskii, V.G., Kas'yanenko, I.I. and Maksimova, L.M. 1983. Physiologically active alcohols from seabuckthorn fruit. *KhimFarm Zh.* 17 (10): 1242-1247.

124. Nichols, J.A. and Katiyar, S.K. 2010. Skin photoprotection by natural polyphenols: Anti-inflammatory, antioxidant and DNA repair mechanism. *Arch Dermatol Res.* 302 (2): 71-90.

125. Nusgens, B.V., Humbert, P. and Rougier, A. 2001. Topically applied vitamin C enhances the mRNA level of collagens I and III, their processing enzymes and tissue inhibitor of matrix metalloproteinase 1 in the human dermis. *Journal of Investigative Dermatology* 116: 853-859.

126. Oomah, B.D. 2003. Seabuckthorn lipids. In: *Seabuckthorn (Hippophae rhamnoides L.): Production and Utilization* (T.S.C. Li and T. Beveridge, Eds.), p. 51-68, NRC Research Press, Ottawa.

127. Padmavathi, B., Upreti, M., Singh, V., Rao, A.R., Singh, R.P. and Rath, P.C. 2005. Chemoprevention by *Hippophae rhamnoides*: Effects on tumorigenesis, phase II and antioxidant enzymes, and IRF-1 transcription factor. *Nutrition and Cancer* 51: 59-67.

128. Panin, G., Strumia, R. and Ursini, F. 2004. Topical alpha-tocopherol acetate in the bulk phase: Eight years of experience in skin treatment. *Annals of the New York Academy of Sciences*: 443-447.

129. Pasonen-Seppanen, S., Suhonen, T.M. and Kirjavainen, M. 2001. Vitamin C enhances differentiation of a continuous keratinocyte cell line (REK) into epidermis with normal stratum corneum ultrastructure and functional permeability barrier. *Histochemistry and Cell Biology* 116: 287-297.

130. Pedriella, P. and Skibsted, L.H. 2002. Antioxidant synergy and regeneration effect of quercetin, (-)-epicatechin, and (+)-catechin on alpha-tocopherol in homogeneous solutions of peroxidating methyl linoleate. *Journal of Agricultural and Food Chemistry* 50: 7138-7144.

131. Pelle, M.T., Crawford, G.H. and James, W.D. 2004. Rosacea: II.Therapy. *Journal of the American Academy of Dermatology* 51 (3): 499-512.

132. Pelletier, G. and Ren, L. 2004. Localization of sex steroid receptors in human skin. *Histology and Histopathology* 19 (2): 629-636.

133. Perricone, N.V. 2001. Treatment of skin damage using conjugated linoleic acid and ascorbyl fatty acid esters. *Official Gazette of the United States Patent and Trademark Office Patents 1251(1) US6296861* .

134. Peterkofsky, B. 1991. Ascorbate requirement for hydroxylation and secretion of procollagen: Relationship to inhibition of collagen synthesis in scurvy. *American Journal of Clinical Nutrition* 54: 1135S-1140S.

135. Phillips, C.L., Combs, S.B. and Pinnell, S.R. 1994. Effects of ascorbic acid on proliferation and collagen synthesis in relation to the donor age of human dermal fibroblasts. *Journal of Investigative Dermatology* 103: 228-232.

136. Pilkington, S.M., Watson, R.E., Nicolaou, A. and Rhodes, L.E. 2011. Omega-3 polyunsaturated fatty acids: Photoprotective macronutrients. *Experimental Dermatology* 20: 537-543.

137. Pinnell, S.R. 2003. Cutaneous photodamage, oxidative stress, and topical antioxidant protection. *Journal of the American Academy of Dermatology* 48: 1-19.

138. Pintea, A., Marpeau, A., Faye, M., Socaciu, C. and Gleizes, M. 2001. Polar lipid and fatty acid distribution in carotenolipoprotein complexes extracted from seabuckthorn fruits. *Phytochemical Analysis* 12: 293-298.

139. Podda, M., Traber, M.G., Weber, C., Yan, L.J. and Packer, L. 1998. UV-irradiation depletes antioxidants and causes oxidative damage in a model of human skin. *Free Radical Biology and Medicine* 24: 55-65.

140. Ponec, M., Weerheim, A. and Kempenaar, J. 1997. The formation of competent barrier lipids in reconstructed human epidermis requires the presence of vitamin C. *Journal of Investigative Dermatology* 109: 348-355.

141. Puglia, C. and Bonina, F. 2008. *In vivo* spectrophotometric evaluation of skin barrier recovery after topical application of soybean phytosterols. *Journal of Cosmetic Science* 59 (3): 217-24.

142. Puupponen-Pimia, R., Nohynek, L., Meier, C., Kahkonen, M., Heinonen, M., Hopia, A. and Oksman-Caldentey, K.M. 2001. Anti-microbial properties of phenolic compounds from berries. *Journal of Applied Microbiology* 90: 494-507.

143. Rana, R.K. 2006. Studies on the therapeutic potential of seabuckthorn (*Hippophae* sp.) seed and pulp oil in the healing of aseptic excisional cutaneous wounds in calves. *Indian Journal of Veterinary Surgery* 27: 1-7.

144. Raschke, T., Koop, U. and Dusing, H.J. 2004. Topical activity of ascorbic acid: From *in vitro* optimization to *in vivo* efficacy. *Skin Pharmacology ans Physiology* 17: 200-206.

145. Ren, L., Yang J., *et al.*, 1992. Observation: Anti-mutation and anti-immune suppresser properties from seabuckthorn seed oil. *Hippophae* 5.

146. Rhie, G., Shin, M.H. and Seo, J.Y. 2001. Aging and photoaging-dependent changes of enzymic and non-enzymic antioxidants in the epidermis and dermis of human skin *in vivo*. *Journal of Investigative Dermatology* 117: 1212-1217.

147. Rice-Evans, C.A., Miller, N.J. and Paganga, G. 1996. Structure-antioxidant activity relationships of flavonoids and phenolic acids. *Free Radical Biology Medicine* 20: 933-956.

148. Rosch, D., Bergmann, M., Knorr, D. and Kroh, L.W. 2003. Structure-antioxidant efficiency relationships of phenolic compounds and their contribution to the antioxidant activity of seabuckthorn juice. *Journal of Agricultural and Food Chemistry* 51: 4233-4239.

149. Rousi, A. 1971. The genus *Hippophae* L.: A taxonomic study. *Annals Botanica Fennic* 8: 177-227.

150. Sabynich, L.V., Sibileva, L.A., Belova, L.S. and Chuchalin, V.S. 1994. Anti-inflammatory properties of thick extract from fruit of pulp of *Hippophae rhamnoides* L. *Rastitel nye, Resursy* 30 (10): 70-74.

151. Saggu, S., Divekar, H.M., Gupta, V., Sawhney, R.C., Banerjee, P.K. and Kumar, R. 2007. Adaptogenic and safety evaluation of seabuckthorn (*Hippophae rhamnoides*) leaf extract: A dose dependent study. *Food and Chemical Toxicology* 45: 609-617.

152. Sauermann, K., Jaspers, S., Koop, U. and Wenck, H. 2004. Topically applied vitamin C increases the density of dermal papillae in aged human skin. *BMC Dermatology* 4: 13-19.

153. Savini, I., Catani, M.V., Rossi A., Duranti, G., Melino, G. and Avigliano, L. 2002. Characterization of keratinocyte differentiation induced by ascorbic acid: Protein kinase C involvement and vitamin C homeostasis. *Journal of Investigative Dermatology* 118: 372-379.

154. Scarpa, M., Stevenato, R. and Viglino, R.A. 1983. Superoxide ion as active intermediate in the autoxidation of ascorbate by molecular oxygen: Effect of superoxide dismutase. *Journal of Biological Chemistry* 258: 6695-6697.

155. Schiffer, R., Neis, M. and Holler, D. 2003. Active influx transport is mediated by members of the organic anion transporting polypeptide family in human epidermal keratinocytes. *Journal of Investigative Dermatology* 120 (2): 285-291.

156. Schroeder, W.R. and Yao, Y. 1995. *Seabuckthorn: A Promising Multipurpose Crop for Saskatchewan*, Supplementary report, p. 95-102, PFRA Shelterbelt Center.

157. Shindo, Y., Witt, E. and Packer, L. 1993. Antioxidant defense mechanisms in murine epidermis and dermis and their responses to ultraviolet light. *Journal of Investigative Dermatology* 100: 260-265.

158. Shindo, Y., Witt, E., Han, D., Epstein, W. and Packer, L. 1994. Enzymic and non-enzymic antioxidants in epidermis and dermis of human skin. *Journal of Investigative Dermatology* (102): 122-124.

159. Shuid, A.N., Anwar, M. and Yusof, A. 2005. The Effects of *Carica papaya* Linn. latex on the healing of burn wounds in rats. *Malaysian Journal of Medicine and Health Sciences* 3: 39-47.

160. Shukla, A., Rasik, A.M. and Patnaik, G. K. 1997. Depletion of reduced glutathione, ascorbic acid, vitamin E and antioxidant defence enzymes in a healing cutaneous wound. *Free Radical Research* 26: 93-101.

161. Shukla, S.K., Chaudhary, P. and Kumar, I.P. 2006. Protection from radiation-induced mitochondrial and genomic DNA damage by an extract of *Hippophae rhamnoides*. *Environmental and Molecular Mutagenesis* 47 (9): 647-656.

162. Sies, H. and Stahl, W. 2004. Nutritional protection against skin damage from sunlight. *Annual Review of Nutrition* 24: 173-200.

163. Singh, V. 2003. Geographical adaptation and distribution of seabuckthorn (*Hippophae* L.) resources. In: *Seabuckthorn (Hippophae L.)-A Multipurpose Wonder Plant. Vol. I. Botany, Harvesting and Processing Technologies* (V. Singh, Ed. in Chief, 2003), p. 21-34, Indus Publishing Company, New Delhi, 518p.

164. Singh, V. 1994. A report on arboreal seabuckthorn (*Hippophae salicifolia*) from dry temperate Himalayas. *Journal of Tree Sciences* 3 (1): 67-68.

165. Singh, V. 2005. Seabuckthorn (*Hippophae* L.) in traditional medicines. In: *Seabuckthorn-A Multipurpose Wonder Plant*. Vol. II. *Biochemistry and Pharamacology* (V.Singh, Editor in Chief), p.505-521, Daya Publishing House, New Delhi, 600 p.

166. Singh, V. and Dogra, K.K. 1996. Characteristics, distribution, biomass, regeneration and nutritional values of seabuckthorn. *Indian Forester* 122 (6): 486-491.

167. Singh, V. and Singh, R.K. 2004. Morphobiochemical variations in seabuckthorn (*Hippophae* L.) populations growing in Lahaul valley, dry temperate Himalayas. *Indian Forester* 130 (6): 663-672.

168. Raina, S.N., Jain, S., Sehgal, D., Kumar, A., Dar, T. H., Bhat, V., Pandey, V., Vaishnavi, S., Bhargav, A., Singh, V., Rani, V., Tandon, R., Tewari, M. and Mahmoudi, A. 2011. Diversity and relationships of multipurpose seabuckthorn (*Hippophae* L.) germplasm from the Indian Himalayas as assessed by AFLP and SAMPL markers. *Genet Resour Crop Evol.* 59 (6):1033-1053.

169. Sorensen, L.T., Toft, B.G. and Rygaard, J. 2010. Effect of smoking, smoking cessation, and nicotine patch on wound dimension, vitamin C, and systemic markers of collagen metabolism. *Surgery* 148: 982-990.

170. St. George, S.D. and Cenkowski, S. 2007. Inuence of harvest time on the quality of oil-based compounds in seabuckthorn (*Hippophae rhamnoides* L. ssp. *sinensis*) seed and fruit. *J. Agric. Food Chem.* 55: 8054-8061.

171. Stahl, W. and Sies, H. 2005. Bioactivity and protective effects of natural carotenoids. *Biochimica et Biophysica Acta* (1740): 101-107.

172. Steiling, H., Longet, K. and Moodycliffe, A. 2007. Sodium-dependent vitamin C transporter isoforms in skin: Distribution, kinetics, and effect of UVB-induced oxidative stress. *Free Radical Biology and Medicine* 43: 752-762.

173. Stewart, M.S., Cameron, G.S. and Pence, B.C. 1996. Antioxidant nutrients protect against UVB-induced oxidative damage to DNA of mouse keratinocytes in culture. *Journal of Investigative Dermatology* 106: 1086-1089.

174. Suryakumar, G. and Gupta, A. 2011. Medicinal and therapeutic potential of seabuckthorn (*Hippophae rhamnoides* L.). *Journal of Ethnopharmacology* 138: 268-278.

175. Tebbe, B., Wu, S., Geilen, C.C., Eberle, J., Kodelja, V. and Orfanos, C.E. 1997. L-ascorbic acid inhibits UVA-induced lipid peroxidation and secretion of IL-1alpha and IL-6 in cultured human keratinocytes *in vitro*. *Journal of Investigative Dermatology* 108: 302-306.

176. Terao, J., Minami, Y. and Bando, N. 2011. Singlet molecular oxygen-quenching activity of carotenoids: Relevance to protection of the skin from photoaging. *Journal of Clinical Biochemistry and Nutrition* (48): 57-62.

177. Thiele, J.J., Hsieh, S.N. and Ekanayake-Mudiyanselage, S. 2005. Vitamin E: Critical review of its current use in cosmetic and clinical dermatology. *Dermatologic Surgery* 31(S1): 805-813.

178. Thiele, J.J., Traber, M.G., Tsang. K., Cross, C.E. and Packer, L. 1997. *In vivo* exposure to ozone depletes vitamins C and E and induces lipid peroxidation in epidermal layers of murine skin. *Free Radical Biology and Medicine* 23: 385-391.

179. Thiele, J.J. and Ekanayake-Mudiyanselage, S. 2007. Vitamin E in human skin: Organ-specific physiology and considerations for its use in dermatology. *Molecular Aspects of Medicine* 28 (5-6): 646-666.

180. Thiem, B. and Goslinska, O. 2004. Antimicrobial Activity of *Rubus chamaemorus* Leaves. *Fitoterapia* 75: 93-95.

181. Traber, M. 2006. Vitamin E. In: *Modern Nutrition in Health and Disease*, 10th Edn. (M.E. Shils, M. Shike, C.A. Ross, B. Cabarell and R. Cousins, Eds.), p. 396-411, Lippincott Williams and Wilkins, Philadelphia.

182. Traikovich, S.S. 1999. Use of topical ascorbic acid and its effects on photodamaged skin topography. *Archives of Otolaryngology-Head and Neck Surgery* 125: 1091-1098.

183. Trommer, H., Wagner, J., Graener, H. and Neubert, R.H. 2001. The examination of skin lipid model systems stressed by ultraviolet irradiation in the presence of transition metal ions. *European Journal of Pharmaceutics and Biopharmaceutics* 51: 207-214.

184. Tsybikova, D.T., Rasputina-Sultumova, D.B., Komissarenko, N.F. and Bolotova, M.N. 1983. In: *Utilization Problems of Medicinal Resources in Siberia and Far East*, Novosibirsk, Russia.

185. Upadhyay, N.K., Kumar R., Mandotra, S.K., Meena, R.N., Siddiqui, M.S., Sawhney, R. C. and Gupta, A. 2009. Safety and healing efficacy of seabuckthorn (*Hippophae rhamnoides* L.) seed oil on burn wounds in rats. *Food and Chemical Toxicology* 47: 1146-1153.

186. Upadhyay, N. K., Kumar, M.S.Y. and Gupta, A. 2010. Antioxidant, cytoprotective and anti-bacterial effects of seabuckthorn (*Hippophae rhamnoides* L.) leaves. *Food and Chemical Toxicology* 48: 3443-3448.

187. Upadhyay, N.K., Kumar, R., Siddiqui, M.S. and Gupta, A. 2011. Mechanism of wound healing activity of *Hippophae rhamnoides* L. leaf extract in experimental burns. *Evidence-Based Complementary and Alternative Medicine*: 659-705.

188. Velioglu, Y.S., Mazza, G., Gao, L. and Oomah, B.D. 1998. Antioxidant activity and total phenolics in selected fruits, vegetables and grain products. *Journal of Agricultural and Food Chemistry* 46: 4113-4117.

189. Vlasov, V.V. 1970. *Hippophae* oil in the treatment of superficial burns of the skin. *Vestn. Dermatol. Venerol.* 44 (6): 69-72.

190. Wang, B., Lin, L., Ni. Q. and Su, C. L. 2011. *Hippophae rhamnoides* Linn. For treatment of diabetes mellitus: A review. *Journal of Medicinal Plants Research* 5: 2599-2607.

191. Wefers, H. and Sies, H. 1988. The protection by ascorbate and glutathione against microsomal lipid peroxidation is dependent on vitamin E. *European Journal of Biochemistry* 174: 353-357.

192. Werninghaus, K. 1995. The role of antioxidants in reducing photodamage. In: *Photodamage. Vol.1* (B. Gilchrest, Ed.) p. 249-258, Blackwell Science, London.

193. Wille, J.J. and Kydonieus, A. 2003. Palmitoleic acid isomer (C16:1delta6) in human skin sebum is effective against gram-positive bacteria. *Skin Pharmacology and Applied Skin Physiology* 16 (3): 176-187.

194. Williams, R.J., Spencer, J.P. and Rice-Evans, C. 2004. Flavonoids: Antioxidants or signalling molecules? *Free Radical Biology Medicine* 36 (7): 838-849.

195. Woolery-Lloyd, H., Baumann, L. and Ikeno, H. 2010. Sodium L-ascorbyl-2-phosphate 5 per cent lotion for the treatment of acne vulgaris: A randomized, double-blind, controlled trial. *Journal of Cosmetic Dermatology* 9: 22-27.

196. Xing, J., Yang, B., Dong, Y., Wang, B., Wang, J. and Kallio, H. 2002. Effects of seabuckthorn (*Hippophae rhamnoides* L.) seed and pulp oils on experimental models of gastric ulcer in rats. *Fitoterapia* 73: 644-650.

197. Xu, M., Sun, S. and Cui, J. 2001. The medicinal research on seabuckthorn. In: *Abstract Book of Int. Workshop on Seabuckthorn*, New Delhi, India.

198. Yang, B. and Kallio, H. 2001. Fatty acid composition of lipids in seabuckthorn (*Hippophae rhamnoides* L.) berries of different origins. *Journal of Agricultural and Food Chemistry* 49: 1939-1947.

199. Yang, B. and Kallio, H. 2002a. Composition and physiological effects of seabuckthorn (*Hippophae*) lipids. *Trends in Food Science and Technology* 13: 160-167.

200. Yang, B. and Kallio, H. 2002b. Supercritical CO_2-extracted seabuckthorn (*Hippophae rhamnoides*) oils as new food ingredients for cardiovascular health. In: *Proceedings of Health Ingredients of Europe*, September 17-19, 2002, Paris, France, 7p.

201. Yang, B. and Kallio, H.P. 2006. Lipophilic components of seabuckthorn (*Hippophae rhamnoides L.*) seeds and berries. In: *Seabuckthorn (Hippophae L.)-A Multipurpose Wonder Plant. Vol. II: Biochemistry and Pharmacology* (V. Singh, Ed. in Chief, 2006), 70-97, Daya Publishing House, New Delhi, 600p.

202. Yang, B., Bongli, A., Pagani, V., Isohanni, T., Von-Knorring, A., Jutila, A. and Judin, V.P. 2009. Effects of oral supplementation and topical application of supercritical CO_2 extracted seabuckthorn oil on skin ageing of female subjects. *Journal of Applied Cosmetology* 27: 1-13.

203. Yang, B., Kalimo, K.O., Tahvonen, R.L., Mattila, L.M., Katajisto, J.K. and Kallio, H.P. 2000. Effect of dietary supplementation with seabuckthorn (*Hippophae rhamnoides*) seed and pulp oils on the fatty acid composition of skin glycerophospholipids of patients with atopic dermatitis. *Journal of Nutritional Biochemistry* 11: 338-340.

204. Yuzhen, Z. and Fuheng, W. 1997. Seabuckthorn flavonoids and their medicinal value. *Hippophae* 10: 39-41.

205. Zadernowski, R., Nowak-Polakowska, H., Lossow, B. and Nesterowicz, J. 1997. Seabuckthorn lipids. *Journal of Food Lipids* 4: 165-172.

206. Zeb, A. 2004. Chemical and nutritional constituents of seabuckthorn juice. *Pakistan Journal of Nutrition* 3: 99-106.

207. Zeb, A. 2006. Anti-carcinogenic potential of lipids from *Hippophae*: Evidence from the recent literature. *Asian Pacific Journal of Cancer Prevention* 7: 32-34.

208. Zhang, Z. and Qiu, D. 1995. Anti-cancer studies from Russia. *Hippophae* 8 (4): 38-40.

209. Zhang, W.L., Zhang, Z.F., Fan, J., Yang, S.Y., Li, Z.M., Dcng, Z.C., Wang, G.L. and Zhang, F.S. 1988. Experimental observation and clinical investigation effects of seabuckthorn oil on acute radio dermatitis. *Hippophae* 1 (1): 27-30.

210. Zhao, Y. 1994. Clinical effects of *Hippophae* seed oil in the treatment of 32 burn cases. *Hippophae* 7: 36-37.

211. Zheng, R.X., Xu, X.D., Tian, Z. and Yang, J.S. 2009. Chemical constituents from the fruits of *Hippophae rhamnoides*. *Natural Product Research* 23: 1451-1456.

212. Ziboh, V.A. and Chapkin, R. S. 1987. Biologic significance of polyunsaturated fatty acids in the skin. *Archieves of Dermatology* 123 (12): 1686-1690.

213. Ziboh, V.A., Cho, Y. and Miller, C.C. 2000. Metabolism of polyunsaturated fatty acids by skin epidermal enzymes: Generation of anti-inflammatory and anti-proliferative metabolites. *American Journal of Clinical Nutrition* 71 (1): 361S-366S.

214. Zou, X. 1997. Multifunctional medicinal super cleansers. *Chinese Patent 1144266.*

Chapter 27

Health Effects of Seabuckthorn Berries: Research and Strategies at the University of Turku, Finland

Heikki P. Kallio and Baoru Yang

Department of Biochemistry and Food Chemistry,
University of Turku, Turku, Finland

SUMMARY

Seabuckthorn has been a target of scientific investigations at the University of Turku since 1980's. In addition to taxonomic, chemical and sensory research of the berries, their health effects have deserved special attention. Nutritional effects of both entire berries and of their oil and ethanol soluble fractions have been investigated. Many of the hypotheses were based on Eastern, especially on Chinese traditions, claims and knowledge. Berries in the Scandinavian countries, also in Finland, both cultivated and wild ones, are commonly regarded as health promoting food ingredients. The major genera are *Vaccinium* (bilberry, lingonberry), *Rubus* (cloudberry, raspberry, arctic bramble), *Ribes* (currants), *Empetrum* (crowberry) and *Hippophae* (seabuckthorn). Seabuckthorn (SB) berries rich in flavonoids, oil-soluble antioxidants and vitamin C were shown to lower concentration of the sensitive CRP in plasma. In addition, consumption of the juice indicated increase of the ratio of HDL cholesterol to LDL cholesterol and elongation of the lag phase of LDL cholesterol oxidation. Berries and especially their ethanol-soluble fraction suppressed the postprandial insulin peak. It was further shown, that the bioavailability of flavonoids was increased by coincide supplementation of seabuckthorn oil. The results of consumption of SB thus indicate possible reduction of the risk of cardiovascular diseases in healthy people. Seabuckthorn seed and pulp oils have been of special interest. The very recent studies showed the unexceptionally high protective, antioxidative effects of SB oils on the isolated DNA *in vitro*. The same was the case with DNA of rat liver homogenate *in vitro*. Whether the positive effects of seabuckthorn oils on dry eyes and atopic skin have the same mechanistic background, is not known. The oils investigated have all been isolated by aseptic CO_2 extraction.

Keywords: *Anti-inflammatory effects, Antioxidative effects, Cardiovascular health, Dry eyes, Hippophaë rhamnoides, Phenolic compounds, Pulp and seed oils and Secondary metabolites.*

Seabuckthorn (*Hippophae* L.): *A Multipurpose Wonder Plant, Vol. 4: Emerging Trends in Research and Technologies* (Virendra Singh, *Editor-in-Chief*, 2014), p. 352–360, Daya Publishing House, New Delhi, India.

INTRODUCTION

Information concerning the nutritional value and medicinal effects of the berries and leaves of seabuckthorn (SB) (genus Hippophae) dates back to the ancient Tibetan Four Books of Pharmacopoeia "rGyud bzi" written by Yu Tuo Yuan Dan Kong Bu in 773-783. The books were revised 300 years later and translated from Tibetan to Chinese in 1987 by Ma Shucai and Lo Dan-Shang (Li and Guo, 1989). More than 300 ways of using seabuckthorn in treating pulmonary, peptic, sputum, inflammatory,

Figure 27.1: Picking of Fruits from the Tibetan-Chinese *H. gyantsensis* by Prof. Baoru Yang and Dr. Sinikka Kallio in Lhasa, Tibet (September, 2011).

gynaecological, etc. diseases were introduced. In addition to berries, fruit and leaf extracts were often used together with other drugs. The originally reported properties were evidently linked to *Hippophae gyantsensis*, nowadays regarded as an independent species and not anymore a subspecies of *H. rhamnoides* (Lian *et al.*, 2000).

The natural area of distribution of seabuckthorn is Eurasia, from China and Mongolia via Middle-East and Middle-Europe up to the Scandinavian countries. The most commonly used berries of the nine *H. rhamnoides* subspecies are *sinensis*, *mongolica* and *rhamnoides*, common native plants in Europe, China and Russia, respectively (Rousi 1971). Scientific and applied research on the berry has been actively conducted during the last decades especially in China, Russia, India, Germany, Sweden and Finland (Singh *et al.*, 2008). The commercial Russian based ssp. *mongolica* varieties and cultivars with big juicy berries are widely distributed also in Europe. In China, Russia, and to some extent in Scandinavia also the best wild clones are used in households and industry as well as in soil conservation in eroded areas (Singh *et al.*, 2003)

The long tradition of exploitation of seabuckthorn in healing and health-maintaining products is, as such, a clear evidence of the positive properties of the plant on human health. But where does the power come from? Most of the medicinal plants in the Qinghai-Tibetan area originate from the open, high-altitude plateau and sandy mountain areas, where they are adapted in extreme weather and climatic conditions (Yang *et al.*, 2011). The native seabuckthorn populations are halo-tolerant and also adapted to dry areas with low annual precipitation. On the coastal areas of Finland and Sweden in Scandinavia seabuckthorn grows, again, next to the sea, often practically roots in the sea water. In the high altitudes of several kilometres in China, India and Southern Europe, the bush is susceptible to high and continuous radiation because there is no shelter of other plants. In fact, competition of other plants, under or above the ground, is close to the only thing seabuckthorn does not tolerate. The plant has also its own root nodules for nitrogen fixation and a sandy soil is enough for survival (Du *et al.*, 1989). Has the causal correlation between the extreme-related chemical composition and the health promoting effects of seabuckthorn been proven? Evidently not yet.

The recent literature from the twenty first century on health effects of seabuckthorn covers antioxidative and anti-inflammatory effects (Chauhan *et al.*, 2007; Chawla *et al.*, 2007; Geetha *et al.*, 2009), apoptosis and proliferation-controlling effects (Padmavathi *et al.*, 2005; Grey *et al.*, 2010) as well as antimicrobial properties (Negi *et al.*, 2005). Also healing of wounds, skin and mucosal membranes including problems related to gastric ulcer (Süleyman *et al.*, 2001), have been reported. Promotion of cardiovascular (Bao and Lou, 2006; Basu *et al.*, 2007; Purushohtman *et al.*, 2008) and liver health (Gao *et al.*, 2003) as well as reduction of risk of type 2 diabetes are also highly evident properties of seabuckthorn. In addition, the progressively increasing number of animal studies has given mainly positive and encouraging results. Thorough reviews of the topics have been written by several authors (Yang and Kallio, 2002a; Singh *et al.*, 2006; Larmo, 2011).

In order to boost scientific co-operation between universities and research institutes around the world, the Ministry of Water Resources in China has established a multinational International Seabuckthorn Association in Beijing. Independently and also as an active party of the Association, the University of Turku has conducted several projects related to seabuckthorn technology, chemistry and health properties. In some cases the projects have been domestic and have been organized together with the Turku University Hospital and other medicinal units of Finnish universities. Some projects have, again, been joined efforts *e.g.* between the University of Turku and Reading University (UK), University of Iceland, Xian Medical University or Beijing University.

The aim of this review is not to repeat what the earlier reviews have shown and proven. The aim is to introduce the strategy and results of the Food Chemistry Unit at the University of Turku. We have combined our knowledge in seabuckthorn chemistry, in health effects via clinical and animal trials as well as in sensory science on seabuckthorn. Many of the ideas can be linked to the traditional Chinese medicine but new dimensions have been reached along with the accumulating compositional information of seabuckthorn.

MATERIALS AND METHODS

Seabuckthorn

Most of the berries investigated by us belong to the three major subspecies, *Hippophae rhamnoides* subsp. *sinensis*, *mongolica* and *rhamnoides*, either wild origin or cultivated varieties or clones. Origin of the raw material is described, if known, in the original publications.

Chemical Analyses and Technology

The review concentrates on clinical, animal and *in vitro* trials of seabuckthorn berries and of its technological fractions. The major products tested have typically been berries or juice, seed oil, pulp oil and ethanol extract of the berries. All the oils investigated were produced by aseptic, supercritical CO_2 extraction (Manninen *et al.*, 1997). The crucial compositional information has been published either as part of each work to be presented or separately in other publications (Yang *et al.*, 2001; Kallio *et al.*, 2002; Yang and Kallio, 2002b; Yang *et al.*, 2006; Yang, 2009; Yang *et al.*, 2009; Yang *et al.*, 2011; Zheng *et al.*, 2011).

Sensory Properties of the Products

Some of the clinical trials conducted did also include sensory evaluations of the test products. Typically, however, investigations related to quantitative sensory profiling of the berries have been published separately (Tiitinen *et al.*, 2005; Tiitinen *et al.*, 2006) and cannot be directly linked to the clinical trials.

Clinical, Animal and *in vitro* Trials

In addition to double blind, randomized cross over studies, also parallel studies and one-dose postprandial studies were conducted, depending on the goal and resources. Especially in small-scale postprandial investigations the test persons may typically act as their own controls. This information is briefly described in the text and in details in the original articles. In case of animal and *in vitro* trials, it is also worth to consult the original publications.

RESULTS AND DISCUSSION

Health Effect of Seabuckthorn Oils

Seed oils of twelve Finnish wild and cultivated berries were isolated by supercritical fluid CO_2 extraction. One of berries was seabuckthorn and also pulp oil from the fruit flesh was isolated and tested. Both seabuckthorn pulp and seed oils showed exceptional antioxidative activities. Investigations of peroxidation inhibition in rat liver microsomes, inhibition of LDL-oxidation (*in vitro*), superoxide anion scavenging activity (by xanthine oxidase), and inhibition of DNA oxidation in rat liver homogenate induced by UV-radiation were conducted. In experiments with rat liver homogenate and purified DNA, seabuckthorn seed oil and pulp oil both were effective in a dose responding manner compared with Trolox and alpha-tocopherol (Yang *et al.*, 2011).

All the oils tested in our clinical nutrition and animal trials have been isolated by CO_2-extraction (Manninen *et al.*, 1997). Composition of the mixed SB oil (combination of seed and pulp oils) supported the idea to test its effectiveness as oral oil supplement to relieve the symptoms of dry eyes. 86 female subjects completed the double blind, randomized parallel study by 2 g oil administration a day for three months. A mixture of medium-chain fatty acid coconut oil and palm kernel oil was used as a placebo (Larmo *et al.*, 2010). The administration of SB oil ended up with reduced redness and burning of the eyes (P=0.04). Also osmolarity of the tears was lower in the SB oil group than in the placebo group (P=0.04) (Larmo *et al.*, 2010). The mechanism of action is unknown but we recognized that there were no differences in the fatty acid compositions in the tear films between the two groups (Jarvinen *et al.*, 2011).

Dry eyes, dry mouth, dry and inflammatory mucosa in the female genital tract and problems in the gastro-intestinal mucosa are typical symptoms of the Sjogren's syndrome. Oral supplementation of SB oils seems to present a way of relieving these symptoms. Especially the scores of itching and dryness feeling as well as discharge were decreased by a 12-week, 3 g/day supplementation (Yang and Erkkola, 2008).

Lipid composition of human skin is effectively buffered. Atopic people were administrated with oral supplementation of mixed SB oil (5 g/day, four months) in a double blind, placebo controlled, parallel study. The subjects got normal diets and normal medication. A hypothesis was that the fatty acid profile in the skin glycerophospholipids would shift towards the normal skin by the SB oil supplementation but this did not happen (Yang *et al.*, 2000). Neither did the skin become less rough during or after the oral SB oil treatment (Yang *et al.*, 2004). However, there was a clear correlation between symptom improvements and increase in proportions of alpha-linolenic acid (18:3n-3) in plasma phospholipids (PL) and neutral lipids (NL) in seabuckthorn seed oil group after one month administration (Yang *et al.*, 1999).

There are many indicators about positive effects of SB oil on cardiovascular health and we decided to test the effect of mixed SB oil on platelet aggregation in 12 healthy, normolipidemic men in a placebo-controlled, randomized, cross-over study. Each subject was his own control. Fractionated coconut oil as the control, a 2 × four-week trial with 5 g oil/d and with a wash-out period of four to eight weeks was conducted. Fasting (12 h) venous blood samples were taken before and after the treatments. ADP and arachidonic as agonists showed SB oil to reduce both the platelet aggregation velocity and the total platelet aggregation (Johansson *et al.*, 2000).

Both pulp oil and seed oil of SB are worth to be tested thoroughly also for improvement of micro circulation in man. The traditional claims of positive effects of SB oils are supported by our investigations on two animal models. In topical mouse auricular and rat mesenteric microcirculation models SB oils typically increased the net capillary intersections and venular or arteriolar calibers (Yang *et al.*, 2008). Whether this mechanism is related to the positive results achieved with several rat model experiments on healing and prevention of peptic ulcer by SB oils (Xing *et al.*, 2002) is not clear.

Health Effects of Seabuckthorn Berries and Juice

The whole berries or puree of seabuckthorn contain a wide variety of cutins, pectins, cellulose, proteins, amino acids, flavonoids, phenolic acids, lignans, vitamin C and other vitamins, sugars, fruit acids and terpenoids but also all the oil-soluble lipophilic components. It is very risky to guess or calculate causal correlations between composition and the effects observed. The first logic question thus is: do we find specified health-promoting effects of the berries or juice?

Even a small quantity of 28 g a day of seabuckthorn puree resulted in significant decrease in the sensitive plasma CRP (P=0.039) in a randomized, parallel, placebo-controlled study of 233 healthy volunteers. The test period lasted for 90 days. Based on diaries, at the same time no effects of SB on common cold symptoms or urinary tract infections were observed (Larmo *et al.*, 2008). Also SB juice has shown mild but positive effects related to cardiovascular health by increasing LDL cholesterol by 20 per cent and by decreasing moderately the susceptibility of LDL oxidation. Twenty healthy male volunteers took either SB juice or placebo over eight weeks. The daily dose of SB juice contained *e.g.* 462 mg vitamin C, 355 mg flavonoids, 3,2 mg alpha-tocopherol and 1 mg beta-carotene. It is worth to be mentioned that a 17 per cent increase in plasma TAG was recognized, even though not statistically significant (Ecclestone *et al.*, 2002a).

Postprandial hyperglycemia and insulin response were shown to be affected by seabuckthorn berry and its ethanol-soluble metabolites (Lehtonen *et al.*, 2010). Ingestion of 40 g of berries suppressed the insulin peak. Removal of lipids from the berries had no effect on the result but removal of the phenolic compounds by ethanol extraction eliminated the positive effect. The ethanol-soluble compounds included *e.g.* the phenolic compounds. Lactose-free yoghurt with added glucose was the base and control. The active ingredients added were either 40 g of SB berries, 32.4 g of CO_2-extracted oil-free berries or 18 g of CO_2-extracted, ethanol-extracted berries. Ten healthy men participated the experiment and acted as their own controls (Lehtonen *et al.*, 2010a). The phenolic compounds, such as flavonoid glycosides, are known to be absorbed from the berries (Lehtonen *et al.*, 2010b). In separate absorption studies it was observed that the bioavailability of flavonoids was increased by coincide supplementation of seabuckthorn oil (Suomela *et al.*, 2006).

CONCLUSION

As long as seabuckthorn is investigated as food of food ingredient the clinical trials should primarily be conducted by using berries, juice, leaves, oils or some other clearly described technological, edible fractions. By simultaneous analysis of both the ingredients and human samples such as tissues, urine or faeces it is possible to approach the question about the most effective ingredients and the target compounds. It is extremely rare that any individual compound alone ends up with an observed positive health promoting effect. Especially we should not neglect the wisdom of the ancient traditions.

REFERENCES

1. Bao, M. and Lou, Y. 2006. Flavonoids from seabuckthorn protects endothelial cells (EA.hy926) from oxidized low-density lipoprotein induced injuries *via* regulation of LOX-1 and eNOS expression. *J. Cardiovasc. Pharmacol.* 48: 834-841.

2. Basu, M., Prasad, R., Jayamurthy, P., Pal, K., Arumughan, C. and Sawhney, R.C. 2007. Anti-atherogenic effects of seabuckthorn (*Hippophae rhamnoides*) seed oil. *Phytomedicine* 14: 770-777.

3. Chauhan, A.S., Negi, P.S. and Ramteke R.S. 2007. Antioxidant and antibacterial activities of aqueous extract of seabuckthorn (*Hippophae rhamnoides*) seeds. *Fitoterapia* 78: 590-592.

4. Chawla, R., Arora, R., Singh, S., Sagar, R.K., Sharma, R.K., Kumar, R., Sharma, A., Gupta, M.L., Singh, S., Prasad, J., Khan, H.A., Swaroop, A., Sinha, A.K., Gupta, A.K., Tripathi, R.P. and Ahuja, P.S. 2007. Radioprotective and antioxidant activity of fractionated extracts of berries of *Hippophae rhamnoides*. *J. Med. Food* 10: 101-109.

5. Du, D., Wang, Y. Yuan F., Li, R. 1989. Study of symbiotic nitrogen fixation of *Hippophae rhamnoides* L. In: *Proceedings of International Symposium on Seabuckthorn (H. rhamnoides L.)*, p. 180-183, October 19-23, 1989, Xian, China.

6. Ecclestone, C., Yang, B., Tahvonen, R., Kallio, H., Rimbach, G.H. and Minihane, A.M. 2002. Effects of an antioxidant-rich juice (seabuckthorn) on risk factors of coronary heart diseases in humans. *J. Nutr. Biochem*. 13: 346-354.

7. Gao, Z.L., Gu, X.H., Cheng, F.T. and Jiang, F.H. 2003. Effect of seabuckthorn on liver fibrosis: A clinical study. *World J. Gastroenterol*. 9: 1615-1617.

8. Geetha, S., Ram, M.S., Sharma, S.K., Ilavazhagan, G., Banerjee, P.K. and Sawhney, R.C. 2009. Cytoprotective and antioxidant activity of seabuckthorn (*Hippophae rhamnoides* L.) flavones against tert-butyl hydroperoxide-induced cytotoxicity in lymphocytes. *J. Med. Food* 12: 151-158.

9. Grey, C., Widen, C., Adlercreutz, P., Rumpunen, K. and Ruan R. 2010. Antiproliferative effects of seabuckthorn (*Hippophae rhamnoides* L.) extracts on human colon and liver cancer cell lines. *Food Chem*. 120: 1004-1010.

10. Jarvinen, R.L., Larmo, P.S., Setala, N.L., Yang B., Engblom, J.R.K., Viitanen, M.H. and Kallio, H.P. 2011. Effects of oral seabuckthorn oil on tear film fatty acids in individuals with dry eye. *Cornea* 30: 1013-1019.

11. Johansson, A., Korte, H., Yang, B., Stanley, J.C. and Kallio, H. 2000. Seabuckthorn berry oil lowers platelet aggregation. *J. Nutr. Biochem*. 11: 491-495.

12. Kallio, H.P., Yang, B. and Peippo, P. 2002. Effects of different origins and harvesting time on vitamin C, tocopherols, and tocotrienols in seabuckthorn (*Hippophae rhamnoides*) berries. *J. Agric. Food Chem*. 50: 6136-6142.

13. Larmo, P. 2011. *The Health Effects of Seabuckthorn Berries and Oil*, Painosalam Oy, Turku, Finland.

14. Larmo, P., Alin, J.A., Salminen, E.K., Kallio, H.P. and Tahvonen, R.L. 2008. Effects of seabuckthorn berries on infections and inflammations: A double-blind, randomized, placebo-controlled trial. *Eur. J. Clin. Nutr*. 62: 1123-1130.

15. Larmo, P.S, Jarvinen, R.L., Setala, N.L., Yang, B., Viitanen, M., Engblom, J.R.K., Tahvonen, R.L. and Kallio, H.P. 2010. Oral seabuckthorn attenuates tear film osmolarity and symptoms in individuals with dry eye. *J. Nutr*. 140: 1462-1468.

16. Lehtonen, H.M., Jarvinen, R., Linderborg, K., Viitanen, M., Venojarvi, M., Alanko, H. and Kallio, H. 2010a. Postprandial hyperglycemia and insulin response are affected by seabuckthorn (*Hippohae rhamnoides* ssp. *turkestanica*) berry and its ethanol-soluble metabolites. *Eur. J. Clin. Nutr*. 64: 1465-1471.

17. Lehtonen, H.M., Lehtinen, O., Suomela, J.P., Viitanen, M. and Kallio, H. 2010b. Flavonol glycosides of seabuckthorn (*Hippophae rhamnoides* ssp. *sinensis*) and lingonberry (*Vaccinium vitis-idaea*) are bioavailable in humans and monoglucuronidated for excretion. *J. Agric. Food Chem*. 58: 620-627.

18. Li, F. and Guo, T. 1989. Application of *Hippophae rhamnoides* L. in Tibetan medicine, p.409-412. In: *Proceedings of the International Symposium on Seabuckthorn* (*H. rhamnoides* L.), Xian, China.

19. Lian, Y.S., Lu, S.G., Xue, S.K. and Chen, X.L. 2000. *Biology and Chemistry of the Genus Hippophae*, Gansu Scientific and Technological Publishing House, Lanzhou, China.

20. Manninen, P., Haivala, E., Sarimo, S. and Kallio, H. 1997. Distribution of microbes in supercritical CO_2 extraction of seabuckthorn (*Hippophae rhamnoides*) oils. *Z. Lebensm. Unters. Forsch*. A204: 202-205.

21. Negi, P.S., Chauhan, A.S., Sadia, G.A., Rohinishree, Y.S. and Ramteke, R.S. 2005. Antioxidant and antibacterial activities of various seabuckthorn (*Hippophe rhamnoides* L.) seed extracts. *Food Chem.* 92: 119-124.

22. Padmavathi, B., Upreti, M., Singh, V., Rao, A.R., Singh, R.P. and Rath, P.C. 2005. Chemoprevention by *Hippophae rhamnoides*: Effect of tumorigenesis, Phase II and antioxidant enzymes, and IRF-1 transcription factor. *Nutr. Cancer* 51: 59.67.

23. Purushohtman, J., Suryamkumar, G., Shukla, D., Malhotra, A.S., Kasiganesan, H., Kumar, R., Sawhney, R.C. and Chami, A. 2008. Modulatory effects of seabuckthorn (*Hippophae rhamnoides* L.) in hypobaric hypoxia induced cerebral vascular injury. *Brain Res. Bull.* 77: 246-252.

24. Rousi, A. 1971. The genus *Hippophae* L.-A taxonomic study. *Ann. Bot. Fennici* 8: 177-227.

25. Singh, V., Kallio, H., Sawhney, R.C., Gupta, R.K., Rongsen, L., Eliseev, I.P., Khabarov, S.N., Korovina, M.A., Skuridin G.M. and Shchapov, N.S. (Eds.) 2003. *Seabuckthorn (Hippophae L.)-A Multipurpose Wonder Plant. Vol. I: Botany, Harvesting and Processing*, Daya Publishing House, New Delhi, India, 518 p.

26. Singh, V., Kallio, H., Yang, B., Morsel, J.T., Sawhney, R.C., Bala, M., Dwevedi, S., Geetha, S., Tyagi, S.P., Li, T.S.C., Rongsen, L., Shunguang, L. and Zubarev, Y. (Eds.) 2008. *Seabuckthorn (Hippophae L.)-A Multipurpose Wonder Plant. Vol. III: Advances in Research and Development*, Daya Publishing House, New Delhi, India, 566p.

27. Singh, V., Yang, B., Kallio, H., Bala, M., Sawhney, R.C., Gupta, R.K., Morsel, J.T., Rongsen, L. and Tolkachev, O.N. (Eds.) 2006. *Seabuckthorn (Hippophae L.)-A Multipurpose Wonder Plant. Vol II: Biochemistry and Pharmacology*, Daya Publishing House, New Delhi, India, 600p.

28. Suleyman, H., Demirezer, L.O., Buyukokuroglu, M.E., Akcay, M.F., Gepdiremen, A., Banoglu, Z.N. and Gocer, F. 2001. Antiulcerogenic affect of *Hippophae rhamnoides* L. *Phytother. Res.* 15: 625-627.

29. Suomela, J.P., Ahotupa, M., Yang, B., Vasankari, T. and Kallio, H. 2006. Absorption of flavonols derived from seabuckthorn (*Hippophae rhamnoides* L.) and their effect on emerging risk factors for cardiovascular disease in humans. *J. Agric. Food Chem.* 54: 7364-7369.

30. Tiitinen, K., Hakala, M. and Kallio, H. 2005. Quality components of seabuckthorn (*Hippophae rhamnoides*) varieties. *J. Agri. Food Chem.* 53: 1692-1699.

31. Tiitinen, K., Hakala, M. and Kallio, H. 2006. Headspace volatiles from frozen berries of seabuckthorn (*Hippophae rhamnoides* L.) varieties. *Eur. Food Res. Technol.* 223: 455-460.

32. Xing, J., Yang, B., Dong, Y., Wang, B., Wang J. and Kallio, H. 2002. Effects of seabuckthorn (*Hippophae rhamnoides* L.) seed and pulp oils on experimental models of gastric ulcer in rats. *Fitoterapia* 73: 644-650.

33. Yang, B. 2009. Sugars, acids, beta-D-glucopyranose and a methyl inositol in seabuckthorn (*Hippophae rhamnoides*) berries. *Food Chem.* 112: 89-97.

34. Yang, B. and Erkkola, R. 2008. Seabuckthorn oils, mucous membranes and Sjogren's syndrome with special reference to latest studies. In: *Seabuckthorn (Hippophae L.)-A Multipurpose Wonder Plant. Vol. III: Advances in Research and Development* (V. Singh, Ed. in Chief, 2008), p. 254-267, Daya Publishing House, New Delhi, India, 566p.

35. Yang, B. and Kallio, H. 2002a. Composition and physiological effects of seabuckthorn (*Hippophae*) lipids. *Trends Food Sci. Technol.* 13: 160-167.

36. Yang, B. and Kallio, H. 2002b. Effects of harvesting time on triacylglycerols and glycreophospholipids of seabuckthorn (*Hippophae rhamnoides* L.) berries of different origins. *J. Food Compos. Anal.* 15: 143-157.

37. Yang, B., Ahotupa, M., Maatta, P. and Kallio, H. 2011. Composition and antioxidative activities of supercritical CO_2-extracted oils from seeds and soft parts of northern berries. *Food Res. Int.* 44: 2009-2017.

38. Yang, B., Halttunen, T., Raimo, O., Price, K. and Kallio, H. 2009. Flavonol glycosides in wild and cultivated berries of three major subspecies of *Hippophae rhamnoides* and changes during harvesting period. *Food Chem.* 115: 657-664.

39. Yang, B., Kalimo, K.O., Tahvonen, R.L., Mattila, L.M., Katajisto, J.K. and Kallio, H.P. 2000. Effect of dietary supplementation with seabuckthorn (*Hippophae rhamnoides*) seed and pulp oils on the fatty acid composition of skin glycerophospholipids of patients with atopic dermatitis. *J. Nutr. Biochem.* 11: 338-340.

40. Yang, B., Kalimo, K.O., Tahvonen, R.L. Mattila, L.M. and Kallio, H.P. 2004. Dietary supplementation with seabuckthorn (*Hippophae rhamnoides*) oils and skin surface roughness of patients with atopic dermatitis. *Global Seabuckthorn Res. Devel.* 1: 44-47.

41. Yang, B., Karlsson, R., Oksman, P. and Kallio, H. 2001. Phytosterols in seabuckthorn (*Hippophae rhamnoides* L.) berries: Identification and effects of different origins and harvesting times. *J. Agric. Food Chem.* 49: 5620-5629.

42. Yang, B., Linko, A.M., Adlecreutz, H. and Kallio, H. 2006. Secoisolariciresinol and matairesinol of seabuckthorn (*Hippophae rhamnoides* L.) berries of different subspecies and harvesting times. *J. Agric. Food Chem.* 54: 8065-8070.

43. Yang, B., Wu, Y., Liu, Q., Wang, B., Kang, J., Wang, J. and Kallio, H. 2008. Supercritical CO_2 extracted seabuckthorn pulp oil and seed oil improve blood microcirculation. In: *Seabuckthorn (Hippophae L.)-A Multipurpose Wonder Plant. Vol. III: Advances in Research and Development* (V. Singh, Ed. in Chief, 2008), p. 268-271, Daya Publishing House, New Delhi, India, 566p.

44. Yang, B., Zheng, J. and Kallio, H. 2011. Influence of origin, harvesting time and weather conditions on content of inositols and methylinositols in seabuckthorn (*Hippohae rhamnoides*) berries. *Food Chem.* 125: 388-396.

45. Zheng, J., Kallio, H., Linderborg, K. and Yang, B. 2011. Sugars, sugar alcohols, fruit acids, and ascorbic acid in wild Chinese seabuckthorn (*Hippophae rhamnoides* ssp. *sinensis*) with special reference to influence of latitude and altitude. *Food Res. Int.* 44: 2018-2026.

Chapter 28

Antioxidant and Anti-carcinogenic Activities of Seabuckthorn (*Hippophae* L.)

Alam Zeb

Department of Biotechnology, University of Malakand, Chakdara, Pakistan

SUMMARY

Seabuckthorn is a medicinal plant, widely grown or cultivated across the different regions of the world. The important and widely used parts of seabuckthorn include whole berries, seeds, leaves, juice and oil. These parts contain different kinds of nutrients and bioactive substances, such as vitamins, carotenoids and phenolic compounds of vast antioxidant potential. This chapter is therefore, focused on the studies of these properties of seabuckthorn during the last five years, 2008-2012. It was observed that during the above period, antioxidant activity was the main focus of research, while limited studies have been carried out with the anti-carcinogenic activity. The antioxidant potential was attributed to the natural compounds present in the respective parts of seabuckthorn, while the antioxidant activity contributed to the anti-carcinogenic potential. It is suggested that anti-carcinogenic studies should be elaborated more in future, so as to find out the complete efficacy of the seabuckthorn.

Keywords: *Seabuckthorn, Antioxidant composition, Antioxidant and Anti-carcinogenic potential.*

INTRODUCTION

Seabuckthorn is a medicinal and aromatic plant, widely praised for its role in environmental conservation and to possess multiple health promoting properties. The seed oil had been used to promote recovery from skin injuries and skin diseases (eczema, burns, and wounds), and the damaging effects of sun on the skin, therapeutic radiation treatments as well as cosmetic laser surgery (Zeb and Khan, 2008; 2009). It had also been used in digestive ulcers and cardiovascular disease (Yang and Kallio, 2002; Zeb, 2004). The clinical trials and scientific studies during the 20th century confirmed medicinal and nutritional values of seabuckthorn. The most important of these studies are the antioxidant and anti-carcinogenic properties. Batool *et al.* (2009) suggested that oral supplementation

Seabuckthorn (Hippophae L.): A Multipurpose Wonder Plant, Vol. 4: Emerging Trends in Research and Technologies (Virendra Singh, *Editor-in-Chief*, 2014), p. 361–369, Daya Publishing House, New Delhi, India.

of the fruit extract from seabuckthorn ssp. *turkestanica* was effective in modifying haloperidol-induced behavioural deficits in rats. It is also observed that increased brain 5-hydroxytryptamine synthesis was due to the extract and the role of serotonin in schizophrenia, seabuckthorn may be used as a nutritive therapy in psychotic patients.

Seabuckthorn is rich in carotenoids, vitamin C, E, and different phenolic compounds. The presence of several of these natural compounds in different seabuckthorn products is the main reason of its antioxidant and anti-carcinogenic activity. This chapter only focused on the studies published in the last five years (2008-2012), regarding the above mentioned properties of seabuckthorn.

ANTIOXIDANTS COMPOSITION

The values of various chemical components of seabuckthorn may vary, depending upon the difference in the variety, methods of analysis and origin and are shown in Table 28.1. The reddish colour of seabuckthorn seed oil is due to the carotenoids, which ranged from 1100-1400 mg/100 g (Zeb and Malook, 2009). The major carotenoids in the seed oils are α-, β-, and γ-carotenes, lycopene, pheophytin, zeaxanthin and β-cryptoxanthin as shown in Figure 28.1 (Giuffrida *et al.*, 2012). Chemat *et al.* (2012) showed that edible oils can be enriched with seabuckthorn carotenoids using ultra-sound assisted extraction. The vitamin C content of the seeds was found to be 35.4 mg/100 g and was higher in fresh seeds/berries (250-333 mg/100 g) (Gutzeit *et al.*, 2008). Sabir *et al.* (2005) showed that seabuckthorn ssp. *turkestanica* contains total plant sterols in the range of 3.3-5.5 per cent. The major sterols in the seed oil were β-sitosterol and 5-avenasterol. Previously, Kallio *et al.* (2002) found that the berries contain up to 290 mg/kg of vitamin E, and may vary with variety and origin.

Table 28.1: Major Antioxidants' Composition in Seabuckthorn

Chemical Component	*Amount*	*Plant Part*	*Reference*
β-Carotene (mg/100 g)	1400	Oil	Zeb and Malook (2009)
Vitamin C (mg/100 g)	35.4	Seeds	Zeb and Malook (2009)
Vitamin E (mg/kg)	290	Berries	Kallio *et al.* (2002)
Phytosterols (mg/100 g)	5.5	Seeds	Sabir *et al.* (2005)
Total Phenolic compounds (mg/kg)	5741	Seeds	Arimboor *et al.* (2008)

Free and bound phenolic acid (PA) composition in different parts of the seabuckthorn ssp. *turkestanica* including seed, using High performance liquid chromatography-diode array detection (HPLC-DAD) were revealed by Arimboor *et al.* (2008). It is found that the total phenolic acid content in the seed kernel (5741 mg/kg) was higher than in the seed coat. The total phenolic acid content in the seed kernel was two-fold higher than that of sesame and cotton seed, and slightly lower than that reported for rapeseed and canola meals, and flax seed. The major phenolic acids present in seed kernel were gallic acid (GA): 3441 mg/kg, protocatechuic acid (ProCA) : 1330 mg/kg, p-hydroxybenzoic acid (p-HBA): 265 mg/kg, vanillic acid (VA): 368 mg/kg, ferulic acid (FA): 156 mg/kg, cinnamic acid (CA): 82 mg/kg, p-coumaric acid (p-CA): 74 mg/kg, and caffeic acid (Caf A): 25 mg/kg. In the seed coat, GA was 230 mg/kg, ProCA: 82 mg/kg, CA: 44 mg/kg, VA: 39 mg/kg, p-HBA: 26 mg/kg, FA: 17 mg/kg and Caf A: 10 mg/kg. ProCA (43 mg/kg) was present as the major phenolic acid of the free phenolic acid fraction, whereas GA (200 mg/kg) was the major constituent in bound phenolic acid fractions from the seed coat.

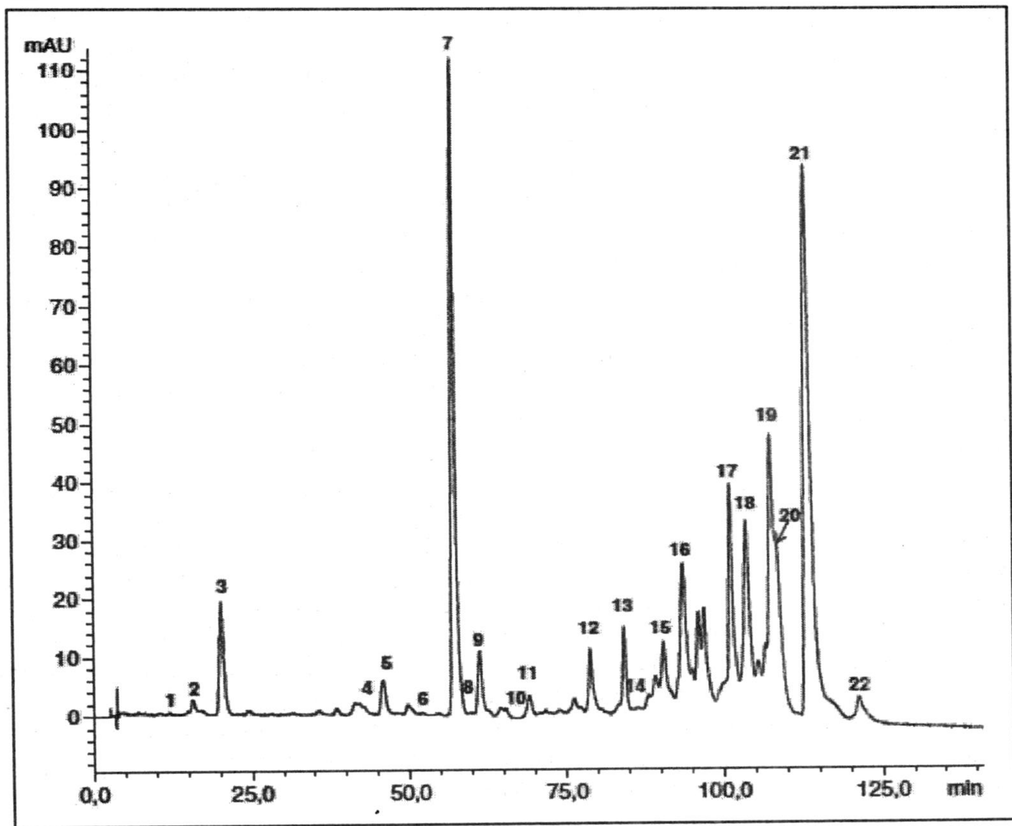

Figure 28.1: Chromatogram (450 nm) of Seabuckthorn Extract (cv. Serbanesti), Using a C30-column Showing Peak 7 is β-carotene.
***Source*: Giuffrida *et al.* (2012).**

ANTIOXIDANT POTENTIAL OF SEABUCKTHORN

Antioxidant activity has been measured in both *in vivo* and *in vitro* studies. Jeong *et al.* (2010) studied the antioxidant and antimicrobial activities from the methanol extracts of the root and stem of *Hippophae rhamnoides* and further partitioned into fractions including hexane, ethyl acetate, butanol, and water. Antioxidant activities of the plant parts were measured by 1,1-dephenyl-2-picryl-hydrazyl (DPPH), superoxide dismutase (SOD)-like activity and ferric thiocyanate (FTC) assays, and compared the results with the standard antioxidants, such as butylated hydroxyl anisole, butylated hydroxyl toluene, α-tocopherol, and ascorbic acid. DPPH assay showed that the overall strong antioxidant activities from the methanol extracts and fractions. The values of SOD activities in hexane fractions of the root and stem were close to the measurement of a reference, ascorbic acid (at 1,000 ppm). The methanol extract of the root in FTC assay showed a remarkable antioxidant and free radical scavenging activity. Similarly, Varshneya *et al.* (2012) studied the 100 per cent methanolic extract (ME), 70 per cent aqua-methanolic extract (AME) and 100 per cent aqueous extract of the seabuckthorn pomace in order to evaluate antioxidant activity. The total phenolic contents in AME were high as compared to other extracts. All the extracts scavenged different *in vitro* radicals in a concentration-dependent manner. The IC_{50} values were lowest in AME for 2, 22 -azino-bis (3-ethylbenzothiazoline-6-sulfonic acid) diammonium salt, 2, 2-diphenyl-1-picrylhydrazyl, superoxide and nitric oxide radicals, while ME had lowest values for hydroxyl radicals. The reducing power of the extracts increased in a dose-dependent manner and was highest in AME.

In animal studies, seabuckthorn products showed significant antioxidant potential. For example, Geetha *et al.* (2008) showed that seabuckthorn leaf extract has significant hepatoprotective effects which might be due to its antioxidant activity and can be developed as a nutraceutical or food supplement against liver diseases. Similarly, Gumustekin *et al.* (2010) studied the effects of vitamin E and seabuckthorn extract on nicotine-induced oxidative stress in rat heart. Increased superoxide dismutase and glutathione S-transferase activities might have taken part in the prevention of nicotine-induced oxidative stress in extract supplemented group in rat heart. It is also suggested that flavonols, such as quercetin and isorahmnetin, tocopherols such as α-tocopherol and β-tocopherol, and carotenoids such as α-carotene and β-carotene, reported to be present in seabuckthorn extracts and may be responsible for the antioxidant activities of this plant extract.

Inhibitory potential of seabuckthorn seed proanthocyanidins against oxidative stress and xanthine oxidase activity was evaluated by Arimboor and Arumughan (2012). It is observed that there is a considerably high antioxidant and xanthine oxidase inhibitory potentials by seed proanthocyanidins. The antioxidant and xanthine oxidase inhibitory capacity evaluation of proanthocyanidin fractions with varying average degree of polymerization showed that proanthocyanidins with lower molecular size were more effective as superoxide anion and hydroxyl radical scavengers and xanthine oxidase inhibitors. Similarly, Kim *et al.* (2012) measured the antioxidant activities of the solvent-dependent seabuckthorn seed extracts using the DPPH and Folin-Ciocalteu assays, and also investigated the efficacy of seabuckthorn seed in protecting skin against UVB-induced damage using cultured human dermal fibroblasts. The findings suggested that the seabuckthorn seed extract may be a potential therapeutic agent for preventing and treating skin photo-aging.

Yeh *et al.* (2012) studied the effects of dietary seabuckthorn on the toxicity of oxidized cholesterol in rats. It was found that seabuckthorn could improve the body weight, decrease the glutathione, superoxide dismutase and catalase level in the liver, increase of liver and kidney weight to body weight ratio, and thiobarbituric acid-reactive substances level in the liver of rats caused by oxidized cholesterol. It also reduced the plasma content of alanine transaminase, aspartate transaminase and alkaline phosphatase indicating that seabuckthorn could protect the hepatotoxicity induced by oxidized cholesterol. These studies showed that oxidative stress can be minimized using extracts of seabuckthorn products. Recently, we have observed that seabuckthorn oil and seed powder has protective role against the toxicity of thermally oxidized lipids. Seabuckthorn was also found to be effective against treating chronic hepatitis induced by oxidized lipids.

ANTI-CARCINOGENIC POTENTIAL OF SEABUCKTHORN

The literature describing the role of seabuckthorn in the prevention and control of cancer is insufficient; however, analysis of the known experimental research information on anti-cancer potential available at present (Zhang, 1989; Mingyu, 1994; Zhang *et al.*, 2005). Most of the work done in this area has been with laboratory animals and the details can be found in review published by Zeb (2006). Early investigations on anti-cancer effects of seabuckthorn have concentrated on extracts from bark (Ambaye *et al.*, 1962), and 5-hydroxy tryptamine (5-HT) has been suggested to be the main responsible component. Preliminary studies have also suggested anti-cancer effects of seabuckthorn oils (Abartiene and Malakhovskis, 1974; 1975; Yang *et al.*, 1989; Zhang *et al.*, 1989; Li *et al.*, 1994; Nersesyan *et al.*, 1990).

Earlier, it has been shown that the inhibition rate of seabuckthorn oil on the cancer cells was not as effective as the positive medicine; for example, the cancer inhibition rate of phosphamide was twice as much as seabuckthorn (Nersesian *et al.*, 2005). A faster recovery of the hemopoietic system after high dose chemotherapy in mice fed the seabuckthorn oil was carried out. The seed oil had been found to enhance non-specific immunity and provided anti-tumor potential in preliminary laboratory studies (Let, 1993). The effects of seabuckthorn on the inhibiting action on the cancer cells and blocking the carcinogenic factors were studied on mice transplanted tumors including sarcoma (S180), lymphatic leukemia (P388) and B16. It was found that both intra-peritoneal injection of seabuckthorn oil and oral administration, inhibited the development of the tumor. Seabuckthorn juice was found to kill the cancer cells of S180 and P388, and inhibit growth of the cell strains of the human gastric carcinoma (SGC7901) and lymphatic leukemia (L1200) (Mingyu *et al.*, 1994). The clinical study had shown that seabuckthorn may be an effective drug for prevention and treatment of liver fibrosis. It had been shown that seabuckthorn inhibits benzo-(α) pyrene-induced forestomach in mouse (Padmavathi *et al.*, 2005).

The studies regarding the anti-carcinogenic potential in human are limited in the recent years than animal models. Grey *et al.* (2010) showed that the anti-proliferative effects of seabuckthorn on both Caco-2 and Hep G2 cells could be associated with the presence of certain compounds, for which a sequential extraction was performed. The compounds were extracted with heptane followed by ethyl acetate, ethanol, water and ethanol:water (1:1) alone. It is found that ethyl acetate fraction was exclusively high of ursolic acid, together with low amounts of other phenolic acids. High levels of phenolic compounds and proanthyocyanidin and little ursolic acid was found to contained in ethanol : water extracts. Strongest inhibitory effect was observed in the ethyl acetate extract for the Caco-2 cells and in the ethanol:water extract for the Hep G2 cells (Figure 28.2). The anti-proliferative effects were dose-dependent in both cases, while in the case of increase in the dose of the ethyl acetate extract was associated with an increase in apoptosis. The authors suggested that the choice of extraction solvent is of considerable importance and that ursolic acid might be more important than the polyphenols in inhibiting the cancer cell proliferation. Bhatia *et al.* (2011) reviewed the anti-cancer potential of several Himalayan medicinal plants. The authors suggested that seabuckthorn has a vast potential of anti-cancer potential in experimental animals.

CONCLUSION AND FUTURE STUDIES

Seabuckthorn has a vast potential of being one of best source of natural antioxidants with anti-carcinogenic potential. Most of the work done in this regard is focused on experimental animals, and limited studies are available regarding human. It was suggested that if a co-operative anti-carcinogen is developed from the lipids, the immunity of cancer patients can probably be strengthened and the side effects caused by chemotherapeutic agents might be decreased (Zeb, 2006). It is therefore, necessary to extract pure effective substances from seabuckthorn and its products, to determine active and appropriate application in human cancer. The ample experimental data showing important ingredients and bioactive substances is vast and continues to increase rapidly with the advent of new technologies. It is possible to conclude that seabuckthorn is a promising plant containing many anti-carcinogenic compounds with a potential beneficial role in experimental animals and possibly in human health. Thus, with the aim to establish whether these compounds from seabuckthorn are really capable to influence positively the incidence and progression of oncogenic reactions at molecular and cellular

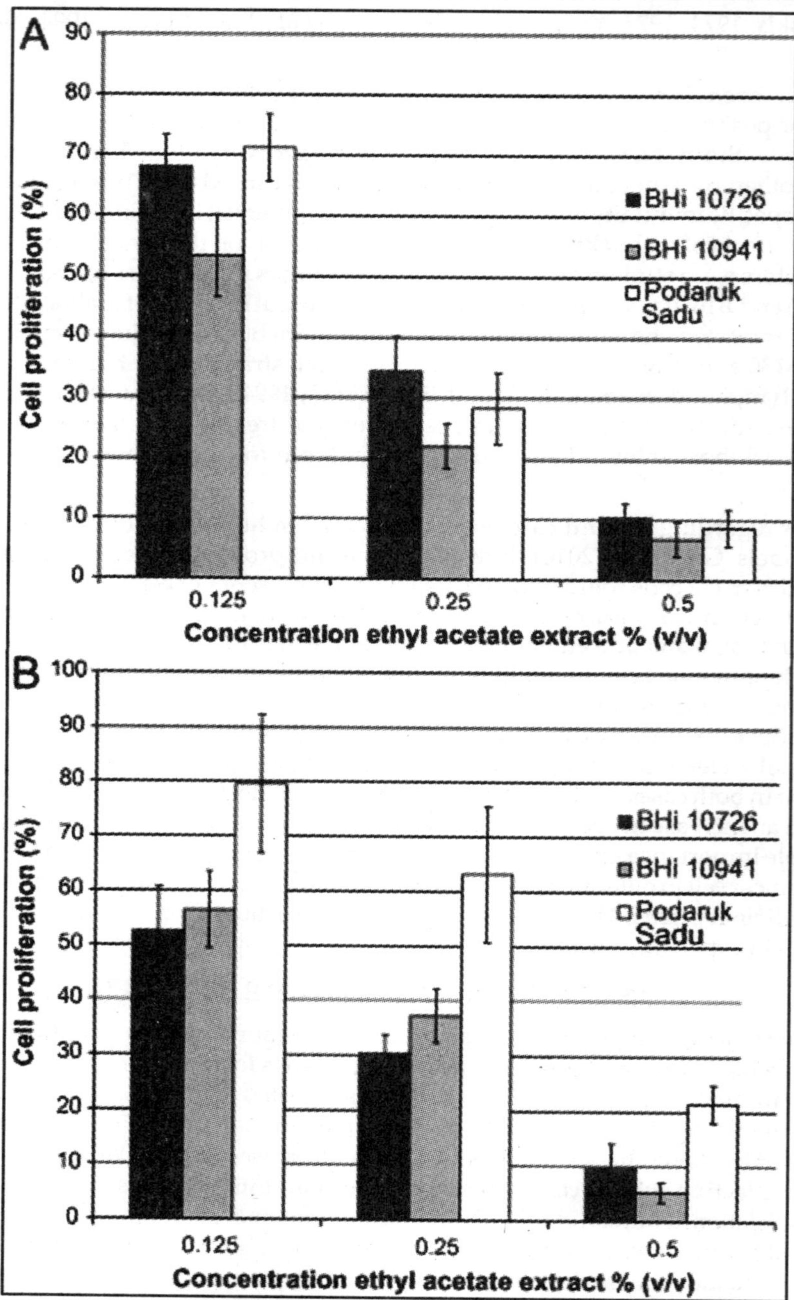

Figure 28.2: Inhibition of Proliferation of Caco-2 Cells (A) and Hep G2 Cells (B) Using Ethyl Acetate Extracts of different Seabuckthorn Species. DMSO (0.5 per cent) in pure medium was used as control.
***Source*: Grey *et al.*, 2010.**

level, a great deal of work in these areas is still necessary. This may include (1) further studies on bioactive compounds from different parts of seabuckthorn and its metabolism in human beings using novel technologies; (2) factors affecting bioavailability; (3) interaction with other therapeutic drugs; (4) dietary variations within and between populations; (5) anti-carcinogenic potential of individual component from seabuckthorn; and (6) molecular or cellular mechanism involve during therapy.

REFERENCES

1. Abartiene, D. and Malakhovskis, A. 1974. Combined action of seabuckthorn oil and anti-tumor preparation on the growth of M-1 sarcoma in rats and their hematological and biochemical indexes. II. Hisphen. *Liet. TSR Mokslu Akad. Darb. Ser. C.* (1): 167-171.

2. Abartiene, D. and Malakhovskis, A. 1975. Combined action of seabuckthorn oil and anti-tumor preparation on the growth of M-1 sarcoma in rats and thier hematological and biochemical indexes. I. Sarcolysine. *Liet. TSR Mokslu Akad. Darb. Ser. C.* (3): 181-186.

3. Ambaye, R.Y., Khanolkar, V.R. and Panse, T.B. 1962. Studies on tumour inhibitory activity of indigenous drugs: Part I. Tumour inhibitory activity of *Hippophae salicifolia* D.Don. *Proc. Ind. Acad. Sci. B* LVI (2): 123-129.

4. Arimboor, R. and Arumughan, C. 2012. Effect of polymerization on antioxidant and xanthine oxidase inhibitory potential of seabuckthorn (*H. rhamnoides*) proanthocyanidins. *Journal of Food Science* 77: C1036-C1041.

5. Arimboor, R., Kumar, K.S. and Arumughan, C. 2008. Simultaneous estimation of phenolic acids in seabuckthorn (*Hippophae rhamnoides*) using RP-HPLC with DAD. *Journal of Pharmaceutical and Biomedical Analysis* 47: 31-38.

6. Batool, F., Shah, A.H., Ahmed, S.D. and Haleem, D.J. 2009. Oral supplementation of seabuckthorn (*Hippophae rhamnoides* L. spp. *turkestanica*) fruit extract modifies haloperidol induced behavioural deficits and increases brain serotonin metabolism. *Journal of Food and Drug Analysis* 17: 257-263.

7. Bhatia, A., Arora, S., Singh, B., Kaur, G. and Nagpal, A. 2011. Anticancer potential of Himalayan plants. *Phytochemistry Reviews* 10: 309-323.

8. Chemat, F., Perino-Issartier, S., Loucif, L., Elmaataoui, M. and Mason, T.J. 2012. Enrichment of edible oil with seabuckthorn by products using ultrasound-assisted extraction. *European Journal of Lipid Science and Technology* 114: 453-460.

9. Geetha, S., Jayamurthy, P., Pal, K., Pandey, S., Kumar, R. and Sawhney, R.C. 2008. Hepatoprotective effects of seabuckthorn (*Hippophae rhamnoides* L.) against carbon tetrachloride induced liver injury in rats. *Journal of the Science of Food and Agriculture* 88: 1592-1597.

10. Giuffrida, E., Pintea, A., Dugo, P., Torre, G., Pop, R.M. and Mondello, L. 2012. Determination of carotenoids and their esters in fruits of seabuckthorn (*Hippophae rhamnoides* L.) by HPLC-DAD-APCI-MS. *Phytochemical Analysis* 23: 267-273.

11. Grey, C., Widen, C., Adlercreutz, P., Rumpunen, K. and Duan, R. 2010. Anti-proliferative effects of seabuckthorn (*Hippophae rhamnoides* L.) extracts on human colon and liver cancer cell lines. *Food Chemistry* 120: 1004-1010.

12. Gumustekin, K., Taysi, S., Alp, M.H., Aktas, O., Oztasan, N., Akcay, F., Suleyman, H., Akar, S., Dane, S. and Gul, M. 2010. Vitamin E and *Hippophae rhamnoides* L. extract reduce nicotine-induced oxidative stress in rat heart. *Cell Biochemistry and Function* 28: 329-333.

13. Gutzeit, D., Baleanu, G., Winterhalter, P. and Jerz, G. 2008. Vitamin C content in seabuckthorn berries (*Hippophae rhamnoides* L. ssp. *rhamnoides*) and related products-A kinetic study on storage stability and the determination of processing effects. *Journal of Food Science* 73: C615-C620.

14. Jeong, J.H., Lee, J.W., Kim, K. S., Kim, J., Han, S. N., Yu, C.Y., Lee, J. K., Kwon, Y. S. and Kim, M.J. 2010. Antioxidant and antimicrobial activities of extracts from a medicinal Plant, seabuckthorn. *Journal of Korean Society of Applied Biological Chemistry* 53: 33-38.

15. Kallio, H., Yang, B. and Peippo, P. 2002. Effects of different origins and harvesting time on vitamin C, tocopherols, and tocotrienols in seabuckthorn (*Hippophae rhamnoides*) berries. *Journal of Agriculture and Food Chemistry* 50: 6136-6142.

16. Kim, H., Cho, H., Seo, Y., Kim, S., Yoon, M.Y., Kang, H., Park, C. and Park, J. 2012. Inhibitory effects of seabuckthorn (*Hippophae rhamnoides* L.) seed on UVB-induced photo-aging in human dermal fibroblasts. *Biotechnology and Bioprocess Engineering* 17: 465-474.

17. Let, Y. 1993. Effects of *Hippophae rhamnoides* juice on immunologic and antitumor functions. *Acta Nutrimenta Sinica* 15: 280-283.

18. Li, J., Wang, X.X. and Zheng, R.L. 1994. Influence of seabuckthorn extracts on syntheses of DNA and protein in cancer cells and plasma cAMP level. *Journal of Lanzhou University (Natural Science)* 33 (4): 548-551.

19. Mingyu, X. 1994. Anti-cancer effects of and direction of research on *Hippophae*. *Hippophae* 7: 41-43.

20. Mingyu, X., Xiaoxuan, S. and Wenxin, T. 1994. Medical research and development on seabuckthorn. *Hippophae* 7: 32-39.

21. Nersesian, A.K., Zilfian, V.N., Kumkumadzhian, V.A. and Proshian, N.V. 2005. Anti-mutagenic properties of seabuckthorn oil. *Genetika* 26: 378-80.

22. Nersesyan, A.K., Zil'fyan, V.N., Kumkumadzhian, V.A. and Proshyan, N.V. 1990. The anti-mutagenic action of seabuckthorn oil. *Genetika (Moscow)* 26 (2): 378-380.

23. Padmavathi, B., Upreti, M., Singh, V., Rao, A.R., Singh, R.P. and Rath, P.C. 2005. Chemoprevention by *Hippophae rhamnoides*: Effects on tumorigenesis, phase II and antioxidant enzymes, and IRF-1 transcription factor. *Nutrition and Cancer* 51: 59-67.

24. Sabir, S.M., Maqsood, H., Ahmed, S.D., Shah, A.H. and Khan, M.Q. 2005. Chemical and nutritional constituents of seabuckthorn (*Hippophae rhamnoides* ssp. *turkestanica*) berries from Pakistan. *Italian Journal of Food Science* 17: 455-462.

25. Varshneya, C., Kant, V. and Mehta, M. 2012. Total phenolic contents and free radical scavenging activities of different extracts of seabuckthorn (*Hippophae rhamnoides*) pomace without seeds. *International Journal of Food Sciences and Nutrition* 63: 153-159.

26. Yang, B. and Kallio, H. 2002. Composition and physiological effects of seabuckthorn (*Hippophae*) lipids. *Trends in Food Science and Technology* 13: 160-167.

27. Yang, J.P., Wang, X., Liu, Y.Y., Li, G.X., Ren, L.F., Jing, J.J., Zhang, H.X., Zhong, C.J. and Shu, R.X. 1989. A preliminary research on the anti-tumor effects of seabuckthorn press residue oil. In: *Proceedings of International Symposium on Seabuckthorn (H. rhamnoides L.)*, p. 382-383, Xian, China.

28. Yeh, Y., Hsieh, Y., Lee, Y. and Shen, Y. 2012. Dietary seabuckthorn (*Hippophae rhamnoides* L.) reduces toxicity of oxidized cholesterol in rats. *e-SPEN Journal* 7: e69-e77.

29. Zeb, A. 2004. Important therapeutic uses of seabuckthorn (*Hippophae*): A review. *Journal of Biological Sciences* 4: 687-693.

30. Zeb, A. 2006. Anti-carcinogenic potential of lipids from seabuckthorn: Evidence from recent literature. *Asian Pacific Journal of Cancer Prevention* 7: 32-35.

31. Zeb, A. and Khan, I. 2008. Composition and medicinal properties of seabuckthorn juice. In: *Seabuckthorn (Hippophae L.)-A Multipurpose Wonder Plant. Vol. III: Advances in Research and Development* (V. Singh, Ed. in Chief, 2008), p.205-213, Daya Publishing House, New Delhi, 566p.

32. Zeb, A. and Khan, I. 2009. Pharmacological applications of seabuckthorn (*Hippophae*). In: *Advances in Phytotherapy Research* (M. Eddouks, Ed.), p. 1-16, Research Signpost, New Delhi, India.

33. Zeb, A. and Malook, I. 2009. Biochemical characterization of seabuckthorn (*Hippophae rhamnoides* L. spp. *turkestanica*) seed. *African Journal of Biotechnology* 8: 1625-1629.

34. Zhang, P. 1989. The anti-cancer activities of *Hippophae* seed oil and its effect on the weight of the immunological organs. *Hippophae* 3: 31-34.

35. Zhang, P., Mao, Y.C., Sun, B., Qian, M. and Qu, W.J. 2005. Changes in apoptosis related genes expression profile in human breast carcinoma cell line Bcap-37 induced by flavonoids from seed residues of *Hippophae rhamnoides* L. *Chinese Journal of Cancer* 24: 454-460.

36. Zhang, P.Z., Ding, X.F., Mao, L.N., Li, D.X. and Li, L.P. 1989. Anti-cancer effects of seabuckthorn juice and seed oil and their effects on immune function. In: *Proceedings of International Symposium on Seabuckthorn (H. rhamnoides L.)*, p. 373-381, Xian, China.

Chapter 29

Herbal Preparation from Seabuckthorn (*Hippophae* L.) Renders Survival Benefit, Protects Haemopoetic and Liver Stem Cells in Whole Body Irradiated Mice

Madhu Bala, Sandhya Tiwari, Jagdish Prasad and Surinder Singh

Radiation Biology Division, Institute of Nuclear Medicine and Allied Sciences,
Brig. S.K. Mazumdar Marg, Delhi – 110 054, India

SUMMARY

The study was planned to evaluate radioprotective effects of a preparation from seabuckthorn (*Hippophae rhamnoides* ssp. *turkestanica*) leaves (code SBL-2), in mice. Whole body ^{60}Co-gamma-irradiation was given without or after intra-peritoneal administration of SBL-2. Haemopoetic stem cells were enumerated using endogenous colony forming assay. Effects on liver cells were studied by RT-PCR followed by DNA sequencing of fibroblast growth factor-inducible14 (Fn14) and Tumor necrosis factor-like weak inducer of apoptosis (TWEAK). Biochemical assays were used to determine thiols, lipid peroxidation, alkaline phosphatase, RBC haemolysis, reducing power and free radical scavenging activity. ^{60}Co-gamma-irradiation (10Gy) caused 100 per cent mortality. SBL-2 treatment before irradiation rendered >90 per cent survival; stimulated hematopoietic stem cells; countered radiation induced changes in (i) mRNA levels of Fn14 and TWEAK; (ii) thiols; (iii) lipid peroxidation; (iv) alkaline phosphatase; and (v) RBC haemolysis. SBL-2 had high reducing power, superoxide radical scavenging activity and hydroxyl radical scavenging activity. This study demonstrated radioprotective potential of SBL-2.

Keywords: *Colony-forming units, Gamma-rays, Hippophae, Stem cells, TWEAK and Fn14.*

INTRODUCTION

Whole body irradiation may happen during radiation accidents as well as during radiotherapy of some tumor. The total body irradiation damages multiple tissues and disrupts many intracellular pathways. Protection to normal tissues from the damaging effects of radiation is, therefore, essential. Development of a drug, which can protect populations exposed to lethal doses of radiation is a global challenge till date and is the subject of extensive research. Synthetic preparations, involving single molecule or small group of molecules, have failed to yield desirable results because of (i) limited protection to some tissues only and/or (ii) unacceptable toxicity (Weiss and Landauer, 2009). More recent efforts include development of radioprotective drug from plants (Nair *et al.*, 2004; Jagetia, 2007; Lee *et al.*, 2010). Our focus is to develop radioprotective preparations from high altitude plants (Goel *et al.*, 2004; Goel and Bala, 2006; Goel *et al.*, 2006). Such plants have multiple-stress tolerant molecules induced by extreme climatic conditions prevalent at high altitude regions. Currently, *Hippophae rhamnoides* L. (seabuckthorn; Family *Elaeagnaceae*), is under active investigations in our laboratory. Seabuckthorn is a hardy, drought- and cold-tolerant plant, found throughout Europe and Asia. In India, it is commonly found in Northwest Indian Himalayas. Traditionally, *Hippophae* is recommended in Indian, Chinese and Tibetan Systems of Medicine for treatment of many ailments *viz.*, cardiac and pulmonary disorders, cough, colds, fever, inflammation, abscesses, toxicity, tumors, and gynecological diseases (Zeb, 2004). Seabuckthorn leaves are pale silvery-green, lanceolate, 3-8 cm long and less than 7 mm broad. The leaves also form constituents of some beverages. High contents of many bioactive molecules *viz.*, vitamin A, C, K and E, organic acids, minerals, carotenoids and phenolic compounds impart a unique ethano-pharmacological importance to *Hippophae* (Bao and Lou, 2006). Earlier, we reported significant radioprotection by alcoholic preparation from berries of *H. rhamnoides* [RH-3, 82 per cent survival (Goel and Bala, 2006)]; and by aqueous preparation from leaves of *H. rhamnoides* [SBL-1, 90 per cent survival (Bala *et al.*, 2009)]. The use of leaves was preferred over berries because of longer shelf life and eco-friendly collection. An ideal radioprotective drug should be potent enough to counter multiple pathologies and offer regeneration/recovery of the tissues damaged by irradiation. In this study, we investigated the protective effects of aqua-alcoholic preparation from *Hippophae* leaves on survival as well as regenerative potential of stem cells in the irradiated mice. Aqua-alcoholic preparation had both polar and non-polar soluble bioactive molecules and was expected to offer better protection than aqueous preparation.

MATERIALS AND METHODS

Extraction and Characterization of SBL-2

Fresh green leaves of seabuckthorn, identified by ethno-botanist and confirmed as *Hippophae rhamnoides* ssp. *turkestanica* (specimen records preserved at Museum, DIHAR, Leh, India), collected from a particular region of Himalayas, were shade dried, powdered and soaked in 70 per cent ethanol. The supernatant was filtered and lyophilized. The dried powder, coded as SBL-2 (drug), was first tested for absence of endotoxins using a Limulus Amebocyte Lysate assay kit (Tsuji and Harrison, 1978). Percentage of ash, moisture, polyphenolics, tannins and proanthocynindines in SBL-2 (Table 29.1) were determined using standard gravimetric methods. HPLC profile of SBL-2 (Figure 29.1) was developed (HPLC system with UV detector, Waters Corporation, USA). Kaempherol and Isorhamnetin were identified on the basis of retention time, co-injections, and spectral matching with standards.

Table 29.1: The Constituents of SBL-2

Constituent	*Value (per cent of total)*
Ash	10.1
Proanthocyanidines	1.1
Polyphenolics	4.5
Moisture	4.0
Tannins	12.0
Unidentified	68.3

Amino acid contents were determined by HPLC using Waters PICO-TAG method (as

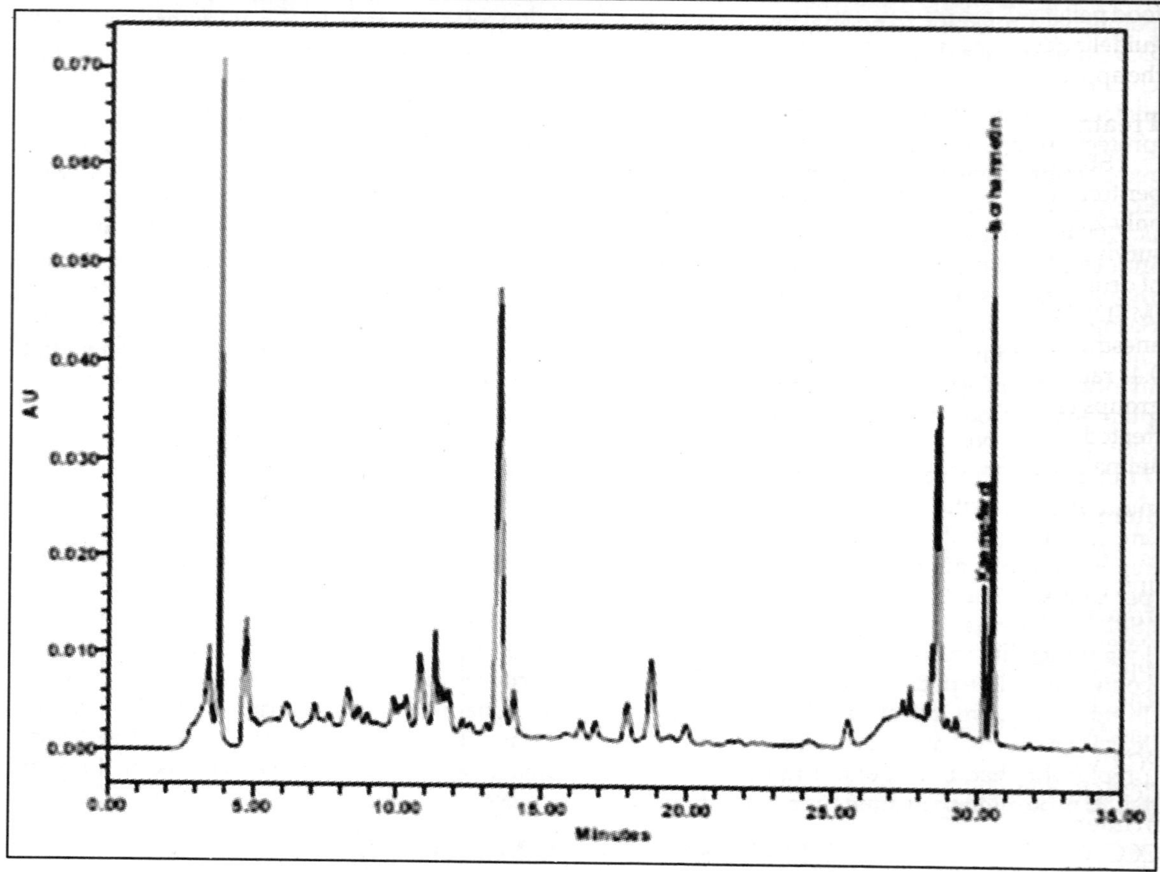

Figure 29.1: The HPLC Profile of SBL-2 Using C-18 Reverse Phase Column and also Shows Kaempherol and Isorhamnetin Identified in SBL-2.

manufacturer's instructions) (Table 29.2). Reducing power was determined using standard potassium ferricyanide reducing assay (Senevirathne *et al.*, 2006). Superoxide radical ($O_2^{\%-}$) scavenging was determined in terms of inhibition of NBT reduction (Nishikimi *et al.*, 1972). Scavenging of hydroxyl radicals (generated by Fenton reaction or by gamma irradiation), was measured by the method Senevirathne *et al.* (2006). Drug concentration rendering 50 per cent inhibition of the maximum inhibition obtained by drug treatment was considered as $IC_{50.}$ Each new batch of SBL-2 was checked for consistency of above mentioned parameters before using for animal experiments.

Table 29.2: Important Amino Acids of SBL-2

Amino Acid	*Content (mmol/mg)*
Aspartic acid	9.653
Arginine	6.720
DL-Phenylalanine	64.64
Lysine	4.480

Animals

Male Strain 'A' inbred mice (6-8 weeks old, weighing 25±2 g), maintained under controlled environment (25±2 °C, RH 60±5 per cent ; 12 h photoperiod) in ventilated cages and fed with standard food pellets (Golden feed, India) were used. All animal experiments were conducted according to the guidelines of Committee for Protection and Care of Small Experimental Animals (CPCSEA) and as per the approved protocol No. CPCSEA 8/GO/a/99/CPCSEA, INMAs Delhi, India.

Treatments

SBL-2 was dissolved in 25 per cent DMSO (Dimethylsulphoxide) and was administered intra-peritoneally (i.p.) to mice [concentration in mg/kg b.w. (body weight)]. Sham treated animals received only 25 per cent DMSO (i.p.). Changes in behaviour, respiration, neuromuscular coordination and survival, manifested within 72 h after drug administration, were recorded. The maximum concentration of drug which yielded no toxic effects was considered as the maximum tolerated drug concentration (MTD). Whole body irradiation to individual mouse was given in perforated plastic bottles, without anesthesia using ^{60}Co-gamma cell (model 220-Atomic Energy Commission, Canada Ltd., dose rate 0.42 rad/sec). Fresh air was continuously circulated to avoid hypoxia. Animals were divided in four groups (6 mice/group) and each experiment was repeated three times. Group I animals were sham treated, non-irradiated; Group II animals received whole body ^{60}Co-gamma-irradiation; Group III animals received drug only; and Group-IV animals received drug prior to ^{60}Co-gamma-irradiation.

Stem Cell Stimulation and other *Ex vivo* Assay

Animals were anesthetized and blood was drawn retro-orbitally. For endogenous colony-forming unit (CFU) assay to enumerate haemopoietic stem cells (HSc), spleen was dissected out on 10th post-irradiation day and fixed in Bouin's solution (Till and McCulloch, 1961). Visible macroscopic colonies were scored. Liver was excised, weighed and homogenized in pre-chilled phosphate buffer saline (PBS, pH 7.2). Both supernatant and liver homogenate were preserved at -80 °C. Protein concentration, RBC haemolysis, thiols, lipid peroxidation and alkaline phosphatase were measured using standardized procedures (Tiwari *et al.*, 2009). RNA extraction, quantification, purification and RT-PCR were performed using kits (Qiagen) (Dwivedi *et al.*, 2008). TWEAK primers (forward, 52-CGAGCTATTGCAGCCCATTAT-32; backward, 52-ACCTGCTTGTGCTCCATCCT-32) and, Fn14 primers (forward, 52- CTAGTTTCCTGGTCTGGAGAAGATG- 32; backward, 52-CCCTCTCCACCAGTCTCCTCTA-32) (Jakubowski *et al.*, 2005), were synthesized from IDT (Coralville). For RT-PCR studies (i-cycler, BioRad, US), reverse transcription was done at 50 °C for 30 minutes; 95 °C for 10 minutes; 30 cycles at 95 °C for 45 seconds and at 58.5 °C for 45 seconds and at 72 °C for 1 minute. Final extension was at 72 °C for 10 minutes followed by cooling at 4 °C. PCR products were quantified by electrophoresis followed by densitometry (Labworks 4.0™, UK). Amplified PCR product was purified using Chromous PCR clean up kit and sequenced (ABI 3130 Genetic Analyzer, Big Dye Terminator version 3.1 Cycle sequencing kit and software 'Sequence analysis_v 5.2').

Statistical Analysis

Multivariate analysis of variance was applied for analyzing statistically important changes in more than one parameter. Post-hoc Student's *t* test was applied for evaluating significance between any two treatments. Value at $p<0.05$ was considered significant. The data was expressed as mean± standard deviation (SD).

RESULTS

MTD, Survival and Body Weight

Single dose of SBL-2 up to 170 mg/kg b.w., was well tolerated by all animals, had no adverse effects and was therefore, considered as MTD; LD_{50} was 190 mg/kg b.w. (Table 29.3). All irradiated (10

Gy) animals, without prior treatment with SBL-2, died within 10 to 12 days (Figure 29.2). One time administration of SBL-2 (20 mg/kg b.w., 30 minute before irradiation, 10 Gy), rendered 92 per cent survival for 30 post-irradiation days and beyond. The interval of 30 minutes between SBL-2 administration and irradiation was most effective, although, SBL-2 administration at increased time intervals (up to 4 h before irradiation, 10 Gy), promoted recovery of radiation induced weight loss (Figure 29.3).

Table 29.3: Tolerance of Single Dose of SBL-2 (observed up to 72 h, n=18)

Dose (mg/kg b.w.)	*Per cent Survival*
50	100.0
100	100.0
120	100.0
140	100.0
160	100.0
170	100.0
180	66.6
190	50.0
200	0.0

Protection to Haemopoietic Stem Cells and Hepatic System

On day 10 after irradiation, endogenous CFU count in spleens were 2.67 (5 Gy), 1.81 (7.5 Gy) and 0.33 (10 Gy). Administration of 20 mg/kg b.w. of SBL-2, 30 min before irradiation rendered significantly ($p < 0.01$) higher CFU counts which at radiation doses 5, 7.5 and 10 Gy were 32.3 ± 1.33, 29.3 ± 0.77 and 17.8 ± 0.65, respectively (Figure 29.4).

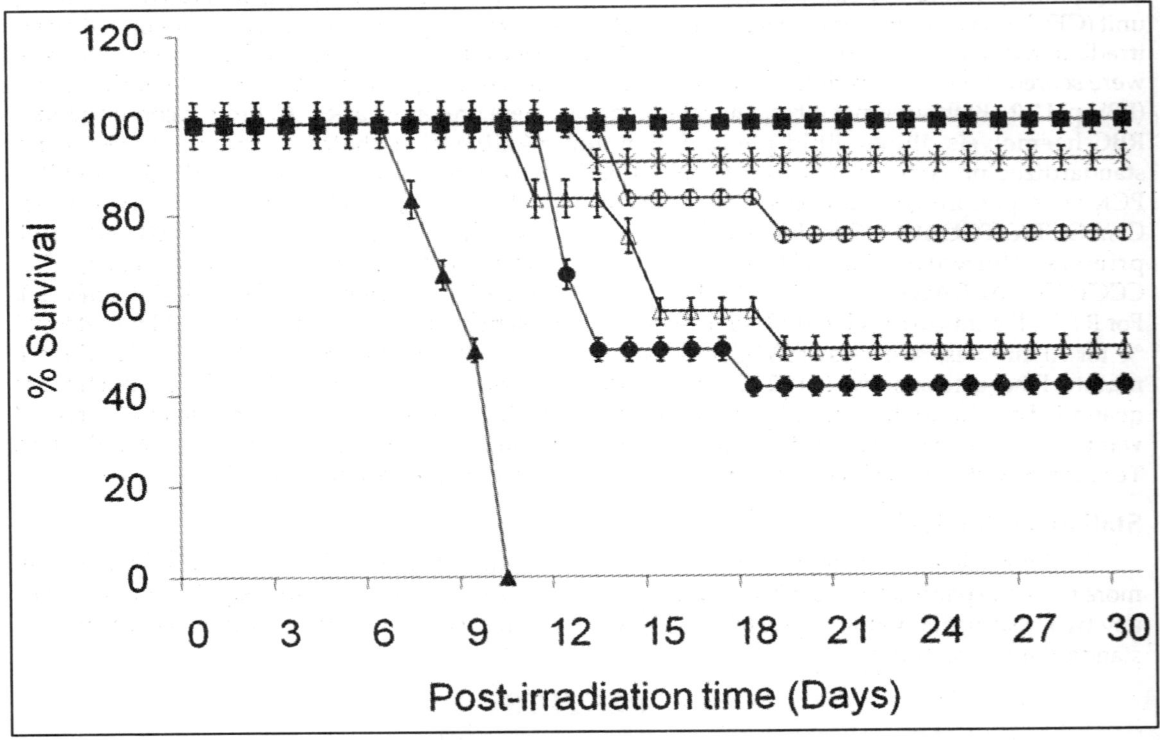

Figure 29.2: The Radioprotective Effects of SBL-2. Test animals were administered intra-peritoneally, with SBL-1, 30 min before exposure to whole body γ-irradiation at 10 Gy; SBL-1 treatments were 0 (s), 10 (l), 20 (S), 30 (O) and 40 (Δ) mg kg^{-1} body weight. The untreated control (□) and drug alone (n) had 100 per cent survival beyond 30 days.

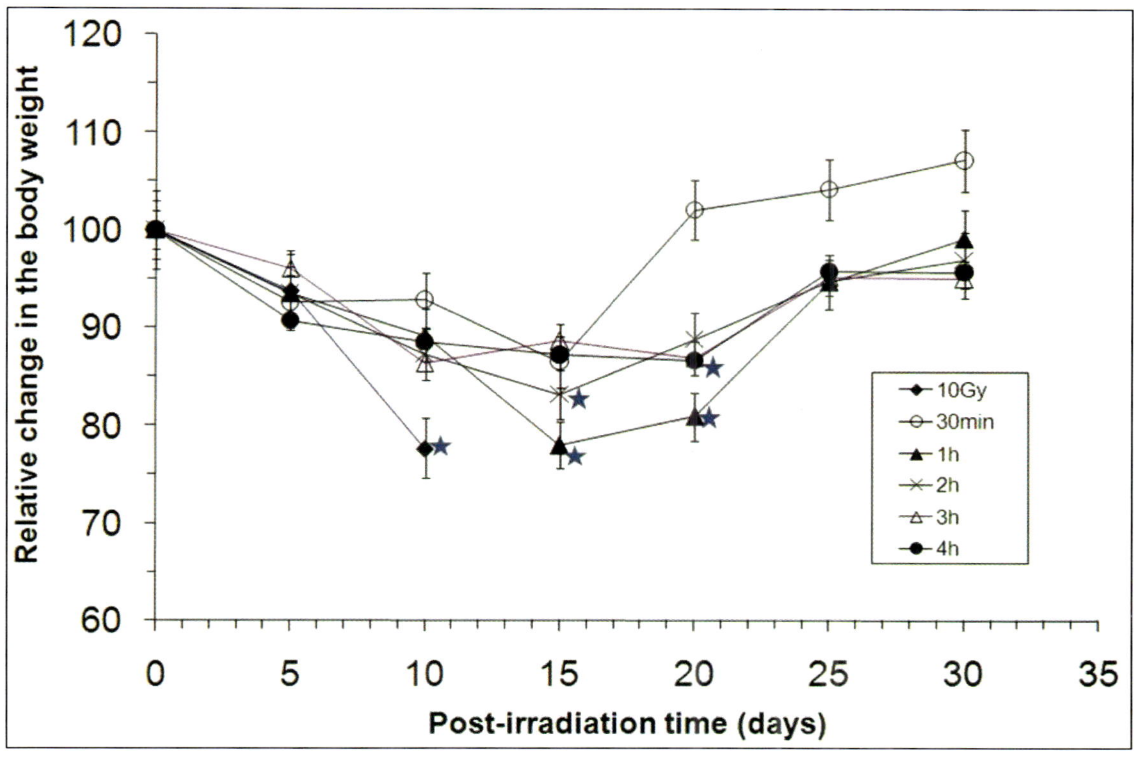

Figure 29.3: Changes in Body Weight of Irradiated Mice Treated with SBL-2 (20 mg/kg b.w.) at different Time Intervals before Irradiation (10 Gy). Significant in comparison to untreated control at $p \leq 0.05$.

The dose reduction factor (DRF) on the basis of CFU counts was approximately 2.4. In comparison to group I, alkaline phosphatase activity of irradiated (10 Gy) animals was significantly increased but in group III and group IV, the activity was same as that in group I. Total thiols content in irradiated (10 Gy) animals was reduced significantly in comparison to group I but not in group III and group IV. The significantly high levels of MDA and therefore, TBARS was observed in irradiated (10 Gy) animals in comparison to group I, suggested high levels of radiation-induced lipid peroxidation. Administration of SBL-2 before irradiation countered the radiation-induced increase in MDA formation (Figure 29.5). Irradiation (7.5 and 10 Gy) increased mRNA levels of Fn14 and TWEAK significantly ($p \leq 0.05$). Treatment with SBL-2, prior to irradiation, countered the radiation induced increase in mRNA of both these genes (Figure 29.6). DNA sequences of all our amplicons, when compared with sequences in NCBI database using BLAST search engine, were found to have identical base pairs (100 per cent) with Fn14 and TWEAK.

In vitro/Ex vivo Studies

SBL-2 showed strong dose dependent reducing power, superoxide radical scavenging and hydroxyl radical ($OH^{\%}$) scavenging ability (Table 29.4). The IC_{50} for superoxide radical scavenging, Fenton reaction and radiation generated $OH^{\%}$ radical scavenging was 115 µg/ml, 97.5 µg/ml and 10

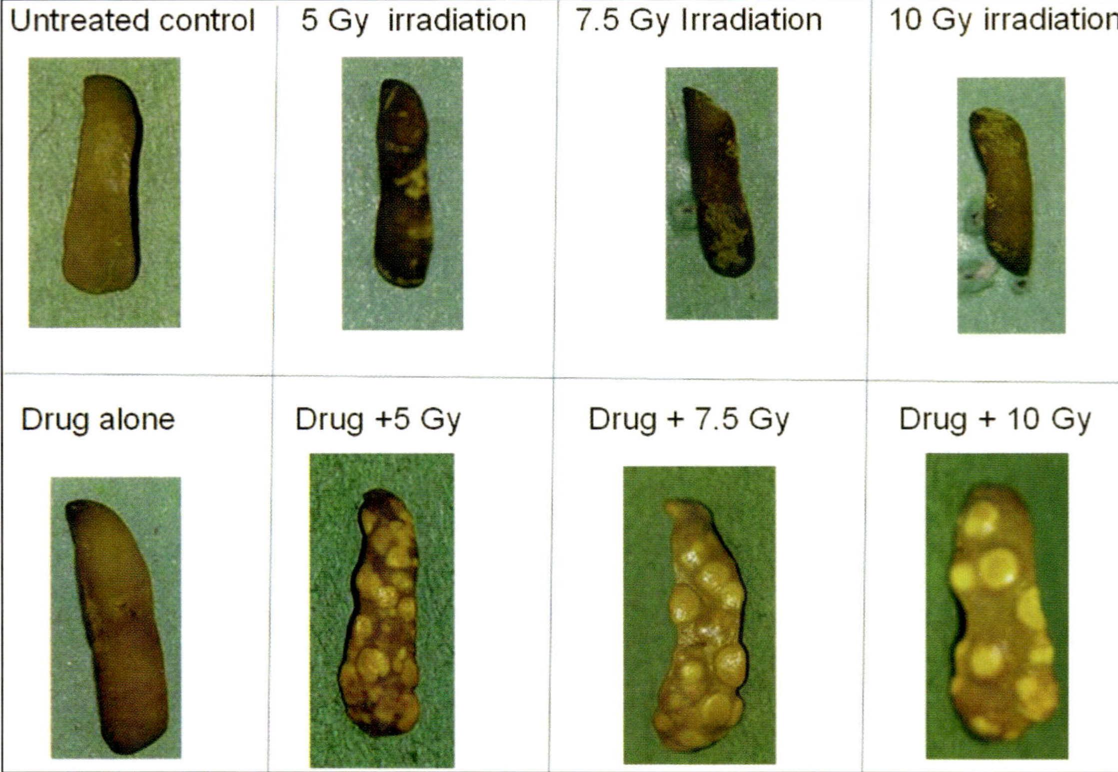

Figure 29.4: Modifying Effect of SBL-2 on Endogenous CFU Counts on Spleens of Irradiated as well as Non-Irradiated Animals.

Table 29.4: Antioxidant Properties of SBL-2, Values are Mean±SD of 6 Replicates

SBL-2	*Fe^{3+} to Fe^{2+} (Ferricyanide)**	*$O_2^{\cdot -}$ Scavenging (NBT Assay)**	*Scavenging OH Radicals Generated by*		*RBC Heamolysis#*	*Lipid Peroxidation (MDA formation)#*
			Irradiation (200 Gy)*	*Fenton Reaction**		
(µg/ml)	*($OD_{700\ nm}$)*	*(Per cent Inhibition)*	*(Per cent)*	*(Per cent)*	*(Per cent Inhibition)*	*nM/mg Protein)*
15.6	0.199± 0.013	1.02±0.023	45.6 ±3.6	11.8±1.6	18.54±5.1	3.6±0.056
31.25	0.627± 0.002	1.4±0.67$	53.3 ±2.06	22.5±0.626	34.5±2.89	2.7±0.053
625	1.1± 0.002	30.0± 4.2$	68.7±1.44$	26.7±0.94	85.9±1.92$	1.70±0.021$
125	1.94± 0.03	40.8±0.6$	73.8±2.1$	39.1±3.14$	90.9±3.5$	1.41±0.021$
250	2.3± 0.015	68.3±3.1$	73±3.1$	45.0±2.2$	92.47±9.5$	1.16±0.025$
500	2.1± 0.014	75.6±1.7$	75.0± 1.8$	60.7±1.24$	92.9±9.7$	1.16±0.025$
10000	2.1± 0.1	78±1.68$	75.1±1.9$	70.0±2.5$	88.7±1.47$	0.92+0.097$

* *In vitro* study, values normalized with reference to the non-drug treated controls.

\# *Ex vivo* study, values normalized with reference to irradiated non-drug treated controls.

$ Significant in comparison to control at $p \leq 0.05$.

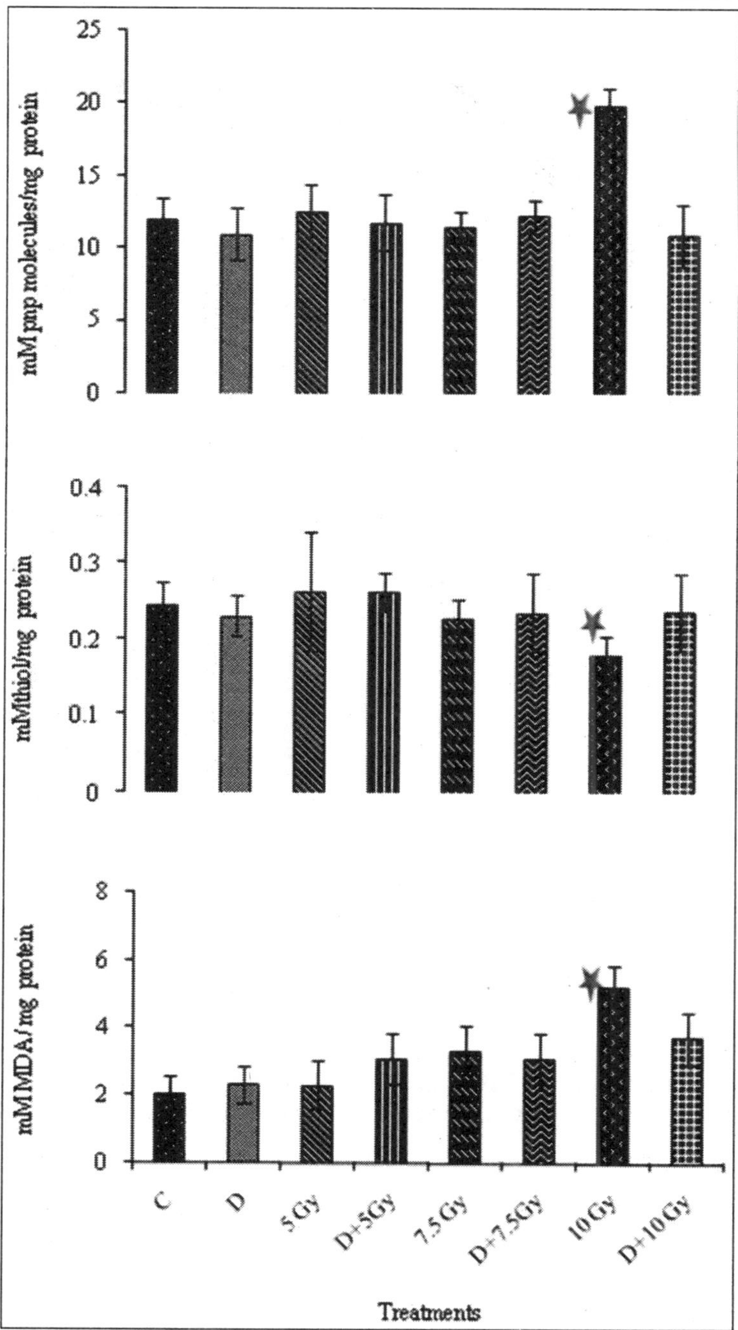

Figure 29.5: ***In vivo*** **Effects of SBL-2 on MDA Formation, Thiol Content and Alkaline Phosphatase Activity in Liver. The data is mean± SD; where six animals were used per treatment. Significant in comparison to untreated control at** $p \leq 0.05$.

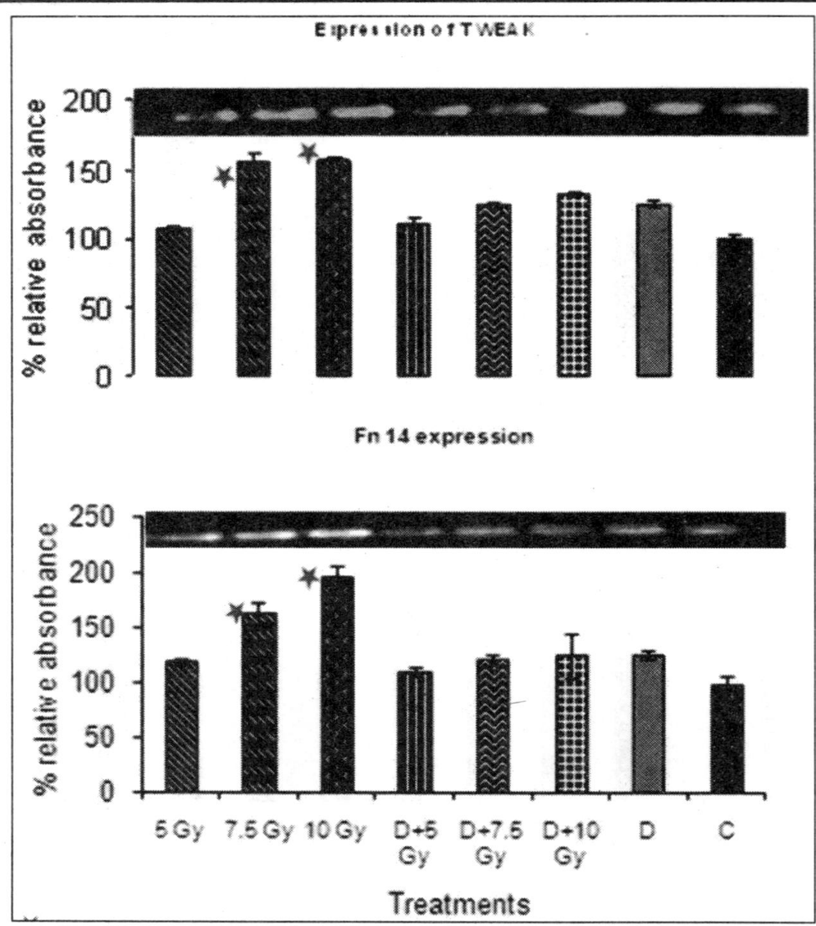

Figure 29.6: Effects of SBL-2 Treatment on mRNA Levels of TWEAK and Fn14 in Liver Cells of Whole Body Irradiated Animals. Significant in comparison to untreated control at $p \leq 0.05$.

μg/ml, respectively. SBL-2 pre-treatment countered the radiation induced MDA formation and IC_{50} was 48 μg/ml. The radiation induced RBC haemolysis was countered by SBL-2 pretreatment and IC_{50} was 35.5 μg/ml. The drug alone did not increase lipid peroxidation or RBC haemolysis.

DISCUSSION

This study demonstrated that single prophylactic dose of SBL-2 (20 mg/kg b.w.) protected 92 per cent of irradiated (10 Gy) mice population (Figure 29.2) and countered radiation induced weight loss (Figure 29.3). Reduced activity of SBL-2 at <20 mg/kg b.w. was due to insufficiency of bioactive molecules. Reduced radioprotective efficacy of SBL-2 at ≥30 mg/kg b.w., indicated adverse effects of excess drug in irradiated animals. Such effects were due to compromised physiology of irradiated animals (Uma Devi *et al.*, 1999; Goel *et al.*, 2004; Goel and Bala, 2006; Goel *et al.*, 2006).

Synthetic radioprotective agents have been unacceptable, because their maximum radioprotective effects occur at doses approaching LD_{50} dose (Nair *et al.*, 2004; Weiss and Landauer, 2009; Lee *et al.*, 2010). In this matter, SBL-2 was considered non-toxic because maximum radioprotective dose (20 mg/kg b.w.) was 9.5 times less than LD_{50} dose (190 mg/kg b.w., Table 29.3). SBL-2, therefore, met with an essential criterion for an ideal radioprotector. The non-toxic effect of *Hippophae* leaves in rats is reported (Saggu *et al.*, 2007). The beneficial properties of SBL-2 could be attributed to high content of antioxidants *viz.*, tannins, polyphenolics (Table 29.1), quercitin and Isorhamnetin (Figure 29.1), as well as some essential amino acids *viz.*, non-polar DL-phenylalanine (DLPA); a base, lysine as well as non-essential amino acids arginine and aspartic acid (Table 29.2). Role of these amino acids in metabolism, wound healing and detoxification of body are known (Nelson and Cox, 2000).

^{60}Co-gamma-irradiation causes massive structural and functional damage, primarily by generating high flux of free radicals and multiple reactive species. Therefore, a radioprotector having broad spectrum antioxidant properties is desirable. Inhibiti-on of TBARS formation by SBL-2 demonstrated its peroxyl radicals scavenging ability and significantly lower IC_{50} for SBL-2 (48 µg/ml) in comparison to SBL-1 (IC_{50} 57 µg/ml) (Bala *et al.*, 2009; Tiwari *et al.*, 2009) and vitamin E (IC_{50} 120 µg/ml) (Sundararajan *et al.*, 2006), demonstrated stronger antioxidant activity of SBL-2 in lipid phase. SBL-2 (IC_{50} 35.5 µg/ml) had better ability to counter radiation induced RBC haemolysis than alpha-tocopherol (IC_{50} 1.6 mg/ml). The scavenging of superoxide radicals by SBL-2 (IC_{50} 115 µg/ml, Table 29.4) was higher than Triphala (antiaging, antioxidant herbal preparation from three Indian medicinal plants, IC_{50} 42 mg/ml) (Naik *et al.*, 2006). Normalizing radiation induced changes in thiols content, lipid peroxidation and alkaline phosphatase, by administration of SBL-2 prior to irradiation (Figure 29.5), suggested that SBL-2 had high antioxidant potential *in vivo*, which restored healthy functioning of liver (Kojima and Sakurada, 1976). Significant increase in HScs in group IV, suggested that SBL-2 acted as regenerative-radioprotective medicine (Figure 29.4). Tissue regeneration is of immense importance for restoration of normal functioning and therefore, for radioprotection (Gokhal *et al.*, 2010). It is suggested that endothelial progenitor cells (EPCs) of bone marrow were mobilized by SBL-2. The EPCs normally home into sites of injury and repair *via* peripheral circulation and give rise to HScs (Burt *et al.*, 2008).

The low constitutive mRNA level of Fn14 (cell surface receptors) in normal mice (Figure 29.6) was in line with other reports (Fausto and Campbell, 2003) and indicated slow hepatocytes turnover. Radiation dose dependent increase in mRNA of Fn14 and TWEAK (a member of TNF super family of cytokines), suggested that oval cells proliferation was augmented after irradiation due to massive injury as in case of hepatocellular carcinoma (Lowes *et al.*, 1999). The 100 per cent match between the sequences of PCR products and NCBI data base, confirmed the amplification of target genes. Inhibition of radiation induced increase in mRNA of Fn14, in group IV animals (Figure 29.6), indicated a check on stimulation of oval cells. This could be due to smaller radiation induced damage in presence of SBL-2, primarily due to antioxidant properties of SBL-2 (Figure 29.5). In this study, we refrained from using internal controls, such as beta-actin, GAPDH or tubulin house keeping gene, because earlier studies showed that irradiation changed levels of all commonly used internal controls (Dwivedi *et al.*, 2008). Therefore, mRNA level of group I samples were used as reference to elucidate the effects of different treatments. Though RT-PCR is semi-quantitative, large quantitative difference between group II and group IV samples suggested protective as well as regenerative effect of SBL-2 on liver and liver stem cells in the mice irradiated with lethal doses of radiation (Figure 29.6). To the best of our knowledge, this is the first report on radioprotective effects of aqua-alcoholic preparation from leaves of *Hippophae rhamnoides*.

CONCLUSION

This study demonstrated that SBL-2 was non-toxic, had high radioprotective potential and could offer survival benefit to 92 per cent of mice population against lethal doses of ^{60}Co-gamma-radiation (10 Gy). SBL-2 acted as a versatile and strong antioxidant, protected and regenerated liver, and haemopoietic stem cells.

ACKNOWLEDGEMENT

This research was funded by DRDO, Government of India. Support from the Director, DIHAR, Leh, in collecting plant material is gratefully acknowledged.

REFERENCES

1. Bala, M., Prasad, J., Singh, S., Tiwari, S. and Sawhney, R.C. 2009. Whole body radioprotective effects of SBL-1: A preparation from leaves of *Hippophae rhamnoides*. *J. Herbs Spices and Med. Plants* 15: 203-215.

2. Bao, M. and Lou, Y. 2006. Flavonoids from seabuckthorn protect endothelial cells (EA.hy926) from oxidized low-density lipoprotein induced injuries *via* regulation of LOX-1 and eNOS expression. *J. Cardiovasc. Pharmacol*. 48: 834-841.

3. Burt, R.K., Loh, Y., Pearce, W., Beohar, N., Barr, W.G., Craig, R., Wen, Y., Rapp, J.A. and Kessler, J. 2008. Clinical applications of blood-derived and marrow-derived stem cells for non-malignant diseases. *J. Am. Med. Ass*. 299: 925-936.

4. Dwivedi, M., Sehgal, N. and Bala, M. 2008.The effects of low dose of ^{60}Co-gamma-radiation on radioresistance, mutagenesis, gene conversion, cell cycle and transcriptome profile in *Saccharomyces cerevisiae*. *Int. J. Low Radiation* 5: 290-309.

5. Fausto, N. and Campbell, J.S. 2003. The role of hepatocytes and oval cells in liver regeneration and repopulation. *Mech. Develop*. 120: 117-130.

6. Goel, H.C. and Bala, M. 2006. Seabuckthorn (*Hippophae rhamnoides* L.) as a radioprotector. In: *Seabuckthorn-A Multipurpose Wonder Plant. Vol. II: Biochemistry and Pharmacology* (V. Singh, Ed. in Chief, 2006), p.419-455, Daya Publishing House, New Delhi, 600p.

7. Goel, H.C., Bala, M., Prasad, J. and Singh, S. 2006. Radioprotection by *Rhdiola imbricata* in mice against whole body lethal irradiation. *J. Medicinal Food* 9: 154-160.

8. Goel, H.C., Prasad, J., Singh, S., Sagar, R.K., Agrawala, P.K., Bala, M., Sinha, A.K. and Dogra, R. 2004. Radioprotective potential of an herbal extract of *Tinospora cordifolia*. *J. Radiat. Res*. 45: 61-68.

9. Gokhal, A., Rwigema, J., Epperly, M.W., Glowacki, J., Wang, H., Wipf, P., Goff, J.P., Dixon, T., Patrene, K. and Greenberger, J.S. 2010. Small molecule GS-nitroxide ameliorates ionizing irradiation-induced delay in bone wound healing in a novel murine model. *In vivo* 24: 377-386.

10. Jagetia, G.C. 2007. Radioprotective potential of plants and herbs against effects of ionizing radiation. *J. Clin. Biochem. Nut*. 40: 74-81.

11. Jakubowski, A., Ambrose, C., Parr, M., Lincecum, J.M., Wang, M.Z., Zheng, T.S., Browning, B., Michaelson, J.S., Baestcher, M., Wang, B., Bissell, D.M. and Burkly, L.C. 2005. TWEAK induces liver progenitor cell proliferation. *J. Clin. Invest*. 115: 2330-2340.

12. Kojima, Y. and Sakurada, T. 1976. Increase in alkaline phosphatase activity in the liver of mice bearing Ehrlich ascites tumor. *Cancer Res*. 36: 23-27.

13. Lee, T.K., O'Brien, K.F., Wang, W., Johnke, R.M., Sheng, C., Benhabib, S.M., Wang, T. and Allison, R.R. 2010. Radioprotective effect of American ginseng on human lymphocytes at 90 minutes post-irradiation: A study of 40 cases. *J. Altern. Complement Med.* 16: 561-567.

14. Lowes, K.N., Brennan, B.A., Yeoh, G.C. and Olynyk, J.K. 1999. Oval cell numbers in human chronic liver diseases are directly related to disease severity. *Am. J. Pathol.* 154: 537-541.

15. Naik, G.H., Priyadarsini, K.I. and Mohan, H. 2006. Free radical scavenging reactions and phytochemical analysis of triphala, an ayurvedic formulation. *Curr. Sci.* 90: 1100-1105.

16. Nair, C.K., Salvi, V., Kagiya, T.V. and Rajagopalan, R. 2004. Relevance of radioprotectors in radiotherapy: Studies with tocopherol monoglucoside. *J. Environ. Pathol. Toxicol. Oncol.* 23: 153-160.

17. Nelson, D.L. and Cox, M.M. 2000. *Principles of Biochemistry* (3rd Ed.), Worth Publishing, New York, 1100p.

18. Nishikimi, M., Appaji, N. and Yagi, K. 1972. The occurrence of superoxide anion in the reaction of reduced phenazine methosulfate and molecular oxygen. *Biochem. Biophys. Res. Comm.* 46: 849-853.

19. Saggu, S., Divekar, H.M., Gupta, V., Sawhney, R.C., Banerjee, P.K. and Kumar, R. 2007. Adaptogenic and safety evaluation of seabuckthorn (*Hippophae rhamnoides*) leaf extract-A dose dependent study. *Food Chem. Toxicol.* 45: 609-617.

20. Senevirathne, M., Kim, S.H., Siriwardhana, N., Ha, J.H., Lee, K.W. and Jeon, Y.J. 2006. Antioxidant potential of *Ecklonia cava* on reactive oxygen species scavenging, metal chelating, reducing power and lipid peroxidation inhibition. *Food Sci. Tech. Int.* 12: 27-38.

21. Sundararajan, R., Haja, N.A., Venkatesan, K., Mukherjee, K., Saha, B.P., Bandyopadhyay, A. and Mukherjee, P.K. 2006. Cystisus scoparious link:A natural antioxidant. *BMC Complement and Altern. Med.* 6: 8-12.

22. Till, J.E.and McCulloch, E.A. 1961. A direct measurement of radiation sensitivity of normal bone marrow. *Radiat. Res.* 14: 213-222.

23. Tiwari, S., Arya, A., Tyagi, S., Prasad, J., Singh, S., Vats, P., Kumar, D., Jain, S.K. and Bala, M. 2009. Anti-mutagenic and radioprotective properties of seabuckthorn leaf (*Hippophae rhamnoides* L.). *J. Med. Plants* 14: 83-89.

24. Tsuji, K. and Harrison, S.J. 1978. Dry-heat destruction of lipo-polysaccharide: Dry-heat destruction kinetics. *Appl. Environ. Microbiol.* 36: 710-741.

25. Uma Devi, P., Ganasoundari, A., Rao, B.S. and Srinivasan, K.K. 1999. *In vivo* radioprotection by *Ocimum* flavonoids-survival of mice. *Radiat. Res* 151: 74-78.

26. Weiss, J.F. and Landauer, M.R. 2009. History and development of radiation-protective agents. *Int. J. Radiat. Biol.* 85: 539-573.

27. Zeb, A. 2004. Important therapeutic uses of seabuckthorn (*Hippophae*): A review. *J. Biol. Sci.* 4: 690-693.

Chapter 30

A Comprehensive Study on Antiviral Polyphenols of Seabuckthorn Leaves

O.P. Sheichenko, V.I. Sheichenko, O.N. Tolkachev, L.D. Shipulina, L.V. Krepkova, S.A. Vichkanova, M.F. Mineeva, V.F. Okhotnikova, T.V. Kachalina and V.A. Bykov

All-Russian Research Institute of Medicinal and Aromatic Plants (VILAR), Grina Str. 7, Moscow 117216, Russia

SUMMARY

Seabuckthorn (*Hippophae rhamnoides* L.) was described during the systematic chemical and biological study of prospective medicinal plants of the Russian flora. Its leaves were shown to contain considerable amounts of biologically active polyphenols (gallo-ellagi tannins) possessing a wide spectrum of antiviral and antimicrobial activity. The original antiviral drug Hiporamin was developed on this basis. New approaches for the chemical composition in the plant material, the fraction of polyphenols, the drug and its pharmaceutical products were elaborated. Large scale methods of the total polyphenols production and the preparative method of individual tannins isolation were worked out. Some results of preclinical and clinical studies of the drug are discussed. The "Hiporamin" drug and its pharmaceutical products are produced at the commercial scale. An original specific enzyme biotest system was created as the method for revealing biologically active substances from the plant materials and forecasting their specific chemotherapeutic and pharmacologic activity.

Keywords: *Seabuckthorn, Hydrolyzing tannins, Hiporamin, NMR, Specific enzyme biotest systems, Toxicological study and clinical trials.*

INTRODUCTION

Seabuckthorn (*Hippophae rhamnoides* L.) is the popular medicinal plant of the Russian flora, which is used since old times in traditional medicine and foods (Sheichenko and Tolkachev, 2006). Numerous forms of the plant grow wildly and improved varieties are cultivated on farm lands at the

Seabuckthorn (*Hippophae* L.): *A Multipurpose Wonder Plant, Vol. 4: Emerging Trends in Research and Technologies* (Virendra Singh, *Editor-in-Chief*, 2014), p. 382–392, Daya Publishing House, New Delhi, India.

Russian territory. All parts of the plants are extensively studied on the chemical composition and biological activities. It was found that polyphenol fraction from seabuckthorn leaves, containing monomeric hydrolyzing gallo-ellagi-tannins (preferably strictinin (I), isostrictinin (II), casuarinin (III), casuarictin (IV) and pedunculagin (V), possesses antiviral activity in respect of wide spectrum of human influenza viruses A and B (*in vitro, in vivo, in ovo*), adenoviruses, paramixoviruses, herpes virus, cytomegalloviruses and HIV viruses etc. (Sheichenko *et al.*, 1987; Yoshida *et al.*, 1991; Sheichenko and Tolkachev, 2006; Krepkova *et al.*, 2008; Shipulina *et al.*, 2008). Specific mechanism of antiviral activity of polyphenols was shown to be related with inhibition of the viral neuraminidase and with the interferon induction action (Shipulina *et al.*, 2005). The drug inhibits gram positive, gram negative and other pathogenic bacteria and fungi, and has interferon induction activity. According to recent report, cyclic polyol quebrachitol (2-O-methyl-L-inositol) (VI) (De Olida *et al.*, 2003; Nobre *et al.*, 2006), possessing antioxidative and cytoprotective action, was also found in the hydrophilic fraction. Quebrachitol is the gastro-protector by the mechanisms involving endogenous nitric oxide release and the activation of K^+ATP channels. Leaf tannin fraction from seabuckthorn by the name Hiporamin was produced at the commercial scale in Russia (Bykov *et al.*, 2005; 2008). The wild forms of seabuckthorn are used for production of the drug. Its breeding varieties and green herbaceous shoots are also suitable as the source for the drug production (Sheichenko *et al.*, 2003). Seabuckthorn tannins inhibit the flu viral neuraminidase (sialidase). N-acetylneuraminic acid-sialic acid (Neu5Ac) (VII) is the important terminal negatively charged glyco-component of glycolipids and glycoproteins in human and animal tissues and saliva, which play an essential role as the immune modulator. It is effective against viral and bacterial infections (influenza viruses A and B, pneumonia, hepatitis, etc.).

MATERIALS AND METHODS

Biochemical Analysis

Total tannin contents and the composition of the tannin fraction from seabuckthorn leaves, the drug and the pharmaceutical forms were determined by the preparative chromatographic method of isolation of the constituents in combination with the chemical and instrumental methods of their identification (NMR and UV spectroscopy, TLC and HPLC) (Sheichenko *et al.*, 2001; Shipulina and Tolkachev, 2006). The procedures of antiviral drug Hiporamin and its pharmaceutical forms production were elaborated.

Clinical Study

The clinical trials of Hiporamin were carried out in 9 medical establishments: Research Institute of Virology RAMN, Research Institute of Transplantology and Artificial Organs, Research Institute of Immunology, Central Research Skin-Venereology Institute, Moscow Sechenov Medical Academy, Russian State Medical Academy of Postgraduate Education. Clinical study were carried out on 626 hospital patients (adults and children) in 6 pharmaceutical forms: Sublingual tablets (0.02), lyophilized powder (sterile) (0.02) for *ex temporal* aqueous solutions (for inhalations and nasal drops), 0.5 per cent ointment for skin and mucous membranes, suppositories (rectal and vaginal) (0.05 for adults), or (0.03 for children) as the antiviral remedy.

RESULTS AND DISCUSSION

Chemical Study and Analytical Determination

The chemical composition of extracts from leaves of seabuckthorn (*Hippophae rhamnoides* L.) (wild forms and varieties in Russia) was extensively studied. They were shown to be rich in

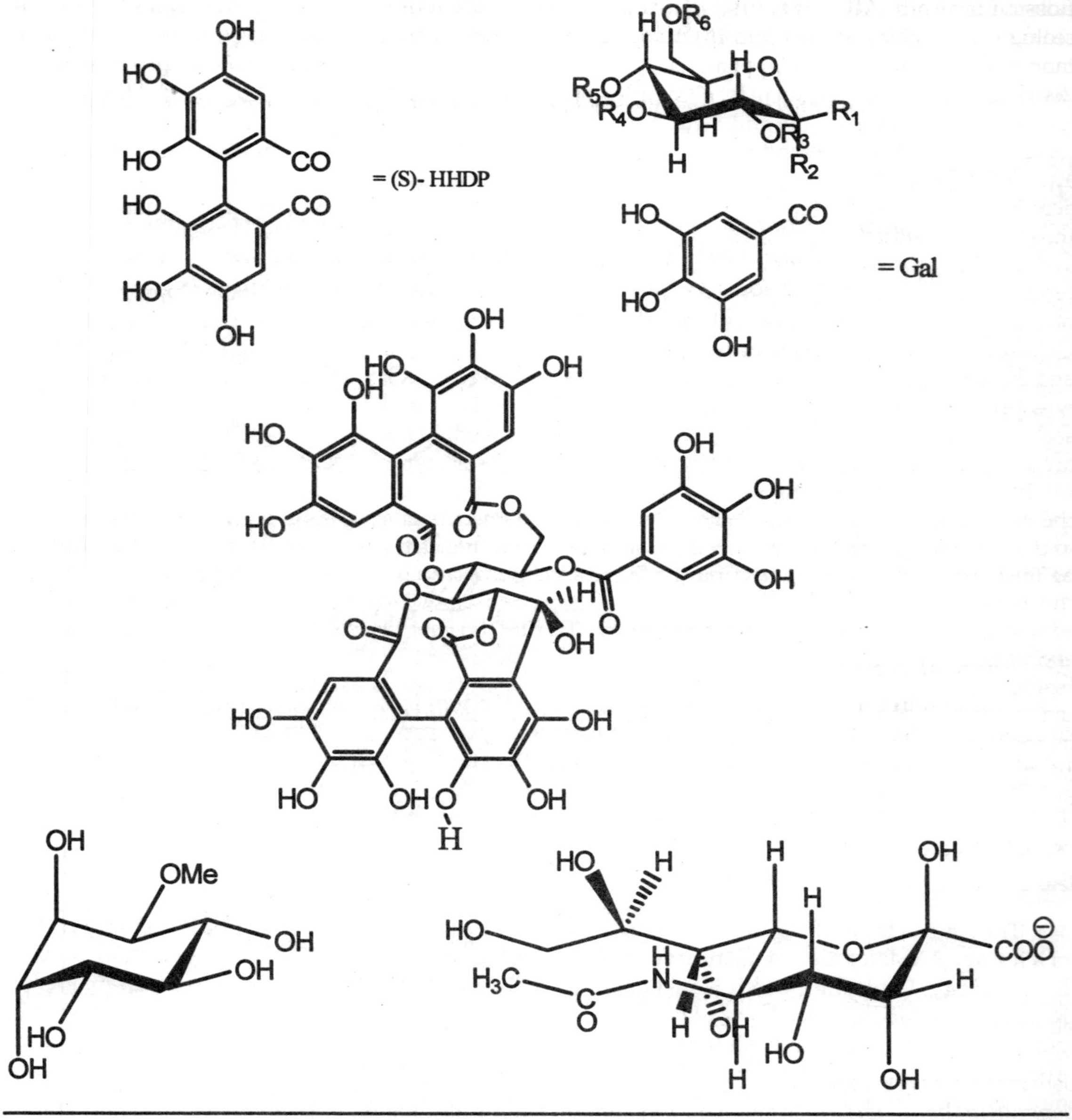

Tannins	R_1	R_2	R_3	R_4	R_5	R_6
Strictinin (I)	OGal	H	H	H	(S)-HHDP	
Isostrictinin (II)	OGal	H	(S)-HHDP		H	H
Casuarictin (IV)	OGal	H	(S)-HHDP		(S)-HHDP	
Pedunculagin (V)	H, OH		(S)-HHDP		(S)-HHDP	

Figure 30.1: Various Types of Polyphenols Fractions Containing Monomeric Hydrolyzing Gallo-ellagi-tannins.

polyphenols-monomeric hydrolysable gallo-ellagi tannins and flavonoids. The structures of the seabuckthorn tannins were elucidated (Sheichenko *et al.*, 2001a) and the chemistry of the seabuckthorn tannins is reported in the previous work (Sheichenko *et al.*, 1987; Yoshida *et al.*, 1991; Sheichenko *et al.*, 2001a; Shipulina and Tolkachev, 2006).

The new methodology of evaluation of the polyphenols in the seabuckthorn leaves, the drug, pharmaceutical forms and by-products using recent techniques were worked out. The UV spectrophotometric method of the total tannin determination, using $AlCl_3$ as the shift agent was elaborated (Sheichenko *et al.*, 2001a). A comparative content of tannins composition was determined by the NMR method. The dynamics of seasonal accumulation of tannins in seabuckthorn leaves were also evaluated by the NMR method (Sheichenko *et al.*, 2001a). The tannin contents in leaves were found maximum during the period of maximal accumulation of biologically active substances (over 20 per cent). The HPLC was also used for the total tannins analysis in the drug Hiporamin (Mozhaiskii *et al.*, 2006). HPLC in combination with the mass-spectrometric detection has helped in study on the composition of the polyphenol fraction from the leaves. Thus, the leaves contained up to 20 per cent gallo-ellagi tannins, among which pedunculagin, casuarinin, casuarictin, isostrictinin, stachyurin, hippophaenin and pterocarinin A were identified (Osipov *et al.*, 2009).

The NMR method was also shown to be suitable for quantitative tannins determination (of total tannin contents and their composition) in the fraction of polyphenols from leaves (Sheichenko *et al.*, 2009). The total tannin fraction according to the ^{1}H NMR spectrum (400 MHz) (the sample of the cultivar "Chuiskaya") has a complicated feature (Figure 30.2), while the basic signals of protons in its spectrum belong to strictinin, isostrictinin, casuarinin, casuarictin and pedunculagin. Intensive signal at 7.11 ppm close to casuarinin signal is obviously belong to stachyurin (its C_1 epimer). Pedunculagin in the spectrum is observed as the double high field signals at 6.29 and 6.30 ppm (two C_1 epimers). Comparative tannin contents in the fraction were calculated on the basic tannin casuarinin (mol.

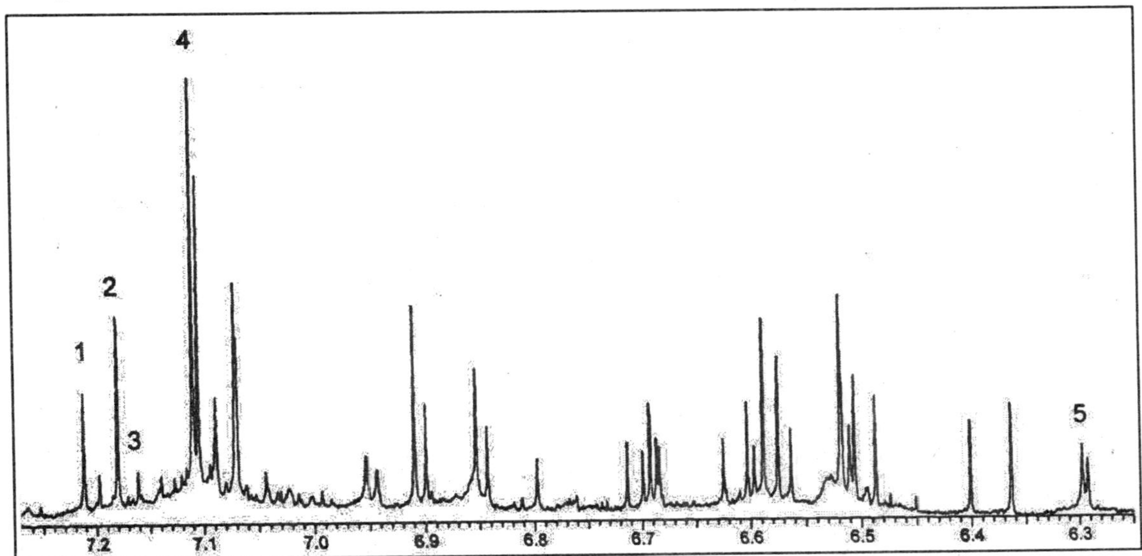

Figure 30.2: The NMR-Spectrum of the Tannin Fraction from Seabuckthorn Leaves. Numbers 1-5 in the spectrum indicated on the diagnostic peaks: (1) Strictinin, (2) Casuarictin, (3) Isostrictinin, (4) Casuarinin, (5) Pedunculagin (scale in ppm).

mass 936.65), using integrals of aromatic protons at (I_t, 6.2-7.3 ppm, 5H) and at (I_c, 3.3-4.3 ppm) for quebrachitol and carbohydrates (mol. mass 180, 6H). An admixture of mineral substances in the drug comes to 4 per cent. The characteristic signals of the galloyl fragment protons in the NMR spectrum were revealed in the aromatic area. Numbers 1-5 belongs to the diagnostic peaks in the spectrum (Figure 30.2).

Equation formula for the tannin contents calculation is represented below, with reappraisal on the basic tannin casuarinin, using the specific integral intensities of aromatic protons in the interval I_t 6.2-7.3 ppm (5H) and carbohydrate protons Ic.

$$X_t \text{ (per cent)} = 100/[0.19 \cdot (I'_c/I'_t) + 1]$$

where, the index $0.19 = \mu_{glucose}/\mu_{casuarinin}$, I'_t-specific intensity integral of the total tannins protons divided on the total protons number in casuarinin molecule ($I_t/5$), I'_c-specific intensity integral of the total carbohydrate protons ($I_c/8$), divided on the protons number in quebrachitol molecule (8f).

The 2H singlets of aromatic protons of the bonded gallic acid fragment at 7.0-7.21 ppm, and of the (S)-hexa-hydroxy-diphenoic and valoneic acids fragments at 6.35-7.05 ppm in the NMR spectra can be clearly seen. Signals of protons of the free gallic acid (as the artifact in technological processes) were not found in the spectrum. A presence of two H_1 signals of glucose in the NMR spectrum of pedunculagin at 6.3 ppm testifies the existence of two of its conformations (α, β). Total tannin contents were determined in a series of the drug samples to be ranged within 62-75.9 per cent. The NMR data are very similar to that produced by the UV spectroscopy method (62.5-74.0 per cent) and comparative data of their analysis is presented in the Table 30.1. Quebrachitol contents were found to be within the range 10-15 per cent.

Table 30.1: Results of NMR and UV Analysis of the Total Tannin Contents in Samples of Commercial Series of the Drug (per cent)

Number of Series	*NMR Method (per cent)*	*UV Method (per cent)*
1	62.0	62.5
2	68.16	70.0
3	73.5	73.5
4	73.9	74.0
5	71.0	71.2
6	73.0	73.6
7	73.0	73.4
8	65.0	64.5

Thus, the ^{1}H NMR spectroscopy may be used for quantitative analysis of the total tannins in the extracts of seabuckthorn leaves and for the determination of individual constituents in the main fraction (in the field of aromatic protons). For tannins identification and their quantitative estimation, there may also be used 2D NMR spectroscopy (H, H COSY) in a region, where the cross peaks of the carbohydrate fragment (glucose) are recorded. There are no spin-spin interaction between aromatic protons of monomeric tannins, what testify to the absence of the cross peaks in their 2D spectra. An appearance of the cross peaks in this interval of the spectrum may indicate the presence of the non-tannin contaminants in the samples analyzed.

Activity Study

Hydrolyzing tannins of seabuckthorn possess a high antiviral activity in respect of various viruses, pathogenic to humans, animals and plants (Influenza, Herpes simplex, cytomegalloviruses, HIV, etc). Virus inhibitory activity of the seabuckthorn drug against the HIV virus in experimental tests exceeds the action of Inosimplex and Timosin. The drug also possesses a high interferon induction activity. The drug inhibited the viral neuraminidase, while casuarinin + casuarictin were equal in activity to the total tannins. The known antiviral drugs Amantadin and Remantadin, on the contrary, are inactive in respect of viral neuraminidase (Shipulina *et al.*, 2000a). Hiporamin is practically non-toxic for experimental animals in acute and chronic tests (mice, white rats and guinea pigs) (Vichkanova *et al.*, 2009). Immuno-enzyme method of analysis has shown that the drug inhibited reproduction of respiratory Syncytium viruses in the cell cultures (Shipulina *et al.*, 2000b). But, the drug does not act on the transcription and translation of viruses. The drug has also inhibited a growth of a number of gram positive, gram negative microorganisms and fungi: *Mycobacterium tuberculosis, Staphyllococcus aureus* 209-P, *Escherichia coli* 844, *Proteus vulgaris* H-3137, *Pseudomonas aeruginosa* 44T, *Microsporum canis, Candida albicans* 1755 etc. (Tolkachev and Shipulina, 2006).

The constituents of Hiporamin drug (casuarinin, casuarictin, hipophaenin B, strictinin and isostrictinin) doses 1000, 100 and 10 μg/ml neutralized influenza viruses, while the total tannins in activity did not exceed any of them. The basic tannin casuarinin is the most potent tannin and it is equal to the total tannins in activity (Shipulina *et al.*, 1997; 2006; 2008). Polyphenol fraction also contains cyclic polyol Quebrachitol (2-O-metyl-L-inositol), possessing antioxidant, cytoprotective and gastro-protective activity. The maximum content of quebrachitol accumulation in the leaves was found during September-October (12-14 per cent).

Specific Enzyme Biotest Systems

Optimization of the phytodrugs creation process is of current importance. In order to solve this problem, original nano-biotechnological tools and specific enzyme biotest systems were used. The latter have allowed revealing *in vitro* principal activity of the plant substances and phytodrugs at a molecular level, correlating with their specified pharmacologic activity. The original specific enzyme biotest systems were created in the institute (VILAR) on the basis of systematic biochemical and pharmacological study, while the correlation results of biochemical study with a definite type of pharmacological activity were confirmed to each created biotest system. These biotest systems possess high sensitivity, selectivity, satisfactory results in reproduction and may be used as the express methods *in vitro* screening study of the plant substances for the antimicrobial and antiviral activities. The above biotest systems were worked out on the basis of the antioxidant defense enzymes NADPH-oxidase, tyrosine-hydroxylase, monooxygenase system of cytochrome P450 and glutathione transferase, acetylcholine esterase, alcohol dehydrogenase and aldehyde dehydrogenase, which allow as a whole to reveal *in vitro* biologically active compounds possessing antioxidant, antimicrobial and antiviral, adaptogenic, immunomodulatory, stress protector, hepatoprotector detoxycation, and anti-alcohol activities. For the comparative determination of the natural products activities, there were estimated rates of enzymatic reactions using enzymes of the antioxidant protection-glutathione reductase, catalase, superoxide dismutase, glutathione-peroxidase, and the standard antimicrobial and antiviral substances as the control, according to the Patent No.: RU 2181891 (2002). The antimicrobial and antiviral substances inhibited the activity of the antioxidant protection enzymes. The method has allowed optimizing the investigations of *in vitro* process during the elaboration and creation of new phytodrugs. The rates of enzymatic reactions were determined spectrophotometrically. Thus, in the presence of Hiporamin, the rates of the enzymatic reactions (counting on the control) come to: For glutathionreductase (28 per cent), catalase (45 per cent), superoxide-dismutase (5 per cent) and glutathione dismutase (41 per cent). Therefore, Hiporamin inhibited all four enzymes of the antioxidant protection in direct contact, producing the maximal effect in respect of superoxide dismutase. The application of the biotest method in a combination with the classical technique of antiviral and antimicrobial activity determination has allowed to reveal new potent substances and to clarify their mode of action using minimal amounts of biologically active samples (Dubinskaya *et al.*, 1997; Mineeva *et al.*, 2003).

Pharmaceutical Forms

Several pharmaceutical forms of Hiporamin for external and internal use were worked out (Dzhavakhyan *et al.*, 2009). Sublingual tablets 0.02 for the treatment of acute respiratory infections (influenza, adenoviral disease), containing auxiliary agents, prolong the action of the drug, and other modifying components; intestinal soluble tablets 0.02 (Okhotnikova and Shevelev, 1999); sublimation of dried powder for *ex temporal* preparation of aqueous solutions 0.1-0.2 (for inhalation and nasal drops); ointment 0.5 per cent for skin and mucous membranes treatment; suppositories rectal and vaginal 0.05 for adults and 0.03 for children, containing sublingual tablets, which are used for the treatment of Influenza for the delivery of the drug directly to blood avoiding the gastrointestinal tract and biliary barrier. During the preclinical study, Hiporamin exceeds the activity of oxolin and sovirax in respect of *Herpes simplex* virus (it was observed that tablets 0.02 and rectal suppository 0.03, ointment 0.5 and gels). Suppositories contain a mixture of hydrogenated plant oils with emulsifiers (up to 5 per cent). A complex of physico-chemical, technological and bio-pharmaceutical methods of studies were tested for their standardization and the storage periods study. Optimal composition of the pharmaceutical forms at the basis of low molecular and polymeric auxiliary substances was established. Regimes of their production were also elucidated. Drug safety study of the soft pharmaceutical forms was done on the rabbits Chinchilla species. Experimental results of the gel and liniment study have shown the prospects for development of new pharmaceutical forms of the drug (Dzhavakhyan, 2006).

In order to clarify a compatibility of the biologically active substances with pharmaceutical bases, new gel forming components, containing occasionally combined hydrophilic polymers (*e.g.* arespol) were also tested. Supramolecular models of the proposed compositions were estimated, using a modification of the computer programs MM2 and Molecular Dynamics (Chem Office, 2004).

Toxicological Study

A detail investigation was carried out on the polyphenol fraction of seabuckthorn leaves, their analytical determination, chemotherapeutic study, and the drug elaboration. Drug safety estimation of the soft pharmaceutical forms (gel and liniment) was realized on the rabbits Chinchilla species (on the skin of the back side of experimental animals of II and IV groups). Animals in the control I and III group got application placebo of the gel and the liniment. Application on the skin of animals for a longer time did not reflect on the morphological composition of the distal blood of rabbits. A number of erythrocytes, thrombocytes, leucocytes, hemoglobin content, hematocrite, the average hemoglobin contents in erythrocytes and their average volumes in animals of II and IV groups were at the level in the control I and III groups and correlated with the physiological level of animals. Various blood parameters studied among the experimental and the control group of animals was found to be statistically at par with each other. The pharmaceutical forms did not produce toxic or irritation effects (Krepkova *et al.*, 2008).

Clinical Trials

The drug was tried on 374 adult patients in case of acute viral infections (influenza A and B, adenoviral infection). The ointment was used in case of *Herpes simplex* and *Varicella zoster* infection. The drug was used for the treatment of Chicken pox on 251 children. There were no general toxic side effects found during I phase of clinical studies. Pharmaceutical forms did not cause local irritation and allergic reactions. Early application of the drug (1-2 days) during acute respiratory infections reduces duration of the fever and toxicosis symptoms. During the second phase of clinical trials, the drug produces statistically positive curative effect in case of uncomplicated influenza during first 24-36 hrs from the beginning of treatment. Hiporamin is also effective in case of tonsillitis. The drug enhances an increase of serum interferon level.

In respect of *Herpes simplex, Varicella zoster* and cytomegaloviral infections 0.5 per cent ointment, rectal and genital suppositories 0.05 produce a market curative effect. All clinics reported a high efficiency and good tolerance during the use of the drug without side effects and contraindications. Hiporamin was permitted for use as a medical and prophylactic antiviral remedy for adults and children for treatment of influenza infection, respiratory syncytial, adenoviral, other acute respiratory viral infections, tonsillitis, cytomegaloviral and other cases (Vichkanova *et al.*, 2009).

CONCLUSION

Results of extensive chemical, analytical, preclinical medico-biological and clinical studies of polyphenols from the fractions of seabuckthorn leaves have revealed their high antiviral and antimicrobial activity in respect of a wide spectrum of pathogenic viruses and microorganisms. A new herbal antiviral drug Hiporamin possessing a unique biological activity was proposed. Original enzyme biotest systems for elucidation of biologically active substances in the plant materials and contaminants in the drug were worked out. A number of pharmaceutical products of the drug were developed. It was shown that the drug possessed a number of advantages over known antiviral drugs used in medicine. Thus, new developments in prospective soft pharmaceutical forms of Hiporamin drug are of a great value. Technological processes of the drug production at the commercial scale were standardized.

REFERENCES

1. Bykov, V.A., Tolkachev, O.N., Morozov, V.I., Sheichenko, O.P., Sheichenko, V.I., Shipulina, L.D. and Krepkova, L.V. 2008. Complex study on seabuckthorn as a source for production of the antiviral drug Hiporhamin. In: *Seabuckthorn (Hippophae L.)-A Multipurpose Wonder Plant. Vol. III: Advances in Research and Developments* (V. Singh, Ed. in Chief, 2008), p. 337-356, Daya Publishing House, Delhi, 566p.
2. Bykov, V.A., Shipulina, L.D., Tolkachev, O.N., Sheichenko, O.P., Sheichenko, V.I., Morozov, V.I. and Val, E.V. 2005. Seabuckthorn leaves as the source for production of the antiviral drug: Organizing and technological aspects. *Practical Phytotherapy, INFI* 4: 6-7. (In Russian)
3. De Olida, T.M., Lemos, T.M., Machado, T.L.G., Rao, L.L. and Santos, F.A. 2003. Quebrachitol-induced gastroprotection against acute gastric lesions: Role of prostaglandins, nitric oxide and K^{+}ATP channels. *Phytomedicine* 15 (5): 327-333.

4. Dubinskaya, V.A., Mineeva, M.F., Krutikova, N.M., Fateeva, T.V., Shipulina, L.D. and Kolkhir, V.K. 1997. Antioxidant enzymes: Possible targets of antimicrobial and antiviral agents. In: *IV Russian National Congress "Man and Drug"*, Moscow, April 8-12, 1997, Abstracts, 43p. (In Russian)

5. Dzhavakhyan, M.A., Okhotnikova, V.F., Kachalina, T.V., Bortnikova, V.V., Tolkachev, O.N. and Krepkova, L.V. 2009. New developments in Hiporamin pharmaceutical forms production and their safety estimation study. In: *Seabuckthorn on the Way Between Science and Industry Interaction.* Proceedings of the 4th International Seabuckthorn Association Conference, ISA-2009, p. 178-180, September 1-6, 2009, Belokurikha, Altai region, Russia, Barnaul.

6. Dzhavakhyan M.A. 2006. *Creation of Composition and the Technology of Pharmaceutical forms of Hiporamin*, Dissertation, Moscow, 119 p. (In Russian)

7. Krepkova, L.V., Bortnikova, V.V., Shkarenkov, A.A., Shipulina, L.D., Tolkachev, O.N. and Borovkova, M.V. 2008. Experimental toxicological study of Hiporamin: An antiviral drug from seabuckthorn. In: *Seabuckthorn (Hippophae L.)-A Multipurpose Wonder Plant. Vol. III: Advances in Research and Developments* (V. Singh, Ed. in Chief, 2008), p. 357-360, Daya Publishing House, Delhi, 566p.

8. Mineeva, M.F., Dubinskaya, V.A., Kolkhir, V.K., Vichkanova, S.A., Shipulina, L.D. and Bykov, V.A. 2003. Enzymatic biotest systems and their uses for the screening of substances with antimicrobial and antiviral activities. In: *Xth Russian National Congress "Man and Drug"*, April 7-11 2003, Abstracts, Moscow, 635 p. (In Russian)

9. Mozhaiskii A.M., Sokol'skaya T.A. and Djavakhyan M.A. 2006. A Study of the possibility of tannin determination in new drugs from seabuckthorn leaves. In: *Research Works of VILAR.* International Conference on the 75 Years of Foundation of the Institute, p. 71-78, Vol. XVII, Moscow. (In Russian)

10. Nobre, Junior, H.V., Cunha, G.M.A., Moraes, M.O., Luciana, M.F.D., Oliveira, R.A., Maia, F.D., Nogueira, M.A.S., Lemos, T.L.G. and Rao, V.S. 2006. Quebrachitol (2-O-methyl-L-inositol) attenuates 6-hydroxydopamine induced cytotoxicity in rat fetal mesencephalic cell cultures. *Food and Chemical Toxicology* 44 (9): 1544-1551.

11. Okhotnikova, V.F. and Shevelev, V.E. 1999. Elaboration of the hiporamin tablets in the intestine soluble capsules. In: *VI Russian National Congress "Man and Drug"*, April 19-23, 1999, Abstracts, Moscow, 453p. (In Russian)

12. Osipov, V.I., Osipova, S.N., Meshkov, A.I., Savina, A.A., Sokol'skaya, T.A. and Bykov, V.A. 2009. Characteristics of phenolic compounds of medicinal plants by the HPLC with mass-spectrometric detection. In: *VII International Symposium on Phenolic Compounds-Fundamental and Applied Aspects*, p. 203-204, Oct 19-21, 2009, Symposium Papers, Moscow. (In Russian)

13. Sheichenko, O.P. and Tolkachev, O.N. 2006. Tannins of seabuckthorn (*Hippophae rhamnoides* L.): Chemical properties and biological action. In: *Seabuckthorn (Hippophae L.)-A Multipurpose Wonder Plant. Vol. II: Biochemistry and Pharmacology* (V. Singh, Ed. in Chief, 2006), p. 197-213, Daya Publishing House, New Delhi, 600p.

14. Sheichenko, O.P., Sheichenko, V.I., Fadeeva, I.I., Zolotarev, B.M. and Tolkachev, O.N. 1987. Tannins from the leaves of *Hippophae rhamnoides* L. *Khim. Prirod. Soedin.* 6: 902-907. (In Russian)

15. Sheichenko, O.P., Sheichenko, V.I., Tolkachev, O.N., Isaev, O.N. and Tsar'kova T.F. 2003. Tannins in green herbaceous shoots of seabuckthorn (*Hippophae rhamnoides* L.). In: *Chemistry, Technology, and Medical Aspects of Natural Compounds.* Proceedings of the International Research Conference, October 8-11, 2003, Almaty, Kazakhstan, 43p. (In Russian)

16. Sheichenko, O.P., Sheichenko, V.I. and Tolkachev, O.N. 2001a. Dynamics of seasonal accumulation of tannins and quebrachitol in leaves of seabuckthorn (*Hippophae rhamnoides* L.). In: *Seabuckthorn-A Resource for Health and Environment in Twenty First Century.* Proceedings of International Workshop on Seabuckthorn, p. 162-166, February 18-21, 2001, New Delhi.

17. Sheichenko, V.I., Semenova, T.S., Tolkachev, O.N. and Sheichenko, O.P. 2001b. Polyphenol fraction from seabuckthorn leaves: Chemical composition. In: *Seabuckthorn-A Resource for Health and Environment in Twenty First Century.* Proceedings of International Workshop on Seabuckthorn, p. 157-161, February 18-21, 2001, New Delhi.

18. Sheichenko, V.I., Sheichenko, O.P., Tolkachev, O.N. and Shipulina, L.D. 2009. On hydrolyzing tannins from seabuckthorn leaves: NMR analysis. In: *"VIIth International Symposium on Phenolic Compounds, Fundamental and Applied Aspects"*, p. 309-310, October 19-23, 2009, Symposium Papers, Moscow.

19. Shipulina, L.D., Fateeva, T.V., Krepkova, L.V., Bortnikova, V.V. and Tolkachev, O.N. 2005. Antiviral drug from seabuckthorn: Data of experimental preclinical studies. *Practical phytotherapy, INFI* 4: 7-16. (In Russian)

20. Shipulina, L.D., Sheichenko, O.P., Sheichenko, V.I., Tolkachev, O.N., Morozov, V.I. and Bykov, V.A. 2008. Seabuckthorn study as the source for production of antiviral drug Hiporamin: Results and prospects. In: *Canada-Russia Agri-Business Knowledge Transfer, Small Fruit Production and Processing Seminar*, July 15, 2008, Manitoba, 36 p.

21. Shipulina, L.D., Leneva, I.A., Fedyakina, I.I., Krepkova, L.V., Bortnikova, V.V. and Tolkachev, O.N. 2005. Hiporamin-Vegetarian antiviral drug: Experimental study and mechanism of its action. In: *Actual problems of innovations with nontraditional natural sources and creation of functional products*. Materials of the 3rd Russian research and practical conference, July 6-7, 2005, RAEN, Moscow, 97p. (In Russian)

22. Shipulina, L.D., Tolkachev, O.N., Sheichenko, O.P., Fateeva, T.V. and Krutikova, N.M. 1997. Correlation analysis of antiviral agents: Hydrolyzable tannins of seabuckthorn. In: *IVth Russian National Congress "Man and Drug"*, April 8-12, 1997, Abstracts, Moscow, 239p. (In Russian)

23. Shipulina, L.D., Tolkachev, O.N., Krepkova, L.V., Bortnikova, V.V. and Shkarenkov, A.A. 2006. Antiviral, antimicrobial and toxicological studies on seabuckthorn (*Hippophae rhamnoides* L.). In: *Seabuckthorn (Hippophae L.)-A Multipurpose Wonder Plant. Vol. II: Biochemistry and Pharmacology* (V. Singh, Ed. in Chief, 2006), p.471-483, Daya Publishing House, New Delhi, 600p.

24. Shipulina, L.D., Leneva, I.A. and Fedyakina, I.T. 2000a. On the mechanism of Hiporamin action. In: *Transactions of VILAR, Chemistry, Technology, Medicine*, p. 250-257, Moscow. (In Russian)

25. Shipulina, L.D., Leneva, I.A. and Fedyakina, I.T. 2000b. Experimental study of hiporamin activity using the method immunoenzymatic analysis in respect of respiratory syncytium virus. In: *Transactions of VILAR, Chemistry, Technology, Medicine*, p. 258-262, Moscow. (In Russian)

26. Tolkachev, O.N. and Shipulina, L.D. 2006. Antiviral polyphenols from seabuckthorn leaves as the source of drug hiporamin. In: *Seabuckthorn-A Resource of Health, a Challenge to Modern Technology*. Proceedings of the 1st Congress of the International Association, p. 90-103, September 14-18, 2003, Berlin, Germany.

27. Vichkanova, S.A., Kolkhir, V.K., Sokol'skaya, T.A., Voskoboinikova, I.V. and Bykov, V.A. 2009. 'Hiporamin-Antiviral drug'. In: *Medicinal Remedies from Plants*, p. 132-143, Moscow, Science Edition, ADRIS, 432 p. (In Russian)

28. Yoshida, T., Tanaka, K., Chen, Xin-Min and Okuda, T. 1991. Tannins from *Hippophae rhamnoides*. *Phytochemistry* 30 (2): 663-666.

Chapter 31

Seabuckthorn (*Hippophae salicifolia* D. Don) Plant Extracts Show Potential Antimicrobial Activity

S.M. Gupta, A.K. Gupta*, Z. Ahmed and Anil Kumar*

Molecular Biology and Genetic Engineering Laboratory,
Defence Institute of Bio-Energy Research (DIBER), DRDO, Haldwani – 263 139, India
*Department of Molecular Biology and Genetic Engineering,
C.B.S and H.G.B. Pant University of Agriculture and Technology, Pantnagar – 263 145, India

SUMMARY

The antimicrobial efficacy of crude leaf and seed extracts, and seed oil of seabuckthorn (*Hippophae salicifolia* D. Don) on two gram-positive bacteria (*B. subtilis* and *B. thuringiensis*), four gram-negative bacteria (*Pseudomonas fluorescens, E. coli, Agrobacterium tumefaciens* and *Acinetobacter junii*) and three fungal cultures (*Rhizopus oryzae, Mucor indicus* and *Tilletia indica*), causing infections/diseases, was investigated by agar disc diffusion method. Kanamycin (1000 µg/ml) and clotrimazole (100 µg/ml) were used as standard drugs for antibacterial and antifungal assays, respectively. The inhibition zones ranged from 7 to 23 mm for antibacterial and from 10 to 27 mm for antifungal activity, respectively. Among all extracts tested, 3 extracts (Dried supernatant of crude leaf extract dissolved in dimethyl sulphoxide, crude seed extract and concentrated crude seed extract) of *H. salicifolia* showed significant and 2 extracts (Dried crude leaf extract dissolved in dimethyl sulphoxide and concentrated supernatant of crude leaf extract) showed intermediate activity against both gram-positive bacteria whereas, no significant activity was observed against all test extracts in gram-negative bacteria, except in case of *Agrobacterium tumefaciens* in which some significant activity was obtained against crude and concentrated seed extract of seabuckthorn. In case of fungal cultures, among all test extracts, only two crude seed extract and concentrated crude seed extract showed significant activity against *Mucor* and *Tilletia* fungus, while in case of *Rhizopus*, no activity was observed against all test extracts. Concentrated crude seed extract of *H. salicifolia* exhibited highest antifungal and antibacterial activity as compared to other extracts tested. These results indicated the possibility of using *H. salicifolia* extracts for medicinal uses and natural food preservation.

Keywords: *Antimicrobial activity, Seabuckthorn (Hippophae salicifolia D. Don), Crude leaf and Seed extracts.*

Seabuckthorn (Hippophae L.): *A Multipurpose Wonder Plant, Vol. 4: Emerging Trends in Research and Technologies* (Virendra Singh, *Editor-in-Chief*, 2014), p. 393–401, Daya Publishing House, New Delhi, India.

INTRODUCTION

In the recent years, there has been an increasing interest in the use of natural bioactive compounds, and some questions concerning the safety of synthetic compounds have encouraged more detailed studies of plant resources (Smid and Gorris, 1999). Infectious diseases caused by bacteria and fungus infection has been a major problem considered for decades that causes spoilage of food products and various diseases in plants, insects and human beings, which leads to significant losses in the crop productivity and health problems worldwide. Microorganisms have developed resistance to many antibiotics due to indiscriminate use of commercial antimicrobial drugs, which has encouraged scientists to search for new antimicrobial substances from various medicinal plants including seabuckthorn.

Hippophae salicifolia D. Don is less explored but higher value species of seabuckthorn grows at low altitude (6500-9000 feet) areas of north-west Himalayan region from Pakistan to Himachal Pradesh, Uttarakhand and north-eastern India (Rousi, 1971; Singh, 1994; Gupta and Ahmed 2010). The main advantage of *Hippophae salicifolia* is that it is mild thorny, contain all essential polyunsaturated fatty acids, particularly omega-3 and omega-6 (Yang and Kallio, 2001; Gupta *et al.*, 2011c; Singh *et al.*, 2011), high quality berries with late maturation character and vitamin C content is the highest among all species of seabuckthorn (Ahmed and Gupta, 2009). Seabuckthorn plant has been used extensively in oriental traditional system of medicine for treatment of asthma, skin diseases, gastric ulcers and lung disorders (Xu *et al.*, 1994). All parts of the plant are considered to be rich source of a large number of bioactive substances like flavonoids (isorhamnetin, quercetin, myricetin, kaempferol and their glycoside compounds), carotenoids (α, β, δ-carotene, lycopene), vitamins (C, E, K), tannins, triterpenes, glycerides of palmitic, stearic and oleic acids and some essential amino acids (Xu, 1991). These seabuckthorn natural products has been found to possess significant anti-microbial, anti-oxidative, anti-inflammatory, immunomodulatory, radio-protective, adaptogenic and tissue regenerative properties (Geetha *et al.*, 2002; Ganju *et al.*, 2005; Negi *et al.*, 2005; Chawla *et al.*, 2007; Saggu *et al.*, 2007). For the past 50 years, several medical preparation of seabuckthorn have been clinically used to treat radiation damage, burn, oral infection, cardiovascular disease, skin disease and many more. Although more than 300 preparations of seabuckthorn have been reported in literature (Rongsen, 1992). Still, there is a need to identify newer potential antimicrobial bioactive compounds from seabuckthorn having better neutraceutical and therapeutic importance. We studied the antimicrobial activity of different *H. salicifolia* extracts against different microorganisms for their potential as a natural preservative and for the development of herbal formulations for future drug development.

MATERIALS AND METHODS

Plant Material and Extraction

H. salicifolia leaves were collected from Field Station of Defence Institute of Bio-Energy Research (DIBER), Auli, Uttarakhand Himalayan State, India, where the plant grows wildly under natural conditions. The fresh leaves (5 g) were ground in porcelain mortar and the leaf paste was squeezed through muslin cloth to give fresh crude leaf extract and 10 µl of the extract was applied on disc 1. The same extract was concentrated (four fold) in centrivap concentrator (Labconco, Biogentek, India) and 10 µl of this was applied on disc 2. The extract was also dried completely in centrivap concentrator and dissolved in 50 µl of Dimethyl sulphoxide (DMSO) by keeping the tubes in the water bath at 37 °C for one hour and cleared by centrifugation at 10,000 × g for 2 min and 10 µl of the supernatant was applied on disc 3. Clarified crude leaf extract (centrifuged at 10000 × g for 10 min) was concentrated four fold in centrivap concentrator and 10 µl of this was applied on disc 4. The remaining aliquot of this supernatant was dried completely in centrivap concentrator and dissolved in 50 µl of DMSO as described earlier and 10 µl of this was applied on disc 5. Standard antibacterial (Kanamycin, 1000 µg/ml) and antifungal (Clotrimazole, 100 µg/ml) drugs were applied separately on disc 6 as positive control. The DMSO solvent was also applied on disc 7 as negative control. Mature seeds (5 g) of *H. salicifolia* were also ground in porcelain mortar using 5 ml DMSO as described earlier to give fresh crude seed extract and 10 µl of the extract were applied on disc 8. The crude seed extract was concentrated (four fold) in centrivap concentrator and 10 µl of this was applied on disc 9. Dried seeds were powdered and successively extracted in a Soxhlet extractor for 8 h at 65 °C with petroleum ether for the extraction of seabuckthorn seed oil following Soxhlet (1879) method and 10 µl of fresh seed oil was applied on disc 10.

Test Microorganisms

Bacterial cultures namely, *B. subtilis* (NCTC 10073), *B. thuringiensis* (MTCC 4714), *P. fluorescens* (NCTC 10038), *E. coli* (NCTC 9002), *A. tumefaciens* (MTCC 609) and *A. junii* (MTCC 1686) and fungal cultures namely, *Rhizopus oryzae* (MTCC 262), *M. indicus* (MTCC 3318) and *T. indica* were obtained from Department of Molecular Biology and Genetic Engineering, G. B. Pant University of Agriculture and Technology, Pantnagar.

Antimicrobial Sensitivity Test

In vitro antimicrobial studies were carried out on different bacterial and fungal cultures against standard antimicrobial drugs (Kanamycin and Clotrimazole) and *H. salicifolia* extracts (leaf, seed extract and seed oil) by agar disc diffusion assay method (Baur *et al.*, 1966). A 100 µl aliquots of bacterial (about 10^6 cells/ml) and fungal cultures (about 10^8 cfu/ml) were evenly spread on to LA (Luria agar, Himedia) and PDA (Potato dextrose agar, Himedia) plates (90 mm diameter) using a sterile glass rod spreader, respectively. The plates were left at room temperature for 15 min to allow the agar surface to dry. The sterile disc (6 mm, Himedia) containing the standard drug and test compounds were placed on the inoculated plates under aseptic conditions. The LA and PDA plates were incubated at 37 °C and 30 °C for 24 to 48 h, respectively. The cleared zones were scored as antimicrobial activity of standard drug and *H. salicifolia* extracts against different tested microorganisms. The minimum inhibitory concentration (MIC100) was reported as the lowest concentration of the drug capable of inhibition of complete growth of the microorganism. The experiments were performed in triplicate and the data were calculated as mean ± SD.

RESULTS

Drug Sensitivity Test

Antimicrobial activity of standard drugs (kanamycin and clotrimazole) against all test microorganisms were measured by disc diffusion assay before conducting the experiments with *H. salicifolia* extracts. The MIC100 values, with respect to kanamycin for *B. subtilis*, *B. thuringiensis*, *Pseudomonas fluorescens*, *E. coli*, *Agrobacterium tumefaciens* and *Acinetobacter junii* were found to be 0.25, 0.5, 1.0, 1.25, 1.25 and 2.0 µg, respectively (Table 31.1). Whereas, the MIC100 values with respect to clotrimazole for *Rhizopus oryzae*, *Mucor indicus* and *Tilletia indica* were found to be 0.1, 0.1 and 20 µg, respectively (Table 31.2). These results indicate that gram-positive bacteria (*Bacillus subtilis* and *Bacillus thuringiensis*) were the most susceptible to the kanamycin drug and having lower MIC values. While, gram-negative bacteria (*Pseudomonas*, *E. coli*, *Agrobacterium tumefaciens* and *Acinetobacter junii*) were found more resistant to kanamycin drug and having higher MIC values in comparison to gram-positive bacteria. On the other hand, among fungal cultures, *Tilletia indica* was found more resistant to

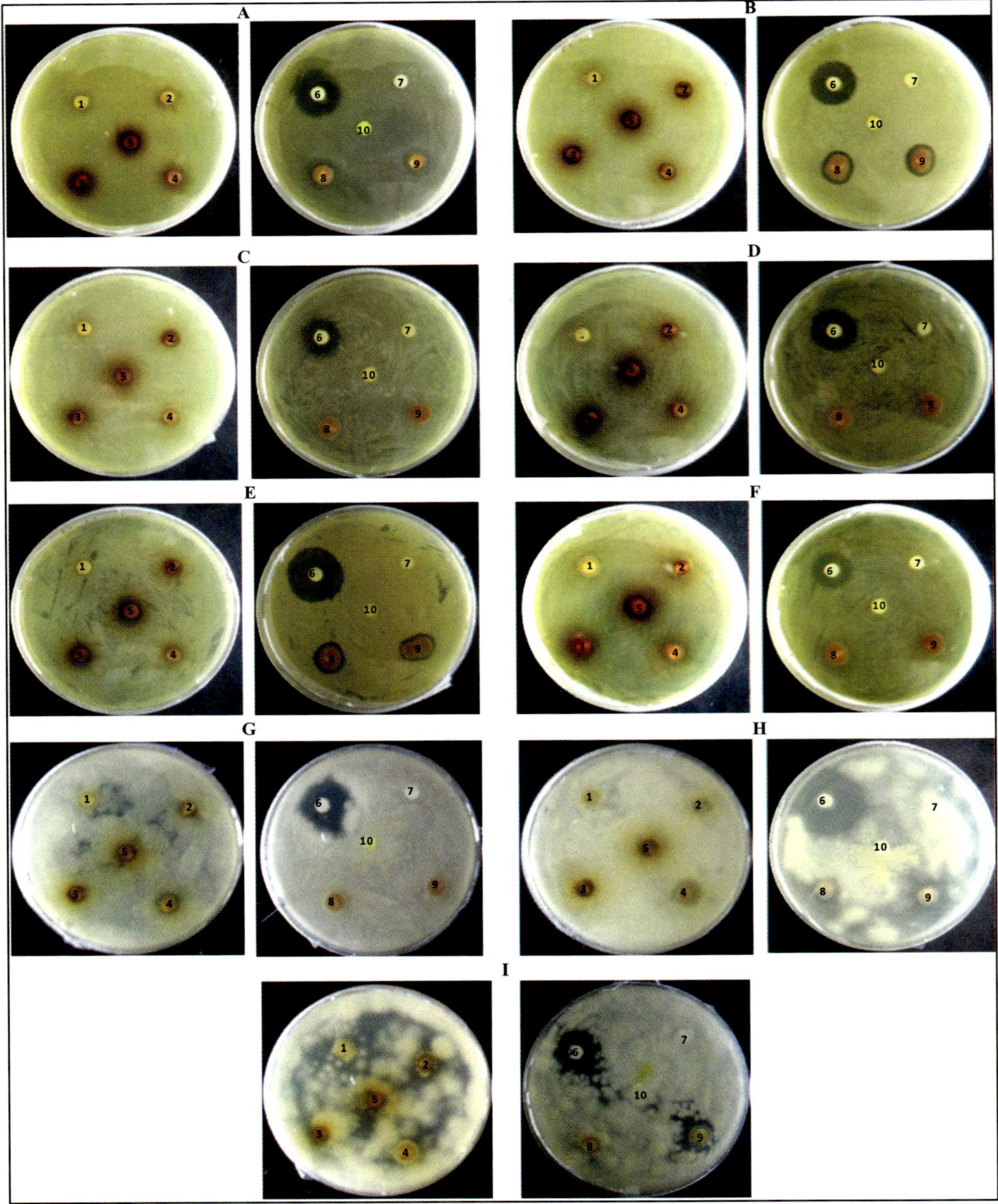

Figure 31.1: Antibacterial and Antifungal Activity in Leaf, Seed Extract and Seed Oil of Seabuckthorn (*Hippophae salicifolia* D. Don) against different bacterial strains. Bacterial strains denoted by A: *Bacillus subtilis*; B: *Bacillus thuringiensis*; C: *Pseudomonas fluorescens*; D: *E. coli*; E: *Agrobacterium tumefaciens*; F: *Acinetobacter junii*. Fungal strains denoted by G: *Rhizopus oryzae*; H: *Mucor indicus*; I: *Tilletia indica*.

more, then it is sensitive. In our experiment, inhibition zones ranged from 7 to 23 mm for antibacterial and from 10 to 27 for antifungal activity, respectively, which clearly indicate the microbicidal activity of seabuckthorn extracts against tested microorganisms.

The reason for higher sensitivity of the gram-positive bacteria in comparison to gram-negative bacteria could be ascribed to the differences between their cell wall compositions. The gram-positive bacteria contain an outer peptidoglycan layer which is an ineffective permeability barrier (Scherrer and Gerhardt, 1971). On the other hand, gram-negative bacteria show higher resistance against seabuckthorn extracts due to the presence of lipopolysaccharides in their outer membrane, which make them inherently resistance to antibiotic, detergent and hydrophilic dyes (Nikaido and Varra, 1985). Similar trend for inhibition of bacterial growth have been observed in earlier studies with seabuckthorn berry extracts (Puupponen-Pimla *et al.*, 2001) and other plant extract (Gupta *et al.*, 2009; Gupta *et al.*, 2011b).

Among different species of seabuckthorn, *H. salicifolia* showed higher total phenolic compounds in the leaf and other reactive phenols in the seed. We proposed that higher antibacterial and antifungal activities of crude seed extract of seabuckthorn may be due to the higher phenolic content. Similarly, Rosch *et al.* (2003) observed that although a few phenolic compounds present in seabuckthorn fruit juice have good antioxidant capacity, their contribution to the antioxidant effect is meagre in comparison to ascorbic acid. Negi *et al.* (2005) also studied the antibacterial activity of chloroform, ethyl acetate, acetone and methanol extracts of seabuckthorn seeds. The crude seed extracts of seabuckthorn have shown good antioxidant and antibacterial activities in various *in-vitro* model systems (Chauhan *et al.*, 2007). The Proanthocyanidins were identified in the defatted crude extract of its seed, which can stimulate phagocytic cells, host-mediated tumour activity, and exerts a wide range of anti-infective actions. Seed oil of seabuckthorn possesses several strong antioxidative and antimicrobial properties, which are due to high content of tocopherols, tocotrienols and carotenoids present in the oil (Chen *et al.*, 1990). In addition, omega-3 and-6 fatty acids present in the seabuckthorn seed oil is known to moderate anti-inflammatory activity of prostaglandins and also critical to vascular health. The seabuckthorn leaves were considered to be rich source of flavonoids (isorhamnetin and quercetin), carotenoids, sterols (free and esterified), triterpenols and isoprenols, which support its medicinal properties, such as antiviral, antimicrobial, anti-cancerous, anti-bronchitis, anti-inflammatory, wound healing, burns, energizer, memory vitalizer and analgesic properties (Shipulina, 2001; Gupta *et al.*, 2005; Upadhyay *et al.*, 2009). Recently, Gupta *et al.* (2011a) have shown the antibacterial and antifungal activity in different extracts of *H. salicifolia*.

CONCLUSION

The microbicidal activity of seabuckthorn extracts is probably due to their ability to form complex with extra cellular and soluble proteins and bacterial cell walls by non-specific forces, such as hydrogen bonding and hydrophobic effects as well as by covalent bond formation. Thus, their mode of antimicrobial actions may be related to their ability to inactive microbial adhesins, enzymes, cell envelope transport proteins, etc. Therefore, the observations of present study suggest that *H. salicifolia* is a potential source of bioactive antimicrobial agents. Further studies are being carried out in our laboratory to assess the practical use of seabuckthorn based bioactive compounds for their potential as a natural preservative and for herbal formulations that can be used as drugs in near future.

REFERENCES

1. Ahmed, Z. and Gupta, S.M. 2009. Seabuckthorn: A source (donar) in molecular breeding. In: *National Conference on Seabuckthorn and Environment: High Altitude Perspectives*. Sep. 25-27, 2009, DIHAR, Leh-Ladakh-194101, Jammu and Kashmir, India.

2. Baur, A.W., Kirby, W.M., Sherris, J.C. and Turck, M. 1966. Antibiotic susceptibility testing by a standard single disk method. *Am. J. Clin. Path.* 45: 493-496.

3. Chauhan, A.S., Negi, P.S. and Ramteke, R.S. 2007. Antioxidant and antibacterial activities of aqueous extract of seabuckthorn (*Hippophae rhamnoides*) seeds. *Fitoterapia* 78: 590-592.

4. Chawla, R., Arora, R., Singh, S., Sagar, R.K., Sharma, R.K. and Kumar, R. 2007. Radioprotective and antioxidant activity of fractionated extracts of berries of *Hippophae rhamnoides*. *J. Med. Food.* 10: 101-109.

5. Chen, Y.D., Jiang, Z.R., Qin, W.I., Ni, M.N., Li, X.L. and He, Y.R.1990. Research on chemical composition and characteristics of seabuckthorn berry and its oil. *Chem. Indus. Forest Prods.* 10 (3): 163-175.

6. Ganju, L., Padwad, Y., Singh, R., Karan, D., Chanda, S., Chopra, M.K., Ganju, L., Padwad, Y., Singh, R., Karan, D., Chanda, S., Chopra, M.K., Bhatnagar, P., Kashyap, R. and Sawhney, R.C. 2005. Anti-inflammatory activity of seabuckthorn (*Hippophae rhamnoides*) leaves. *Int. Immunopharmcol.* 5: 1675-1684.

7. Geetha, S., Sai Ram, M., Singh, V., Ilavazhagan, G. and Sawhney, R.C. 2002. Antioxidant and immunomodulatory properties of seabuckthorn (*Hippophae rhamnoides* L.)-An *in vitro* study. *J. Ethnopharmacology* 79: 373-378.

8. Gupta, A., Kumar, R., Pal, K., Banerjee, P.K. and Sawhney, R.C. 2005. A preclinical study of the effects of seabuckthorn (*Hippophae rhamnoides* L.) leaf extract on cutaneous wound healing in albino rats. *Int. J. Lower Extrem. Wounds* 4: 88-92.

9. Gupta, S., Kumar, N. and Gupta, S.M. 2009. Antibacterial and antifungal activity in extract and oil of *Piper betle* landrace bangla mahoba. *J. Adv. Zoo.* 30 (1): 16-20.

10. Gupta, S.M. and Ahmed, Z. 2010. Seabuckthorn (*Hippophae salicifolia* L.) plant: As source donor of cold tolerant genes for improving high altitude agriculture during cold stress. *Res. Environ. Life Sci.* 3 (3): 105-112.

11. Gupta, S.M., Arif, M. and Ahmed, Z. 2011b. Antimicrobial activity in leaf, seed extract and seed oil of *Jatropha curcas* L. plant. *J. Appl. Natural Sci.* 3 (1): 102-105.

12. Gupta, S.M., Gupta, A.K., Ahmed, Z. and Kumar, A. 2011a. Antibacterial and antifungal activity in leaf, seed extract and seed oil of seabuckthorn (*Hippophae salicifolia* D. Don) plant. *J. Plant Pathol. Microbiol.* 2: 105. *doi: 104172/2157-7471.1000105*.

13. Gupta, S.M., Pandey, P., Grover, A. and Ahmed, Z. 2011c. Breaking seed dormancy in *Hippophae salicifolia*, a high value medicinal plant. *Physiol. Mol. Biol. Plants* 17 (4): 403-406.

14. Negi, P.S., Chauhan, A.S., Sadia, G.A., Rohinishree, Y S. and Ramteke, R.S. 2005. Antioxidant and antibacterial activities of various seabuckthorn (*Hippophae rhamnoides* L.) seed extracts. *Food Chem.* 92: 119-124.

15. Nikaido, H. and Vaara, M. 1985. Molecular basis of bacterial outer membrane permeability. *Microbiological Reviews* 49: 1-32.

16. Puupponen-Pimia, R., Nohynek, L., Meier, C., Kahkonen, M., Heinonen, M., Hopia, A. and Oksman-Caldentey, K.M. 2001. Antimicrobial properties of phenolic compounds from berries. *J. Appl. Microbiol.* 90: 494-507.

17. Rongsen, L. 1992. *Seabuckthorn: A Multipurpose Plant Species of Fragile Mountains*. Occasional paper No.20, ICIMOD, Katmandu, Nepal, 62p.

18. Rosch, D., Bergmann, M., Knorr, D. and Kroh, L.W. 2003. Structure-antioxidant efficiency relationships of phenolic compounds and their contribution to the antioxidant activity of seabuckthorn juice. *J. Agric. Food Chem.* 51: 4233-4239.

19. Rousi, A. 1971. The genus *Hippophae* L.-A taxonomic study. *Ann. Bot. Fenn.* 8: 177-227.

20. Saggu, S., Diveker, H.M., Gupta, V., Sawhney, R.C., Banerjee, P.K. and Kumar, R. 2007. Adaptogenic and safety evaluation of seabuckthorn (*Hippophae rhamnoides* L.) leaf extract: A dose dependent study. *Food Chem. Toxicol.* 45: 609-617.

21. Scherrer, R. and Gerhardt, P. 1971. Molecular sieving by the *Bacillus megatrium* cell wall and protoplast. *J. Bacteriol.* 107: 718-735.

22. Shipulina, L.D. 2001. A study of antiviral activity and other biological properties of hiporamin-A new antiviral drug from seabuckthorn leaves. In: *Seabuckthorn-A Resource for Health and Environment in Twenty First Century*. Proceeding of International Workshop on Seabuckthorn (V. Singh and P.K. Khosla, Eds.), p. 212-213, February18-21, 2001, New Delhi.

23. Singh, V., Gupta, R.K., Arumughan, C., Sawhney, S.C., Rana, R.K., Singh, A., Lal, M., Devi, R. and Sharma, N. 2011. Biochemical evaluation of *Hippophae salicifolia* and *H. mongolica* as horticultural crops in dry temperate Himalayas. In: *Emerging Trends on R and D on Health Protection and Environmental Conservation*. Proceedings of National Conference on Seabuckthorn, p.75-92, December 1-3, 2011, CSK HPKV, Palampur, HP, 324 p.

24. Singh, V. 1994. A report on arboreal seabuckthorn (*Hippophae salicifolia*) from dry temperate Himalayas. *Journal of Tree Sciences* 3 (1): 67-68.

25. Smid, E.J. and Gorris, L.G.M. 1999. Natural antimicrobials for food preservation. *In: Handbook of Food Preservation* (M.S. Rahman, Ed.), p. 285-308, Marcel Dekker, New York. 809p.

26. Soxhlet, F. 1879. Die genichtsanalytische bestimmung des miichfettes. *Dingler Poly Technisches J.* 232: 461.

27. Upadhyay, N.K., Kumar, R., Siddiqui, M.S. and Gupta, A. 2009. Mechanism of wound-healing activity of *Hippophae rhamnoides* L. leaf extract in experimental burns. *eCAM* : 1-9.

28. Xu, M. 1991. Present conditions and future research on seabuckthorn medicinal use. *J. Water Soil Conserv. China* 5: 38.

29. Xu, M.Y., Sun, X.X. and Tong, W.X. 1994. Medical research and development of seabuckthorn. *Hippophae* 7: 32-40.

30. Yang, B. and Kallio, H. 2001. Fatty acid composition of lipids in seabuckthorn (*Hippophae rhamnoides* L.) berries of various origins. *J. Agric. Food Chem.* 49 (4): 1939-1947.

Chapter 32

Eco-climatic Imprint and Gender Bias in the *In vitro* Antioxidant Activity and Total Polyphenols in the Leaves of *Hippophae rhamnoides* and Related Species from Trans-Himalaya

Kapil Sharma, Brajinder Singh, Brij Lal and Harsh Pratap Singh

Institute of Himalayan Bioresource Technology (CSIR), Palampur – 176 062, Himachal Pradesh, India

SUMMARY

The leaves of male and female plants of *Hippophae rhamnoides* and related species from trans-Himalayan region were evaluated for *in vitro* antioxidant activity expressed as trolox equivalent antioxidant capacity (TEAC) by two different bioassays: 1,12-Diphenyl-2-picrylhydrazyl (DPPH) and 2, 22- azino-bis (3-ethylbenzothiazoline-6-sulfonic acid) (ABTS) radical scavenging assays. The 50 per cent aqueous ethanol extracts of female plants of *Hippophae rhamnoides* were found to possess strong antioxidant activity (34.33 to 50.76 and 43.44 to 62.94 mM trolox equivalent/g) with ABTS and DPPH assays, respectively. In general, leaves of female plants of *Hippophae* species were found to have higher antioxidant capacity than from male plants of the same species and amongst the species, *H. rhamnoides* being superior to all. The TEAC measured by ABTS and DPPH methods of leaf extracts was found to be in the following order: *H. rhamnoides* ♀ > *H. rhamnoides* ♂ > *H. salicifolia* ♀ > *H. salicifolia* ♂ > *H. tibetana* ♀ > *H. tibetana* ♂. The chemical composition of extracts studied in terms of the total polyphenol contents (TPC) and total flavonoid contents (TFC) was found to be in the range of 14.1 to 82.0 mg/g gallic acid equivalent and 6.2 to 39.3 mg/g quercetin equivalent, respectively, and showed positive correlation with $TEAC_{ABTS}$ and $TEAC_{DPPH}$, suggested polyphenols in the extracts were responsible for the antioxidant activities.

Keywords: *Hippophae rhamnoides, Antioxidant activity, TEAC, Phenolics and flavonoids.*

Seabuckthorn (*Hippophae* L.): *A Multipurpose Wonder Plant, Vol. 4: Emerging Trends in Research and Technologies* (Virendra Singh, *Editor-in-Chief,* 2014), p. 402–411, Daya Publishing House, New Delhi, India.

INTRODUCTION

There is a growing interest in the plant derived antioxidants as stress reliever and as daily dietary supplements in human nutrition. These are accumulated as secondary metabolites in all parts of plant especially in foliage. The genus *Hippophae* is a rich source of dietary antioxidants and traditionally, being used as forage for animals and also taken in the form of tea made from it in high altitude trans-Himalayan region (Xing, 2003).

Hippophae rhamnoides L. commonly known as seabuckthorn, is a dioecious, wind pollinated and wild shrub growing at high altitudes (2100-4500 m asl) in harsh climatic conditions, and also have distinct male and female population (Polunin and Stainton, 1984; Yoshida *et al.*, 1991). The species has restricted ecological adaptation in trans-Himalayan belt found growing along river sides and river beds, also on rocky mountain slopes, where soil moisture is seen. The three *Hippophae* species, *H. rhamnoides* (HR), *H. salicifolia* (HS) and *H. tibetana* (HT) are recognized world over. The trans-Himalayan region of Indian subcontinent particularly Lahaul-Spiti region of Himachal Pradesh and Leh-Ladakh region of Jammu and Kashmir, is the prime dwelling place of *H. rhamnoides*. It is found growing in all the geoclimatic zones in these regions including Lahaul-Spiti and Ladakh (Singh, 2003), whereas the other two species, namely *H. salicifolia* and *H. tibetana*, it seems that they are more geo-climatic specific, restricted only in certain localized patches in Lahaul and Spiti valleys. The climatic condition of Leh-Ladakh is not different than prevalent in Lahaul and Spiti (Chaurasia and Singh, 1996).

The plant has been recognized as a versatile nutraceutical crop with diverse uses, from controlling soil erosion to being a source of horse fodder, nutritious foods, drugs and skin-care products (Fan *et al.*, 2007). Many medicinal effects of seabuckthorn against flu, cardiovascular diseases, mucosal injuries and skin disorders (Xiao, 1980; Beveridge *et al.*, 1999) have been suggested to be due to the high contents of antioxidant substances present in this plant. The leaves are the richest source of bioactive flavonoids and are known to have considerable medicinal values (Kallio *et al.*, 2002). The leaves are also rich in vitamin A, vitamin C and amino acids. The leaves of plant are processed into herbal tea for use as refreshing drink by the local populace of high altitude regions.

In earlier studies, the antioxidant and immunomodulatory properties of *Hippophae rhamnoides* leaves from trans-Himalayan region has been reported (Geetha *et al.*, 2002; 2003) but scientific studies on the antioxidant activity of leaves of *H. rhamnoides* and related species with special reference to male and female plants of trans-Himalayan region are lacking. The present investigation was undertaken to evaluate the antioxidant properties of leaves of male and female plant of *H. rhamnoides* and related species for its use as a source of natural antioxidant and in the formulation of herbal teas.

MATERIALS AND METHODS

Chemicals

DPPH (1, 12- Diphenyl-2-picrylhydrazyl), ABTS [2, 22- azino-bis (3-ethylbenzothiazoline-6-sulfonic acid)], trolox, gallic acid, ascorbic acid, Folin-Ciocalteu's phenol reagent and quercetin were purchased from Sigma-Aldrich Chemie (Steinheim, Germany). All other solvents and chemicals were of analytical grade and obtained from Merck, Mumbai (India).

Leaves Sample Collection

The leaves of male and female plants of *H. rhamnoides* were collected from Lahaul-Spiti and Leh-Ladakh region, whereas the leaves of *H. salicifolia* and *H. tibetana* were procured from Lahaul-Spiti region of trans-Himalaya in the month of September, 2006 and voucher specimens collected were identified, processed and deposited in the Herbarium (PLP) of Institute of Himalayan Bioresource Technology, Palampur, India. About 1.0 kg leaves of all the species, *i.e. H. rhamnoides* from Darena (PLP 6521), Tinno (PLP 6468), Kishori (PLP 6593) of Lahaul-Spiti and Leh (PLP 6489) of Ladakh region; *H. salicifolia* from two different locations *viz.*, Kirting (PLP 6496) and Keylong (PLP 4064) and *H. tibetana* from Takcha (PLP 4334) and Losar (PLP 4335) of Lahaul-Spiti region were collected and pooled separately. After harvest, the leaves were air dried under shade (temperature, 27 ± 2 °C; and relative humidity, 34 ± 3 per cent and stored at an ambient temperature (27 °C).

Extraction of Plant Material

About 5 g of powdered plant material was sonicated with 50 ml of 50 per cent aqueous ethanol in an ultrasonicator bath (Elma Ultrasonic, Germany) at a controlled temperature (30 ± 5 °C) for 40-80 min. An extraction time of 60 min was taken as optimum on mass yield basis. The extracts were filtered and concentrated to dryness under vacuum (temperature, 40-45 °C) and then subjected to lyophilization using Savant (Savant, USA) vacuum concentrator until a constant weight was obtained.

Determination of Total Polyphenol Contents

Total polyphenol contents (TPC) in samples were measured by using Folin-Ciocalteu's method (Swain and Hillis 1959; Lin *et al.*, 1999). For preparation of a calibration curve, 20, 40, 60, 80, and 100 µl aliquots of aqueous gallic acid (0.2 per cent) were mixed with 0.5 ml of 1N Folin-Ciocalteu's phenol reagent and 1.0 ml of 35 per cent Na_2CO_3 in a 25 ml volumetric flask and the solution was made to 25 ml in distilled water. The absorbance was read at 730 nm by using spectrophotometer after 35 min of incubation at an ambient temperature. The total polyphenol contents of the sample were expressed as mg of gallic acid equivalent (GAE)/g of plant material. A total of 50 µl of aqueous ethanol extracts of samples were mixed with the same reagent as described above, and after 35 min of incubation, the absorption was measured at 730 nm for determination of total polyphenol contents.

Determination of Total Flavonoid Content

Total flavonoid content (TFC) in the samples was measured by using a modified colorimetric method (Kosalec *et al.*, 2004). For preparation of calibration curve, 0.5 ml of 12.5, 25.0, 50.0, 80.0 and 100 µg/ml quercetin in 80 per cent ethanol (v/v) was mixed with 1.5 ml 95 per cent ethanol (v/v), 0.1 ml 10 per cent aluminum chloride (w/v), 0.1ml of 1 mol/l potassium acetate and 2.8 ml water in 5.0 ml volumetric flask. After incubation at room temperature for 30 min, the absorbance of the reaction mixture relative to blank was measured at 415 nm by using spectrophotometer. The total flavonoid content of samples was expressed as mg quercetin/g dry plant material. Aliquots (0.5 ml) of aqueous ethanol extracts were mixed with the same reagent as described above and after 30 minutes, the absorption was measured at 415 nm for determination of total flavonoids.

EVALUATION OF ANTIOXIDANT ACTIVITY

ABTS Radical Cation Scavenging Activity

ABTS radical cation decolourisation test is a spectrophotometric method widely used for assessment of antioxidant activity of various substances. The experiment was carried out using an improved ABTS decolourisation assay (Re *et al.*, 1999). In brief, ABTS was dissolved in water to 7 mM concentration. ABTS radical cation was produced by reacting ABTS stock solution with 2.45 mM potassium persulphate (final concentration) and allowed the mixture to stand in the dark at room temperature for 16 h before use. The radical was stable in this form for more than 2 days when stored in the dark at room temperature. For the study of phenolic compounds and food extracts, the $ABTS^{\cdot+}$

was diluted with ethanol to an absorbance of 0.70 ± (0.02) at 734 nm and equilibrated at 30 °C. After addition of 2.0 ml of diluted $ABTS^{\cdot+}$ Solution (A = 0.700 ± 0.020) to 50 µl of antioxidant compounds, the absorbance reading was taken exactly after 4 min, appropriate solvent blank was run in each assay. The percentage inhibition of absorbance at 734 nm is calculated and plotted as a function of concentration of antioxidants.

DPPH Radical Scavenging Activity

Radical scavenging activity of leaf extract against stable DPPH· was also determined spectrophotometrically. When DPPH· reacts with an antioxidant compound which can donate hydrogen, is reduced. The change in colour (from deep violet to yellow) was measured at 517 nm. The radical scavenging activity of extracts was measured by slightly modified method of Brand-Williams *et al.* (1995). 0.1 ml extract was added to 2.9 ml of the 100 µM DPPH·solution prepared in 80 per cent aqueous ethanol. The mixture was shaken vigorously and allowed to stand at 23 °C in dark for 30 min and decrease in absorbance of the resulting solution was monitored at 517 nm against a blank consisted of 0.1 ml of 50 per cent aqueous ethanol and 2.9 ml of DPPH·solution. All measurements were done in triplicate.

Trolox Equivalent Antioxidant Capacity (TEAC) of *Hippophae* Samples

Trolox standard curves that relate the concentration of trolox and the amount of absorbance reduction caused by it were obtained using the ABTS and DPPH assays (Kim *et al.*, 2002; Miliauskas *et al.*, 2004). The absorbance reduction at 734 and 517 nm of trolox was measured at various concentrations by ABTS and DPPH assay, respectively. For preparation of calibration curve of trolox, 50 µl of aqueous ethanol solutions of above (final concentration 0-30 µM) were mixed with 2.0 ml of diluted $ABTS^{\cdot+}$ solution (A = 0.700 ± 0.020) and the absorbance was read exactly 4 minute after initial mixing at 734 nm. In case of DPPH, for preparation of calibration curve of trolox, 100 µl of aqueous ethanol solution of above (final concentration 0-30 µM) was mixed with 2.9 ml of 100 µM DPPH· prepared in 80 per cent aqueous ethanol. The solutions were shaken vigorously and allowed to stand at 23 °C in dark for 30 min. The decrease in absorbance of the resulting solutions was monitored at 517 nm at 30 min. For samples, 50 µl of extracts were assayed as above.

The radical scavenging activity was calculated by the following formula:

$$\text{Per cent Inhibition} = [(A_B - A_A)/A_B] \times 100$$

where, A_B-absorption of blank sample and A_A-absorption of tested extract solution.

The calculation of TEAC of each compound at the various concentrations levels was made using trolox standard curves (Wang *et al.*, 1996; Van den Berg *et al.*, 1999). For *Hippophae* leaf samples, properly diluted extracts were assayed by both ABTS and DPPH models.

Statistical Analysis

Results were expressed as mean ± standard error. The correlation between total phenolics, flavonoid contents and antioxidant activity was established by regression analysis. All determinations were performed in triplicate.

RESULTS AND DISCUSSION

Total Polyphenol and Flavonoid Contents

Phenolic compounds are commonly found in plants and have several biological activities including antioxidant properties. The total polyphenol contents in aqueous ethanol extracts were determined from regression equation of calibration curve ($y = 0.0042x + 0.0224$, $R^2 = 0.9983$) and expressed in gallic acid equivalent between 14.1 to 82.0 mg GAE/g of dry plant material (Table 32.1). The highest amount was found in the leaf extracts of female plants of *H. rhamnoides* collected from Spiti valley. The general trend in descending order as follows: HR (Spiti) ♀> HR (Spiti) ♂> HR (Leh) ♀> HR (Leh) ♂> HS (Spiti) ♀> HS (Spiti) ♂> HT (Spiti) ♀> HT (Spiti) ♂. The total phenolics in the leaves of female plants of *H. rhamnoides* collected from Spiti region was 1.12 times more as compared to female plants from Leh, and was 1.82 and 4.24 times more than female plants of *H. salicifolia* and *H. tibetana* from Spiti region. On the other hand, total phenolics in the leaves of male plants of *H. rhamnoides* collected from Spiti region was 1.18 times more as compared to male plants from Leh was 2.35 and 5.08 times more than male plants of *H. salicifolia* and *H tibetana* from Spiti region (Table 32.1).

Table 32.1: Total Phenolics, Flavonoids and Antioxidant Activity in the Extracts of *Hippophae* Leaves Samples

Region	Species	Female			Male		
		TPC (mg gallic acid equivalent/g)	TFC (mg quercetin equivalent QE/g)	TEAC (mM trolox equivalent/ g)	TPC (mg gallic acid equivalent/g)	TFC (mg quercetin equivalent QE/g)	TEAC (mM trolox equivalent/ g)
Spiti	*H. rhamnoides*	82.0 ± 0.03	39.3 ± 0.00	50.76 ± 0.42[a] 62.94 ± 3.00[b]	74.8 ± 0.03	36.3 ± 0.00	45.37 ± 0.42[a] 58.69 ± 0.86[b]
	H. salicifolia	44.9 ± 0.78	21.6 ± 0.05	16.69 ± 0.14[a] 21.98 ± 0.28[b]	31.8 ± 0.79	13.1 ± 0.04	9.36 ± 0.10[a] 12.83 ± 0.21[b]
	H. tibetana	19.3 ± 0.19	11.0 ± 0.14	7.63 ± 0.07[a] 10.67 ± 0.80[b]	14.1 ± 0.19	6.2 ± 0.16	5.75 ± 0.11[a] 8.33 ± 0.64[b]
Leh	*H. rhamnoides*	73.0 ± 0.13	30.7 ± 0.00	34.33 ± 0.63[a] 43.44 ± 0.74[b]	63.1 ± 0.06	28.8 ± 0.02	32.83 ± 0.63[a] 41.44 ± 0.22[b]
	H. salicifolia[c]	n.d.	n.d.	n.d.	n.d.	n.d.	n.d.
	H. tibetana[c]	n.d.	n.d.	n.d.	n.d.	n.d.	n.d.

a: TEAC assayed by ABTS method.

b: TEAC assayed by DPPH method; and

c: Plant species were not available in the region for sample collection.

n.d.: Not determined.

Similarly, total flavonoids in aqueous ethanol extracts were determined from regression equation of calibration curve ($y=0.005x-0.0069$, $R^2=0.999$) and expressed in quercetin equivalent between 6.2 to 39.3 mg QE/g of dry plant material (Table 32.1). The highest flavonoids contents were found in the extract of female plants of *H. rhamnoides* leaves collected from Spiti valley and followed the same trend as observed in case of total phenolics. For the extraction purpose, ethanol along with water was taken as solvent since aqueous alcohols are most widely used in antioxidant extraction work (Spigno and DeFaver, 2007).

TEAC of *Hippophae* Leaves Samples

Keeping in view, the effectiveness of antioxidant study, it has recently been recommended to employ at least two different *in vitro* methods due to the differences between various free radical-scavenging assay systems (Schlesier *et al.*, 2002; Aruoma, 2003). Thus, the extracts were subjected to two different antioxidant bioassays employing $ABTS^{\cdot+}$ and $DPPH^{\cdot}$ (expressed in mM/g). In this study, the antioxidant activity was expressed as mM trolox equivalents per gram of plant material on the dry basis, as it is a more meaningful and descriptive expression than assays that express antioxidant activity as the percentage decrease in absorbance. As such the results provide a direct comparison of the antioxidant activity with trolox. Till now, to the best of our knowledge, there is no such report available which could highlight the comparative evaluation of antioxidant potential in the leaves of male and female plants of *Hippophae rhamnoides* and related species from trans-Himalayan region. The results of ABTS and DPPH inhibition by standard antioxidant trolox are summarized in Table 32.1.

As evident from Table 32.1, the TEAC values for the extracts had wide variability and ranged between 5.75-50.76 mM trolox/g by ABTS assay and 8.33-62.94 mM trolox/g by DPPH assay. The TEAC assayed by both ABTS and DPPH methods for leaves of female plants of *H. rhamnoides* from Spiti region was 1.45 times more as compared to leaves of female plants of Leh. Similarly, the TEAC of male plants of Spiti region was 1.4 times more than Leh region. The leaves of female plants of Spiti valley possessed the higher antioxidant capacity followed by male plants of Spiti and female and male plants of Leh, respectively. The gender bias observed in the estimated antioxidant values in *H. rhamnoides* of trans-Himalayan region was further confirmed by similar trends observed in the values in other related species of the plant, *i.e. H. tibetana* and *H. salicifolia*. This could not be confirmed for Leh region because the other two species could not be found in this region during survey and sample collection work. However, it could be safely inferred from the results (Table 32.1) that the trend of the analyses as observed in the Spiti region for the three species could not be greatly different for the Leh region as well as both the regions are part of the larger trans-Himalayan belt.

Furthermore, the locale specific gender bias observed in the estimated antioxidant values in *H. rhamnoides* in Lahaul-Spiti region were further confirmed by similar trends observed in the values in other related species of the plant, *i.e. H. tibetana* and *H. salicifolia* (Table 32.1). The male-female ratio of antioxidant values for *H. rhamnoides* (Spiti) was found 1.1, 0.95 for *H. rhamnoides* (Leh), 1.78 for *H. salicifolia* (Spiti) and 1.32 for *H. tibetana* (Spiti). These values suggested species specific indices for *Hippophae* plant, whereas indices obtained by dividing the antioxidant values of *H. rhamnoides* (Spiti) female plant taken as a unit to other female species values like 0.64 for *H. rhamnoides* (Leh), 0.32 for *H. salicifolia* (Spiti) and 0.15 for *H. tibetana* (Spiti) indicate true antioxidant potential of related species of the region (Table 32.1). However, this could not be confirmed for Ladakh region because of non-availability of other species due to geo-climatic factors. The TEAC assayed by ABTS method for leaves of female plants of *H. rhamnoides* from Spiti region was 3.0 and 6.7 times more than *H. salicifolia* and *H. tibetana*, respectively, while the TEAC assayed by DPPH method for leaves of female plants of *H. rhamnoides* from Spiti region was 2.86 and 5.9 times more than *H. salicifolia* and *H. tibetana*. On the other hand, TEAC assayed by ABTS method for male plants of *H. rhamnoides* from Spiti region was 4.8 and 7.8 times more than *H. salicifolia* and *H. tibetana* whereas TEAC assayed by DPPH method for male plants of *H. rhamnoides* from Spiti region was 4.6 and 7 times more than *H. salicifolia* and *H. tibetana*. A linear relationship was established between total polyphenols, flavonoids and TEAC values and it was observed that antioxidant activity increased proportionally to the polyphenol and flavonoids contents. It will be relevant to mention here that earlier reports (Gao *et al.*, 2000; Ruberto *et al.*, 2007) have demonstrated the correlation between phenolic content of plant to their antioxidant power. In this study, the contents of polyphenols and flavonoids in the extracts positively correlate with their antiradical activity confirming participation to the radical scavenging activity of the leaves of *Hippophae*

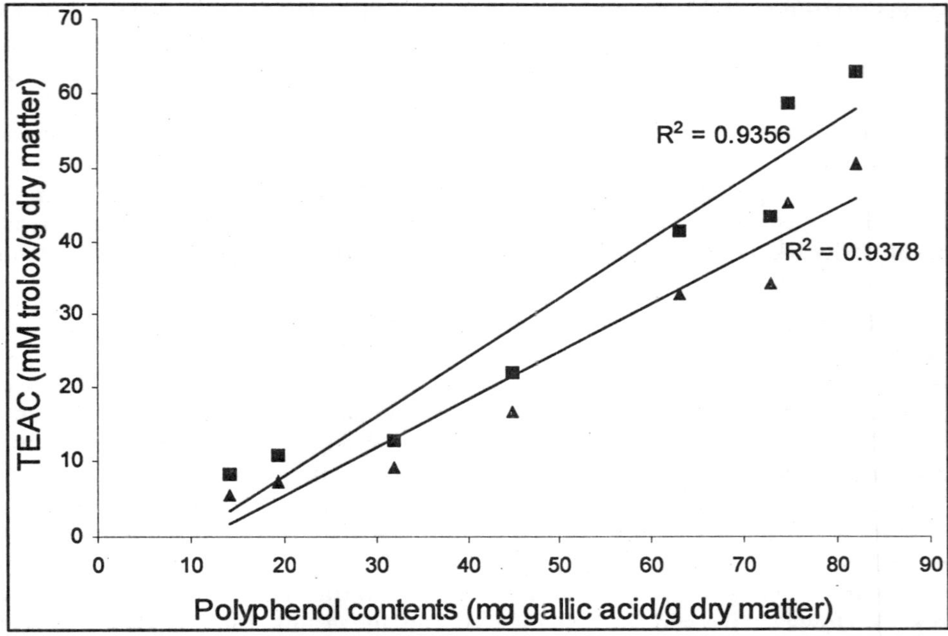

Figure 32.1: Correlation between Total Polyphenol Contents and TEAC Measured in different *Hippophae* Leaves Samples. ■: DPPH; ▲: ABTS.

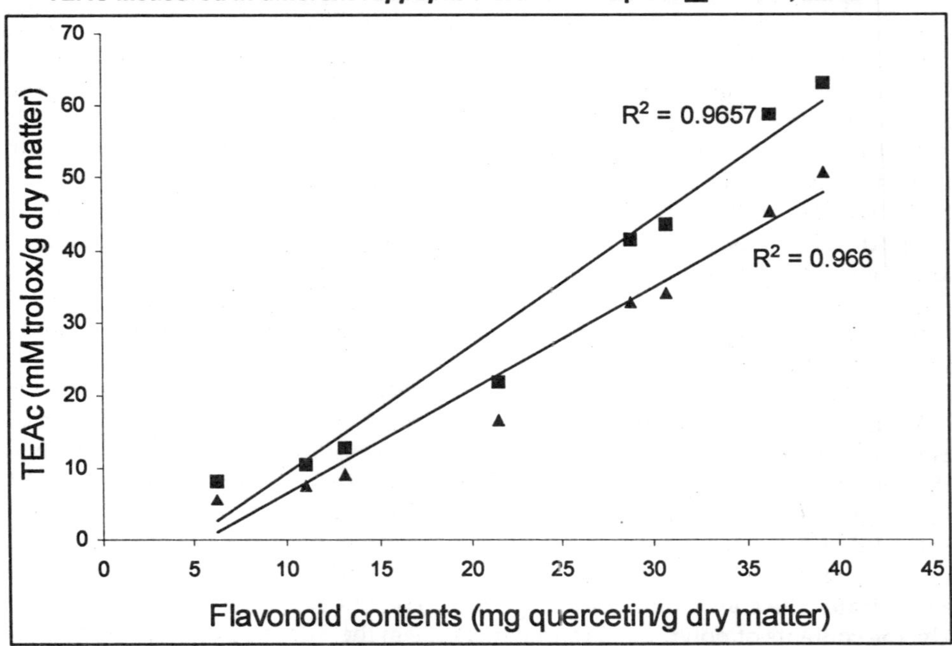

Figure 32.2: Correlation between Total Flavonoid Contents and TEAC Measured in different *Hippophae* Leaves Samples. ■: DPPH; ▲: ABTS.

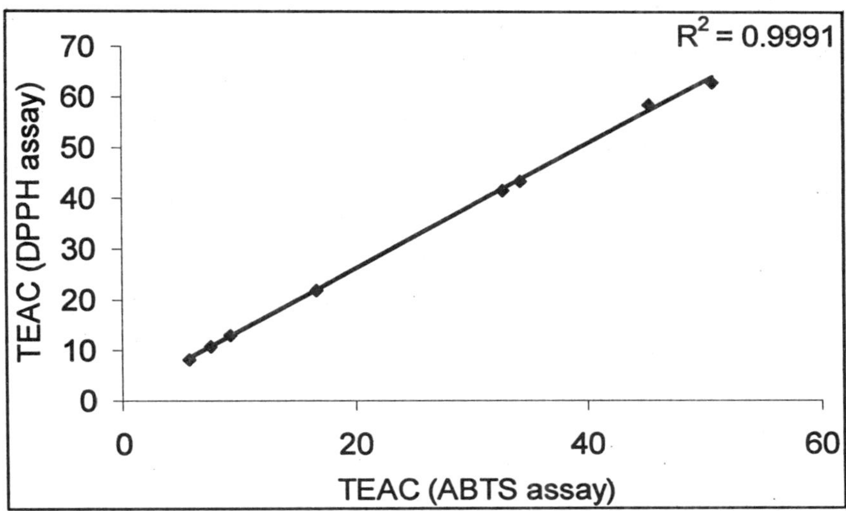

Figure 32.3: Correlation between ABTS and DPPH Assay in *Hippophae* Leaves Samples.

extracts (Figures 32.1 and 32.2) which in general was observed in other plants as well (Javanmardi *et al.*, 2003; Minoggio *et al.*, 2003). Both the assay methods (ABTS and DPPH) for determining the antioxidant activity in the extracts of the *Hippophae* leaves samples were in high conformity to the results expressed (Figure 32.3).

CONCLUSION

The results of this study showed that phenolics are the major contributors towards the total antioxidant activity of leaves of *Hippophae* species. The high correlation of the phenolic and flavonoid content with the antioxidant activity in aqueous ethanol extract suggested that either of them could be used as an index of the antioxidant activity of leaves of *Hippophae* species. Moreover, the leaves of female plants of *Hippophae* species in general, and *H. rhamnoides* collected from Spiti valley of trans-Himalayan region in particular, are good source of natural antioxidants for medicinal and commercial use.

ACKNOWLEDGEMENT

We acknowledge financial support from Defence Research and Development Organization (DRDO), India, for funding the Project entitled "*Hippophae* Based Herbal Tea". Kapil Sharma is indebted to Council of Scientific and Industrial Research, New Delhi, India for the award of Senior Research Fellowship. Thanks to Director, IHBT, for providing research facilities, encouragement and permission to publish this work.

REFERENCES

1. Aruoma, O.I. 2003. Methodological considerations for characterizing potential antioxidant actions of bioactive components in plant foods. *Mutat. Res.* 523/524: 9-20.
2. Beveridge, T., Li, T.S.C. and Oomah, B.D. 1999. Seabuckthorn products: Manufacture and composition. *J. Agric. Food Chem.* 47: 480-3488.

3. Brand-Williams, W., Cuvelier, M. E. and Berse, C. 1995. Use of free radical method to evaluate antioxidant activity. *Lebensm. Wiss. Technol.* 28: 25-30.

4. Chaurasia, O.P. and Singh, B. 1996. *Cold Desert Plants-Leh Valley.* Oxford University Press, New Delhi, 146p.

5. Fan, J., Ding, X. and Gu, W. 2007. Radical-scavenging proanthocyanidins from seabuckthorn seed. *Food Chem.*102: 168-177.

6. Gao, X.Q., Ohlander, M., Jeppsson, N., Bjork, L. and Trajkovski, V. 2000. Changes in antioxidant effects and their relationship to phytonutrients in fruits of seabuckthorn (*Hippophae rhamnoides* L.) during maturation. *J. Agric. Food. Chem.*48: 1485-1490.

7. Geetha, S., Ram, M.S., Mongia, S.S., Singh, V., Ilavazhagan, G. and Sawhney, R.C. 2003. Evaluation of antioxidant activity of leaf extract of seabuckthorn (*Hippophae rhamnoides*) on chromium (VI) induced oxidative stress in albino rats. *J. Ethnopharmacol.* 87: 247-251.

8. Geetha, S., Ram, M.S., Singh, V., Ilavazhagan, G. and Sawhney, R.C.2002. Antioxidant and immunomodulatory properties of seabuckthorn (*Hippophae rhamnoides*)-An *in vitro* study. *J. Ethnopharmacol.* 79: 373-378.

9. Javanmardi, J., Stushnoff, C., Locke, E. and Vivanco, J.M. 2003. Antioxidant activity and total phenolic content of Iranian *Ocimum* accessions. *Food Chem.* 83: 47-50.

10. Kallio, H., Yang, B.R. and Peippo, P.J. 2002. Effects of different origins and harvesting time on vitamin C, tocopherols, and tocotrienols in seabuckthorn (*Hippophae rhamnoides*) berries. *J. Agric. Food Chem.* 50: 6136-6142.

11. Kim, D.O., Lee, K.W., Lee, H.J. and Lee, C.Y. 2002. Vitamin C equivalent antioxidant capacity (VCEAC) of phenolic phytochemicals. *J. Agric. Food Chem.* 50: 3713-3717.

12. Kosalec, I., Bakmaz, M., Pepeljnjak, S. and Vladimir-Knezevic, S. 2004. Quantitative analysis of the flavonoids in raw propolis from northern Croatia. *Acta Pharmaceut.* 54: 65-72.

13. Lin, Y.L., Juan, I.M., Chen, Y.L., Liang, Y.C. and Lin, J.K. 1999. Composition of polyphenols in fresh tea leaves and associations of their oxygen-radical absorbing capacity with antiproliferative actions in fibroblast cells. *J. Agric. Food Chem.*44: 1387-1394.

14. Miliauskas, G., Venskutoins, P.R. and Vanbeek, T.A. 2004. Screening of radical scavenging activity of some medicinal and aromatic plant extracts. *Food Chem.* 85: 231-237.

15. Minoggio, M., Bramati, L., Simonetti, P., Gardana, C., Iemoli, L., Santangelo, E., Mauri, P.L., Spigno, P., Soressi, G.P. and Pietta, P.G. 2003. Polyphenols pattern and antioxidant activity of different tomato lines and cultivars. *Ann. Nutr. Metab.* 47: 64-69.

16. Polunin, O. and Stainton, A. 1984. *Flowers of the Himalaya.* p. 357-358, Oxford University Press, New Delhi, 545p.

17. Re, R., Pellegrini, N., Proteggente, A., Pannala, A., Yang, M. and Rice Evans, C. 1999. Antioxidant activity applying an improved ABTS radical cation decolourisation Assay. *Free radical Biol.* 26: 1231-1237.

18. Ruberto, G., Renda, A., Daquino, C., Amico, V., Spatafora, C., Tringali, C. and Tommasi, N.D. 2007. Polyphenol constituents and antioxidant activity of grape pomace extracts from five Sicilian red grape cultivars. *Food Chem.* 100: 203-210.

19. Schlesier, K., Harwat, M., Bohm, V. and Bitsch, R. 2002. Assessment of antioxidant activity by using different *in vitro* methods. *Free Radical Res.* 36: 177-187.

20. Singh, V. 2003. Geographical adaptation and distribution of seabuckthorn (*Hippophae* L.) resources. In: *A Multipurpose Wonder Plant-Botany.* Vol. I. *Botany, Harvesting and Processing Technologies* (V.Singh, Editor in Chief), p.21-34, Indus Publishing Company, New Delhi, 518p.

21. Spigno, G. and DeFaver, D.M. 2007. Antioxidants from grape stalks and marc: Influence of extraction procedure on yield, purity and antioxidant power of the extracts. *J. Food Eng.* 78: 793-801.

22. Swain, T. and Hillis, E. 1959. The phenolic constituents of *Prunus domestica* L. The quantitative analysis of phenolic constituents. *J. Sci. Food Agric.* 10: 63-68.

23. Van den Berg, R., Haenen, G.R., Van den Berg, H. and Bast, A. 1999. Applicability of an improved trolox equivalent antioxidant capacity (TEAC) assay for evaluation of antioxidant capacity measurement of mixtures. *Food Chem.* 66: 511-517.

24. Wang, H., Cao, G. and Prior, R.L. 1996. Total antioxidant capacity of fruits. *J. Agric. Food Chem.* 44: 701-705.

25. Xiao, Z. 1980. Chemical study on the flavonoids of *Hippophae rhamnoides*. *Acad. J. Sichuan. Med. Univ.* **11:** 174-176. (In Chinese)

26. Xing, C. 2003. Health protection and processing technology of seabuckthorn tea. In: *Seabuckthorn (Hippophae L.)-A Multipurpose Wonder Plant. Vol. I: Botany, Harvesting and Processing Technologies* (V. Singh, Ed. in Chief, 2003), p.475-478, Indus Publishing Company, New Delhi, 518p.

27. Yoshida, T., Tanaka, K. and Chen, X. 1991. Tannins from *Hippophae rhamnoides*. *Photochemistry* 30: 663-666.

Chapter 33

Antioxidant Efficiency of Seabuckthorn (*Hippophae rhamnoides* L.) Extract in Ehrlich Ascites Carcinoma (EAC) of Balb/c Mice Model

Mustafa Koray Gumustas, Aliye Aras Perk, Berna Yavuz, Nensi Cavak and Mehmet Bona

Department of Medical Biochemistry, Cerrahpasa Medical Faculty, Istanbul University, Istanbul, Turkey
Department of Biology, Faculty of Science, Istanbul University, Istanbul, Turkey

SUMMARY

Seabuckthorn, which is a unique and valuable plant, has recently gained worldwide attention, mainly due to its ecological, medicinal and nutritional potential. The aim of the present study was to evaluate the antioxidant activity of seabuckthorn against EAC cells by measuring primer antioxidant enzyme superoxide dismutase (SOD) and lipid peroxidation products malondialdehyde (MDA). From the results, it is clear that seabuckthorn has antioxidant efficiency in ascetic fluid and also on EAC cells. Seabuckthorn application into abdominal has also antioxidative effects on general circulation and possibly on whole body.

Keywords: *Seabuckthorn, Ehrlich ascites tumor, Antioxidant and Lipid peroxidation.*

INTRODUCTION

Large number of studies has revealed that a regular consumption of fruits and vegetables provide good protection against cancer in many body parts and some of the foods and herbs contain a host of cancer-protective phytochemicals (Hung *et al.*, 2004). These beneficial compounds alter metabolic pathways and hormonal actions that are associated with the development of cancer, stimulate the immune system and the antioxidant activity (Craig, 1999).The majority of the world's population in developing countries still relies on herbal medicines to meet their health needs in cases when synthetic

Seabuckthorn (*Hippophae* L.): *A Multipurpose Wonder Plant, Vol. 4: Emerging Trends in Research and Technologies* (Virendra Singh, *Editor-in-Chief*, 2014), p. 412–417, Daya Publishing House, New Delhi, India.

medicine could not relieve patients who suffer from painful illnesses like cancer (Nadkarni, 2005). In the modern system of medicine, several chemotherapeutic agents have been developed as a result of screening of the medicinal plants in various parts of world (Radha *et al.*, 2008). So, there is a growing interest in the pharmacological evaluation of medicinal plants.

Seabuckthorn belongs to the family *Elaeagnaceae*, is a unique and valuable plant has recently gained worldwide attention, mainly for its ecological, medicinal and nutritional potential. According to the last records, Lian *et al.* (2003) described a new subspecies of seabuckthorn (*Hippophae rhamnoides* ssp. *wolongensis*) (Lain *et al.*, 2000) and therefore, this genus has 7 species and 9 subspecies (Sun *et al.*, 2003). Seabuckthorn has an extremely wide distribution but fragmented in Europe and Asia, from China, India, Mongolia, Russia, Kazakhstan, Turkey, Romania, Switzerland, France to Britain and north to Finland, Norway and Sweden (Rousi, 1971; Rongsen, 1997; Bartish *et al.*, 2000; Bartish *et al.*, 2002; Singh, 2003). This plant has been distributed over Turkey, mainly in northern and eastern regions from the sea level to high elevations of about 3000 m (Mc Kean, 1982).

Almost all parts of seabuckthorn contains vitamins (A, C, E, K, riboflavin, folic acid), mineral elements, organic acids, free amino acids, monosaccharides sugars, large amount of carotenoids (α, β, δ-carotene, lycopene), different flavonoids (quercetin, isorhamnetin, myricetin, kaempferol), fatty acids, triacylglycerol, glycerophospholipids, phytosterols (ergosterol, stigmasterol, lansterol, amyrins), zeaxanthin esters, phenolic compounds and alpha-tocopherol acids (Yao and Tigerstedt, 1992; Beveridge *et al.*, 1999; Jeppson and Gao, 2000; Weller and Breithaupt, 2001; Yang *et al.*, 2001; Kallio *et al.*, 2002; Yang and Kallio, 2002; Luhua *et al.*, 2004; Fu *et al.*, 2005; Pintea *et al.*, 2005; Vaher *et al.*, 2005; Zhang and Cui, 2005).

Like seabuckthorn, these compounds are of our interest from the chemical point of view, and also because many of them possesses biological and therapeutic activities including antioxidant, cardiovascular, cancer therapy, healing, anti-inflammation, antiradiation effect, treatment of gastrointestinal ulcers, as a liver protective agent, antioxidant, platelet aggregation and immunomodulator (Patel *et al.*, 2012).

For these reasons, the search for antioxidants as cancer chemopreventive agents is a continued process. Another aspect of antioxidant administration in cancer patients is that these could affect antineoplastic efficacy or the development of side effects of anticancer drugs (Conklin, 2004). A number of studies indicate that plant derived natural products like seabuckthorn, which possess diverse pharmacological properties and have antioxidant activity (Geetha *et al.*, 2002; Negi *et al.*, 2005). The aim of the present study was to evaluate the antioxidant activity of seabuckthorn against EAC cells by measuring primer antioxidant enzyme superoxide dismutase (SOD), and lipid peroxidation products malondialdehyde (MDA).

MATERIALS AND METHODS

Seabuckthorn Extraction

This study used commercial seabuckthorn seed oil extracted by using supercritical carbon dioxide. The seed oil of *Hippophae rhamnoides* L. were obtained from Inner Mongolia Yuhangren Hi-Tech Industrial Co., Ltd.

Animals

Fifteen male Balb/c mice (3-4 month) were obtained from Experimental Animal Research Center, Cerrahpasa Medical Faculty, Istanbul University. They were used throughout the study and housed in iron cages in a controlled environment (temperature (25±2) °C and 12 h dark/light cycle) with standard laboratory diet and water *ad libitum*. Experiments were carried out in accordance with Ethical Committee Guidelines laid down by the local committee regarding the care and use of animals for experimental procedures.

Tumor Cell Line

A line of EAC obtained from the Faculty of Pharmacy, Istanbul University. The tumor line was maintained in male Balb/c mice once by interperitoneal injection of 1.5×10^6 cells/mouse. Such developed tumor is characterized by its moderate rapid growth which could not kill the animal due to accumulation of ascites before injection using the bright line heamocytometer and dilutions were made by physiological saline and desired numbers of cells were injected in a volume of 0.5 ml.

Experimental Design

Mice were divided into three groups to carry out the experimental study. First group was control, second group was the animals once interperitoneal injected EAC and received diet and water *ad libitum* like control group. Third group was the animal once interperitoneal injected EAC and diet-water *ad libitum* like group two but interperitoneal injected seabuckthorn extract daily 10 Unit for three weeks.

At the end of three weeks, all groups were sacrificed under ether anesthesia, the accumulation of ascithes ejected, whole blood samples were collected from the heart of the animals and 5 ml liquid of ascithes mai was collected separately.

In both experimental groups, SOD activity of abdominal region tumor fluid and tumor cells (pellet) were determined by the method (Sun *et al.*, 1988). In both experimental groups and control group, MDA levels of abdominal region tumor fluid, tumor cells and plasma were determined by the method of Budge and Aust (1978). In EAC cells, protein contents were determined by the method of Lowry *et al.* (1951).

RESULTS AND DISCUSSION

At the end of the experiment period (3 weeks) in both experimental groups, a regular and rapid increase in ascetic fluid volume and weight were observed in all mice. Ascetic fluid is the direct nutritional source for tumor growth, because it meets the nutritional requirements for tumor cells (Shimizu *et al.*, 2004). These observations showed that in both experimental groups all mice were successfully inoculated with tumor cells and growth.

In EAC group, both in ascetic fluid and EAC cell oxidative stress parameters were significantly high when compared with EAC threaded with SBT group. It is because MDA as an oxidative stress marker was decreased when EAC group was treated with seabuckthorn ($p<0.001$). Although, the most important primer antioxidant enzyme SOD activity was found decreased in ascetic fluid ($p<0.001$), same enzyme activity was found increased in EAC cells ($p<0.001$) (Tables 33.1 and 33.2). From these results, we can conclude that increased SOD activity in EAC cell is more important to decrease oxidative stress and growth of EAC cells. It is clear that seabuckthorn has antioxidant efficiency in ascetic fluid and also in EAC cells.

In both ascetic fluid and EAC cells, a significant decrease of MDA was observed and showed that EAC+SBT group might be the result of direct injection of SBT into abdominal ascetic fluid compartment and its reflection to EAC cells. On the other hand, a significant decrease of SOD activity of ascetic fluid in EAC+SBT group, but increase of SOD activity in the same group EAC cells that seems to be contradictory, but SOD expression and activation may be the result of cell response of EAC cells (Table 33.1). So, the high SOD activity of EAC cells and antioxidant efficiency of SBT might be the result of decreased MDA levels when EAC cells treated with SBT (Table 33.2).

Table 33.1: Results and Comparison of Ascetic Fluid MDA Levels and SOD Activity of Mice in EAC and EAC Treated with SBT Groups

	EAC	*EAC + SBT*
MDA (nmol/mg protein)	93.27 ± 5.63	26.61 ± 4.18*
SOD (U/mg protein)	63.64 ± 3.54	32.64 ± 3.35*

*p<0.001; against EAC group.

Table 33.2: Results and Comparison of EAC Cells MDA Levels and SOD Activity of Mice in EAC and EAC Treated with SBT Groups

	EAC	*EAC + SBT*
MDA (nmol/mg protein)	220.31 ± 13.91	121.59 ± 5.24*
SOD (U/mg protein)	69.46 ± 5.20	101.05 ± 5.10*

*p<0.001; against EAC group.

When plasma EAC group compared with controls, high levels of EAC plasma MDA levels were found and EAC also has an influenced oxidative stress on general circulation (p=0.001). But, seabuckthorn application on EAC showed that plasma oxidative stress was found decreased by MDA levels (p<0.001). From these results, SBT application into abdominal also has an antioxidative effects on general circulation and possibly on whole body. These effects might be the result of the antioxidant vitamins and specific beneficial biomolecules combinations of SBT (Table 33.3).

Table 33.3: Results and Comparison of Plasma MDA Levels of Mice in Control, EAC and EAC Treated with SBT Groups

	Control	*EAC*	*EAC + SBT*
MDA (nmol/mL)	16.73 ± 1.31	44.73 ± 2.98[b]	26.75 ± 3.30[a,*]

*p<0.001; against EAC group; [a]p=0.001; [b]p<0.001; against control group.

CONCLUSION

From the present study, the results showed that increased SOD activity in EAC cell is more important to decrease oxidative stress and growth of EAC cells. It is clear that seabuckthorn has antioxidant efficiency in ascetic fluid and also on EAC cells. Also, seabuckthorn oil application on EAC showed that plasma oxidative stress was decreased in abdominal and has an antioxidative effects on general circulation, and possibly on whole body. These effects might be the result of the antioxidant vitamins and specific beneficial biomolecules combinations of seabuckthorn.

REFERENCES

1. Bartish, I.V., Jeppsson, N. and Bartish, G.I. 2000. Inter-and intraspecific genetic variation in *Hippophae* (*Elaeagnaceae*) investigated by RAPD markers. *Plant Syst. Evol.* 225 (1-4): 85-101.

2. Bartish, I.V., Jeppsson, N. and Nybom, H. 2002. Phylogeny of *Hippophae* (*Elaeagnaceae*) inferred from parsimony analysis of chloroplast DNA and morphology. *Systematic Botany* 27 (1): 41-54.

3. Beveridge, T., Li, T.S., Oomah, B.D. and Smith, A. 1999. Seabuckthorn products- Manufacture and composition. *J. Agric. Food Chem.* 47: 3480-3488.

4. Buege, J.A. and Aust, S.D. 1978. Microsomal lipid peroxidation. *Methods Enzymol.* 12: 302-310.

5. Conklin, K.A. 2004. Cancer chemotherapy and antioxidants. *J. Nutr.* 134: 3201S-3204S.

6. Craig, W.J. 1999. Nutrition and Wellness. *A Vegetarian Way to Better Health Golden Harvest books*, Berrian Springs.

7. Fu, S.C., Hui. C.W., Li, L.C., Cheuk, Y.C., Qin, L., Gao, J. and Chan, K.M. 2005. Total flavones of *Hippophae rhamnoides* promote early restoration of ultimate stress of healing patellar tendon in a rat model. *Med. Eng. Phys.* 27 (4): 313-321.

8. Geetha, S., Sai Ram, M., Singh, V., Ilavazhagan, G. and Sawhney, R.C. 2002. Antioxidant and immunomodulatory properties of seabuckthorn: An *in vitro* study. *J. Ethnopharmacol.* 79: 373-378.

9. Hung, H.C., Joshipura, K.J., Jiang, R., Hu, F.B., Hunter, D., Smith-Warner, S.A., Colditz, G. A., Rosner, B., Spiegelman, D. and Willett, W.C. 2004. Fruit and vegetable intake and risk of major chronic disease. *Journal National Cancer Institute* 96 (21): 1577-1584.

10. Jeppsson, N. and Gao, X.Q. 2000. Changes in the contents of kaempferol, quercetin and L-ascorbic acid in seabuckthorn berries during maturation. *Agri. Food Sci. Finland* 9: 17-22.

11. Kallio, H., Yang, B., Peippo, P., Tahvonen, R. and Pan, R. 2002. Triacylglycerols, glycerophospholipids, tocopherols, and tocotrienols in berries and seeds of two subspecies (ssp. *sinensis* and *mongolica*) of seabuckthorn (*Hippophae rhamnoides*). *J. Agric. Food Chem.* 50: 3004-3009.

12. Lian, Y.S., Chen, X.L., Sun, K. and Ma, R. 2003. A new subspecies of *Hippophae* (*Elaeagnaceae*) from China. *Novon* 13 (2): 200-202.

13. Lian, Y.S., Lu, S.G., Xue, S.K. and Chen, X.L. 2000. Biology and chemistry of the genus *Hippophae*. *Lanzhou, Gansu Science Technology Press*: 1-226. (In Chinese)

14. Lowry, O.H., Rosenbrough, N.J., Farr, A.L. and Randal, R.J. 1951. Protein measurement with folin-phenol regent. *J. Biol. Chem.* 193: 265.

15. Luhua, Z., Ying, T., Zhengyu, Z. and Guangji, W. 2004. Determination of alpha-tocopherol in the traditional Chinese medicinal preparation seabuckthorn oil capsule by non-aqueous reversed phase-HPLC. *Chem. Pharm. Bull.* 52: 150-152.

16. Mc Kean, D.R. 1982. *Flora of Turkey and the East Aegean Islands. Vol. 7.* (P.H. Davis, Ed.), p. 532-534, Edinburg, 947p.

17. Nadkarni, A.K. 2005. *Indian Materia Medica*, p. 972-973, Popular Prakashan, Bombay.

18. Negi, P.S., Chauhan, A.S., Sadia, G.A., Rohinishree, Y.S. and Ramteke, R.S. 2005. Antioxidant and antibacterial activities of various seabuckthorn seed extracts. *Food Chem.* 92: 119-124.

19. Patel, C.A., Divakar, K., Santani, D., Solanki, H.K. and Thakkar, J.H. 2012. Remedial prospective of *Hippophae rhamnoides* Linn. (seabuckthorn). *ISRN Pharmacology, doi:10.5402/2012/436857*.

20. Pintea, A., Varga, A., Stepnowski, P., Socaciu, C., Culea, M. and Diehl, H. A. 2005. Chromatographic analysis of carotenol fatty acid esters in *Physalis alkekengi* and *Hippophae rhamnoides*. *Phytochem. Anal.* (16): 188-195.

21. Radha, R., Kavimani, S. and Ravichandran, V. 2008. Antitumor activity of methanolic extract of *Plumeria alba* L. leaves against Dalton lymphoma ascites in mice. *Int. J. Health Res.* 1: 79-85.

22. Rongsen, L. 1997. Eco-geographical distribution of seabuckthorn and prospects of International cooperation. In: *Worldwide Research and Development Seabuckthorn* (S. Lu, M. Li, J. Hu and S. Liu, Eds.), p.11-22, Beijing, Chinese Science and Technology Press, China.

23. Rousi, A. 1971. The genus *Hippophae* L.-A taxonomic study. *Ann. Bot. Fenn.* 8: 177-227.

24. Shimizu, M., Azuma, C., Taniguchi, T. and Murayama, T. 2004. Expression of cytosolic phospholipase A2α in murine C12 cells, a variant of L929 cells, induced arachidonic acid release in response to phorbol myristate acetate and Ca^{2+} ionophores, but not tumor necrosis factor-α. *J. Pharm. Sci.* 96: 324-332.

25. Singh, V. 2003. Geographical adaptation and distribution of seabuckthorn (*Hippophae* L) resources. In: *Seabuckthorn (Hippophae L.)-A Multipurpose Wonder Plant. Vol. I: Botany, Harvesting and Processing Technologies* (V. Singh, Ed. in Chief, 2003), p.21-34, Indus Publishing Company, New Delhi, 518p.

26. Sun, K., Ma, R.J. and Chen, X.L. 2003. Hybrid origin of the diploid species *Hippophae goniocarpa* evidenced by the internal transcribed spacers (ITS) of nuclear rDNA. *Belg. J. Bot.* 136 (1): 91-96.

27. Sun, Y., Oberly, L.W. and Ying, L. 1988. A simple method for clinical asset of superoxide dismutase. *Clin. Chem.* 34: 479-500.

28. Vaher, M., Ehala, S. and Kaljurand, M. 2005. On-column capillary electrophoretic monitoring of rapid reaction kinetics for determination of the antioxidative potential of various bioactive phenols. *Electrophoresis* 26: 990-1000.

29. Weller, P. and Breithaupt, D.E. 2001. Identification and quantification of zeaxanthin esters in plants using liquid chromatography mass spectrometry. *J. Agric. Food Chem.* 51: 7044-7049.

30. Yang, B., Karlsson, R.M., Oksman, P.H. and Kallio, H.P. 2001. Phytosterols in seabuckthorn (*Hippophae rhamnoides* L.) berries: Identification and effects of different origins and harvesting times. *J. Agric. Food Chem.* 49: 5620-5629.

31. Yang, B.R. and Kallio, H. 2002. Composition and physiological effects of seabuckthorn (*Hippophae*) lipids. *Trends Food Sci. Tech.* 13 (5): 160-167.

32. Yao, Y. and Tigerstedt, P. 1992. Variation of vitamin C concentration between and within natural seabuckthorn (*Hippophae rhamnoides* L.) populations. *Acta Agriculturae Scandinavica* 42: 12-17.

33. Zhang, Q. and Cui, H. 2005. Simultaneous determination of quercetin, kaempferol, and isorhamnetin in phytopharmaceuticals of *Hippophae rhamnoides* L. by high-performance liquid chromatography with chemiluminescence detection. *J. Sep. Sci.* 28: 1171-1178.

Chapter 34

Anti-stress and Adaptogenic Potential of Seabuckthorn (*Hippophae rhamnoides* L.)

Priyanka Sharma, Ravi Kumar, Rajendra Prasad and Geetha Suryakumar

Defence Institute of Physiology and Allied Sciences,
Lucknow Road, Timarpur, Delhi – 110 054, India

SUMMARY

Seabuckthorn is a unique and valuable plant currently being investigated for its medicinal values all over the world. All parts of the plant are considered to be a good source of a large number of bioactive substances. The present study was undertaken to evaluate anti-stress and adaptogenic activity of five different extracts of seabuckthorn fruit and seed oil. The dose dependent adaptogenic study of these extracts was carried out at different doses in the range of 25-500 mg/kg body weight. The extract was administered orally, 30 min prior to cold (5° C)-hypoxia (428 mm Hg)-restraint (C-H-R) exposure for the fruit extracts and as overnight oral dose for the seed oil. The extracts were evaluated for tolerance to cold and hypoxia on the basis of the delay in the time taken to attain rectal temperature T_{rec} 23 °C during C-H-R exposure and faster recovery to T_{rec} 37 °C. After administration of fruit extracts, it was observed that there was 15-20 per cent resistance to C-H-R induced hypothermia observed by a delay in attaining rectal temperature 23 °C and about 20 per cent faster recovery back to T_{rec} 37 °C. Out of the four extracts, the fresh fruit serial alcoholic extract was found to have the most potent adaptogenic activity and its maximal effective dose was found to be 100 mg/kg body weight. The seed oil was also found to be a potent adaptogen as it provided almost 50 per cent resistance during the Cold-Hypoxia-Restraint stress exposure and about 17 per cent faster recovery back at 37°C. Hence, seabuckthorn can be developed as a nutraceutical or health supplement to enhance resistance to environmental stress conditions.

Keywords: *Seabuckthorn, Adaptogen, Stress, Hypoxia, Resistance, Seed oil and Cold.*

INTRODUCTION

The exposure to high altitude is a well known environmental stress with physiological and metabolic consequences. High altitude conditions are emphasized by diminished ambient oxygen

Seabuckthorn (*Hippophae* L.): *A Multipurpose Wonder Plant, Vol. 4: Emerging Trends in Research and Technologies* (Virendra Singh, *Editor-in-Chief*, 2014), p. 418–426, Daya Publishing House, New Delhi, India.

pressure, decreased temperature, lower humidity, and increased ultraviolet radiation. Hypobaric hypoxia generated due to low barometric pressure, leads to a shift in the redox status thereby producing reactive oxygen species. Mechanistically, low oxygen availability at high altitude decreases mitochondrial respiration, leading to the accumulation of reducing equivalents in the cellular electron transport chain causing partial reduction of oxygen, which in turn generates superoxide radicals (Mohanraj *et al.*, 1998). Hypobaric hypoxia, the primary stressor at high altitude, when accompanied with low temperature induces complex biochemical and molecular responses, arbitrated through the amplified free radical generation.

In recent years, supplementation studies, using vitamins, carotenoids, antioxidants and other biomolecules from plants have been rising for combating the oxidative stress caused by high altitude (Pfeiffer *et al.*, 1999; Schmidt *et al.*, 2002; Subudhi *et al.*, 2006). A number of biological active extracts derived from herbal plants growing at high altitudes are reported to enhance stress tolerance and helps in faster adaptation (Saggu *et al.*, 2007). Under stressful environment, supplementation with various nutrients and/or single and polyherbal preparations has been demonstrated to enhance stress tolerance, besides curtailing stress induced disorders (Kumar *et al.*, 2002). The herbal formulations which can enhance physical endurance, mental function and non-specific resistance of the body against stress have been termed as adaptogens (Brekhman, 1980). Adaptogens have a wide range of therapeutic activity with minimal alteration of body function and manifest its adaptogenic action only under conditions of challenge to the system.

Seabuckthorn (*Hippophae rhamnoides* L., *Elaeagnaceae*) is a unique and valuable plant, currently being investigated for its medicinal values all over the world. This shrub has drawn a great attention worldwide because of its amazing ability to acclimatize in extremes of environment. A wide spectrum of pharmacological effects of seabuckthorn have been recently reported, including antioxidant, immunomodulatory, anti-atherogenic, anti-stress, hepatoprotective, radioprotective and tissue repair (Geetha *et al.*, 2002a; 2002b; Goel *et al.*, 2002; Gupta *et al.*, 2005; Basu *et al.*, 2007; Chawla *et al.*, 2007; Saggu *et al.*, 2007; Geetha *et al.*, 2008; Upadhyay *et al.*, 2009; 2011). The word *Hippophae* has been derived from a Latin word '*Hippo*' meaning 'horse' and '*Phaos*' which means 'shine'. In Greece, seabuckthorn leaves and twigs were used to feed animals which resulted in weight gain and shining coat, especially in horses. It has a rich history of use in treating numerous medical conditions. Many of its pharmacological effects have been recorded in classics, such as Sibu Yidian from the Tang Dynasty and Jing Zhu Ben Cao from the Qing Dynasty. It was used as a medicinal plant in Tibet as early as 900 A.D. For centuries, the people of central and south-eastern Asia have used seabuckthorn as an agent of traditional medicine to prevent various ailments. It has been considered to be one of the most important valuable bioresource, being locally used for centuries as fuel, fodder, small timber, food and medicine. Every part of the plant, such as fruit, leaf, twig, root and bark has been traditionally used as medicine, nutritional supplement, firewood and fence. The seabuckthorn berries were used as a source of herbal medicines, health food and natural skin care in Europe and Asia.

The adaptogenic activity of seabuckthorn leaf extract was studied in rats using a passive Cold-Hypoxia-Restraint (C-H-R) animal model (Ramachandran *et al.*, 1990). In this multi-stress model, the animals were restrained and put in a decompression chamber maintained at 5°C and low oxygen tension of 428 mm Hg pressure. The colonic temperature of the rats was monitored continuously on Iso-thermex recorder every minute. When the rats attained a rectal temperature of 23°C, it was taken out of the chamber and allowed to recover to normal colonic temperature at atmospheric pressure in a room maintained at 32±1°C. Rectal temperature represents deep body temperature and is an indirect measure of physical and mental performance capacity, the metabolic status of the animal and hypothalamic and higher brain functions. Its maintenance during adverse stress conditions is taken as an indicator of tolerance of the organism.

Seabuckthorn leaf aqueous extract at a dose of 100 mg/kg BW, had significant anti-stress and adaptogenic activity (Saggu *et al.*, 2007). It is reported that hypoxia, cold and restraint stress during C-H-R exposure produce oxidative stress (Bhaumik *et al.*, 1995; Simon, 1996) and skeletal muscle fatigue (Barclay and Hansel, 1991). Supplementation with seabuckthorn leaf extract reduced the oxidative stress in liver and muscle of rats during C-H-R exposure and post-stress recovery (Saggu and Kumar, 2007a). During severe stressful exposure to C-H-R and post stress recovery, the aerobic metabolism as well as hexose monophosphate pathway is suppressed. The single and five doses of seabuckthorn leaf extract treatment restricted the decrease or better maintained tissue glycogen and enzyme activities such as hexokinase, phosphofructokinase, citrate synthase and glucose-6-phosphate dehydrogenase in blood, liver and muscle during C-H-R exposure. The studies suggested that seabuckthorn leaf aqueous extract treatment caused a trend for shifting anaerobic metabolism to aerobic during multiple stresses C-H-R exposure and post stress recovery (Saggu and Kumar, 2007b; 2008). The alcoholic leaf extract of seabuckthorn was also observed to have potent antioxidant and adaptogenic activity (Geetha *et al.*, 2006).

The fruits of seabuckthorn are a rich source of many bioactive substances like vitamin A, C, E, K, carotenoids (β-carotene, lycopene), flavonoids (isorhamnetin, quercetin, kaempferol and glycosides), organic acids, micro and macronutrient (Chen *et al.*, 1990; Yao and Tigerstedt, 1992; Yang and Kallio, 2001). The extract of whole fruit, oil from its pulp and seed were found to possess immunomodulatory, antioxidant and anti-bacterial activity (Geetha *et al.*, 2002b; Negi *et al.*, 2005), effective in the treatment of various diseases, such as gastric ulcers (Suleyman *et al.*, 2001; Xing and Yang, 2002), skin disorders (Yang *et al.*, 2000), coronary heart disease (Eccleston and Baoru, 2002) and wound healing (Gupta *et al.*, 2005). Aqueous fruit extract of seabuckthorn have been shown to have adaptogenic activity (Tulswani *et al.*, 2010). The HPLC analysis showed the presence of rutin and ascorbic acid in extract and the phenolic content of the extract was found to be 48.4 mg/g of gallic acid equivalent, which may be responsible for the observed adaptogenic activity. The present study was aimed to evaluate the anti-stress and adaptogenic activity of the different extracts of seabuckthorn fruit (fresh and dry) and seed-oil using cold-hypoxia- restraint stress animal model. The animals were evaluated for stress endurance on the basis of delay in the time taken to attain rectal temperature (T_{rec}) 23 °C during C-H-R exposure and faster recovery to (T_{rec}) 37 °C.

MATERIALS AND METHODS

Extract Preparation

Five different extracts of seabuckthorn were prepared (Table 34.1). The fresh fruits of seabuckthorn were cleaned and washed thoroughly with distilled water and dried under shade in a clean and dust free environments, then seeds were removed and dried pulp was crushed using laboratory blender. The crushed fruits (fresh fruit pulp in case of fresh fruits extracts) were soaked in distilled water (1:5 w/v) at room temperature (25 ± 2°C) to obtain the aqueous extract using cold percolation method. After 24 hours, the supernatant was decanted and the residue was resoaked in fresh water. The process was repeated four times for the complete extraction. The supernatants were pooled by passing through a muslin cloth and stored in amber coloured bottles and centrifuged at 8000 g, 4°C. The supernatant obtained after centrifugation was then dried under reduced pressure by using Buchi Rotavapor R-124 (Buchi Labortechnik AG, Postfach, CH-9230 Flawil/Schweiz, Switzerland) at 35-40°C to obtain vacuum dried aqueous extract and stored at -20°C in an airtight glass tube for further studies. The crude yield of the vacuum dried extract was determined gravimetrically by weighing the dried residues obtained from the known volumes of extract. For direct alcoholic extract preparation, 70 per cent alcohol was used instead of water. For serial alcoholic extract preparation the pulp was first soaked in distilled water for four days and then in 70 per cent alcohol for another four days. The rest of the procedure is same as mentioned above. In another study, seabuckthorn seed oil is used to determine its adaptogenic potential in rats using C-H-R animal model.

Table 34.1: Dose Dependent Adaptogenic Potential of Seabuckthorn Fruit Extracts

Extract Code	*Name of the Extract*	*Maximal Effective Dose (mg/kg body weight)*	$^{a}T_1$	$^{b}T_2$
–	Control	–	67±1.7	153±1.8
A	Seabuckthorn dry fruit alcoholic extract (serial)	200	68±2.5	131±3.9
B	Seabuckthorn dry fruit alcoholic extract (direct)	300	81±2.9	115±4.2
C	Seabuckthornfresh fruit aqueous extract	300	79±4	113±4
D	Seabuckthorn fresh fruit alcoholic extract (serial)	100	85±4.2	114±5.2

*: Values are mean ± SE of six rats in each group and expressed as percentage change with respect to control.

a: T_1: Time taken (in min) to attain T_{rec} 23 °C; b: T2: Time taken (in min) to attain T_{rec} 37 °C.

EXPERIMENTAL DESIGN FOR EVALUATION OF ADAPTOGENIC ACTIVITY

Experimental Animals

The study was approved by the Animal Ethical Committee of our Institute. Sprague-Dawley inbred male rats, 12-14 weeks old, weighing 150±15 g, from the animal colony of the Defence Institute of Physiology and Allied Sciences (Delhi, India), were used for the study. The rats were maintained inside polypropylene cages under a controlled environment in the Institute's animal house at 22±1°C, 55 per cent ± 1 per cent humidity and 12 h light-dark cycle. The rats had free access to standard animal food pellets and water.

Simulated High Altitude Animal Exposure Chambers

The animal chamber has two units: One unit is Animal Decompression Chamber also called as Cold-Hypoxia-Restraint (C-H-R) multiple stress animal model where the rats kept in restrainers, are exposed to cold (5pC) and hypoxia (428 mm Hg) equivalent to an altitude of 4,572 m and the air flow is 2 l min^{-1}. Another unit is Animal Conditioning Chamber which is maintained at 32±1pC and normal atmospheric pressure.

DOSE DEPENDENT ADAPTOGENIC STUDIES

Total 156 animals were used in the study divided in five groups, A-E designated according to the seabuckthorn extract being analyzed: (A) Dry fruit alcoholic extract (serial); (B) Dry fruit alcoholic extract (direct); (C) Fresh fruit aqueous extract; (D) Fresh fruit alcoholic extract (serial); and (E) Seed oil. Each group had 12 to 36 rats depending upon the dose being studied (Table 34.2, Figures 34.1A-D). For each dose, six rats were used. In each group, one of the groups of six rats served as control, administered with distilled water (0.5 ml for 150 g of rat) and other groups were administered with extract at different doses. The extracts as oral dose were administered to the overnight fasted rats 30 minutes prior to exposure in the C-H-R animal model. In group E, seed oil was administered orally (0.5

Figure 34.1: Adaptogenic Potential of Seabuckthorn Fruit Extracts in Rats at the Doses Ranging from 25-500 mg/kg Body Weight p.o. 30 Minutes Prior to Exposure in Cold-Hypoxia-Restraint Animal Model.

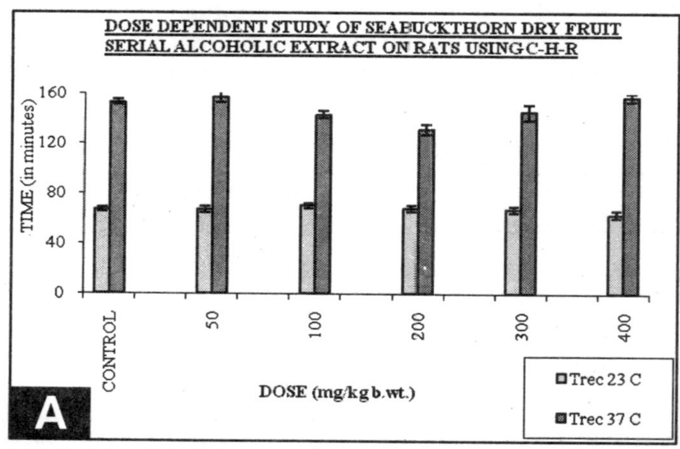

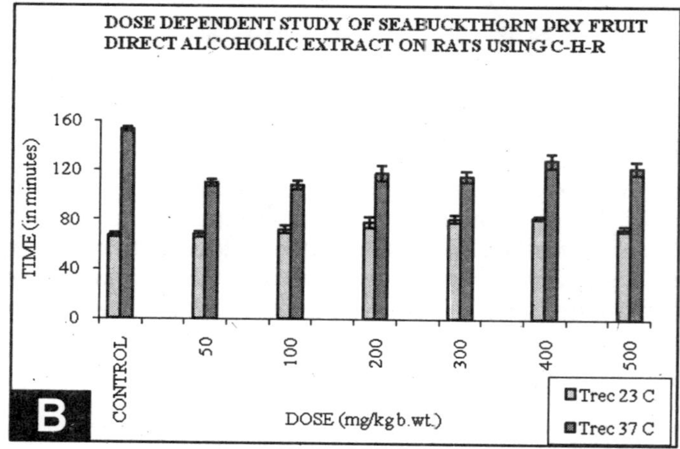

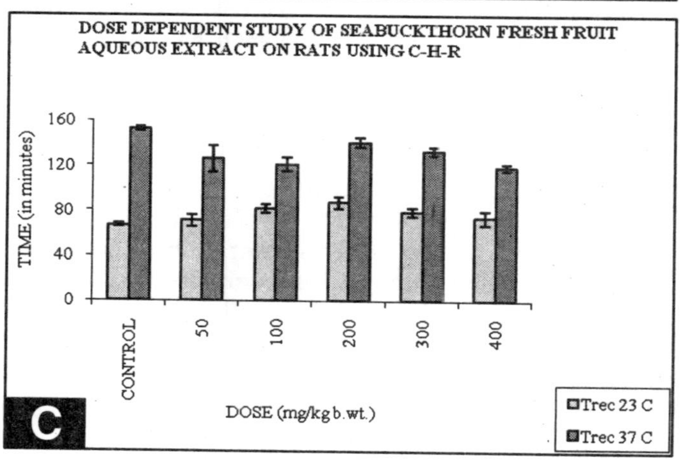

Contd...

Figure 34.1–*Contd...*

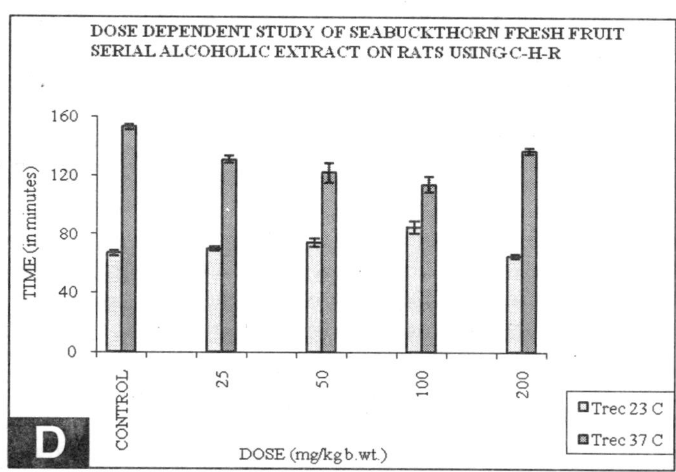

ml for 150 g of rat) 12 hours prior to exposure in the C-H-R animal model. The rectal probe was inserted 2 cm past the rectum of the rat and retained there with the help of adhesive plaster. The rectal temperature of the rats was monitored continuously, every minute by Iso-thermex temperature recorder. When the rectal temperature reaches 23 pC, the rats were taken out and kept in conditioning chamber still in restrained state and allowed to recover to normal rectal temperature of 37 pC. The rectal temperature of 23 pC was taken as termination point of the C-H-R exposure, as a further fall in the rectal temperature resulted in high mortality. The time taken (min) to reach termination point, *i.e* rectal temperature at 23pC (T_{rec} 23 pC) and recovery to 37 pC (T_{rec} 37 pC) were used as a measure of endurance (Ramachandran *et al.*, 1990).

Table 34.2: Adaptogenic Potential of Seabuckthorn Seed Oil

Extract Code	*Name of the Extract*	*Dose*	$^{a}T_{1}$	$^{b}T_{2}$
E	Seed-oil	0.5ml/150 g of rat	102	128

* Values are mean ± SE of six rats in each group and expressed as percentage change with respect to control.

a: T_1: Time taken (in min) to attain T_{rec} 23 °C; b: T_2: Time taken (in min) to attain T_{rec} 37 °C.

RESULTS AND DISCUSSION

Figure 34.1 shows the dose dependent adaptogenic potential of different seabuckthorn fruit extracts. In all the studies, the doses administered were two doses below and two doses above the maximal adaptogenic dose (Table 34.1). It was observed that adaptogenic substances show a bell shaped pattern of anti-stress activity, *i.e.* higher doses of minimum effective dose of adaptogenic substance do not provide increase activity instead it starts falling (Figure 34.1) (Sharma *et al.*, 2012). For the adaptogenic activity of the fruit extracts, there was 15-20 per cent delay in the time taken to attain T_{rec} 23° C and similarly, about 20 per cent faster recovery was observed, *i.e.* time taken to attain T_{rec} 37° C. Among the four seabuckthorn fruit extracts, the fresh fruit serial alcoholic extract was found to have potent adaptogenic activity and its maximal effective dose was 100 mg/kg body weight. Many medicinal effects of seabuckthorn berries, such as improving blood circulation and functions of digestive system have been reported (Yang and Kallio, 2002).

Seabuckthorn seed oil was found to be a better adaptogen as it provided almost 50 per cent resistance during the first phase of the exposure and about 17 per cent faster recovery to 37° C. Earlier studies have reported the protective effects of seed oil against liver injury (Cheng, 1992), atopic dermatitis (Yang *et al.*, 2000) and atherosclerosis (Basu *et al.*, 2007). Seabuckthorn seed oil also has protective effects against ischemic cerebral infarction in rats (Cheng, 2003). Seabuckthorn seed oil is one of the most versatile natural oil and is rich in bioactive substances like carotenoids, tocopherols, omega-3 and omega-6 fatty acids, and phytosterols, which have a potent antioxidant activity and anti-inflammatory activity. The tocopherols and carotenoids, which are present in higher levels in seabuckthorn seed oil (Yang and Kallio, 2002), have been reported to contain antioxidant activity (Sergio and Robert, 1999; Clair *et al.*, 2002), which might be responsible for adaptogenic activity.

CONCLUSION

Seabuckthorn fruit and seed oil were found to have potent anti-stress and adaptogenic activity. The detailed studies on adaptogenic properties and phytochemical analysis are in progress to develop the extract as a nutraceutical to enhance performance under environmental stress conditions.

REFERENCES

1. Barclay, J.K. and Hansel, M. 1991. Free radicals may contribute to oxidative skeletal muscle fatigue. *Canadian Journal of Physiology and Pharmacology* 69: 279-284.
2. Basu, M., Prasad, R., Jayamurthy, P., Pal, K., Arumughan, C. and Sawhney, R.C. 2007. Anti-atherogenic effects of seabuckthorn (*Hippophae rhamnoides*) seed oil. *Phytomedicine* 14: 770-777.
3. Bhaumik, G., Srivastava, K.K., Selvamurthy, W. and Purkayastha, S.S. 1995. The role of free radicals in cold injuries. *International Journal of Biometeorology* 3: 171-175.
4. Brekhman, I.I. 1980. *Man and Biologically Active Substances*, Pergamon Press, Oxford, 89p.
5. Chawla, R., Arora, R., Singh, S., Sagar, R.K., Sharma, R.K., Kumar, R., Sharma, A., Gupta, M.L., Singh, S., Prasad, J., Khan, H.A., Swaroop, A., Sinha, A.K., Gupta, A.K., Tripathi, R.P. and Ahuja, P.S. 2007. Radioprotective and antioxidant activity of fractionated extracts of berries of *Hippophae rhamnoides*. *Journal of Medicinal Food* 10: 101-109.
6. Chen, Y., Jiang, Z., Qin, W., Ni, X. and He, Y. 1990. Chemical composition and characteristics of seabuckthorn fruit and its oil. *Chemistry and Industry of Forest Products* 10: 163-175.
7. Cheng, T.J. 1992. Protective action of seed oil of *Hippophae rhamnoides* L. (HR) against experimental liver injury in mice. *Zhonghua Yu Fang Yi Xue Za Zhi* 4: 227–229.
8. Cheng, T.J., Wang, Y.B., Gao, L.P., Sun, Y.F. and Zhang, J. 2003. The protection of seed oil of *Hippophae rhamnoides* on ischemic cerebral infarction in rats. *Zhongguo Zhong Yao Za Zhi* 28: 548-550.
9. Clair, E., Yang, B., Raija, T., Heikki, K., Gerald, H.R. and Anne, M.M. 2002. Effects of an antioxidant-rich juice (seabuckthorn) on risk factors for coronary heart disease in humans. *Journal of Nutritional Biochemistry* 13: 346-354.
10. Eccleston, C. and Baoru, Y. 2002. Effects of an antioxidant-rich juice (seabuckthorn) on risk factors for coronary heart disease in humans. *Journal of Nutritional Biochemistry* 13: 346-354.

11. Geetha, S., Jayamurthy, P., Pal, K., Pandey, S. and Sawhney, R.C. 2008. Hepatoprotective activity of seabuckthorn (*Hippophae rhamnoides* L.) against carbon tetrachloride induced hepatic damage in rats. *Journal of Science of Food and Agriculture* 88: 1592-1597.

12. Geetha, S., Sai Ram, M., Grover, S.K., Divekar, H.M., Singh, V., Ilavazhagan, G. and Sawhney, R.C. 2006. Seabuckthorn in curtailing oxidative cold and hypoxic stresses. In: *Seabuckthorn (Hippophae L.)-A Multipurpose Wonder Plant. Vol. II: Biochemistry and Pharmacology* (V. Singh, Ed. in Chief, 2006), p. 413-418, Daya Publishing House, New Delhi, 600p.

13. Geetha, S., Sai Ram, M., Singh, V., Ilavazhagan, G. and Sawhney, R.C. 2002a. Antioxidant and immunomodulatory properties of seabuckthorn (*Hippophae rhamnoides*)-An *in vitro* study. *Journal of Ethnopharmacology* 79: 373–378.

14. Geetha, S., Sai Ram, M., Singh, V., Ilavazhagan, G. and Sawhney, R.C. 2002b. Effect of seabuckthorn against sodium nitroprusside induced oxidative stress in murine macrophages. *Biomedicine and Pharmacotherapy* 56: 463-467.

15. Goel, H.C., Prasad, J., Singh, S., Sagar, R., Prem Kumar, I. and Sinha, A.K. 2002. Radioprotection by a herbal preparation of *Hippophae rhamnoides* RH-3, against whole body lethal irradiation in mice. *Phytomedicine* 9: 135-143.

16. Gupta, A., Kumar, R., Pal, K., Banerjee, P.K. and Sawhney, R.C. 2005. A preclinical study of the effects of seabuckthorn (*Hippophae rhamnoides* L.) leaf extract on cutaneous wound healing in albino rats. *International Journal of Lower Extremity Wounds* 4: 88-92.

17. Kumar, R., Divekar, H.M., Gupta, V. and Srivastava, K.K. 2002. Anti stress and adaptogenic activity of lecithin supplementation. *Journal of Alternative Complementary Medicine* 8: 487-492.

18. Mohanraj, P., Merola, A.J., Wright, V.P. and Clanton, T.L. 1998. Antioxidants protect rat diaphragmatic muscle function under hypoxic conditions. *Journal of Applied Physiology* 84: 1960-1966.

19. Negi, P.S., Chauhan, A.S., Sadia, G.A., Rohinishree, Y.S. and Ramteke, R.S. 2005. Antioxidant and antibacterial activities of various seabuckthorn (*Hippophae rhamnoides* L.) seed extracts. *Food Chemistry* 92: 119-124.

20. Pfeiffer, J.M., Askew, E.W., Roberts, D.E., Wood, S.M., Benson, J.E., Johnson, S.C. and Freedman, M.S. 1999. Effect of antioxidant supplementation on urine and blood markers of oxidative stress during extended moderate-altitude training, *Wilderness and Environmental Medicine* 10: 66-74.

21. Ramachandran, U., Divekar, H.M., Grover, S.K. and Srivastava, K.K. 1990. New experimental model for evaluation of adaptogenic products. *Journal of Ethnopharmacology* 29: 275-281.

22. Saggu, S. and Kumar, R. 2007a. Modulatory effect of seabuckthorn leaf extract on oxidative stress parameters in rats during exposure to cold, hypoxia and restraint (C-H-R) stress and post stress recovery. *Journal of Pharmacy and Pharmacology* 59: 1739-1745.

23. Saggu, S. and Kumar, R. 2007b. Possible mechanism of adaptogenic activity of seabuckthorn (*Hippophae rhamnoides*) during exposure to cold, hypoxia and restraint (C-H-R) stress induced hypothermia and post stress recovery in rats. *Food and Chemical Toxicology* 45: 2426-2433.

24. Saggu, S. and Kumar, R. 2008. Effect of seabuckthorn leaf extracts on circulating energy fuels, lipid peroxidation and antioxidant parameters in rats during exposure to cold, hypoxia and restraint (C-H-R) stress and post stress recovery in rats. *Phytomedicine* 15: 437-446.

25. Saggu, S., Divekar, H.M., Gupta, V., Sawhney, R.C., Banerjee, P.K. and Kumar, R. 2007. Adaptogenic and safety evaluation of seabuckthorn (*Hippophae rhamnoides*) leaf extract-A dose dependent study. *Food and Chemical Toxicology* 45: 609-617.

26. Schmidt, M.C., Askew, E.W., Roberts, D.E., Prior, R.L., Ensign, W.Y. and Hesslink, R.E. 2002. Oxidative stress in humans training in a cold, moderate altitude environment and their response to a phytochemical antioxidant supplement. *Wilderness and Environmental Medicine* 13: 94-105.

27. Sergio, A.R.P. and Robert, M.R. 1999. β-carotene and other carotenoids as antioxidants. *Journal of American College of Nutrition* 18: 426-433.

28. Sharma, P., Kirar, V., Meena, D.K., Suryakumar, G. and Misra, K. 2012. Adaptogenic activity of *Valeriana wallichii* using cold, hypoxia and restraint multiple stress animal model. *Biomedicine and Aging Pathology* 2: 198-205.

29. Simon, S. E. 1996. Oxidative stress at high altitude and effects of vitamin E. *In: Nutritional Needs in Cold and in High Altitude Environments*, Washington DC, National Academy Press.

30. Subudhi, A.W., Jacobs, K.A., Hagobia, T.A., Fattor, J.A., Muza, S.R., Fulco, C.S., Cymermon, A. and Friedlander, A.L. 2006. Changes in ventilatory threshold at high altitude- Effect of antioxidants. *Medicine Science Sports and Exercise* 38: 1425-1431.

31. Suleyman, H. and Demirezer, L.O. 2001. Anti-ulcerogenic effect of *Hippophae rhamnoides* L. *Phytotherapy Research* 15: 625-627.

32. Tulsawani, R., Sharma, P., Divekar, H.M., Meena, R.N., Singh, M. and Kumar, R. 2010. Supplementation of fruit extract of *Hippophae rhamnoides* speeds adaptation to simulated high altitude stressors in rats. *Journal of Complementary and Integrative Medicine* 7 (1): Article 30.

33. Upadhyay, N.K., Kumar, R., Mandotra, S.K., Meena, R.N., Siddiqui, M.S., Sawhney, R.C. and Gupta, A. 2009. Safety and wound healing efficacy of seabuckthorn (*Hippophae rhamnoides* L.) seed oil in experimental rats. *Food and Chemical Toxicology* 47: 1146-1153.

34. Upadhyay, N.K., Kumar, R., Siddiqui, M.S. and Gupta, A. 2011. Mechanism of wound healing activity of *Hippophae rhamnoides* L. leaf extract in experiment burns. *Evidence based Complementary and Alternative Medicine. doi:10.1093/ecam/nep 189.*

35. Xing, J. and Yang, B. 2002. Effects of seabuckthorn (*Hippophae rhamnoides* L.) seed and pulp oils on experimental models of gastric ulcer in rats. *Fitoterapia* 73: 644-650.

36. Yang, B. and Kallio, H. 2002. Composition and physiological effects of seabuckthorn (*Hippophae*) lipids. *Trends in Food and Science Technology* 13: 160-167.

37. Yang, B. and Kallio, H.P. 2001. Fatty acid composition of lipids in seabuckthorn (*Hippophae rhamnoides* L.) berries of different origins. *Journal of Agricultural Food Chemistry* 49: 1939-1947.

38. Yang, B., Kalimo, K.O., Tahvonen, R.L., Mattila, L.M., Katajisto, J.K. and Kallio, H.P. 2000. Effect of dietary supplementation with seabuckthorn (*Hippophae rhamnoides*) seed and pulp oils on the fatty acid composition of skin glycerophospholipids of patients with atopic dermatitis. *Journal of Nutritional Biochemistry* 11: 338-340.

39. Yao, Y. and Tigerstedt, P. 1992. Variation of vitamin C concentration between and within natural seabuckthorn (*Hippophae rhamnoides* L.) populations. *Acta Agriculturae Scandinavica* 42: 12-17.

Chapter 35

Anti-oxidative and Anti-inflammatory Activities of *Hippophae rhamnoides* and *Hippophae salicifolia*: An *In vitro* and *In vivo* Comparative Study

Rajkumar Tulsawani, Manimaran Manickam, Koganti Praveen, Kshipra Misra and Virendra Singh*

Defence Institute of Physiology and Allied Science, Delhi – 110 054, India
*CSK Himachal Pradesh Agricultural University, Palampur – 176 062, Himachal Pradesh, India

SUMMARY

The present study was undertaken to compare anti-stress, anti-hypoxic and anti-inflammatory activities of aqueous extracts of two high yielding selections of seabuckthorn (*Hippophae rhamnoides and Hippophae salicifolia)* cultivated in Lahaul, Himachal Himalayas. The per cent yield of *Hippophae rhamnoides* was found better than *Hippophae salicifolia*. However, in terms of phenolic content, flavonoid content and antioxidant activity, the extract of *Hippophae salicifolia* was richer than *Hippophae rhamnoides*. Both the extracts of *Hippophae rhamnoides* and *Hippophae salicifolia* demonstrated the resistance against hypoxia driven HT22 cell death and reversed oxidative stress. However, among the both of extracts evaluated, the protective efficacy extract of *Hippophae salicifolia* was found better than of extract of *Hippophae rhamnoides*. In separate *in vivo* studies, rats supplemented with aqueous extracts of *Hippophae rhamnoides* and *Hippophae salicifolia* gain resistance to oxidative stress was evident by restoration of stress markers, such as MDA, GSH, GPx and SOD and minimized inflammatory response (TNF-α, IL6 and NF-κB) associated with exposure to extreme acute hypobaric hypoxia. Overall, the findings from these *in vitro* and *in vivo* studies confirmed that *Hippophae salicifolia* extract had better anti-stress, anti-hypoxic and anti-inflammatory activity than *Hippophae rhamnoides*.

***Keywords:** Hippophae rhamnoides, Hippophae salicifolia, Hypoxia, Oxidative stress and Inflammation.*

INTRODUCTION

The genus *Hippophae* is a wild thorny, deciduous shrub growing at high altitude (2500-4000 m) in adverse climatic conditions and is native of Europe and Asia (Rousi, 1971). In India, the genus is

Seabuckthorn (Hippophae L.): *A Multipurpose Wonder Plant, Vol. 4: Emerging Trends in Research and Technologies* (Virendra Singh, *Editor-in-Chief,* 2014), p. 427–441, Daya Publishing House, New Delhi, India.

widely distributed in Himachal Pradesh, Uttarkhand, Jammu and Kashmir, Sikkim and Arunachal Pradesh state (Dhyani *et al.*, 2010). Out of the seven species of *Hippophae* found all over the world, three species are reported in India *viz.*, *H. rhamnoides* ssp. *turkestania* Rousi, *H. salicifolia* and *H. tibetana* (Singh, 2003). Among various species, most of the scientific studies have been conducted on *H. rhamnoides* (Li and Beveridge, 2003).

Numerous medicinal effects of *H. rhamnoides* are well documented and include beneficial effects in cardiovascular diseases, mucosal injuries, skin disorders, cytoprotective and anti-oxidative, wound healing and adaptogenic activity have been documented (Beveridge *et al.*, 1999; Eccleston *et al.*, 2002; Gupta *et al.*, 2005; Narayanan *et al.*, 2005; Saggu *et al.*, 2007; Saggu and Kumar, 2008; Tulsawani *et al.*, 2010a). The medicinal properties of *Hippophae rhamnoides* leaves extract are attributed due to the presence of high contents of flavonoids, tannins and triterpenes (Kallio *et al.*, 2002). The major flavonoids found in the leaves of seabuckthorn include quercetin, keampferol, isorhamnetin, catechin and rutin (Zu *et al.*, 2006). *Hippophae salicifolia* D. Don is a deciduous tree, restricted to the Himalayan region and is reported best species among all the species of *Hippophae* in terms of high quality fruit, high yield and less thorns (Rongsen *et al.*, 2001; Rattan and Tomar, 2011). However, the scientific literature available for *Hippophae salicifolia* is limited (Singh, 1994; Singh *et al.*, 2011).

It is documented that *Hippophae* exhibit variations in its morphological structure depending upon the species and microclimate variation. *Hippophae* species represent hydrophytic to xerophytic features (Rongsen, 1992; Ghaffar, 1997). Therefore, it is reasonably expected that the composition and properties of *Hippophae* from different regions may vary and further bioactive compounds isolation from plants depends on the extraction method and growing conditions. In the present study, two species of *Hippophae, i.e. H. rhamnoides* and *H. salicifolia* cultivated in Lahaul were used. The advance and controlled accelerated solvent extraction technique was utilized for extract preparation under pressurized conditions, which prevent the chance of microbial contamination and exposure of bio-active compounds for longer duration at room temperature in traditional cold percolation method. Subcritical fluid extraction (SCFE) is useful technique to replace the traditional extraction methods and provides higher extracts yields (Rodriguez *et al.*, 2006).

Many products based on *Hippophae* species are available in markets, such as jam, pickle, blended/ spiced beverages and cosmetics. The availability of raw material for the commercialization of products at large scale has to be warranted. In the present study, anti-stress, anti-hypoxic and anti-inflammatory activities of aqueous extract of *Hippophae salicifolia* were carried out and compared with *Hippophae rhamnoides* cultivated in same regional conditions.

MATERIALS AND METHODS

Plant Material

Hippophae salicifolia and Hippophae rhamnoides leaves cultivated in Lahaul, Himachal Pradesh, were used in the present study. Prof. Virendra Singh carried out the ethanobotanical identification of the plant characterized and the voucher specimen (DIP-HIP-SF/2011 and DIP-HIP-SR/2011) of the plant material is preserved in the herbarium. The fresh leaves of *Hippophae* mentioned above, were cleaned, washed thoroughly with water and re-washed with distilled water. The washed fresh leaves were dried under shade in a clean, dust free environment and crushed using laboratory blender.

Extract Preparation

For the preparation of sub critical water extracts, an Accelerated Solvent Extraction system ASE 350 equipped with a solvent controller unit from Dionex Corporation (Sunnyvale, CA, USA) was used. The extractions were carried out in triplicate using water as a solvent. Individual extractions were performed using the sample at room temperature for 15 min extraction time. All extractions were performed in 33 ml extraction cells, containing 2 g of sample. Between extractions, a rinse of the complete system was made in order to overcome any extract carryover. Finally, the supernatant solution was lyophilized (Allied frost, India) and the dried extract was stored at 4 °C for the further studies.

Determination of Total Phenol Content

The total phenolic content in extract was estimated using Folin-Ciocalteu reagent (FCR) based assay (Singleton and Rossi, 1965). 20 µl of stock solution (1 mg/ml) of the extract, 80 µl of water and 500 µl of FCR were taken in the test tube. After 5 min incubation in the dark at room temperature, 400 µl of 7.5 per cent sodium carbonate solution was added. The mixture was incubated in the dark for 30 min at room temperature and absorbance of the colour developed was read at 765 nm. Total phenols (mg/g) in the extract were expressed as gallic acid equivalent (GAE), using standard curve prepared from gallic acid (0.1 mg/ml) solution.

Determination of Total Flavonoid Content

Total flavonoid content was estimated by the aluminum chloride colorimetric assay as described by Zhishen *et al.* (1999), using rutin as a standard. To the 1 ml of extract (1 mg/ml), 4 ml distilled water and subsequently, 0.3 ml of 5 per cent $NaNO_2$ solution was added. After 5 min, 0.3 ml of 10 per cent $AlCl_3$ solution was added and allowed to stand for 5 min, then 0.2 ml of 4 per cent NaOH solution was added to the mixture and the volume was adjusted up to 10 ml with distilled water. The absorbance of the mixture was read at 510 nm. Total flavonoid content (mg/g) in the extract was expressed as rutin equivalent.

FRAP (Ferric Reducing Antioxidant Power) Assay

The FRAP assay was estimated using method described by Benzie and Strain (1996). The fresh FRAP working solution was prepared by mixing 25 ml of 300 mM acetate buffer (3.1 g $C_2H_3NaO_2.3H_2O$ and 16 ml $C_2H_4O_2$, pH 3.6), 2.5 ml of 10 mM TPTZ (2, 4, 6-tripyridyl-s-triazine in 40 mM HCl), and 2.5 ml of 20 mM $FeCl_3.6H_2O$ solution which was warmed at 37 °C prior to use. 150 µl of extract (1 mg/ml) was allowed to react with 2850 µl of the FRAP solution for 30 min in dark and the absorbance of the colour developed was read at 593 nm. Results are expressed as mg Trolox equivalent/g of extract using standard curve prepared from Trolox solution.

Cell Culture

HT22 cells (a kind gift of ***Dr. Dave Schubert, Salk Institute,*** San Diego, CA, USA), a murine immortalized hippocampal neuronal cell line was maintained in DMEM medium containing 10 per cent FBS and 1X antibiotic-antimycotic solution. The cells were then plated in a density of 1×10^4 cells/ml in 96 well plates or 1×10^6 cells/ml in 60 mm plates or 3×10^6 cells/ml in 100 mm plates and maintained for 24 h following treatments. The viability of the cells was determined by MTT assay. The cells were exposed to 0.5 per cent of O_2 for 24 h in presence or absence of *Hippophae* extracts.

Mitochondrial Integrity (MTT Assay)

The mitochondrial integrity was measured by the reduction of mitochondrial enzyme succinate dehydrogenase by MTT. The MTT reduction was carried out by the method of Mosman (1983). Briefly, MTT was dissolved in culture medium at a concentration of 0.5 mg/ml and filtered to remove a small amount of insoluble residue. The MTT containing medium was added to each well in a volume of 0.2

ml and incubated for three hours. Thereafter, the supernatant was removed and a solution of 0.4N HCl-isopropanol (1:24 v/v) was added to extract and solubilize the formazon. After 30 min of incubation at room temperature, the absorbance of the formazon was read at wavelength of 570 nm on a plate reader (Power Wave XS2, BioTek, USA).

Assay for Intracellular Redox State

Intracellular redox state levels were measured using the fluorescent dye, H2-DCFH-DA. Briefly, cells were washed once with HBSS and incubated in the same buffer containing 5-10 µg of DCFH-DA for 30 min at 37 °C. Intracellular fluorescence was detected with excitation at 485 nm and emission at 530 nm (Cary Eclipse, Varian, USA).

Bioassays

Further, at the end of exposure period, the HT22 cells were suspended in 0.1M phosphate buffer (pH 7.4) and sonicated. The homogenate was then centrifuged at 10,000 rpm for 10 min and supernatant was collected for the analysis of reduced glutathione (GSH), glutathione peroxidase (GPx) and superoxide dismutase (SOD). The cellular GSH content was measured by the method described elsewhere (Beutler *et al.*, 1963). The activities of GPX and SOD were measured by commercial diagnostic kits. The values were expressed as percent change over controls.

Animals

The male Sprague-Dawley rats (200-250 g) bred in the animal facility of Defence Institute of Physiology and Allied Sciences (DIPAS) were used for all experiments. The animals were maintained on a bedding of rice husk in polypropylene cages under controlled environment in the Institute's animal house at 22 ± 1°C, 55 per cent ± 10 per cent humidity and 12 h light-dark cycle. The animals had access to standard rodent pellet feed and water *ad libitum*. The study has the approval of the Institutional Ethical Committee on Animal Experimentations (IAEC) and the experiments were performed in accordance with the regulations specified by the IAEC and conformed to national guidelines on the care and use of laboratory animals, India.

Animal Exposure

The animals were randomly subdivided into following experimental group of six animals each (i) Control (ii) Hypobaric hypoxia 6 h (iii) Hypobaric hypoxia 6 h + *Hippophae salicifolia* extract (12.5-100 mg/kg) (iv) Hypobaric hypoxia 6 h + *Hippophae rhamnoides* extract (12.5-1,250 mg/kg). All the groups of animals were exposed to a simulated altitude of 25,000 feet in a decompression chamber where temperature and humidity were maintained at 28 °C and 55-60 per cent, respectively, for 6 h in the absence or presence of *Hippophae rhamnoides* extract except group one, *i.e.* control animals.

Biochemical Analysis

After hypoxic exposure, the rats were sacrificed and heart was dissected out after perfusion with 20 ml sterile PBS (pH 7.4) to remove the blood and stored at -80°C for further analysis. Later, the heart was homogenized in 0.1 M phosphate buffer (pH 7.4) to obtain 10 per cent homogenate (w/v). The homogenate was then centrifuged at 10,000 rpm for 30 min at 4°C and supernatant was collected for the analysis of reduced glutathione (GSH), glutathione peroxidase (GPx) and superoxide dismutase (SOD). The cellular GSH content was measured by the method of Beutler *et al.* (1963). The activities of GPx and SOD were measured by commercial diagnostic kits. Lipid peroxidation was measured by estimating malondialdehyde (MDA) formed by thiobarbituric acid (TBA) reaction and measuring the absorbance at 531 nm (Ohkawa *et al.*, 1979). 1,1,3,3-tetraethoxy propane was used as standard and the levels of lipid peroxides were expressed as nmol MDA/g tissue.

Protein Expression Studies

Quantitative Elisa kits were used for studying expression of NF-kB, TNF-α and IL-6 following the manufactures instructions (R&D Systems, USA; Invitrogen, USA). Total tissue lysates were prepared using 0.1 M phosphate buffer (pH 7.4) containing protease inhibitor cocktails added just prior to extraction (Sigma, St. Louis, MO). In general, equal amounts of protein samples were loaded onto each specific antibody coated wells and incubated. After incubation, the Elisa plates were further incubated with horseradish peroxidase-conjugated secondary antibodies following washing. The colour was generated by incubating the Elisa wells with TMB substrate and colour generation was terminated by stop solution. The yellow colour generated for the samples was read at 450 nm against the specific standards provided in the kit.

Statistical Analysis

The results were expressed as mean ± SE. The data was analyzed by one-way ANOVA followed by Dunnet's test for comparing control and the various groups, using GraphPad software. The statistical significance was estimated at the 5 per cent level.

RESULTS AND DISCUSSION

CHEMICAL STANDARDIZATION OF *HIPPOPHAE RHAMNOIDES* LEAF EXTRACT

Intake of *Hippophae* species herbal or dietary supplements may help to maintain an adequate antioxidant status in the body as they contain a wide variety of antioxidant compounds including phenolics (phenol and polyphenols) and flavonoids (Kallio *et al.*, 2002; Rababah *et al.*, 2004; Lotito and Frei, 2006). The per cent yield of *Hippophae salicifolia* and *Hippophae rhamnoides* was found to be 29.2 ± 0.52 and 40.4 ± 1.02 (Figure 35.1A). In the present study, it was found that per cent yield of *Hippophae rhamnoides* was more than *Hippophae salicifolia*. However, in terms of phenolics, flavonoid content and antioxidant activity in chemical assay of aqueous leaf extract of *Hippophae salicifolia* was better than *Hippophae rhamnoides* (Figures 35.1B and 1C). The total phenolic and flavonoid content in the *Hippophae salicifolia* and *Hippophae rhamnoides* aqueous leaf extract was found to be 35.3 ± 0.52 GAE/g extract and 23.1 ± 0.61 GAE/g extract and 17.8 ± 0.36 and 12.8 ± 0.18, respectively (Figure 35.1B). Further, the extract showed antioxidant activity of *Hippophae salicifolia* and *Hippophae rhamnoides* was 120.2 ± 0.42 and 91.7 ± 0.45 µM TE/g extract and 123.9 ± 0.84 and 78.5 ± 0.78 as evident by FRAP and DPPH assays, respectively (Figure 35.1B).

The major flavonoids found in the leaves of *Hippophae rhamnoides* include keampferol and isorhamnetin (Zu *et al.*, 2006). RP-HPLC analysis revealed the presence of isorhamnetin and keampferol in the water extract of *Hippophae salicifolia* and *Hippophae rhamnoides* (Figure 2). The active content of isorhamnetin and keampferol ranged in between 14-15 µg/ml and 19-20 µg/ml in both extracts, respectively. The active content of isorhamnetin and keampferol was more or less similar in both the extracts. It is reported that *Hippophae rhamnoides* has high content of flavonoids (Kallio *et al.*, 2002).

Figure 35.1: Per cent Yield (A), Total Phenols and Flavonoid Content (B), and Antioxidant Activity of Aqueous Extract (C) of *Hippophae salicifolia* and *Hippophae rhamnoides*.

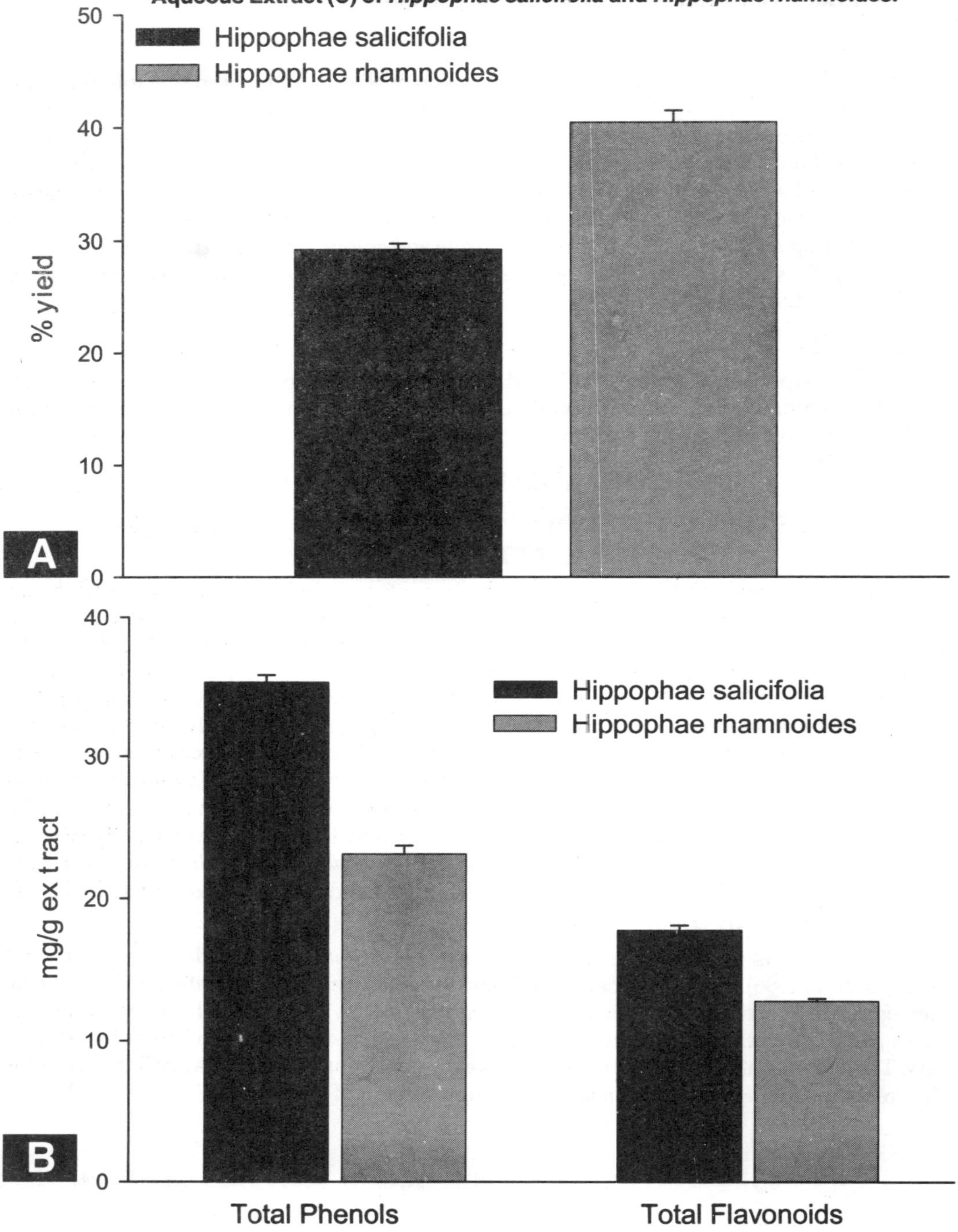

A
% yield
0
10
20
30
40
50
Hippophae salicifolia
Hippophae rhamnoides

B
mg/g extract
0
10
20
30
40
Hippophae salicifolia
Hippophae rhamnoides
Total Phenols
Total Flavonoids

Contd...

Figure 35.1–*Contd...*

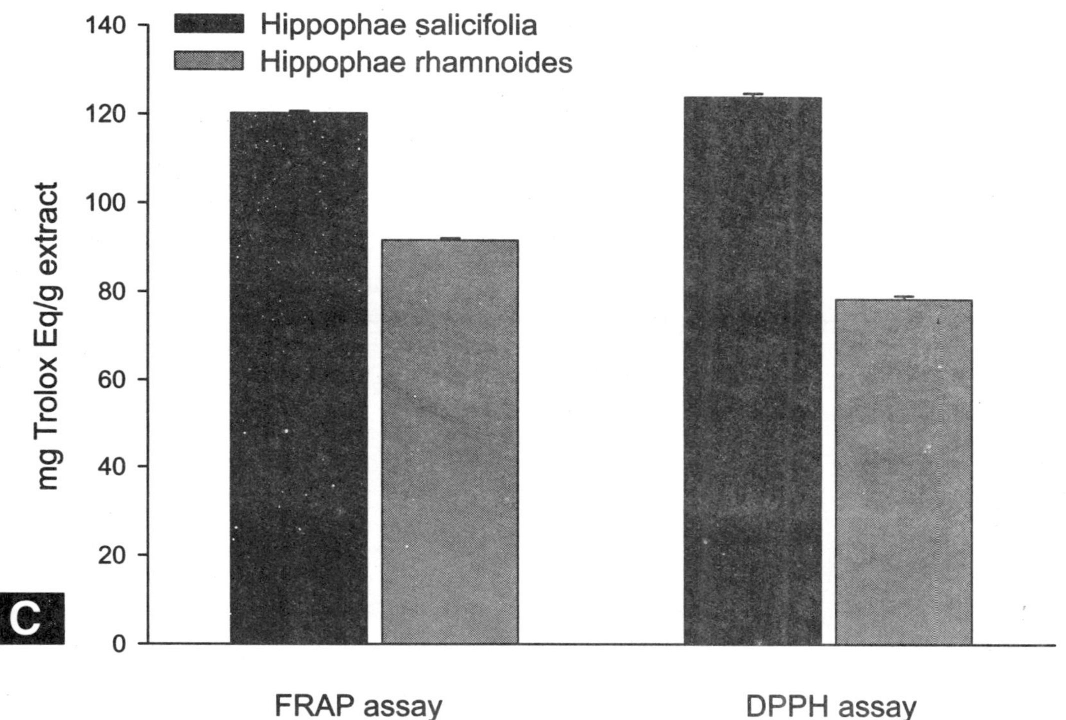

ANTI-HYPOXIA AND ANTI-STRESS EFFECT OF *HIPPOPHAE SALICIFOLIA* AND *HIPPOPHAE RHAMNOIDES IN VITRO* AND *IN VIVO*

The manner in which chronic hypoxia causes cell or tissue damage or cell death has been described in several recent reports documenting that the hypoxia evokes overproduction of ROS, which are a major culprit in cell damage (Kaur *et al.*, 2008). Figure 35.3 depicts protective effects of *Hippophae salicifolia* and *Hippophae rhamnoides* in HT22 cells exposed to hypoxia for 24 h. The cells exposed to hypoxia in the absence of extracts showed 32.6 per cent cell death and had higher levels of reactive oxygen species (Figure 35.3A-C; dark grey bars). The treatment of cells with aqueous extract from both the species showed dose dependent cyto-protection; however, in terms of protection against oxidative stress *Hippophae salicifolia* was found to be better by 6-11 per cent in cell survival assay and 12-14 per cent in preventing ROS generation after hypoxia exposure at various concentrations (25-200 µg/ml) of extract (Figure 35.3 A-C; light grey bars).

The hypoxia exposure is associated with reduced expression and activity of survival molecules and antioxidants (Tulsawani *et al.*, 2010b). The cellular antioxidants defense network comprising the well known endogenous enzymes superoxide dismutase (SOD), and glutathione peroxidase (GSH-Px), as well as non-enzymatic antioxidants reduced glutathione (GSH). The cell exposed to hypoxia had lesser GSH content and decreased GPx activity with increase in its SOD activity as compared to unexposed cells (Figure 35.3C-E; dark grey bars). All such changes observed in hypoxic cells were prevented by treatment of extracts obtained from both species of *Hippophae* (Figure 35.3A-C; light grey

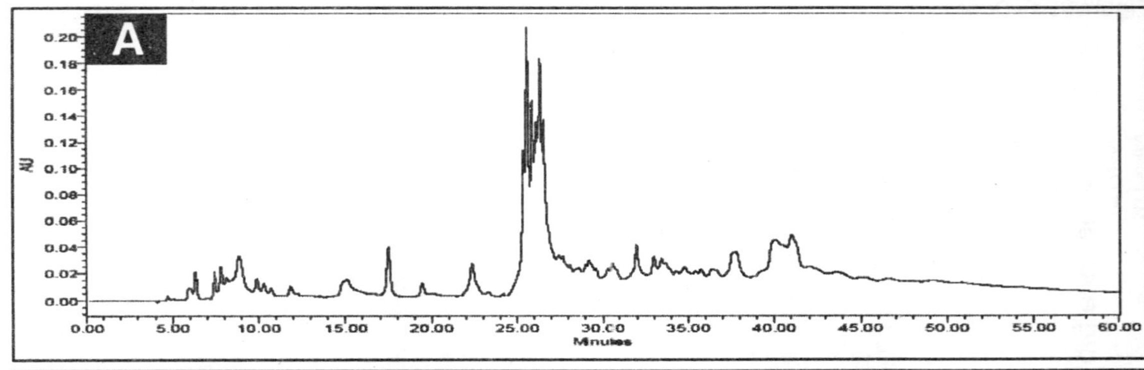

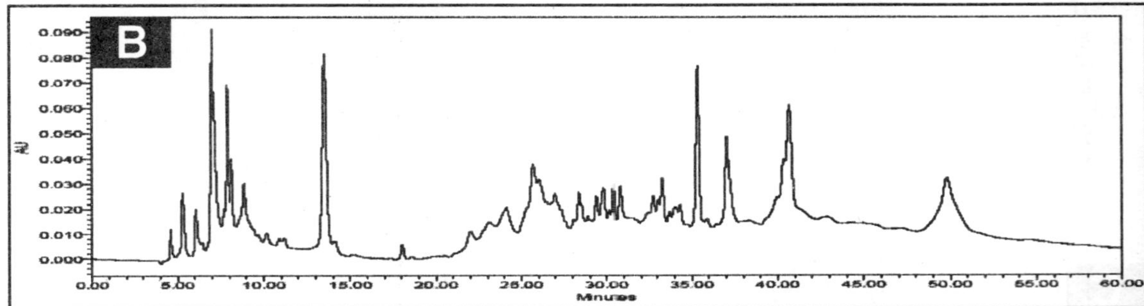

Figure 35.2: RP-HPLC Chromatogram of Aqueous Extract of *Hippophae salicifolia* (A) and *Hippophae rhamnoides* (B).

bars). Among both the species, the protection offered by *Hippophae salicifolia* was more than *Hippophae rhamnoides* in restoring GSH, GPx and SOD activities by 3-8 per cent, 2-7 per cent and 4-9 per cent, respectively, at various concentrations of extracts ranging from 25-200 µg/ml.

Further in the *in vivo* studies, rats exposed to hypobaric hypoxia showed an increase in plasma MDA levels with decrease in content of GSH and activities of GPX and SOD (Figure 35.4 A-D; dark grey bars). The findings are well in agreement with earlier published reports (Tribble *et al.*, 1988; Nakanishi *et al.*, 1995; Lluis *et al.*, 2005). The treatment of rats with extract of *Hippophae rhamnoides* and *Hippophae salicifolia* prevented such changes induced by hypobaric hypoxic, evidencing anti-stress and anti-hypoxic effects. The anti-stress and anti-hypoxic activity of *Hippophae salicifolia* was found to be better in comparison to the *Hippophae rhamnoides* by restoring MDA, GSH, GPx and SOD levels by 7-19 per cent, 4-19 per cent, 4-16 per cent and 4-9 per cent, respectively, after treatment of rats at various doses (12.5-100 mg/kg) of extract.

ANTI-INFLAMMATORY EFFECTS OF *HIPPOPHAE SALICIFOLIA* AND *HIPPOPHAE RHAMNOIDES IN VITRO* AND *IN VIVO*

The oxidative stress induces various stress responsive transcriptional factors, such as NF-kB and activator protein-1 (AP-1) through which it triggers inflammatory response by inducing IL-6, TNF-α, Intercellular Adhesion Molecule-1 (ICAM-1) and Vascular Cell Adhesion Molecule-1 (VCAM-1) (Gudrun *et al.*, 2004). NF-kB plays an important role in inflammatory responses and cell growth

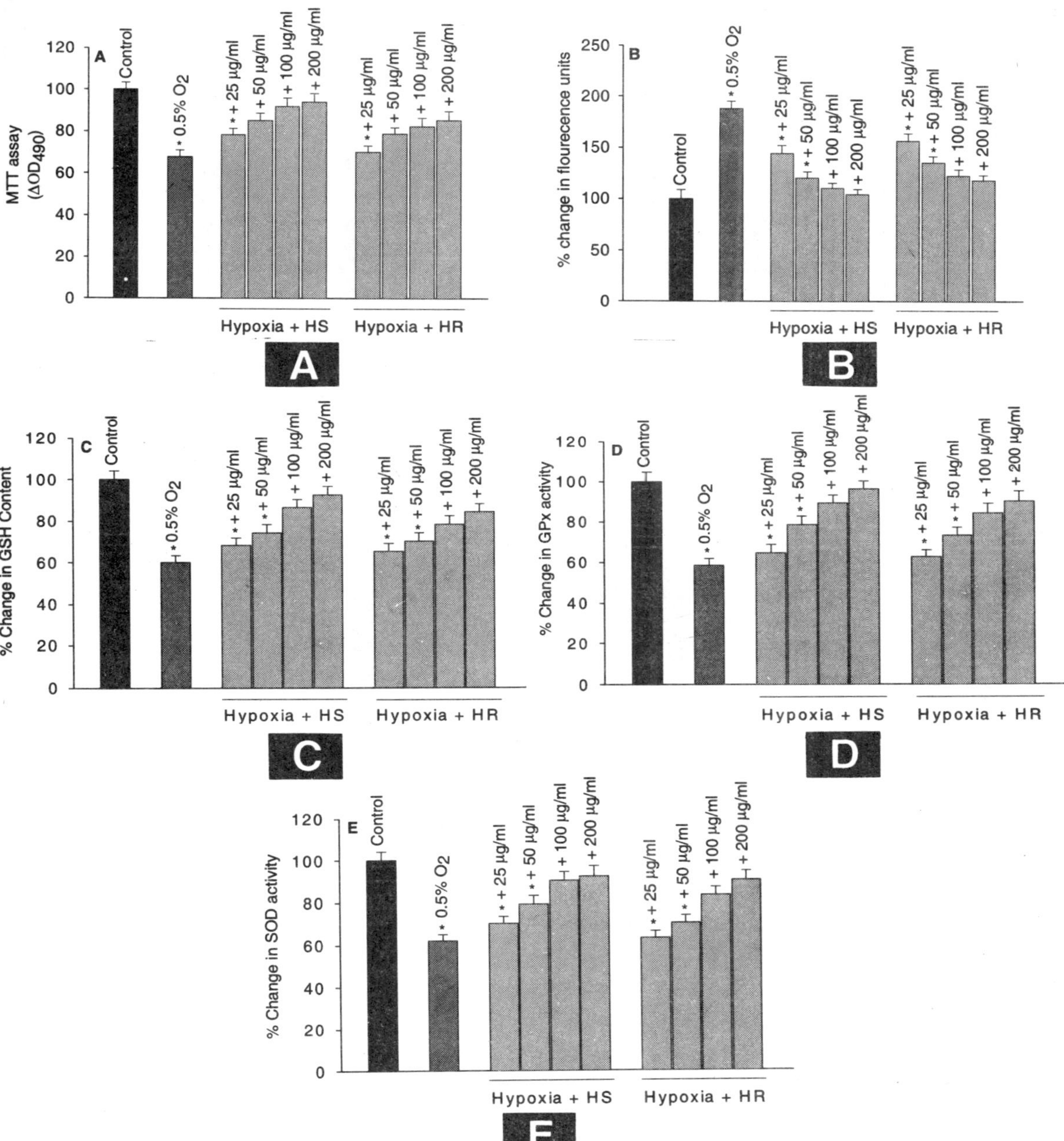

Figure 35.3: Effect of Aqueous Extracts of *Hippophae salicifolia* and *Hippophae rhamnoides* on Cell Viability and antioxidant Status of Cells Exposed to hypoxia for 24 h. Treatment of cells with aqueous extracts of *Hippophae salicifolia* and *Hippophae rhamnoides* reduced cell death (A), ROS levels (B) and prevented decrease in GSH content (C) and activities of GPx (D) and SOD (E) after hypoxia exposure.

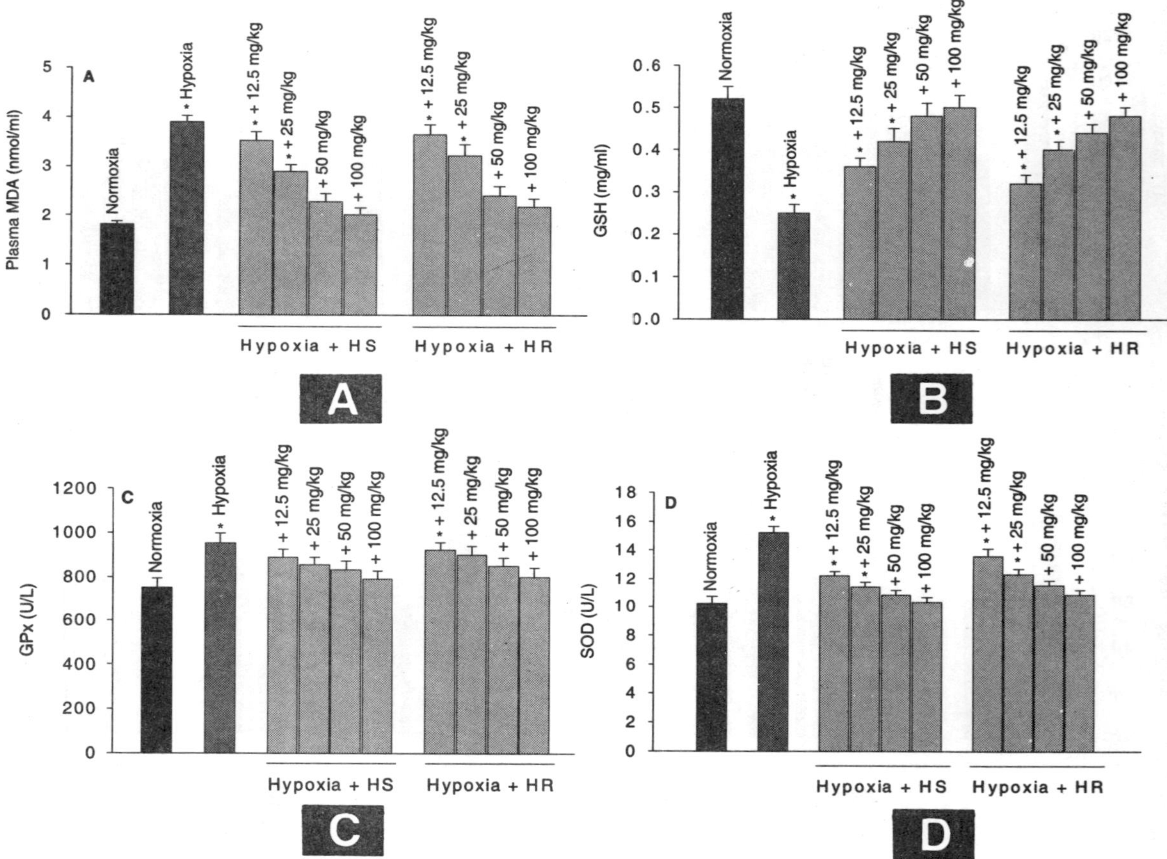

Figure 35.4: Effects of Aqueous Extracts of *Hippophae salicifolia* and *Hippophae rhamnoides* on Antioxidant Status of Rats Exposed to Hypobaric Hypoxia for 6 h. Pre-treatment (-30 min) of Rats with aqueous extracts of *Hippophae salicifolia* and *Hippophae rhamnoides* prevented increase in plasma MDA levels (A) and further minimized decrease in plasma GSH content (B) and activities of GPx (C) and SOD (D) after hypoxia exposure.

regulation and it is involved in the expression of many cytokines and acute phase proteins (Lenardo and Baltimore, 1989).

Figure 35.5 and 35.6 depicts the effect of hypoxia on NF-κB expression and levels of pro-inflammatory cytokines in absence or presence of *Hippophae salicifolia* and *Hippophae rhamnoides* extract. NF-kB activation is usually associated with the higher levels of reaction oxygen species observed after hypoxia exposure and play a key role in many inflammatory diseases. As expected, the expression of NF-kB was found to be increased in cells exposed to hypoxia (Figure 35.5A; dark grey bar) by specific ELISA assay. It is reported that hypoxia induced oxidative stress is usually associated with NF-kB activation (Sarada *et al.*, 2008). The treatment with extracts from both the species of *Hippophae* de-activated NF-kB response of hypoxia (Figure 35.5A; red and blue bar). Concomitant with this, higher levels of pro-inflammatory markers, such as TNF-α and IL-6 were observed after exposure of

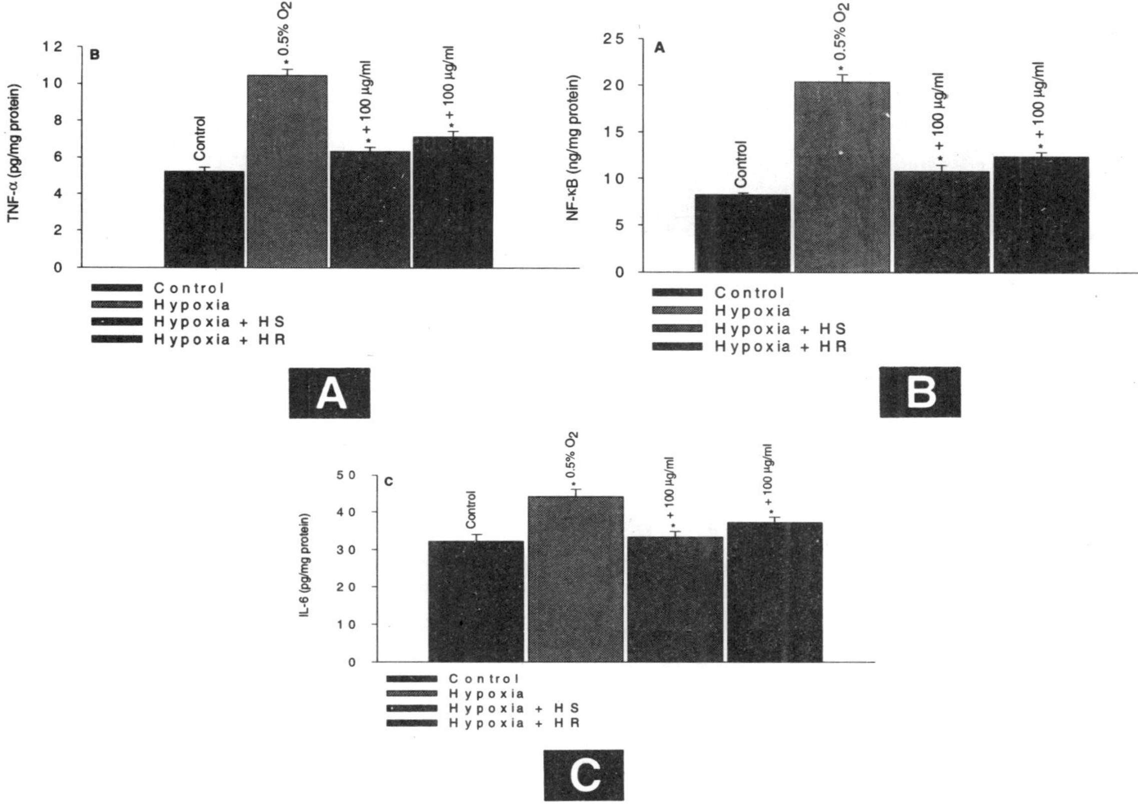

Figure 35.5: Anti-inflammatory Effect of *Hippophae salicifolia* and *Hippophae rhamnoides In vitro*. HT22 cells exposed to hypoxia had higher expressions of NF-kB (A), TNF-α (B) and IL6 (C) grey bars. Treatment of cells with aqueous extracts of *Hippophae salicifolia* and *Hippophae rhamnoides* minimized above mentioned pro-inflammatory markers (red and blue bars).

HT22 cells to 0.5 per cent O_2 for 24 h (Figure 35.5B and 35.5C; dark grey bar). It is documented that experimental rats subjected to hypobaric hypoxia showed changes in pro-inflammatory cytokines, such as IL-6 and TNF- α as compared with control (Sarada *et al.*, 2008). Both TNF-α and IL-6 are pleiotropic cytokines with varying effects. TNF-α may help in maintaining homeostasis during stressful situations or may cause devastating effects depending on the degree of expression. In the present study, the activation of pro-inflammatory markers could be prevented by the treatment of extracts from both species by *Hippophae salicifolia* and *Hippophae rhamnoides* (Figures 35.5B and 35.5C; red and blue bar). In terms of anti-inflammatory response, the extract of *Hippophae salicifolia* was better than that of *Hippophae rhamnoides* by 19.4 per cent, 15.4 per cent and 11.8 per cent in preventing an increase in NF-kB, TNF-α and IL-6 levels, respectively, after hypoxia exposure. More or less, similar findings were observed *in vivo* studies in which rats supplementation of 100 mg/kg extracts from both the species of *Hippophae* prevented rise in plasma NF-kB, TNF-α and IL6 levels after hypoxia exposure for 6 h (Figure 35.6 A-C).

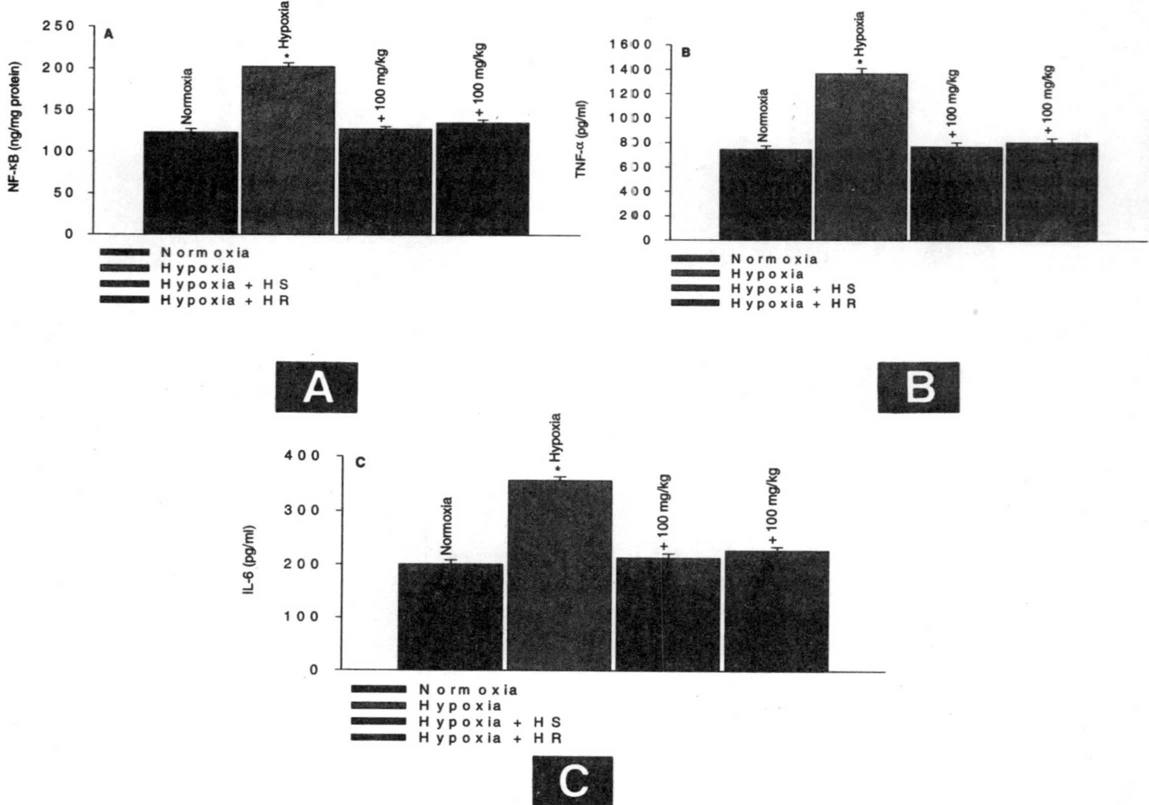

Figure 35.6: Anti-inflammatory Effect of *Hippophae salicifolia* and *Hippophae rhamnoides In vivo*. Pre-treatment (-30 min) of rats with aqueous extracts of *Hippophae salicifolia* and *Hippophae rhamnoides* prevented an increase in pro-inflammatory cytokines, such as NF-κB, TNF-alpha and IL6 associated with hypoxia exposure (red and blue bars).

CONCLUSION

In nutshell, the present study indicates that aqueous extract of *Hippophae salicifolia* has better anti-stress, anti-hypoxic and anti-inflammatory effects as compared to *Hippophae rhamnoides*, as both cultivated in Lahaul valley of cold desert area of Himachal Pradesh, India.

REFERENCES

1. Benzie, I.F. and Strain, J.J. 1996. The ferric reducing ability of plasma (FRAP) as a measure of antioxidant power-The FRAP assay. *Anal. Biochem.* 239: 70-76.
2. Beutler, E., Duron, O. and Kelly, B.M. 1963. Improved method for the determination of blood glutathione. *J. Lab. Clin. Med.* 61: 882-888.
3. Beveridge, T., Li, T.S., Oomah, B.D. and Smith, A. 1999. Seabuckthorn products-Manufacture and composition. *J. Agric. Food Chem.* 47: 3480-3488.

4. Dhyani, D., Maikhuri, R.K., Misra, S. and Rao, K.S. 2010. Endorsing the declining indigenous ethnobotanical knowledge system of seabuckthorn in Central Himalaya, India. *J. Ethnopharmacol.* 127: 329-334.

5. Eccleston, C., Baoru, Y., Tahvonen, R., Kallio, H., Rimbach, G.H. and Minihane, A.M. 2002. Effect of an antioxidant rich juice (seabuckthorn) on risk factors for coronary heart disease in humans. *J. Nutr. Biochem.* 13: 346-354.

6. Ghaffar, A. 1997. Seabuckthorn (*Hippophae rhamnoides*)-A multipurpose plant. In: *Proceedings of the International Seminar on Farming System Research in the Context of Food Security*, p.115-120, University of Arid Agriculture, Rawalpindi, Pakistan.

7. Gudrun, R., Michal, T. and Berhned, H. 2004. Quercetin protects against linolenic acid induced porcine endothelial cell dysfunction. *J. Nutr.* 134: 771-775.

8. Gupta, A., Kumar, R., Pal, K., Banerjee, P.K. and Sawhney, R.C. 2005. A preclinical study of the effects of seabuckthorn (*Hippophae rhamnoides* L.) leaf extract on cutaneous wound healing in albino rats. *Int. J. Low Extrem. Wounds* 4: 88-92.

9. Kallio, H., Yang, B., and Peippo, P. 2002. Effects of different origins and harvesting time on vitamin C, tocopherols and tocotrienols in seabuckthorn (*Hippophae rhamnoides* L.) berries. *J. Agric. Food Chem.* 50: 6136-6142.

10. Kaur, C., Foulds, W.S. and Ling, E.A. 2008. Hypoxia-ischemia and retinal ganglion cell damage. *Clin. Ophthalmol.* 2: 879-889.

11. Lenardo, M.J. and Baltimore, D. 1989. NF-kB: A pleiotropic mediator of inducible and tissue-specific gene control. *Cell* 58: 227-229.

12. Li, T.S.C. and Beveridge, H.J. 2003. In: *Seabuckthorn (Hippophae rhamnoides)-Production and Utilization*, NRC Research Press, Ottawa, Ontariopp, 133p.

13. Lluis, J.M., Morales, A., Blasco, C., Colell, A., Mari, M., Garcia-Ruiz, C. and Fernandez-Checa, J.C. 2005. Critical role of mitochondrial glutathione in the survival of hepatocytes during hypoxia. *J. Biol. Chem.* 280: 3224-3232.

14. Lotito, S.B. and Frei, B. 2006. Consumption of flavonoid-rich foods and increased plasma antioxidant capacity in humans: Cause, consequence, or epiphenomenon? *Free Radic. Biol. Med.* 41: 1727-1746.

15. Mosman, T. 1983. Rapid colorimetric assay for cellular growth and survival-Application to proliferation and cytotoxicity assays. *J. Immnunol. Meth.* 65: 55-63.

16. Nakanishi, K., Tajima, F., Nakamura, A., Yagura, S., Ookawara, T., Yamashita, H., Suzuki, K., Taniguchi, N. and Ohno, H. 1995. Effects of hypobaric hypoxia on antioxidant enzymes in rats. *J. Physiol.* 489: 869-879.

17. Narayanan, S., Ruma, D., Gitika, B., Sharma, S.K., Pauline, T., Ram, M.S., Ilavazhagan, G., Sawhney, R.C., Kumar, D. and Banerjee, P.K. 2005. Antioxidant activities of seabuckthorn (*Hippophae rhamnoides*) during hypoxia induced oxidative stress in glial cells. *Mol. Cell Biochem.* 278: 9-14.

18. Ohkawa, H., Ohishi, N. and Yagi, K. 1979. Assay of lipid peroxides in animal tissues by TBA reaction. *Anal. Biochem.* 95: 351-358.

19. Rababah, T.M., Hettiarachchy, N.S. and Horax, R. 2004. Total phenolics and antioxidant activities of fenugreek, green tea, black tea, grape seed, ginger, rosemary, gotu kola, and ginkgo extracts, vitamin E, and tert-butylhydroquinone. *J. Agric. Food Chem.* 52: 5183-5186.

20. Rattan, V., and Tomar, A. 2011. Effect of different temperatures on the seed germination of willow leaf seabuckthorn (*Hippophae salicifolia*). In: *Souvenir of National Conference on Forest Biodiversity by Uttar Pradesh State Biodiversity Board*, p.56-61, Lucknow.

21. Rodrýguez-Meizoso, I., Marin, F.R., Herrero, M., Senorans, F.J., Reglero, G., Cifuentes, A. and Ibanez, E. 2006. Subcritical water extraction of nutraceuticals with antioxidant activity from oregano. Chemical and functional characterization. *J. Pharm. Biomed. Anal.* 41: 1560-1565.

22. Rongsen, L. 1992. *Seabuckthorn-A Multipurpose Plant Species for Fragile Mountains*, Occasional paper, No. 20, ICIMOD, Kathmandu, 62p.

23. Rongsen, L., Cao, Y., and Lu, S. 2001. Introduction of seabuckthorn (*Hippophae salicifolia*) from high altitude to low altitude area. In: *Proceedings of International Workshop on Seabuckthorn*, p.89-90.

24. Rousi, A. 1971. The genus *Hippophae* L.-A taxonomic study. *Annals Bot.* 8: 177-227.

25. Saggu, S. and Kumar, R. 2008. Effect of seabuckthorn (*Hippophae rhamnoides*) leaf aqueous and ethanol extracts on avoidance learning during stressful endurance performance of rats-A dose dependent study. *Phytother. Res.* 22: 1183-1187.

26. Saggu, S., Divekar, H.M., Gupta, V., Sawhney, R.C., Banerjee, P.K. and Kumar, R. 2007. Adaptogenic and safety evaluation of seabuckthorn (*Hippophae rhamnoides*)-A dose dependent study. *Food Chem. Toxicol.* 45: 609-617.

27. Sarada, S., Himadri, P., Mishra, C., Geetali, P., Ram, M.S. and Ilavazhagan, G. 2008. Role of oxidative stress and NFkB in hypoxia-induced pulmonary edema. *Exp. Biol. Med. (Maywood)* 233: 1088-1098.

28. Singh, V. 1994. A report on arboreal seabuckthorn (*Hippophae salicifolia*) from dry temperate Himalayas. *Journal of Tree Sciences* 3 (1): 67-68.

29. Singh, V. 2003. Geographical adaptation and distribution of seabuckthorn (*Hippophae* L.) resources. In: *Seabuckthorn (Hippophae L.)-A Multipurpose Wonder Plant. Vol. I: Botany, Harvesting and Processing Technologies* (V. Singh, Ed. in Chief, 2003), p. 21-34, Indus Publishing Company, New Delhi, 518p.

30. Singh, V., Gupta, R.K., Arumughan, C., Sawhney, S.C., Rana, R.K., Singh, A., Lal, M., Devi, R. and Sharma, N. 2011. Biochemical evaluation of *Hippophae salicifolia* and *H. mongolica* as horticultural crops in dry temperate Himalayas. In: *Proceedings of National Conference on Seabuckthorn: Emerging Trends on R&D on Health Protection and Environmental Conservation*, p. 75-92, CSK HPKV, Palampur, HP, 324p.

31. Singleton, V.L. and Rossi, J.A. 1965. Colorimetric determination of total phenolic with phosho-molybidic-phospho-tungstic acid reagents. *Am. J. Enol. Vitic.* 16: 144-158.

32. Tribble, D.L., Jones, D.P. and Edmondson, D.E. 1998. Effect of hypoxia on tert-butylhydroperoxide-induced oxidative injury in hepatocytes. *Mol. Pharmacol.* 34: 413-420.

33. Tulsawani, R., Kelly, L.S., Fatma, N., Chhunchha, B., Kubo, E., Kumar, A. and Singh, D.P. 2010b. Neuroprotective effect of peroxiredoxin 6 against hypoxia-induced retinal ganglion cell damage. *BMC Neuroscience* 5: 11:125.

34. Tulsawani, R.K., Sharma, P., Divekar, H.M., Meena, R.N., Singh, M. and Kumar, R. 2010a. Supplementation of fruit extract of *Hippophae rhamnoides* speeds adaptation to simulated high altitude stressors in rats. *J. Complement Integr. Med.* 7: Article 30.

35. Zhishen, J., Mengcheng, T. and Jianming, W. 1999. The determination of flavonoid content in mulberry and their scavenging effects on superoxide radicals. *Food Chemistry* 64: 555-559.

36. Zu, Y., Li, C., Fu, Y. and Zhao, C. 2006. Simultaneous determination of catechin, rutin, quercetin kaempferol and isorhamnetin in the extract of seabuckthorn (*Hippophae rhamnoides* L.) leaves by RP-HPLC with DAD. *J. Pharm. Biomed. Anal.* 41: 714-719.

Chapter 36

Phenol Content and Antioxidant Activity of the Ethanolic and Aqueous Extracts of Seabuckthorn (*Hippophae rhamnoides* L.) Vegetative Parts

E. Sne, D. Seglina, R. Galoburda* and L. Krasnova

Latvia State Institute of Fruit Growing, Graudu 1, Dobele, LV-3701, Latvia
*Latvia University of Agriculture, Liela 2, Jelgava, LV-3001, Latvia

SUMMARY

Seabuckthorn (*Hippophae rhamnoides* L.) has been highly evaluated because of its nutritional, medical and cosmetic properties. Mostly berries are studied and used as a raw material in processing, although other parts of plant contain biologically active compounds. Therefore, the aim of the study was to evaluate the content of total phenols and total flavonoids and evaluate the antioxidant capacity by DPPH, ABTS, and FRAP assays of seabuckthorn vegetative parts depending on harvest time. Seabuckthorn leaves and shoots were collected in May, June, July, and October months. The total phenol and flavonoid content and antioxidant activity by FRAP, DPPH and ABTS assays of ethanolic and water extracts were investigated. The obtained results showed significant differences between the collected samples of seabuckthorn vegetative parts. Leaves and shoots harvested in October reached the highest results in several categories: Total phenol (127.71 mg/g) and flavonoid (77.43 mg/g) content, and antioxidant activity of water extracts by ABTS (53.26 per cent), FRAP (181.58 mg/g) and DPPH (338.29 mg/g) assays. The present study proved that due to biologically active compounds and their biological activity, seabuckthorn vegetative parts could be used as raw material not only in pharmacological and cosmetic industries, but also for food processing.

Keywords: *Seabuckthorn, Leaves, Shoots, Phenols, Flavonoids, FRAP, DPPH and ABTS.*

INTRODUCTION

Seabuckthorn (*Hippophae rhamnoides* L.) is widespread in Europe and Asia (Singh, 2003), but currently is being cultivated in various parts of the world, because of its nutritional and pharmacological value. Different parts have been used in traditional medicine in various countries of the world (Guliyev *et al.*, 2004; Upadhyay *et al.*, 2010), mostly in India and China. Although fruits of seabuckthorn are considered to be the most valuable because of their potential health benefits, other parts are also rich source of different bioactive compounds (Prabhu *et al.*, 2011).

Leaves and their extracts possess high antioxidant activity (Tolkachev and Sheichenko, 2006), which is mainly attributed to the presence of polyphenols. Phenolics in seabuckthorn leaves are represented by phenolic acids (gallic, caffeic, p-coumaric, p-hydroxybenzoic, ferulic, vanillic, and cinnamic), tannins and the largest group-flavonoids. Quercetin, catechin, rutin, isorhamnetin, kaempferol, and astragalin and their glucoside compounds have been isolated (Singh, 2006; Zu *et al.*, 2006; Arimboor *et al.*, 2008; Kim *et al.*, 2011). Flavonoids are strong radical scavengers and antioxidants, and are believed to be preventive against cancer, cardiovascular diseases, atherochlerosis (Singh, 2006). Antitumor and antiviral properties of flavonoids also have been reported (Arimboor *et al.*, 2008). Other compounds found in seabuckthorn leaves are carotenoids, different minerals, vitamins (C and E), proteins, triterpenes and isoprenols, and free and esterified sterols have been found in leaves (Singh, 2006; Badaluta *et al.*, 2010). According to Guliyev *et al.* (2004), the extracts of seabuckthorn leaves and bark are used in traditional folk medicine against colitis, diarrhoea, gastrointestinal and dermatological and other problems. Industrially, leaves have been used in production of pharmacological preparations, cosmetics and as a component of tea.

For the last few years, leaf extracts have been studied widely due to their high levels of antioxidants and medical properties (Upadhyay *et al.*, 2010; Kumar *et al.*, 2011; Lee *et al.*, 2011; Michel *et al.*, 2012). Kim *et al.* (2011) also suggested that seabuckthorn leaf extracts have a potential as natural compounds, for example, they could be used as food additives. Extracts could be eventual food stabilizers-natural antioxidants having a great potential for replacement of synthetic ones like butylated hydroxyanisole (BHA) and butylated hydroxytoluene (BHT), which due to their toxicity, cancerogenesis and overall potential health risks are seriously criticized (Dudonne, 2009; Kumar *et al.*, 2011). Addition of leaf extracts to the food products would not only prolong their storage time, but also increase their nutritional value. It has been stated that increased consumption of food rich in radical scavengers and antioxidants decreases the risk of cardiovascular diseases and cancer (Singh, 2006).

Shunguang *et al.* (2003) reported that interest in seabuckthorn leaf utilization has increased. Together with new, tender non-woody shoots they are promising stock material not only for forage and medicinal products, but also in food processing due to their valuable chemical composition and great productivity. Therefore, the aim of the present study was to detect the content of total phenols and total flavonoids and to evaluate the antioxidant capacity by DPPH, ABTS, and FRAP assays of seabuckthorn vegetative parts depending on harvest time.

MATERIALS AND METHODS

CHEMICALS USED

1, 1-diphenyl-2-picrylhydrazyl (DPPH), 3,4,5-trihydroxybenzoic acid (gallic acid), rutin, petroleum ether, Folin-ciocalteu reagent were purchased from Sigma. 6-hydroxy-2,5, 7,8-tetramethylchroman-2-carboxylic acid (Trolox), $FeCl_3.6H_2O$, F (II) ($FeSO_4.7H_2O$) were obtained from Aldrich. 2,2-azino-bis-3-ethylbenzthiazoline-6-sulphonic acid (ABTS) and ethanol (96 per cent) was bought from Fluka and SIA Jaunpagasts Plus, respectively.

Plant Material

The samples for the investigation were young, green and non-woody shoots, and leaves collected at four different harvests from the same field in southern part of Latvia in 2010. Samples from female trees were harvested in May, June, July and October months. They were collected only in spring and early summer in order to impede berry formation, and in autumn, when berries have been harvested.

Preparation of Extracts

Before experiments, samples were frozen and lyophilized (FreeZone, Labconco, USA) at temperature -50 ± 1 °C and pressure 0.063 ± 0.01 mbar. After drying, they were ground into a fine powder by mill (Foss, Germany).

Ethanolic Extract for Total Phenols and Total Flavonoids Assay

Extracts were prepared by taking about 0.1 g of freeze dried ground leaves, added 5 ml of water and mixed. Then, added 50 ml of 70 per cent ethanol and continuously shaking for 2 h. Afterwards, the extract was filtered by vacuum filter using glass filter and placed into 100 ml volumetric flask. Sediment was rinsed 3 times by 10 ml of 80 per cent ethanol. Volumetric flask was filled up till mark by water and shaken well. Half of the prepared extract (50 ml) was placed in a dividing funnel, after 20 ml of petroleum ether was added and mixed for 2 minutes. The funnel was left in dark place till mixture became differentiated where water-ethanol part was used for further analysis (Singleton *et al.*, 1999).

Ethanolic and Water Extracts for Antiradical Activity and Ferric Reducing Power Assays

Extracts were prepared by soaking about 1 g of freeze dried ground leaves in 50 ml of 70 per cent ethanol or distilled water at room temperature (22 ± 2 °C) for 24 h continuously shaking. Then extract was filtered through a vacuum filter using glass filter and centrifuged for 5 min at 9000 g (Eppendorf Centrifuge 5804R, Eppendorf AG, Germany).

Total Phenol Content

The concentration of total phenol in the ethanolic extracts was determined using Folin-ciocalteu reagent using method described by Singleton *et al.* (1999). Briefly, 1 ml of extract was added to 4 ml 7.5 per cent Na_2CO_3 and 5 ml Folin-ciocalteu solution (1 part FCR diluted with 10 parts of water). Then mixture was kept in dark place for 30 min. The absorbance was measured at 760 nm using spectrophotometer UV-1650 PC (Shimadzu, Japan). The calibration curve was made using gallic acid solution as standard with concentrations from 0 to 0.06 mg/ml. Results were expressed as mg of gallic acid equivalents (GAE) per 1 g of dry weight (DW).

Total Flavonoid Content

Total flavonoids were determined using the method of Kumar *et al.* (2011) and Upadhyay *et al.* (2010). Ethanolic extract (1 ml) was mixed with 2 ml of water and 0.15 ml 5 per cent $NaNO_2$ and left to react for 6 min. Then 0.15 ml of 10 per cent $AlCl_3$ solution was added, mixed and allowed to react for another 6 min. Afterwards, 2 ml of 4 per cent NaOH was added to the mixture and shaken vigorously. The absorbance was read against blank after 15 min at 510 nm using spectrophotometer UV-1650 PC. The calibration curve (0-0.6 mg/ml) was prepared using a standard solution of rutin. The result was expressed as mg of rutin equivalents (RUE) per 1 g of dry weight (DW).

Antioxidant Activity

FRAP Assay

The ferric reducing antioxidant potential (FRAP) assay was determined by method as described previously by Benzie and Strain (1996) with some modifications. The experiment was performed by adding 2.9 ml of FRAP reagent to 0.1 ml of the leaf extract (water or ethanolic) and left to react for 10 min in dark place at room temperature 22 ± 2 °C. Absorbance was measured at wavelength 593 nm using spectrophotometer UV-1650 PC. The standard curve was made in a range from 0 to 0.1 mg/ml. Results were expressed as mg of trolox equivalents (TE) per 1 g of dry weight (DW).

DPPH Assay

The free radical scavenging activity of samples for 1,1-diphenyl-2-picrylhydrazyl (DPPH) radical was detected according to Brand-Williams *et al.* (1995) with slight modifications. The analysis was done by taking 0.1 ml of leaf extract (water or ethanolic) and by mixing with 2.9 ml of a DPPH solution. Mixture was left at room temperature in dark place for 30 min. The adjusted absorbance for DPPH solution has to be 0.900 at wavelength 517 nm. The absorbance of final solution was measured at 517 nm using spectrophotometer UV-1650 PC. The trolox calibration curve was done between 0 and 0.1 mg/ml. Results were expressed as mg of trolox equivalents (TE) per 1 g of dry weight (DW).

ABTS Assay

2, 2-azino-bis-3-ethylbenzthiazoline-6-sulfonic acid (ABTS) radical scavenging capacity assay was carried out according to Re *et al.* (1999) with some modifications. The scavenging activity was detected by taking 3 ml of ABTS solution and added to 0.03 ml of extract (water or ethanolic). Then mixture was kept in dark place at room temperature for 10 min. The absorbance of ABTS work solution has to reach 0.700 ± 0.001 at 734 nm. The optical density was read at wavelength 734 nm using spectrophotometer UV-1650 PC.

The activity of the extracts in scavenging ABTS radical was calculated as follows:

$$\%\ \text{Inhibition} = \frac{\text{Absorbance of control} - \text{Absorbance of sample}}{\text{Absorbance of control}} \times 100$$

Results were expressed as per cent of scavenged radicals.

Statistical Analysis

All data were expressed as means ± standard deviations at least of triplicate measurements and analyzed by SPSS for Windows (version 17.0) to test any significant differences among various parameters. One-way analysis of variance (ANOVA) and Duncan's multiple range tests were used. Values of $p < 0.05$ were considered statistically significant ($\alpha = 0.05$).

RESULTS

Total phenol and total flavonoid content were statistically differed ($p < 0.05$) between analysed samples regarding to the harvest time (Table 36.1). The content of total phenols varied from 100.28 to 127.71 mg/g and total flavonoid content from 43.22 to 77.43 mg/g. The highest values of both above mentioned parameters were found in seabuckthorn vegetative parts harvested in October, while the lowest in samples harvested in May.

Table 36.1: Total Phenol and Total Flavonoid Content of Seabuckthorn Vegetative Parts

Harvesting Month	*Total Phenol Content (mg GAE/g)*	*Total Flavonoid Content (mg RUE/g)*
May	100.28 ± 0.16[d]	43.22 ± 3.81[c]
June	114.69 ± 0.15[c]	61.62 ± 5.20[b]
July	120.27 ± 0.10[b]	46.38 ± 1.15[c]
October	127.71 ± 0.11[a]	77.43 ± 2.22[a]

Note: Values marked with the same letters in the column are not significantly different at $p < 0.05$.

GAE: Gallic acid equivalent, RUE: Rutin equivalent.

The ferric reducing antioxidant potential (FRAP) of analysed seabuckthorn young shoots and leaves were statistically different ($p < 0.05$) by harvesting time (Figure 36.1). The highest value of water extracts was detected in samples collected in October (181.58 mg/g) and it was about 2.5 times higher than the samples collected in May (71.58 mg/g). Meanwhile, the opposite tendency was observed in ethanolic extracts, when the highest level of FRAP potential between all samples was found in leaves and shoots in May (167.19 mg/g), and lowest in October ones (145.27 mg/g).

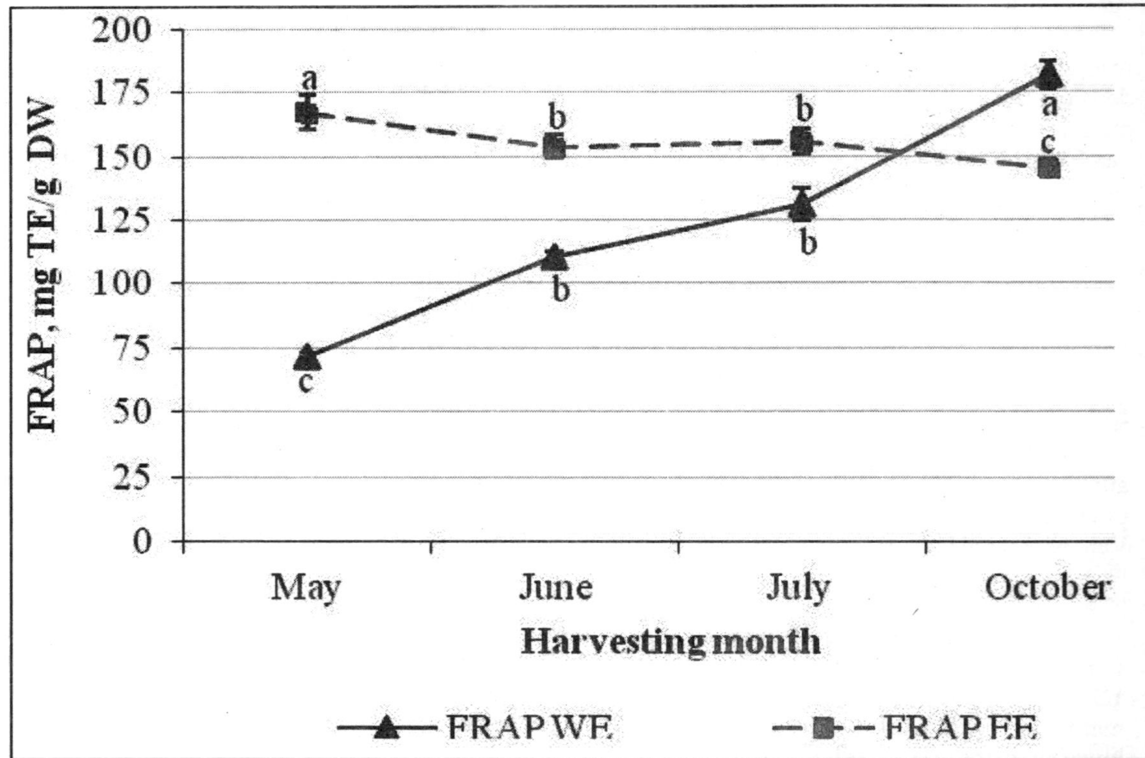

Figure 36.1: The Ferric Reducing Antioxidant Power (FRAP) of Seabuckthorn Vegetative Parts. (Values marked with the same letters of each kind of extract are not significantly different at p<0.05. TE: Trolox equivalent; WE: Water extract; EE: Ethanolic extract.)

Antioxidant activity evaluation was done using two radicals: 1,1-diphenyl-2-picrylhydrazyl (DPPH) and 2,2-azino-bis-3-ethylbenzthiazoline-6-sulphonic acid (ABTS). The values of DPPH radical scavenging activity of analysed samples differed significantly ($p < 0.05$) depending on harvest time (Figure 36.2). In DPPH assay, the water extracts of samples collected in October were found more than 2.5 times higher (338.29 mg/g) as compared to those harvested in May (129.71 mg/g).

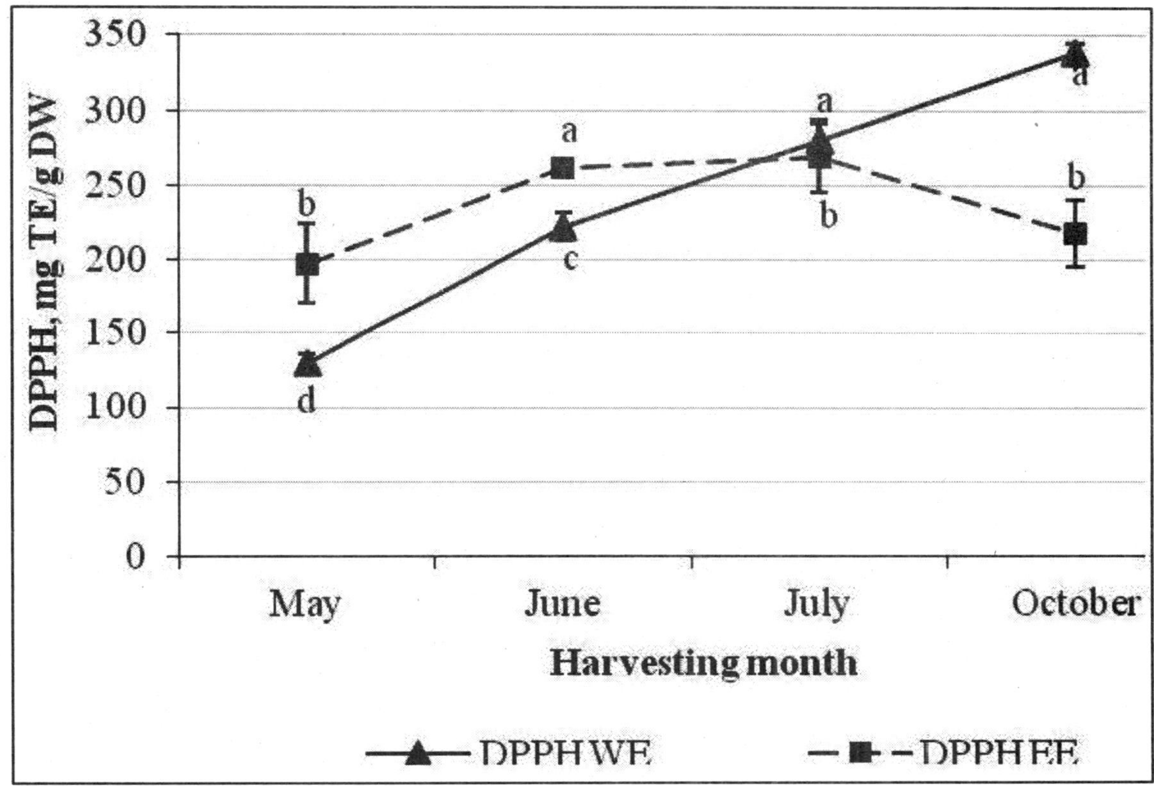

Figure 36.2: The DPPH Radical Scavenging Activity of Seabuckthorn Vegetative Parts.
(Values marked with the same letters of each kind of extract are not significantly different at p<0.05. TE:Trolox equivalent; WE: Water extract; EE: Ethanolic extract.)

Radical scavenging activities of ethanolic extracts of seabuckthorn leaves and shoots proved to be the highest in those samples which were harvested in the summer months-June and July (262.03 and 269.43 mg/g, respectively). The ABTS assay values of analysed samples of vegetative parts differed significantly ($p < 0.05$) depending on harvest time (Figure 36.3). All ethanolic extracts of seabuckthorn vegetative parts possessed very high radical scavenging potential by ABTS assay, ranging from 80.62 per cent (May samples) to even 99.26 per cent (October samples). On the other hand, water extracts had lower scavenging activity compared to ethanolic ones and could be ranked as follows: 23.68 per cent (May) < 35.07 per cent (June) < 42.55 per cent (July) < 53.29 per cent (October).

Pearson correlation test was used to detect any interconnection between total phenol and total flavonoid content, FRAP, DPPH and ABTS activity of water and ethanolic extracts in seabuckthorn vegetative parts (Table 36.2).

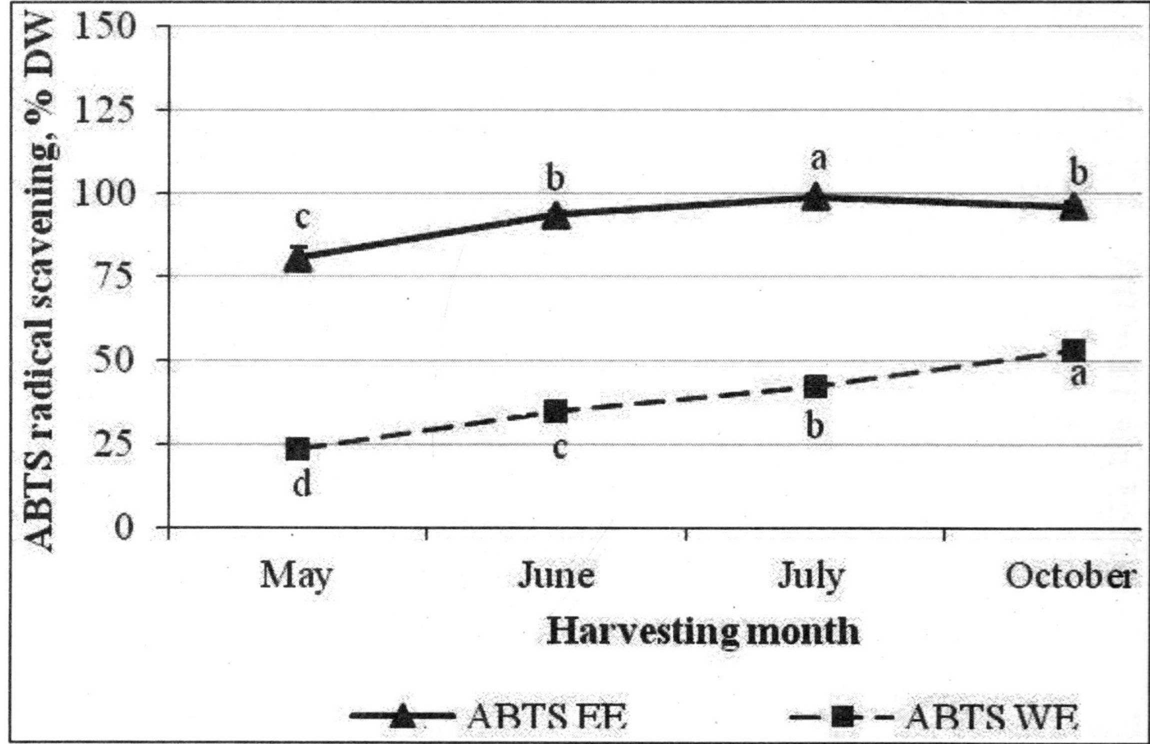

Figure 36.3: The ABTS Radical Scavenging Activity of Seabuckthorn Vegetative Parts. (Values marked with the same letters of each kind of extract are not significantly different at p<0.05. WE: Water extract; EE: Ethanolic extract.)

Table 36.2: The Correlations between Tested Parameters of Seabuckthorn Vegetative Parts

	TPC	*TFC*	*$ABTS_{WE}$*	*$ABTS_{EE}$*	*$DPPH_{WE}$*	*$DPPH_{EE}$*	*$FRAP_{WE}$*	*$FRAP_{EE}$*
TPC	1	0.721*	0.979*	0.889*	0.991*	0.359	0.964*	–0.837*
TFC		1	0.757*	0.443	0.704*	-0.014	0.795*	–0.744*
$ABTS_{WE}$			1	0.815*	0.989*	0.227	0.991*	–0.834*
$ABTS_{EE}$				1	0.865*	0.655	0.758*	–0.699*
$DPPH_{WE}$					1	0.291	0.976*	–0.827*
$DPPH_{EE}$						1	0.137	-0.255
$FRAP_{-WE}$							1	–0.826*
$FRAP_{EE}$								1

* Correlation is significant at $\alpha = 0.01$ level (2-tailed). TPC: Total phenol content; TFC: Total flavonoid content; WE: Water extract; EE: Ethanolic extract.

The very strong and positive correlation was observed between activities of water extracts (ABTS $_{WE}$, DPPH $_{WE}$ and FRAP $_{WE}$) and TPC. The strong correlation was detected between TFC and TPC, and also all activities of water extracts. Ethanolic extracts on FRAP assays showed interesting tendency-almost all correlation coefficients were strong but negative.

DISCUSSION

Phenolics are bioactive compounds that are considered to provide beneficial effect on human health due to their natural antioxidant properties (Rivas-Arreola *et al.*, 2010). They work as scavengers of free harmful radicals which act to human's body (Boonchum *et al.*, 2011), deteriorating immune system and increasing risk of potential health problems and illnesses. Also, flavonoids as a part of phenolics possess antioxidant properties and are reported as compounds that have protective qualities against coronary and cardiovascular diseases (Singh, 2006; Rivas-Arreola *et al.*, 2010).

In this study, total phenol and total flavonoid contents of seabuckthorn vegetative parts harvested at different time were evaluated. Total phenol content varied from 100.28 to 127.71 mg/g and total flavonoid content from 43.22 to 77.43 mg/g. These results are higher than the ones obtained by Upadhyay *et al.* (2010) in alcoholic extract of seabuckthorn leaves collected in north-west Himalayas. The content of phenols and flavonoids reached to 56.28 mg/g and 20.76 mg/g, respectively. It is less than Singh (2006) reported in leaves collected from west Pamirs where flavonoid content was from 31 to 123.8 mg/g of dry weight. Kumar *et al.* (2011) was exploring various extraction solvents and methods at different temperatures for phenol and flavonoid detection in leaves collected from west Himalayas. Between water, ethanolic and subcritical water extraction (hot water under pressure), the highest result of total phenol was given by subcritical water extraction at 150 °C, reaching 93.72 mg/g. The same extraction method proved to be the best also for flavonoid extraction-66.40 mg/g and 66.03 mg/g by extraction at 200 °C and 150 °C, respectively. The overall differences of phenolic content could be explained not only by analytical method used but also genetic, seasonal, and agronomic variations (Erturk *et al.*, 2010), different age of plant, harvesting time, and storage conditions. High total phenol content in the current study could be reached due to the geographical position. Kondakova *et al.* (2009) reported that fruits of berry plants grown in colder climate and having shorter vegetation period have higher level of these compounds, probably such observation could be attributed to other parts of plant as well.

Seabuckthorn berries grown in Turkey contained only from 21.31 to 55.38 mg/g total phenol of dry weight (Ercisli *et al.*, 2007). According to Bal *et al.* (2011), flavonoid content in berries reached to 3.54 mg/g of fresh weight, accordingly it would be 23.60 mg/g of dry weight (if moisture content is 85 per cent), but it is still very small amount as compared to leaves. Young leaves of other berry plants like blackberry and strawberry contained from 74.0 to 91.6 mg/g and 55.2 mg/g total phenols of dry weight, respectively (Wang and Lin, 2000).

Antioxidant properties of different natural products and plants have been widely investigated. Due to the different nature of bioactive substances and complexity of the oxidation-antioxidation process, antioxidant activity should be evaluated by various methods using different solvents (Rivas-Arreola *et al.*, 2010; Upadhyay *et al.*, 2010). Soluble phenolic compounds could be extracted using water and other polar organic solvents, such as ethanol, methanol, acetone and their mixture with water (Escribano-Bailon and Santos-Buelga, 2003; Tsao, 2010).

In this study, the antioxidant activity of water and ethanol extracts of seabuckthorn vegetative parts was evaluated by three assays: Ferric reducing antioxidant power (FRAP), ABTS and DPPH radical scavenging activity. Assays are based on stable free radical bleaching (ABTS and DPPH) and Fe^{3+} reduction to Fe^{2+} (FRAP) by bioactive compounds in extracts. There is no particular solvent which could provide extraction of all bioactive compounds, but commonly used solvents in plant extraction for soluble phenolics are water and its mixtures with ethanol, methanol and acetone (Escribano-Bailon and Santos-Buelga, 2003; Sun and Ho, 2005). Ethanol and water are considered as low toxic solvents with high extraction yield (Franco *et al.*, 2008; Dudonne *et al.*, 2009; Michel *et al.*, 2012). Furthermore, polarity of solvent can be controlled using mixtures of both of these solvents (Franco *et al.*, 2008). Therefore, in the current study, water and water-ethanol (30:70 w/w) mixture was used for antioxidant activity tests.

The present results showed that almost all ethanolic extracts of seabuckthorn vegetative parts possesses higher antioxidant activity (FRAP, ABTS, and DPPH) as compared to those prepared by using water as a solvent. These results are in accordance with the findings of other authors that the highest yields are obtained by ethanol and methanol and their mixtures with water (Franco *et al.*, 2008). It could be explained by the solubility of low polar and liposoluble antioxidants, for example, carotenoids, tocopherols, comparing to water extraction (Franco *et al.*, 2008).

Interestingly, the water extracts of parts collected in October for FRAP and DPPH assays recorded higher antioxidant activity than ethanolic extracts. These results are similar to those of Boonchum *et al.* (2011) results, obtained on seaweed extracts and could be explained by the presence of attached sugars (forming glycosides), which provide better water solubility of phenolics (Escribano-Bailon and Santos-Buelga, 2003).

Also, the remarkable increase of total flavonoid content in autumn vegetative parts has to be taken into consideration (Table 36.1). Seabuckthorn leaves are reported to contain not only free flavonoids, but also in a bonded form as glucosides (Bekker and Glushenkova, 2001; Arimboor *et al.*, 2008). Flavonoids are phenolic compounds that probably exhibiting the strongest antioxidant activity during growing season, vegetative parts have accumulated flavonoids in a form of glucosides, reaching the maximum after berry harvesting.

Regarding to the harvesting time, the results varied and were uneven. However, sample collected in October was standing out, reaching the highest results in several categories: Total phenol and flavonoid contents and antioxidant activity by ABTS, FRAP and DPPH assays of water extracts. Erturk *et al.* (2010) in their study of tea leaves observed similar tendency. Results showed that there were significant differences in the content of phenols and antioxidant activity between harvests (May, July, September) and it was rising, where the highest level was reached on third harvest time, *i.e.* in autumn. Total phenols of young shoots of *Camellia sinensis* (a raw material for green and black tea) clones varied between 33.0 to 90.27 mg/g of dry weight in May, from 83.00 to 122.09 mg/g in July and from 111.18 to 291.18 mg/g in September (Erturk *et al.*, 2010). Meanwhile, in this study, the content of phenols was 100.28 mg/g in May, 120.27 mg/g in July and 127.71 mg/g in the beginning of October. Such difference in the content of total phenols and flavonoids could be explained by ecological parameters. According to Erturk *et al.* (2010), sunlight and longer time of daylight has a possible positive effect on formation of phenolics in tea shoots. However, it is also reported (Erturk *et al.*, 2010) that in stress conditions (irregular rainfall, high fluctuations of temperature, UV irradiation, pathogens, etc.), plants produce more phenolic compounds.

Therefore, high content of phenols and flavonoids and their provided antioxidant properties could be used as desirable characters in the selection of seabuckthorn leaf and young shoot extracts for processing in food industry. According to the information collected by Singh (2006), for processing seabuckthorn leaves could be collected during all growing season.

CONCLUSION

Although,there were significant differences ($p < 0.05$) in chemical properties depending on harvesting time, obtained results showed uneven variations, therefore, it is not possible to declare the best time of seabuckthorn vegetative part collection. The highest level of total phenol and flavonoid contents, and antioxidant activities of water extracts was detected in October, while parts collected in July showed the second highest result in total phenol content, but the highest in antioxidant activities of ethanolic extracts. These two were followed by June samples, where total flavonoid content was high together with good results of DPPH activity in ethanolic extract.

The study showed negative and/or no correlation between water and ethanolic extracts within the same antioxidant assay, what could be explained by the presence of different antioxidant compounds and their solubility properties. In addition, already small amount of extracts were very active with regard to antioxidant potential, and therefore, it is suggested to test the activity more extensively concerning different methods, parameters, solvents, and ratios between extracts and solvents.

Hence, the detailed research on various parts of seabuckthorn is needed in particular phenolic and other antioxidant activity possessing compounds, and their changes during all growing season. Also, extraction technique, solvent type and impact of various process parameters on the chemical composition of extract should be examined. Still, the present study showed that vegetative parts of seabuckthorn plant due to high content of phenolic compounds and their possessed antioxidant properties have potential natural source for pharmacological, cosmetic and food processing industries.

ACKNOWLEDGEMENT

Research and publication was done in framework of ERAF project "*Use of Seabuckthorn Vegetative parts for Development of Food Products with High Antioxidant activity*" (2010/0246/2DP2.1.1.0/10/APIA/VIAA/159).

REFERENCES

1. Arimboor, R., Kumar, S.K. and Arumughan, C. 2008. Simultaneous estimation of phenolic acids in seabuckthorn (*Hippophae rhamnoides*) using RP-HPLC with DAD. *Journal of Pharmaceutical and Biomedical Analysis* 47: 31-38.
2. Badaluta, N., Zamfirache, M.M., Oprica, L., Olteanu, Z. and Ghiorghita, G. 2010. The monthly dynamics of the protein biosynthesis in the leaves harvested from *Hippophae rhamnoides* L. varieties. *Analele Stiintifice Ale Universitatii Sectiunea Ii A. Genetica si Biologie Moleculara* XI (4): 123-129. (In Romanian)
3. Bal, L.M., Meda, V., Naik, S.N. and Satya, S. 2011. Seabuckthorn berries: A potential source of valuable nutrients for nutraceuticals and cosmoceuticals. *Food Research International* 44: 1718-1727. *doi:10.1016/j.foodres.2011.03.002.*
4. Bekker, N.P. and Glushenkova, A.I. 2001. Components of certain species of the *Elaeagnaceae* family. *Chemistry of Natural Compounds* 37 (2): 97-116.
5. Benzie, I.F. and Strain, J.J. 1996. The ferric reducing ability of plasma (FRAP) as a measure of "Antioxidant Power": The FRAP assay. *Analytical Biochemistry* 239: 70-76.
6. Boonchum, W., Peerapornpisal, Y., Vacharapiyasophon, P., Pekkoh, J., Pumas, C., Jamjai, U., Amornlerdpison, D., Noiraksar, T. and Kanjanapothi, D. 2011. Antioxidant activity of some seaweed from the gulf of Thailand. *International Journal of Agriculture and Biology* 13 (1): 95-99.

7. Brand-Williams, W., Cuvelier, M.E. and Berset, C. 1995. Use of a free radical method to evaluate antioxidant activity. *Lebensmittel Wissenschaft und Technologie-Food Science and Technology* 28: 25-30.

8. Dudonne, S., Vitrac, X., Coutiere, P., Woillez, M. and Merillon, J.M. 2009. Comparative study of antioxidant properties and total phenolic content of 30 plant extracts of industrial interest using DPPH, ABTS, FRAP, SOD, and ORAC assays. *Journal of Agricultural and Food Chemistry* 57: 1768-1774.

9. Ercisli, S., Orhan, E., Ozdemir, O. and Sengul, M. 2007. The genotypic effects on the chemical composition and antioxidant activity of seabuckthorn (*Hippophae rhamnoides* L.) berries grown in Turkey. *Scientia Horticulturae* 115: 27–33.

10. Erturk, Y., Ercisli, S., Sengul, M., Eser, Z., Haznedar, A. and Turan, M. 2010. Seasonal variation of total phenolic, antioxidant activity and minerals in fresh tea shoots (*Camelia sinensis* var. *sinensis*). *Pakistan Journal of Pharmacological Science* 23: 69-74.

11. Escribano-Bailon, M.T. and Santos-Buelga, C. 2003. Polyphenols extraction from foods. In: *Methods in Polyphenol Analysis* (C. Santos-Buelga and G. Williamson, Ed.), p.1-16, Royal Society of Chemistry, Cambridge, UK, 380p.

12. Franco, D., Sineiro, J., Rubilar, M., Sanchez, M., Jerez, M., Pinelo, M., Costoya, N. and Nunez, M.J. 2008. Polyphenols from plant materials: Extraction and antioxidant power. *Electronic Journal of Environmental, Agricultural and Food Chemistry* 7 (8): 3210-3216.

13. Guliyev, V.B., Gul, M. and Yildirim, A. 2004. *Hippophae rhamnoides* L.: Chromatographic methods to determine chemical composition, use in traditional medicine and pharmacological effects. *Journal of Chromatography* 812: 291-307.

14. Kim, J.S., Kwon, Y.S., Sa, Y.J. and Kim, M.J. 2011. Isolation and identification of seabuckthorn (*Hippophae rhamnoides*) phenolics with antioxidant activity and α-glucosidase inhibitory effect. *Journal of Agricultural and Food Chemistry* 59 (1): 138-144.

15. Kondakova, V., Tsvetkov, I., Batchvarova, I.R., Badjakov, I., Dzhambazova, T. and Slavov, S. 2009. Phenol compounds: Qualitative index in small fruits. Biotechnology and *Biotechnological Equipment* 23 (4): 1444-1448.

16. Kumar, Y.M.S., Dutta, R., Prasad, D. and Misra, K. 2011. Subcritical water extraction of antioxidant compounds from seabuckthorn (*Hippophae rhamnoides*) leaves for the comparative evaluation of antioxidant activity. *Food chemistry* 127: 1309-1316.

17. Lee, H.I., Kim, M.S., Park, S.K., Seo, K.I., Kim, H.J., Kim, M.J., Choi, M.S. and Lee, M.K. 2011. Anti-visceral obesity and antioxidant effects of powdered seabuckthorn (Hippophae rhamnoides L.) leaf tea in diet-induced obese mice. *Food and Chemical Toxicology* 49 (9): 2370-2376.

18. Michel, T., Destandau, E., Le Floch, G., Lucchesi, M.E. and Elfakir, C. 2012. Antimicrobial, antioxidant and phytochemical investigations of seabuckthorn (*Hippophae rhamnoides* L.) leaf, stem, root and seed. *Food Chemistry* 131: 754-760.

19. Prabhu, R.A, Prasad, G.A., Manivannan, J., Elavarasan, S. and Balaji, S. 2011. Review on curative assets of seabuckthorn. *Journal of Pharmacy Research* 4 (1): 164-166.

20. Re, R., Pellegrini, N., Proteggente, A., Pannala, A., Yang, M. and Rice-Evans, C. 1999. Antioxidant activity applying and improved ABTS radical cation decolourization assay. *Free Radical Biology and Medicine* 26: 1231-1237.

21. Rivas-Arreola, M.J., Rocha-Guzman, N.E., Gallegos-Infante, J.A., Gonzalez-Laredo, R.F., Rosales-Castro, M., Bacon, J.R., Cao (Tsao), R., Proulx, A. and Intriago-Ortega, P. 2010. Antioxidant activity of oak (*Quercus*) leaves infusions against free radicals and their cardioprotective potential. *Pakistan Journal of Biological Sciences* 13 (11): 537-545.

22. Shunguang, L., Zhengping, J., Xiufeng, W., Keqin, W. and Dengming, L. 2003. Seabuckthorn breeding for the purpose of leaves utilization. In: *Proceedings of the 1st Congress of the International Seabuckthorn Association*, p.36-46, September 14-18, 2003, Berlin, Germany.

23. Singh, V. 2003. Geographical adaptation and distribution of seabuckthorn (*Hippophae* L.) resources. In: *Seabuckthorn (Hippophae L.)-A Multipurpose Wonder Plant. Vol. I. Botany, Harvesting and Processing Technologies* (V. Singh, Ed. in Chief, 2003), p. 21-34, Indus Publishing Company, New Delhi, 518p.

24. Singh, V. 2006. Free radicals, diseases, antioxidants and antioxidant properties of seabuckthorn (*Hippophae rhamnoides* L.). In: *Seabuckthorn (Hippophae L.)-A Multipurpose Wonder Plant. Vol. II: Biochemistry and Pharmacology* (V. Singh, Ed. in Chief, 2006), p. 3-69, Daya Publishing House, New Delhi, 600p.

25. Singleton,V.L., Orthofer, R. and Lamuela-Raventos, R.M. 1999. Analysis of total phenols and other oxidation substrates and antioxidants by means of Folin-ciocalteau reagent. *Methods in Enzymology* 299:152-178.

26. Sun, T. and Ho, C.T. 2005. Antioxidant activities of buckwheat extracts. *Food Chemistry* 90 (4): 743-749.

27. Tolkachev, O.N. and Sheichenko, O.P. 2006. Flavonoids of seabuckthorn (*Hippophae rhamnoides* L.): Chemistry and Pharmacology. In: *Seabuckthorn (Hippophae L.)-A Multipurpose Wonder Plant. Vol. II: Biochemistry and Pharmacology* (V. Singh, Ed. in Chief, 2006), p. 159-167, Daya Publishing House, New Delhi, 600p.

28. Tsao, R. 2010. Chemistry and biochemistry of dietary polyphenols. *Nutrients* 2: 1231-1246.

29. Upadhyay, N.K., Kumar, Y.M.S. and Gupta, A. 2010. Antioxidant, cytoprotective and anti-bacterial effects of seabuckthorn (*Hippophae rhamnoides* L.) leaves. *Food and Chemical Toxicology* 48: 3443-3448.

30. Wang, S.Y. and Lin, H.S. 2000. Antioxidant activity in fruits and leaves of blackberry, raspberry, and strawberry varies with cultivar and developmental stage. *Journal of Agricultural and Food Chemistry* 48: 140-146.

31. Zu, Y., Li, C., Fu, Y. and Zhao, C. 2006. Simultaneous determination of catechin, rutin, quercetin, kaempferol and isorhamnetin in the extract of seabuckthorn (*Hippophae rhamnoides* L.) leaves by RP-HPLC with DAD. *Journal of Pharmaceutical and Biomedical Analysis* 41: 714-719.

Chapter 37

Antimicrobial Properties of Seabuckthorn (*Hippophae* L.)

Rajesh Chahota and Mandeep Sharma

Department of Veterinary Microbiology, DGCN COVAS,
CSK Himachal Pradesh Agricultural University, Palampur, India

SUMMARY

Seabuckthorn (*Hippophae* L.), has recently gained interest for its nutritional and medicinal values. Ancient literature offers abundant information about medicinal and health beneûts of seeds, oils, leaves and berries of *H. rhamnoides* and other *Hippophae* species. In recent years, research impetus towards studying its antimicrobial properties is gaining momentum, as antimicrobial resistance to antimicrobial agents has become an increasingly important and pressing global problem due to rampart use of antibiotics and antimicrobials. Seabuckthorn seeds oil, leaf extract, stem, roots, bark and berries are now observed to have antimicrobial properties, and those can have application in medical, pharmaceutical and food industry. Different parts of seabuckthorn are rich in various biomolecules including polyphenolic compounds/avonoids, phytosterols, proanthocyanidins and other compounds with antimicrobial actions. The antioxidant potential of the leaves of *H. rhamnoides* are correlated to avonoids and phenolic acids derivatives present in it. Lately, antimicrobial effects of seabuckthorn extracts/oils has been tested by many research groups and confirmed by various *in-vitro* and *in-vivo* studies. Further studies are needed for the characterization of individual compounds to elucidate the underlying mechanisms, bioactive properties and the existence of possible synergism among other compounds present in seabuckthorn. These medicinal properties of seabuckthorn can effectively replace the modern medicine for treating various health related issues in future.

Keywords: *Seabuckthorn, Biomolecules, Seed oil, In vitro and In vivo antimicrobial studies.*

INTRODUCTION

The plants are the oldest resources of useful drugs that mankind has used since time memorial. The history of the medicine and pharmacy begins with Hippocrates, the father of medicine, who

Seabuckthorn (Hippophae L.): A Multipurpose Wonder Plant, Vol. 4: Emerging Trends in Research and Technologies (Virendra Singh, *Editor-in-Chief*, 2014), p. 454–464, Daya Publishing House, New Delhi, India.

mentioned nearly 400 plant samples as medicinal substance. India has rich diversity of medicinally important herbal flora. Globally, efforts have been made to find out active phytochemicals as alternative medicine for treatment of various ailments in human and animals. This trend of using herbal formulations is increasing in recent years, not only for therapeutic purpose but also as prophylactic and health promotive agents.

Seabuckthorn (*Hippophae* L.) belonging to family *Elaeagnaceae*, a unique and versatile plant has recently gained worldwide attention, mainly for its medicinal and nutritional potential. Seabuckthorn has been used in traditional Chinese medicine since the Tang Dynasty, going back more than 1000 years. In-depth survey and documentation of indigenous ethnobotanical knowledge of seabuckthorn reveal that this plant was traditionally utilized by local people of Asia, Nordic countries and the Baltic region in multidimensional aspects of food, fuel, medicine, veterinary, agricultural tools and bio-fencing (Yang *et al.*, 2000; Dhyani *et al.*, 2010). The current research is now beginning to understand and support the traditional uses of seabuckthorn. A wide spectrum of pharmacological effects of SBT have been recently reported, including antioxidant, immunomodulatory, antiatherogenic, antistress, hepatoprotective, radioprotective and tissue repair (Geetha *et al.*, 2002; Goel *et al.*, 2002; Suleyman *et al.*, 2002; Xing *et al.*, 2002; Gao *et al.*, 2003; Gupta *et al.*, 2006; Basu *et al.*, 2007; Saggu and Kumar, 2007; Upadhyay *et al.*, 2010, 2011).

The present chapter will be focusing specifically on the antimicrobial properties of SBT plant as well as different methodologies used for analysis of these properties and its significance in modern medical and veterinary sciences owing to its medicinal as well as pharmacological activities.

MEDICINAL USES OF SEABUCKTHORN

The medicinal and nutritional value of seabuckthorn for human was recorded as far back as the 8th century in the Tibetan classic "*rGyud Bzi*" and in Sibu Yidian from the Tang Dynasty and Jing Zhu Ben Cao from the Qing Dynasty. The medicinal plant has also been mentioned in the writings of ancient Greek as proven remedy for horses. The genus name Hippophae is classical Latin for "*shining horse*," a name that was given in ancient times after it was found that feeding the leaves to horses improved their health and made their hair shiny. The use of seabuckthorn has primarily been limited to human medicine. The scarce efforts have been made to exploit the therapeutic potential of seabuckthorn in animal diseases.

In recent studies, seabuckthorn with modern scientific techniques has been proven to be effective against various ailments of infectious and metabolic nature. Its juices, oil and leaf extract is said to have properties to combat the modern age life style diseases. Seabuckthorn has got tremendous potential for therapeutic utilization in human and veterinary medicine, as it is reported to possess many bioactive substances notably lipids, fatty acids, vitamins (both fat soluble and water soluble), flavonoids, tannins, phenols, steroids, 5-hydroxytryptamine etc. Besides, antimicrobial and antioxidant activities seabuckthorn also helps in controlling infection by strengthen non-specific immunity as has been observed in laboratory mice and guinea pigs by Zhong *et al.* (1989) with the compound extracts of *Hippophae* that strengthen the phagocytic function of the mouse macrophagocyte, increased serum-lysozyme in mice and alexin in guinea pigs. Thus, seabuckthorn along with directly affecting bacterial, viral and mycotic agent through various phytochemicals, may also acts as immunostimulant for combating the infections in humans and animals. Clinical tests on medicinal uses were first initiated in Russia during the 1950s.

The most important pharmacological functions attributed to seabuckthorn oil are: anti-inflammatory, antimicrobial, pain relief, and promoting regeneration of tissues. More than ten different drugs have been developed from seabuckthorn in Asia and Europe and are available in different forms such as liquids, powders, plasters, films, pastes, pills, liniments, suppositories and aerosols. A new drug preparation, Hiporamin, a purified tannin fraction from seabuckthorn has been found to possess wide spectrum antiviral and antibacterial activity (Shipulina *et. al.*, 2005).

MEDICINALLY IMPORTANT PARTS OF SEABUCKTHORN PLANT

Seabuckthorn is a thorny nitrogen-ûxing deciduous shrub of cold arid region native to Asia and Europe. In India, it grows in high altitude regions of Himachal Pradesh, Uttrakhand, and some North-Eastern States. In Himachal Pradesh, three species of seabuckthorn namely *H. rhamnoides*, *H. salicifolia* and *H. tibetana* have been identified, out of which *H. rhamnoides* is the predominant species. *Hippophae rhamnoides* has had been used by Chinese and Tibetan people as a traditional medicine since time immemorial. It is currently domesticated in several parts of the world due to its nutritional and medicinal values (Rousi, 1971; Li, 2003). Recent work has shown that there is variation in availability of useful bioactive compounds among the prevalent species of seabuckthorn (Sharma *et al.*, 2011). All parts of this plant are considered to be a good source of large number of bioactive substances but the following components of the plant have been studied experimentally and found to have bioactive molecules in high concentrations.

Fruits/Berries

Traditionally, fruits or berries of seabuckthorn owing to its hemostatic and anti-inammatory effects were added to prescriptions in pulmonary gastrointestinal, cardiac, blood and metabolic disorders (Gammerman, 1982; Bazaron and Aseeva, 1984; Badareva, 1985) in Indian and Tibetan medicine. At present time, few studies have demonstrated that *H. rhamnoides* L. has anti-microbial and anti-tumoral effects. Phenolic compounds of *H. rhamnoides* L. berry inhibit the growth of Gram-negative bacteria (Puupponen-Pimia *et al.*, 2001).

Leaves

The extracts of *H. rhamnoides* L. branches and leaves are used to treat colitis and enterocolitis in humans and animals in Mongolia (Tsybikova *et al.*, 1983). Branches and leaves are also used in the treatment of diarrhea (Vereshchagin *et al.*, 1959). Leaves are used in gastrointestinal and dermatologic disorders (Tsybikova *et al.*, 1983). In recent studies, both aqueous and methanolic extracts of the seabuckthorn leaves have been experimentally tested against different pathogenic and non-pathogenic bacteria. In one of such study, aqueous and hydroalcoholic leaf extracts of seabuckthorn showed growth inhibiting effect against *Bacillus cereus*, *Pseudomonas aeruginosa*, *Staphylococcus aureus* and *Enterococcus faecalis* (Upadhyay *et al.*, 2010). The seabuckthorn leaf extract also showed a signiûcant anti-dengue activity when evaluated against Dengue virus type-2 infected blood-derived human macrophages with a decrease and increase in TNF-α and IFN-γ, respectively (Jain *et al.*, 2008). Recently, leaf extract of *H. salicifolia* was found to have intermediate activity against both gram-positive bacteria (Gupta *et al.*, 2011). However, in one of the study, *Salmonella typhimurium* strain was found to be resistant against all tested crude extracts of seabuckthorn (*H. rhamnoides* L.) derived from pomace, seeds and leaves (Arora *et al.*, 2012). Comparative antimicrobial activities of leaf extracts of *H. rhamnoides* and *H. salicifolia* species of seabuckthorn against bacterial pathogens associated with skin and wound infections of livestock was done and better antimicrobial effects of leaf extract of *H. salicifolia* was observed (Sharma *et al.*, 2011). Another study on effects of seabuckthorn leaf extract on dermal fungi revealed that the inhibitory effect of seabuckthorn leaf extract at 5 per cent concentration was of almost 80 per cent in 12 h of reading as compared to the standard drugs used as positive control against fungal isolates (Verma *et al.*, 2013).

Stem

Antimicrobial activity of the extracts and fractions from the stem of seabuckthorn were evaluated by determining the minimum inhibitory concentration (MIC) values and superior antimicrobial activity was observed as compared to known antimicrobial agents like catechin, ketoconazol, and mycostantin (Jeong *et al.*, 2010). This study indicates that the plant stem contain a variety of compounds contributing to antimicrobial activity, which could be used for food additives and the development of useful natural compounds. While in another study, the stem of seabuckthorn was found less effective antimicrobial organ (Michael *et al.*, 2012).

Root

The extracts of the root also showed enhance antimicrobial activity when compared with other antimicrobial agents (Jeong *et al.*, 2010), which indicates that the plant root also have an array of compounds contributing to antimicrobial activities. Extract from root showed 63 ± 6 per cent inhibition for *Enterococcus durans* in another study (Michael *et al.*, 2012).

Seeds and Seed Oil

The oil from seeds and fruits is used in the treatment of eczema, lupus erythematosus, chronic wounds that are difcult to heal, inammatory diseases, erosion of the cervix uteri, in the treatment of burns and frozen parts of the body (Chopik *et al.*, 1983). Also, its berry oil is reported to treat skin disease and thrombosis (Cheng *et al.*, 2003). Oil extracts are used externally in dermatologic diseases such as eczema, psoriasis, lupus erythematosus, and chronic dermatoses and in ophthalmology for treatment of keratitis, trachoma, injuries or burns of eye-lid, conjunctivitis (Guliyev *et al.*, 2004).

The aqueous extract of seabuckthorn seeds has been screened for antibacterial and antioxidant activities. The extract was also found to possess antibacterial activity against *Listeria monocytogenes* and *Yersinia enterocolitica* (Chauhan *et al.*, 2007). The antimicrobial effect of the extract implicates its potential for natural preservation. It has also been shown that seabuckthorn oil had effects on anathrepis, eliminating inflammation and easing pain, promoting immune function and strengthening body resistance (Mingyu *et al.*, 1993). Seed extract of *H. salicifolia* also showed significant activity against both gram-positive bacteria, whereas no activity was observed against gram-negative bacteria except in case of *Agrobacterium tumefaciens* and in case of fungal cultures, seed extract also showed significant activity against *Mucor* and *Tilletia* fungi, while in case of *Rhizopus*, no activity was observed (Gupta *et al.*, 2011). Effects of seabuckthorn seed oil has also been evaluated against drug resistant *Staphylococcus aureus* field strains isolated from the purulent wounds and the animals skin infection and noteworthy growth inhibitions was observed (Chahota *et al.*, 2010).

Extracts from seabuckthorn seeds inhibited the growth of *Bacillus cereus, Bacillus coagulans, Bacillus subtilis, Listeria monocytogenes* and *Yersinia enterocolitica* (Puupponen-Pimia *et al.*, 2001; Negi *et al.*, 2005). Ethanol extracts of seabuckthorn inhibited the growth of *Helicobacter pylori* at a MIC around 60 mcg/ml (Li *et al.*, 2005). Effects of seabuckthorn different types of seed extracts on growth of different bacteria was also studies and it was found that methanolic extract was found to be the most effective, followed by chloroform and acetone extracts (Negi *et al.*, 2005). Higher resistance of Gram-negative bacteria to external agents has been earlier reported, and it is attributed to the presence of lipopolysaccharides in their outer membranes, which make them inherently resistant to antibiotics, detergent and hydrophilic dyes (Nikaido and Vaara, 1985). The reason for higher sensitivity of the Gram-positive bacteria than Gram-negative bacteria could be ascribed to the differences between their cell wall compositions. The Gram-positive bacteria contain an outer peptidoglycan layer, which is an ineffective permeability barrier (Scherrer and Gerhardt, 1971). Similar trend for inhibition of bacterial growth have been observed with seabuckthorn berry extracts.

Pomace and Bark

In one of the study, moderate antibacterial property of aqueous and methanolic extracts of seabuckthorn pomace, was observed against some microorganisms like *E. coli, E. faecalis* and *S. aureus* (Arora *et al.*, 2012). The bark of seabuckthorn has been used for treating diarrhoea in traditional medicine.

ANTIMICROBIAL PHYTOCHEMICALS IN SEABUCKTHORN

Since time immemorial, man has used various parts of plants in the treatment and prevention of various ailments (Tanaka *et al.*, 2002). In recent years, secondary plant metabolites (phytochemicals), previously with unknown pharmacological activities, have been extensively investigated as a source of medicinal agents (Krishnaraju *et al.*, 2005). Thus, it is anticipated that phytochemicals with adequate antibacterial efficacy may be used for the treatment of bacterial infections (Balandrin *et al.*, 1985). The antimicrobial activity of plant extracts may reside in a variety of different components, including aldehyde and phenolic compounds (Lai and Roy, 2004). Naturally occurring combinations of these compounds can be synergistic and often result in extracts having greater antimicrobial activity than the purified individual constituents (Delaquis *et al.*, 2002).

Adequate presence of different bioactive substances in various parts of seabuckthorn contributes enormously to its various microbiological/therapeutic properties like antibacterial, antifungal, immunostimulant, anti-inflammatory, antiradiation, antioxidant, hepatoprotective, cytoprotective, anti-cancerous, anti-artherosclerosis, anti-diabetic, anti-ulcer and wound healing actions. Following classes of phytochemicals are reported to play important role in antimicrobial properties in seabuckthorn most probably in combinations:

Phenolic/Polyphenolic Compounds

Flavonoids are ubiquitous in photosynthesising cells and are commonly found in fruit, vegetables, nuts, seeds, stems and owers of plants. For centuries, preparations containing these compounds as the principal physiologically active constituents have been used to treat human diseases. Increasingly, this class of natural products is becoming the subject of anti-infective research, and many groups have isolated and identied the structures of avonoids possessing antifungal, antiviral and antibacterial activity. It is reported that phenolic compounds from the berries of seabuckthorn inhibit the growth of gram-negative but not gram-positive bacteria. Seabuckthorn fruits (berries) have a complex composition, containing a mixture of different components including different phenolic compounds. The bioactive components vary with fruit maturity, fruit size, species, geographic locations, climate and methods of extraction (Zeb, 2004; Leskinen *et al.*, 2010). The berries are also a rich source of compounds like multiple vitamins (C and E), carotenoids (β-carotene, lycopene, lutein and zeaxanthin), avonoids (isorhamnetin, quercetin, isorhamnetin-3-β- D-glucoside, isorhamnetin-3-β-D-glucosaminide, kaempferol) organic acids, amino acids, micro and macronutrients (Kallio *et al.*, 2002; Yang and Kallio, 2002).

Moreover, several groups have demonstrated synergy between active avonoids as well as between avonoids and existing chemotherapeutics. The reports of activity in the eld of antibacterial avonoid research are widely conicting, probably owing to inter-and intra-assay variation in susceptibility testing. However, several high-quality investigations have examined the relationship between avonoid structure and antibacterial activity and these are in close agreement. In addition, numerous research groups have sought to elucidate the antibacterial mechanisms of action of selected avonoids. The activity of quercetin, for example, has been at least partially attributed to inhibition of DNA gyrase. It has also been proposed that sophoraavone G and (")-epigallocatechin gallate inhibit cytoplasmic membrane function, and that licochalcones A and C inhibit energy metabolism. Other avonoids whose mechanisms of action have been investigated include robinetin, myricetin, apigenin and rutin (Cushnie and Lamb, 2005).

Phytosterols and Fatty Acids

Phytosterol present in vegetable oil are considered to have antibacterial and antiinflammatory properties (Beveridge *et al.*, 2002). In seabuckthorn, sitosterol and avenasterol are quantitatively the most important phytosterols (Li *et al.*, 2007). Sitosterol was identied as the major constituent of phytosterols in the seabuckthorn oil (Moersel *et al.*, 2005) but their specific antimicrobial action has not been studied yet. In addition to the plethora of antioxidants, the berries of seabuckthorn are also rich in fatty acids (saturated 13.7 per cent and 86.3 per cent unsaturated) including palmitic acid, oleic acid (omega-9), palmitoleic acid (omega-7), linoleic acid (omega-6), and linolenic acid (omega-3); and phytosterols. Many bioactive compounds were isolated from the berries of seabuckthorn such as hippophae cerebroside, oleanolic acid, ursolic acid, 19-alpha-hydroxyursolic acid, dulcioic acid, 5-hydroxymethyl-2-furancarbox-aldehyde, cirsiumaldehyde, octacosanoic acid, palmitic acid and 1-O-hexadecanolenin (Zheng *et al.*, 2009).

Antimicrobial Assays

For studying the antimicrobial properties of seabuckthorn parts, various methods are used depending upon the nature of the seabuckthorn product. Mostly methanolic or ethanolic extracts are obtained from powdered leaves, seed or stem. After drying of leaves at 37 °C for 3 to 5 days grinding is recommended without exposure to sunlight to prevent the loss of active compounds. Alcoholic extraction methods include hot extraction with soxlet apparatus for 8 hours in at 60 °C and cold extraction method by keeping alcoholic and powdered suspension 24 hours on orbital shaker at 25-30°C.

The oil from seed can also be obtained by cold press method or CO_2 extraction methods. There is variation in desired contents obtained by different methods. Here is the brief account of various techniques used to check susceptibility of various bacterial, mycotic agents against seabuckthorn products:

1) *In-vitro* Studies

For different methods listed below standard protocols are used.

1. Disc diffusion test: Inhibition of microbial growth can be tested using the sterile paper disc diffusion method by adding predetermined quantity of extracted materials. This method is preferred for leaf and other water-soluble extracts, as it needs to diffuse agar gel.
2. Agar gel dilution method: In this method seabuckthorn oil in mixed with low concentrated slurry and mixed them with agar gel so that oil is spread throughout the agar for better contact with test microorganisms.
3. Agar diffusion method: In this method, holes are punched in the agar gel and test material (preferably water soluble) is allowed to diffuse to surrounding to inhibit microbial growth.

4. Minimum inhibitory concentration: MIC was determined by the dilution (both micro and macro) methods using tubes or 96 well microtitre plate. In this method bacterial growth inhibition is determined either by counting bacterial colony forming units (cfu) or by measuring optical density (OD) of the suspension.

2) *In-vivo* Studies

For testing the effects of seabuckthorn products/extracts, different animals species are used. Effects of leaf extract on wound healing in rabbit (Verma *et al.*, 2012); in calves (Mahajan *et al.*, 2005) and flavone from seabuckthorn in rats (Gupta *et al.*, 2006). Effects of seabuckthorn seed oil on healing of gastric ulcers have also been tested on dogs (Varshney *et al.*, 2005). All these studies concluded that seabuckthorn leaf extracts and seed oil has beneficial effects on wound healing on cutaneous and gastric mucosa.

CONCLUSION

Development of drug resistance by different bacterial pathogens against the available drugs/ antibiotics is becoming the global problem. Finding new antimicrobial drugs to treat such infections is a great challenge to mankind. Exploration of different phytochemicals for their possible antimicrobial properties can provide answer for future drug development. Seabuckthorn is a versatile plant with multiple health promoting and antimicrobial properties. Its leaves, fruits, seeds and roots are reported to have bioactive molecules with antimicrobial activities. Limited studies worldwide have shown that various seabuckthorn products exhibited antibacterial properties against Gram negative and Gram positive pathogenic and non-pathogenic bacteria. Besides antibacterial activities, its antiviral and antimycotic properties have also been reported. Thus, seabuckthorn plant and its products can be examined in detail for futuristic drug development, thus addressing the problem of emerging drug resistance.

REFERENCES

1. Arora, R., Mundra, S., Yadav, A., Srivastava, R.B. and Stobdan, T. 2012. Antimicrobial activity of seed, pomace and leaf extracts of seabuckthorn (*Hippophae rhamnoides* L.) against foodborne and food spoilage pathogens. *African Journal of Biotechnology* 11(45): 10424-10430.
2. Badareva, B.B. 1985. *Monument of Tibetian medicine-"Dzeytsxar migch-jan"*, Novosibirsk. (In Russian)
3. Balandrin, M.F, Kjocke, A.J., Wurtele, E. Bllinger, W.H. 1985. Natural plant chemicals: Sources of industrial and mechanical materials. *Science* 228: 1154-1160.
4. Basu, M., Prasad, R., Jayamurthy, P., Pal, K., Arumughan, C. and Sawhney, R.C. 2007. Anti-atherogenic effects of seabuckthorn (*Hippophae rhamnoides*) seed oil. *Phytomedicine* 14: 770-777.
5. Bazaron, E.G. and Aseeva, T.A. 1984. *Thesis of Indo-Tibetian medicine-"Vaydurya-onbo"*, Novosibirsk, 118p. (In Russian)
6. Beveridge, T.H.J., Li, T.S.C. and Drover, J.C.G. 2002. Phytosterol content in American ginseng seed oil. *Journal of Agricultural and Food Chemistry* 50 (4): 744-750.
7. Guliyev, V.B., Gul, M. and Yildirima, A. 2004. *Hippophae rhamnoides* L.: Chromatographic methods to determine chemical composition use in traditional medicine and pharmacological effects. *Journal of Chromatography B* 812: 291-307.
8. Chahota, R., Palial, A., Verma, H. and Sharma, M. 2010. Evaluating antimicrobial properties of seabuckthorn (*Hippophae rhamnoides* L.) seed oil against strains of *Staphylococcus* spp. commonly

involved in animal skin infections. In: *Proceedings of Annual Conference of Indian Association for the Advancement of Veterinary Research (IAAVR)*, College of Veterinary Science and Animal Husbandry, March 11-12, 2010, MP, India.

9. Chauhan, A.S., Negi, P.S. and Ramteke, R.S. 2007. Antioxidant and antibacterial activities of aqueous extract of seabuckthorn (*Hippophae rhamnoides*) seeds. *Fitoterapia* 78: 590-592.
10. Cheng J.Y., Kondo K., Suzuki Y., Ikeda Y., Meng X.S. and Umemura K. 2003. Inhibitory effects of total flavones of *Hippophae rhamnoides* L. on thrombosis in mouse femoral artery and *in vitro* platelet aggregation. *Life Sci.*72: 2263-2271.
11. Chopik, V.I., Dudchenko, L.G. and Krasnova, A.N. 1983. *Wild Medicinal Plants of Ukrain Kiev*. (In Russian)
12. Cushnie T.P.T. and Lamb, A.J. 2005. Antimicrobial activity of avonoids. *International Journal of Antimicrobial Agents* 26: 343-356.
13. Delaquis, P.J., Stanich, K., Girard, B. and Mazza, G. 2002. Antimicrobial activity of individual and mixed fractions of dill, cilantro, coriander and eucalyptus essential oils. *Int. J. Food Microbiol.* 74: 101-109.
14. Dhyani, D., Maikhuri, R.K., Misra, S. and Rao, K.S. 2010. Endorsing the declining indigenous ethnobotanical knowledge system of seabuckthorn in Central Himalaya. *Indian Journal of Ethnopharmacology* 127: 329-334.
15. Gao, Z.L., Gu, X.H., Cheng, F.T. and Jiang, F.H. 2003. Effect of seabuckthorn on liver brosis: A clinical study. *World Journal of Gastroenterology* 9: 1615-1617.
16. Gammerman, A.F. 1982. *Materials to Learn the Resources of Traditional System of Indo-Tibetian Medicine*, Novosibirsk, 53p. (In Russian)
17. Geetha, S., Sai Ram, M., Singh, V., Ilavazhagan, G. and Sawhney, R.C.2002. Antioxidant and immunomodulatory properties of seabuckthorn (*Hippophae rhamnoides*)-An *in vitro* study. *Journal of Ethnopharmacology* 79: 373-378.
18. Goel, H.C., Prasad, J., Singh, S., Sagar, R., Prem Kumar, I. and Sinha, A.K. 2002. Radioprotection by a herbal preparation of *Hippophae rhamnoides* RH-3, against whole body lethal irradiation in mice. *Phytomedicine* 9: 135-143.
19. Gupta, A., Kumar, R., Pal, K., Singh, V., Banerjee, P.K. and Sawhney R.C. 2006. Influence of (*Hippophae rhamnoides*) flavone on dermal wound healing in rats. *Molecular and Cellular Biochemistry* 290: 193-198.
20. Gupta, S.M., Gupta, A.K., Ahmed, Z. and Kumar, A. 2011. Antibacterial an antifungal activity in leaf, seed extract and seed oil of seabuckthorn (*Hippopha salicifolia* D. Don) Plant. *J. Plant Pathol. Microbiol.* 2:105.
21. Verma, H., Chahota, R., Palial, A., Patil, R.D., Sharma, A., Kuradey, N.P. and Sharma, M. 2012. Effect of seabuckthorn (*Hippophae* L.) leaf extract and seed oil on infected cutaneous wound healing process in rabbit model. *Ind. J. Anim. Sciences*. 82 (1): 58-60.
22. Jain, M., Ganju, L., Katiyal, A., Padwad, Y., Mishra, K.P., Chanda, S., Karan, D., Yogendra, K.M. and Sawhney, R.C. 2008. Effect of *Hippophae rhamnoides* leaf extract against Dengue virus infection in human blood-derived macrophages. *Phytomedicine* 15: 793-799.

23. Jeong J.H., Lee, J.W., Kim, K.S., Kim, J., Han, S.N., Yu, C.Y., Lee, J.K., Kwon, Y.S. and Kim, M.J. 2010. Antioxidant and antimicrobial activities of extracts from a medicinal plant, seabuckthorn. *J. Korean Soc. Appl. Biol. Chem.* 53(1): 33-38.

24. Kallio, H., Yang, B. and Peippo, P. 2002. Effects of different origins and harvesting time on vitamin C, tocopherols and tocotrienols in seabuckthorn (*Hippophae rhamnoides* L.) berries. *Journal of Agriculture and Food Chemistry* 50: 6136-6142.

25. Krishnaraju, A.V., Rao, T.V.N., Sundararaju, D., Vanisree M., Tsay H.B., Subbaraju G.V. 2005. Assessment of bioactivity of Indian medicinal plants using Brine shrimp (*Artemia salina*) lethality assay. *Int. J. Appl. Sci. Eng.* 2: 125-134.

26. Lai, P.K. and Roy, J. 2004. Antimicrobial and chemopreventive properties of herbs and spices. *Curr. Med. Chem.* 11: 1451-1460.

27. Larmo, P., Alin, J., Salminen, E., Kallio, H. and Tahvonen, R. 2008. Effects of sea buckthorn berries on infections and inflammation: A double-blind, randomized, placebo-controlled trial. *Eur. J. Clin. Nutr.* 62 (9): 1123-1130.

28. Padwad, Y., Ganju, L., Jain, M., Chanda, S., Karan, D., Banerjee, P.K. and Chand Sawhney, R.C. 2005. Effect of leaf extract of seabuckthorn on lipopolysaccharide induced inflammatory response in murine macrophages. *Int. Immunopharmacol*. 6 (1): 46-52.

29. Leskinen, H.M., Suomela, J.P., Yang, B. and Kallio, H.P. 2010. Regioisomer compositions of vaccenic and oleic acid containing triacylglycerols in seabuckthorn (*Hippophae rhamnoides*) pulp oils: Inuence of origin and weather conditions. *Journal of Agriculture and Food Chemistry* 58: 537-545.

30. Li, T.S.C. 2003.Taxonomy, natural distribution and botany. *In*: *Seabuckthorn (Hippophae rhamnoides L.): Production and Utilization* (T.S.C. Li and T. Beveridge, (Eds.), p.7-11, NRC Research Press, Ottawa, ON.

31. Li, T.S.C., Beveridge, T.H.J and Drover, J.C.G. 2007. Phytosterol content of seabuckthorn (*Hippophae rhamnoides* L.) seed oil: Extraction and identification. *Food Chemistry* 10: 1633-1639.

32. Li. Y., Xu, C., Zhang, Q., Liu, J.Y. and Tan, R.X. 2005. *In vitro* anti-*Helicobacter pylori* action of 30 Chinese herbal medicines used to treat ulcer diseases. *J Ethnopharmacol*. 98: 329-333.

33. Mahajan, A., Sharma, S.K., Asrani, R.K., Varshney, A.C., Tyagi, S.P. and Amit, K. 2005. Efficacy of seabuckthorn (*Hippophae rhamnoides*) in the healing of infected cutaneous wounds in calves. *Indian Journal of Veterinary Surgery* 26: 31-33.

34. Michel, T., Destandau, E., Floch, G.L., Lucchesi, M.E. and Elfakir, C. 2012. Antimicrobial, antioxidant and phytochemical investigations of seabuckthorn (*Hippophae rhamnoides* L.) leaf, stem, root and seed. *Food Chemistry* 131: 754-760.

35. Mingyu, X. 1993. A brief report on an anti-bacterial experiment using seabuckthorn oil. *Hippophae* 6 (2): 28-29.

36. Moersel, T., Heilscher, K. and Morsel, C., 2005. Quality parameters of Seabuckthorn (*Hippophae rhamnoides* L.) oil. In: *Seabuckthorn (Hippophae L.)-A Multipurpose Wonder Plant. Vol. II: Biochemistry and Pharmacology* (V. Singh, Ed. in Chief, 2006), p.337-341, Daya Publishing House, New Delhi, 600p.

37. Negi, P., Chauhan, A., Sadia, G., Rohinishree, Y. and Ramteke, R. 2005. Antioxidant and antibacterial activities of various seabuckthorn (*Hippophae rhamnoides* L.) seed extracts. *Food Chem.* 92: 119-124.

38. Nikaido, H. and Vaara, M. 1985. Molecular basis of bacterial outer membrane permeability. *Microbiological Reviews* 49: 1-32.

39. Padwad, Y., Ganju, L., Jain, M., Chanda, S., Karan, D., Banerjee, P.K. and Chand Sawhney, R.C. 2005. Effect of leaf extract of seabuckthorn on lipopolysaccharide induced inflammatory response in murine macrophages. *Int. Immunopharmacol*. 6 (1): 46-52.

40. Puupponen-Pimia, R., Nohynek, L. and Meier, C. 2001. Antimicrobial properties of phenolic compounds from berries. *J. Appl. Microbiol*. 90: 494-507.

41. Rousi, A. 1971. The genus *Hippophae* L., a taxonomic study. *Annals Botanica Fennici* 8, 177-227.

42. Saggu, S. and Kumar, R. 2007. Possible mechanism of adaptogenic activity of seabuckthorn (*Hippophae rhamnoides*) during exposure to cold, hypoxia and restraint (C–H–R) stress induced hypothermia and post stress recovery in rats. *Food and Chemical Toxicology* 45: 2426-2433.

43. Scherrer, R. and Gerhardt, P. 1971. Molecular sieving by the *Bacillus megatrium* cell wall and protoplast. *Journal of Bacteriology* 107: 718-735.

44. Sharma, M., Chahota, R., Palial, A. and Singh, V. 2011. Comparative antimicrobial activities of leaf extracts of *Hippophae rhamnoides* and *Hippophae salicifolia* species of seabuckthorn against bacterial pathogens associated with skin and wound infections of livestock of Himachal Pradesh. In: *Proceedings of Nation Seabuckthorn Conference*, p.136-140, HPKV, Palampur, HP, India.

45. Shipulina, L.D., Tolkachev, O.N., Krepkova, L.V., Bortnikova, V.V. and Shkarenkov, A.A. 2005. Anti-viral anti-microbial and toxicological studies on Seabuckthorn (*Hippophae rhamnoides*). In: *Seabuckthorn (Hippophae L.)-A Multipurpose Wonder Plant. Vol. I: Botany, Harvesting and Processing Technologies* (V. Singh, Ed. in Chief, 2003), p.471–483, Indus Publishing Company, New Delhi, 518p.

46. Suleyman, H., Demirezer, L.O., Buyukokuroglu, M.E., Akcay, M.F., Gepdiremen, A., Banoglu, Z.N. and Gocer, F. 2001. Antiulcerogenic effect of *Hippophae rhamnoides*. *Phytotherapy Research* 33: 77-81.

47. Tanaka, H., Sato, M. and Fujiwara, S. 2002. Antibacterial activity of isoflavonoids isolated from *Erythrina variegata* against methicillin-resistant *Staphylococcus aureus*. *Lett Appl Microbiol* 35: 494-498.

48. Tsybikova, D.T., Rasputina-Sultumova, D.B., Komissarenko, N.F. and Bolotova, M.N. 1983. Utilization problems of medicinal resources in Siberia and Far East, Novosibirsk. (In Russian)

49. Upadhyay, N.K., Kumar, M.S.Y. and Gupta, A., 2010. Antioxidant, cytoprotective and anti-bacterial effects of seabuckthorn (*Hippophae rhamnoides* L.) leaves. *Food and Chemical Toxicology* 48: 3443-3448.

50. Upadhyay, N.K., Kumar, R., Siddiqui, M.S. and Gupta, A. 2011. Mechanism of wound healing activity of *Hippophae rhamnoides* L. leaf extract in experimental burns. *Evidence based Complementary and Alternative Medicine.doi*:10.1093/ecam/nep 189.

51. Varshney, A.C., Tyagi, S.P., Kumar, A., and Singh, V. 2005. Studies for the evaluation of anti-inflammatory activities of seabuckthorn (*Hippophae rhamnoides*) oil in canine. *Global Seabuckthron Research* and *Development* **2**: 7-10.

52. Vereshchagin, B.A., Sobolevskaya, K.A. and Yakubov, A.N. 1959. *Useful Plants of Western Siberia.* (In Russian)

53. Verma, H., Sharma, M., Chahota, R. and Palial, A. 2013. Assessment of antimycotic activity of seabuckthorn (*Hippophae rhamnoides*) leaf extracts against common fungi associated with skin dermatitis. *Vet. World* 6 (4): 205-208.

54. Yang, B., Kalimo, K.O., Tahvonen, R.L., Mattila, L.M., Katajisto, J.K., Kallio, H.P.2000. Effect of dietary supplementation with seabuckthorn (*Hippophae rhamnoides*) seed and pulp oils on the fatty acid composition of skin glycerophospholipids of patients with atopic dermatitis. *J. Nutr. Biochem*. 11: 338-340.

55. Yang, B., Kalimo, K.O., Tahvonen, R.L., Mattila, L.M., Katajisto, J.K. and Kallio, H.P. 2000. Effect of dietary supplementation with sea buckthorn (*Hippophae rhamnoides*) seed and pulp oils on the fatty acid composition of skin glycerophospho-lipids of patients with atopic dermatitis. *Journal of Nutritional Biochemistry* 11: 338-340.

56. Yang, B. and Kallio, H. 2002. Supercritical CO_2-extracted seabuckthorn (*Hippophae rhamnoides*) oils as new food ingredients for cardiovascular health. In: *Proceedings of Health Ingredients of Europe*, Paris, France, 7p.

57. Zhong Fei *et al*., 1989. Study on the immunopharmacology of the components extracted from *Hippophas rhamnoides* L. In: *Proceedings of the International Symposium on Seabuckthorn*, 368-370, Xi'an, China.

58. Xing, J., Yang, B., Dong, Y., Wang, B., Wang, J. and Kallio, P.H. 2002. Effects of seabuckthorn (*Hippophae rhamnoides* L.) seed and pulp oils on experimental models of gastric ulcer in rats. *Fitoterapia* 73: 644-650.

59. Zeb, A. 2004. Important therapeutic uses of seabuckthorn (*Hippophae*): A review. *Journal of Biological Sciences* 4: 687-693.

60. Zheng, R.X., Xu, X.D., Tian, Z. and Yang, J.S. 2009. Chemical constituents from the fruits of *Hippophae rhamnoides*. *Natural Product Research* 23: 1451-1456.

Chapter 38

Seabuckthorn Regulation of Homocystein Impairment Mediated Neurodegeneration with Special Reference to Senile Dementia of Alzheimer's Type (SDAT)

Aruna Agrawal, Sushma Tewari*, Shivapriya Shivakumar*, K. Ilango***, R.G. Singh** and G.P. Dubey****

Faculty of Ayurveda; *Department of Nephrology, Institute of Medical Sciences,
**National Facility for Tribal and Herbal Medicine, IMS,
Banaras Hindu University, Varanasi, U.P., India
***Interdisciplinary School of Indian System of Medicine, SRM University, Chennai, T.N., India

SUMMARY

Neurodegenerative disorders particularly Alzheimer's and Parkinson's disease is affecting millions of individuals worldwide. These disorders gradually occur in old age. Neuropsychological loss is significantly observed in age-associated cognitive decline, multi-infarct dementia, Parkinson's disease, etc. In clinical presentation characteristically, neurodegenerative deficits are associated with impaired concentration, poor attention, over arousal, changes in behavior, apprehensiveness and perceptual abnormalities. Hyperhomocysteinemia (HHcy) has been implicated as a risk factor for neurodegenerative disorders. There is evidence indicating HHcy increases oxidative stress, DNA damage, triggering of apoptosis and excitotoxicity, the all important mechanisms involved in neurodegeneration. Folic acid supplementation has a protective effect on homocystein (Hcy) induced oxidative stress by reducing intracellular superoxide levels and to a lesser extent quenching hydrogen peroxide. Hippocampal neurons from Alzheimer's disease patients have shown higher level of Hcy, triggering the population of cell division proteins resulting in pathology or death in cell division does not occur, or cell division proteins are markers of failure of DNA repair in the presence of HHcy. Thus two major mechanisms for neuronal damage in Alzheimer's disease are the accumulation of amyloid-h-peptide (Ah) and the abnormal phosphorylation of tau protein. Hcy promotes both pathways. Seabuckthorn contains maximum essential micronutrients particularly the folic acid regulates the Hcy concentration and

Seabuckthorn (Hippophae L.): A Multipurpose Wonder Plant, Vol. 4: Emerging Trends in Research and Technologies (Virendra Singh, *Editor-in-Chief*, 2014), p. 465–478, Daya Publishing House, New Delhi, India.

prevents neuronal damage. One hundred twenty four adult volunteers of both sexes with age range of 55-70 years, were selected from various localities of Chennai, Delhi, and Varanasi cities showing high concentration of Hcy (measured in fasting plasma samples with ELISA kit methods), impaired oxidative stress markers with cognitive deficits. A placebo controlled procedure was adopted. 76 subjects were treated with Seabuckthorn extract in effective doses for 6 months and 48 patients were kept on placebo to serve as control. The beneficial role of Seabuckthorn was evaluated on Hcy, oxidative stress markers, *i.e.* TBARS, SOD, Catalase and glutathione including inflammatory cytokines IL-6 and TNF-α following laboratory standard method. A significant correlation was observed between Hcy and advancing age, and females showed much higher Hcy levels than men volunteers. The reduction in Hcy level following 6 months of Seabuckthorn treatment prevented the Hcy mediated neurodegeneration, improved memory, attention and behavioural disturbances and also with its significant synergistic role by reducing oxidative injury and vascular inflammation involved with neurodegeneration particularly among SDAT patients. The clinical trial was conducted strictly as per prescribed standard ethical norms.

Keywords: *Homocystein, Cognitive deficit, Oxidative stress markers, IL-6, TNF-α, SDAT, and Seabuckthorn.*

INTRODUCTION

Hcy has been found to be elevated in several neurodegenerative diseases such as Alzheimer's disease and Parkinson's disease (Clarke *et al.*, 1998; Morris, 2003; O'suilleabhain *et al.*, 2004). An Hcy serum concentration increase with age and enhances the incidence of neurodegenerative disease (Obeid *et al.*, 2006). Elevated levels of Hcy is correlated with cognitive impairment and finally with dementia (Boxer *et al.*, 2005). Additionally, a relationship between Hcy and brain atrophy in elderly individuals has been found (Sachdev *et al.*, 2002). Methylmalonic acid (MMA) is closely linked to Hcy metabolism. It has been demonstrated in vitro that MMA is neurotoxic (Narasimhan *et al.*, 1996; Kruman *et al.*, 2002).

Higher levels of Hcy are considered to be a risk factor for a number of pathologies. Common causes of HHcy include renal impairment, induced by several pathologies (Arnadottir *et al.*, 1996), *i.e.* deficiency of vitamins involved in Hcy metabolism (McNulty *et al.*, 2008), overloading of dietary methionine or excessive accumulation of reactive oxygen species (ROS) in tissues (Boldyrev *et al.*, 2007), as well as deficiency of enzymes which promote Hcy metabolism. The term HHcy is used when the Hcy level exceeds 15μmol/litre. A level of 15-30μmol/litre is considered mild, 30-100μmol/litre moderate and more than 100μmol/litre to severe homocysteinemia (Welch *et al.*, 1998; Warren, 2002). Genetic studies, animal experiments as well as cell culture studies have demonstrated the involvement of neuronal cell death in neurodegenerative diseases; Duan *et al.*, 2002; Ho *et al.*, 2003; de Lau *et al.*, 2005) involving several possible mode of actions (Loscalzo 2002; Sachdev 2005).

Regular alcohol consumption reduces folate and vit. B_{12} status and increases Hcy levels in blood of healthy persons. Homocysteinuria is usually accompanied by thromoboembolia, progressive cardiovascular diseases, mental deficiency (Stein *et al.*, 1998) and neurodegenerative diseases (Clark *et al.*, 1998; McCaddon *et al.*, 1998; Hanbey and Elkelboon, 1999; Morris 2003; Boldyrev *et al.*, 2007; Kim *et al.*, 2008). Severe HHcy results in brain convulsions and dementia (Stein *et al.*, 2002), as well as may contribute to Alzheimer's disease (Seshadri *et al.*, 2002). Recently, it has been found that genetically manifesting deficit of methylenetetrahydrofolate reductase and coherent accumulation of Hcy in blood is a risk factor of Alzheimer's disease (Kim *et al.*, 2008). In Alzheimer's disease patient's, plasma and brain Hcy levels were found to be increased significantly (McCaddon *et al.*, 1998; Eto *et al.*, 2002). It was found similar in cerebrospinal fluid (Isobe *et al.*, 2006). HHcy is considered as a predictive factor for cognitive impairment in the aged (Isobe *et al.*, 2006; Nilsson *et al.*, 2002) and for Alzheimer's disease (Adunsky *et al.*, 2005).

Hcy may also play a provocative role in neurotoxicity of amyloid β peptides and the presence of amyloid-β in the brain of Alzheimer's disease patients, was shown to be exacerbated by homocysteic acid (HcA) (Hasegawa *et al.*, 2005). Both Hcy and homocysteic acid may increase sensitivity to toxic effects of amyloid-β peptides directly affecting their conformation and inducing the formation of neurotoxic β-fibrils (Hasegawa *et al.*, 2005; Agnati *et al.*, 2007).

The toxic effect of Hcy and its derivatives on the cerebellum granule cells has been described by various workers and it was suggested to have action via glutamate receptors (Kim *et al.*, 1996; Shi *et al.*, 2003). The most expected subject or candidate could be glutamate receptors activated by N-methyl-D aspartic acid (NMDA), the so-called NMDA receptors of their activation results in a significant rise in intracellular Ca^{++} and severe accumulation of reactive oxygen species (ROS) inside the neurons (Boldyrev *et al.*, 2004). The increase in ROS levels could be an immediate reason of the toxic effect for both superoxide dismutase and catalase proved to be strongly protective (Kim *et al.*, 1996; Lipton *et al.*, 1997). This phenomena cause damage of neuronal membrane. Homocysteic acid a derivative of Hcy is considered to be an endogenic neurotoxin (Do *et al.*, 2004), and may have much stronger neurotoxic effect than Hcy (Gortz *et al.*, 2004; Boldyrev, 2005). The toxic effect of Hcy will aggravate the development of disease revealing neurological symptoms.

Dis-circulatory encephalopathy (DE) is characteristic of chronic sub-acute expansion accompanied by increased levels of blood Hcy (Perry *et al.*, 1995). Parkinson's disease (PD) which is a chronic neurodegenerative disease characterized by progressive loss of dopaminergic neurons in the substantia nigra pars compacta is a disease of multi-factorial nature including both environmental and genetic factors, which result in development of oxidative stress in specific areas of brain. One of the factors accompanying PD development is pronounced HHcy (Duan *et al.*, 2002).

HHcy is often considered in the light of stroke, Alzheimer's disease, Parkinson's disease, schizophrenia, depression, renal disease, leukemia and diabetes (Godfrey *et al.*, 1990; Hultberg *et al.*, 1993; Kishi *et al.*, 2000; Yasui *et al.*, 2000; Blandini *et al.*, 2001; Madonna *et al.*, 2002; Kishi *et al.*, 2003; Pniewski *et al.*, 2003; Parnetti *et al.*, 2004; DeLeuis *et al.*, 2005). A strong negative correlation between Hcy level and concentrations of folate and vitamins B_6 and B_{12} in body fluids has also been reported (Bissoli *et al.*, 2002; Guerra Shinoharo *et al.*, 2002).

Elevated levels of Hcy increase the vulnerability of hippocampal neurons to excite toxic and oxidative injury in cell culture and in animal experimentation (Kruman *et al.*, 2002). A direct neurotoxicity of Hcy to neuronal cells has been demonstrated (Ho *et al.*, 2003). The mechanism like excite-toxicity by stimulating NMDA receptors, increased production of beta-amyloid aggregates, promotion of neuronal apoptosis and an increased oxidative stress are explained in Hcy induced neurotoxicity (Ho PI *et al.*, 2003; Sachdev 2005; Loscalzo, 2002; Sung *et al.*, 2002). Further, the elevated Hcy levels are also responsible for increased risk of arteriosclerosis (Obeid and Herrmann, 2006).

Excitotoxicity is an attractive target for neuro-protective efforts because it is implicated directly or indirectly in the pathophysiology of a wide variety of acute and chronic neurodegenerative disorders including Alzheimer's disease, PD, Huntington's disease, amyotrophic lateral sclerosis, and multiple sclerosis (Salinska *et al.*, 2005). These conditions are caused by different specific mechanisms but they may share a final common pathway, leading to neuronal injury that involves overstimulation of glutamate receptors especially the NMDA subtype (Chen and Lipton, 2006). Acute brain disorders such as stroke, CNS trauma and epilepsy include a component of excitotoxicity.

The traditional causative treatment of HHcy, designed to reduce the levels of Hcy in human body fluids involves supplementation of the diet with folic acid and/or vitamins B_6 and B_{12}; however in some cases this treatment is unsuccessful (Scott, 2001) keeping the facts in view the plant seabuckthorn which is a rich source of vit. C, B_6, B_{12} and mainly contains folic acid has been evaluated for its therapeutic potential in Hcy impairment mediated neurodegeneration particularly in Senile Dementia of Alzheimer's disease through its Hcy reducing property.

MATERIALS AND METHODS

One hundred twenty four (124) volunteers were recruited from field study of three cities Chennai, Delhi and Varanasi. The volunteers showing elevated Hcy level, with decline cognitive function, oxidative injury and higher level of inflammatory markers IL-6 and TNF-α were enrolled for clinical trial of seabuckthorn extract. Each potential participant of present study was thoroughly informed about the study design, signed an informed consent and underwent an initial laboratory and clinical examination. Out of total screened 334 candidates only 124 volunteers between the age group of 55-70 years of both sexes were included under this clinical trial program as these cases exhibited neuro-psychophysiological status of SDAT.

The study was approved by Institute Ethical Committee and designed as placebo control trial study. Each selected volunteer visited institute minimum three times, *i.e.* before starting treatment (initial visit), after 3 months and six months follow up recording of the parameters. On each visit clinical examination along with blood pressure measurement were carried out and blood samples were taken for estimation of parameters under investigation. 76 subjects of both sexes (43 males and 33 females) were treated with seabuckthorn extract (450 mg morning and 450 mg evening) for six months. Similar to test drug capsules, placebo capsules were also prepared and administered to 48 cases falling in selection criteria to compare the results.

Fasting venus blood samples were collected, stored and processed as per standard procedure for the estimation of oxidative Hcy and scavenging enzymes like lipid peroxidation (TBARS method, Ohkawa *et al.*, 1979). Total glutathione (Ellman *et al.*, 1961), glutathione peroxidase was measured using spectrophotometer. Hcy, TNF-α, interleukin-6 were measured by ELISA kits. Memory span scores and attention span were determined following sophisticated computerized electronic devices.

Table 38.1: Scoring Parameters of Memory of Patients

Complaints	*Normal control*	*SDAT*
Attention span	79.55	32.22
Initiation/preservation	23.55	10.04
Construction	21.37	8.93
Conceptualization	2.55	1.45
Memory	21.18	9.79
Total Score	10.91	5.87

Diagnosis of SDAT was carried out among the selected subjects only those who exhibited more than three of mentioned complaints were kept in senile dementia of Alzheimer's type group and were assessed on various neuropsychophysiological parameters. Dementia rating scale-2 (Jurica, 2001) has been used.

Preparation of Drug

The hydro-methanolic (50:50) extract of Seabuckthorn (225 mg fruits; 225 mg leaves) were concealed in 500 mg empty capsules with the help of additional additive. The trial drug capsules including placebo capsules were provided by Tulsi Ayurvedic Products and Research Pvt. Ltd., Bhadohi for conducting clinical trial. The values of parameters obtained at initial level were compared with the values of 6 months in both treatment groups. Standard statistical procedures were applied to obtain significance level.

RESULTS AND DISCUSSION

It is observed that Hcy enhances with advancing age as compared to Hcy level determined in the younger age group (40-50 years) of both sex. Results show that at the initial level elevated Hcy level is estimated that has shown association with neuropsychophysiological and inflammatory cytokines IL-6 and TNF-α function responsible for neurodegenerative disorders, declined significantly following 6 months of Seabuckthorn treatment.

Daily administration of test drug for a period of six months was found to be associated with a significant decrease in Hcy concentration and oxidative stress markers that lead to improvement in cognitive function particularly in memory and attention span of treated subjects of SDAT. A retard in inflammatory biomarkers IL-6 and TNF-α under test drug treatment indicated neuro-protective and neuromodulatory activity of the drug (Tables 38.2–38.10).

HHcy is one of the serious risk factors for neurodegenerative disease as well as much other pathology in which various manifestations of oxidative stress are involved. The molecular basis of toxicity of Hcy and its derivatives consists in over-activation of the NMDA class specific glutamate receptors. A wide distribution of these receptors in a number of tissues including the brain, heart and immune competent cells make Hcy toxic for many tissues. Thus a satisfactory use of naturally occurring compounds to protect body from Hcy toxicity might find a favourable application as noticed in the present clinical trial results.

It is demonstrated that reduction of Hcy with folic acid supplementation effectively improved the cognitive function which is known to decline with advancing age (Durga *et al.*, 2007).The ripe fruits of *Hippophae rhamnoides* contain malic acid, oxalic acid, phospholipid, vitamin C, A, B complex, E and K. Other nutrients like fat, protein, organic acids, and flavonoids are also found (Zhang *et al.*, 1989; Zhang *et al.*, 2001).

A number of studies have evaluated the role of HHcy in neurodegenerative disease (Allain *et al.*, 1995; Rogers *et al.*, 2003; Valkovic *et al.*, 2005; Isobe *et al.*, 2006; Zoccolella *et al.*, 2008; Valentino *et al.*, 2010). In the present study the Hcy concentrations along with level of oxidative stress markers like TBARS, glutathione, SOD, GPx with IL-6, TNF-α inflammatory cytokines were estimated in selected SDAT individuals. The volunteers showing altered levels of these markers along with cognitive deficits were treated with high altitude plant Seabuckthorn extract for its beneficial effect. As this plant is a rich source of vitamin C, B6, B12 along with folic acid there is marked reduction in Hcy concentration among SDAT cases exhibiting higher concentration of Hcy before treatment (Agrawal *et al.*, 2001; 2002; 2008; 2009; Dubey *et al.*, 2002; 2003; 2004; 2005; 2008; 2010). Further various studies are in support that seabuckthorn has antioxidant, anti-inflammatory, immunomodulatory effects with down regulation of Hcy in neurodegenerative diseases particularly in SDAT and PD patients (Dubey *et al.*, 2008; Agrawal *et al.*, 2009). Hcy has been associated with levodopa treatment can be explained the

Table 38.2: Pattern of Homocysteine, IL-6 and TNF-α Measured in Relation to Age

Age (Years)	*Homocysteine (mmol/L)*	*95 per cent Confidence Interval of the Limit of the Mean*	*IL-6 (pg/ml)*	*95 per cent Confidence Interval of the Limit of the Mean*	*TNF-α (pg/ml)*	*95 per cent Confidence Interval of the Limit of the Mean*
20-30 (N=61)	8.259±2.436	7.636 -8.882	0.423±0.092	0.4-0.446	201.840±39.450	191.738-211.942
31-40 (N=48)	11.684±3.112	10.786-12.582	0.982±0.108	0.951-1.013	411.982±84.702	381.53- 436.434
41-50 (N=50)	14.902±3.756	13.84-15.964	1.665±0.201	1.609-1.721	561.302±102.22	532.396- 590.208
51-60 (N=63)	21.384±4.901	20.119-22.649	2.191±0.204	2.139-2.243	739.42±113.752	710.046-768.794
61-70 (N=65)	28.773±6.313	27.207- 30.339	3.464±0.287	3.393-3.353	982.62±188.904	935.158-029.482

Table 38.3: Relationship between Homocystein, Memory Performance and Attention Span Determined in Relation to Age

Age (Years)	*Homocystein (mmol/L)*	*95 per cent Confidence Interval of the Limit of the Mean*	*Short Term Memory (Score)*	*95 per cent Confidence Interval of the Limit of the Mean*	*Long Term Memory (Score)*	*95 per cent Confidence Interval of the Limit of the Mean*	*Attention Span (Score)*	*95 per cent Confidence Interval of the Limit of the Mean*
20-30 (N=61)	8.259±2.436	7.636 -8.882	9.310±1.046	9.043- 9.577	8.346±1.306	8.012-8.680	16.452±3.142	15.648- 17.256
31-40 (N=48)	11.684±3.112	10.786-12.582	8.452±1.316	8.073-8.831	6.894±1.212	6.545- 7.243	14.229±2.998	13.797-14.661
41-50 (N=50)	14.902±3.756	13.84-15.964	8.116±1.332	7.74-8.492	6.142±1.314	5.771-6.513	12.901±2.542	12.183- 13.619
51-60 (N=63)	21.384±4.901	20.119-22.649	5.851±1.049	5.581 -6.121	5.252±1.087	5.112-5.392	8.669±1.755	8.216-9.122
61-70 (N=65)	28.773±6.313	27.207- 30.339	5.112±1.101	4.839-5.385	3.882±0.985	3.638-4.126	7.412±1.806	6.964-7.860

hylation of levodopa *via* catechol-O-methyltransferase, present in brain and peripheral tissues, releases S-adenosylhomocysteine from levodopa which is subsequently hydrolyzed by S-adenosylhomocysteine hydrolase to Hcy (dos Santos *et al.*, 2009). In the present investigation, perhaps this pathway is regularized by the treatment of Seabuckthorn extract resulting in decline of Hcy concentration.

Table 38.4: Decline in Hcy Concentration among SDAT cases following seabuckthorn extract

Treatment Groups	*No. of Cases*	*Homocystein (nmol/l)*			*Comp. Initial vs After 6 Months*
		Initial	*After 3 Months*	*After 6 Months*	
Placebo treated	48	42.986±4.129	44.039±3.775	40.793±5.102	P<0.05
Seabuckthorn treatment	76	48.329±3.978	41.783±4.802	37.481±3.886	P<0.001

Note: Normal range: 5-15 nmol/l.

Table 38.5: Beneficial Role of Seabuckthorn on Oxidative Stress Markers among SDAT Cases

Treatment Groups	*Parameters*	*Initial*	*After 3 Months*	*After 6 Months*	*Comp. Initial vs After 6 Months*
Placebo treated (N=48)	SOD (per cent inhibition)	1518.68±113.72	1435.42±118.36	1342.45±104.98	p<0.001
	GPx(IU/ml)	68.322±9.75	69.641±13.75	72.136±15.38	p>0.05
	GSH (mg GSH/ mg protein)	1.683±0.112	1.524±0.098	1.382±0.102	p<0.001
	TBARS (nM TBARS/g tissue)	3.972±0.136	3.887±0.201	3.994±0.188	p>0.05
Seabuckthorn treatment (N=76)	SOD (per cent inhibition)	1608.93±201.48	1413.35±156.98	1228.90±177.35	p<0.001
	GPx(IU/ml)	67.940±12.482	61.394±13.016	58.829±8.902	p<0.001
	GSH (mg GSH/ mg protein)	1.394±0.102	2.202±0.201	2.304±1.223	P<0.001
	TBARS (nM TBARS/g tissue)	3.884±0.166	3.406±0.293	2.852±0.284	p<0.001

Table 38.6: Improvement in Short-term Memory Performance after Seabuckthorn Extract Treatment among SDAT Cases

Treatment Groups	*No. of Cases*	*Short Term Memory (Score)*			*Comp. Initial vs After 6 Months*
		Initial	*After 3 Months*	*After 6 Months*	
Placebo treated	48	6.394±1.014	6.286±1.002	5.891±0.984	p>0.05
Seabuckthorn treatment	76	5.974±0.872	6.342±0.639	6.486±0.752	p<0.001

Note: Normal range-10 (Score).

Table 38.7: Improvement in Long Term Memory Scores after Seabuckthorn Extract Treatment in SDAT Cases

Treatment Groups	No. of Cases	Long Term Memory (Score)			Comp. Initial vs After 6 Months
		Initial	After 3 Months	After 6 Months	
Placebo treated	48	3.882±0.112	3.509±0.092	3.112±0.089	p<0.001
Seabuckthorn treatment	76	3.110±0.121	3.813±0.110	4.118±0.116	p<0.001

Note: Normal range-10 (Score).

Table 38.8: Improved Attention Span Scores after Seabuckthorn Treatment in SDAT Cases

Treatment Groups	No. of Cases	Attention Span (Score)			Comp. Initial vs After 6 Months
		Initial	After 3 Months	After 6 Months	
Placebo treated	48	4.752±1.002	4.801±0.124	4.592±0.216	p>0.05
Seabuckthorn treatment	76	4.489±0.169	4.973±0.224	5.098±0.132	p<0.001

Note: Normal range-10 (Score)

Table 38.9: Reduction in Interleukin-6 following Seabuckthorn Treatment in SDAT Cases

Treatment Groups	No. of Cases	IL-6 (pg/ml)			Comp. Initial vs After 6 Months
		Initial	After 3 Months	After 6 Months	
Placebo treated	48	1.392±0.041	1.514±0.068	1.486±0.094	p<0.001
Seabuckthorn treatment	76	1.832±0.088	1.298±0.098	1.106±0.089	p<0.001

Note: Normal range: ≤1 (pg/ml)

Table 38.10: Retard in TNF-α Concentration following Seabuckthorn Treatment in SDAT Cases

Treatment Groups	No. of Cases	TNF-α (pg/ml)			Comp. Initial vs After 6 Months
		Initial	After 3 Months	After 6 Months	
Placebo treated	48	934.72±88.73	1002.73±91.86	988.64±87.36	p<0.01
Seabuckthorn treatment	76	1016.82±112.90	886.35±89.32	759.22±94.03	p<0.001

Note: Normal range: 25-800 (pg/ml).

As pointed out elevated Hcy increases the risk of neurodegenerative diseases (Obeid and Herrmann, 2006), rather there is association of more rapid disease progression for PD patients, or other neurodegenerative conditions, and also an association of HHcy with progression of cognitive impairment including dementia has been established (Boxer *et al.*, 2005). In the present series of investigation various mechanism of action are proposed to be involved for the manner in which Seabuckthorn has shown its protective action against neurodegeneration. Reactive oxygen and nitrogen species may cause chemical alterations of biomolecules and the redox state of the whole organism is maintained by a complex of mechanisms like scavenging enzyme activities like superoxide dismutase, SOD, glutathione peroxidase, GPx reflects antioxidant defense status which has been proven by several studies (Agrawal *et al.*, 2008; 2009; Dubey *et al.*, 2001; 2008; 2010). Glutathione is one of the most abundant intracellular anti-oxidant found to be increased under treatment of Seabuckthorn extract. The reaction with thiobarbituric acid, thiobarbuturic acid reactive substances (TBARS) is most widely used that mainly originates in lipid peroxidation (Lefevre *et al.*, 1998), decreased following Seabuckthorn treatment. As the drug Seabuckthorn has been proven as a potent antioxidant agent, it acted through above mechanism by reducing oxidative stress markers.

CONCLUSION

The neurodegenerative disease SDAT investigated in present study have shown different pathological mechanisms suggesting that a number of subjects will develop dementia during the course of process involved with them. It is speculated that since the Hcy has been reported to be elevated in Alzheimer's disease patients it might be a risk factor for progression of dementia, SDAT/AD in near future among selected subjects. As the test drug Seabuckthorn has shown promising results in reducing elevated Hcy and improvement in oxidative stress markers and vascular inflammation, it may be proposed as a effective adjuvant therapy for the prevention and management of Hcy mediated neurodegenerative conditions particularly senile dementia of Alzheimer's type.

REFERENCES

1. Adunsky, A., Arinzon, Z., Fidelman, Z., Krasniansky, I., Arad, M. and Gepstein, R. 2005. Plasma homocysteine levels and cognitive status in long-term stay geriatric patients: A cross-sectional study. *Arch. Geront. Geriatr.* 40: 129-138.
2. Agarwal, A., Tiwari, S., Rajamanickam, G.V. and Dubey, G.P. 2009. Antioxidant effect of seabuckthorn. In: *Seabuckthorn (Hippophae spp.): The Golden Bush*, p.303-335, Satish Serial Publishing House, Delhi.
3. Agrawal, A., Dixit, S.P. and Dubey, G.P. 2002. Medhya Rasayana Property of *Hippophae rhamnoides* in age related cognitive deficits. In: *Ayurvedic conference on Rasayana, Rashtriya Ayurveda Vidyapeeth (National Academy of Ayurveda), Ministry of Health and Family Welfare*, Dhanwantari Bhawan, New Delhi.
4. Agrawal, A., Khuntia, B.B. and Dubey, G.P. 2001. Seabuckthorn modulates the mental health status and its effect on immunosenescence among the elderly-A double blind placebo control study. In. *Proceeding of International Workshop on Seabuckthorn* (V. Singh and P.K. Khosla, Eds.), p. 214-220, February 18-21, 2001, Delhi, 214p.
5. Agrawal, A., Khuntia, B.B. and Dubey, G.P. 2001. Seabuckthorn modulates the mental health status and Immunosenescence among the elderly-A double blind placebo control study. *Indian J. of Gerontology*: 183-212.
6. Agrawal, A., Sharma, A., Rajamanickam, G.V. and Dubey, G.P. 2008. Age related neuro-chemical and neuro-psychological modulation by *Hippophae rhamnoides* with special reference to senile dementia. *Journal of Research in Ayurveda and Siddha* XXIX (1): 123-145.
7. Agrawal, A., Tewari, S. and Dubey, G.P. 2001. Role of seabuckthorn (*Hippophae rhamnoides*) in the management of age related memory disorder. In: *Proceeding of International Workshop on Seabuckthorn* (V. Singh and P.K. Khosla, Eds.), p. 225-230 New Delhi.

8. Agrawal, A., Vasanthi, H.R., Attrey, D.P. and Dubey, G.P. 2004. An ayurvedic drug therapeutic strategy in the prevention and management of neuropsychiatric problems among elderly. In: *Seminar on Hippophae rhamnoides-A resource for Environment, Health and Economy*, Chanakya Bhawan, New Delhi.

9. Allain, P., Le, B.A., Cordillet, E., Le, Q.L., Bagheri, H. and Montastruc, J.L. 1995. Sulphate and cysteine levels in the plasma of patients with Parkinson's disease. *Neurotoxicology* 16: 527-529.

10. Arnadottir, M., Hultberg, B., Vladov, V., Nilsson-Ehle, P. and Thysell, H. 1996. Hyperhomocysteinemia in cyclosporine-treated renal-transplant recipients. *Transplantation* 61 (3): 509-512.

11. Bissoli, L., Di Francesco, V., Ballarin, A., Mandragona, R., Trespidi, R., Brocco, G., Caruso, B., Bosello, O. and Zamboni, M. 2002. Effect of vegetarian diet on homocysteine levels. *Ann. Nutr. Metab* 46:73-79.

12. Blandini, F., Fancellu, R., Martignoni, E., Mangiagalli, A., Pacchetti, C., Samuele, A. and Nappi, G. 2001. Plasma homocysteine and l-dopa metabolism in patients with Parkinson disease. *Clin.Chem.* 47: 1102-1104.

13. Boldyrev, A.A. and Johnson, P. 2007. Homocysteine and its derivatives as possible modulators of neuronal and non-neuronal cell glutamate receptors in Alzheimer's disease. *J. Alzheimer's Dis.* 11: 219-228.

14. Boldyrev, A.A., Carpenter, D.O. and Johnson, P. 2005. Emerging evidence for a similar role of glutamate receptors in the nervous and immune systems. *J. Neurochem*. 95: 913-918.

15. Boldyrev, A.A., Kazey, V.I., Leinsoo, T.A., Mashkina, A.P., Tyulina, O.V., Johnson, P., Tuneva, J.O., Chittur, S. and Carpenter, D.O. 2004. Rodent lymphocytes express functionally active glutamate receptors. *Biochem. Biophys. Res. Commun*. 324: 133-139.

16. Boxer, A.L., Kramer, J.H., Johnston, K., Goldman, J., Finley, R. and Miller, B.L. 2005. Executive dysfunction in hyperhomocystinemia responds to homocysteine-lowering treatment. *Neurology* 64: 1431-1434.

17. Chen, H.S. and Lipton, S.A. 2006. The chemical biology of clinically tolerated NMDA receptor antagonists. *J. Neurochem*. 97: 1611-1626.

18. Clarke, R., Smith, A.D., Jobst, K.A., Refsum, H., Sutton, L. and Ueland, P.M. 1998. Folate, vitamin B_{12}, and serum total homocysteine levels in confirmed Alzheimer disease. *Arch. Neurol*. 55: 1449-1455.

19. de Lau, L.M., Koudstaal, P.J., van Meurs, J.B., Uitterlinden, A.G., Hofman, A. and Breteler, M.M. 2005. Methylene tetrahydrofolate reductase C677T genotype and PD. *Ann. Neurol*. 57: 927-930.

20. DeLuis, D.A., Fernandez, N., Arranz, M.L., Aller, R., Izaola, O. and Romero, E. 2005. Total homocysteine levels relation with chronic complications of diabetes, body composition, and other cardiovascular risk factors in a population of patients with diabetes mellitus type-2. *J. Diabetes Complications* 19: 42-46.

21. Do, K.Q., Benz, B., Binns, K.E., Eaton, S.A. and Salt, T.E. 2004. Release of homocysteic acid from rat thalamus following stimulation of somatosensory afferents in vivo: feasibility of glial participation in synaptic transmission. *Neuroscience* 124: 387-393.

22. dos Santos, E.F., Busanello, E.N., Miglioranza, A., Zanatta, A., Barchak, A.G., Vargas, CR, Saute, J., Rosa, C., Carrion, M.J., Camargo, D., Dalbem, A., da Costa, J.C., de Sousa Miguel, S.R., de Mello Rieder, C.R. and Wajner, M. 2009. Evidence that folic acid deficiency is a major determinant of hyperhomocysteinemia in Parkinson's disease. *Metab. Brain Dis*. 24: 257-269.

23. Duan, W., Ladenheim, B., Cutler, R.G., Kruman, I.I., Cadet, J.L. and Mattson, M.P. 2002: Dietary folate deficiency and elevated homocysteine levels endanger dopaminergic neurons in models of Parkinson's disease. *J. Neurochem*. 80: 101-110.

24. Dubey, G.P. and Agrawal, A. 2003. Memory enhancing potentials of *Hippophae rhamnoides* (*Hippophae rhamnoides*)-A placebo controlled study. In: *Congress of the International Hippophae rhamnoides Association-ISA*, Berlin, Technical University, Berlin.

25. Dubey, G.P., Agrawal, A., Rajamanickam, G.V. and Lavekar, G.S., 2008: Brain ageing and ayurveda. In: *Central Council for Research in Ayurveda and Siddha*, Deptt. of AYUSH, Ministry of Health and Family Welfare, Govt. of India, New Delhi.

26. Dubey, G.P., Agrawal, A. and Dixit, S.P. 2001. Neuro-physiological adaptation following stress under effect of *Hippophae rhamnoides*. In: *Proceeding of International Workshop on Seabuckthorn* (V. Singh and P.K. Khosla, Eds.), p.199-204, February18-21, 2001, New Delhi.

27. Dubey, G.P., Agrawal, A. and Dixit, S.P. 2002. Role of seabuckthorn (*Hippophae rhamnoides*) in the maintenance of cardiovascular homeostasis following cold stress. *Journal of Natural Remedies* 3: 136-140.

28. Dubey, G.P., Agrawal, A. and Rastogi, M. 2010. Antioxidant and Neuro-modulatory activity of *Hippophae rhamnoides*. In: *Emerging Trends in Production to consumption*. National Conference on Seabuckthorn, Krishi Vishvavidyalaya, February 16-18, 2010, Palampur, Himachal Pradesh.

29. Dubey, G.P., Agrawal, A., Wilson, E., Vyas, N. and Rajamanickam, G.V. 2005. Novel markers of atherothrombotic risk-Therapeutic benefits. In: *Second International Conference of the International Society for Ayurveda and Health and International seminar on Chronic Diseases and its Management by Complementary and Alternative Medicine (CAM)*, Dept. of Medicinal Chemistry, IMS, BHU and University of Connecticut, School of Medicine, USA.

30. Dubey, G.P., Attreya, D.P. and Agrawal, A. 2004. Multi-dimensional therapeutic potential of *Hippophae rhamnoides* with special reference to memory disorder. In: *Seminar on Hippophae rhamnoides, a Resource for Environmental, Health and Economy*, Defence Research and Development Organization Chanakya Bhawan, New Delhi.

31. Durga, J., van Boxtel, M.P., Schouten, E.G., Kok, F.J., Jolles, J., Katan, M.B. and Verhoef, P. 2007. Effect of 3-year folic acid supplementation on cognitive function in older adults in the FACIT trial: A randomized, double blind, controlled trial. *The Lancet*. 369: 208-216.

32. Ellman, G.L., Courtney, K.D., Andres, V.J. and Featherstone, R.M. 1961. A new and rapid colorimetric determination of acetylcholinesterase activity. *Biodiem. Plutrmac*. 7: 88-89.

33. Eto, K., Asada, A., Arima, K., Makifuchi, T. and Kimura, H. 2002. Brain hydrogen sulfide is severely decreased in Alzheimer's disease. *Biochem. Biophys. Res. Commun*. 293: 1485-1488.

34. Godfrey, P.S., Toone, B.K., Carney, M.W., Flynn, T.G., Bottiglieri, T., Laundy, M., Chanarin, I. and Reynolds, E.H. 1990. Enhancement of recovery from psychiatric illness by methylfolate. *The Lancet* 336: 392-395.

35. Gortz, P., Hoinke, A., Fleischer, W., Otto, F., Schwahn, B., Wendel, U. and Siebler, M.J. 2004. Implications for hyperhomocysteinemia: Not homocysteine but its oxidized forms strongly inhibit neuronal network activity. *Neurol. Sci.* 218: 109-114.

36. Guerra-Shinohara, E.M., Paiva, A.A., Rondo, P.H., Yamasaki, K., Terzi, C.A. and D'Almeida, V. 2002. Relationship between total homocystein and folate levels in pregnant womenand their newborn babies according to maternal serum levels of vitamin B12. *BJOG* 109: 784-791.

37. Hasegawa, T., Ukai, W., Jo, D.G., Xu, X., Mattson, M.P., Nakagawa, M., Araki, W., Saito, T. and Yamada, T. 2005. Homocysteic acid induces intraneuronal accumulation of neurotoxic Abeta42: Implications for the pathogenesis of Alzheimer's disease. *J. Neurosci. Res.* 80: 869-876.

38. Ho. P.I., Ashline, D., Dhitavat, S., Ortiz, D., Collins, S.C., Shea, T.B. and Rogers, E. 2003: Folate deprivation induces neurodegeneration: Roles of oxidative stress and increased homocysteine. *Neurobiol. Dis.* 14: 32-42.

39. Hultberg, B., Andersson, A. and Sterner, G. 1993. Plasma homocysteine in renal failure. *Clin. Nephrol.* 40: 230-235.

40. Isobe, C., Murata, T., Sato, C. and Terayama, Y. 2006. Increase of total homocysteine concentration in cerebrospinal fluid in patients with Alzheimer's disease and Parkinson's disease. *Life Sci.* 77: 1836-1843.

41. Jurica, P.J., Leittern, C.L. and Mattis, S. 2001. Dementia rating scale-2. *Professional Manual,* Psychological Assessment Resources, Inc., Florida.

42. Kim, J.M., Steward, R., Kim, S.W., Yang, S.J., Shin, I.S., Shin, H.Y. and Yoon, J.S. 2008. Methylenetetrahydrofolate reductase gene and risk of Alzheimer's disease in Koreans. *Int. J. Geriatr. Psychiatry* 23: 454-459.

43. Kim, W.K., and Pae, Y.S. 1996. Involvement of N-methyl-D-aspartate receptor and free radical homocysteine-mediated toxicity on rats cerebellar granule cells in culture. *Neurosci. Lett.* 216, 117-120.

44. Kishi, S., Griener, J., Cheng, C., Das, S., Cook, E.H., Pei, D., Hudson, M., Rubnitz, J., Sandlund, J.T., Pui, C.H. and Relling, M.V. 2003. Homocysteine, pharmacogenetics, and neurotoxicity in children with leukemia. *J. Clin.Oncol.* 21: 3084-3091.

45. Kishi, T., Tanaka, Y. and Ueda, K. 2000. Evidence for hypomethylation in two children with acute lymphoblastic leukemia and leukoencephalopathy. *Cancer* 89: 925-931.

46. Kruman, I.I., Kumaravel, T.S., Lohani, A., Pedersen, W.A., Cutler, R.G., Kruman, Y., Haughey, N., Lee, J., Evans, M. and Mattson, M.P. 2002. Folic acid deficiency and homocysteine impair DNA repair in hippocampal neurons and sensitize them to amyloid toxicity in experimental models of Alzheimer's disease. *J. Neurosci.* 22: 1752-1762.

47. Lefevre, G., Beljean-Leymarie, M., Beyerle, F., Bonnefont-Rousselot, D., Cristol, J.P., Therond, P. and Torreilles, J. 1998. Evaluation of lipid peroxidation by measuring thiobarbituric acid reactive substances (In French). *Ann. Biol.Clin. (Paris)* 56: 305-319.

48. Lipton, S.A., Kim, W.K., Choi, Y.B., Kumar, S., D'Emilia, D.M., Rayudu, P.V., Arnelle, D.R. and Stamler J.S. 1997. Neurotoxicity associated with dual actions of homocysteine at the N-methyl-D-aspartate receptor. *Proc. Natl. Acad. Sci.* 94: 5923-5928.

49. Loscalzo, J. 2002. Homocysteine and dementias. *N. Engl. J. Med.* 346: 466-468.

50. Madonna, P., de Stefano, V., Coppola, A., Cirillo, F., Cerbone, A.M., Orefice, G. and Di Minno, G. 2002. Hyperhomocysteinemia and other inherited prothrombotic conditions in young adults with a history of ischemic stroke. *Stroke* 33: 51-56.

51. McCaddon, A., Davies, G., Hudson, P., Tandy, S. and Cattell, H., 1998. Total serum homocysteine in senile dementia of Alzheimer type. *Int. J. Gen. Psych.* 13: 235-239.

52. McNulty, H., Pentieva, K., Hoey L. and Ward, M. 2008. Homocysteine, B-vitamins and CVD. *Proc. Nutr. Soc.* 67 (2): 232-237.

53. Morris, M.S. 2003. Homocysteine and Alzheimer's disease. *The Lancet Neurol.* 2: 425-428.

54. Narasimhan, P., Sklar, R., Murrell, M., Swanson, R.A. and Sharp, F.R. 1996. Methylmalonyl-CoA mutase induction by cerebral ischemia and neurotoxicity of the mitochondrial toxin methylmalonic acid. *J. Neurosci.* 16: 7336-7346.

55. Nilsson, K., Gustafson, L. and Hultberg, B. 2002. Relation between plasma homocysteine and Alzheimer's disease. *Dement. Geriatr. Cogn. Disord.* 14: 7-12.

56. O'Suilleabhain, P. and Diaz-Arrastia, R. 2004. Levodopa elevates homocysteine: Is this a problem? *Arch. Neurol.* 61: 633-634.

57. Obeid, R. and Herrmann, W. 2006. Mechanisms of homocysteine neurotoxicity in neurodegenerative diseases with special reference to dementia. *FEBS Lett.* 580: 2994-3005.

58. Ohkawa, H., Ohishi, N. and Yagi, K. 1979. Assay for lipid peroxides in animal tissues by thiobarbituric acid reaction. *Annals of Biochemistry* (95): 351-358.

59. Parnetti, L., Caso, V., Santucci, A., Corea, F., Lanari, A., Floridi, A., Conte, C. and Bottiglieri, T. 2004. Mild hyperhomocysteinemia is a risk factor in all etiological subtypes of stroke. *Neurol. Sci.* 25: 13-27.

60. Perry, I.J., Refsum, H., Morris, R.W., Ebrahim, S.B., Ueland, P.M. and Shaper, A.G. 1995. Prospective study of serum total homocysteine concentration and risk of stroke in middle-aged British men. *The Lancet* 346: 1395.

61. Pniewski, J., Chodakowska-Zebrowska, M., Wozniak, R., Stepien, K. and Stafiej, A. 2003. Plasma homocysteine level and the course of ischemic stroke. *Acta Neurobiol. Exp. (Wars)* 63: 127-130.

62. Rogers, J.D., Sanchez-Saffon, A., Frol, A.B. and Diaz-Arrastia, R. 2003. Elevated plasma homocysteine levels in patients treated with levodopa: Association with vascular disease. *Arch. Neurol.* 60: 59-64.

63. Sachdev, P.S. 2005. Homocysteine and brain atrophy. *Prog. Neuropsychopharmacol. Biol. Psychiatry* 29: 1152-1161.

64. Sachdev, P.S., Valenzuela, M., Wang, X.L., Looi, J.C. and Brodaty, H. 2002. Relationship between plasma homocysteine levels and brain atrophy in healthy elderly individuals. *Neurology* 58: 1539-1541.

65. Salinska, E., Danysz, W. and Lazarewicz, J.W. 2005. The role of excitotoxicity in neurodegeneration. *Folia Neuropathol.* 43: 322-339.

66. Scott, J.M. 2001. Modification of hyperhomocysteinemia. In: *Homocysteine in Health and Disease* (R. Carmel and D.W. Jacobsen, Eds.), p. 267-276, Cambridge University Press.

67. Seshadri, S., Beiser, A., Selhub, J., Jacques, P.F., Rosenberg, I.H., D'Agostino, R.B., Wilson, P.W.F. and Wolf, P.A. 2002. Plasma homocysteine as a risk factor for dementia and Alzheimer's disease. *N. Engl. J. Med*. 346: 476-483.

68. Shi, Q., Savage, J.E., Hufeisen, S.J., Rauser, L., Grajkowska, E., Ernsberger, P., Wroblewski, J.T., Nadeau, J.H. and Roth, B.L. 2003. L-homocysteine sulphinic acid and other acidic homocysteine derivatives are potent and selective metabotropic glutamate receptor agonists. J. *Pharmacol. Exp. Ther*. 305: 131-142.

69. Stein, J. and McBride, P. 1998. Hyperhomocysteinemia and atherosclerotic vascular disease: pathophysiology, screening and treatment. *Arch. Intern. Med*. 158: 1301-1306.

70. Stein, J.H., Bushara, M., Bushara, K., McBride, P.E., Jorenby, D.E. and Fiore, M.C. 2002. Smoking cessation, but not smoking reduction, reduces plasma homocysteine levels. *ClinCardiol*. 25 (1): 23-26.

71. Sung, J.J., Kim, H.J., Choi-Kwon, S., Lee, J., Kim, M. and Lee, K.W. 2002. Homocysteine induces oxidative cytotoxicity in Cu, Zn-superoxide dismutase mutant motor neuronal cell. *NeuroReport* 13: 377-381.

72. Valentino, F., Bivona, G., Butera, D., Paladino, P., Fazzari, M., Piccoli, T., Ciaccio, M. and La Bella, V. 2010. Elevated cerebrospinal fluid and plasma homocysteine levels in ALS. *Eur. J. Neurol*. 17: 84-89.

73. Valkovic, P., Benetin, J., Blazicek, P., Valkovicova, L., Gmitterova, K. and Kukumberg, P. 2005. Reduced plasma homocysteine levels in levodopa/entacapone treated Parkinson patients. *Parkinsonism Relat. Disord*. 11: 253-256.

74. Warren, C.J. 2002. Emergent cardiovascular risk factor: Homocysteine. *Prog. Cardiovasc. Nurs*. 17 (1): 35-41.

75. Welch, G. and Loscalo, J. 1998. Homocysteine and atherosclerosis. *New Engl. J. Med*. 338 (15): 1042-1050.

76. Yasui, K., Kowa, H., Nakaso, K., Takeshima, T. and Nakashima, K. 2000. Plasma homocysteine and MTHFR c677T genotype in levodopa-treated patients with PD. *Neurology* 55: 437-440.

77. Zhang, W.J., Yan, J., Duo, B., Ren and Guo, J. 1989. Preliminary study of biochemical constitutes of berry of seabuckthorn growing in Shanxi Province and their changing trend. In: *Proceeding of International symposium on seabuckthorn* (*H. rhamnoides L.*), Xian, China.

78. Zhang, X., Zhang, M., Gao, Z., Wang, J. and Wang, Z. 2001. Effect of total flavones of *Hippophae rhamnoides* L. on sympathetic activity in hypertension. *Hua Xi yiKe da XueBao* 32: 547-550.

79. Zoccolella, S., Simone, I.L., Lamberti, P., Samarelli, V., Tortelli, R., Serlenga, L. and Logroscino, G. 2008. Elevated plasma homocysteine levels in patients with amyotrophic lateral sclerosis. *Neurology* 70: 222-225.

Chapter 39

Improvement in Insulin Sensitivity Following Treatment with Seabuckthorn (*Hippophae* L.) Extract

G.P. Dubey, A. Agrawal*, S. Sandhiya*, K. Ilango*** and R.G. Singh****

National Facility for Tribal and Herbal Medicine, IMS
*Faculty of Ayurveda, Institute of Medical Sciences
**Department of Nephrology, National Facility for Tribal and Herbal Medicine, IMS, Banaras Hindu University, Varanasi, U.P., India
*** Interdisciplinary School of Indian System of Medicine, SRM University, Chennai, India

SUMMARY

Adiponectin is an adipokine profoundly expressed in adipose tissues and directly sensitizes the body to insulin. Hypoadiponectinemia manifesting due to genetic as well as environmental factors, appears to play an important role in insulin resistance associated with type-2 diabetes mellitus and metabolic syndrome that are linked to obesity. Insulin resistance is a key feature of these diseases, and any defect in the insulin signaling cascade can cause insulin resistance. A decrease in plasma adiponectin levels preceded the onset of diabetes, decreased insulin sensitivity causes obesity, metabolic syndrome and independently associated with cardiovascular disease and hypertension. Keeping the above facts in view 96 diagnosed cases of insulin resistance, associated with obesity or metabolic syndrome provided the data of clinical trial on Seabuckthorn evaluated for its adiponectin enhancing property. Adiponectin estimation was performed by utilizing ELISA kit methods adopting standard procedure. The conventional drug Orlistat were prescribed in a group of 54 cases of both sexes to compare the results. Test drug Seabuckthorn extract was administered to 96 selected patients along with conventional drug Orlistat for 3 months. The results revealed beneficial role of Seabuckthorn by enhancing adiponectin concentration with a decrease in leptin level. The test drug also exerted improvement in body energy balance, improved vascular inflammation and prevented atherosclerosis through its adiponectin increasing effect. The patients of Orlistat treatment group could not responded positively to enhance adiponectin content and assessed to be more at risk of development of cardiovascular disorders.

Seabuckthorn (*Hippophae* L.): *A Multipurpose Wonder Plant, Vol. 4: Emerging Trends in Research and Technologies* (Virendra Singh, *Editor-in-Chief*, 2014), p. 479–487, Daya Publishing House, New Delhi, India.

It is concluded that there is improvement in insulin sensitivity due to increase in circulating adiponectin and insulin responsiveness following treatment with seabuckthorn extract. All the selected patients had given informed written consent to participate in clinical trial program, and ethical guidelines were followed to conduct the study.

Keywords: *Adiponectin, Leptin, CRP, Obesity, Metabolic syndrome and Seabuckthorn.*

INTRODUCTION

Insulin resistance, a pathological state where insulin action is impaired in target tissues including liver, skeletal, muscle and adipose tissue is a central abnormality of the metabolic syndrome is strongly associated with intra-abdominal fat. The increasing incidence of obesity, leading to metabolic complications has become a major public health problem. Adipose tissue an endocrine organ plays an important role in modulation of insulin sensitivity, inflammation and vascular thrombosis (Stefan and Stumvoll, 2002; Diez and Iglesias, 2003). Adipose tissue secretes adipokines markedly influence lipid contents and insulin metabolism, oxidative stress and cardiovascular integrity (Lowell, 1999).

Hypoadiponectinemia is caused due to genetic as well as environmental factors ultimately manifesting obesity, insulin resistance, type-2 diabetes mellitus and metabolic syndrome that are mainly linked with obesity. A decrease in circulating levels of adiponectin due to various genetic or environmental factors contributes in development of diabetes and metabolic syndrome. Further, it is reported that plasma adiponectin levels are also decreased in obese animals that frequently developed type-2 diabetes (Hotta *et al.*, 2001), and correlated inversely with insulin resistance (Arita *et al.*, 1999; Ryo *et al.*, 2004; Yatagai *et al.*, 2003; Yamamoto *et al.*, 2004). Thus, hypoadiponectinemia has been demonstrated to be independently associated with the metabolic syndrome (Matsushita *et al.*, 2006), in particular and also playing a role in cardiovascular disease, hypertension etc. (Kumada *et al.*, 2003; Pischon *et al.*, 2004; Ouchi *et al.*, 2003; Adamczak *et al.*, 2003).

Adiponectin an anti-inflammatory adipokine have a role in regulation of energy balance and peripheral tissue lipid metabolism (Berg *et al.*, 2002). The molecular mechanism involved with insulin sensitivity increases fatty acid oxidation and glucose uptake via activation of AMP-activated protein kinase (AMPK) and peroxisome proliferator-activated receptors-α (PPAR-α) responsible for glucose metabolism and insulin sensitivity (Kadowaki *et al.*, 2003; Yamauchi and Kamon, 2002). Leptin is produced by white adipose tissue regulates body weight, insulin sensitivity, metabolism and reproductive function (Oral *et al.*, 2002).

An association between adipokines (adiponectin and leptin) and insulin resistance has been noted in diabetes and metabolic states (Weyer *et al.*, 2001; Tschritter *et al.*, 2003; Yamamoto *et al.*, 2002). It is reported by various researchers that adiponectin play a direct role in determination of insulin mediated glucose uptake (Yamamoto *et al.*, 2002; Yamauchi *et al.*, 2001; Berg *et al.*, 2001; Kubota *et al.*,. 2002; Berg, 2002). Insulin resistance and obesity are associated with lower plasma adiponectin concentration. Further, adiponectin concentrations are significantly associated with body fat (Nemet *et al.*, 2003; Arita *et al.*, 1999; Yang *et al.*, 2001; Comuzzie *et al.*, 2001). It is pointed out that insulin resistance and hyperinsulinemia are linked with obesity and both decline with weight loss (Olefsky *et al.*, 1974; McLaughlin *et al.*, 2001a, 2001b). Thus improving insulin resistance or reduction of insulin levels through application of pharmacologic or non-pharmacologic measures may increase adiponectin concentration. As studied that both obese and non-obese can be insulin sensitive as well as insulin resistant (Olefsky *et al.*, 1974; McLaughlin *et al.*, 2001a, 2001b, Ferrannini *et al.*, 1997), therefore estimation of plasma adiponectin levels in metabolic disorders particularly in diabetes, obesity, metabolic syndrome will be of clinical significance and understanding the association of insulin resistance with level of adiponectin will be helpful in developing a therapeutic modality having potential in enhancing the adiponectin concentration through a specific mechanism like weight control mechanism improving insulin sensitivity or improving glucose metabolism etc. In the recent past several researches have shown therapeutic potentials of the plant Seabuckthorn as an anti-oxidant, anti-inflammatory, immunomodulatory, homocystein lowering effects. Therefore in the present study this drug is evaluated for its adiponectin enhancing property among insulin resistant subjects of obesity and metabolic syndrome.

MATERIAL AND METHODS

A clinical trial was conducted in selected obesity and metabolic syndrome cases of both sexes showing insulin sensitivity with low adiponectin concentration to validate the beneficial role of Seabuckthorn on adiponectin, leptin and anthropometric indices particularly body mass index (BMI).For this study 53 obesity cases and 43 metabolic syndrome cases were selected following certain standard inclusion and exclusion criteria. The standard extract of Seabuckthorn was obtained from Venkatesh Food Industries, Chhindwara, Madhya Pradesh and evaluated on BMI, fasting serum insulin, adiponectin and leptin levels. For the estimation of parameters venus blood sample was collected in a plain vial in sterile condition. The serum for adiponectin, leptin and insulin was stored at 20°C till the analysis.

Serum insulin was estimated by sandwich ELISA method using insulin assay kit provided for research purpose. Adiponectin estimation was done using Human ELISA commercial kit and leptin was also measured through kit by ELISA method.

All the cases were asked to complete a personal health and medical history questionnaire, which served as a screening tool of selected patients. The ethics committee approved the clinical trial procedure and study protocols, which were fully explained to all subjects. A written consent form was signed by each subject. The consent form explained the details of the clinical trial program. A baseline reading of parameters was carried out and selected cases were advised for treatment schedule of both group, *i.e.* Orlistat conventional treatment group and another group where the Orlistat was given along with Seabuckthorn extract in two divided doses (450 mg morning and 450mg evening). A successive follow up study was conducted at the interval of 6weeks and 12 weeks of the parameters *i.e.* fasting serum insulin, adiponectin, leptin, CRP and BMI. The blood glucose level, lipid profile was also estimated in both the treatment groups. On statistical application, values are presented as mean±SD and paired 't' test is applied for obtaining significance level.

RESULTS AND DISCUSSION

The results of present study showed an increase in serum adiponectin levels among subjects who underwent 3 months of Seabuckthorn treatment. The result revealed concurrent improvement in BMI, increase in serum adiponectin level with a reduction in leptin concentration with increased insulin sensitivity.

In the present clinical trial a therapeutic strategy for the treatment of insulin resistance in obesity and metabolic syndrome patients exerted upregulation of adiponectin concentration with a down regulation of leptin level and decrease in BMI, serum insulin and CRP that proves the drug Seabuckthorn is an adiponectin receptor agonist. The drug has been proven to improve systemic insulin sensitivity through its adiponectin enhancing property. Table shows improvement in various targeting markers, *i.e.* anthropometric indices particularly BMI, serum insulin level, decrease in leptin level with significant increase in adiponectin concentration. The inflammatory marker CRP also decreased following treatment suggesting improved cerebrovascular inflammation, indicating reduced cerebrovascular or cardiac events among insulin sensitive individuals (Tables 39.1–39.4).

Table 39.1: Effect of Test Drug Seabuckthorn on BMI among Obesity and Metabolic Syndrome Cases

Treatment Groups	*Clinical Groups*	*No. of Cases*	*BMI*			*Comp. Initial vs. After 12 Weeks*
			Initial	*After 6 Weeks*	*After 12 Weeks*	
Orlistat treated group	Obesity	53	34.398±4.802	32.948±3.716	31.859±4.084	P<0.01
	Metabolic syndrome	43	31.936±3.702	30.741±2.984	30.224±2.885	P>0.05
Orlistat + Seabuckthorn	Obesity	28	35.198±4.732	33.561±3.905	31.904±2.887	P<0.05
	Metabolic syndrome	26	32.316±3.116	28.410±3.104	28.292±2.972	P<0.05

Normal range: 18-24.

Table 39.2: Adiponectin Enhancing Effect of Seabuckthorn among Cases of Obesity and Metabolic Syndrome

Treatment Groups	*Clinical Groups*	*No. of Cases*	*Adiponectin (μg/ml)*			*Comp. Initial vs. After 12 Weeks*
			Initial	*After 6 Weeks*	*After 12 Weeks*	
Orlistat treated group	Obesity	53	6.931±1.801	6.390±1.710	7.103±1.061	P>0.05
	Metabolic syndrome	43	9.016±2.042	8.793±1.894	9.831±2.139	P>0.05
Orlistat + Seabuckthorn	Obesity	28	5.358±0.826	8.221±1.804	10.314±2.114	P<0.001
	Metabolic syndrome	26	8.904±1.986	11.093±2.751	13.885±2.139	P<0.01

Normal range: 5-30 μg/ml.

Table 39.3: Decline in Leptin Concentration following Seabuckthorn Treatment in Obesity and Metabolic Syndrome

Treatment Groups	*Clinical Groups*	*No. of Cases*	*Leptin (μgm/L)*			*Comp. Initial vs. After 12 Weeks*
			Initial	*After 6 Weeks*	*After 12 Weeks*	
Orlistat treated group	Obesity	53	41.792±6.351	38.756±7.094	37.816±5.902	P<0.01
	Metabolic syndrome	43	38.684±4.752	36.902±6.315	34.812±5.539	P<0.01
Orlistat + Seabuckthorn	Obesity	28	44.271±7.124	38.369±6.883	33.711±6.359	P<0.001
	Metabolic syndrome	26	36.889±3.758	31.049±4.065	28.426±5.228	P<0.001

Normal range: 5-12 μgm/l.

Table 39.4: Reduction in C-reactive Protein following Treatment with Seabuckthorn Extract in Obesity and Metabolic Syndrome Cases

Treatment Groups	*Clinical Groups*	*No. of Cases*	*C-reactive Protein (mg/L)*			*Comp. Initial vs. After 12 Weeks*
			Initial	*After 6 Weeks*	*After 12 Weeks*	
Orlistat treated group	Obesity	53	3.109±0.932	2.896±0.751	2.748±0.663	P>0.05
	Metabolic syndrome	43	3.881±0.904	3.436±0.886	3.198±0.894	P<0.01
Orlistat + Seabuckthorn	Obesity	28	3.836±0.885	2.693±0.902	2.104±0.794	P<0.001
	Metabolic syndrome	26	4.423±1.065	3.610±1.002	2.879±0.805	P<0.001

Normal range: 1-3 mg/L.

The mechanism based studies suggested that insulin sensitizing action of adiponectin stimulates AMPK phosphorylation and activation in the liver (Yamauchi *et al.*, 2002). The drug Seabuckthorn acted through an adipocyte derived anti-diabetic hormone having adiponectin enhancing activity through activating AMPK, thereby regulates glucose metabolism and improves insulin sensitivity. Due to adiponectin enhancing property, the drug Seabuckthorn extract perhaps increased fatty acid combustion and energy consumption via PPAR-α and PPAR-γ activation which cause decrease in triglycerides content in the liver and skeletal muscle and increases the insulin sensitivity.

The therapeutic strategy for the prevention and management of insulin resistance in type-2 diabetes mellitus, obesity, metabolic syndrome including cardiovascular disease may include the upregulation of plasma adiponectin levels. A number of therapeutic agents are known to improve systemic insulin sensitivity of varying etiology through enhancing glucose disposal in skeletal muscle and suppressing gluconeogenesis in the liver. The drug Seabuckthorn is proposed to ameliorate insulin resistance through binding and activation of PPAR-γ in adipose tissue thereby promoting adipocyte differentiation and increasing the number of small adipocytes that are more sensitive to insulin secretion. Plasma adiponectin levels have been observed to be enhanced after test drug treatment. Perhaps the drug may have effect on adiponectin through generating small adipocytes that markedly secretes adiponectin. Further, the Seabuckthorn may also directly facilitate the generation of high molecular weight adiponectin. Since, adiponectin is an insulin sensitizing adipokine, it is speculated that the action of test drug increase insulin sensitivity by increasing adiponectin levels.

Studies have indicated that adiponectin deficient ob/ob mice when investigated, it is proposed to observe whether the PPAR-γ agonist therapeutic agent is capable of ameliorating insulin sensitivity in the absence of adequate adiponectin level or not (Kubota *et al.*, 2002; 2006). It was noticed that severity of insulin resistance and diabetes was significantly reduced in association with significant upregulation of serum adiponectin level in experimental animals. The amelioration of insulin resistance attributed to decreased glucose production and increased AMPK levels in the liver but not to increased glucose uptake in skeletal muscle. The test drug Seabuckthorn seems to increase adiponectin levels through activation of adiponectin gene transcription without stimulating adipocyte differentiation. Thereby enhances AMPK activation, decreasing gluconeogenesis in the liver and improving insulin resistance and blood glucose levels. The result obtained out of present study suggests that increase in adiponectin concentration can ameliorate the obesity associated insulin resistance. However, a larger sample size based studies are required to determine the physiological and patho-physiological roles of adiponectin through modulation of various adiponectin genes for the treatment of insulin resistance, metabolic syndrome and type-2 diabetes mellitus.

It is apparent from the present study that adiponectin levels are closely linked with insulin concentrations and insulin resistance. Further, it was noticed that adiponectin levels in insulin resistant subjects were uniformly low in addition to only obese persons. These results demonstrate a clear and definite relationship between insulin resistance and plasma adiponectin concentrations. Several studies have indicated that insulin resistance is closely related to adiponectin levels even after adopting suitable measures for the management of obesity (Weyer *et al.*, 2001; Tschritter *et al.*, 2003). The present study extend the findings that there is independent association of insulin resistance with adiponectin level and insulin may have direct effects on adiponectin concentrations and the drug Seabuckthorn is able in elevating the adiponectin levels indicating that higher levels of insulin in insulin resistant subjects may up-regulate levels of adiponectin.

Seabuckthorn through its adiponectin enhancing effects improved obesity-linked insulin resistance and diabetes by enhancing glucose disposal in skeletal muscle and suppressing gluconeogenesis in liver. Further independently of adiponectin, it is speculated that Seabuckthorn decreased adipocyte size, serum FFA levels and expression of TNF-α and resistin, thus contributing to amelioration of insulin sensitivity in skeletal muscle.

As pointed out adiponectin concentrations are inversely correlated with the degree of adiposity. The present data demonstrated that adiponectin levels are related to degree of insulin resistance/ hyperinsulinemia independent of weight, body mass index or waist circumference as measures of obesity. The possibility that differences in degree of adiposity or fat distribution if measured more accurately and precisely, will be helpful in determining the course complication and ultimate prognosis of patients with insulin resistance and hypoadiponectinemia may be predisposed to develop more cardiovascular risk factors than those with obesity alone (McLaughlin *et al.*, 2001a; 2002).

It is proposed that various mechanism have been observed to be responsible for improvement in insulin sensitivity that includes (1) increased post-receptor insulin signaling; (2) increased glucose transporter protein and mRNA; (3) increased muscle glucose delivery and changes in muscle composition; and (4) restoring insulin sensitivity by weight management that might be mediated mainly by mechanism other than adiponectin, by AMP-activated protein kinase pathway discussed earlier. Thus, it is concluded that Seabuckthorn improved various body composition, insulin sensitivity and enhanced adiponectin level among insulin resistance patients of obesity and metabolic syndrome.

CONCLUSION

It is concluded that adiponectin exerts a potent insulin-sensitizing role through binding of adiponectin receptors, leading to activation of AMPK, PPAR-α and presumably some other signaling pathways. Thus estimation of adiponectin is a good predictor marker for type-2 diabetes and metabolic syndrome. Increase in adiponectin levels by effective therapeutic measures is proven to be beneficial in the prevention and management of obesity, diabetes and metabolic syndrome. From the present study it is established that there is significant positive correlation between adiponectin, leptin with BMI and insulin resistance. A significant direct correlation between circulating leptin levels and insulin resistance markers is noticed. Seabuckthorn has potentiality to enhance adiponectin concentration that enhances/improves insulin sensitivity of varying etiology.

REFERENCES

1. Adamczak, M. 2003. Decreased plasma adiponectin concentration in patients with essential hypertension. *Am. J. Hypertens*. 16: 72-75.

2. Arita, Y. 1999. Paradoxical decrease of an adipose-specific protein, adiponectin, in obesity. *Biochem. Biophys. Res. Commun*. 257: 79-83.

3. Arita, Y., Kihara S., Ouchi, N., Takahashi, M., Maeda, K., Miyagawa, J., Hotta, K., Shimomura, I., Nakamura, T., Miyaoka, K., Kuriyama, H., Nishida, M., Yamashita, S., Okubo K., Matsubara, K., Muraguchi, M., Ohmoto, Y., Funahashi, T. and Matsuzawa, Y. 1999. Paradoxical decrease of an adipose-specic protein, adiponectin, in obesity. *Biochem Biophys Res Commun* 257: 79-83.

4. Berg, A.H., Combs, T.P. and Scherer, P.E. 2002. ACRP30/adiponectin: An adipokine regulating glucose and lipid metabolism. *Trends Endocrinol. Metab.* 13:84-89.

5. Berg, A.H., Combs, T.P., Du, X., Brownlee, M. and Scherer, P.E. 2001. The adipocyteşecreted protein Acrp30 enhances hepatic insulin action. *Nat. Med.* 7: 947-953.

6. Comuzzie, A.G., Funahashi, T., Sonnenberg, G., Martin, L.J., Jacob, H.J., Black, A.E., Maas, D., Takahashi, M., Kihara, S., Tanaka, S., Matsuzawa, Y., Blangero, J., Cohen, D. and Kissebah, A. 2001. The genetic basis of plasma variation in adiponectin, a global endophenotype for obesity and the metabolic syndrome. *J. Clin. Endocrinol. Metab.* 86: 4321-4325.

7. Diez, J.J. and Iglesias, P. 2003. Role of novel adipocyte derived hormone adiponectin in human disease. *Eur. J. Clin. Endocrinol.*148: 293-300.

8. Ferrannini, E., Natali, A., Bell, P., Cavallo-Perin, P., Lalic, N. and Mingrone, G. 1997. Insulin resistance and hypersecretion in obesity: European group for the study of insulin resistance (EGIR). *J. Clin. Invest*. 100: 1166-1173.

9. Hotta, K., Funahashi, T., Bodkin, N.L., Ortmeyer, H.K., Arita, Y., Hansen, B.C. and Matsuzawa, Y. 2001. Circulating concentrations of the adipocyte protein adiponectin are decreased in parallel with reduced insulin sensitivity during the progression to type 2 diabetes in rhesus monkeys. *Diabetes* 50: 1126-1133.

10. Kadowaki, T., Hara, K., Yamauchi, T., Terauchi, Y., Tobe, K. and Nagai, R. 2003. Molecular mechanisms of insulin resistance and obesity. *Exp. Biol. Med*. 228: 1111-1117.

11. Kubota, N., Terauchi, Y., Kubota, T., Kumagai, H., Itoh, S., Satoh, H., Yano, W., Ogata, H., Tokuyama, K., Takamoto, I., Mineyama, T., Ishikawa, M., Moroi, M., Sugi, K., Yamauchi, T., Ueki, K., Tobe, K., Noda, T., Nagai, R., and Kadowaki, T. 2006. Pioglitazone ameliorates insulin resistance and diabetes by both adiponectin-dependent and -independent pathways. *J. Biol. Chem*. 281: 8748-8755.

12. Kubota, N., Terauchi, Y., Yamauchi, T., Kubota, T., Moroi, M., Matsui, J., Eto, K., Yamashita, T., Kamon, J., Satoh, H., Yano, W., Froguel, P., Nagai, R., Kimura, S., Kadowaki, T. and Noda, T. 2002. Disruption of adiponectin causes insulin resistance and neointimal formation. *J. Biol. Chem*. 277: 25863-25866.

13. Kubota, N., Terauchi, Y., Yamauchi, T., Kubota, T., Moroi, M., Matsui, J., Eto, K., Yamashita, T., Kamon, J., Satoh, H., Yano, W., Froguel, P., Nagai, R., Kimura, S., Kadowaki, T. and Noda, T. 2002. Disruption of adiponectin causes insulin resistance and neointimal formation. *J. Biol. Chem*. 277: 25863-25866.

14. Kumada, M., Kihara, S., Sumitsuji, S., Kawamoto, T., Matsumoto, S., Ouchi, N., Arita, Y., Okamoto, Y., Shimomura, I., Hiraoka, H., Nakamura, T., Funahashi, T. and Matsuzawa, Y. 2003. Association of hypoadiponectinemia with coronary artery disease in men. *Arterioscler. Thromb. Vasc. Biol.* 23: 85-89.

15. Lowell, B.B. 1999. PPARc: An essential regulator of adipogenesis and modulator of fat cell function. *Cell* 99: 239-242.

16. Matsushita, K., *et al.*, 2006. Comparison of circulating adiponectin and proinflammatory markers regarding their association with metabolic syndrome in Japanese men. *Arterioscler. Thromb. Vasc. Biol.* 26: 871-876.

17. McLaughlin, T., Abbasi, F., Kim, H.S., Lamendola, C., Schaaf, P. and Reaven, G. 2001a. Relationship between insulin resistance, weight loss, and coronary heart disease risk in healthy, obese women. *Metabolism* 50: 795-800.

18. McLaughlin, T., Abbasi, F., Lamendola, C., Kim, H.S. and Reaven, G.M. 2001b. Metabolic changes following sibutramine-assisted weight loss in obese individuals: role of plasma free fatty acids in the insulin resistance of obesity. *Metabolism* 50: 819-824.

19. McLaughlin, T., Abbasi, F., Lamendola, C., Liang, L.G., Schaaf, P. and Reaven, P. 2002. Differentiation between obesity and insulin resistance in the association with C-reactive protein. *Circulation*106: 2908-2912.

20. Nemet, D., Wang, P., Funahashi, T., Matsuzawa, Y., Tanaka, S., Engelman, L. and Cooper, D.M. 2003. Adipocytokines, body composition, and tness in children. *Pediatr. Res.* 53: 148-152.

21. Olefsky, J., Reaven, G.M. and Farquhar, J.W. 1974. Effects of weight reduction on obesity: Studies of lipid and carbohydrate metabolism in normal and hyperlipoproteinemic subjects. *J. Clin. Invest.* 53: 64-76.

22. Oral, E.A., Simha, V., Ruiz, E., Andewelt, A., Premkumar, A., Snell, P., Wagner, A.J., DePaoli, A.M., Reitman, M.L., Taylor, S.I., Gorden, P. and Garg, A. 2002. Leptin-replacement therapy for lipodystrophy. *N. Engl. J. Med.* 346: 570-578.

23. Ouchi, N., Ohishi, M., Kihara, S., Funahashi, T., Nakamura, T., Nagaretani, H., Kumada, M., Ohashi, K., Okamoto, Y., Nishizawa H., Kishida, K., Maeda, N., Nagasawa, A., Kobayashi, H., Hiraoka, H., Komai, N., Kaibe, M., Rakugi, H., Ogihara, T. and Matsuzawa, Y. 2003. Association of hypoadiponectinemia with impaired vasoreactivity. *Hypertension* 42: 231-234.

24. Pischon, C.J., Girman, G.S. and Hotamisligil, G.S. 2004. Plasma adiponectin levels and risk of myocardial infarction in men. *JAMA* 291: 1730-1737.

25. Ryo, M., Nakamura, T., Kihara, S., Kumada, M., Shibazaki, S., Takahashi, M., Nagai, M., Matsuzawa, Y. and Funahashi, T. 2004. Adiponectin as a biomarker of the metabolic syndrome. *Circ. J.* 68: 975-981.

26. Stefan, N. and Stumvoll, M. 2002.Adiponectin: Its role in metabolism and beyond. *Horm. Metab. Res.* 43: 469-474.

27. Tschritter, O., Fritsche, A., Thamer, C., Haap, M., Shirkavand, F., Rahe, S., Staiger, H., Maerker, E., Haring, H. and Stumvoll, M. 2003. Plasma adiponectin concentrations predict insulin sensitivity of both glucose and lipid metabolism. *Diabetes* 52: 239-243.

28. Weyer, C., Funahashi, T., Tanaka, S., Hotta, K., Matsuzawa, Y., Pratley, R.E. and Tataranni, P.A. 2001. Hypoadiponectinemia in obesity and type 2 diabetes: close association with insulin resistance and hyperinsulinemia. *J. Clin. Endocrinol. Metab.* 86: 1930-1935.

29. Yamamoto, Y., Hirose, H., Saito, I., Nishikai, K. and Saruta, T. 2004. Adiponectin, an adipocytederived protein, predicts future insulin-resistance: Two-year follow-up study in Japanese population. *J. Clin. Endocrinol. Metab.* 89:87-90.

30. Yamamoto, Y., Hirose, H., Saito, I., Tomita, M., Taniyama, M., Matsubara, K., Okazaki, Y., Ishii, T., Nishikai, K. and Saruta, T. 2002. Correlation of the adipocytederived protein adiponectin with insulin resistance index and serum high-density lipoprotein-cholesterol, independent of body mass index, in the Japanese population. *Clin. Sci.* (*Lond.*) 103: 137-142.

31. Yamauchi, T. and Kamon, J. 2002. Adiponectin stimulates glucose utilization and fatty acid oxidation by activating AMP-kinase. *Nat. Med.* 8: 1288-1295.

32. Yamauchi, T., Kamon, J., Waki, H., Terauchi, Y., Kubota, N., Hara, K., Mori, Y., Ide, T., Murakami, K, Tsuboyama-Kasaoka, N., Ezaki, O., Akanuma, Y., Gavrilova, O., Vinson, C., Reitman, M.L., Kagechika, H., Shudo, K., Yoda, M., Nakano, Y., Tobe, K., Nagai, R., Kimura, S., Tomita, M., Froguel, P. and Kadowaki, T. 2001. The fat-derived hormone adiponectin reverses insulin resistance associated with both lipoatrophy and obesity. *Nat. Med.* 7: 941-946.

33. Yang, W.S., Lee, W.J., Funahashi, T., Tanaka, S., Matsuzawa, Y., Chao, C.L., Chen, C.L., Tai, T.Y. and Chuang, L.M. 2001. Weight reduction increases plasma levels of an adipose-derived anti-inammatory protein, adiponectin. *J. Clin. Endocrinol. Metab.* 86: 3815-3819.

34. Yatagai, T., Nagasaka, S., Taniguchi, A., Fukushima, M., Nakamura, T., Kuroe, A., Nakai. Y. and Ishibashi, S. 2003. Hypoadiponectinemia is associated with visceral fat accumulation and insulin resistance in Japanese men with type 2 diabetes mellitus. *Metabolism* 52: 1274-1278.

PART VII: PROCESSING TECHNOLOGIES

Chapter 40

Bioactivity of a Non-traditional Seabuckthorn (*Hippophae rhamnoides* L.) Wine

Gargi Dey and Bharti Negi

Department of Biotechnology and Bioinformatics,
Jaypee University of Information Technology, Waknaghat, Solan – 173 234, H.P., India

SUMMARY

Indian Himalayas region is very well recognized for the rich biodiversity as its inhabitants are largely dependent on this for their wealth, and long-term nutrition and health. The species of *Hippophae*, commonly known as seabuckthorn, is very rich in phytonutrients including flavonoids and has a potential to provide healthy foods. The indigenous uses of *Hippophae* sp. in the tribal communities have been in juice, jam and wine preparation among which wine has the highest cost benefit. This work relates to development of a non-traditional fermented product of seabuckthorn (*Hippophae rhamnoides* L.) with *Saccharomyces cerevisiae* and *Issatchenkia orientalis*. The most significant effect of co-fermentation on physico-chemical characteristics was the reduction of malic acid content to almost half, which is highly desirable attribute for acidic musts like seabuckthorn. The aim of the present study was to evaluate the effects of seabuckthorn wine against high-cholesterol diet (HCD) induced hypercholesterolemia in male LACA mice. For this, the animals were divided into following three groups: Group 1 animals served as control; group 2 animals were fed on high-cholesterol diet (HCD); and group 3 animals were fed on HCD with seabuckthorn wine. Administration of HCD resulted in significant increase in total cholesterol, triglycerides, HDL-cholesterol (HDL-C) and LDL-cholesterol (LDL-C) as compared to control group. The administration of seabuckthorn wine was able to bring down the total cholesterol content and LDL-cholesterol in group 3 as compared to group 2. The seabuckthorn wine supplementation also exhibited a significant increase in HDL-C/LDL-C ratio as compared to high-cholesterol diet-fed mice. These studies provide an important evidence of bioactivity of the non-traditional seabuckthorn wine against high cholesterol diet.

Keywords: *Seabuckthorn, Co-fermentation, Wine, Lipid profile and High-cholesterol diet.*

Seabuckthorn (Hippophae L.)*: A Multipurpose Wonder Plant, Vol. 4: Emerging Trends in Research and Technologies* (Virendra Singh, *Editor-in-Chief,* 2014), p. 491–499, Daya Publishing House, New Delhi, India.

INTRODUCTION

The chronic inflammation disease, atherosclerosis, is a typical multi-factorial disease most often dependent on risk factors, such as hypercholesterolemia, diabetes and hypertension. Several previous studies (Waddington *et al.*, 2004; Corder *et al.*, 2006; Forester and Waterhouse, 2009) suggested that moderate red wine intake decreases cardiac mortality due to atherosclerosis. The mechanistic studies indicate that most beneficial effects of drinking are attributable to flavonoids that are present in the fruit pulps and juices. In this context, fruit wines from blueberries, blackcurrant, cranberries have been reported to have comparable or even higher flavonoid and phenolic contents (Vuorinen *et al.*, 2000; Pinhero and Paliyath, 2001; Dey *et al.*, 2009; Arozarena *et al.*, 2012; Mudnic *et al.*, 2012). The Indian Himalayas region is very well recognized for the rich biodiversity as its inhabitants are largely dependent on this for their wealth, and long-term nutrition and health. The species of *Hippophae*, commonly known as seabuckthorn, is very rich in phytonutrients including flavonoids and has a potential to provide health foods. These berries contain essential oils and vitamin C, proteins (especially, globulins and albumins), carotene, fatty acids (saturated and unsaturated), free amino acids, flavonoids and vitamin E (Guliyer *et al.*, 2004; Zheng *et al.*, 2010; Suryakumar and Gupta, 2011). A wide spectrum of pharmacological effects of seabuckthorn have been recently reported including antioxidant, immunomodulatory, anti-atherogenic, anti-stress, hepatoprotective, radioprotective and tissue repair (Suleyman *et al.*, 2001; Geetha *et al.*, 2002; Basu *et al.*, 2007; Chawla *et al.*, 2007; Upadhyay *et al.*, 2011). *In vitro* assays evidenced distinct cytoprotective properties of seabuckthorn berry extracts, which could be attributed to their antioxidant activity (Raffo *et al.*, 2004).

The indigenous uses of seabuckthorn in the tribal communities have been in juice, jam and wine preparation. Interestingly, the cost benefit analysis of the seabuckthorn products developed showed that seabuckthorn wine fetched the highest net profit of Rs 140/litre as compared to Rs 70/litre for juice and Rs 10-15/kg of fresh fruits (Singh *et al.*, 2011). With the appropriate value-addition and adequate proof of concept of health claims of traditional or non-traditional wine products, these can be in great demand by local consumers and tourists.

In our laboratory, we have explored the strains *Saccharomyces cerevisiae* (MTCC 10421) and *Issatchenkia orientalis* (MTCC 6191) for the development of a quality wine from seabuckthorn juice. In this study, we present the effect of co-fermentation on the physico-chemical characteristics and more interestingly, the *in vivo* effects of seabuckthorn wine on serum levels of lipids in diet-induced hypercholesterolemic mice.

MATERIALS AND METHODS

Co-fermentation Process

Fresh seabuckthorn berries were collected from Spiti, Himachal Pradesh and juice extraction from the berry was carried out using pressing technique according to Bump (1989). Extracted juice was thermally treated at 85 °C for 30 min and stored at 4 °C for further use. The co-fermentation of seabuckthorn juice was carried out by the method of Dey and Negi (2012) using two yeast strains, *Saccharomyces cerevisiae* (MTCC 10421) and *Issatchenkia orientalis* (MTCC 6191) which were obtained from Institute of Microbial Technology (IMTECH), Chandigarh, India (Figures 40.1 and 40.2).

Physico-chemical and Organic Acid Analysis

Reducing sugar was estimated by the method of Miller (1959). The ethanol content and titratable acidity were estimated by the methods described by Joshi (1997).

Figure 40.1: Bioreactor for Co-fermentation Set Up.

Figure 40.2: Seabuckthorn Wine.

Lactic, acetic, citric and malic acids were determined using high performance liquid chromatography (HPLC) according to Ong *et al*. (2006). An Aminex HPX-87H column (300 × 7.8 mm, Bio-Rad Laboratory, Richmond, California, U.S.A.), was used. Sulphuric acid (0.009 N) filtered through a 0.45 μm membrane filter (Millex) was used as a mobile phase at a flow rate of 0.6 ml/min. The detection device was an ultraviolet-visible detector set at 220 nm with running time of 15 min.

Ascorbic acid was estimated by the method of Raffo *et al.* (2004). The chromatographic column was Prevail C18 (5 µm, 250 × 4.6 mm) and the ascorbic acid was eluted isocratically with 25 mM KH_2PO_4 (pH 2.8) at a flow rate of 1 ml/min and the eluate was monitored at 214 nm.

In-vivo Study

Male LACA mice (25-30 g body weight, 8-10 weeks of age) were obtained from the Central Animal House, Panjab University, Chandigarh, India. The animals were kept in plastic cages under hygienic conditions in the animal rooms of department with 12/12 h dark-light cycle. They were allowed free access to standard pellet diet (Hindustan Lever Pvt. Ltd. India) and water *ad libitum*. Mice were acclimatized to the animal room conditions for a week before initiating the study. Necessary approvals (CAH/131) for animal studies were obtained from the Institutional Ethics Committee, Panjab University, Chandigarh. The animals care and handling was done according to the guidelines set by the World Health Organization (WHO), Geneva, Switzerland and the Indian National Science Academy (INSA), New Delhi, India. The treatment protocol was for 15 days. Animals were divided into three groups (six animals in each group) for the study. Group 1 animals served as control. These animals were fed on normal chow diet. Group 2 animals were fed on high-cholesterol diet (HCD) (2 per cent cholesterol with normal chow diet). Group 3 animals were fed on HCD with seabuckthorn wine. 500 µl (per 25 g body weight of animal) of seabuckthorn wine was given daily through oral administration for 15 days.

Body weights of mice and their daily food and fluid intake were recorded on alternate days. After completion of treatment period, blood was drawn from the retro-orbital sinus of the ether anaesthetized overnight (12 h) fasting mice. Blood was collected into a clean, sterile, labelled centrifuge tubes without an anticoagulant. After clotting, the serum was separated by centrifugation at 1500 g for 10 min.

Lipid Profile Analysis

The serum lipid profile analysis was performed using colorimetric kits to estimate the total cholesterol, triglycerides, high-density lipoprotein cholesterol (HDL-C) and low-density lipoprotein cholesterol (LDL-C) levels. Total cholesterol level was estimated using CHOD-POD (cholesterol oxidase/peroxidase phosphotungstate method) based kit obtained from Human Diagnostics, Germany. Triglyceride levels and HDL-C were estimated using commercial available assay kits (Sigma-Aldrich Co., USA). LDL-C was calculated by the Friedewald formula (Friedewald *et al.*, 1972).

Statistical Analysis

All data were presented as the average of triplicate experiments with standard deviation. Results were statistically interpreted with one-way analysis of variance (ANOVA) followed by Tukey's post hoc analysis to locate the significant differences indicated with ANOVA. The data for ANOVA were analyzed at different significance level using statistical package MSTAT/Minitab (Minitab Inc. USA, Version 13, 2004 for Windows®).

RESULTS

Fermentation is a potent tool in the production of quality wines with balance physico-chemical characteristics, especially acids and residual sugar. In this context, we developed a co-fermented product from seabuckthorn (*Hippophae rhamnoides* L.) berries with low alcohol content.

Effect of Co-fermentation on Physico-chemical Properties and Organic Acids

The physico-chemical characteristics of seabuckthorn juice and co-fermented seabuckthorn wine are presented in Table 40.1. The reducing sugar content was 1109 mg/100 ml in seabuckthorn juice which decreased to 800 mg/100ml after the co-fermentation. The alcoholic content of the wine was 8.6 per cent (v/v). The pH variation as a result of fermentation of seabuckthorn juice was from 2.50 to 3.89. A significant decrease was observed in the titratable acidity of seabuckthorn juice from 923 mg/100 ml to 230 mg/100 ml after the co-fermentation.

Table 40.1: Physico-chemical Characteristics of Seabuckthorn Juice and Wine

	Reducing Sugar (mg/100 ml)	*pH*	*Alcohol per cent (v/v)*	*Titratable Acidity (mg/100 ml)*
Seabuckthorn juice	1109 ± 17. 18	2.5 ± 0.03	–	923 ± 15.12
Seabuckthorn wine	800 ± 15.30[a]	3.89 ± 0.01[a]	8.6 ± 0.56	230 ± 3.70[a]

a: The values indicate a significant difference ($p \leq 0.001$) between seabuckthorn juice and co-fermented seabuckthorn product based on one-way ANOVA followed by Tukey's post hoc analysis.

Further, the effect of co-fermentation on organic acids (acetic, lactic, malic, citric and ascorbic acid) contents in seabuckthorn wine was estimated using HPLC (Table 40.2). The malic acid content decreased from 2862 to 1287 mg/100 ml and ascorbic acid from 20.56 to 17.68 mg/100 ml. However, no significant variation was observed in the content of other organic acids. The co-fermentation resulted in a seabuckthorn product of low alcohol content.

Table 40.2: Comparative Study of Organic Acids of Seabuckthorn Juice and Wine

Organic Acids (mg/100 ml)	*Seabuckthorn Juice*	*Seabuckthorn Wine*
Acetic acid	11 ± 0.05	8 ± 0.09
Lactic acid	46 ± 0.14	51 ± 0.17
Malic acid	2862 ± 20.45	1287 ± 10.81[a]
Citric acid	252 ± 4.87	247 ± 6.49
Ascorbic acid	20.56 ± 0.54	17.68 ± 0.22[a]

a: The variation in the values are found to be significantly different at the level of $p \leq 0.001$ based on one-way ANOVA followed by Tukey's post hoc analysis.

Effect of Seabuckthorn Wine Administration on Lipid Profile

We were more interested in evaluating the bioactivity of this fermented seabuckthorn product and hence, we examined the impact of seabuckthorn wine administration on high-cholesterol diet induced hypercholesterolemia in male LACA mice. For this study, animals were divided into following three groups: Group 1 animals served as control; group 2 animals were fed on high-cholesterol diet (HCD); and group 3 animals were fed on HCD with seabuckthorn wine. The final body weight of mice in group 2 (35.17 g) were significantly higher ($p \leq 0.001$) than that of the group 1 (27.14 g) and group 3 (27.78 g). The oral administration of seabuckthorn wine along with high-cholesterol diet did not lead to any change in body weights of mice.

To evaluate the extent of hypercholesterolemia, lipid profile was analyzed in all the three groups (Table 40.3). Mice were fasted overnight (12 h) before blood collection. Administration of high-cholesterol diet (HCD) resulted in increase in total cholesterol, triglycerides, HDL-cholesterol (HDL-C) and LDL-cholesterol (LDL-C) significantly ($p \leq 0.001$) in group 2 as compared to group 1 (control). On administration of seabuckthorn wine, total cholesterol content decreased significantly ($p \leq 0.001$) in group 3 as compared to group 2. Also, the LDL-cholesterol showed a significant decrease ($p \leq 0.001$) to 15.16 mg/dl in group 3 from 44.18 mg/dl in group 2. On the other hand, the triglycerides, HDL-cholesterol and VLDL-cholesterol did not show any remarkable change. The ratio of HDL-cholesterol to LDL-cholesterol depicted a noteworthy decrease in group 2 (HDL-C/LDL-C: 1.50) as compared to group 1 (HDL-C/LDL-C: 3.82), while on the oral administration of seabuckthorn wine along with HCD (group 3), this ratio was 4.46.

Table 40.3: Effect of Seabuckthorn Wine Administration on Serum Lipid Profile

	Group 1 (Control Mice)	*Group 2 (High-cholesterol diet treated mice)*	*Group 3 (High-cholesterol diet + Seabuckthorn wine treated mice)*
Total Cholesterol (mg/dl)	76.26 ± 3.04	130.45 ± 5.59[a]	102.56 ± 2.80[a,b]
Triglycerides (mg/dl)	80.89 ± 4.25	100.02 ± 7.90[a]	99.22 ± 6.55[a]
HDL Cholesterol (mg/dl)	47.62 ± 2.47	66.27 ± 4.46[a]	67.56 ± 2.55[a]
LDL Cholesterol (mg/dl)	12.46 ± 1.65	44.18 ± 3.30[a]	15.16 ± 1.51[b]
VLDL (mg/dl)	16.18 ± 1.21	20.00 ± 1.46[a]	19.84 ± 1.31
HDL/LDL	3.82	1.50	4.46

Effect of seabuckthorn wine on diet induced hypercholesterolemia in blood serum.

a: Comparison of group 2 and 3 *vs.* control group (group 1). b: Comparison of group 3 vs. group 2.

DISCUSSION

Seabuckthorn berries have many therapeutically useful compounds. But, the main drawback of these berries is its very high malic acid content and low pH, which has limited the proper post-harvest utilization of these berries for value added products. We standardized and developed a method for production of low alcohol fermented product using co-fermentation with *S. cerevisiae* (MTCC 10421) and *I. orientalis* (MTCC 6191). The main rationale for using *I. orientalis* was that it is an acidophilus yeast with an ability to degrade malic acid (Hong *et al.*, 2010) as *Saccharomyces* sp. which are used for wine production cannot reduce malic acid proficiently (Volschenk *et al.*, 2001).

Sugar is known as the most important raw material during fermentation process which dictates the final concentration of alcohol in wine. The production of alcohol in co-fermented seabuckthorn product was only 8.6 per cent (v/v) in spite of a significant decrease (28 per cent) in the content of reducing sugar (Table 40.1). Our result corroborates the study of Kim *et al.* (2008) who reported the low alcohol production rate (7.8 per cent, v/v) after the co-fermentation of grape must by *S. cerevisiae* and *I. orientalis*. The content of reducing sugar indicated that the final product of seabuckthorn was of semi dry type with pH of 3.89, because as per Conde *et al.* (2007), technically dry wines have sugar concentration of less than 400 mg/100 ml. The process developed in our laboratory for seabuckthorn wine resulted in a table wine, as the alcohol contents of such wines are usually 7 to 16 per cent of alcohol (Joshi, 1997).

The seabuckthorn juice contained a high level of malic acid content (2862 mg/100 ml) which would negatively affect the wine quality (Table 40.2). Malic acid is one of the major organic acids contributes to the acidic taste of wine and serves as a substrate for contaminating wine after bottling (Henick-Kling, 1993). Thus, it was important to remove excess malic acid from the juice to guarantee the quality of the final product (Pretorius, 2000). In the present study, we noticed that *I. orientalis* degraded about 55 per cent of malic acid in seabuckthorn juice. The co-fermentation also caused a 14 per cent decrease in ascorbic acid. However, other organic acids (acetic, lactic and citric acid) were not significantly influenced by the co-fermentation. Similar results have been reported by previous studies where an attempt was made to degrade the malic acid in wine through the *I. orientalis* (Seo *et al.*, 2007; Kim *et al.*, 2008; Hong *et al.*, 2010).

In order to evaluate the effects of the non-traditional seabuckthorn wine on serum levels of lipids, we used diet-induced hypercholesterolemic mice. The triglycerides in serum increased by 24 per cent in the HCD group as compared to the control group. Supplementation of seabuckthorn wine did not lower the serum triglyceride levels significantly. Interestingly, however, there was the decrease of serum total cholesterol by 21 per cent as a result of seabuckthorn wine administration as compared to the HCD treated group. More than 60 per cent reduction in LDL-cholesterol was observed in group 3 (HCD + seabuckthorn wine) in comparison to group 2 mice. A significant increase (197 per cent) in the HDL-C/LDL-C ratio was also observed in mice fed with seabuckthorn wine in comparison to HCD group. It may be concluded that the increase in HDL/LDL-cholesterol ratio is more significant due to the decrease in LDL-cholesterol concentration. Considering that high LDL-cholesterol levels is correlated to high atherogenic process (Ho *et al.*, 2003), the effect of seabuckthorn wine on reduction of LDL-cholesterol levels is an important finding of our study. The various *in vivo* studies reported in past by other workers on the effects of different seabuckthorn products on the lipid profile are not conclusive. Our results are similar to that of a recent study where the administration of a seabuckthorn product (seabuckthorn tea) showed a reduction in the plasma total cholesterol in high fat diet-induced obese mice in 42 days, but did not affect the plasma HDL-cholesterol concentration (Lee *et al.*, 2011).On the other hand, Larmo *et al.* (2009) did not detect any changes in total serum, LDL and HDL-cholesterol in healthy adults after consumption of seabuckthorn berries. Since our studies indicated that seabuckthorn wine reduced the LDL-cholesterol levels, it may be expected that the dietary interventions with this wine are likely to prove beneficial.

CONCLUSION

The present study provided a valuable insight into the potential bioactivity of seabuckthorn wine. It suggested that inclusion of seabuckthorn wine in diet may be an effective way of lowering serum cholesterol level, however, further studies on the mechanism of action of the wine needs to be investigated.

ACKNOWLEDGEMENT

The authors are grateful to Prof. Mohinder P. Bansal (Department of Biophysics, Panjab University, Chandigarh, India) under whose supervision the animal studies were carried out. We also thank the laboratory technicians of Panjab University for providing assistance during the conduct of *in vivo* studies.

REFERENCES

1. Arozarena, I., Ortiz, J., Hermosin-Gutierrez, I., Urretavizcaya, I., Salivatierra, S., Cordova, I., Marin-Arroyo, M.R., Noriega, M. and Novarro, M. 2012. Colour, ellagitannins, anthocyanins and antioxidant activity of Andean blackberry wine. *J. Agric. Food Chem*. 60 (30): 7463-7473.
2. Basu, M., Prasad, R., Jayamurthy, P., Pal, K., Arumughan, C. and Sawhney, R.C. 2007. Anti-atherogenic effects of seabuckthorn (*Hippophae rhamnoides*) seed oil. *Phytomedicine* 14: 770-777.
3. Bump, V.L. 1989. Apple pressing and juice extraction. In: *Processed Apple Products* (D.L. Downing, Ed.) p.53-82, Avi, Van Nostrand Reinhold, New York, 210p.
4. Chawla, R., Arora, R., Singh, S., Sagar, R.K., Sharma, K., Kumar, R., Sharma, A., Gupta, M.L., Singh, S., Prasad, J., Khan, H.A., Swaroop, A., Sinha, A. K., Gupta, A. K., Tripath, R.P. and Ahuja, P.S. 2007. Radioprotective and antioxidant activity of fractionated extracts of berries of *Hippophae rhamnoides*. *J. Med. Food* 10:101-109.
5. Conde, C., Silva, P., Fontes, N., Dias, A.C.P., Tavares, R.M., Sousa, M.J., Agasse, A., Delrot, S. and Geros, H. 2007. Biochemical changes throughout grape berry development and fruit and wine quality. *Food* 1 (1): 1-22.
6. Corder, R., Mullen, W., Khan, N.Q., Marks, S.C., Wood, *E.G.*, Carrier, M.J. and Crozier, A. 2006. Oenology: Red wine procyanidins and vascular health. *Nature* 444: 566-568.
7. Dey, G., Negi, B. and Gandhi, A. 2009. Can fruit wines be considered as functional foods-An overview. *Natural Products Radiance* 8: 314-322.
8. Dey, G. and Negi, B. 2012. "Novel method for producing seabuckthorn wine and wine thereof" Application no. 499/DEL/2010, *Indian Patent*, Publication Date (U/S11A) 10/02/2012.
9. Forester, S.C. and Waterhouse, A.L. 2009. Metabolites are key to understanding health effects of wine polyphenolics. *Journal of Nutrition* 139 (9): 18245-18315.
10. Friedewald, W.T., Levy, R.I. and Fredrickson, D.S. 1972. Estimation of the concentration of low-density lipoprotein cholesterol in plasma, without use of the preparative ultracentrifuge. *Clin.Chem.* 18: 499-502.

11. Geetha, S., Sai Ram, M., Singh, V., Ilavazhagan, G. and Sawhney, R.C. 2002. Antioxidant and immunomodulatory properties of seabuckthorn (*Hippophae rhamnoides*)-An *in vitro* study. *J. Ethnopharmacol.* 79: 373-378.

12. Guliyer, V.B., Gul, M. and Yildirim, A. 2004. *Hippophae rhamnoides* L: Chromatographic methods to determine composition and use in traditional medicine and pharmacological effects. *J. Chromatogr. B* 812: 291-307.

13. Henick-Kling, T. 1993. Malolactic fermentation. In: *Wine Microbiology and Biotechnology* (G.H. Fleet, Ed.), p.289-326, Harwood Academic, Church, Switzerland, 513p.

14. Ho, H.M., Leung, L.K., Chan, F. L., Huang, Y. and. Chen, Z.Y. 2003. Soy leaf lowers the ratio of non-HDL to HDL cholesterol in hamsters. *J. Agric. Food Chem.* 51: 4554-4558.

15. Hong, S.K., Lee, H.J., Park, H.J., Hong, Y.A., Rhee, I.K., Lee, W.H., Choi, S.W., Lee, O.S. and Park, H.D. 2010. Degradation of malic acid in wine by immobilized *Issatchenkia orientalis* cells with oriental oak charcoal and alginate. *Lett. Appl. Microbiol.* 50: 522-529.

16. Joshi, V.K. 1997. *Fruit Wine* (2nd Ed.), India: Directorate of Extension Education, Dr. Y S Parmar University of Horticulture and Forestry, Nauni, Solan (HP), 255p.

17. Kim, D.H., Hong, Y.A. and Park, H.D. 2008. Co-fermentation of grape must by *Issatchenkia orientalis* and *Saccharomyces cerevisiae* reduces the malic acid content in wine. *Biotechol. Lett.* 30: 1633-1638.

18. Larmo, P.S., Yang, B., Hurme, S.A.M., Alin, J.A., Kallio, H., Salminen, E.K. and Tahvonen, R.L. 2009. Effect of a low dose of seabuckthorn berries on circulating concentrations of cholesterol, triacylglycerols, and flavonols in healthy adults. *Eur. J. Nutr.* 48: 277-282.

19. Lee, H.I., Kim, M.S., Lee, K.M., Park, S.K., Seo, K. L., Kim, H. J., Kim, M.J., Choi, M.S. and Lee, M.K. 2011. Anti-visceral obesity and antioxidant effects of powdered seabuckthorn (*Hippophae rhamnoides* L.) leaf tea in diet-induced obese mice. *Food Chem. Toxicol.* 49: 2370-2376.

20. Miller, G.L. 1959. Use of dinitrosalicylic acid reagent for determination of reducing sugar. *Anal. Chem.* 31: 426-428.

21. Mudnic, I., Budimir, D., Modun, D., Gunjaaca, G., Generalic, I., Skroza, D., Katalinic, V., Ljubenkov, I. and Bobaln, M. 2012. Antioxidant and vasodilatory effects of blackberry and grape wines. *J. Med. Food* 15 (3): 315-321.

22. Ong, L., Henriksson, A. and Shah, N.P. 2006. Development of probiotic Cheddar cheese containing *Lactobacillus acidophilus*, *Lb. casei*, *Lb. Paracasei* and *Bifidobacterium* spp. and the influence of these bacteria on proteolytic patterns and production of organic acid. *Int. Dairy J.* 16: 446-456.

23. Pinhero, R.G. and Paliyath, G. 2001. Antioxidant and calmodulin inhibitory activities of phenolic compounds in fruit wines and its biotechnological implications. *Food Biotechnology* 15 (3): 179-192.

24. Pretorius, I.S. 2000. Tailoring wine yeast for the new miilennium: Novel approaches to the ancient art of wine making. *Yeast* 16: 675-729.

25. Raffo, A., Paoletti, F. and Antonelli, M. 2004. Changes in sugar, organic acid, flavonol and carotenoid composition during ripening of berries of three seabuckthorn (*Hippophae rhamnoides* L.) cultivars. *Eur. Food Res. Technol.* 219: 360-368.

26. Seo, S.H., Rhee, C.H. and Park, H.D. 2007. Degradation of malic acid by *Issatchenkia orientalis* KMBL 5774, an acidophilic yeast strain isolated from Korean grape wine pomace. *J. Microbiol.* 45: 521-527.

27. Singh, A., Samant, S.S., Lal, M., Sharma, P., Butola, J.S. and Marpa, S. 2011. Assessment, mapping and harnessing, economic, potential of *Hippophae* sp. for the socio-economic development of tribal communities in Himachal Pradesh, India. In: *Proceedings of National Conference on Seabuckthorn* (V. Singh, Ed. in Chief), p. 260-269, Dec 1-3, 2011, CSK HPKV, Palampur, 324p.

28. Suleyman, H., Demirezer L.O., Buyukokuroglu, M.E., Akcay, M.F., Gepdiremen, A., Banoglu, Z.N. and Gocer, F. 2001. Antiulcerogenic effect of *Hippophae rhamnoides*. *Phytotherapy Res.* 33: 77-81.

29. Suryakumar, G. and Gupta, A. 2011. Medicinal and therapeutic potential of seabuckthorn (*Hippophae rhamnoides* L.). *J. of Ethnopharmacol.* 138: 268-278.

30. Upadhyay, N.K., Kumar, R., Siddiqui, M.S., and Gupta, A. 2011. Mechanism of wound healing activity of *Hippophae rhamnoides* L. leaf extract in experimental burns. *Evid. Based Complement Alternat. Med. doi:10.1093/ecam/nep 189.*

31. Volschenk, H., Viljoen-Bloom, M., Subden, R.E. and van Vuuren. H.J.J. 2001. Malo-ethanolic fermentation in grape must by recombinant strains of *Saccharomyces cerevisiae*. *Yeast* 18: 963-970.

32. Vuorinen, H., Maatta, K. and Torronen. 2000. Content of flavonols, myrecetin, quercitin and kaempferol in Finnish berry wines. *J. Agric. Food Chem.* 48: 2675-2680.

33. Waddington, E., Puddey, I.B. and Croft, K.D. 2004. Red wine polyphenolic compounds inhibit atherosclerosis in apolipoprotein E-deficient mice independently of effects on lipid peroxidation. *Am. J. Clin. Nutr.* 79: 54-61.

34. Zheng, J., Kallio, H., Linderborg, K. and Yang, B. 2010. Sugars, sugar alcohols, fruit acids, and ascorbic acid in wild Chinese seabuckthorn (*Hippophae rhamnoides* ssp. *sinensis*) with special reference to influence of latitude and altitude. *Food Res. International. doi:10.1016/j.foodres.2010.10.007.*

Chapter 41

Formulation and Characterization of Seabuckthorn Oil Emulsion

Snezana Cupara[1], Milica Ninkovic[2], Ivana Arsic[3], Vanja Tadic[4], Milan Knezevic[1], and Slobodan Jankovic[1]

[1] Faculty of Medical Sciences, University of Kragujevac, Kragujevac, Serbia
[2] Institute of Medical Research, Military Medical Academy, Belgrade, Serbia
[3] Medical Faculty, University of Nis, Serbia
[4] Institute for Medicinal Plant Research "Dr Josif Pancic", Belgrade, Serbia

SUMMARY

Seabuckthorn oil (*Hippophae rhamnoides* L.) has been medicinally effective in external application, but the application is unsuitable due to its liquid, lipophilic and highly coloured nature. These difficulties has been overcome by formulation of semisolid emulsion with seabuckthorn oil. The formulation of emulsion with seabuckthorn oil and combination of emulsifiers that form liquid crystals has been proposed and two different quantities of seabuckthorn oil were incorporated. The samples of semisolid emulsion were prepared with 10 per cent and 40 per cent of seabuckthorn oil. The wound healing potential and suitability for topical use were evaluated. The wound healing potential was evaluated on excised model animal during 4, 8, and 14 days and histo-pathological examination of tissue specimens was done. The accelerated wound healing was noted in the group treated 8 days by sample containing lower concentration of seabuckthorn oil. The samples of seabuckthorn emulsion were orange, semisolid, shiny, easily spread on skin, and the smear on glass plate was homogen. There was absence of smell and the emulsion was oil/water type. These results suggested that emulsion with enhanced structure by liquid crystal is suitable vehicle for seabuckthorn oil, attaining the optimum wound healing activity at lower concentration of the oil. The constituents and method used in the proposed formulation were adequate for preparation of semisolid emulsion with seabuckthorn oil for external use.

***Keywords:** Seabuckthorn, Oil emulsion and Wound healing potential.*

Seabuckthorn (*Hippophae* L.)*: A Multipurpose Wonder Plant, Vol. 4: Emerging Trends in Research and Technologies* (Virendra Singh, *Editor-in-Chief,* 2014), p. 500–506, Daya Publishing House, New Delhi, India.

INTRODUCTION

Hippophae rhamnoides L. (seabuckthorn) is a bushy tree, naturally found in north of Europe, near the Black sea, on Caucaz, Alps and Carpathians and in delta of Danube, cold and high altitude regions of China and Himalayas (Singh, 2003). It is a cultivated in Germany and Russia, and was identified in Serbia (Rousi, 1971; Saric, 1989; Tucakov, 1990; Brad, 2002). The berries of this plant have been researched due to their pharmacological effects (Xing *et al.*, 2002; Yang and Kallio, 2002a; Zeb, 2006; Upadhyay *et al.*, 2009; Majewska and Gendaszewska-Darmach, 2011; Suryakumar and Gupta, 2011). They are round, fleshy, predominantly of orange colour (Beveridge *et al.*, 1999) and collected from September to late December. The content of vitamin C and fatty oil in berries depends on harvesting time (Yang and Kallio, 2002b). The chemical content of oil is rather unusual for plant kingdom-saturated and unsaturated fatty acids (palmitic, palmitoleic, oleic, linoleic, linolenic, myristic, stearic), accompanied by vitamin A and E, beta carotene, sterols, etc. (Beveridge *et al.*, 1999; Yang and Kallio, 2002a; Cupara *et al.*, 2010). Seabuckthorn oil is pharmacologically active in both external and internal application (Xing *et al.*, 2002; Yang and Kallio, 2002a; Suryakumar and Gupta, 2011). Externally applied seabuckthorn oil has tissue-regenerative, anti-inflammatory, antioxidant and antibacterial properties in wounds, burns and atopic dermatitis (Yang and Kallio, 2002a; Majewska and Gendaszewska-Darmach, 2011). The topical use of seabuckthorn oil in burns stimulates proliferation of fibroblasts, collagen synthesis, expression of specific matrix metaloproteinazas and angiogenesis. These effects are linked to high content of unsaturated omega-3 and omega-6 fatty acids, carotenoids and tocopherols in oil (Upadhyay *et al.*, 2009; Majewska and Gendaszewska-Darmach, 2011).

We obtained oil from fruits of seabuckthorn from natural forest (Cupara *et al.*, 2010). It was orange, viscous, lipophilic liquid and its external application was difficult in a reproducible manner. The obstacles were dripping, leaking and formation of occlusive film on skin surface that prevents absorption of coloured active substances. We proposed to formulate a semisolid oil/water topical emulsion of suitable pH for external application, containing a phase of liquid crystals. The emulsions containing liquid crystal structure possess enhanced characteristics for topical use and could be formed by use of non-ionic emulsifiers, such as polyoxyethylene stearyl ether. Liquid crystals possess characteristic feature, birefringence, similar to real crystals and could be detected by polarized light microscopy (Muller-Goymann, 1984; Nesseem, 2001; Muller-Goymann, 2004). Since, there was no data recorded for seabuckthorn oil ever being used within liquid crystal structure emulsion, we were attempted to formulate it and assess the effect of such chemically enhanced structure of emulsion containing seabuckthorn oil may achieve. In order to evaluate its pharmacological effect on wound healing, we used the excision model on *Wistar* rats and also the formulation was proposed to characterize by estimating its suitability for external use (organoleptic properties, type, pH value).

MATERIALS AND METHODS

We have used the ingredients showing in Table 41.1.

Seabuckthorn oil was prepared *ex tempore* (Cupara *et al.*, 2010). All substances were divided in two groups for inner/oily and outer/watery phase based on lipophilic and hydrophilic affinity and were merged together. Semi-solid emulsion containing 10 per cent and 40 per cent of seabuckthorn oil was prepared by standard emuslifying method (Cupara *et al.*, 2012). The samples were packed in tubes and sealed immediately after the preparation. Organoleptic properties were evaluated visually, such as colour, smell, brightness, consistency and homogeneity. Homogeneity was tested by smearing it on a glass plate to observe trapped air or other inhomogeneous parts. The type of emulsion was determined by conductometric method. pH values of samples were measured 48 h after the preparation and after 90 days at room and on different storage temperature (Cupara *et al.*, 2012). the liquid crystal structure of the samples was examined under the polarized light microscope (X600, Leica DMR, Germany) (Cupara *et al.*, 2011).

Table 41.1: Substances Used for Preparation of Semisolid Emulsion with Seabuckthorn Oil

Substance/INCI	*Brand Name/Producer*
Heksadecyl (cetyl) alcohol	Lanette 16®, Cognis, Germany
Oktadecyl (stearyl) alcohol	Lanette 18®, Cognis, Germany
Polyoxyethylene(20) stearyl ether (steareth-21)	Brij® 721P, Croda, SAD
Polyoksyletilene(2)stearyl ether, (steareth-2)	Brij® 72, Croda, SAD
Polyoksipropylene glykol(15)stearyl ether	Arlamol E®, Croda, SAD
Methyl hydroksibenzoat	Nipagin®, Nipa, SAD
Aqua purificata	Zorkapharma, Serbia
Propylene glycol	Zorkapharma, Serbia
Seabuckthorn oil	*ex tempore*

The wound healing effect was evaluated by excision wound model on *Wistar* rats, allowing monitoring of wound contraction by planimetric measurements. It was performed for both samples of emulsion, *i.e.* with 40 per cent of seabuckthorn oil (ESBT 40) and emulsion with 10 per cent of seabuckthorn oil (ESBT 10) in comparison to the control group that received no treatment and group that topically received only seabuckthorn oil (SBT oil). The outcome was expressed as percentage of wound healing (Cupara *et al.*, 2011). The emulsion samples were applied in non-sterile conditions once daily, during 4, 8, and 14 days. The granulation tissue specimens for histo-pathological examination were excised and fixed in buffered formaldehyde solution (4 per cent, pH 7.6). No animal showed signs of infection during the study. The microscopic histo-pathological analysis was performed using hematoxylin-eosin and Van Gieson staining colouring methods (Carl Zeiss, Axioskop 40, Germany).

Statistical Analysis

The statistical analysis was done by one way analysis of variance (ANOVA) followed by LSD test using SPSS package. Significance was noted at $p< 0.05$ value.

RESULTS

Both samples of emulsion containing 40 per cent and 10 per cent of seabuckthorn oil were semisolid, homogen and possesed the enhanced structure with liquid crystals (Cupara *et al.*, 2011). These could be easily spread, with shiny surface, no smell and of orange colour. Sample containing 40 per cent of seabuckthorn had more intensive orange colour. Both samples showed to be non sticky and formed mild occlusive film on the skin after application. Organoleptic characteristics remained unchanged 30, 90 and 365 days after the preparation. The colour of samples diminished after 365 days. The change of colour intensity was more evident in sample with 40 per cent of seabuckthorn oil. Results of electrical conductivity, which were for both samples higher of 50 µS/cm, confirmed that oil in water type of emulsion was formed. pH values of both samples has remained within limits for external use during the time of experiment (Cupara *et al.*, 2012). Increased wound healing was noted after 8 days in the group where sample ESB 10 was applied in comparison to the control group (p≤0.01), SBT oil group (p<0.05) and ESBT 40 group (p<0.05) (Figure 41.1).

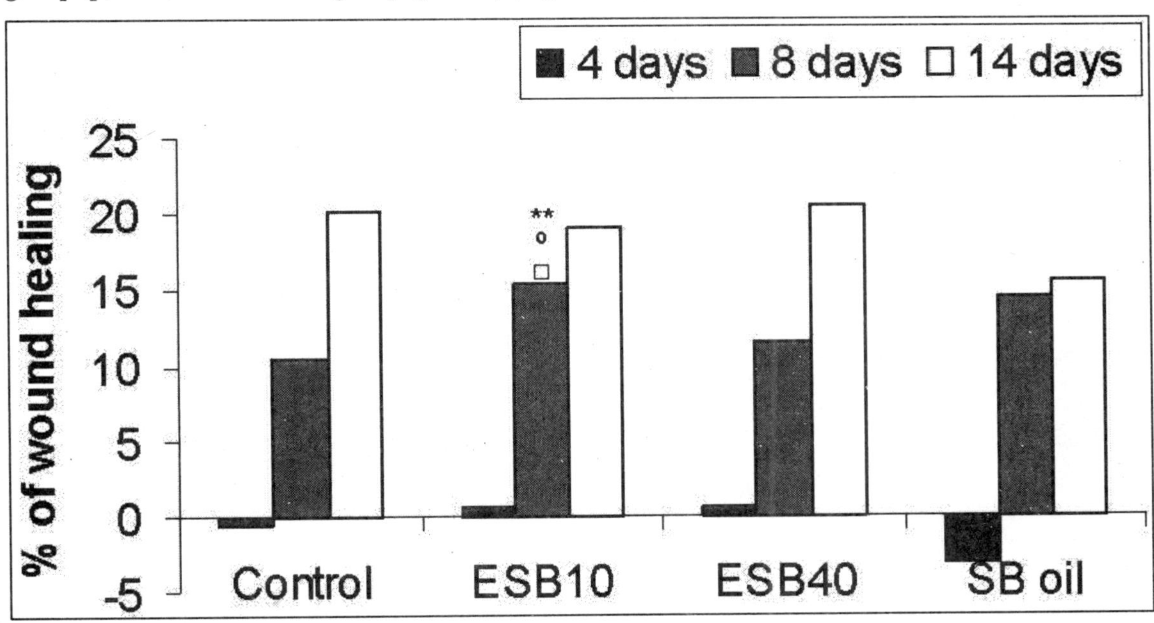

Figure 41.1: Percentage of Wound Healing for Control and Treated Animals with Seabuckthorn Oil and Emulsions with SB Oil after 4, 8, and 14 days.

****High statistical significance in comparison to the control p≤0.01; °-Statistically significant in comparison to the SB oil p<0.05; and %-Statistically significant in comparison to the ESB 40 p<0.05.**

DISCUSSION

Seabuckthorn oil shows anti-inflammatory and epitelization stimulating properties, which makes it ideal for incoroporation in a semisolid topical formulation (Upadhyay *et al.*, 2009; Majewska and Gendaszewska-Darmach, 2011). Seabucktorn oil used in this reasearch was obtained from sponatneous flora and its organoleptic properties, yield and chemical content were in accordance with literature data (Cupara *et al.*, 2010). We have used it in the formulation of semisolid emulsion, producing samples in which two different quantities of seabuckthorn oil were succesfully incorporated.

The pharmacological effects of proposed formulation were evaluated by wound healing animal model (Figure 41.2). The formulation showed higher wound healing potential then seabuckthorn oil. We suppose that unique combination of the ingredients allowed sinergistic activity-seabuckthorn oil hastened epitalization, while the enhanced structure of liquid crystals in emulsion contributed to the healing by providing proper hydration. Fibroblast proliferation were seen in treated groups by emulsion samples. Significant accelerated wound healing, accompanied by increased collagen deposition and subsequent shrinkage of wound area, occurred in our study after the treatment by emulsion with liquid crystals loaded with 10 per cent of seabuckthorn oil. This implied that ingredients and their quantities used in this investigation formed emulsion as suitable vehicle for incorporation of different quantity of seabuckthorn oil, which promoted wound healing with lower concentration of the oil (Cupara *et al.*, 2011).

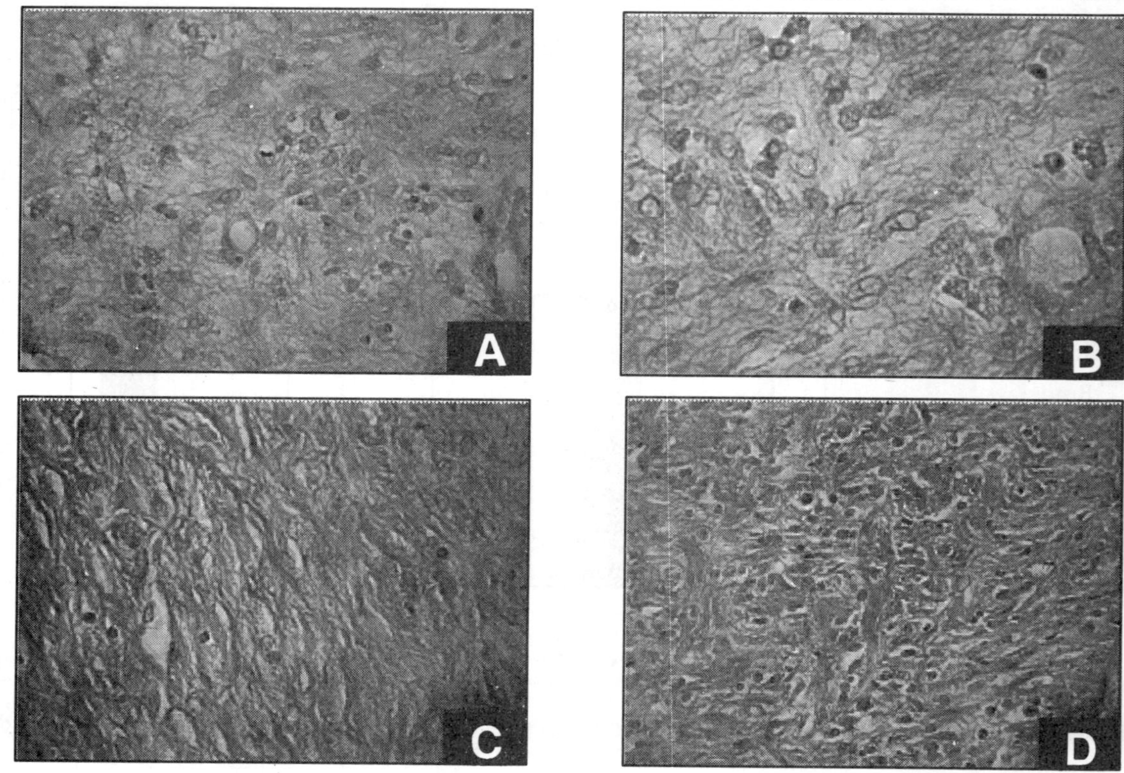

Figure 41.2: Histological Studies of the Granulation Tissue Samples after 8 Days (ESB 10 and ESB 40) Showed Cellular Proliferation.

A) Control group-Increased number of inflammatory cells, few macrophages and small amount of collagen fibers (Van Gieson X200); B) SBT oil treated group-Mild response of inflammatory cells, more macrophages (yellow stained) and moderate deposition of collagen in wavy manner, diffusely distributed (Van Gieson X200); C) ESB 10 treated group-Increased deposition of collagen (Van Gieson X200); and D) ESB 40 treated group-Increased deposition of collagen (Van Gieson X200).

Studying organoleptic characteristics and pH properties of this emulsion helped us determine its adequacy for topical use. Our study showed that proposed formulation in both its varieties (samples with 10 per cent and 40 per cent of seabuckthorn oil) possesed acceptable organoleptic and tactile properties for external application. Samples were naturally coloured and consistency was in accordance with desired type of emulsion-oil in water. Absence of smell links the formulation to medicinal use. Emulsion was homogenous with no trapped air. Diminished colour of both samples after 12 months, indicated possible need of caroten quantification over the time. It seems to us that external application of proposed formulation should not affect natural pH value of the skin and weaken the skin barrier function, since pH values of both samples were appropriate for topical use (Cupara *et al.*, 2012).

CONCLUSION

Our study showed that seabuckthorn oil could be successfully incorporated in enhanced structure emulsion, which contributed to pharmacological activity of seabuckthorn oil. The choice of components used in emulsion formulation influenced quantity of active ingredient needed for optimal treatment. Based on these results we proposed that inadeqacies in skin application of liquid seabuckthorn oil which could be overcome by formation of semisolid emulsion with seabuckthorn oil, that shows wound healing potential and is suitable for skin application.

ACKNOWLEDGEMENT

The authors would also like to express gratitude to the Ministry of Education and Science of the Republic of Serbia, for Grant OI N°175014.

REFERENCES

1. Beveridge, T., Li, T.S.C., Oomah, B.D. and Smith, A. 1999. Seabuckthorn products: Manufacture and composition. *J. Agric. Food Chem.* 47: 3480-3488.
2. Brad, I. 2002. *Catina alba*, p.18-20, Editura Tehnica, Bucuresti.
3. Cupara, S., Jankovic, S., Arsic, I., Tadic, V. and Jacevic, V. 2012. Characterization of seabuckthorn oil emulsion. *Mil. Med. Sci. Lett.* (*Voj. Zdrav. Listy*) 81 (2): 56-60.
4. Cupara, S., Ninkovic, M., Knezevic, M., Vuckovic, I. and Jankovic, S. 2011. Wound healing potential of liquid crystal structure emulsion with seabuckthorn oil. *HealthMED*5: 1218-1223.
5. Cupara, S., Sobajic, S., Tadic, V., Arsic, I., Djordjevic, S., Runjaic-Antic, D. and Djordjevic, B. 2010. Dry seabuckthorn berries (*Hippophae rhamnoides* L.): Fatty acid and carotene content in pericarp and seed oil. *HealthMED* 4: 788-791.
6. Majewska, I. and Gendaszewska-Darmach, E. 2011. Proangiogenic activity of plant extracts in accelerating wound healing: A new face of old phytomedicines. *Acta Biochimica Polonica* 58 (4): 449-460.
7. Muller-Goymann, C.C. 1984. Liquid crystals in emulsions, creams and gels containing ethoxylated alcohols as surfactants. *Pharm. Res.* 154.
8. Muller-Goymann, C.C. 2004. Physicochemical characterization of colloidal drug delivery systems, such as reverse micelles, vesicles, liquid crystals and nanoparticles for topical administration. *Eur. J. Pharm. Biopharm.* 58: 343-356.
9. Nesseem, D.I. 2001. Formulation and evaluation of itraconazole *via* liquid crystal for topical delivery system. *Journal of Pharmaceutical and Biomedical Analysis* 26: 387-399.
10. Rousi, A. 1971.The genus *Hippophae* L.: A taxonomic study. *Annales Botanici. Fennici.* 8: 177-227.
11. Saric, M. 1989. *Lekovite biljke SR Srbije.* Srpska akademija nauka i umetnosti, p. 307-309, Beograd.
12. Singh, V. 2003. Geographical adaptation and distribution of seabuckthorn (*Hippophae* L.) resources. In: *Seabuckthorn (Hippophae L.)-A Multipurpose Wonder Plant-Botany. Vol. I. Botany, Harvesting and Processing Technologies* (V. Singh, Ed. in Chief, 2003), p. 21-34, Indus Publishing Company, New Delhi, 518p.
13. Suryakumar, G. and Gupta, A. 2011. Medicinal and therapeutic potential of seabuckthorn (*Hippophae rhamnoides* L.). *J. Ethnopharmacol.* 138: 268-278.
14. Tucakov, J. 1990. *Lecenje biljem-fitoterapija.* p. 532, Rad, Beograd.
15. Upadhyay, N.K., Kumar, R., Mandotra, S.K., Meena, R.N., Siddiqui, M.S., Sawhney, R.C. and Gupta, A. 2009. Safety and healing efficacy of seabuckthorn (*Hippophae rhamnoides* L.) seed oil on burn wounds in rats. *Food Chem. Toxicol.* 47 (6): 1146.

16. Xing, J., Yang, B., Dong, Y., Wang, B., Wang, J. and Kallio, H.P. 2002. Effects of seabuckthorn (*Hippophae rhamnoides* L.) seed and pulp oils on experimental models of gastric ulcer in rats. *Fitoterapia* 73: 644-650.

17. Yang, B. and Kallio, H. 2002a. Composition and physiological effects of seabuckthorn (*Hippophae*) lipids. *Trends in Food Science and Technology* 13: 160-167.

18. Yang, B. and Kallio, H. 2002b. Effects of harvesting time on triacylglycerols and gycerophospholipids of seabuckthorn (*Hippophae rhamnoides* L.) berries of different origins. *Journal of Food Composition and Analysis* 15: 143-157.

19. Zeb, A. 2006. Anticarcinogenic potential of lipids from *Hippophae*: Evidence from the recent literature. *Asian Pac. J. Cancer Prev.* 7 (1): 32-35.

Chapter 42

Effect of Spices on the Quality of Seabuckthorn Nectar

Dalija Seglina, Anita Olsteine, Inta Krasnova, Karina Juhnevicca and Elga Sne

Latvia State Institute of Fruit Growing, Garndu Str.1
Dobele, LV-3701, Lativa

SUMMARY

Seabuckthorn (*Hippophae rhamnoides* L.) is a multipurpose plant, due to its typical biochemical composition of the fruits, mainly used in medicine, cosmetics and food industry. Seabuckthorn beverages can be used as functional foods in our daily diet as their consumption is known to be beneficial for the human body.

One of the problems faced by the consumer is the short shelf life of the fresh beverage. Many commonly consumed spices possess antioxidant and antimicrobial activities, which are linked to their flavonoids and tannin content. In the present study, qualitative characteristics (microbiology and sensory properties, contents of vitamin C and E, total carotenoids, phenolic compounds, and antiradical activity), of the fresh seabuckthorn nectar were estimated during storage to evaluate the effect of used spices (ginger, cinnamon, and cloves) in the preparation of the nectar.

Using hedonic rating scale model, the nectar along with the spice mixture was rated as the best reaching highest flavour on the 16th day of storage. The addition of spice mixture to the seabuckthorn nectar can extend its shelf life by five days as compared with the control due to reduction of microbial contamination. Results showed that manufacturers can successfully use a mixture of spices (cinnamon, ginger, and cloves) to improve the quality characteristics and prolong the shelf life of fresh seabuckthorn nectar.

Keywords: *Cinnamon, Cloves, Ginger, Shelf life, Seabuckthorn nectar and Quality.*

INTRODUCTION

Seabuckthorn from a little known wild shrub has evolved into a well established fruit crop. Fresh seabuckthorn fruits contain significant amount of vitamin C, E, and B, carotenoids, organic acids,

Seabuckthorn (Hippophae L.): A Multipurpose Wonder Plant, Vol. 4: Emerging Trends in Research and Technologies (Virendra Singh, *Editor-in-Chief*, 2014), p. 507–517, Daya Publishing House, New Delhi, India.

polyunsaturated fatty acids, phenolic compounds and essential elements. It has been used in traditional medicines in Tibet and Indian Himalayas (Singh, 2006a). The positive effects of seabuckthorn for prevention of various diseases have been reported in a large number of scientific studies. It is considered as one of the most important plants found in nature, containing many and various biologically active substances known to prevent the formation of free radicals. The fruits and juice of seabuckthorn have high antioxidative capacity, as reported by many researchers (Suryakumar *et al.*, 2001; Allman *et al.*, 2005; Gao, 2005; Singh, 2006b). The experiments conducted on animals as well as clinical investigations on human volunteers carried out in different countries, allows considering the processed products of seabuckthorn fruits as a functional food (Varshney, 2003; Singh, 2006a).

The beverages are assumed as the optimal food for use in any human diet. One of the most popular seabuckthorn processed products is juice. And due to its high acidity, seabuckthorn juice is widely used in beverage production. Traditionally, the shelf-life stability of juices and beverages has been achieved by thermal processing. However, thermal pasteurization usually reduces the product quality and freshness. During the last couple of decades, some non-thermal pasteurization methods have been proposed, including high hydrostatic pressure, high pressure homogenization, pulsed electric field and ultrasound. These emerging techniques seem to have the potential to provide safe fruit juices with prolonged shelf-life comparable to the fresh juice (Rupasinghe and Yu, 2012). Apart from thermal pasteurization, some chemical preservatives including potassium sorbate and sodium benzoate are also widely used for the extension of the shelf-life of fruit juices and beverages. Many safe and environmental friendly preservatives of natural origin are available for the manufacturer to satisfy the consumer's demand. One of them is the use of natural antimicrobials of plant origin, including plant extracts (essential oils, aldehydes, and esters), herbs and spices. The main elements of these antimicrobials are phenolic compounds, such as caffeic, cinnamic, ferulic, and gallic acids, oleuropein, thymol, and eugenol (Singh *et al.*, 2007). Fruit juices and beverages contain water, sugars, organic acids, vitamins, and trace elements which provide an ideal environment for microbial spoilage. The extracts of cinnamon and clove plants characterized as great natural antimicrobials are generally recognized as safe (GRAS) additives (Rupasinghe and Yu, 2012).

The recent trends show that the consumer preference towards unpasteurized seabuckthorn juice or beverages has increased due to its nutritive values. However, storage time was the prime limiting factor for the shelf life of this product. A lot of work has been done on the development, pasteurization and nutritional characterization of seabuckthorn juice, but there are no reports on specific physical, chemical, sensory and microbiological properties of fresh seabuckthorn nectar with spices during shelf life.

MATERIALS AND METHODS

EXPERIMENTAL DESIGN

The experiment was carried out at the Experimental Processing Laboratory of Latvia State Institute of Fruit-Growing in 2010. The object of study was fresh seabuckthorn nectar produced by "Satori Alpha" Ltd. (Latvia). The composition of investigated product: Water, puree of seabuckthorn fruits, mixture of glucose, sucrose and fructose. Seabuckthorn fruit puree content in nectar is not less than 42 per cent. The current storage time of fresh seabuckthorn nectar at temperature 4±2 °C is 14 days.

For shelf life extension of seabuckthorn nectar, dried spices (cinnamon, ginger, cloves) were purchased from the market. Cinnamon and cloves were ground by mill IKA® (Germany), while ginger was already in a powder form (originated from China). Seabuckthorn nectar samples were prepared as: Control sample without spices; sample with ginger powder (70 g per litre) and sample with mixture of dried powdered spices (50 g per litre, 46.7 per cent ginger, 33.3 per cent cinnamon, 20 per cent cloves).

STORAGE AND ANALYSIS OF SAMPLES

The samples were stored for 21 days in a Commercial Freezer/Cooler camera at the temperature 4.0±0.5 °C, controlled by MINILog Gresinger electronic (Germany). The analysis of the nectar was done at the sample preparation day and at 14 16, 19 and 21 days of storage. Samples was estimated according to Latvian Standards and analyzed in triplicate. The following parameters in nectar were determined as follows:

Microbial Analysis

Collected samples were processed for microbial load by standard plate count technique. Microbial load was calculated according to the following formula:

$$\text{CFU/ml} = \frac{\text{Number of Colonies} \times \text{Dilution Factor}}{\text{Volume of Sample used}}$$

The samples for microbial testing were prepared according to Latvian standard (LVS EN ISO 6887-5:2011). Determination of the total bacteria count of microorganisms (TBC)-***mesophilic aerobic and facultative anaerobic*** was according to the Latvian standard (LVS EN ISO 4833:2003) and determination of yeasts and moulds was done according to the Latvian standard (LVS ISO 21527: 2008).

Vitamin C

The content of ascorbic acid (vitamin C) (mg/100 g) was determined by the HPLC method (LVS EN ISO 14130:2003).

Vitamin E and β-carotene

The contents of vitamin E and β-carotene (mg/100 g) in samples were determined by a modified method. The method is based on extraction of vitamin E and carotenoids with petroleum benzene (Seglina *et al.*, 2008).

Phenolic Compounds

The total content of phenols (mg/100 g) was determined by the Folin-ciocalteu method, using a spectrometer UV-1650-PC (Shimadzu, Japan) at a wavelength 765 nm (Singleton *et al.*, 1999).

Radical Scavenging Activity

Radical scavenging activity and presence of hydrogen donors in juice and beverage samples were examined by reduction of 2, 2-diphenyl-1-picrylhydrazyl (DPPH) in methanol according to Brand-Williams *et al.* (1995) with minor modifications. The reaction mixture containing 100 μl juice extract (10 mg/ml) in methanol mixed with 2900 μl of DPPH (1×10^{-4} M) ethanol solution. DPPH radical was determined from the difference in absorbance with or without the sample (control) and the percent of inhibition was calculated using following equation:

$$\% \text{ Inhibition} = \frac{(ac - at)}{ac} \times 100$$

where,

ac-absorbance of control and at-absorbance of test sample.

Sensory Analysis

Seabuckthorn nectar was evaluated using the hedonic rating and line scale according to ISO 4121:1987 by 20 trained panellists aged 25 to 70 years. Hedonic rating test was done using a scale, in which 9 for like extremely; 5 for neither like nor dislike and 1 is for dislike extremely. The taste, aroma, sourness, sweetness, and colour of nectar were recorded using a line scale.

Statistical Analysis

The data processing was carried out by General Linear Model procedure using SPSS 15 software package and MS Excel variance analysis. *Sheffe* criterion was used in the clarification of significant differences ($p<0.05$) among studied samples.

RESULTS AND DISCUSSION

Microbiology

The microflora of seabuckthorn nectar after adding spices (cinnamon, ginger and cloves) varied significantly during storage ($p<0.05$). The changes in total bacteria count; yeast and mould colony forming units (CFU) are represented in Table 42.1. Total bacteria count (TBC) during the investigation period was higher in the control sample as compared to the samples with ginger and spice mixture. The CFU of TBC in the sample with spice mixture increased on an average 19 times from the beginning of study till storage day 16, in the sample with ginger it raised 25 times, while in the control sample

Table 42.1: Microbial Flora (Total bacteria count, yeasts and moulds CFU) in Seabuckthorn Nectar at different Storage Time (Days)

Sample	*Storage Time, Days*				
	0	*14*	*16*	*19*	*21*
			In total bacteria count (CFU/ml)		
Control	3.69±0.59	5.48±0.59	7.57±0.25	9.65±1.47	10.47±1.34
Ginger	3.64±0.73	4.57±0.28	7.43±0.04	9.71±1.28	9.72±1.77
Mixture of spices	2.83±0.28	4.25±0.56	5.79±0.79	7.90±1.59	9.20±1.73
			In yeasts (CFU/ml)		
Control	2.56±0.61	5.63±0.22	7.59±0.42	9.87±0.39	10.76±0.96
Ginger	1.10±1.23	4.76±1.02	7.54±1.09	9.57±1.12	9.64±1.15
Mixture of spices	0±0.75	2.83±1.63	5.61±1.2	6.45±1.73	8.42±1.39
			In moulds (CFU/ml)		
Control	3±0.00	5±0.00	5.2±0.90	197±1.02	354±15.08
Ginger	6.7±0.00	7.2±0.00	7.6±1.12	210±8.82	287±12.02
Mixture of spices	0±0.00	0±0.00	3±1.33	37±3.33	120±11.55

raised 48 times. After three more days of storage (day 19) in the sample with spice mixture, TBC yielded slightly lower as compared to the control sample, while in the sample with ginger, it was higher as compared to the other samples. At the end of study, all samples had a large number of micro-organisms therefore, they were not recommended for use in diet.

The yeast development throughout the study period was more intense in the control sample. From the beginning to the 16th storage day, the sample with a mixture of spices yeast colony forming units had grown the least, while in the sample with ginger and in the control sample, their amount increased significantly ($p<0.05$). The sample with spice mixture after 19 days of storage had almost four times lower amount of yeast CFU than the control sample ($p<0.05$).

There were no moulds in nectar sample with spice mixture from the beginning and till storage day 14. In nectar sample with ginger at the beginning of the experiment, 6.7 CFU of mould were found which reached 7.2 units at 14th day. Meanwhile, in the control sample 3.0 CFU were found, and they grew up to 5.0 CFU during the same storage period. After 19 days of storage, the amount of mould CFU was five times lower ($p<0.05$) in sample with spice mixture.

To ensure the microbial safety of seabuckthorn juice and beverage, most of the companies traditionally use the product sterilization or pasteurization (Bates *et al.*, 2001). The spices, such as sage, rosemary, clove, coriander, garlic, and onion were listed as potential antimicrobials for use in food industry (Deans and Ritchie, 1987). The ground cinnamon (0.3 per cent) could inhibit the growth of *Staphylococcus aureus*, *Y. enterocolitica* and *Salmonella typhimurium* in apple juice (Yuste and Fung, 2004). Linchong (2003) reported that 10 per cent ginger extract was the optimum concentration to reduce microbial population and still produce the product with acceptance from panellists. Kinnow mandarin juice blended with pomegranate juice and 3 per cent ginger juice was lower in bacterial, mould and yeast population during six month storage period as compared to the control sample without ginger juice (Bhardwaj and Mukherjee, 2011). The essential oils of species and active compounds in them (carvacrol, cinnamaldehyde, eugenol, citral geraniol, and D-limonene) have been identified and used for juice stabilization (Nassar *et al.*, 2007; Bevilacqua *et al.*, 2010; Campaniello *et al.*, 2010).

In our study, the mixture of spices (ginger, cinnamon, and clove) showed better antimicrobial properties than using ginger alone. In addition, the microbiological parameters of ginger powder were tested and from the results attained, it was concluded that the raw material is contaminated with large amount of TBC, yeast and mould colony forming units and therefore, sterilization of spice is recommended before nectar production.

Chemical Properties

The changes in total phenolic compounds, carotenoids, vitamin C and E, and antiradical activity are presented in Table 42.2. Significantly higher total phenolic content was observed in the sample with spice mixture as compared to the control sample during the storage for 19 days ($p<0.05$). On the sample preparation day and after 16 days of storage the content of phenolic compounds in nectar with spice mixture were 97.4 and 96.0 mg/100 g, respectively. The control sample and sample with ginger during the first four days of storage contained lower amount of phenolic compounds (86.0 and 87.9 mg/100 g, respectively). After 19 days of storage, decrease in phenolic content was observed for all samples.

Table 42.2: Contents of Total Phenolics and Carotenoids, Vitamin C and E, and DPPH Radical-scavenging Activity in Seabuckthorn Nectar

Sample	*Storage Time, Days*				
	0	*14*	*16*	*19*	*21*
			Total phenolics (mg/100 g)		
Control	87.29±0.68^{b}	87.29±0.53^{b}	87.99±1.04^{b}	87.79±0.93^{b}	84.82±1.11^{bc}
Ginger	86.03±0.77^{b}	86.23±1.06^{b}	86.48±0.76^{b}	85.98±0.54^{b}	84.57±1.23^{bc}
Mixture of spices	97.42±0.77^{a}	95.96±0.56^{a}	96.01±0.77^{a}	89.00±0.70^{a}	87.24±0.62^{b}
			Vitamin C (mg/100 g)		
Control	29.38±0.89^{a}	29.38±0.58^{a}	29.16±0.74^{a}	28.95±0.21^{a}	28.29±0.39^{a}
Ginger	28.97±0.23^{a}	28.97±0.38^{a}	28.75±0.57^{a}	28.32±0.78^{a}	28.29±0.75^{a}
Mixture of spices	27.57±0.41^{a}	27.57±0.63^{a}	27.34±0.54^{a}	26.96±0.31^{a}	26.72±0.51^{b}
			Carotenoids (mg/100 g)		
Control	42.2±0.34^{a}	41.4±0.89^{b}	41.6±0.92^{b}	40.9±0.33^{b}	38.5±0.21^{b}
Ginger	41.7±0.17^{a}	43.8±0.38^{a}	43.4±0.31^{a}	42.8±0.29^{a}	41.7±0.18^{a}
Mixture of spices	41.4±0.68^{a}	44.3±0.15^{a}	44.0±0.39^{a}	43.2±0.43^{a}	41.1±0.25^{a}
			Vitamin E (mg/100 g)		
Control	14.00±0.16^{a}	13.41±0.29^{a}	13.63±0.18^{b}	11.00±0.22^{b}	9.90±0.44^{b}
Ginger	13.86±0.06^{a}	13.87±0.12^{a}	13.59±0.25^{b}	11.21±0.36^{b}	10.91±0.39^{a}
Mixture of spices	13.89±0.11^{a}	13.95±0.14^{a}	14.16±0.23^{a}	11.74±0.11^{a}	10.88±0.56^{a}
			DPPH radical-scavenging activity (per cent inhibition)		
Control	24.0±1.20^{a}	23.8±0.44^{b}	23.1±0.65^{b}	22.4±0.24^{b}	21.4±0.22^{b}
Ginger	26.3±1.09^{a}	26.3±0.49^{a}	25.0±0.61^{a}	24.5±0.60^{a}	24.0±1.00^{a}
Mixture of spices	24.8±0.93^{a}	26.7±0.70^{a}	26.0±0.60^{a}	25.3±0.78^{a}	24.6±1.18^{a}

Note: Values in the same column with different superscript letters (a, b, c) differ significantly ($p < 0.05$).

The content of vitamin C in seabuckthorn nectar is an important indicator, which characterizes the quality of product. During storage, the content of vitamin C decreased in all samples, and added spices had significant effect on it ($p>0.05$). In the sample with ginger the content of vitamin C decreased slightly in the all storage time as compared to the control sample.

A considerable decrease of vitamin C was observed in the sample with spice mixture as compared with other samples during all storage time. Overall, in the entire storage time, there was no significant differences ($p=0.09$) between samples with ginger and control samples. The nectar with spice mixture contained a slightly smaller amount of vitamin C as compared with other samples. The proposed hypothesis is that the high polyphenol content of spices (especially cloves) has a negative impact on the stability of ascorbic acid in nectar, causing faster oxidation processes after 21 days of storage.

Plant pigments (carotenoids) ensure colour intensity from yellow to red in seabuckthorn fruits. Initially, the content of β-carotene in the nectar samples was 41.4-42.2 mg/100 g. Between storage days 14 and 19, the amount of carotenoids in samples with ginger and spice mixture slightly increased in comparison to the control sample and did not substantially change after 19 days of storage ($p=0.51$). The decrease in carotenoids was observed in all studied nectar samples after 19 days storage.

Initially, the content of vitamin E in seabuckthorn nectar was on an average 14.0 mg/100 g. No significant differences in vitamin E was observed between samples within 14 days of storage (p=0.65). After 19 days of storage, the amount of vitamin E decreased by 16-20 per cent in all nectar samples.

The highest DPPH free-radical scavenging activity at the beginning of experiment was found in nectar with ginger, and it decreased during three week storage period. At the day 14, the increase of antioxidant activity (1.1 per cent) was detected in samples with spice mixture, while in the sample with ginger it remained unchanged, when compared to the control sample, in which downward trend was observed. Generally, the study showed that the highest antioxidant activity was found in the sample with spice mixture. After 21 of storage days, DPPH free-radical scavenging activity was only 1 per cent less as compared with initially, while in the control sample and sample with ginger, it was lower (10.8 per cent and 8.8 per cent, respectively). It is related to antioxidant activity differences of spices and their changes during the storage time.

The DPPH radical scavenging assay is commonly employed to evaluate the ability of antioxidants to scavenge free-radicals. The use of the DPPH free-radical is advantageous in evaluating antioxidant effectiveness because it is more stable than hydroxyl or superoxide radicals (Siriwardhana and Shahidi, 2002). The extract of clove buds at higher concentration (400 mg/ml) have remarkable inhibition of DPPH (45 to 93 per cent) in comparison with 400 mg/ml synthetic antioxidants, such as butylated hydroxyl toluene (95 per cent) (Nassar *et al.*, 2007). In another study, it is mentioned that cinnamon leaf and clove buds in 10 mg/ml of essential oil concentration showed 96.7 per cent and 96.1 per cent respectively, while ginger only 18.4 per cent of the DPPH free-radical scavenging activity (Wang *et al.*, 2009). In the concentration of 1 mg/ml, the total phenolic amounts in cinnamon leaf and clove buds essential oils were 420 and 480 mg/g of GAE, respectively, whereas total phenolic compounds in the ginger sample was up to 30 times less (16 mg/g of GAE). The phenolic compounds represent a substantial portion of spice antioxidants, reaching 88-100 mg/g for clove and 20-30 mg/g for ginger (Suhaj and Horvathova, 2007).

It can be concluded from this study that seabuckthorn nectar with spice mixture contained the highest amount of polyphenols which positively influenced the DPPH free-radical scavenging activity as compared with other samples. The significant antimicrobial and antioxidant activities of spice mixture suggested that it could serve as a source of compounds with preservative phenomenon.

Sensory Properties

Acceptability of seabuckthorn nectar during storage by hedonic and line scale are represented in Table 42.3. According to hedonic scale, the highest sensory evaluation ("like moderately") was for nectar with ginger at the sample preparation day, while the sample with spice mixture was estimated with the lowest-"like slightly". This could be because the particles of added spices did not diffuse well into the nectar and therefore, the taste of each spice was perceptible individually. In particular, the strong clove flavour was dominating. The ginger taste was separately observed, and its fresh, cool taste and odour was considered pleasant by panellists. The control sample was estimated by the lowest values within a range of "like slightly" and "like moderately" between storage days 14 and 19, whereas seabuckthorn nectar samples with ginger and spice mixture during this period were evaluated with slightly higher sensory scores. Acceptability of all nectar samples decreased after the storage day 16. In general, after 14 days of storage till end of investigation, the sample with spice mixture was assessed as the best.

Table 42.3: Sensory Evaluation of Seabuckthorn Nectar

Sample	*Storage Time, Days*				
	0	*14*	*16*	*19*	*21*
			*Hedonic scale		
Control	6.58	7.00	7.05	6.65	6.65
Ginger	7.16	7.05	7.45	6.95	6.70
Mixture of spices	6.42	7.37	7.47	7.05	6.90
			*Line scale (taste)		
Control	7.17	7.61	7.50	7.60	7.87
Ginger	8.25	7.90	8.49	8.01	7.56
Mixture of spices	8.97	8.44	7.91	8.59	8.72
			Line scale (aroma)		
Control	6.71	5.73	7.08	5.77	6.61
Ginger	7.27	7.04	7.17	6.04	6.72
Mixture of spices	8.11	8.07	8.07	7.46	7.73
			Line scale (sourness)		
Control	6.65	6.06	6.58	6.89	6.99
Ginger	5.91	6.95	6.46	6.77	6.84
Mixture of spices	6.22	5.97	5.98	6.73	5.96
			Line scale (sweetness)		
Control	6.27	5.60	6.45	5.42	5.90
Ginger	6.41	5.17	6.19	5.47	5.40
Mixture of spices	6.83	5.65	5.78	5.69	7.02
			Line scale (colour)		
Control	9.28	9.75	8.77	9.26	9.93
Ginger	9.38	9.59	9.12	9.33	10.08
Mixture of spices	9.21	9.43	8.70	8.96	9.00

* Consumer Acceptance Testing (n=20); 9 point hedonic scale-9=like extremely; 8=like very much; 7=like moderately; 6=like slightly; 5=neither like nor dislike; 4=dislike slightly; 3=dislike moderately; 2=dislike very much and 1=dislike extremely.

In order to evaluate and justify the addition of spices to seabuckthorn nectar, line scale method was used. This method allows to gain more detailed description (taste, aroma, sweetness, acidity, and colour) of the product liking degree. The best taste was evaluated for sample with spice mixture during the study. The acceptable taste of nectar was provided by biologically active substances in spices. For comparison, the taste and aroma of sample with ginger during storage increased gradually. The dry ginger powder swells and diffuses slowly in the nectar, so the best taste was established at the day 19. During the study, the sample with spice mixture was characterized by stronger aroma as compared with the control sample and sample with ginger. Using hedonic scale, seabuckthorn nectar sample with ginger was evaluated almost equally as sample with spice mixture at the storage day 16. The highest sourness was noted for control sample and for sample with spice mixture in the first day of investigation, while for sample with ginger this quality was estimated with less point. Interestingly, the sample with spice mixture was assessed by lower sourness after storage day 14. Probably, it could be explained by anaesthetic properties of clove oil which makes impossible to taste the nectar sourness. In addition, high sweetness was evaluated for sample with spice mixture at the preparation day. Probably, the warm flavour of cinnamon affected taste buds on the tongue, which led to feel the greater sweetness. In comparison with control, sample with ginger was found to be sweeter. Results showed that added ginger slightly intensified the sweet taste of nectar. During the study, the higher colour was assessed for sample with ginger as compared to the control sample and sample with spice mixture and could be explained by the yellow colour of ginger powder which intensifies the orange colour of seabuckthorn nectar. Whereas, the colour for sample with spice mixture was evaluated as the lowest, probably it was due to dark colour of cinnamon and cloves.

CONCLUSION

The micro-flora of seabuckthorn nectar after adding spices (cinnamon, ginger and cloves) varied significantly during storage. The total bacteria count in sample with spice mixture decreased by 27.2 per cent as compared with control sample after 19 days of storage. Using a mixture of spices, the shelf life of manufactured seabuckthorn nectar can be extended by five days by reducing yeast and mould colony forming units. Significantly, higher content of total phenolics was recorded in the sample with spice mixture and only after day 19, it decreased by 7 per cent. The content of carotenoids and vitamin E were stable in seabuckthorn nectar samples between storage days 14 and 16. During the study, the content of vitamin C decreased by 2.8 per cent in sample with spice mixture, and by 1.4 per cent sample with ginger. Using hedonic rating scale model, the nectar with the spice mixture was rated as the best and the highest flavour this nectar reached on the 16th day of storage. Results showed that manufacturers can successfully use a mixture of spices (cinnamon, ginger, and cloves) to improve the quality characteristics and to prolong shelf life of non pasteurized seabuckthorn nectar.

ACKNOWLEDGEMENT

This research was supported by the Latvia State Research Program No. 2010.10-4/VPP5/4.

REFERENCES

1. Allman, S., Strass, G., Kranl, K., Frank, T., Bitsch, I., Bitsch, R. and Netzel, M. 2006. Bioactivity and bioactive compounds of seabuckthorn fruit juice. In: *Seabuckthorn (Hippophae L.)-A Multipurpose Wonder Plant. Vol. II: Biochemistry and Pharmacology* (V. Singh, Ed. in Chief, 2006), p. 402-412, Daya Publishing House, New Delhi, 600p.
2. Bates, R.P., Morris, J.R. and Crandall, P.G. 2001. *Principles and Practices of Small-and Medium Scale Fruit Juice Processing*, FAO, Rome. *http://www.fao.org/docrep/005/y2515e/y2515e00.htm*
3. Bevilacqua, A. Corbo, M.R. and Sinigaglia, M. 2010. *In vitro* evaluation of the antimicrobial activity of eugenol, limonene and citrus extract against bacteria and yeasts, representative of the spoiling microflora of fruit juices. *Journal of Food Protection* 73: 888-894.
4. Bhardwaj, R.L. and Mukherjee, S. 2011. Effects of fruit juice blending ratios on kinnow juice preservation at ambient storage condition. *African Journal of Food Science* 5 (5): 281-286.
5. Brand-Williams, W., Cuvelier, M.E. and Berset, C. 1995. Use of a free radical method to evaluate antioxidant activity. *Lebensm-Wiss. U. Tehnol.* 28: 25-30.

6. Campaniello, D., Corbo, M.R. and Sinigaglia, M. 2010. Antifungal activity of eugenol against *Penicillium, Aspergillus* and *Fusarium* species. *Journal of Food Protection* 73: 1124-1128.
7. Deans, S.G. and Ritchie, G. 1987. Antibacterial properties of plant essential oils. *International Journal of Food Microbiology* 5: 165-180.
8. Gao, X. 2006. Antioxidants and antioxidative activity of seabuckthorn (*Hippophae* L.). In: *Seabuckthorn (Hippophae L.)-A Multipurpose Wonder Plant. Vol. II: Biochemistry and Pharmacology* (V. Singh, Ed. in Chief, 2006), p. 390-401, Daya Publishing House, New Delhi, 600p.
9. ISO 4121-1987. *Sensory analysis-Methodology-Evaluation of Food Products by methods using Scales.*
10. Linchong, S. 2003. Preservation of orange juice by combined application of mild heat and ginger extract. *Food Journal* 33 (3): 211-220.
11. LVS EN ISO 14130:2003. *Foodstuffs-Determination of vitamin C by HPLC.*
12. LVS EN ISO 4833:2003. *Microbiology of food and animal feeding stuffs-Horizontal method for the enumeration of m/y-colony count tech. at 30 °C.*
13. LVS EN ISO 21527-1:2008. *Microbiology of food and animal feeding stuffs-Horizontal method for the enumeration of yeast and mould-Part I: Colony count technique in products with water activity less than or equal to 0.95.*
14. LVS EN ISO 6887-5:2011. *Microbiology of food and animal feeding stuffs-Preparation of test samples, initial suspension and decimal dilutions for microbiological examination. Part 5: Specific rules for the preparation of milk and milk products (ISO 6887-5:2010).*
15. Nassar, M.I., Gaara, A.H., El-Ghorab, A.H., Farrag, A.R.H., Shen, H., Huq, E. and Mabry, T.J. 2007. Chemical constituents of clove (*Syzygium aromaticum*, Fam. *Myrtaceae*) and their antioxidant activity. *Revista Latinoamericana de Quimica* 35 (3): 47-57.
16. Rupasinghe, H.P.V. and Yu, L.J. 2012. Emerging preservation methods for fruit juices and beverages. In: *Food Additive* (Y. El-Samragy, Ed.), ISBN: 978-953-51-0067-6, InTech, 266p.
17. Seglina, D., Karklina, D., Gailite, I., Ruisa, S., Krasnova, I. and Heidemane, G. 2008. Changes of biochemical compounds in seabuckthorn marc during storage. In: *Proceedings of the International Scientific Conference 3rd Baltic Conference on Food Science and Technology, FOODBALT-2008*, p. 109-114, Latvia University of Agriculture, Jelgava.
18. Singh, G., Maurya, S., deLampasona, M.P. and Catalan Cesar, A.N. 2007. A comparison of chemical, antioxidant and antimicrobial studies of cinnamon leaf and bark volatile oils, oleoresins and their constituents. *Food and Chemical Toxicology* 45 (9): 1650-1661.
19. Singh, V. 2006a. Seabuckthorn (*Hippophae L.*) in traditional medicines. In: *Seabuckthorn (Hippophae L.)-A Multipurpose Wonder Plant. Vol. II: Biochemistry and Pharmacology* (V. Singh, Ed. in Chief, 2006), p. 505-521, Daya Publishing House, New Delhi, 600p.
20. Singh, V. 2006b. Free radicals, diseases, antioxidants and antioxidant properties of seabuckthorn (*Hippophae rhamnoides* L.). In: *Seabuckthorn (Hippophae L.)-A Multipurpose Wonder Plant. Vol. II: Biochemistry and Pharmacology* (V. Singh, Ed. in Chief, 2006), p. 3-69, Daya Publishing House, New Delhi, 600p.
21. Singleton, V.L., Orthofer, R. and Lamuela-Raventos, R.M. 1999. Analysis of total phenols and other oxidation substrates and antioxidants by means of Folin-ciocalteu reagent. *Methods in Enzymology* 299:152-178.

22. Siriwardhana, S.S.K.W and Shahidi, F. 2002. Antiradical activity of extracts of almond and its byproducts. *Journal American Oil Chemistry Society* 79: 903-908.

23. Suhaj, M. and Horvithovi, J. 2007. Changes in antioxidant activity induced by irradiation of clove (*Syzygium aromaticum*) and ginger (*Zingiber officinale*). *Journal of Food and Nutrition Research* 46 (3): 112-122.

24. Suryakumar, G., Sai, R., Singh, V., Havazhagan, G. and Sawhney, R.C. 2001. Antioxidant and immunomodulatory properties of seabuckthorn (*Hippophae rhamnoides*): An *in vitro* study. *Journal of Ethnopharmacology* 79: 373-378.

25. Varshney, A.C. 2003. Therapeutic evaluation of seabuckthorn oil in cutaneous burn wound healing in bovine: A clinico-haematological study. In: *Seabuckthorn-A Resource of Health, a Challenge to Modern Technology*, Proceedings of 1st Congress of the International Seabuckthorn Association, p. 104-107, Berlin, Germany.

26. Wang, H.F., Wang, Y.K. and Yih, K.H. 2009. DPPH free-radical scavenging ability, total phenolic content, and chemical composition analysis of forty-five kinds of essential oils. *International Journal of Cosmetic Science* 31 (6): 475-476.

27. Yuste, J. and Fung, D.Y. 2004. Inactivation of *Salmonella typhimurium* and *Escherichia coli* O157:H7 in apple juice by a combination of nisin and cinnamon. *Journal of Food Protection* 67: 371-377.

PART VIII:
ENVIRONMENTAL CONSERVATION

Chapter 43

Status of Soil Nutrients under Seabuckthorn (*Hippophae rhamnoides*) Vegetation in Lahaul Valley, Himachal Himalayas

Alpy Sharma, Virendra Singh, Manohar Lal, Ashok Singh and S.P. Dixit*

Department of Biology and Environmental Sciences,
*Department of Soil Sciences,
CSK Himachal Pradesh Agricultural University, Palampur – 176 062, Himachal Pradesh, India

SUMMARY

The present study has been carried out to determine the soil nutrients status under the seabuckthorn vegetation in relation to the altitude and in comparison to other tree species in Lahaul valley, Himachal Pradesh, India. Soil samples under represented sites were collected along with information of their geographical locations and then analysed for the soil characteristics. Under the seabuckthorn vegetation, soil pH was neutral to mild alkaline (7.2 to 7.6) in the top soil layer (0-10 cm) and increased with increasing soil depth, however, it decreased with an increase in altitude. Soil organic carbon ranged from 1.95 to 2.30 per cent. Soil available N, P, K and S contents also varied from 271 to 305 kg/ha, 41 to 59 kg/ha, 105 to 201 kg/ha and 21 to 34 kg/ha, respectively, in the soil top layer and increased with an increase in altitude and decreased with increasing soil depth. Soil available micronutrients such as Fe, Zn, Cu and Mn varied from 32.6 to 58.8 mg/kg, 2.0 to 4.4 mg/kg, 1.8 to 3.6 mg/kg and 3.1 to 12.8 mg/kg, respectively, in the soil top layer and also decreased with increasing soil depth. However, available micronutrients such as Zn and Mn decreased with an increase in altitude. As compared to other tree species such as willow, poplar, juniper and control (wasteland), soil nutrients status under seabuckthorn vegetation was significantly higher. As seabuckthorn plays an important role in improving the soil fertility, therefore, it is the most suitable species for the environmental conservation of the fragile ecosystem of cold deserts of the Himachal Himalayas.

***Keywords:** Seabuckthorn, Soil nutrient status, Soil fertility and Lahaul and Spiti.*

Seabuckthorn (Hippophae *L.*): *A Multipurpose Wonder Plant, Vol. 4: Emerging Trends in Research and Technologies* (Virendra Singh, *Editor-in-Chief*, 2014), p. 521–532, Daya Publishing House, New Delhi, India.

INTRODUCTION

The fragile ecosystem of Himalayas is facing an alarming environmental degradation at a frightening rate. A huge transects of Himalayas are turning into barren lands as a result of irreversible process of soil degradation. This is due to the deforestation and indiscriminate and unscientific management practices which result in various types of land degradation or soil fertility problems. Keeping these factors in view, the management of these natural resources in general, has become an indispensable need of time. Therefore, management and maintenance of soil fertility is central to the development of sustainable agriculture (Sehgal and Abrol, 1994).

In contrast to the hot deserts of India, the soils of cold arid regions of high hill Himalayas such as Lahaul have received very little attention due to very harsh outlook for life. Lahaul is situated on the western side of district Lahaul-Spiti in Himachal Pradesh. The region is characterized by high elevation barrier and dissected land, skeletal soils capes, seismic disturbance, peculiar natural drainage, snow covered mountains and glaciers and irregular valleys. The soil of high altitude regions of Lahaul mostly occurs on steeply sloping lands. Therefore, soils of Lahaul and neighbouring high hills and low rainfall regions have received little or no attention, practically nothing or very little is known about the soils, especially their fertility status.

Lahaul falls under semi-arid and arid temperate zone of Himachal Himalayas supports limited growth of vegetation due to the cold and dry climatic conditions. Thus, for ensuring better environmental conservation and ecological conditions in this region an appraisal of soils supporting seabuckthorn forests are of a paramount importance. Seabuckthorn is a prominent species of temperate forests in the cold desert areas of Himachal Himalayas. It is one of the most important shrubs belonging to the family *Elaeagnaceae*, existing between 2500 m and 4500 m above sea level in high mountainous transect of the Pradesh (Singh, 2003). Being an important source of anti-oxidant, it has extraordinary economic benefits, as it has a high potential in health, food, cosmetic and drug industries. This plant also has an outstanding ability to conserve water and moisture, apart from fixing atmospheric nitrogen. Recent studies indicate that the seabuckthorn plays an important role in improving the soil fertility (Rongsen, 1992). It is, therefore, imperative to study soils under seabuckthorn vegetation for their better management and conservation. If the seabuckthorn vegetation is properly managed and effectively conserved, it will help in sustaining the soil productivity.

Keeping in view, the importance of soils supporting seabuckthorn forests in the Lahaul valley of Himachal Pradesh, an attempt has been made to study the soil nutrients status under seabuckthorn population in comparison with other three important forest species in relation to different geographical and climatic variables in Lahaul, Himachal Pradesh.

MATERIALS AND METHODS

Study Area

The present study was conducted in the Lahaul valley region. The Lahaul valley (31° 44′ 57″- 32 59′ 57″N, 76° 46′29″-78° 41′34″ E) is situated towards western side of district and covers an area of 6244 km, is a part of the proposed Cold Desert Biosphere Reserve. The altitude ranges from 2400 to 6517 m above sea level. The climate varies from dry temperate to alpine and it is enclosed between the high mountains. The track is narrow, steep with mountain peaks rising to an average of 5480 m and highest peak reaching at 6400 m. The rainfall varies from 100 mm to 500 mm, depending the altitude and year. The temperature during winter may go down to -20C. vegetation is scarce particularly in higher altitude areas. The study was conducted in the region of Lahaul, where the species of the seabuckthorn (*Hippophae rhamnoides*), juniper (*Juniperus macropoda*), willow (*Salix daphanoides*) and poplar (*Populus nigra*) grow on marginal lands in the region. The soil samples were collected from the represented sites under different vegetation (Singh and Awasthi, 1995).

The study was conducted on the represented sites of the Lahaul valley. At the time of the collection of soil samples, information on altitude, different latitudes, on various aspects and habitats with reference to each spot was procured from various government departments (Forest, university etc.) for making details about the site.

In order to meet the objective of present work, soil samples were collected under seabuckthorn (*Hippophae rhamnoides*), juniper (*Juniperus macropoda*), willow (*Salix dephanoides*) and poplar (*Populus nigra*) vegetation. The soil samples were collected under seabuckthorn vegetation from the different locations such as Kukumseri, Jahlma, Keylong, (semi-arid), Gemour, Jispa and Darcha (arid region) with increasing altitude (Table 43.1). As there was no poplar stand in the Bhaga valley (Gemour, Jispa and Darcha villages), therefore, soil analysis work was confined to only seabuckthorn in Bhaga valley.

Samples Collection

Soil samples were collected from the represented sites under different vegetations. The soil samples were drawn from 0-10 cm, 10-20 cm and 20-30 cm depth of soil from 10 random locations in each stand with the help of a soil auger and were thoroughly mixed together weighing about 200 g. All the samples were brought to the laboratory, air dried, and ground with pestle mortar and passed through 2 mm sieve.

Table 43.1: Comparison of Sites of Collection of Soil Samples from different Species

Species	*Site*	*Latitude*	*Longitude*	*Altitude (m asl)*	*Aspect*	*Habitat*
Seabuckthorn	Kukumseri	32°40'00.1"N	76°39'09.2"E	2630	S-W	Riverside
	Jahlma	3237'58.1"N	76°51'38.3"E	2781	S-W	Slope
	Keylong	32°34'34.9"N	77°01'01.0"E	3176	S-W	Slope
	Gemour	32°30'20.7"N	77°11'05.4"E	3250	S-W	Marshy
	Jispa	32°36'38.9"N	077°08'46.7"E	3280	S-W	Slope
	Darcha	32°40'07.6"N	7712'50.3E	3316	S-W	Riverside
Poplar	Kukumseri	32°40'00.1"N	76°39'09.2"E	2630	S-W	Slope
	Jahlma	3237'58.1"N	76°51'38.3"E	2781	S-W	Roadside
	Keylong	32°34'34.9"N	77°01'01.0"E	3176	S-W	Slope
Juniper	Kukumseri	32°40'00.1"N	76°39'09.2"E	2630	S-W	Dry slope
	Jahlma	3237'58.1"N	76°51'38.3"E	2781	S-W	Dry slope
	Keylong	32°34'34.9"N	77°01'01.0"E	3176	S-W	Slope
Willow	Kukumseri	32°40'00.1"N	76°39'09.2"E	2630	S-W	Grassland
	Jahlma	3237'58.1"N	76°51'38.3"E	2781	S-W	Slope
	Keylong	32°34'34.9"N	77°01'01.0"E	3176	S-W	Slope
Wasteland	Kukumseri	32°40'00.1"N	76°39'09.2"E	2630	S-W	Riverside
	Jahlma	3237'58.1"N	76°51'38.3"E	2781	S-W	Roadside
	Keylong	32°34'34.9"N	77°01'01.0"E	3176	S-W	Habitation

The soil samples after proper processing were analysed for soil pH, organic carbon, available N, P, K, S, Fe, Zn, Cu and Mn by the standard laboratory methods.

Soil pH

Soil pH was measured by the pH meter electrode measured in a 1:2:5 soil: water ratio following the method of Jackson (1967).

Organic carbon

It was estimated in the soil by the Rapid titration method (Walkley and Black, 1934).

Available N

It was determined in the soil by the Alkaline potassium permanganate method (Subbiah and Asija, 1956).

Available P

It was estimated in the soil by the by the alkali extraction. The alkali solution of 0.5 N Na HCO_3 (pH 8.5) is used and known as Olsen's extractant (Olsen *et al.*, 1954).

Available K

It was determined in the soil by the Neutral normal Ammonium acetate method (Black, 1968).

Available S

It was estimated in the soil by the 0.15 percent $CaCl_2$ extractable method (William and Steinberg, 1959).

Available Micronutrients (Fe, Mn, Zn, and Cu)

It was determined in the soil by the DTPA method in the Atomic absorption spectrophotometer (Lindsay and Norvell, 1978).

RESULTS AND DISCUSSION

Soil pH under seabuckthorn vegetation was generally neutral to mild alkaline in reaction. In top soil layer (0-10 cm), soil pH varied from 7.2 to 7.6. The soil pH increased with increasing soil depth and decreased with increasing altitude (Figure 43.1). It was maximum (7.7) at high altitude Kukumseri (2630 m) and minimum (7.2) at lowest altitude Darcha (3316 m) (Table 43.2). As compared with the other tree species, seabuckthorn had the maximum in the soil top layer (pH 7.6) and minimum in the juniper (pH 7.1). Overall, soil pH was higher under willow (pH 7.7) and was lower for juniper (pH 7.2) (Tables 43.3). Under low rainfall conditions, pH values are high and rich in the exchangeable bases (Sharma and Kanwar, 2010).

Organic carbon varied from 1.95 to 2.30 per cent in the soil top layer 0-10 cm and decreased with increasing soil depth and increased with increasing altitude (Figure 43.2), which was due to the addition of leaf litter annually and slow decomposition of organic residues under low temperature conditions (Minhas *et al.*, 1997). It was maximum (2.27 per cent) at high altitude Darcha (3316 m) and minimum (1.91 per cent) at lowest altitude Kukumseri (2630 m) (Table 43.2). As compared with the other tree species, organic carbon accumulated maximum (2.05 per cent) under seabuckthorn and minimum (0.86 per cent) for control (wasteland) in the soil top layer (0-10 cm). Overall, organic carbon was significantly ($p \leq 0.05$) higher under seabuckthorn (2.02 per cent) (Tables 43.2 and 43.3). Soil

Table 43.2: Effect of the Seabuckthorn (*Hippophae rhamnoides*) and Depth on the Soil pH, Organic Carbon and other Soil available Nutrients

Locations	*pH*	*OC (per cent)*	*N (kg/ha)*	*P (kg/ha)*	*K (kg/ha)*	*S (kg/ha)*	*Fe (mg/kg)*	*Zn (mg/kg)*	*Cu (mg/kg)*	*Mn (mg/kg)*
Kukumseri	7.7	1.91	266	37	100	19	29.8	4.3	1.5	11.6
Jahlma	7.6	2.06	282	38	111	24	31.6	3.9	1.7	10.5
Keylong	7.5	2.09	288	41	156	25	37.3	3.5	2.4	7.4
Jispa	7.4	2.13	291	46	161	25	42.5	2.9	2.6	7.3
Gamour	7.3	2.22	295	49	165	27	46.4	2.2	3.3	3.6
Darcha	7.2	2.27	299	55	195	30	55.1	1.5	3.5	3.0
SE±	0.024	0.006	1.046	0.665	0.791	0.760	0.253	0.076	0.052	0.112
CD (P£0.05)	0.069	0.017	3.000	1.907	2.270	2.181	0.725	0.219	0.150	0.334
Depth (cm)										
D1 (0-10)	7.4	2.14	293	49	155	29	43.5	3.3	2.9	8.0
D2 (10-20)	7.5	2.12	288	44	148	25	40.6	3.0	2.5	7.3
D3 (20-30)	7.5	2.08	280	40	142	21	37.3	2.8	2.0	6.4
SE (m)±	0.017	0.004	0.740	0.470	0.560	0.538	0.179	0.054	0.037	0.082
CD ($P \leq 0.05$)	0.049	0.012	2.121	1.348	1.605	1.542	0.513	0.154	0.106	0.236

Table 43.3: Effect of the different Species and Depth on the soil pH, Organic Carbon and Other Available Nutrients

Species	*pH*	*OC (per cent)*	*N (kg/ha)*	*P (kg/ha)*	*K (kg/ha)*	*S (kg/ha)*	*Fe (mg/kg)*	*Zn (mg/kg)*	*Cu (mg/kg)*	*Mn (mg/kg)*
Seabuckthorn	7.6	2.02	279	47	122	23	32.9	2.2	1.9	4.6
Willow	7.7	1.55	208	32	224	22	26.2	1.9	1.7	3.9
Poplar	7.6	1.19	234	31	233	14	18.8	1.5	1.5	4.3
Juniper	7.2	1.07	182	34	163	16	12.9	1.3	1.6	4.6
Wasteland	7.6	0.86	136	31	249	14	12.8	0.8	1.2	1.4
Se(m)±	0.013	0.005	0.571	0.321	1.987	0.316	0.258	0.044	0.037	0.059
CD (P£0.05)	0.037	0.014	1.606	0.903	1.401	0.888	0.729	0.125	0.103	0.165
Depth (cm)										
D1 (0-10)	7.5	1.41	214	38	204	21	22.8	1.7	1.8	4.2
D2 (10-20)	7.5	1.37	208	35	198	18	20.5	1.5	1.6	4.0
D3 (20-30)	7.6	1.34	202	32	192	15	18.9	1.3	1.3	3.1
Se(m)±	0.010	0.004	0.442	0.249	1.987	0.245	0.201	0.034	0.029	0.046
CD (P£0.05)	0.029	0.011	1.244	0.699	2.426	0.688	0.564	0.096	0.080	0.128

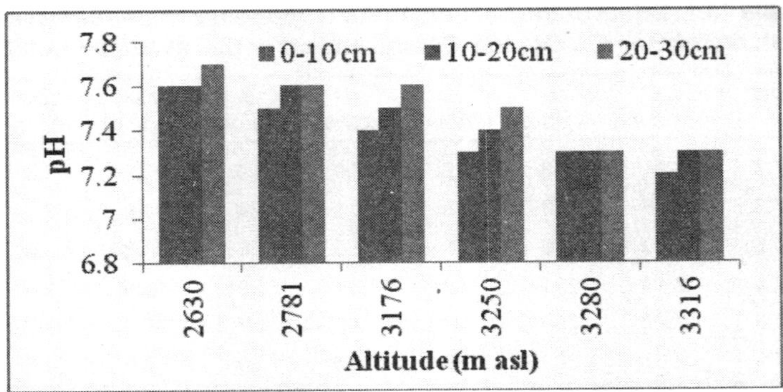

Figure 43.1: Variation in pH with Altitude and Depth under Seabuckthorn in Lahaul.

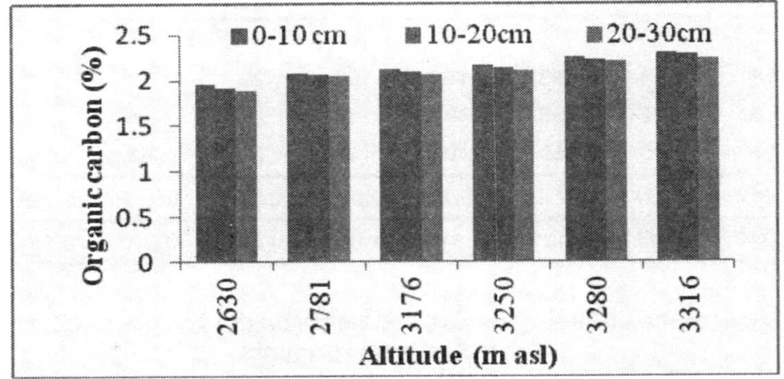

Figure 43.2: Variation in Organic Carbon with Altitude and Depth under Seabuckthorn in Lahaul.

organic carbon decreased with an increase in soil depth in all the species. Seabuckthorn plant produces greater biomass both above and underground through its metabolism which result in high amount of organic carbon (Quanzhong *et al.*, 1989).

Soil available N under seabuckthorn vegetation was low to medium. It varied from 271 to 305 kg/ha in the soil top layer (0-10 cm). It decreased with increasing soil depth and increased with increasing altitude (Figure 43.3). This was due to the decrease in the temperature with an increase in altitude which retards the rate of the decomposition of the plant residues as reported by Biddapa and Rao (1973). Available N was maximum (299 kg/ha) at high altitude Darcha (3316 m) and minimum (266 kg/ha) at Kukumseri (2630 m) (Table 43.2). As compared with the other tree species, seabuckthorn had maximum soil available N (286 kg/ha) and minimum (143 kg/ha) in wasteland in the soil top layer (0-10 cm) (Tables 43.2 and 43.3). Overall, soil available N was significantly ($p \leq 0.05$) maximum (279 kg/ha) under seabuckthorn vegetation. Soil available N decreased with an increase in soil depth in all the species. Seabuckthorn plant coupled with efficient nitrogen fixation and nitrogen-fixing species have a function to enrich N in soil (Rong *et al.*, 2010).

Soil available P under seabuckthorn vegetation was generally high. In top soil layer (0-10 cm), soil P varied from 41 to 59 kg/ha. Soil P increased with increasing altitude however decreased with increasing soil depths (Figure 43.4). It was maximum (55 kg/ha) at high altitude Darcha (3316 m) and minimum (37 kg/ha) at low altitude Kukumseri (2630 m) (Table 43.2). As compared with the other tree

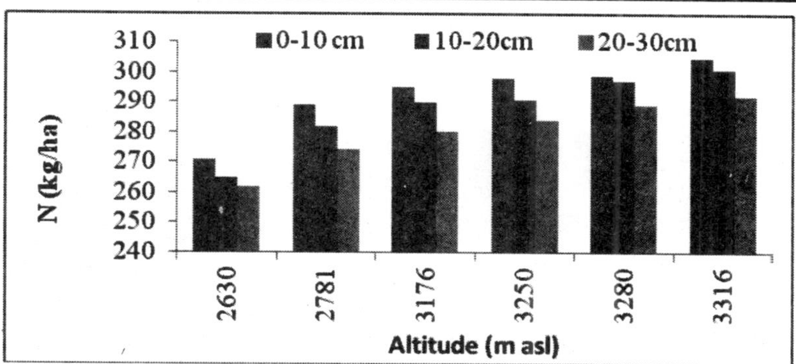

Figure 43.3: Variation in N (kg/ha) with Altitude and Depth Under Seabuckthorn in Lahaul.

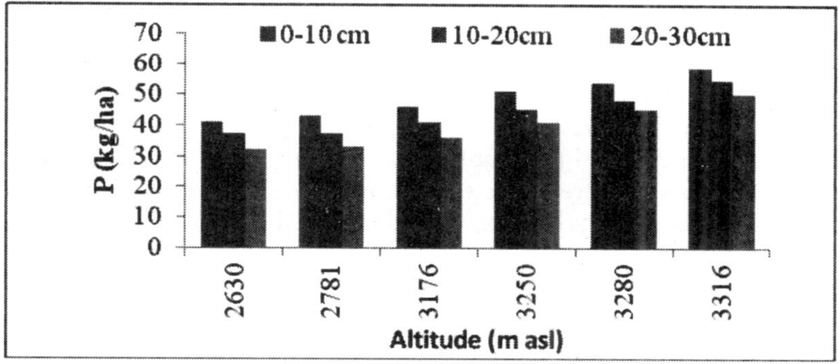

Figure 43.4: Variation in P (kg/ha) with Altitude and Depth Under Seabuckthorn in Lahaul.

species, seabuckthorn had maximum soil available P (52 kg/ha) in the soil top layer (0-10 cm) and minimum in the wasteland (34 kg/ha). Soil available P decreased with increase in soil depth in all the cases. Overall, soil available P was significantly ($p \leq 0.05$) maximum under seabuckthorn vegetation (47 kg/ha) and minimum for poplar (31 kg/ha) which was statistically at par with the wasteland (Table 43.3). Fertility status with respect to available P depends upon the amount of organic material added to the soil. Therefore, high organic matter soil contains high phosphorus contents (Dongale, 1993).

Soil available K under seabuckthorn was low to medium. In soil top layer (0-10 cm) under seabuckthorn, soil available K varied from 105 to 201 kg/ha. Available K increased with increasing altitude, but decreased with increasing soil depth (Figure 43.5). Available K was maximum (195 kg/ha) at higher altitude Darcha (3316 m) and minimum (100 kg/ha) at lower altitude Kukumseri (2630 m). Grieve *et al.* (1990) found decreasing intensity of weathering and nutrient decomposition with increasing altitude due to which with increasing altitude potassium content also increased. As compared with other tree species, soil available K was maximum (254 kg/ha) under control (wasteland) in the soil top layer (0-10 cm) and minimum in the seabuckthorn (129 kg/ha). Overall, soil available K was significantly ($p \leq 0.05$) maximum under control *i.e.* wasteland (249 kg/ha) and minimum (122 kg/ha) for seabuckthorn (Table 43.3). Due to the presence of high organic carbon under the seabuckthorn vegetation, it contained low amount of the potassium (Mishra *et al.*, 2009). Higher concentration of the

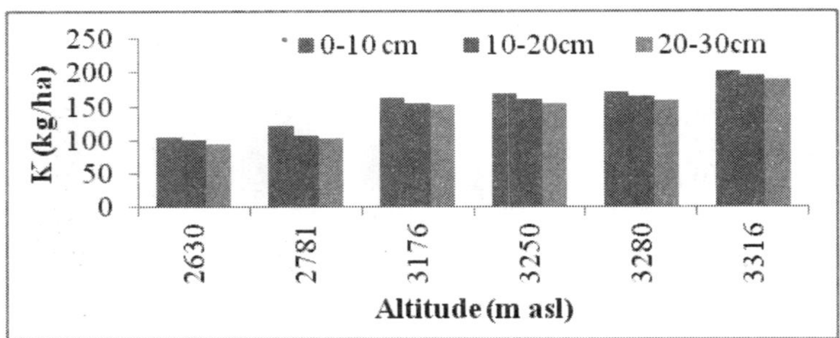

Figure 43.5: Variation in K (kg/ha) with Altitude and Depth Under Seabuckthorn in Lahaul.

potassium in the surface might be due to the action of the plant roots in transporting potassium to the surface and due to the addition of the plant residue (Black, 1968).

Soil available S was generally high. In the soil top layer (0-10 cm), soil available S varied from 21 to 34 kg/ha under seabuckthorn vegetation. Soil available S increased with altitude and decreased with soil depth (Figure 43.6). Soil available S was maximum (30 kg/ha) at higher altitude Darcha (3316 m) and minimum (19 kg/ha) at lower altitude Kukumseri (2630 m) under seabuckthorn (Table 43.2). As compared with the other tree species, soil available S was maximum (26 kg/ha) under seabuckthorn and minimum (16 kg/ha) for wasteland in the soil top layer (0-10 cm). Overall, seabuckthorn had maximum available S in the soil which was significantly ($p \leq 0.05$) higher (23 kg/ha), as presented in Table 43.2. Sulphur added to soils by both root and leaf litters (David *et al.*, 1987).

Soil available Fe under seabuckthorn vegetation was generally in the sufficient range at different places in Lahaul. In the soil top layer (0-10 cm), soil available Fe varied from 32.6 to 58.8 mg/kg. Soil available Fe in all cases, increased with increasing altitude and decreased with increasing soil depth (Figure 43.7). The surface soils contained high amount of the available Fe due to the regular addition of plant residues (Jalali *et al.*, 1989). Soil available Fe was maximum (55.1 mg/kg) at higher altitude Darcha (3316 m) and minimum (29.8 mg/kg) at lower altitude Kukumseri (2630 m) (Table 43.2). Lahiri and Chakravati (1989) also found that high altitude soils to be high in the available Fe due to slow decomposition of nutrients with increasing altitude. As compared with the other tree species, the soil

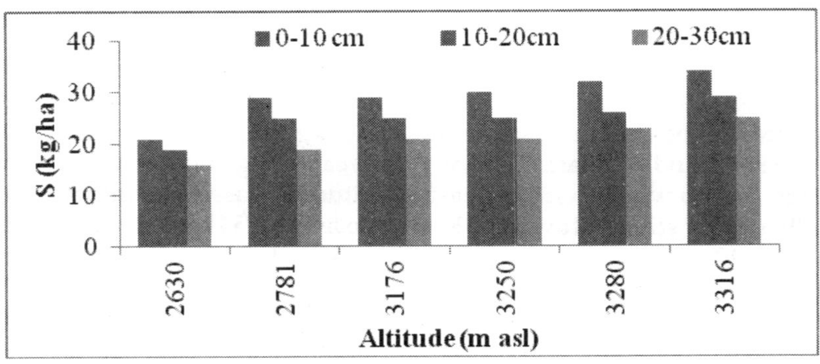

Figure 43.6: Variation in S (kg/ha) with Altitude and Depth Under Seabuckthorn in Lahaul.

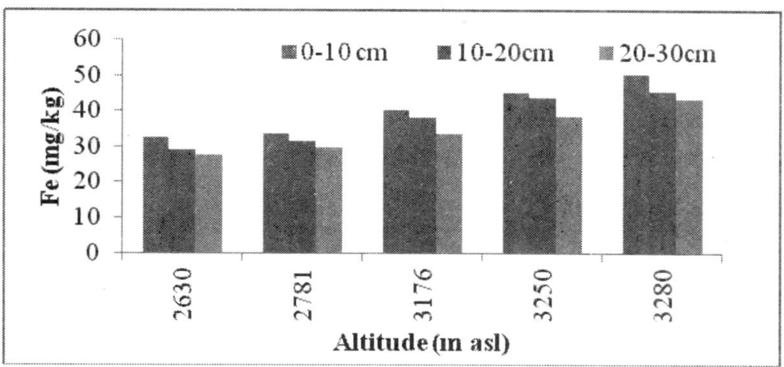

Figure 43.7: Variation in Fe (mg/kg) with Altitude and Depth Under Seabuckthorn in Lahaul.

available Fe was maximum (35.4 mg/kg) in the soil top layer (0-10 cm) under seabuckthorn and minimum (14.7 mg/kg) for the control (wasteland) (Tables 43.2 and 43.3). Overall, seabuckthorn had maximum (32.9 mg/kg) available Fe in the soil which was significantly ($p \leq 0.05$) higher than the other species (Table 43.2).

Soil available Zn was generally in the sufficient range. Soil available Zn varied from 2.0 to 4.4 mg/kg in the soil top layer (0-10 cm). Soil available Zn in all cases decreased with increasing soil depth and with altitude (Figure 43.8). Soil available Zn was maximum (4.3 mg/kg) at lower altitude Kukumseri (2630 m) and minimum (1.5 mg/kg) at higher altitude Darcha (3316 m) under seabuckthorn (Table 43.2). Lower contents of the Zn in the high altitude soils might be due to the strong sorption of the Zn by the organic matter (Lahiri and Chakarvarti, 1989). As compared with the other tree species, seabuckthorn had maximum available Zn (2.0 mg/kg) and minimum (0.6 mg/kg) for the control (wasteland) in the soil top layer (0-10 cm). Overall, it was higher under seabuckthorn (2.2 mg/kg) and is shown in Table 43.2.

Soil available Cu was mainly in the sufficient range. In the soil top layer (0-10 cm), it varied from 1.8 to 3.6 mg/kg. Soil available Cu increased with altitude and decreased with increasing soil depth (Figure 43.9). It was maximum (1.5 mg/kg) at higher altitude Darcha (3316 m) and minimum (3.4 mg/kg) at lower altitude Kukumseri (2630 m). As compared with the other tree species, seabuckthorn had

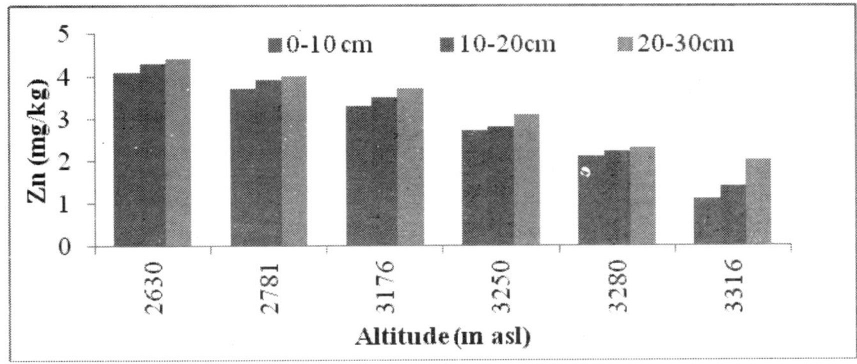

Figure 43.8: Variation in Zn (mg/kg) with Altitude and Depth Under Seabuckthorn in Lahaul.

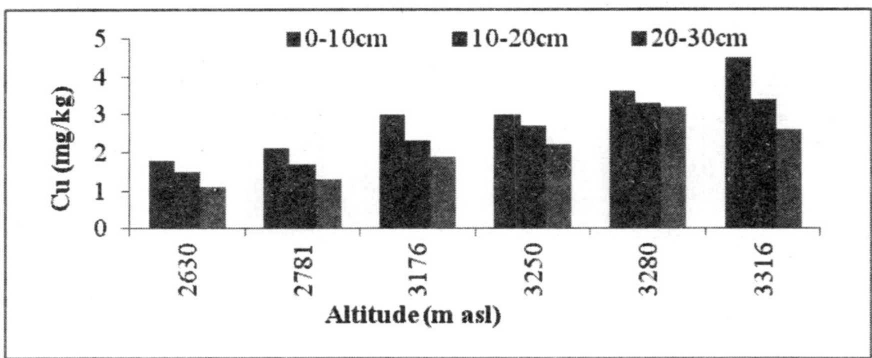

Figure 43.9: Variation in Cu (mg/kg) with Altitude and Depth Under Seabuckthorn in Lahaul.

maximum available Cu (2.3 mg/kg) and minimum (1.4 mg/kg) for control (wasteland) in the soil top layer (0-10 cm) and is presented in Table 43.3.

However overall soil available Cu was maximum (1.9 mg/kg Cu) in the seabuckthorn soil which was significantly higher ($p \leq 0.05$) than the other species (Table 43.3). Soil Cu concentration was higher due to higher inputs of organic matter (Singh and Sharma, 2004).

Soil available Mn was in the sufficient range. In general soil available Mn varied from 3.1 to 12.8 mg/kg in the soil top layer (0-10 cm) (Table 43.1). Soil available Mn decreased with altitude and soil depth (Figure 43.9). It was maximum (11.6 mg/kg) at lower altitude Kukumseri (2630 m) and minimum (3.0 mg/kg) at higher altitude Darcha (3316 m) under seabuckthorn. This is due to the fact that organic matter formed insoluble complexes with the manganese (Jalali *et al.*, 1989). As compared with the other tree species, soil available Mn was maximum of 3.7 mg/kg under seabuckthorn and minimum of 1.3 mg/kg for control (wasteland) in the soil top layer (0-10 cm). Overall, it was significantly ($p \leq 0.05$) maximum for the seabuckthorn (4.6 mg/kg) which was statistically at par with the juniper and a minimum for the control (1.4 mg/kg) (Table 43.3).

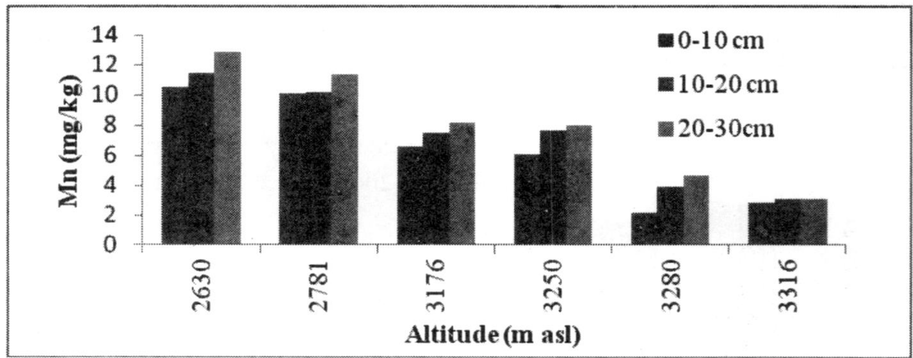

Figure 43.9: Variation in Mn (mg/kg) with Altitude and Depth Under Seabuckthorn in Lahaul.

CONCLUSION

Seabuckthorn soils accumulated organic carbon, available N, P, Fe, Zn, Cu and Mn, which were significantly ($p \leq 0.05$) higher than the soils of poplar, juniper, willow and baseland. It was also found that contents of soil nutrients increased with altitude and decreased with soil depths under seabuckthorn. Seabuckthorn habitat conditions were most suitable for the natural growth of the seabuckthorn forest at Jahlma, Keylong, Gemour and Jispa. The valleys were broader higher, there were larger duration of the sun hours, more exposure to the sun, better availability of moisture and soil nutrients, and other better habitat conditions, nitrogen fixing ability of seabuckthorn, which also helped in the growth of other vegetation around the seabuckthorn. Consequently, there is more accumulation of the nitrogen, organic carbon and other minerals under seabuckthorn as compared to other species. From the present study, it has been concluded that seabuckthorn is the most promising species of the region for the improvement of the soil fertility and improvement. Therefore, it is the most suitable species for the afforestation and environmental conservation of the fragile ecosystem of the region.

REFERENCES

1. Black, C.A. 1968. *Soil Plant Relationship*. Indian Reprint, Willey Eastern (Pvt.) Limited, New Delhi.
2. Biddapa, C.C. and Rao, V. 1973. Altitudinal pattern of soil characteristics under forest in eastern Himalayan region. *Journal of Indian Society of Soil Science* 25 (1): 21-42.
3. David, M.B., Mitchell, M.J. and Scott, T.J. 1987. Importance of biological processes in the S budget of a northern ecosystem. *Biology and Fertility of Soils* 52 (1): 58-64.
4. Dongale, J.M.1993. Depth distribution of different forms of phosphorus in lateritic soils of coastal region. *Journal of Indian Society of Soil Science* 4 (1): 62-66.
5. Grieve, I.C., Proctor, J. and Cousins, S.A. 1990. Soil variation with altitude on Volcan Barva, Costa Rica. *CATENA* 17 (1): 525-534.
6. Jackson, M.L. 1967. *Soil Chemical Analysis*. p. 219-221, Prentice-Hall of India, LTD., New Delhi.
7. Jalali, V.K., Talib, A.R. and Takkar, P.N. 1989. Distribution of micronutrients in some benchmark soils of Kashmir at different altitudes. *Journal of Indian Society of Soil Science* 37 (5): 465-469.
8. Lahiri, T. and Chakravarti, S.K. 1989. Characteristics of some soils of Sikkim at various altitudes. *Journal of the Indian Society of Soil Science* 37: 451-454.
9. Lindsay, W.L. and Norvell, W.A. 1978. Development of DTPA SOIL test for zinc, iron, manganese and copper. *Soil Science Society of American Journal* 42 (1): 421-428.
10. Minhas, R.S., Minhas, H. and Verma, S.D. 1997. Soil characterization in relation to forest vegetation in the wet temperate zone of Himachal Pradesh. *Journal of Indian Society of Soil Science* 45 (1): 146-151.
11. Mishra, S., Dhyani, D. and Maikhuri, R.K. 2009. Understanding the role of seabuckthorn in soil fertility improvement in central Himalaya, India. In: Proceedings of ISA Conference on *"Seabuckthorn on the Way Between Science and Industry Interaction*, Russia.
12. Olsen, S.R., Cole, C.V., Watanbe, F.S. and Dean, L.A. 1954. Estimation of available Phosphorus in soils by extraction with sodium carbonate. *USDA Circ.* 939 (6): 19-33.
13. Quanzhong, L. 1989. Multiple effects of artificial common seabuckthorn (*Hippophae rhamnoides L*) forests in western Liaoning. In: *Proceedings of the International Symposium on Seabuckthorn (H. rhamnoides)*, p.288-297, China.
14. Rongsen, L. 1992. *Seabuckthorn-A Multipurpose Plant Species for Fragile Mountains*. Occasional Paper No. 20, ICIMOD, Kathmandu, Nepal, 62p.

15. Rong, M., De-Hui, Zeng, Gui-Yan, Ai Dan, Y., Lu-Jun, L. and Yun-Xia, L. 2010. Soil microbiological and chemical effects of a nitrogen fixing shrub in poplar in semi-arid regions of north-east China. *European Journal of Soil Biology* 46 (5): 325-329.
16. Subbiah, B.V. and Asija, G.L. 1956. A rapid procedure for estimation of the available nitrogen in soils. *Current Science* 25 (1): 259-260.
17. Singh, R.P. and Gupta, M.K. 1990. Soil and vegetation study of Lahaul and Spiti cold desert of western Himalayas. *Indian Forester* 116 (10): 785-790.
18. Sehgal, J.L. and Abrol, I. P. 1994. *Soil degradation in India: Status and Impact*. Oxford and IBH Publishing Co. Pvt. Ltd., New Delhi, India, 80p.
19. Singh, V., Singh, B. and Awasthi, C.P. 1995. Distribution, taxonomy and nutritional values of seabuckthorn growing in dry temperate Himalayas. In: *Proceeding of International Workshop on Seabuckthorn*, p.52-59, Dec.12-17, 1995, Beijing, China, 206p.
20. Singh, V. 2003. Geographical adaptation and distribution of seabuckthorn (*Hippophae* L) resources. In: *Seabuckthorn (Hippophae L.)-A Multipurpose Wonder Plant. Vol. I: Botany, Harvesting and Processing Technologies* (V. Singh, Ed. in Chief, 2003), p.21-34, Indus Publishing Company, New Delhi, 518p.
21. Singh, V. and Sharma, R.K. 2004. Morpho-biochemical variations in seabuckthorn (*Hippophae* L.) populations growing in Lahaul Valley, dry temperate Himalayas. *Indian Forester* 130 (6): 663-672.
22. Sharma, V. and Kanwar, B.B. 2010. Copper status and its relation with soil properties in pea growing soils of high hills dry temperate zone of the Himachal Pradesh. *Indian Journal of Agricultural Research* 44(1): 32-37.
23. Tapesh, L. and Chakravarti, S.K. 1989. Characteristics of some soils of Sikkim at various altitudes. *Journal of Indian Society of Soil Science* 37 (1): 451-454.
24. Walkley, A. and Black, C.A. 1934. An examination of wet acid method for determining soil organic matter and a proposed modification of the chromic acid titration method. *Soil Science* 37 (4): 29-38.
25. William, C.H. and Steinbergs, A. 1959. Soil sulphur fraction as the chemical indices of available sulphur in some Australian soil. *Australian Journal of Agricultural Research* 10 (1): 340-352.

Chapter 44

Integrated Conservation Techniques for Prevention of Soil Erosion in Soft Rock Area, China

Bi Cifen[1], TaiYuanlin[1], Wang Fugui[2], Qiao Wanglin[3], Li Gui[3] and Hu Cunsheng[3]

[1]International Seabuckthorn Development Center,
Yellow River Conservancy Commission, Ministry of Water Resources, China
[2]Bureau of the Upper and Middle Reaches of Yellow River,
Yellow River Conservancy Commission, Ministry of Water Resources, China
[3]Erdos City Water Science Institute, Inner Mongolia

SUMMARY

The present study has provided information on the integrated soil and water conservation techniques for soil erosion prevention in soft rock region, given a full and systematic treatment model, and also the quantitative relations of the model. The model can separate coarse sediment from fine sediment and retain them in all mini-gullies and all these factors can provide better basis for sustainable development in the region.

***Keywords:** Soil conservation, Soft rock, Soil erosion, Gully and sediment, Loess plateau, Yellow river, China.*

INTRODUCTION

Soft rock region is the core area of bedrock sediment formation area in Loess plateau, China, which is an important part of coarse sediment release in Yellow River. The main characteristic of soft rock is that it is as hard as stone when it is dried and soft as mud when it is wet, and it is weathering when wind is blowing. Because of strong frost-frozen action, loose layer on gully slope is 5-10 cm (Ainai, 2000). It is hard for people to live in the area, local farmers describe the kind of rock as arsenic and called it soft rock. The land form is fragmented, vegetation is rare, climate is dry, precipitation is less and lot of wind-sand storm and so, it was hard to solve the problems. For long time, it couldn't be treated thoroughly. Since the 1980s, the government has implemented a key dam construction for soil

Seabuckthorn (*Hippophae* L.): *A Multipurpose Wonder Plant, Vol. 4: Emerging Trends in Research and Technologies* (Virendra Singh, *Editor-in-Chief*, 2014), p. 533–536, Daya Publishing House, New Delhi, India.

and water conservation and ecological development of seabuckthorn. This has become a better method for the control of accelerating local soil erosion. The chapter focuses on the integrated gully treatment technical system for rapid ecological rehabilitation, and also land and water development, at the same time seeking a operational model for soil erosion control in the soft rock area radically.

INTEGRATED SOIL AND WATER CONSERVATION TECHNIQUES

The practices of several decades have proved that any single soil and water conservation technique could not change the soil erosion situation. So, there were varied techniques carried out so as to control soil erosion by aiming at soil erosion characteristics in the region.

Soil Erosion Process

There are eight months for non-runoff soil erosion, such as frost-frozen and weathering. According to the investigation, it was estimated that non-runoff sediment yield in ravines is about 90 per cent of the total. Sediment discharge concentrates in several rainstorms during flood season, especially in the first rainstorm of a year. Here, rainstorm was usually in large quantity for a short time. Because, the gully bed ratio is in between 1/10 and 1/100, strand flow will erode gully bed and banks after transporting materials piled-up in the gully.

The eroded sediment yield on slope land was less than that in gully and is the bed of Erdos lake. After thousands of years erosion, large-sized grain remained including coarse grains covered up fine grains, so sediment yield on slope land is less and wind erosion was more on cultivated land.

INTEGRATED SOIL AND WATER CONSERVATION TECHNIQUES FOR SOIL EROSION PREVENTION

On the basis of soil erosion process in the region, it was found that soil erosion occurs mainly in the developing gully head. In order to control sediment yield source essentially, one should first control gully head and gully bank expansion, as these are the key parts for soil erosion prevention. Based on the decade research on the experiment of plant "flexible dam" in soft rock region, we succeeded that seabuckthorn can be used as materials for dam works. As long as seabuckthorn is planted crossing the flowing direction with certain distance between rows and individual plant in gully, then "flexible dam" system is arranged. The dam system was established within 3 years. This can either retain sediment or recover ecological environment. You will benefit from such dam system for ten years. The study has found that the "flexible dam" techniques for preventing gully head erosion.

In order to promote replacing slope farmland by trees and shrubs, recovering natural grassland, one must seek for high-yield farmland for replacing slope farmland. So, it must use flat gully bed in the third and the fourth gully, and build high-yield farmland through building warping dams, key dams and techniques of artificial wetland and artificial bank land.

To order to build mini-reservoir in gully mouth for flood control, "flexible dam" system techniques, rigid warping dam techniques, key dam and mini-reservoir techniques have formed a technical system for soil erosion prevention of quick gully ecological recovery, sediment retention, water storage and land regulation in small watershed (Cefen *et al.*, 2001). But, a key technique of the system starts from "flexible dam", which is the principal part of watershed ecological rehabilitation.

SUSTAINABLE DEVELOPMENT GULLY TREATMENT MODEL

By taking plant "flexible dam" as the principal part of sediment retention works, dam land, artificial wetland and artificial bank land as the main components of basic farmland, a key dam as support and mini-reservoir as guarantee, the formula will be formed that coarse sediment was retained in gullies, fine sediment was retained in artificial bank land and dam land, artificial wetland and dam land produced between dams. This can increase runoff infiltration and mini-reservoir was storing all runoff not infiltrating (Cifen and Guifen, 1999; Cifen and Cifen, 2001). So, the objectives were achieved, such as water and sediment was treated separately, water and sediment is balanced, ecology is balanced so as to get sustainable development. The following models are given in terms of a 20 km² small watershed.

Basic model: $F_d+W_d+K_d+V_d$ (1)

Expanded model: $F_d+W_{dU}+W_{dL}+K_d+K_{dU}+K_{dL}+V_d$ (2)

where, F_d-Plant "flexible dam" system in the fourth and fifth small gully of Yellow River; W_{dU}-Warping dam in the upper reaches of the fourth gully in accordance with the controlled area; W_{Dl}-Warping dam in the lower reaches of the fourth gully; K_d-Key dam in the upper reaches of the third gully of Yellow River, K_{Du}-Sediment retention from parts between dams, runoff retention from main gully; K_{Dl}-Runoff retention from part between dams, sediment retention from uncontrolled area; and V_d-Mini-reservoir at outlet of the third gully of Yellow River.

NUMBER OF RIGID DAM AND FLEXIBLE DAM FOR THE INTEGRATED GULLY TREATMENT MODEL IN THE FORM OF SUSTAINABLE DEVELOPMENT

The above mentioned integrated sustainable development treatment model in the soft rock region can be explained from the example collocation of rigid dam and "flexible dam" in Xizhao gully of Wulanmulun river.

Table 44.1: The Collocating Dam Number in a Small Watershed with the Area of 20 km²

Works Name	*Mini-Reservoir* V_d	*Key Dam* K_d	*Warping Dam* W_d	*Flexible Dam* F_d
Quantity	1	3	9	117

From the Table 44.1, it is found that the relationship between dams as:

$$V_d = 1 \quad (3)$$

$$K_d = 3V_d \quad (4)$$

$$W_d = 3K_d \quad (5)$$

$$F_d = 13W_d = 117V_d \quad (6)$$

At present, these formulas can be used as guidance to practical action. In the soft rock region, they can be used as planning and construction regulations in gully treatment. These formulas also indicates that a better collocation ratio for integrated soil and water conservation techniques in preventing soil erosion in gullies under the function of "flexible dam". The collocation ratio expresses "flexible dam" to be at the position of principal part, *i.e.* it relies on "flexible dam" to rehabilitate gully head ecological environment and coordinate the proper ratio of gully works after retaining coarse sediment produced in gully head section. Meanwhile, it expresses the time sequence that effect along gully bed.

According to the above formulas, it was tried to do planning of soil and water conservation works in 510 small watersheds in Huangfu River with the total area of 3240 km² (Zhengping, 1992). The base small watershed area is 22.8 km² and the planned dam number was:

$$V_d = 142$$

$$K_d = 3V_d = 426$$

$$W_d = 3K_d = 1278$$

$$F_d = 13W_d = 16614$$

CONCLUSION

The present study has provided the integrated soil and water conservation techniques for soil erosion prevention in soft rock region, given a full and systematic treatment model, and also the quantitative relations of the model. The model can separate coarse sediment from fine sediment and retain them in all mini-gullies. It not only prevent the soil erosion radically, eliminate coarse sediment danger to lower reaches and reservoir in the main channel of Yellow River, but also can replace slope farmland with basic farmland in gullies so as to promote, to replace slope farmland with trees and shrubs. All of these factors can provide better basis for sustainable development in the region.

REFERENCES

1. Ainai, M. 2000. *Geography Science and Geographical Information Science*, August 2000, Wuhan Publishing House.
2. Cifen, B. and Cifen, B. 2001. Discussion on treatment countermeasure in bedrock area on Loess plateau. *Journal of Sediment Research* 4.
3. Cifen, B. and Guifen, L. 1999. Soil erosion control investigation and experimental site selection by seabuckthorn. In: *Bureau of the Upper and Middle Reaches of Yellow River*, February, 1999.
4. Cifen, B., Fugui, W. and Guangyao, Z. 2001. To develop artificial wetland in gullies, to improve ecology in bedrock erosion area. *Soil and Water Conservation in China* 5.
5. Zhengping, J. 1992. *Soil Erosion Systematic Model and Treatment Model in Huangfu River*, Sea and Ocean Publishing House.

Chapter 45

Multiple Functions of Seabuckthorn in the Sustainable use of Water and Land Resources in Soft Rock Area of China

Bi Cifen, Tai Yuanlin and Wang Fugui*

International Seabuckthorn Development Center,
Yellow River Conservancy Commission, Ministry of Water Resources, China
*Bureau of the Upper and Middle Reaches of Yellow River,
Yellow River Conservancy Commission, No. 1, West Hansen Road,
Ministry of Water Resources, Xi'an 710043, China

SUMMARY

Seabuckthorn is a kind of structural material for conservation of soil and water for sustainable use in soft rock region. The coordinative results formed four artificial ecological systems, a core industry and a sustainable development management system, *i.e.* gully artificial forest ecological system, artificial wetland system, artificial lake system, slope grassland system, seabuckthorn industry development system and sustainable development and management system. Each aspect contains a series of sub-system. It also proved that the fact is the coordinative function by seabuckthorn for the suggestion of "*Taking seabuckthorn resource development as a breakthrough for accelerating treatment of soil loss in Loess plateau*", China.

Keywords: *Soft rock, Coordinative functions, Flexible dam, Seabuckthorn, Loess plateau and China.*

INTRODUCTION

The soft rock region of China with the geographical area of 11.7 thousand km^2, located in the boundary area of Shaanxi and Inner Mongolia provinces, where rock is composed of mud rock, sandstone and sand shale. The mountain is bare, slope is steep and flow is torrential. It is called the most serious soil erosion area in the world and now, it is involved in the poverty and environmental degeneration situation. This has threatened human being's life and social development. How to

Seabuckthorn (Hippophae L.): A Multipurpose Wonder Plant, Vol. 4: Emerging Trends in Research and Technologies (Virendra Singh, *Editor-in-Chief*, 2014), p. 537–541, Daya Publishing House, New Delhi, India.

change the environment into better way is an important problem concerned by governments and scholars. In the course of practice and near ten-year scientific experiment in this area, solution was found. Qian Zhengying, the former minister of MWR who understood and rescued seabuckthorn, as natural and human activities destroyed the ecological environment of Loess plateau, consequently local tree species of seabuckthorn disappeared over time. She put forward scientific suggestion of *"Taking seabuckthorn resource development as a breakthrough for accelerating treatment on Loess plateau"* on November 16th of 1985 after scientific investigation and study by scientists (Zhengying, 1986). It was started first time for seabuckthorn rehabilitation and development in Loess plateau and it became a milestone of artificial seabuckthorn resource construction in Loess plateau. In 16 years, seabuckthorn planting was done in over 19 provinces in northeast China, northwest China, north China and southwest China. It has become the pioneer tree species for soil erosion prevention and quick ecological recovery. Since, the successful experiment of seabuckthorn plantation on steep slope land in the soft rock region, 66,700 ha artificial seabuckthorn forest was raised, despite serious drought lasted from 1997 to 2001 (Jinrong *et al.*, 1997). Drought caused the incidence of insect pest and led seabuckthorn mortality in certain areas (Zhangyi, 2002), but did not affect the soil and water conservation programme. The experimental research on *"Plant Flexible Dam" in Soft Rock Region* during 1997 and 2001 was carried out as follows: 1. The experimental data proved that the "Flexible dam" could hold up big-size sediment from gullies and large-size sediment and small size sediment naturally for the first time. Seabuckthorn can grow continuously with high growth speed. The buried seabuckthorn root system can sprout into new horizontal roots and new shoots, and grow in the horizontal zone so as to increase "Flexible dam" body; 2. The research studies proved that seabuckthorn can be cheap dam material in the region and possesses the value of ecological rehabilitation, sediment retention and soil conservation.

The crux of Yellow River is sediment. The complex problem will become simple while materials for sediment control were arranged. It was also pointed out that the fact of the scientific suggestion mentioned above is the coordinative function of "Flexible dam" system for conservation of sand, soil and water in branch gullies, and the ecological and economic values of seabuckthorn. On the basis of report of *"Comprehensive Soil and Water Conservation Model for Soil Erosion Prevention in Soft Rock Region"* published by the experts, and report on *"Agricultural Planting and Management Model of Seabuckthorn Ecology"* by Zhou Zhangyi, the study analyzed the important role of seabuckthorn "Flexible dam" system in multiple uses and gully head ecological rehabilitation as well as the four aspects of coordinative relationship.

TWO MODELS WERE DEVELOPED

MODEL 1

The Comprehensive Soil and Water Conservation Model for the control of soil erosion in soft rock region, in short, is a comprehensive gully treatment model.

a. Comprehensive Gully Treatment Model

To arrange flexible dam, warping dam, key dam and mini-reservoir starting from gully head in a small watershed with an area of 20 km².

b. Numerical Relation of the Model

$$\text{Model: } m=F_d+W_d+K_d+V_d \quad (1)$$

The relation is:

$$V_d = 1 \quad (2)$$

$$K_d = 3V_d \quad (3)$$

$$W_d = 3K_d \quad (4)$$

$$F_d = 13W_d = 39K_d = 117V_d \quad (5)$$

MODEL 2

Agricultural Planting and Management Model of seabuckthorn ecology, in short, is an agricultural planting and management model. The model taking into account seabuckthorn planting in the first year, harvesting leaves and fruits after 2-3 years, harvesting branches after 7-8 years from bad planting site, cutting the stem for rejuvenation after 15 years.

The two models should be used simultaneously. The former one established a good foundation for utilizing sediment, soil and water resources, and the later one settled a good base for seabuckthorn entering into market as a new kind of "crop". All of these will make a whole foundation for harmonious living between human beings and the outside environment.

RELATION OF SOIL EROSION PREVENTION ON THE BASIS OF COORDINATING WATER AND SOIL IN GULLY CHANNEL

To Coordinate Sediment, Soil and Water Resources

Seabuckthorn "flexible dam" system shows the dualism along flowing direction. The "Flexible dam" system by mixing with warping dam, key dam and mini-reservoir formed the systematic gully engineering works for regulating sand, soil and water. From formula (Cefen, 2001), one can see that quantity of "Flexible dam" is nine times as warping dam and twenty seven times as key dam. As a result, they formed artificial seabuckthorn forest in gully head section, artificial wetland in middle section and artificial lake near gully outlet. This can create better soil and water resources in gully channel so as to establish good basis for commercial cultivation, and agricultural and animal husbandry development.

To Coordinate Gully Channel and Slope Land

As long as there is good land and water resources, the relationship between gully channel and slope land will be coordinated so as to promote development of land by planting trees, shrubs and grasses, and to develop and protect natural grassland on slope land. According to present investigation, income from gully land is 3-5 times higher than that in slope land after gully treatment. In addition, there are natural grasses and shrubs, as the area belongs to desert pasture zone. Like caragana, seabuckthorn and other grasses can be planted on bare soft rock and on soil covered soft rock. The sandy willow, sand sagebrush and seabuckthorn can be planted on sand covered soft rock. Therefore, we can select suitable shrubs for planting contour plant hedge. Grasses can be planted between hedges to prevent sand storm, control of desertification and improving slope land grass ecology.

To Coordinate Ecology and Economy

After the ordinate and perpendicular regulation, this will settle good base for seabuckthorn planting planning and development. During the 9-years of seabuckthorn development, it can make green products for twenty first century and become the raw material for industries of medicines and healthcare products. It can also make great economic benefit. So, the seabuckthorn industry will become the potential core industry for the poverty elimination. Artificial seabuckthorn forest ecological system is formed in the gully head section. Basic farmland locates in gully channel and this presents *"Loess plateau ecological and economic science is on the basis of agri-ecological economy, and by taking forestry ecological economy as a principal unit"* (Cifen, 1999).

To Coordinate Human beings and Environment

Through the three kinds of coordinative relations and presented above, there will form a relative balance between ecology and economy. After obtaining economic income, farmers can provide input again for enlarging production. Thus, it will be developed as stable ecological environment with great economic results. It also ensures to meet the local people's demands. It is also evaluated that the whole development course will be increasing helicity cycle so as to reach the sustainable development and utilization.

SEABUCKTHORN FLEXIBLE DAM SYSTEM

COORDINATION IN RELATION TO HUMAN BEINGS AND ENVIRONMENT

Figure 45.1 shows the relations coordinated by seabuckthorn "flexible dam" in a small watershed with an area of 20 km². This is a macro-frame diagram. Each aspect contains some sub-systems.

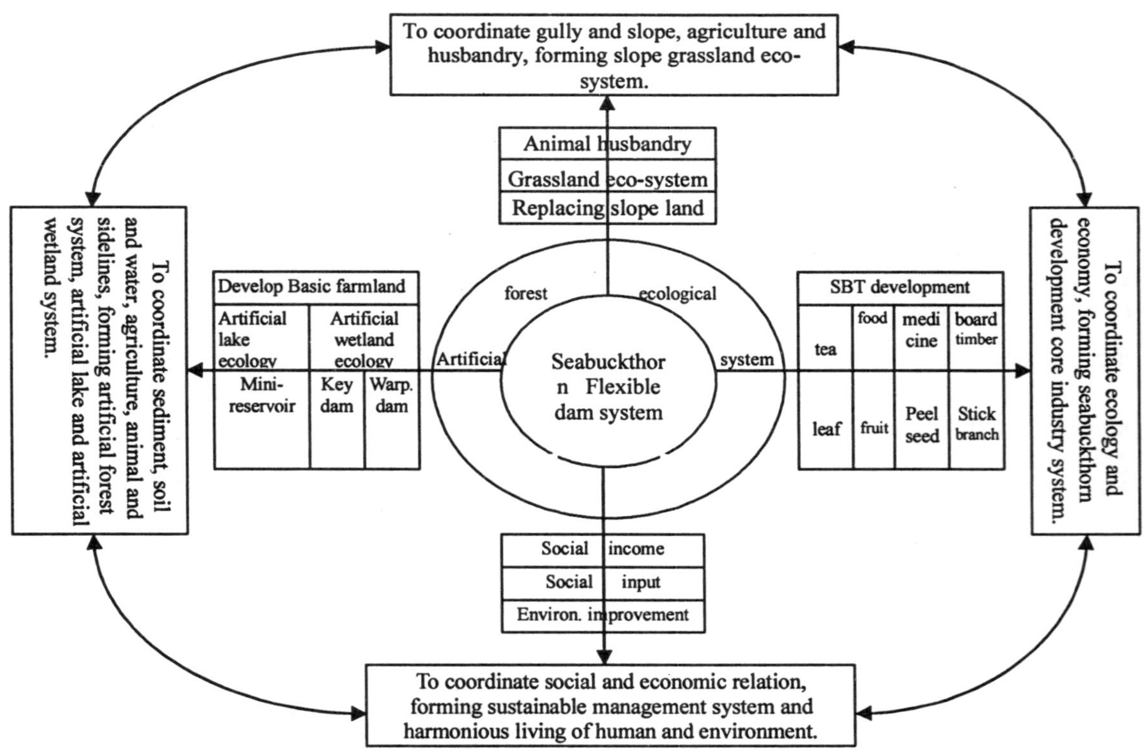

Figure 45.1: Seabuckthorn "Flexible Dam" System.

From the Figure 45.1, one can see that good cycle was formed from "flexible dam" coordination to gully sediment, soil and water.

CONCLUSION

Seabuckthorn presents a kind of structural material for sustainable use of soil and water in soft rock region. The integrated collocation of gully treatment techniques not only prevents soil erosion, but also regulate sediment, soil [and water, and reflect the dominant function of sediment retention in gully head section and principal function in local ecological recovery. Therefore, it formed the harmonious relations between gully and slope land, between ecology and economy, between economy and society, between human beings and environment. The coordinative results formed four artificial ecological systems, a core industry and a sustainable development management system, *i.e.* gully artificial forest ecological system, artificial wetland system, artificial lake system, slope grassland system, seabuckthorn industry development system and sustainable development and management system. Seabuckthorn "Flexible dam" system is a macro-frame diagram. Each aspect contains a series of sub-systems. This system accords with the sustainable development strategic plan in the outline of *'Agenda of the Twenty First Century'*. It also proved that the fact is the coordinative function by seabuckthorn for the suggestion of *"Taking seabuckthorn resource development as a breakthrough for accelerating treatment on Loess plateau"*.

REFERENCES

1. Cifen, B. 1999. Discussion on research content of eco-economy for soil and water conservation in Loess plateau. *Soil and Water Conservation in China* 4.
2. Cifen, B. 2001. Discussion on treatment counter measure in bedrock area on Loess plateau. *Journal of Sediment Research* 4: 1-5.
3. Jinrong, G., Liping, F., *et al.*, 1997. Experimental report on steep slopeland treatment and use in soft rock region. In: *Bureau of Inner Mongolia Water Conservancy*, Zhunger Banner Soil and Water Conservation Experiment Station.
4. Zhangyi, Z. 2002. Investigation report of large-area seabuckthorn mortality reasons and strategy. *Hippophae* 2.
5. Zhengying, Q. 1986. By taking seabuckthorn resource as a breakthrough for accelerating treatment on Loess plateau. *Soil and Water Conservation Technical Information* 4.

PART IX:
MISCELLANEOUS

Chapter 46

Seabuckthorn Residue (Cake) Feeding to the Poultry Birds and Cattle Calves for Nutrients Utilization and Production

V.K. Sharma, A. Sharma, D. Wadhwa and Virendra Singh*

Department of Animal Nutrition, College of Veterinary and Animal Sciences
*Department of Biology and Environmental Sciences, College of Basic Sciences,
H.P. Agricultural University, Palampur – 176 062, H.P., India

SUMMARY

Balanced and effective feeding are the most important prerequisites for economic production of livestock and poultry. Therefore, minimization of feed cost would be a great achievement for livestock and poultry farmers if the unconventional but potential ingredient like seabuckthorn (*Hippophae* L.) cake would be introduced as the replacer of major protein sources, as nutritionally, seabuckthorn (SBT) cake is very rich in proteins, fat, amino acids, vitamins and minerals. Very little systematic and scientific work has been conducted to study the effects of feeding SBT cake to the livestock and poultry. Hence, the present study was envisaged to see the effect of replacement of crude protein (CP) of conventional feed of layers and calves, and SBT cake to evaluate biological performances, nutrient utilization, rumen metabolites and cost benefit ratio by feeding SBT cake.

Three experiments were conducted to study the effect of SBT cake on layers and calves production. First trial was conducted in 360 BV-300 layer chicks which were offered conventional rations in control (S_0) group and replacing 10, 20, 30, 40 and 50 per cent CP by SBT cake CP. The second experiment was aimed to study the effect of SBT cake on growth of calves where 18 calves of 4-5 months age were divided in 3 treatment groups having 6 animals in each group. The conventional calf grower ration (CGR) was offered in control group (SBT_0) and replacing 50 per cent and 75 per cent CP of the control group with the CP of SBT cake in SBT_{50} and SBT_{75} groups, respectively. One more trial was conducted as a third experiment, aimed to observe the effect of SBT cake on rumen metabolites and rumen microbial population in adult cattle. Three adult cattle (fistulated) were maintained on three rations with a switch over design. The crude protein contents of the basal ration (T_1) were replaced with SBT cake CP at 50 per cent (T_2) and 75 per cent (T_3) CP in the adult cattle ration.

Seabuckthorn (Hippophae L.): A Multipurpose Wonder Plant, Vol. 4: Emerging Trends in Research and Technologies (Virendra Singh, *Editor-in-Chief*, 2014), p. 545–575, Daya Publishing House, New Delhi, India.

The egg production, egg mass and body weight gain of layers were higher in S_{20} group where 20 per cent CP of conventional concentrate was replaced with the CP of the SBT cake. FCR in respect of egg mass as well as biological performance of the layer was found better in S_{20} treatment group. It was noticed that, as the level of SBT cake increased beyond 20 per cent, the biological performance of the layers declined though the quality traits of eggs were not affected by replacement of SBT cake at any level. It was revealed that replacement of CP of conventional layer feed with SBT cake CP up to 20 per cent level was economic for layer production. Higher body weight gain and better balance of nutrient in calves was observed in SBT_{50} group where 50 per cent CP of the traditional CGR was replaced with the SBT cake CP. Better feed conversion ratio (FCR) of calves was found in SBT_{50} group, hence, can be considered as the highest dose of replacement with SBT cake CP in CGR. Higher microbial population in older calves was found in T_2 group, where 50 per cent CP of the traditional adult calves ration was replaced with the CP of SBT cake. Better rumen metabolites were also found in the T_2 group.

In the present investigations, it can be concluded that 20 per cent CP of the traditional concentrate ration could be replaced with the CP of SBT cake in layer birds for a viable and cost effective egg production, whereas, 50 per cent CP of the traditional calves ration could be replaced with the CP of SBT cake in young as well as older calves to improve the nutrient utilization efficiency for sustainable livestock production.

Keywords: *Seabuckthorn cake, Crude protein, Poultry bird, Egg production, Cattle calves, and Rations.*

INTRODUCTION

Livestock and poultry are the major components of agriculture in most of the developing countries in the world that play an important role in their economy. The economy of an agricultural based country mostly depends upon its agricultural sectors, such as livestock, poultry, fisheries and crops which is an integral part of farming systems of the country. Livestock, poultry and their products provide direct cash income, as animals and domestic birds are a living bank for many farmers and are critical to agricultural intensification by providing milk, meat, egg, power, fuel and manure for our daily intake.

Reports regarding the availability of fodder in the country project an alarming gap between demand and supply. The available fodder can meet the demand of only 46.7 per cent of the total livestock. In the year 2000, the deficiency in green and dry fodder in India was 61.10 per cent and 21.93 per cent, respectively. There is a deficit of dry, green fodder as well as of concentrates for livestock feeding and more than 20 per cent livestock population do not get minimum dry matter requirement in many parts of India. In such situation, it is necessary to bridge this gap, and so alternative feeds for livestock must be discovered to meet the requirements. This can be possible by exploiting the NCFR (Non-conventional feed resources) of feed by the use of some available plants and other agriculture byproducts as feed for livestock.

One of such plant which can be exploited as NCFR is "Seabuckthorn" (*Hippophae*), which grows in high altitude areas of cold deserts of Himalayas (Singh *et al.*, 1995; Singh and Dogra, 1996). Rich fodder values of seabuckthorn have been studied by Singh *et al.* (1999). As poultry is one of the fastest growing segments of Indian agriculture, and balanced and effective feeding are the most important requisites to a superior germplasm for economic production. Hence, the inclusion of seabuckthorn leaves, cake and its industrial byproducts (concentrate mixture) will be of economic relief for the poultry and dairy farmers. Feed is only one of the important items for the livestock and poultry production that occupies about 70 to 80 per cent of the total recurring costs.

Furthermore, seabuckthorn fruit contains high quality of pulp and seed oils, which have high contents of peculiar types of fatty acids like linolenic and palmitolic acids (Yang *et al.*, 2006) and other phyto-nutrients, which have many medicinal properties like protection from cardiovascular, gastric ulcer, skin diseases and cancer (Suryakumar and Gupta, 2012). However, scientific validations of these properties of Indian forms of seabuckthorn are needed for the development of useful preparations for eventual utilization by industries. Preliminary study done at CSKHPKV showed the potential of seabuckthorn oil for prevention of gastric ulcer in dogs and feed formulations to the cattle for growth and milk production and poultry for meat production. Further, the appropriate processing technologies have been standardized for many selected seabuckthorn products, which are otherwise popular in International markets like China, Russia and Europe. Therefore, there is a need for the development of variety of food products, veterinary and neutraceutical products, which will add value to seabuckthorn and increase its demand in the market.

On the basis of available information regarding seabuckthorn, a very little systematic and scientific work has been carried out to study the effects of feeding seabuckthorn cake to the cattle, poultry and other animals to evaluate its nutrient utilization for various productive functions. Keeping in view the usefulness of seabuckthorn cake, it will be of great importance to evaluate its nutrient utilization in calves for growth and poultry for egg production. Feeding seabuckthorn cake can meet some amount of the nutrient availability to the layers and calves for growth and egg production. So, an attempt was made under this study to see the effect of replacement of crude protein of conventional feed of layer and calf grower rations with crude protein of seabuckthorn cake to minimize the feed cost and to study the nutrient utilization with the following objectives: To evaluate the biological performances of commercial layers and calves by feeding Seabuckthorn cake (*Hippophae* L.) and study the nutrient availability after the replacement of crude protein of conventional concentrate with seabuckthorn cakes at different levels in layers and calves and feed conversion ratio for layers and calves production.

MATERIALS AND METHODS

The present study was carried out in the metabolic stall of the Department of Animal Nutrition, College of Veterinary and Animal Sciences (COVAS) and in the Livestock Farm of CHK Sarwan Kumar Himachal Pradesh Agricultural University (CSK HPKV), Palampur, Kangra (H.P.), India. To achieve the objectives of research programme, the research work was divided into three experiments: Experiment I was conducted in poultry layers from day-old chicks to 40 weeks of age including 24 weeks of egg laying of the layers, Experiment II was conducted in the calves for 8 months to study the growth pattern and nutrient utilization and Experiment III was conducted with adult calves for 2 months to see the rumen fermentation pattern.

Experimental Plan

The seabuckthorn cake was manufactured by separating the seed from the SBT pomace procured from the Lahaul and Spiti Seabuckthorn Cooperative Society Ltd., Keylong (HP), a NGO of farmers in seabuckthorn growing area of Himachal Pradesh. The oil was extracted from the seed with the help of mechanical press and the left out SBT cake was used for conducting the experiments on the poultry layer for egg production, cross bred Jersey calves for growth and adult calves for rumen fermentation studies.

EXPERIMENT I

Under this experiment for evaluating the replacement of seabuckthorn cake CP in the basal ration of poultry for egg production, 360 one day-old layer chicks were collected from the hatchery of M/S Venketshawara Hatcheries Pvt. Ltd., Nalagrah, HP near Chandigarh. These chicks were divided into 6 groups with four replications, each consisting of 15 chicks. The crude protein (CP) content of the basal ration (control, S_0) was replaced with SBT cake CP at 10, 20, 30, 40 and 50 per cent levels in S_{10},

S_{20}, S_{30}, S_{40} and S_{50} treatment group, respectively, in different phases of growth and production in layers according to the BIS (1992) standard. The whole period of experiment was conducted keeping in view the following phases as per BIS standard.

Analytical Procedures

The proximate principles of ingredients, feeds (layer starter, grower and finisher), left over feed and the seabuckthorn cake was done as per the method given in AOAC (1985). The data obtained from the different experiments was analyzed as per the standard methods of statistical analysis (Snedecor and Cochran, 1967) using Software GraphPad InStat tm 1990-1994, by applying One Way Analysis of Variance. The differences between the treatment means were tested by applying Duncan's (1955) multiple range test (DMRT).

Physical Composition of Feed

The physical composition of different experimental rations on dry matter basis and experimental layout are presented in Tables 46.1 and 46.2, respectively.

Table 46.1: Physical Composition of Rations

		Layer Starter Ration				
Concentrate	100.00	92.10	84.20	76.30	68.40	60.50
Seabuckthorn cake	0.00	07.90	15.80	23.70	31.60	39.50
		Layer Grower Ration				
Concentrate	100.00	93.10	86.20	79.30	72.40	65.50
Seabuckthorn cake	00.00	06.90	13.80	20.70	27.60	34.50
		Layer Finisher Ration				
Concentrate	100.00	93.50	87.00	80.50	74.00	67.50
Seabuckthorn cake	00.00	06.50	13.00	19.50	26.00	32.50

Table 46.2: Experimental Layout

Treatment No.	*Treatment Detail*	*Abbreviation Used*	*Supplementation Level*		
			Chick Phase (Starter ration, 0-8 wks)	*Growing Chick Phase (Grower ration, 8-20 wks)*	*Laying Chick Phase (Layer Ration, 20 wks onwards)*
1	Control	S_0	Basal diet	Basal diet	Basal diet
2	10 per cent CP of Control feed replaced with CP of SBT cake	S_{10}	7.9 kg of SBT cake	6.9 kg of SBT cake	6.5 kg of SBT cake
3	20 per cent CP of Control feed replaced with CP of SBT cake	S_{20}	15.8 kg of SBT cake	13.8 kg of SBT cake	13 kg of SBT cake
4	30 per cent CP of Control feed replaced with CP of SBT cake	S_{30}	23.7 kg of SBT cake	20.7 kg of SBT cake	19.5 kg of SBT cake
5	40 per cent CP of Control feed replaced with CP of SBT cake	S_{40}	31.6 kg of SBT cake	27.6 kg of SBT cake	26.0 kg of SBT cake
6	50 per cent CP of Control feed replaced with CP of SBT cake	S_{50}	39.5 kg of SBT cake	34.5 kg of SBT cake	32.5 kg of SBT cake

Note: CP: Crude protein; SBT: Seabuckthorn; wks: Weeks.

Chemical Composition of Feed Ingredients

The chemical compositions of various feed ingredients used for formulating different rations during these trials have been presented in the Table 46.3.

Table 46.3: Chemical Composition of Feed Ingredients (per cent DM Basis)

Ingredients	*DM*	*CP*	*CF*	*EE*	*NFE*	*TA*	*Ca*	*P*	*ME*
Maize	94.64	09.78	02.50	03.50	82.21	02.01	00.35	00.28	3261
Sunflower cake	93.59	28.50	13.00	04.45	44.95	09.10	00.20	01.00	2721
Molasses	72.56	02.05	00.00	00.00	85.45	12.50	01.10	00.18	2854
Soya meal	89.03	49.70	05.40	02.36	35.69	06.85	00.29	00.75	2959
Fishmeal	91.38	42.27	05.72	04.21	34.30	13.50	03.73	02.43	2806
GNC	93.30	35.50	08.84	08.50	38.14	09.02	00.20	00.30	3027
Mustard cake	94.66	32.10	09.50	05.86	45.76	06.78	08.00	03.50	2971
Rice Polish	91.90	11.40	06.20	13.50	57.83	11.07	0.08	1.40	3250
DORB	89.18	16.87	15.00	00.60	56.53	11.00	00.07	01.50	2434
SBT Cake	90.00	26.00	14.00	04.50	53.00	02.50	00.50	02.70	2907

Note: Each value was the mean of duplicate determinations.

The metabolizable energy (ME) values of maize, sunflower cake, molasses, soya meal, fish meal, GNC, mustard cake, rice polish and DORB were calculated as 3261, 2721, 2854, 2959, 2806, 3027, 2971, 3250 and 2434 kcal/kg, respectively. These nutritional values of the feed ingredients found closer with the findings of other researchers (Gupta, 2003; Ambatkar, 2009; Sharma and Ambatkar, 2009; Guleria, 2010; Shama, 2011. The value of CP and ME of the ingredients were also corroborated with the standard value (BIS, 1992).

Chemical Composition of Seabuckthorn Cake

The chemical composition of the SBT cake used as crude protein replacer in different experimental rations has been given in Table 46.3. The DM and CP of SBT cake were found as 90.00 and 26.00 per cent, respectively. The Energy value of the SBT cake was calculated as 2907 kcal/kg. The CF, EE, TA and NFE content of SBT cake were found as 14.00, 4.50, 2.50 and 53.00 per cent, respectively. The Ca and P content were 0.50 and 2.70 per cent, respectively. The CF content of cake was found to be very high at 14.00 than required in poultry (BIS, 1992). The calcium and phosphorus content was ideal and their ratio was also within prescribed limits (BIS, 1992). These values of SBT cake was corroborated with the findings of Guleria (2010) and Shama (2011).

Chemical Composition of Layer Rations

Data regarding the chemical composition of layer rations of different phases (chick, grower and layer) have been presented in Tables 46.4–46.6. As far as possible, the ration formulated was kept iso-nitrogenous and iso-caloric. The little difference in the values of different nutrients in the different treatments groups was due to the varying physical composition of feed ingredients while formulating the rations.

The data regarding the overall average gain in live weight and FCR during 1st to 17th weeks of age of BV-300 layer chicks is presented in Table 46.7. The average live weight gain in chicks for S_0, S_{10}, S_{20},

S_{30}, S_{40} and S_{50} treatment groups was 990.69±9.07, 976.71±16.41, 999.83±10.81, 926.17±7.14, 908.01±4.29 and 915.76±14.56 g, respectively. The values of weight gain ranged from 908.01±4.29 to 999.83±10.81 g being highest in S_{20} and lowest in S_{40} treatment group. The result of statistical analysis of the data revealed the significant ($p<0.05$) effect of treatment. The mean comparison indicated that CP replacement from control feed with higher level of SBT cake (after 30 per cent) had the lower average daily gain but that of the lower rate (up to S_{20}) had the positive impact on average gain in weight of layer chicks.

Table 46.4: Chemical Composition of Layer Chick Ration (per cent DM basis)

Trait	DM	CP	CF	EE	NFE	TA	AIA	Ca	P	Lys	Meth	Na	ME
S_0	95.02	20.68	03.77	03.40	60.01	05.66	03.59	00.98	00.55	00.56	00.77	00.25	2878
S_{10}	95.10	20.80	03.56	03.80	56.39	05.92	03.93	01.03	00.73	00.91	00.79	00.29	2860
S_{20}	95.40	20.44	03.46	04.40	53.67	05.95	04.38	01.04	00.91	01.29	00.78	00.31	2858
S_{30}	95.00	20.36	03.27	04.80	50.09	06.27	04.85	01.10	01.08	01.65	00.86	00.35	2832
S_{40}	95.30	20.38	03.09	05.30	46.78	06.55	05.24	00.98	01.26	02.00	00.97	00.38	2818
S_{50}	95.03	20.50	02.53	03.70	42.65	06.99	05.87	00.98	01.44	02.36	01.19	00.42	2784

Table 46.5: Chemical Composition of Layer Grower Ration (per cent DM basis)

Trait	DM	CP	CF	EE	NFE	TA	AIA	Ca	P	Lys	Meth	Na	ME
S_0	95.20	18.20	03.40	03.20	61.00	05.30	03.30	01.60	00.50	00.50	00.90	00.20	2800
S_{10}	95.30	18.50	03.30	03.60	59.00	05.80	03.90	01.20	00.70	00.80	00.90	00.23	2830
S_{20}	95.33	18.50	03.20	04.00	56.00	06.10	04.40	01.20	00.80	01.10	01.00	00.20	2810
S_{30}	95.00	18.60	03.10	04.50	52.00	06.50	04.90	01.10	00.90	01.40	01.10	00.20	2790
S_{40}	95.05	18.40	02.90	04.90	50.00	06.70	05.30	01.10	01.00	01.80	01.20	00.30	2770
S_{50}	95.09	18.50	02.70	05.20	47.00	07.00	05.60	01.10	01.10	02.00	1.30	00.30	2750

Table 46.6: Chemical Composition of Layer Ration (per cent DM basis)

Trait	DM	CP	CF	EE	NFE	TA	AIA	Ca	P	Lys	Meth	Na	ME
S_0	95.00	17.80	03.60	03.50	57.00	04.80	02.70	03.30	00.80	00.80	00.80	00.04	2600
S_{10}	95.08	17.60	03.50	03.90	55.00	05.30	03.30	03.70	01.00	01.10	00.90	00.07	2600
S_{20}	95.11	17.60	03.40	04.20	52.00	05.70	03.80	03.90	01.00	01.30	01.00	00.10	2600
S_{30}	95.06	17.70	03.20	04.70	50.00	05.80	04.20	03.90	01.10	01.50	01.00	00.13	2620
S_{40}	95.02	17.50	03.20	05.30	48.00	06.30	04.80	03.80	01.20	01.70	01.20	00.16	2620
S_{50}	95.08	17.80	03.10	05.50	45.00	06.60	05.20	03.90	01.40	02.00	01.30	00.19	2610

Note: Each value was the mean of duplicate determinations.

DM: Dry matter; CP: Crude protein; CF: Crude fibre; EE: Ether extract; NFE: Nitrogen free extract TA: total ash; AIA: Acid insoluble ash, Ca: Calcium; P: Phosphorus; Lys: Lysine; Meth: Methonine; Na: Sodium and ME: Metabolizable energy.

Biological Performance of Layers

The FCR in chicks for S_0, S_{10}, S_{20}, S_{30}, S_{40} and S_{50} treatment groups was found as 3.83±0.02, 3.94±0.08, 3.73±0.04, 4.21±0.05, 4.34±0.04, 4.09±0.07, respectively. The values of FCR ranged from 3.73±0.04 to

Table 46.7: Overall Biological Performance of Layer during 1st To 17th Weeks of Age

Traits	Treatment Groups						Level of Significance
	S_0	S_{10}	S_{20}	S_{30}	S_{40}	S_{50}	
Average initial weight (g)	38.00 ± 0.03	38.00 ± 0.03	38.00 ± 0.03	38.00 ± 0.03	38.00 ± 0.03	38.00 ± 0.03	NS
Average final weight (g)	$1028.70^{a}\pm9.07$	$1014.70^{a}\pm16.44$	$1037.80^{a}\pm10.79$	$964.17^{b}\pm7.12$	$946.01^{b}\pm4.30$	$953.76^{b}\pm14.58$	*
Average gain in weight (g)	$990.69^{a}\pm9.07$	$976.71^{a}\pm16.41$	$999.83^{a}\pm10.81$	$926.17^{b}\pm7.14$	$908.01^{b}\pm4.29$	$915.76^{b}\pm14.56$	*
Feed intake (g)	$3788.80^{abe}\pm17.84$	$3840.30^{bc}\pm15.44$	$3731.30^{e}\pm10.08$	$3900.00^{cd}\pm33.17$	$3941.50^{d}\pm17.10$	$3745.00^{e}\pm25.33$	*
FCR	$3.83^{a}\pm0.02$	$3.94^{ed}\pm0.08$	$3.73^{a}\pm0.04$	$4.21^{bc}\pm0.05$	$4.34^{c}\pm0.04$	$4.09^{bd}\pm0.07$	*

Note: Values bearing different superscripts in the same row differ significantly; *p<0.05, **p<0.01, ***p<0.001, NS: Non-significant.

Table 46.8: Biological Performance of Layers

Particular	Treatment Groups						Level of Significance
	S_0	S_{10}	S_{20}	S_{30}	S_{40}	S_{50}	
DMI (g/hen/day)	$80.24^{a}\pm1.99$	$80.84^{a}\pm4.34$	$80.45^{a}\pm3.96$	$116.68^{bc}\pm5.55$	$124.55^{c}\pm5.78$	$86.58^{ab}\pm4.70$	***
DMI per dozen egg (g)	$1554.26^{ab}\pm73.23$	$1368.63^{a}\pm65.76$	$1372.76^{a}\pm53.77$	$2049.69^{bc}\pm102.9$	$2671.59^{c}\pm150.2$	$1968.08^{b}\pm180.5$	*
Egg produ-ction (per cent)	$74.61^{ab}\pm4.21$	$75.08^{a}\pm2.22$	$77.22^{a}\pm1.63$	$66.17^{b}\pm3.34$	$65.15^{b}\pm4.19$	$54.77^{c}\pm3.70$	**
No. of egg/Fortnight	11.21	11.26	11.58	9.93	9.77	8.22	—
Egg weight (g)	52.59 ± 0.69	52.85 ± 0.62	52.84 ± 0.70	51.50 ± 0.73	52.32 ± 0.60	51.96 ± 0.91	NS
Egg mass (g/hen/d)	$32.58^{ab}\pm2.08$	$37.46^{a}\pm0.96$	$37.16^{a}\pm0.84$	$35.18^{ab}\pm1.62$	$29.27^{bc}\pm2.03$	$27.43^{c}\pm1.77$	**
Feed Efficiency	$2.46^{ab}\pm0.11$	$2.16^{a}\pm0.09$	$2.16^{a}\pm0.08$	$3.32^{c}\pm0.16$	$4.26^{d}\pm0.66$	$3.16^{bc}\pm0.26$	*
Weight per dozen egg (g)	631.08 ± 3.04	634.20 ± 5.04	634.08 ± 6.87	618.00 ± 5.34	627.84 ± 5.41	623.52 ± 11.11	NS
Livability (per cent)	98.33	96.66	100	100	98.33	96.66	—

Note: Values bearing different superscripts in a same row differ significantly (*p<0.05, **p<0.01, ***p<0.001). NS: Non-significant.

4.34±0.04 being highest in S_{40} and lowest in S_{20} treatment group. The result of statistical analysis of the data revealed the significant ($p<0.05$) effect of treatment at higher level of inclusion. The mean comparison indicated best FCR in S_{20} group as compared to the S_{30}, S_{40} and S_{50} treatment groups. It indicated that CP replacement from control feed up to 20 per cent level of SBT cake CP had the positive impact on FCR of layer chicks during 1st to 17th weeks of age of the layers.

These results corroborated with the findings of Gupta (2003); Simsek and Kilic (2006). The findings of current research found higher than that of *de* Witt *et al.* (2009) and lower than the findings of Kaistha (1996); Katoch (1996).

Biological Performance of Layers During 18th To 32nd Weeks of Age

The experimental layer chicks started laying at 18th week of age. The data regarding the biological performances of layers at the age of 18th to 32 weeks have been presented in the Table 46.8.

Body Weight Changes of Layers at Different Ages

The body weight at different ages of the layer chicks has been presented in Table 46.9 and depicted in graphical representation in Figure 46.1. Average weight at 3rd day was 38.00±0.03 in all the treatment groups. The average weight at the 2nd week was found as 63.73±0.71, 61.93±0.75, 63.14±0.60, 57.87±0.66, 55.18±0.98 and 52.23±0.76 g for S_0, S_{10}, S_{20}, S_{30}, S_{40} and S_{50} treatment groups, respectively. The average weight at 2nd week varied from 52.23±0.76 g (S_{50}) to 63.73±0.71g (S_0) found to be statistically significant ($p<0.05$) among the different groups. Body weight of chicks at 4th week of age was found as 180.25±1.10, 175.29±1.90, 173.22±1.69, 156.35±1.80, 145.52±1.08 and 144.25±2.28 g for S_0, S_{10}, S_{20}, S_{30}, S_{40} and S_{50} treatment groups, respectively. The weight at 4th week varied from 144.25±2.28 g (S_{50}) to 180.25±1.10 g (S_0) found to be statistically significant ($p<0.05$). Body weight at 6th week of age of chicks was found as 342.48±6.15, 334.80±10.96, 321.12±10.87, 291.78±4.37, 290.14±2.75 and 288.18±5.25 g for S_0, S_{10}, S_{20}, S_{30}, S_{40} and S_{50} treatment groups, respectively. The weight at 6th week varied from

Table 46.9: Body Weight Changes of Layers at different Ages (g/Bird)

Attributes	*Treatment Groups*						*Level of Significance*
	S_0	S_{10}	S_{20}	S_{30}	S_{40}	S_{50}	
3rd day	38.00±0.03	38.00±0.03	38.00±0.03	38.00±0.03	38.00±0.03	38.00±0.03	NS
2nd week	63.73^a±0.71	61.93^a±0.75	63.14^a±0.60	57.87^b±0.66	55.18^{bc}±0.98	52.23^c±0.76	*
4th week	180.25^a±1.10	175.29^a±1.90	173.22^a±1.69	156.35^b±1.80	145.52^c±1.08	144.25^c±2.28	**
6th week	342.48^a±6.15	334.80^a±10.96	321.12^a±10.87	291.78^b±4.37	290.14^b±2.75	288.18^b±5.25	**
8th week	436.52^a±3.97	418.65^a±17.10	420.70^a±11.31	356.00^b±9.44	363.74^b±2.93	366.35^b±16.22	*
10th week	677.00^a±11.71	668.00^a±10.03	657.00^a±11.90	581.17^b±11.00	597.50^b±9.06	578.42^b±5.15	**
14th week	854.05±9.04	852.75±32.48	879.32±16.73	818.83±4.66	804.52±7.23	824.22±16.23	NS
17th week	1028.70^a±9.07	1014.70^a±16.44	1037.80^a±10.79	964.17^b±7.12	946.01^b±4.30	953.76^b±14.58	*
22nd week	1203.33^a±10.48	1176.67^a±4.77	1196.34^a±5.29	1109.50^b±12.65	1087.50^b±2.62	1083.30^b±13.03	***
32nd week	1250.30^a±12.73	1252.40^a±15.25	1259.70^a±27.32	1203.40^{ab}±12.48	1183.10^{ab}±30.66	1160.80^b±11.63	*
40th week	1335.20^{ab}±24.29	1400.00^a±20.09	1378.80^a±13.31	1296.20^{bc}±07.81	1278.20^{bc}±16.17	1238.30^c±11.47	**

Note: Values bearing different superscripts in a same row within the same week differ significantly (*$p<0.05$, **$p<0.01$, ***$p<0.001$). NS= Non-significant.

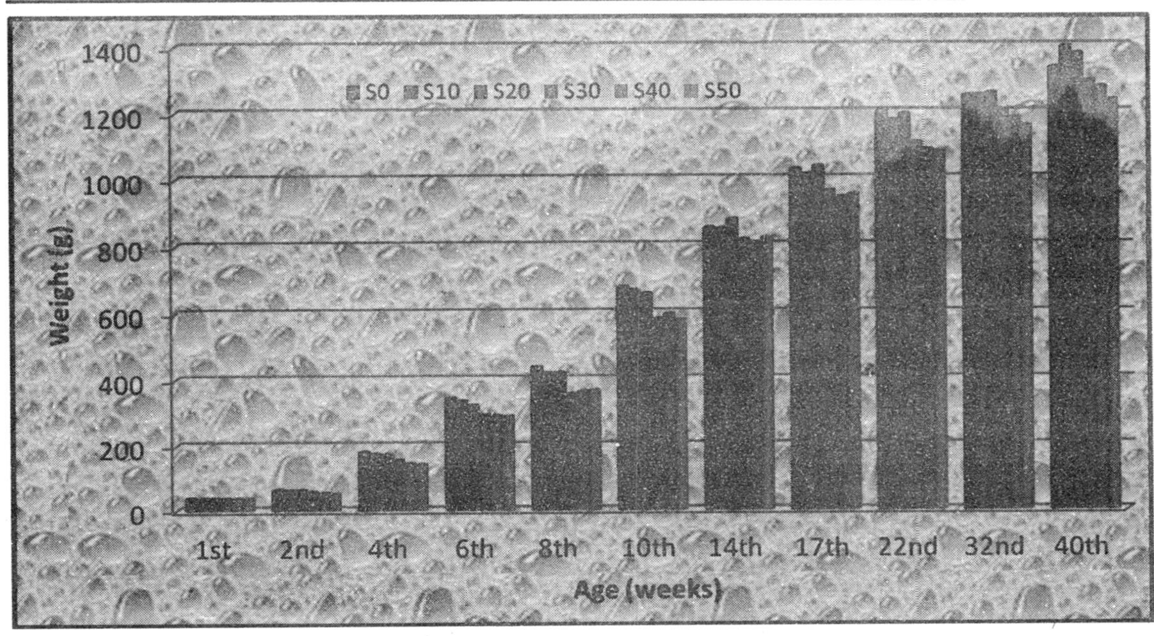

Figure 46.1: Body Weight of Layer at Different Ages.

288.18±5.25 g (S_{50}) to 342.48±6.15 g (S_0) and found to be highly significant ($p<0.01$). Body weight at 8th week was found as 436.52±3.97, 418.65±17.10, 420.70±11.31, 356.00±9.44, 363.74±2.93 and 366.35±16.22 g for S_0, S_{10}, S_{20}, S_{30}, S_{40} and S_{50} treatment groups, respectively. The data in this age varied from 356.00±9.44 g (S_{30}) to 436.52±3.97 g (S_0) found to be significant ($p<0.05$). The Body weight of chicks at 10th week of age was found as 677.00±11.71, 668.00±10.03, 657.00±11.09, 581.17±11.00, 597.50±9.06 and 578.42±5.15 g for S_0, S_{10}, S_{20}, S_{30}, S_{40} and S_{50} treatment groups, respectively. The weight at 10th week varied from 578.42±5.15 g (S_{50}) to 677.00±11.71 g (S_0) and found to be highly significant ($p<0.01$). The body weight at 14th week of age was found as 854.05±9.04, 852.75±32.48, 879.32±16.73, 818.834.66, 804.52±7.23 and 824.22±16.23 g for S_0, S_{10}, S_{20}, S_{30}, S_{40} and S_{50} groups, respectively found to be non-significant ($p>0.05$). The body weight at 17th week of age was found as 1028.70±9.07, 1014.70±16.44, 1037.80±10.79, 964.17±7.12, 946.01±4.30 and 953.76±14.58 g for S_0, S_{10}, S_{20}, S_{30}, S_{40} and S_{50} treatment groups, respectively. The weight varied from 946.01±4.30 (S_{40}) to1037.80±10.79 (S_{20}) g and found to be significant ($P<0.05$). The body weight of layers at 22nd week of age was found as 1203.33±10.48, 1176.67±4.77, 1196.34±5.29, 1109.50±12.65, 1087.50±2.62 and 1083.30±13.03 g for S_0, S_{10}, S_{20}, S_{30}, S_{40} and S_{50} groups, respectively. The values ranged from 1083.30±13.03 g (S_{50}) to 1203.33±10.48 g (S_0), found to be extremely significant ($p>0.001$). The body weight of layers at 32nd week of age was found as 1250.33±12.73, 1252.40±15.25, 1259.70±27.32, 1203.40±12.48, 1183.10±30.36 and 1160.80±11.63 g for S_0, S_{10}, S_{20}, S_{30}, S_{40} and S_{50} groups, respectively. The value ranged from 1160.80±11.63 (S_{50}) to 259.70±27.32 being highest in S_{20} and lowest in S_{50} group, found to be significant ($p>0.05$). Finally, the body weight of layers at 40th week of age was found as 1335.20±24.29, 1400.00±20.09, 1378.80±13.31, 1296.20±7.81, 1278.20±16.17 and 1238.30±11.47 g for S_0, S_{10}, S_{20}, S_{30}, S_{40} and S_{50} treatment groups, respectively. The data ranged from 1238.30±11.47 to 1400.00±20.09 g being highest in S_{10} and lowest in S_{50} g and found to be highly significant ($p<0.01$).

Table 46.10: Fortnightly Egg Production from 20th to 31st Weeks of Age (per cent)

Duration	Treatment Groups						Level of Significance
	S_0	S_{10}	S_{20}	S_{30}	S_{40}	S_{50}	
1st Fortnight (20-21 week)	$55.71^{a}\pm2.63$	$58.79^{a}\pm3.07$	$60.22^{a}\pm4.05$	$42.67^{b}\pm3.27$	$48.37^{ab}\pm3.03$	$42.11^{b}\pm1.75$	*
2nd Fortnight (22-23 week)	$75.83^{a}\pm1.50$	$77.79^{a}\pm1.37$	$81.38^{a}\pm1.68$	$68.31^{b}\pm1.78$	$65.22^{bc}\pm1.82$	$60.39^{c}\pm1.95$	*
3rd Fortnight (24-25 week)	$85.33^{a}\pm2.00$	$75.56^{b}\pm1.24$	$77.89^{ab}\pm1.39$	$66.00^{cd}\pm1.95$	$68.67^{bc}\pm2.89$	$58.70^{d}\pm2.24$	*
4th Fortnight (26-27 week)	$78.07^{a}\pm1.65$	$82.56^{a}\pm1.68$	$83.11^{a}\pm1.24$	$70.22^{b}\pm2.64$	$70.74^{b}\pm2.05$	$47.19^{c}\pm0.42$	*
5th Fortnight (28-29 week)	$81.21^{a}\pm1.99$	$81.78^{a}\pm1.29$	$81.19^{a}\pm1.32$	$78.19^{a}\pm2.37$	$74.20^{a}\pm2.23$	$61.62^{b}\pm2.37$	*
6th Fortnight (30-31 week)	$71.53^{ab}\pm2.20$	$73.52^{a}\pm1.69$	$79.50^{a}\pm1.78$	$71.65^{ab}\pm2.29$	$63.70^{bc}\pm2.74$	$58.58^{c}\pm2.46$	*
Overall	$74.61^{a}\pm4.24$	$75.00^{a}\pm3.54$	$77.22^{a}\pm3.48$	$66.17^{ab}\pm4.99$	$65.15^{ab}\pm3.69$	$54.77^{b}\pm3.30$	*

Note: Values bearing different superscripts in a same row within the same week differ significantly (*p<0.05, **p<0.01, ***p<0.001).

From Figure 46.1, it was found that the body weight of layers in S_0, S_{10} and S_{20} group attained parallel weight up to the age of 4th week. From 6th to 8th weeks of age the layers belonged to S_0 group attained higher body weight followed by S_{10} and S_{20} groups. Whereas, on 14th to 32nd weeks of age of layers the S_{20} group attained higher body weight than control (S_0) group. It may be due to the well utilization of SBT cake up to 20 per cent level by the layers. Finally, at 40th week of age the layer belonged to S_{10} group showed highest body weight followed by S_{20}, S_0, S_{30}, S_{40} and S_{50} groups in the descending order.

The result of weight gain was lower than the findings of Gupta (2003) and higher than that of Bhatt (1993). However, the finding of the current study was corroborated with the findings of Fayeye *et al.* (2005) in respect to body weight changes.

Egg Production of Layers

The egg laying was started at 18th week of age onwards of the layers. The average fort-nightly egg production has been given in Table 46.10 and depicted in Figure 46.2 up to 8 months of age. The overall per cent egg production in S_0, S_{10}, S_{20}, S_{30}, S_{40} and S_{50} treatment groups up to 31 weeks age were found as 74.61±4.21, 75.00±2.22, 77.22±1.63, 66.17±3.34, 65.15±4.19 and 54.77±3.70 per cent, respectively. The egg production (per cent) was significantly highest ($p<0.01$) in S_{20} group (77.22±1.63) and minimum in S_{50} (54.77±3.70) group. The statistical comparison of the result of average per cent of egg production revealed the positive effect of SBT cake CP up to 20 per cent level of inclusion.

The egg production at 1st fortnight for S_0, S_{10}, S_{20}, S_{30}, S_{40} and S_{50} treatment groups was found as 55.71±2.63, 58.79±3.07, 60.22±4.05, 42.67±3.27, 48.37±3.03 and 42.11±1.75 per cent, respectively. The values ranged from 42.11±1.75 (S_{50}) to 60.22±4.05 (S_{20}) and found to be significant ($p<0.01$). The egg production at 2nd fortnight for S_0, S_{10}, S_{20}, S_{30}, S_{40} and S_{50} treatment groups were 75.83±1.50, 77.79±1.37, 81.38±2.07, 68.31±1.80, 65.22±1.91 and 60.39±2.06 per cent, respectively. The values were significant ($p<0.05$), highest in S_0 and lowest in S_{50} group. The egg production at 3rd fortnight for S_0, S_{10}, S_{20}, S_{30}, S_{40} and S_{50} treatment groups were 85.33±2.00, 75.56±1.35, 77.79±2.28, 66.00±2.15, 68.67±2.89 and 58.70±2.28 per cent, respectively. The values ranged from 58.70±2.28 (S_{50}) to 85.33±2.00 (S_c) and found to be significant ($p<0.05$). The egg production at 4th fortnight for S_0, S_{10}, S_{20}, S_{30}, S_{40} and S_{50} treatment groups were 78.07±1.65, 82.56±1.68, 83.11±1.24, 70.22±2.64, 70.74±2.05 and 47.19±0.42 per cent, respectively. The values ranged from 47.19±0.42 (S_{50}) to 83.11±1.24 (S_{20}) and found to be significant ($p<0.05$). The egg production at 5th fortnight for S_0, S_{10}, S_{20}, S_{30}, S_{40} and S_{50} treatment groups were 81.21±1.99, 82.78±1.29, 81.19±1.32, 78.19±2.37, 74.20±2.23 and 61.62±2.40 per cent, respectively. The values ranged from 61.62±2.40 (S_{50}) to 82.78±1.29 (S_{10}) and found to be significant ($p<0.05$). The egg production at 6th fortnight for S_0, S_{10}, S_{20}, S_{30}, S_{40} and S_{50} treatment groups were 71.53±2.20, 73.52±1.69, 79.50±1.80, 71.65±2.29, 63.70±2.77 and 58.58±2.50 per cent, respectively. The values ranged from 58.58±2.50 (S_{50}) to 79.50±1.80 (S_{20}) and found to be significant ($p<0.05$). Finally, the overall egg production from 1st to 6th fortnight for S_0, S_{10}, S_{20}, S_{30}, S_{40} and S_{50} treatment groups were 74.61±4.24, 75.00±3.54, 77.22±3.48, 66.17±4.99, 65.15±3.69 and 54.77±3.30 per cent, respectively. The values ranged from 54.77±3.30 (S_{50}) to 77.22±3.48 (S_{20}) and found to be significant ($p<0.05$). From the above data, it was found that the best production was up to S_{20} groups and after that the production declined. The lowest production was always found in S_{50} group.

From Figure 46.2, it was found that the peak egg production was found in between 3rd to 5th fortnight. After that the production of all the treatment groups declined gradually. The average production up to 6th fortnightly was found highest in S_{20} group followed by S_{10} and S_0 groups. The production of S_{30} to S_{50} was found lower at most of the times. It may be due the fact that the inclusion

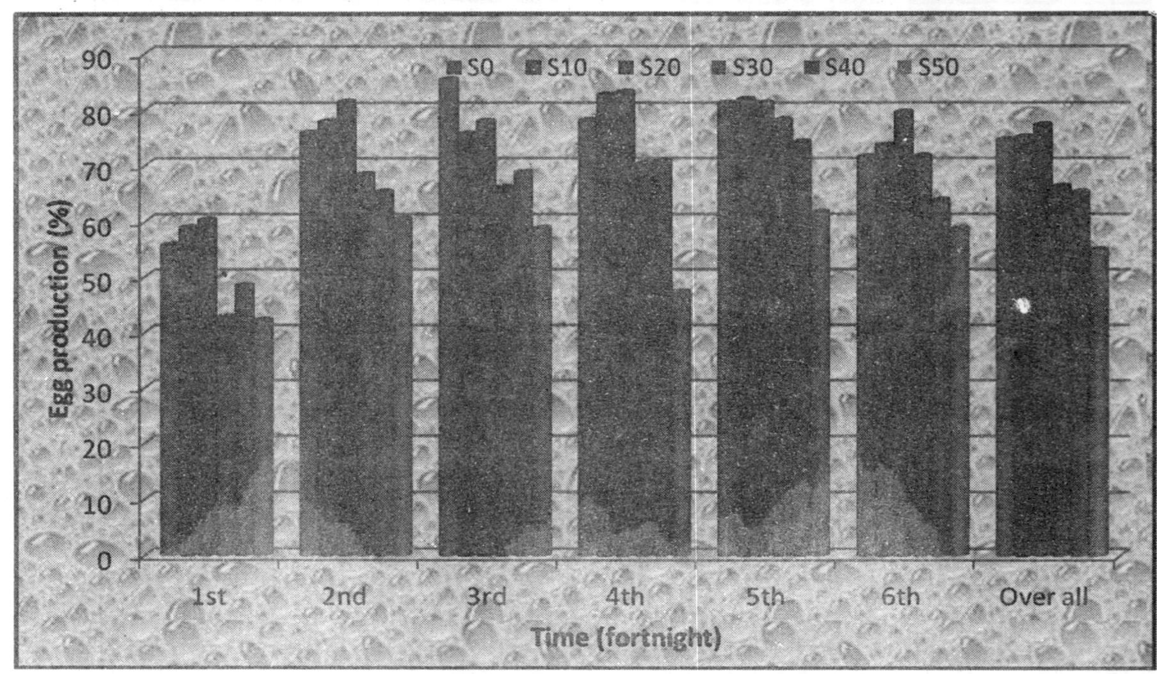

Figure 46.2: Fortnightly Egg Production (per cent).

of CP of SBT (>20 per cent) cannot be used for layer poultry and SBT cake more inclusion was having higher percentage of phosphorous leads to imbalance of nutrient availability and lowered the egg production.

Feed Conversion Ratio (FCR)/Feed Efficiency

The feed efficiency of the layers was calculated during the digestibility trial of the layers. The feed efficiency for S_0, S_{10}, S_{20}, S_{30}, S_{40} and S_{50} treatment groups were 2.46±0.11, 2.16±0.09, 2.16±0.08, 3.32±0.16, 4.26±0.66 and 3.16±0.26, respectively. The values ranged from 2.16±0.08 to 4.26±0.66 being lowest in S_{20} and highest in S_{50} treatment group. The statistical analysis of the data showed significant ($p<0.05$) effect of S_{10} and S_{20} treatment over all other treatments except S_0 group. The comparison of the data revealed that the inclusion of CP of SBT cake in traditional layer ration up to 20 per cent level had the positive feed efficiency. The result regarding feed efficiency was in close agreement with the findings of Bhatt (1993), but higher than that of *de* Witt *et al.* (2009).

COST BENEFIT RATIO OF LAYER PRODUCTION AT 31ST WEEK OF AGE

For Single Egg Production

The cost benefit ratio of single egg was also calculated and found that the profit (Rs.) per egg for S_0, S_{10}, S_{20}, S_{30}, S_{40} and S_{50} treatment groups were Re. 00.95, 01.15, 01.10, 00.09, -00.88 and 00.07, respectively. From the numerical value, it was found that the profit was higher in S_{10} group followed by S_{20}, S_0, S_{30}, S_{50} and S_{40} groups (Table 46.11).

Table 46.11: Economics of Egg Production

Attributes	*Treatment Groups*					
	S_0	S_{10}	S_{20}	S_{30}	S_{40}	S_{50}
DMI (g) per single egg	129.52	114.05	114.40	170.81	222.63	164.01
Cost of feed (Rs/kg)	15.00	15.4	15.78	16.18	16.56	16.96
Cost of Production (Rs./egg)	02.05	01.85	01.90	02.91	03.88	02.93
Market value (Rs./egg)	03.00	03.00	03.00	03.00	03.00	03.00
Profit (Rs./egg)	00.95	01.15	01.10	00.09	–00.88	00.07

For Meat and Egg Production

As the initial price, labour, electricity and other fixed costs were similar in all the treatments group except feed cost and production cost in respect of meat and egg production, the cost benefit ratio was calculated on the basis of feed and production only. The total profit (Rs.) for S_0, S_{10}, S_{20}, S_{30}, S_{40} and S_{50} treatment groups was 169.12, 166.59, 170.51, 84.07, 64.62 and 89.43 per bird for 12 weeks (Table 46.12). From the data, it was found that the profit was higher in S_{20} group followed by S_0, S_{10}, S_{50}, S_{30} and S_{40} treatment group, respectively.

Table 46.12: Economics of Layer Production from 20th to 31st Weeks of Age

Attributes	*Treatment Groups*					
	S_0	S_{10}	S_{20}	S_{30}	S_{40}	S_{50}
Body weight	1250.30	1252.40	1259.70	1203.40	1133.10	1160.80
Meat Cost @Rs.70	87.52	87.67	88.18	84.24	82.82	81.26
Avg. Egg production (per cent)	74.61	75.00	77.22	66.17	65.15	54.77
Egg Cost @Rs. 3	188.03	189.00	194.58	166.76	164.18	138.01
Total Income (Rs./bird)	275.55	276.67	282.76	250.99	247.00	219.26
Dry Matter Intake (g/b/d)	80.24	80.84	80.45	116.68	124.55	86.58
Feed Intake (g/b/d)	84.46	85.09	84.68	122.82	131.11	91.14
Feed consumption (kg)	07.09	07.15	07.11	10.32	11.01	07.66
Feed Cost (Rs/kg)	15.00	15.4	15.78	16.18	16.56	16.96
Total Feed Cost (Rs.)	106.42	110.08	112.25	166.93	182.37	129.84
Profit (Rs./bird)	169.12	166.59	170.51	84.07	64.62	89.43
Egg mass g/hen/day	32.58	37.46	37.16	35.18	29.27	27.43
FCR (Feed/egg mass)	02.59	02.27	02.28	03.49	04.48	03.32

Blood Serum Minerals Analysis

The blood sample was collected at the last day of the digestibility trial of the layers. The calcium content in blood serum was found as 28.87±1.14, 26.90±1.57, 23.91±1.13, 19.54±0.43, 19.21±1.76 and 14.92±1.49 mg/dl for S_0, S_{10}, S_{20}, S_{30}, S_{40} and S_{50} treatment groups, respectively (Table 46.13). The calcium level ranged from 14.92±1.49 to 28.87±1.14 mg/dl being highest in S_0 group and lowest in S_{50} group found to be extremely significant ($p<0.001$) among them. Blood serum Magnesium was found as 4.28±0.54, 4.60±0.06, 4.30±0.09, 4.10±0.05, 3.97±0.30 and 3.47±0.11 mg/dl for S_0, S_{10}, S_{20}, S_{30}, S_{40} and S_{50} treatment groups, respectively. The statistical analysis of the data for Magnesium level in the blood serum found to be non-significant. Blood serum iron level was found as 0.24±0.03, 0.26±0.04, 0.24±0.03, 0.29±0.02, 0.30±0.02, 0.28±0 mg/dl in S_0, S_{10}, S_{20}, S_{30}, S_{40} and S_{50} treatment groups, respectively and the iron level in the blood serum found to be non-significant. Zinc content in blood serum was found as 0.36±0.07, 0.28±0.01, 0.23±0.01, 0.21±0.04, 0.20±0.03, 0.10±0.03 mg/dl in S_0, S_{10}, S_{20}, S_{30}, S_{40} and S_{50} treatment groups, respectively. The zinc content varied from 0.10±0.03 to 0.36±0.07 mg/dl being highest in S_0 and lowest in S_{50} treatment groups and found to be significant ($p<0.05$). Blood serum copper level was found as 27.08±0.68, 31.68±2.54, 38.13±4.50, 36.03±2.99, 23.98±1.23 and 18.10±3.88 mg/dl for S_0, S_{10}, S_{20}, S_{30}, S_{40} and S_{50} treatment groups, respectively. The copper content varied from 18.10±3.88 to 38.13±4.50 mg/dl being highest in S_{20} and lowest in S_{50} treatment group. The statistical analysis of the data found significant ($p<0.05$) effect of treatment in S_{20} group. Blood serum cobalt content was found as 14.75±1.80, 15.75±1.60, 18.50±0.65, 19.25±0.25, 20.50±0.87 and 17.25±0.63 µg/dl for S_0, S_{10}, S_{20}, S_{30}, S_{40} and S_{50} treatment groups, respectively. The cobalt content varied from 14.75±1.80 to 20.50±0.87 µg/dl being highest in S_{40} and lowest in S_0 treatment groups and was significant ($p<0.05$). Blood serum manganese content was found as 6.00±0.91, 11.25±1.38, 9.00±1.68, 4.50±1.44, 3.75±1.11, 3.25±1.32 µg/dl for S_0, S_{10}, S_{20}, S_{30}, S_{40} and S_{50} treatment groups, respectively. The manganese content varied from 3.25±1.32 to 11.25±1.38 µg/dl being highest in S_{10} and lowest in S_{50} treatment groups and found to be significant ($p<0.05$). Blood serum sodium was found as 121.44±1.53, 124.56±2.25, 123.72±1.28, 124.46±2.40, 122.88±1.48 and 127.88±2.05 mEq/L for S_0, S_{10}, S_{20}, S_{30}, S_{40} and S_{50} groups, respectively and the sodium level in the blood serum is non-significant ($p>0.05$). Blood serum potassium was found as 3.72±0.26, 3.63±0.18, 3.81±0.21, 3.48±0.23, 3.59±0.23 and 4.09±0.16 mEq/L for S_0, S_{10}, S_{20}, S_{30}, S_{40} and S_{50} groups, respectively and the its level in the blood serum was found to be non-significant.

Table 46.13: Blood Serum Minerals Level at the Last Day of Metabolism Trial of Layers

Particulars	*Treatment Groups*						*Level of Significance*
	S_0	S_{10}	S_{20}	S_{30}	S_{40}	S_{50}	
Ca (mg/dl)	28.87^{a}±1.14	26.90^{a}± 1.57	23.91^{ab}±1.13	19.54^{bc}±0.43	19.21^{bc}±1.76	14.92^{c}±1.49	***
Mg (mg/dl)	4.28±0.54	4.60±0.06	4.30±0.09	4.10±0.05	3.97±0.30	3.47±0.11	NS
Fe (mg/dl)	0.24±0.03	0.26±0.04	0.24±0.03	0.29±0.02	0.30±0.02	0.28±0.02	NS
Zn (mg/dl)	0.36^{a}±0.07	0.28^{a}±0.01	0.23^{ab}±0.01	0.21^{ab}±0.04	0.20^{ab}±0.03	0.10^{b}±0.03	*
Cu (µg/dl)	27.08^{abc}±0.68	31.68^{ab}±2.54	38.13^{a}±4.50	36.03^{ab}±2.99	23.98^{bc}±1.23	18.10^{c}±3.88	*
Co (µg/dl)	14.75^{a}±1.80	15.75^{ab}±1.60	18.50^{ab}±0.65	19.25^{ab}±0.25	20.50^{b}±0.87	17.25^{ab}±0.63	*
Mn (µg/dl)	6.00^{ab}±0.91	11.25^{a}±1.38	9.00^{ab}±1.68	4.50^{b}±1.44	3.75^{b}±1.11	3.25^{b}±1.32	*
Na (mEq/l)	121.44±1.53	124.56±2.25	123.72±1.28	124.46±2.40	122.88±1.48	127.88±2.05	NS
K (mEq/l)	3.72±0.26	3.63±0.18	3.81±0.21	3.48±0.23	3.59±0.23	4.09±0.16	NS

Note: Values bearing different superscripts in the same row differ significantly ($p<0.05$); NS: Non-significant.

The comparison of the data from statistical analysis revealed the calcium level in the blood serum of layer decreased sharply with the increase of SBT cake in the feed, though the average production was highest in S_{20} group. As the SBT cake is containing more phosphorous, it might have some positive balance in S_{20} group. The iron level of the blood serum of layers was also found more for the

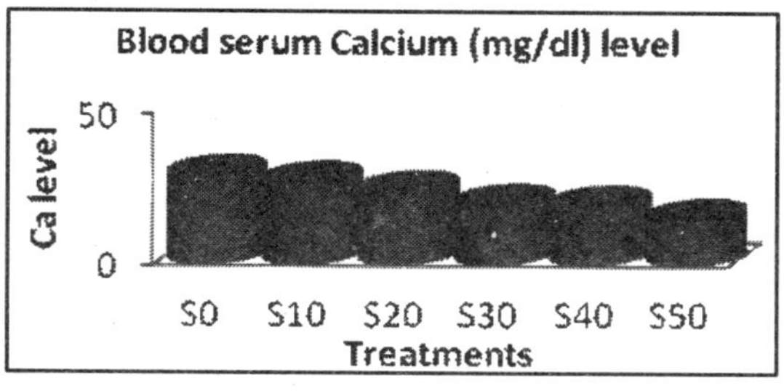

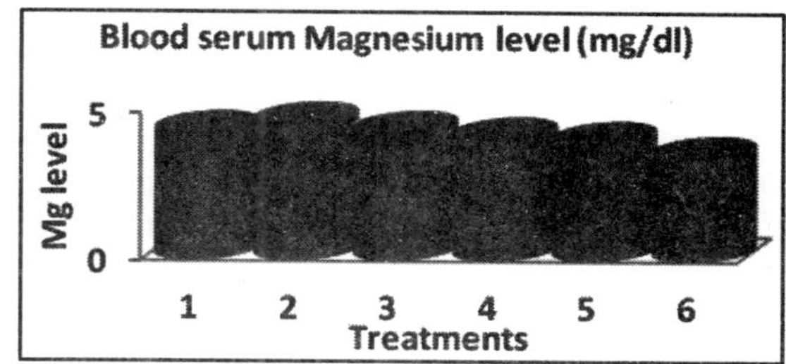

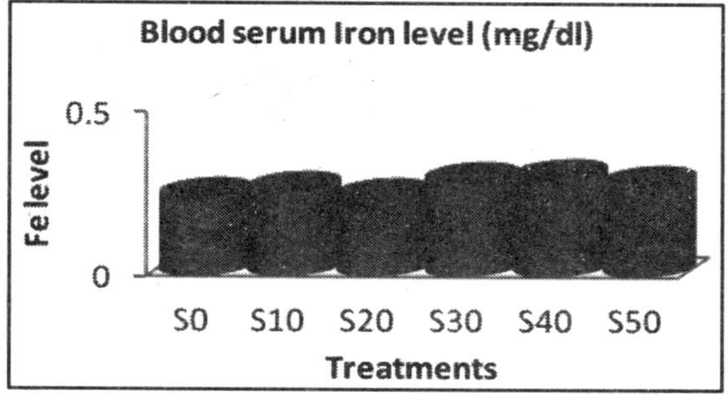

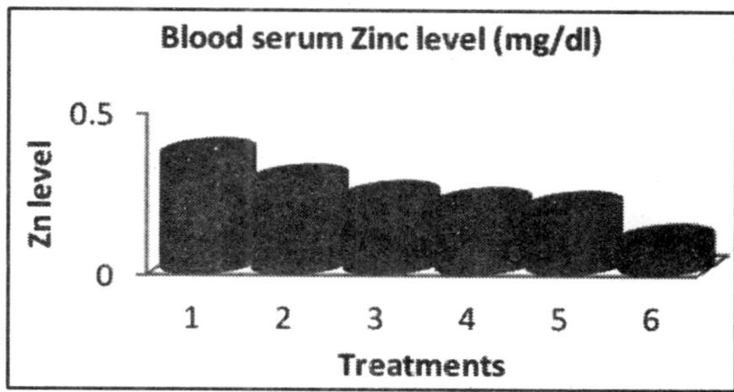

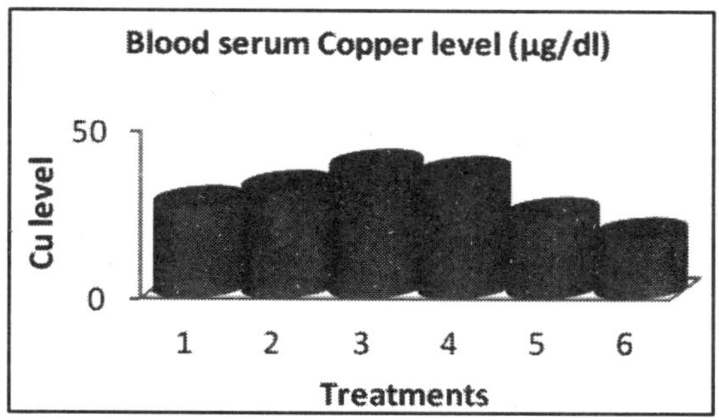

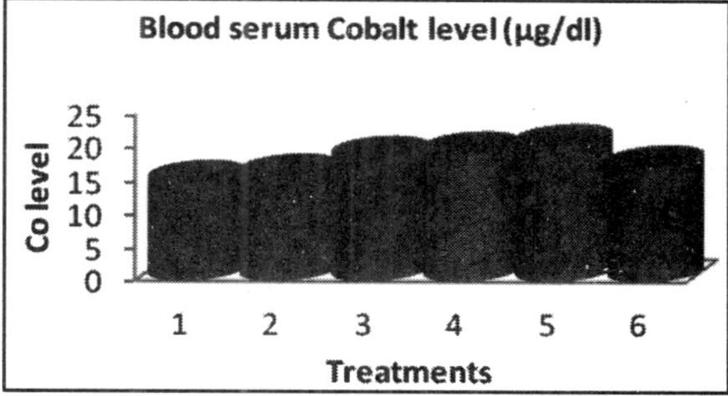

Figure 46.3: Blood Serum Minerals Level at Last Day of Digestibility Trial.

increased level of SBT cake in the rations. This may be due to higher iron value of SBT cake. The absorption of iron is greatly influenced by various constituents of the diet, such as ascorbic acid, which promote absorption (Henry and Miller, 1995). As the SBT cake contained ascorbic acid, it has helped in the increased level of iron in blood of layers as per the increased level of SBT cake in the layer ration. A small increase in dietary calcium reduced the availability of zinc (Wedekind, *et al.*, 1994). The level of zinc decreased as the level of addition of SBT cake in rations. Duodenum is the major site of zinc absorption (Davies, 1980). Mucosal induction of the metal-binding protein metallothionein limits zinc absorption at high intakes (Cusins, 1996). The SBT cake had the higher level of phosphorous, so with increased level of SBT cake in rations, the feed and phosphorous intakes were higher simultaneously, hence, both the condition may be the reason for decreased zinc level in blood with the increased level of SBT cake in ration. The copper level in blood was slightly higher than the normal level though, it was having decreasing trend with the increased level of SBT cake in diets. The toxicity of copper to chicks is increased when the diet is low in calcium and based on casein-gelatin rather than maize-soybean (Leach *et al.*, 1990). As the calcium and copper level in blood was in decreased trend and the production of egg was higher in S_{20} group, it may be due to the optimum balance of calcium and copper up to group S_{20} and thereafter due to the low level of calcium, the copper toxicity occurs and the production fallen down. The blood cobalt level of layers increased with the increased level of SBT cake in ration. For optimum production of layer, a positive correlation in between cobalt and Vit. B_{12} is necessary (Underwood and Suttle, 1999). In this study, the optimum production was found in S_{20} group and the production was fell down thereafter, this may be due to the improper balance of Vit. B_{12} and cobalt after the inclusion of S_{20} group level of SBT cake.

EXPERIMENT II

For carrying out this experiment to evaluate the effect of seabuckthorn cake on calves for growth, 18 calves of 4-5 months of age were divided randomly into three treatment groups in a randomized block design (RBD). The control (SBT_C) group was traditional calf grower ration (CGR) and crude protein content of the control (SBT_C) group was replaced with crude protein of SBT cake at 50 per cent and 75 per cent in two experimental groups SBT_{50} and SBT_{75}, respectively.

Selection of the Animal

Eighteen crossbred Jersey calves of age 4-5 months were procured from the University Livestock Farm of CSK Himachal Pradesh Krishi Vishvavidyalaya, Palampur, Kangra (HP) of almost similar body weight, breed, sex and size dividing them in 3 groups of six calves in each group and replacing 50 and 75 per cent of CP of control group (SBT_c) into two experimental groups SBT_{50} and SBT_{75}, respectively (Table 46.14).

Table 46.14: Experimental Plan

Particulars of the Treatment	*No. of Calves*	*Abbreviations*	*Supplementation Level (per cent)*	
			Basal Diet	*Seabuckthorn Cake*
Control	6	SBT_C	100	–
50 per cent Crude protein of Control feed replaced with crude protein of Seabuckthorn cake	6	SBT_{50}	64.40	35.60
75 per cent Crude protein of Control feed replaced with crude protein of Seabuckthorn cake	6	SBT_{75}	43.75	56.25

GROWTH OF CALVES AND NUTRIENT UTILIZATION

Body Weight Record

The individual body weight of the calves was taken for first three consecutive days in the morning before feeding. The average body weight was computed of each animal and grouped and after that fortnightly weight of the calves was recorded. Body weight changes of the calves were calculated. The effect of inclusion of SBT cake on growth of crossbred calves was studied having six (6) calves in each group for the period of 7 months.

CHEMICAL COMPOSITION OF FEED INGREDIENTS

The various feed ingredients used for formulating different rations during feeding experiments were analyzed individually including SBT cake for their proximate composition and cell wall constituents..The data for the different rations formulated were chemically analyzed and the values have been presented in the Table 46.15.

Table 46.15: Chemical Composition of Calf Grower Ration (per cent DM basis)

Particulars	*Treatment Groups*		
	SBT_C	SBT_{50}	SBT_{75}
DM	95.50	94.98	95.05
ME (kcal/kg)	2924	2873	2863
CP	18.00	18.20	18.50
CF	05.10	03.80	03.50
EE	05.60	06.70	07.30
NFE	61.00	45.00	38.00
TA	06.60	08.00	08.50
AIA	04.70	06.30	07.00
Ca	01.50	02.10	02.80
P	00.90	01.30	01.60
NDF	39.39	49.33	50.55
ADF	29.85	36.02	32.77
Hemicellulose	9.54	13.31	17.78
Cellulose	13.93	09.29	05.09

NDF: Neutral detergent fibre; ADF: Acid detergent fibre.

Proximate Composition

As far as possible, the rations formulated were kept iso-nitrogenous and iso-caloric and the difference in the same ration of different groups was due to the formulation of rations and variations in physical composition in different constituents of ration and particularly of SBT cake chemical composition.

Feed Consumption

The body weight and daily intake of DM in all treatment groups was recorded during the entire period of study. Average DMI was calculated as kg/d, kg/kg $W^{0.75}$ ($W^{0.75}$ =Metabolic body weight) and kg/100 kg body weight and has been presented in Table 46.16.

The average values of DMI (kg/d) in SBT_C, SBT_{50} and SBT_{75} were found to be 2.93±0.25, 2.99±0.15 and 2.19±0.11, respectively (Table 46.16). Whereas, the mean values of DMI expressed as kg/100 kg body weight were found to be 3.37±0.48, 3.12±0.13 and 3.11±0.21 in SBT_C, SBT_{50} and SBT_{75} treatment groups, respectively. The average DMI (g/kg $W^{0.75}$) were found to be 102.77±13.11, 97.63±3.79 and 90.14±5.65 for SBT_C, SBT_{50} and SBT_{75} treatment groups, respectively.

The average values obtained for DMI (kg/d) ranged from 2.19±0.11 to 2.99±0.1 being highest in SBT_{50} and lowest in SBT_{75} treatment groups. The results of statistical analysis of data revealed significant ($p<0.05$) effect of SBT_C and SBT_{50} over SBT_{75} treatment groups on DMI (kg/d). Whereas, the DMI in terms of kg per 100 kg body weight and g/kg $W^{0.75}$ found to be non-significant.

The results of DMI in respect of kg per 100 kg body weight as well as g/kg $W^{0.75}$ were in close agreement with the findings of Shihag *et al.* (2008) and Singh *et al.* (2009) but higher than that of Pandya *et al.* (2009) and Ikhar *et al.* (2011).

Table 16: Body weight and dry matter intake

Attributes	*Treatment*			*Level of Significance*
	SBT_C	SBT_{50}	SBT_{75}	
Body weight (kg)	$89.13^{ab}\pm6.85$	$95.99^{a}\pm4.11$	$70.72^{b}\pm2.21$	*
Metabolic Body size ($W^{0.75}$)	$28.96^{a}\pm1.65$	$30.65^{a}\pm0.98$	$24.38^{b}\pm0.57$	*
DMI (kg/d)	$2.93^{a}\pm0.25$	$2.99^{a}\pm0.15$	$2.19^{b}\pm0.11$	*
DMI (kg/100kg BW)	3.37±0.48	3.12±0.13	3.11±0.21	NS
DMI (g/kg $W^{0.75}$)	102.77±13.11	97.63±3.79	90.14±5.65	NS
OMI (kg/d)	$2.65^{a}\pm0.07$	$2.58^{a}\pm0.12$	$1.94^{b}\pm0.05$	***

Note: Means bearing different superscripts in a same row differ significantly ($p<0.05$). OMI: Organic matter intake.

The nutrient intake from different feed was calculated as total DMI from roughage and concentrate along with CP intake from these respective ingredients and values. The mean value of DMI from roughage and concentrate as well as CP intake found to be significant ($p<0.05$) in SBT_C and SBT_{50} over SBT_{75} treatment groups. The statistical mean revealed that total CP intake from both the sources were higher up to the 50 per cent level of inclusion of SBT cake CP as compared to its higher level of 75 per cent of inclusion (Table 46.17).

These values in respect of CP, DMI (roughage and concentrate) were significantly lower ($p<0.001$) in case of SBT_{75} treatment group as compared to SBT_C and SBT_{50} treatment groups. The total CP intake was significantly higher ($p<0.001$) in SBT50 treatment group than the other two treatment groups. The data revealed the positive effect of SBT cake CP inclusion in CGR up to 50 per cent level.

Table 46.17: Nutrient Intake (kg/d) and Plane of Nutrition of different Rations in Calves

Attributes	*Treatment*			*Level of Significance*
	SBT_C	SBT_{50}	SBT_{75}	
DMI (roughage)	$1.85^{a}\pm0.13$	$1.92^{a}\pm0.07$	$1.24^{b}\pm0.08$	***
DMI (concentrate)	$1.07^{a}\pm0.02$	$1.07^{a}\pm0.02$	$0.95^{b}\pm0.05$	*
CP (roughage)	$0.22^{a}\pm0.01$	$0.22^{a}\pm0.01$	$0.17^{b}\pm0.01$	**
CP (concentrate)	0.12±0.00	0.13±0.01	0.10±0.00	NS
Total CP intake	$0.30^{ab}\pm0.01$	$0.35^{a}\pm0.01$	$0.26^{b}\pm0.01$	***
Plane of nutrition/nutritional worth				
DCP intake (g/d)	$194.06^{a}\pm12.43$	$244.14^{b}\pm13.97$	$158.49^{a}\pm8.43$	***
DCP per cent in ration	11.69	12.50	11.11	—
TDN intake (kg/d)	$1.60^{a}\pm0.11$	$1.67^{a}\pm0.07$	$1.04^{b}\pm0.07$	***
TDN per cent in ration	57.32	48.27	41.32	—

abc: Means bearing different superscripts in a row differ significantly (**$p<0.01$, ***$p<0.001$).

DCP: Digestible crude protein; TDN: Total digestible nutrients.

Body Weight Gain

The weight gain is very reliable measure of performance of animals on the production ration, hence the body weight of experimental animals were recorded at monthly interval and has been presented in Table 46.18 and the body weight gain trend has been depicted in Figure 46.4, respectively.

Table 46.18: Body Weight Gain and Feed Efficiency of different Rations

Parameter	*Treatment*			*Level of Significance*
	SBT_C	SBT_{50}	SBT_{75}	
Initial BW (kg)	$85.60^{ab}\pm6.37$	$91.63^{a}\pm3.98$	$68.31^{b}\pm2.18$	*
Final BW (kg)	$89.13^{ab}\pm6.85$	$95.99^{a}\pm4.11$	$70.72^{b}\pm2.21$	*
Total weight gain (kg)	$3.53^{ab}\pm0.55$	$4.36^{a}\pm0.25$	$2.41^{b}\pm0.41$	*
ADG (kg/d)	$0.71^{ab}\pm0.11$	$0.87^{a}\pm0.05$	$0.48^{b}\pm0.08$	*
Total DMI (kg/d)	$2.93^{a}\pm0.25$	$2.99^{a}\pm0.15$	$2.19^{b}\pm0.11$	*
DMI (kg/kg gain)	4.62±1.08	4.48±0.20	5.36±1.60	NS
Feed cost (Rs./kg)	15.00	16.96	17.94	—
Cost (Rs./kg gain)	68.67	64.76	90.95	—

Note: Means bearing different superscripts in same row differ significantly ($p<0.05$). BW: Body weight.

The average daily gain was found to be 0.71±0.11, 0.87±0.05 and 0.48±0.08 (kg/d) for SBT_C, SBT_{50} and SBT_{75} treatment groups, respectively. These values ranged from 0.48±0.08 to 0.87±0.05 being highest in SBT_{50} and lowest in SBT_{75} treatment groups. The result of statistical analysis of the data revealed the significantly higher ($p<0.05$) average daily gain in SBT_{50} over SBT_{75} and control (SBT_C) treatment groups. The comparison of means indicated that CP replacement of control feed with higher

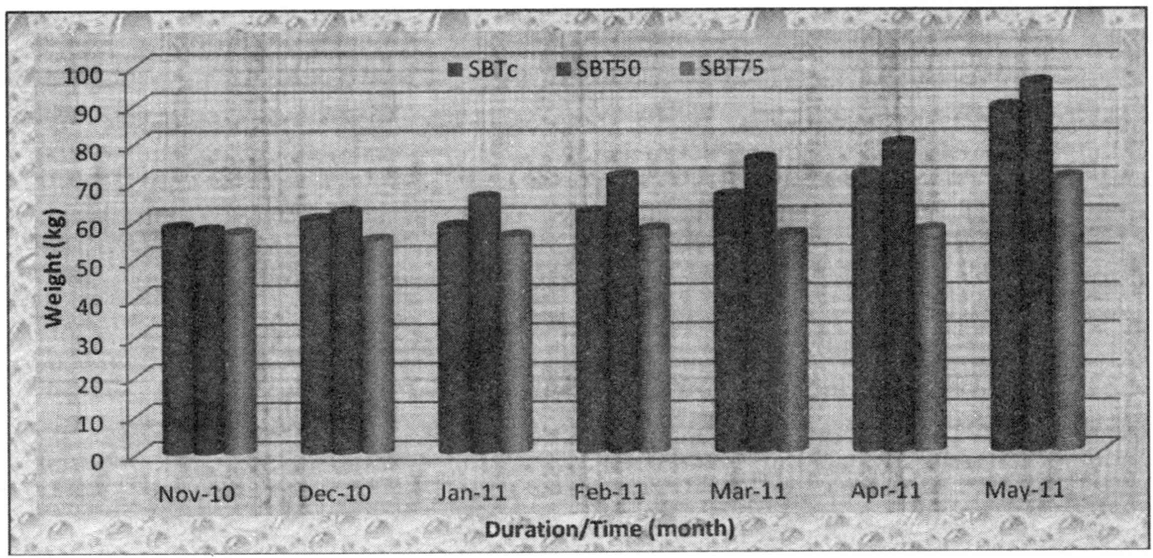

Figure 46.4: Monthly Body Weight (kg) of Calves.

level of SBT cake had the lower average daily gain but that of the lower rate (SBT_{50}) had the positive impact on ADG of calves.

From the above results, it can be concluded that the experimental calves of SBT_{50} group had higher average daily gain (kg/d) as compared to SBT_{75} treatment group. Higher DCP and TDN intake and nutrient digestibility in respect to CP and EE may be the reason for significantly higher weight gain in SBT_{50} group. The results in respect of ADG (kg/d) and DMI (kg/kg gain) were supported by the findings of Mandal *et al.* (2008), Singh *et al.* (2009) but higher than that of Pandya *et al.* (2009) and Ikhar *et al.* (2011).

Feed Conversion Ratio (FCR) and Cost Benefit Ratio

The FCR was calculated as DMI (kg/kg gain). These values were 4.62±1.08, 4.48±0.20 and 5.36±1.60 for SBT_C, SBT_{50} and SBT_{75} treatment groups, respectively. The values ranged from 4.48±0.20 to 5.36±1.60 found lowest in SBT_{50} and highest in SBT_{75} treatment groups. The result of statistical analysis revealed non-significant ($p>0.05$) effect of treatment. The lowest FCR was found in SBT_{50} treatment group in which 50 per cent CP was replaced with the SBT cake of CGR.

The cost benefit ratio of CGR was done on the basis of the production performances of calves. As all the managemental cost like purchasing of calves, electricity costs, labour charges and other costs were fixed to all the treatment groups and only the feed cost varies. So, cost (Rs/kg gain) was calculated on the basis of feed consumption only. The cost in term of Rs. per kg gain was found as 68.67, 64.76 and 90.95 for SBT_C, SBT_{50} and SBT_{75} treatment groups, respectively (Table 46.18). These values revealed that the better performance in respect of lower feed consumption was found in SBT_{50} group of treatment. The result in respect of cost (Rs./kg gain) was in close relation with the findings of Ikhar *et al.* (2011) but higher than that of Pandya *et al.* (2009).

NITROGEN BALANCE

The balance of nitrogen was calculated from the total intake, excretion through faeces and urine. From Table 46.19, it can be seen that the mean values of nitrogen intake (g/d) were found to be 47.21±2.07, 56.30±2.36 and 41.81±1.19 in SBT_C, SBT_{50} and SBT_{75} treatment groups, respectively. The overall mean values of nitrogen balance in calves for SBT_C, SBT_{50} and SBT_{75} treatment groups were found to be 24.17±2.21, 33.31±2.33 and 21.82±1.44 g/d, respectively. Whereas, the overall nitrogen balance as per cent of absorbed were found to be 49.79±3.44, 58.11±2.03 and 51.40±2.22 in SBT_C, SBT_{50} and SBT_{75} treatment groups, respectively.

Table 46.19: Nitrogen Balance of different Rations

Parameter	*Treatment*			*Level of Significance*
	SBT_C	*SBT_{50}*	*SBT_{75}*	
Intake (g/d)	47.21^b±2.07	56.30^a±2.36	41.81^b±1.19	***
Faecal outgo (g/d)	16.17±0.76	17.24±0.35	16.45±0.37	NS
Urinary outgo (g/d)	6.88^a±0.55	5.75^a±0.39	3.54^b±0.34	***
Balance (g/d)	24.17^b±2.21	33.31^a±2.33	21.82^b±1.44	***
Retained (per cent of absorbed)	49.79±3.44	58.11±2.03	51.40±2.22	NS

Note: Means bearing different superscripts in a same row differ significantly ($p<0.001$). NS: Non-significant.

The statistical analysis of data showed significant ($p<0.001$) effect of SBT cake CP on SBT_{50} as compared to SBT_C and SBT_{75} treatment groups on nitrogen intake (g/d) and balance of nitrogen (g/d). This was due to the improved gross activity of rumen micro flora, alteration in numbers and species of micro-organism in rumen, increased TVFA concentration on inclusion of SBT cake in the rations of the calves. Whereas, the balance of nitrogen as per cent of absorbed showed non-significant effect of SBT cake CP over that of control group. This was due to the improved digestibility of CP on the respective group. However, all the animals irrespective of treatment found to be in positive nitrogen balance. This study also revealed that there were higher significant N-balances available for various activities in case of SBT cake 50 per cent fed group as compared to other two groups.

Higher value of balance of nitrogen (g/d) than the current study was observed by Pandya *et al.* (2009), and lower value was observed by Kumar *et al.* (2008). However, the result in respect of balance of nitrogen (g/d) supported by the findings of Mandal *et al.* (2008) and Mudgal *et al.* (2008).

FEED EFFICIENCY

The final body weight (kg), total weight gain (kg) as well as average daily gain (g/d) was significantly ($p<0.05$) higher in SBT_{50} group as compared to SBT_{75} group (Table 46.20). The variation of DMI (kg/kg gain) was non-significant. The comparison revealed that SBT_{50} treatment showed better performance as compared to SBT_{75} group.

Table 46.20: Feed Efficiency of different Rations during Entire Experimental Period

Parameter	*Treatment*			*Level of Significance*
	SBT_C	*SBT_{50}*	*SBT_{75}*	
Initial BW (kg)	58.15±4.36	57.43±3.28	56.65±3.93	NS
Final BW (kg)	89.13[ab]±6.85	95.99[a]±4.11	70.72[b]±2.21	*
Total weight gain (kg)	30.98[ab]±8.66	38.57[a]±1.40	14.07[b]±2.97	*
ADG (g/d)	172.11[ab]±48.09	214.27[a]±7.76	78.15[b]±16.50	*
Total DMI (kg/d)	2.93[a]±0.25	2.99[a]±0.15	2.19[b]±0.11	*
DMI (kg/kg gain)	23.67±9.49	13.96±0.34	32.05±7.09	NS

Means bearing different superscripts in a same row differ significantly ($p<0.05$).

ESTIMATION OF BLOOD SERUM MINERALS LEVELS

The blood serum minerals levels of calves at the last day of digestibility cum metabolism trial conducted have been presented in Table 46.21. The Ca content in blood serum was found as 8.50±0.38, 8.15±0.56 and 6.98±0.64 mg/dl for SBT_C, SBT_{50} and SBT_{75} groups, respectively being higher in SBT_C and lower in S_{75} group, found to be non-significant ($p>0.05$) among them but decreasing in trend with the increased level of SBT cake in rations. The P content in blood serum found to be 4.58±0.09, 4.46±0.18 and 4.68±0.05 mg/dl for SBT_C, SBT_{50} and SBT_{75} groups, respectively. The maximum P value was observed in case of SBT_{75} and minimum in SBT_{50} group.

Blood serum Mg was found as 1.91±0.21, 2.29±0.35 and 2.23±0.18 mg/dl for SBT_C, SBT_{50} and SBT_{75} groups, respectively and the Mg level in the blood serum found to be non-significant effect of treatments. Blood serum Fe was found as 30.37±1.61, 27.94±2.07 and 29.93±0.74 mg/dl for SBT_C, SBT_{50} and SBT_{75} treatment groups, respectively and the Fe level in the blood serum found to be non-significant. Zn content in blood serum was found as 10.61±0.24, 11.39±1.64 and 10.43±0.95 mg/dl for SBT_C, SBT_{50} and SBT_{75} groups, respectively and the Zn level in the blood serum found to be non-significant. Blood serum Cu was 8.59±0.70, 7.22±0.38 and 8.08±0.30 mg/dl for SBT_C, SBT_{50} and SBT_{75} treatment groups, respectively, found to be non-significant ($p>0.05$) effect of treatments. Blood serum cobalt content was found as 5.75±0.25, 5.50±0.65 and 4.50±0.29 µg/dl for SBT_C, SBT_{50} and SBT_{75} groups, respectively which was found in decreasing in trend with the increased level CP of SBT cake in CGR. The statistical analysis of the data revealed non-significant effect of treatment on blood serum cobalt. Blood serum Mn content was found as 10.00±0.00, 15.00±2.89 and 12.50±2.50 µg/dl for SBT_C, SBT_{50} and SBT_{75} groups, respectively and the Mn level in the blood serum found to be non-significant. Blood serum Na was found as 74.17±3.25, 76.77±5.01 and 76.71±4.81 mEq/L for SBT_C, SBT_{50} and SBT_{75} groups, respectively and found to be non-significant. Blood serum K was found as 3.14±0.23, 3.10±0.16 and 3.22±0.40 mEq/L for SBT_C, SBT_{50} and SBT_{75} groups, respectively and the its level in the blood serum found to be non-significant.

Table 46.21: Blood Serum Minerals Level at the End of Metabolism Trial of Calves

Attributes	*Treatment*			*Level of Significance*
	SBT_C	*SBT_{50}*	*SBT_{75}*	
Ca (mg/dl)	8.50±0.38	8.15±0.56	6.98±0.64	NS
P (mg/dl)	4.58±0.09	4.46±0.18	4.68±0.05	NS
Mg (mg/dl)	1.91±0.21	2.29±0.35	2.23±0.18	NS
Fe (mg/dl)	30.37±1.61	27.94±2.07	29.93±0.74	NS
Zn (mg/dl)	10.61±0.24	11.39±1.64	10.43±0.95	NS
Cu (mg/dl)	8.59±0.70	7.22±0.38	8.08±0.30	NS
Co (µg/dl)	5.75±0.25	5.50±0.65	4.50±0.29	NS
Mn (µg/dl)	10.00±0.00	15.00±2.89	12.50±2.50	NS
Na (mEq/l)	74.17±3.25	76.77±5.01	76.71±4.81	NS
K (mEq/l)	3.14±0.23	3.10±0.16	3.22±0.40	NS

Note: Means bearing different superscripts in same row differ significantly ($p<0.05$).

The Ca level in the blood serum of calves decreased sharply with the increment of SBT cake of the feed beyond 50 per cent and though the average production was highest in SBT_{50} group. As the SBT cake is containing more P, the blood P level was increased for the increased level of SBT cake which may have some positive balance in SBT_{50} treatment group. The iron level of the blood serum of calves found in increasing trend for the increased level of CP of the SBT cake (Figure 46.5). It may be due to higher Fe value of SBT cake. The absorption of Fe is greatly influenced by various constituents of the diet, such as ascorbic acid, which promote absorption (Henry and Miller, 1995).

As the SBT cake contained ascorbic acid, it may help in the increased level of Fe in blood of calves according to the increment level of SBT cake in the calves ration. The SBT cake had the higher level of P, so with increment level of SBT cake in rations, the FI was higher and simultaneously P intakes were also higher, hence both the condition may be the reason for decreased trend of Zn in blood with the increment level of SBT cake in diet. The Cu level in the blood was in opposite trend of Zn. Though, the Ca was in opposite trend with Cu level in the blood, production of calves was higher in SBT_{50} group, it may be due to the optimum balance of Ca and Cu up to that particular level (SBT_{50}). For optimum

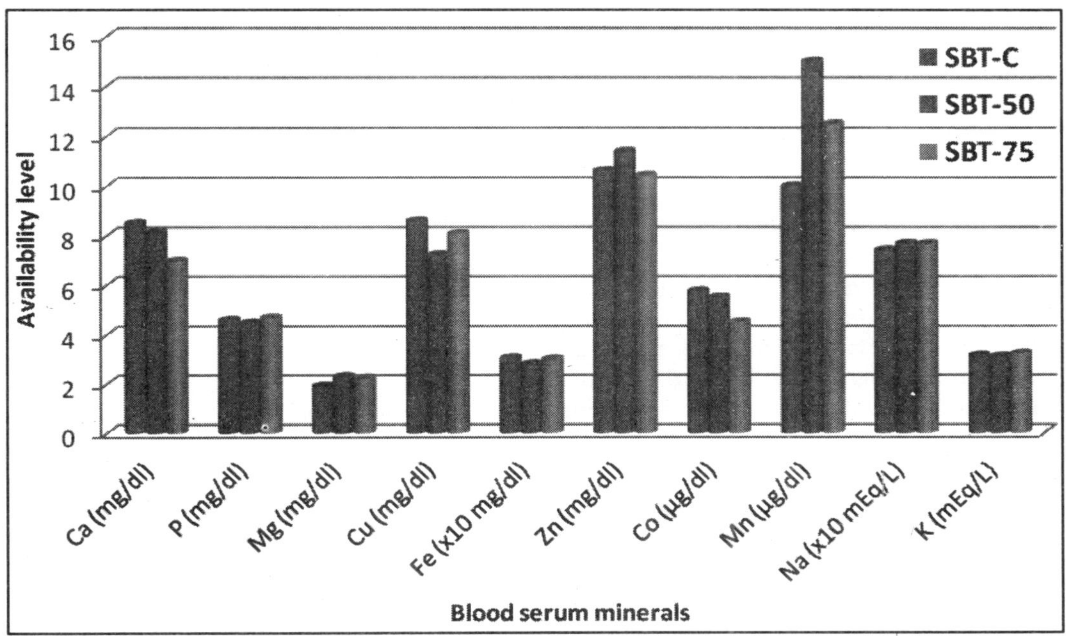

Figure 46.5: Blood Serum Minerals Level at the End of Metabolism Trial of Calves.

production of calves a positive correlation in between Co and Vit-B12 is necessary (Underwood and Suttle, 1999). In this study, the optimum production was found in SBT_{50} group and the production was fell down thereafter, this may be due to the improper balance of Vit. B_{12} and Co in SBT_{75} group.

EXPERIMENT–III

RUMEN FERMENTATION STUDY OF ADULT CALVES

To study the rumen fermentation, rumen liquor samples was collected from the experimental animals at 2, 4, 8, 12, 16, 20 and 24 h post feeding (concentrate was offered at 10.00 AM, once only) to assess the replacement effect of CP of adult calf ration with the CP of Seabuckthorn cake on rumen functioning.

pH of the Rumen Liquor

The variations in rumen metabolites of adult male calves at different hours of feeding have been presented in Table 46.22. The pH of the rumen liquor varied from 6.10±0.06 to 7.27±0.13, 6.10±0.06 to 7.37±0.03, and 6.80±0.06 to 7.30±0.17 in T_1, T_2 and T_3 group, respectively. The highest and lowest pH values were recorded at 12 h and 24 h post feeding in all the treatments. The overall pH value for T_1, T_2 and T_3 was 6.60±0.16, 6.60±0.17and 7.04±0.07, respectively, found to be non-significant. Mahouchi *et al.* (2003) reported lowest value of rumen pH in sheep at 1-2 h post feeding. In case of buffalo, the pH of the rumen represented significantly lower levels at 4 h post feeding either in presence or absence of protozoa (Chaudhary *et al.*, 2000; Kamra *et al.*, 2000). Grazing on either natural grassland or silvi-pasture system maintained a rumen pH of 7.30 to 7.96 (Samanta *et al.*, 1998). The other experiment conducted by Samanta *et al.* (2005) showed that the cattle's grazing on natural pasture, the rumen pH was always highest above 7.0 irrespective of post feeding intervals. In the present investigation, T_3 group was highest irrespective of post feeding intervals as compared to other all groups. It might be due to higher secretion of alkalizing agents through saliva.

Total Volatile Fatty acids (TVFA)

Total volatile fatty acids (TVFA) concentration of rumen liquor varied and was 18.33±0.88 to 57.00±0.58, 47.33±0.33 to 71.67±1.20, and 21.00±1.00 to 68.33±0.88 mEq/dl and reached to the peak levels at 24 h, 4 h, and 16 h of post feeding in T_1, T_2 and T_3 diet, respectively. The overall value of TVFA was 36.05±5.52, 62.19±2.99 and51.24±6.15 for T_1, T_2 and T_3 treatment, respectively, showed highly significant ($p<0.01$) effect of T_2 over T_1. This might be due to the sufficient availability of fermentable carbohydrates for their rapid and faster metabolism into volatile fatty acids at those particular feeding intervals and particular (T_2) treatment. Irrespective of post feeding intervals, the rumen TVFA concentration was significantly ($p<0.05$) maximum in T_2 followed by T_3 and T_1while taking consideration of the ration only. Hence the animals on SBT cake based ration led towards higher TVFA production. By virtue of huge microbial population, ruminants ferment the plant complex biomolecules into volatile fatty acids for deriving the energy. Though the concentration of VFA was highly dependent on the nature and type of ration; however, it generally varied from 60 to 120 mmol/ litre (Houtert, 1993). In case of buffalo, Santra *et al.* (1996) reported the highest TVFA concentration (12.65 m mol/dl Strained rumen liquor (SRL)) at 4 h post feeding. Devasena *et al.* (2004) recorded that the TVFA concentration ranges from 68.8 to 101.8 in sheep and goat, whereas Pandya *et al.* (2009) reported that the TVFA concentration in SRL varies from 74 to 126 mEq/l in adult calves.

Ammonia Nitrogen (NH_3-N)

The rumen ammonia concentration represents the balance between the degradation of nitrogenous compounds (including protein and non-protein nitrogenous compounds) and uptake by the microbes for the synthesis of microbial protein. In the present experiment, the rumen ammonia nitrogen (NH_3-N) concentration ranged from 14.00±0.58 to 30.67±0.67, 14.33±0.33 to 30.00±1.16, and 14.33±0.88 to 28.00±0.58 mg/dl in T_1, T_2 and T_3 treatments, respectively. Furthermore, the overall value of NH3-N was 20.71±2.51, 21.10±2.43 and 21.37±2.02, found to be non-significant (Table 46.22). It was higher than the minimum values (5-7 mg/dl) for optimum microbial activity (Satter and Slyter, 1974) in SBT ration. The average range of NH_3-N was corroborated with the findings of Samanta *et al.* (2006) and Devasena *et al.* (2004). The rumen NH_3-N concentration peaked ($p<0.05$) at 4 h post feeding in T_1;2 h post feeding in T_2 and T_3 treatment which was corroborated with the findings of Dahiya *et al.* (2011) in the calves.

Total Nitrogen (Total-N)

The total-N concentration in rumen liquor ranged from 117.26±6.92 to 247.93±7.06, 115.90±4.78 to 226.12±4.94, and 95.20±5.83 to 196.59±9.24 mg/dl in treatment T_1, T_2 and T_3, respectively. Highest concentration was observed at 12 h, 16 h, and 12 h post feeding in T_1, T_2 and T_3 treatment, respectively and lowest N-concentration was found at 24 h of post feeding in T_1 and T_2 treatments. Whereas, the overall value of total-N was found 188.46±17.37, 174.59±14.33, 159.25±11.85 in T_1, T_2 and T_3 treatment, respectively. This might be due to slow degradation of dietary protein and hydrolysis of NPN substances in the rumen. The subsequent changes of total-N concentration were observed as a zigzag manner might be due to changes in the rumen volume through inflow of saliva as well as outflow of digesta (Grubb and Dehority, 1975). The total-N concentration in the rumen liquor corroborated with the findings of Samanta *et al.* (2006), Rekha *et al.* (2006) and Lailer *et al.* (2011).

Table 46.22: Effect of Diet and Time on Various Rumen Metabolites in Adult Calves

Rumen metabolite	Treatment	Hours Post Feeding							
		2 h	4 h	8 h	12 h	16 h	20 h	24 h	Overall
pH	T_1	6.83^{a}±0.09	6.63^{a}±0.09	6.17^{a}±0.09	6.10^{a}±0.06	6.30^{a}±0.06	6.90^{a}±0.06	7.27±0.13	6.60±0.16
	T_2	6.73^{a}±0.07	6.73^{a}±0.07	6.80^{b}±0.06	6.10^{a}±0.06	6.10^{a}±0.06	6.40^{b}±0.06	7.37±0.03	6.60±0.17
	T_3	7.23^{b}±0.12	7.17^{b}±0.09	7.00^{b}±0.00	6.80^{b}±0.06	6.93^{b}±0.03	6.83^{a}±0.03	7.30±0.17	7.04±0.07
	SL	*	*	***	***	***	**	NS	NS
TVFA (mEq/l)	T_1	44.00^{a}±0.58	30.67^{a}±0.67	29.67^{a}±0.33	22.33^{a}±0.33	18.33^{a}±0.88	50.33^{a}±0.33	57.00^{a}±0.58	36.05^{a}±5.52
	T_2	51.00^{b}±0.58	71.67^{b}±1.20	63.67^{b}±0.33	65.00^{b}±0.00	59.67^{b}±0.88	70.67^{b}±0.67	53.67^{b}±0.33	62.19^{b}±2.99
	T_3	41.00^{c}±0.58	38.00^{c}±0.58	65.00^{b}±0.58	66.67^{b}±0.88	68.33^{c}±0.88	27.00^{c}±0.00	52.67^{b}±0.33	51.24^{ab}±6.15
	SL	*	**	***	***	**	***	**	**
NH_3-N (mg/dl)	T_1	26.00^{a}±0.58	30.67^{a}±0.67	24.00^{a}±0.58	21.67^{a}±0.88	14.33±0.33	14.00±0.58	14.33^{a}±0.33	20.71±2.51
	T_2	30.00^{b}±1.16	20.67^{b}±0.67	29.67^{b}±0.33	16.33^{b}±0.33	16.00±0.58	14.33±0.33	20.67^{b}±0.67	21.10±2.43
	T_3	28.00^{ab}±0.58	23.67^{c}±0.33	26.00^{a}±0.58	22.33^{a}±0.33	14.33±0.33	14.33±0.88	21.00^{b}±0.58	21.37±2.02
	SL	*	*	**	**	NS	NS	***	NS
Total-N (mg/dl)	T_1	139.41^{a}±4.86	186.67^{a}±9.33	205.76^{a}±11.95	247.93^{a}±7.06	224.00^{a}±3.23	198.21±6.19	117.26±6.92	188.46±17.37
	T_2	152.90^{a}±3.90	146.62^{b}±4.58	201.52^{a}±2.84	196.51^{b}±8.05	226.12^{a}±4.94	182.59±2.74	115.90±4.78	174.59±14.33
	T_3	119.38^{b}±2.55	173.86^{ab}±6.79	155.70^{b}±5.36	196.59^{b}±9.24	176.40^{b}±4.28	178.52±9.22	114.29±7.89	159.25±11.85
	SL	*	*	*	*	***	NS	NS	NS
TCA-N (mg/dl)	T_1	119.89±11.76	130.33^{a}±4.07	157.57^{a}±6.29	179.20^{a}±5.60	159.18^{a}±3.61	154.59^{a}±2.74	108.35^{a}±2.49	144.16±9.52
	T_2	103.77±2.89	107.33^{b}±4.67	130.92^{b}±2.53	124.90^{b}±3.90	173.68^{a}±4.20	158.24^{a}±2.87	85.95^{b}±2.39	126.40±11.74
	T_3	89.85±6.80	119.30^{ab}±1.96	120.40^{b}±4.28	135.59^{b}±2.28	133.81^{b}±2.15	132.34^{b}±6.31	69.58^{c}±1.67	114.41±9.54
	SL	NS	*	*	***	**	*	**	NS
Soluble-N or NPN (mg/dl)	T_1	19.51^{a}±8.50	56.34±11.91	48.19^{a}±6.97	68.73±2.06	64.82^{a}±3.61	43.61^{a}±4.30	12.58^{a}±4.47	44.83±8.16
	T_2	49.13^{b}±5.98	39.28±8.01	70.60^{b}±2.20	71.61±4.31	52.44^{b}±0.76	24.35^{b}±0.98	29.95^{ab}±3.31	48.19±6.99
	T_3	29.53^{ab}±4.62	54.56±6.14	35.30^{a}±1.96	61.01±7.97	42.59^{b}±2.21	46.18^{a}±3.72	44.71^{b}±8.16	44.84±4.05
	SL	*	NS	*	NS	*	*	*	NS

Note: Values bearing different superscripts in a column within the same hour differ significantly; *$p<0.05$, **$p<0.01$, ***$p<0.001$; NS: Non-significant; SL: Level of significance; TCA: Trichloride amino acid.

Table 46.23: Effect of Ration and Time on Various Rumen Microbes In Adult Calves

Rumen metabolite	Treatment	Hours Post Feeding							
		2 h	4 h	8 h	12 h	16 h	20 h	24 h	Overall
Bacteria	T_1	1.60^{a}±0.06	1.80^{a}±0.06	2.10^{a}±0.06	1.30^{a}±0.06	6.13^{a}±0.13	1.43^{a}±0.03	0.93^{a}±0.07	2.21^{b}±0.70
(Cell x 10^{10})	T_2	10.27^{b}±0.27	7.13^{b}±0.09	9.67^{b}±0.03	9.40^{b}±0.21	9.17^{b}±0.17	6.03^{b}±0.03	11.13^{b}±0.13	8.97^{a}±0.67
	T_3	6.20^{c}±0.10	3.10^{c}±0.06	3.10^{c}±0.06	5.10^{c}±0.10	3.40^{c}±0.21	4.03^{c}±0.03	2.80^{c}±0.06	3.96^{b}±0.47
	SL	***	***	*	***	***	***	***	***
Protozoal	T_1	++++	++++	++++	++++	+++	+++	++	3.43+
Motility#	T_2	++++	++++	++++	+++	+++	+++	+++	3.43+
	T_3	++++	++++	++++	++++	++++	++++	++++	4+
Protozoa	T_1	1.33^{a}±0.03	2.20^{a}±0.10	1.37^{a}±0.07	2.77^{a}±0.03	1.53^{a}±0.03	2.13^{a}±0.03	1.87^{a}±0.03	1.89±0.20
(Cell x 10^{6})	T_2	1.60^{b}±0.06	1.93^{a}±0.03	2.10^{b}±0.06	2.70^{a}±0.00	2.60^{b}±0.06	2.37^{b}±0.03	1.63^{b}±0.03	2.13±0.18
	T_3	1.87^{c}±0.03	4.13^{b}±0.07	2.80^{c}±0.06	2.03^{b}±0.03	4.30^{c}±0.15	4.27^{c}±0.03	2.87^{c}±0.03	3.18±0.40
	SL	*	***	***	***	***	**	**	NS

Values bearing different superscripts in a column within the same hour differ significantly; *p<0.05, **p<0.01, ***p<0.001.

NS= Non-significant; SL: Level of significance.

Protozoal Motility: ++++: Very rapid movement, whole mass is moving; +++: Rapid movement, very large population of protozoa showing their motility; and ++: Moderate movement, less number of protozoa is moving moderately.

TCA Precipitable Nitrogen (TCA-N)

The TCA-N concentration ranged from 108.35±2.49 to 179.20±5.60, 85.95±2.39 to 173.68±4.20, and 69.58±1.67 to 135.59±2.28 mg/dl, with the highest concentration recorded at 12 h, 16 h, and 12 h of post feeding in T_1, T_2, and T_3 treatments, respectively. The lowest concentration was observed at 24 h of post feeding in all the treatments. Whereas, the overall value of the TCA-N was 144.16±9.52, 126.40±11.74 and 114.41±9.54 in T_1, T_2, and T_3 treatments, respectively. The concentration of TCA-N in rumen liquor was higher in this study than the previous studies, which might be due to the optimum proportion of N and energy provided by SBT cake required for microbial protein synthesis.

Soluble Nitrogen (Soluble-N)/NPN

The Soluble-N concentration ranged from 12.58±4.47 to 68.73±2.06, 26.22±4.06 to 71.61±4.31, and 19.13±5.14 to 61.01±7.97 mg/dl, in T_1, T_2, and T_3 diet, respectively with the highest concentration recorded at 12 h of post feeding in all the treatment groups. The lowest concentration was recorded at 24 h, 20 h and 2 h post feeding in T_1, T_2 and T_3 treatments, respectively. The overall value to NPN was 44.83±8.16, 48.19±6.99 and 44.84±4.05 in T_1, T_2 and T_3 treatments, respectively (Table 22). The NPN concentration in SBT (T_2 and T_3) ration group was found always higher in plea of few cases than the control group. It might be due to the synchronized availability of energy and microbial protein in the rumen. These results supported by the findings of Lailer *et al.* (2011) and Pandya *et al.* (2009).

RUMEN MICROBIAL POPULATION

Rumen Bacteria

The microbial population in case of rumen bacteria (cells x 10^{10}) ranged from 0.93±0.07 to 6.13±0.13, 6.03±0.03 to 11.13±0.13, and 2.80±0.06 to 6.20±0.10 per ml of rumen liquor in T_1, T_2 and T_3 ration, respectively. The bacteria population attained peak level at 16 h, 2 h and 2 h of post feeding and lowest values was found at 24 h, 20 h and 24 h of post feeding in T_1, T_2, and T_3 treatment, respectively (Table 46.23). Moreover, the overall number of rumen protozoa was 2.21±0.70, 8.97±0.67 and 3.96±0.47 and found to be significant ($p<0.01$). The population of bacteria was higher both in T_2 and T_3 diet might be due to the positive balance of protein and energy in rumen where SBT cake was added. These results were similar as found by Rekha *et al.* (2006), Thakur *et al.* (2006) and Chandanshive *et al.* (2007) (Table 46.23).

Rumen Protozoa

The rumen protozoal population (cells x 10^6) ranged from 1.33±0.03 to 2.77±0.03, 1.53±0.03 to 2.70±0.00, and 1.87±0.03 to 4.27±0.03 per ml of rumen liquor in T_1, T_2 and T_3 treatment groups, respectively, being highest at 12 h of post feeding in T_1 and T_2 diet, and 20 h of post feeding in T_3 treatment. Moreover, the overall number of rumen protozoa found 1.89±0.20, 2.13±0.18 and 3.18±0.40, which were non-significant. The concentration of protozoal population in the SRL was supported by the Murugeswari *et al.* (2007).

CONCLUSION

In the present study, the egg production, egg mass and body weight gain of layers were higher in S_{20} group where 20 per cent CP of conventional concentrate was replaced with the CP of the SBT cake. Higher body weight gain and better balance of nutrient in calves was observed in SBT_{50} group where 50 per cent CP of the traditional CGR was replaced with the SBT cake CP. Higher microbial population in adult calves was found in T_2 group, where 50 per cent CP of the traditional adult calves ration was replaced with the CP of SBT cake.

REFERENCES

1. Ambatkar, S.K. 2009. Evaluation of seabuckthorn leaves (*Hippophae* L.) as protein replacer for efficient broiler production. *M.VSc. Thesis*, Department of Animal Nutrition, CSK Himachal Pradesh Krishi Vishvavidyalaya, Palampur, India, 80p.
2. AOAC. 1985. Official methods of analysis (14th edn.). *Association of Official Analytical Chemists* 24: 868-878.
3. Bhatt, R.S. 1993. Application of some probiotics in broilers and egg production. *PhD. Thesis*, Department of Animal Nutrition, CSK Himachal Pradesh Krishi Vishvidyalaya, Palampur, India, 40 p.
4. BIS.1992. Poultry feed specification. IS: 1374 (4th revision). In: *Bureau of Indian Standards*, Manak Bhavan, 9 Bahadur Zafar Marg, New Delhi, India.
5. Chandansive, S.V., Mani, V., Kaur, H. and Kewalramani, N. 2007. Zinc supplementation to meliorate adverse effects of cadmium on rumen fermentation in cattle. *Indian Journal of Animal Nutrition* 24 (2): 67-71.
6. Chaudhary, L.C., Kamra, D.N., Singh, R., Agarwal, N. and Pathak, N.N. 2000. Effect of feeding bentonite on rumen fermentation, enzyme activity and protozoal population in adult buffaloes (*Bubalus bubalis*). *Buffalo Journal* 1: 73-79.
7. Cusins, R.B. 1996. Zinc. *Present Knowledge in Nutrition* (7th Edn.) (L.J. Filer and E.F. Ziegler, Eds.). International Life Science Institute-Nutrition Foundation, Washington, DC, 540p.
8. Dahiya, S.S., Lailer, P.C., Lal, D. and Chauhan, T.R. 2011. Rumen fermentation pattern in buffaloes fed complete feed blocks. *Indian Journal of Animal Nutrition* 28 (2):177-180.
9. Davies, N.T. 1980. Studies on the absorption of zinc by rat intestine. *British Journal of Nutrition* 43: 189-203.
10. *de* Witt, F.H., Kuleile, N.P., van der Merwe, H.J. and Fair, M.D. 2009. Effect of limestone particle size on egg production and eggshell quality of hens during late production. *South African Journal of Animal Science* 39 (Supplement 1).
11. Devasena, B., Haritha, K., Ravi, A. and Prasad, R. 2004. Evaluation of fruit wastes in small ruminants. *Indian Journal of Animal Nutrition* 21(2): 85-88.
12. Duncan, D.B. 1955. Multiple range and F-tests. *Biometrics* 11:1-42.
13. Fayeye, T.R., Adeshiyan, A.B. and Olugbami, A.A. 2005. Egg traits, hatchability and early growth performance of the Fulani-ecotype chicken. *Livestock Research for Rural Development* 17 (8): *www.cipav.org.co/lrrd.*
14. Grubb, J.A. and Dehority, B.A. 1975. Effect of an abrupt change in ration from all roughages to high concentrate diet upon rumen microbial number in sheep. *Applied Microbiology* 30: 404-415.
15. Guleria, R. 2010. Evaluation of seabuckthorn (*Hippophae* L.) cake as a protein source in broiler ration. *M.VSc. Thesis*, Department of Animal Nutrition, CSK Himachal Pradesh Krishi Vishvavidyalaya, Palampur, India, 58p.
16. Gupta, M. 2003. Effect of combinations of different strains of microbes isolated from indigenous sources on the performance of chicken. *M.VSc. Thesis*, Department of Animal Nutrition, CSK Himachal Pradesh Krishi Vishvavidyalaya, Palampur, India, 86 p.

17. Henry, P.R. and Miller, E.R. 1995. Iron bioavailability. In: *Bioavailability of Nutrients for Animals* (C.B. Ammerman, D.H. Baker and A.J. Lewis, Eds.), p.169-200, Academic Press, New York.

18. Houtert, M.F.J. 1993. The production and metabolism of volatile fatty acids by ruminants fed roughages-A review. *Animal Feed Science and Technology* 43: 189-225.

19. Ikhar, U.V., Singh, S.K. and Mudgal, V. 2011. Effect of different feeding regiemen on growth and reproductive performance of shahiwal heifers. *Indian Journal of Animal Nutrition* 28 (2): 181-184.

20. Kaistha, M. 1996. Application of some locally raised bacterial strains for economic poultry production. *M.VSc. Thesis,* CSK Himachal Pradesh Krishi Vishvavidyalaya, Palampur, India.

21. Kamra, D.N., Singh, R., Chaudhary, L.C., Agarwal, N. and Pathak, N.N. 2000. Soapnut as natural defaunating agent: Its effect on rumen fermentation and *in sacco* degradability of jowar hay in buffaloes. *Buffalo Journal* 1: 99-104.

22. Katoch, S. 1996. Biological effects of locally isolated microbial strains in chicken. *M.VSc. Thesis,* CSK Himachal Pradesh Agricultural University, Palampur, India.

23. Kumar, V., Makkar, H.P.S. and Becker, K. 2008. Detoxification of *Jatropha curcas* seed meal and its utilization as a protein source in fish diet. In: *25th European Society for Comparative Physiology and Biochemistry* (*ESCPB*) *Congress,* September 7-11, 2008, Ravenna, Italy.

24. Lailer, P.C., Dahya, S.S., Lal, D. and Chauhan, T.R. 2011. Rumen fermentation pattern in buffaloes fed completes feed blocks. *Indian Journal of Animal Nutrition* 2: 177-180.

25. Leach, R.M., Rosenblum, J.C.I., Amman, M.J. and Burdette, J. 1990. Broiler chicks fed low calcium diets. 2. Increased sensitivity to copper toxicity. *Poultry science* 69: 1905-1910.

26. Mahouchi, M., Haddad, L., Kayouli, C., Thewis, A. and Beckers, Y. 2003. Effects of nature of nitrogen supplement on voluntary intake, rumen parameters and ruminal degradation of dry matter in sheep fed oat silage based diets. *Small Ruminant Research* 48: 181-187.

27. Mandal, G.P., Dass, R.S. and Garg, A.K. 2008. Effect of supplemental zinc on performance, nutrient utilization and trace mineral metabolism in crossbred calves. *Indian Journal of Animal Nutrition* 25 (3): 211-214.

28. Mudgal, V., Garg, A.K. and Dass, R.S. 2008. Effect of zinc, copper and selenium supplementation on growth rate and nutrient utilization in buffalo (*Bubalus bubalis*) calves. *Indian Journal of Animal Nutrition* 25 (3): 272-277.

29. Muruggeswari, R., Balakrishnan, V. and Vijayakumar, R. 2007. Effect of processing Tapioca leaves on rumen fermentation. *Asian Journal of Animal and Veterinary Advances* 2 (3): 115-124.

30. Pandya, P.R., Desai, M.C., Patel, G.R., Talpada, P.M., Pande, M.B. and Parnerkar, S. 2009. Economical rearing of crossbred calves on complete feeds, based on sugarcane bagasse and non conventional feeds. *Indian Journal of Animal Nutrition* 26 (3): 211-215.

31. Rekha, N.A., Prasad, J.R., Ramana, J.V. and Ravi, A. 2006. Evaluation of groundnut haulms based complete rations with or without yeast culture in sheep. *Indian Journal of Animal Nutrition.* 23 (1): 1-4.

32. Samanta, A.K., Singh, K.K., Das, M.M. and Kundu, S.S. 2006. Rumen fermentation pattern in crossbred cattle maintained on complete feed block. *Indian Journal of Animal Nutrition* 23 (1): 50-52.

33. Samanta, A.K., Singh, K.K., Das, M.M., Pailan, G.H., Verma, N.C. and Rai, S. 2005. Influence of roughage source on rumen metabolites and microbial population with special reference to anaerobic fungus in cattle. *Animal Nutrition and Feed Technology* 5: 179-193.

34. Samanta, A.K., Upadhyay, V.S. and Pathak, P.S. 1998. Effect of silvipasture system on rumen fermentation and growth in crossbred heifers. *Indian Journal of Animal Nutrition* 15: 18-20.

35. Santra, A., Kamra, D.N. and Pathak, N.N. 1996. Influence of ciliate protozoa on biochemical changes and hydrolytic enzymes in the rumen of Murrah buffalo (*Bubalus bubalis*). *Buffalo Journal* 1: 95-100.

36. Satter, L.D. and Slyter, L.L. 1974. Effect of ammonia concentration on rumen microbial protein production *in vitro*. *British Journal of Nutrition* 32: 199-208.

37. Shama, S. 2011. Effects of probiotics supplementation on the utilization of seabuckthorn cake in broiler production. *M.VSc. Thesis*, Department of Animal nutrition, CSK Himachal Pradesh Krishi Vishvavidyalaya, Palampur, HP. India.

38. Sharma, V.K. and Ambatkar, S.K. 2009. Seabuckthorn leaves as a protein replacer for the poultry broiler production. In: *13th Biennial Animal Nutrition Conference*, 225 p.

39. Shihag, Z.S., Sajjan, S. and Dahiya, D.S. 2008. Effect of higher plane of nutrition on the performance of pregnant crossbred heifers. *Indian Journal of Animal Nutrition* 25 (3): 206-210.

40. Simsek, E. and Kilic, I. 2006. Building environment and interaction of population density and position and their relationship to layer performance. *International Journal of Poultry Science* 5 (9): 856-862.

41. Singh, B., Chandan, B.K. and Gupta, D.K. 2003. Adaptogenic activity of a novel with anolide-free aqueous fraction from the roots of *Withania somnifera* Dun. II. *Phytotherapy Research* 17: 531-536.

42. Singh, V. and Dogra, K.K. 1996. Characteristics, distribution, biomass, regeneration and nutritional values of seabuckthorn. *Indian Forester* 122 (6): 486-491.

43. Singh, V., Gupta, R.K., Uppal, R., Singh, B. and Lata, S. 2005. Morpho-biochemical variations in the seeds of different biotyes of seabuckthorn (*Hippopahe* L.). *Non-timber Forest products* 12 (1): 25-29.

44. Singh, V., Singh, R.K. and Singh, B. 1999. Fodder values of foliage of seabuckthorn (*Hippopahe rhamnoides* ssp. *turkestanica*) growing in Lahaul valley. In: *Proc. of International Workshop on Seabuckthorn*, p. 55-59, August 29-September 2, 1999, China Science and Technology Press, Beijing, China, 97p.

45. Singh, V., Singh, B. and Awasthi, C.P. 1995. Distribution, taxonomy and nutritional values of seabuckthorn growing in dry temperate Himalayas. In: *Proceeding of International Workshop on Seabuckthorn*, p.52-59, Dec.12-17, 1995, Beijing, China, 206p.

46. Snedecor, G.W. and Cochran, W.G. 1967. *Statistical Methods* (8th Edn.), Oxford and IBH Publishing Company, Calcutta.

47. Suryakumar, G. And Gupta, A. 2012. Medicinal and therapeutic potential of seabuckthorn (*Hippophae rhamnoides* L.). *Journal of Ethnopharmacology* 138 (2011) 268– 278.

48. Thakur, S.S., Tomar, S.K. and Malik, R. 2006. Effect of feeding total mixed rations on the performance of buffalo calves. *Indian Journal of Animal Nutrition* 23 (1): 5-9.

49. Underwood, E.J. and Suttle, N.F. 1999. The mineral nutrition of livestock. *CABI Publishing*, New York, USA.

50. Wedekind, K.J., Collings, G., Hancock, J. and Titgemeyer, E. 1994. The bioavailability of zinc-methionine relative to zinc sulphate is affected by calcium level. *Poultry Science* 73 (suppl.1): 114.

51. Yang, B. and Kallio, H. 2006. Lipophilic components of seabuckthorn (*Hippophae rhamnoides* L.) seeds and berries. In: *Seabuckthorn (Hippophae L.)-A Multipurpose Wonder Plant. Vol. II: Biochemistry and Pharmacology* (V. Singh, Ed. in Chief, 2006), p.70-97, Daya Publishing House, New Delhi, 600p.

Chapter 47

Trials on Protective Agents against the Major Pests of Seabuckthorn

Lubov Shamanskaya

The Lisavenko Research Institute of Horticulture for Siberia, Barnaul, Russian Academy of Agricultural Sciences, Barnaul, Russia

SUMMARY

The present study presents the information about seabuckthorn flies (*Rhagoletis batava obscuriosa* Kol.) life cycles, and control measures by the chemical agent "Actellic" and organic agent "Phitoverm" against pests of seabuckthorn. This study reveals the techniques of the possible agent consumption dose reduction, providing the best plant growth, avoiding harvest losses and receiving toxic residues-free seabuckthorn produces.

Keywords: *Seabuckthorn, Pests, Seabuckthorn fly, Agent, Efficiency, Productivity, Profitability, and Toxicological assessment.*

INTRODUCTION

Seabuckthorn is the major garden plant in the industrial horticulture sphere in Siberia. High ecological adaptation, heavy fruiting and a wealth of nutrients in fruits promotes the cultivation of seabuckthorn throughout many countries. At the same time, together with the promotion of seabuckthorn to other regions of Russian Federation and abroad, the spread of pests with their habitat increases. Among the pests, seabuckthorn fly (*Rhagoletis batava obscuriosa* Kol.) is the most malicious problem.

It is very difficult to fight against this pest, which results in fall in production of seabuckthorn fruits. The search of safer ways for use against seabuckthorn fly affirmed promise usage of fungal agents like "Boverin" (Kalvish *et al.*, 1986) and preparations based on predatory nematodes (Berger and Danilov, 1998). However, the long term practice of using against seabuckthorn fly confirmed that the application of chemical pesticides reliably protects the harvest, but often entailing to shown consequences (Goreeva and Prokofiev, 1996; Shamanskaya, 2008).

Seabuckthorn (*Hippophae* L.): *A Multipurpose Wonder Plant, Vol. 4: Emerging Trends in Research and Technologies* (Virendra Singh, *Editor-in-Chief*, 2014), p. 576–583, Daya Publishing House, New Delhi, India.

High efficiency against this pest demonstrates biological agent "Fitoverm" developed in Scientific-biological center "Pharmbiomed" (Moscow), which provides the production of ecologically pure crops (Shamanskaya, 2010). However, the high price of "Fitoverm" is an obstacle for the active promotion of this kind of agents.

The aim of our further researches lies in the development of economically and ecologically viable measures of seabuckthorn protection against seabuckthorn fly.

MATERIALS AND METHODS

We conducted tests on agent against seabuckthorn fly in experiments on the small plots of the ground with agent once-and double-sprayings usage. Besides, the major pest seabuckthorn fly, other pests like brown tail seabuckthorn moth (*Gelechia hippophaeella* Schak.), seabuckthorn gall mite (*Aceria hippophaena* Nal.) and leaf mite (*Vasates tibialis* Liro.), also affected the growth and development of young plants.

We calculated expected yield loss of seabuckthorn plants and products (Shamanskaya, 2006). The plants were sprayed with the backpack sprayer "Kwazar" with drain of 0.5-0.6 litre per 1 tree. The controlled (non-treated) plants were not treated. The age of the plants was 3-4 years old (the first and the second year of fruiting). The total experimental plot area was 0.7 hectares square. The experiments were done in triplicate.

We determined the efficiency of the chemical against seabuckthorn fly by viewing under a microscope MBS-10 in 100 damaged fruits. Simultaneously, their related malicious objects were accounted.

Biochemical analyses of the fruit was carried out according to standard techniques in the laboratory of biochemistry of the Lisavenko Research Institute of Horticulture for Siberia, Toxicological analyses were conducted by the method of gas-liquid chromatography in the Central Scientific and Production Veterinary Radiological Laboratory (Barnaul city, Altai). The toxicity of the fruits have been done by biological agent "Fitoverm", at Scientific-Biological Center "Pharmbiomed" by the method of fluorescent high performance liquid chromatography.

RESULTS

The techniques required against pest were standardized on the basis of the knowledge of their vital life cycle stages. Seabuckthorn fly has complex life cycle and the basic phases of development are shown in Figure 47.1.

The pest overwinters in cask-shaped, straw yellow colour and false pupa in the soil 10 cm below surface under the canopy of damaged trees. The flights of adult insects start from 9-22 June, depending on the weather conditions in Altai territory. Departure of the flights takes place at the end of June, which was marked in the years of cold and prolonged spring. Flies breeding lasts for 2-3 weeks and the mass breeding lasts for about a week and usually occurs in the first decade of July. The flies feed on the juices of seabuckthorn fruits, piercing into their hull by ovipositor. Flights, mating and oviposition occur only in warm weather, but when rainy and cool weather occur adult insects formed, hides under lumps of soil, fallen leaves and grass. Females lay eggs under the hull of green fruits, placing them one by one. The fecundity of one female reaches to 200 eggs.

A larva hatching starts in the I-II weeks of July and ends in the II-III weeks of August. A larva feeds on the flesh of seabuckthorn fruit during 3 weeks, making long winding passages under the hull; fruits remain intact and become transparent. Usually, each fruit is devoured by one larva, sometimes

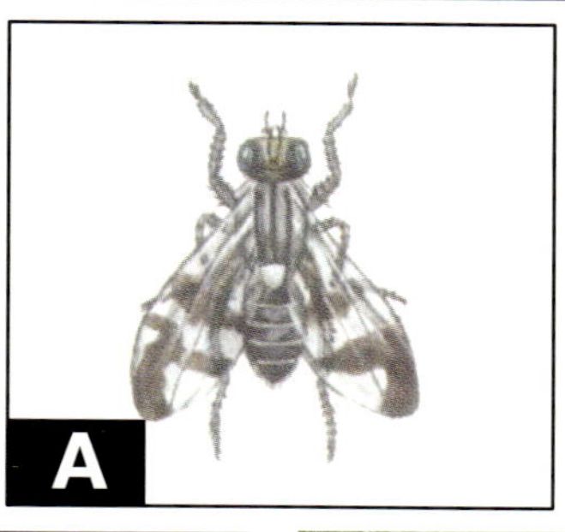

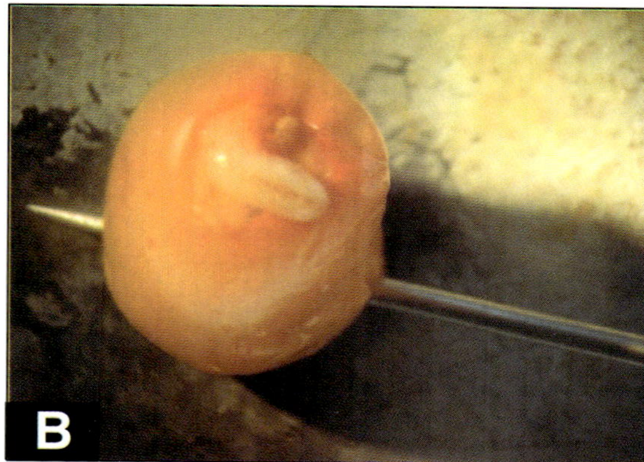

Figure 47.1
A: The adult insect; B: Larva; and C: False pupa.

two and rarely three. One individual larva damages 4-5 fruits. After feeding, larva comes out from fallen fruit and starts digging in the ground, and then forms a false pupa, in which it hibernates.

The initial indications of the fruit damage by seabuckthorn fly are orange spots, and small dents appearance on the green fruit hull. Then, the fruits are gradually deformed, become darker and dry, while remaining hung on the branches. The number of seabuckthorn flies varies considerably from year to year. In our experiments, protective measures were taken against damages caused at the larval phase. Preliminary studies have shown the possibility of reducing the consumption of chemical and biological agents by their complex usage together with biologically active supplement that have been developed by us. The effectiveness of this method for seabuckthorn protection was tested on a large scale with an expected crop loss in sort of a plant "Chuyskaya" 100 per cent in 2011 and 35 per cent in 2012.

A chemical agent "Actellic" and biological agent "Fitoverm" caused the death of seabuckthorn flies larvae at the level of 96.8-100.0 per cent in 2011 (Table 47.1).

The safety of the crop in the treatment of full-dose or low-dose of "Actellic" amounted to 100 per cent (Figure 47.2B).

The effectiveness against other related malicious pests was estimated from 95.4-100.0 per cent (brown tail seabuckthorn moth) to 69.1-100 per cent (mites). The most convincing acaricidal effect showed by agent "Fitoverm" usage in full- and half-dosages, which led to the mites death at the level

Figure 47.2: Effectiveness of the Treatment Against Seabuckthorn Flies. A-Control; and B-Treatment with "Actellic" (0.04 per cent), in combination with bioactive supplement.

Table 47.1: Effectiveness of Agents Studied against Seabuckthorn Flies Carried on 2011.07.10 and 2011.07.17 in Sort "Chuyskaya"

Variants of the Experience	*Number of Treatment*	*Death of Larvae (per cent)*	
		13.07	*26.07*
Control	–	5.6	2.4
Actellic-0.1 per cent	1	100.0	100.0
Actellic-0.04 per cent + bioactive supplement	1	93.5	100.0
Fitoverm (1 per cent e.c.)-0.3 per cent	2	89.4	96.8
Fitoverm (1 per cent e.c.)-0.15 per cent + bioactive additive	2	91.8	96.3
HCP_{05}	–	1.1	0.8

of 97.2-100.0 per cent . The agents application also affects the sprout recovery from the damage caused by brown tail seabuckthorn moth caterpillars (Figure 47.3), as well as for shoots and leaves set free from the mites (Figure 47.4).

Figure 47.3: The Effectiveness of the Treatment Against *Gelechia hippophaella*. A-Control; and B-Treatment with "Actellic" (0.04 per cent), in combination with bioactive supplement.

Figure 47.4: The Effectiveness of the Treatment Against Gall Mites. A-Control; and B-Treatment with "Fitoverm" (0.15 per cent), in combination with bioactive supplement.

The best additional plant growth was achieved by usage of the agents with combination of bioactive supplement during experiments held in vegetation period. The value of total increment in that moment reached to 21.8-23.3 m, as compared to length of 20.5-20.7 m, using the total dose of agents but without the supplement. The increment in control (non-treated) plants growth per 1 plant was 18.1 m.

In 2011, the first year of seabuckthorn fruiting was failed, because plenty of berries were consumed by birds. In 2012, the treatment of plants was carried out according to the same scheme (as in 2011) and similar results in effectiveness have received.

It was estimated that the crop yield in controlled (non-treated) plants had been found to be lowest (1.8 tones/hectare). The crop protection from seabuckthorn fly by chemical agent "Actellic" and biological agent "Fitoverm" in full dosage, and reduced dosage provided a yield increase to 3.2-4.5 tones/hectare (Table 47.2).

Biologically active supplement contributed to the increase in weight of the fruit in the drought year of 2012 from 53.1-54.5 g against 51.0-51.4 g of each fruit when using agents in the full dosage and 50.1 g in control plants.

Moreover, the biologically active supplement in combination with "Fitoverm" contributed to the improvement of contents of carotenoids and oils. There was an increase in content of carotenoids, oil, dry soluble substances, sugars and vitamin C after treatment with "Actellic" in combination with bioactive supplement.

Table 47.2: Yield and Weight of the Seabuckthorn Fruits in different Techniques of Treatment in Sort "Chuyskaya" in 2008

Variants of the Experience	*Number of Treatments*	*Weight of 100 Fruits (g)*	*Yield of Crop*		*Increase to Control (per cent)*
			kg/bush	*t/ha*	
Control	–	50.1	1.5	1.8	–
Actellic-0.1 per cent (standard)	1	51.0	3.2	3.9	116
Actellic-0.04 per cent + bioactive additive	1	53.1	3.6	4.5	150
Fitoverm (1 per cent e.c.)- 0.3 per cent	2	51.4	2.6	3.2	77
Fitoverm (1 per cent e.c.)-0.15 per cent + bioactive additive	2	54.5	3.5	4.3	138
HCP_{05}	–	0.7	0.8	0.98	–

The results of the toxicological assessment of crops showed an absence of toxic residues in the use of low consumption dose of "Actellic" in combination with biologically active supplement as well as in the application by "Fitoverm" in full- and half-dose.

DISCUSSION

Protection of seabuckthorn by chemical agent "Actellic" at a single spraying and biological agent "Fitoverm" with the double treatments at full and reduced dosages provided a high efficiency in not only against the seabuckthorn fly, but also against seabuckthorn moth and mite, activity of which restrain the plants growth and development. The best growth of the plants treated with techniques in 2011 in increasing the productivity of seabuckthorn in 2012. Maximum increase of the crop (138-150 per cent) was obtained by treatment with agents "Fitoverm" and "Actellic" at low dosages in combination with biologically active supplement. That results of using supplement indicate of its appropriate additional, favourable effect on the growth and protection of plants. These treatment techniques provide obtaining produces of better quality than other.

The use of the agent "Actellic" at reduced dosages in combination with biologically active supplement provides a fruit without toxical residues, while the full dosage of "Actellic" usage gives the toxicity of the fruit at 0.043 mg/kg. Biological agent "Fitoverm" in full-and half-dosages allows receiving ecologically pure products.

The calculations of economic efficiency showed not high profitability level of production at 33.0-72.3 per cent . The higher cost of labour for manual harvesting and incomplete stage of the plants fruiting restricted more high levels of profitability.

The highest profitability of seabuckthorn fruits production (72.3 per cent) was obtained techniques of using low-dosages agent "Actellic" in combination with supplement. When having used expensive agent "Fitoverm" at half dosage in combination with the supplement, profitability of the fruit production was reached to 62.5 per cent . When having used agent "Fitoverm" in full dosage, profitability of fruit production was below the target (controlled plantation). Consequently, application of the biological agent "Fitoverm" for the protection of seabuckthorn against seabuckthorn fly is economically justified only in the reduction of its consumption doubled on the base of integrated complex usage with bioactive supplement.

CONCLUSION

Environmentally, ecologically and economically justified protection of seabuckthorn against the most malicious pest seabuckthorn fly and other related species: Brown tail seabuckthorn moth, seabuckthorn gall mite and leaf mite by using chemical agent "Actellic" and biological agent "Fitoverm" at low dosages in combination with biologically active supplement, which have been developed by us. These techniques increase the crop yield by 138-150 per cent and profitability by 62.5-72.3 per cent . Biologically active supplement promotes better and safe plants growth, increase the weight of the fruits, and improve their quality and producing toxic residues-free seabuckthorn produces.

REFERENCES

1. Berger, L. and Danilov, L. 1998. Prospects of application of entomo-pathogenic nematodes against seabuckthorn fly. In: *Report of 3th International Symposium on Seabuckhorn*. p. 94-95, Ulan-Ude.
2. Goreeva, T. and Prokofiev, M. 1996. The problems of protection of plants against pests and diseases in the gardens of Siberia. In: *Scientific Bases of Horticulture in Siberia*. p.111-114, Novosibirsk.
3. Kalvish, T., Sanzhimiputova, R., Sharapov, V. and Khodyrev, V.P. 1986. In: *Microbiological means of Protection of Plants*. p. 110-113, Novosibirsk.
4. Shamanskaya, L. 2006. Phytosanitary optimization of fruit and berry plantations in the conditions of the Altai region. In: *Monograph*. Barnaul, Russia, 235 p.
5. Shamanskaya, L. 2008. Perspective directions in the protection of seabuckthorn plantations from seabuckthorn fly (*Rhagoletis bataba obscuriosa* Backdoor. Kol.). In: *Modern Trends of Development of Industrial Horticulture, Materials of the International Scientifically-Practical Conference*, 75th Anniversary of the Lisavenko Research Institute of Horticulture for Siberia, p. 332-338, Barnaul, Russia.
6. Shamanskaya, L. 2010. The biological method of protection of seabuckthorn from seabuckthorn fly (*Rhagoletis batava obscuriosa* Backdoor. kol.). *Achievements of Science and Technology of the Agroindustrial Complex* 6: 48-50.

Chapter 48

Development of Seabuckthorn Fruit Harvesting Tools and Devices

D.K. Vatsa and Virendra Singh*

Department of Agricultural Engineering,
*Department of Biology and Environmental Sciences,
CSK Himachal Pradesh Agricultural University, Palampur – 176 062, H.P., India

SUMMARY

Seabuckthorn fruit harvesting tools and devices were developed under the NAIP Scheme of Indian Council of Agricultural Research by CSK Himachal Pradesh Agricultural University, Palampur. Seabuckthorn harvesting comb, shaker, power run branch shaker and handle with clip were standardized for the two India seabuckthorn species, *i.e. Hippophae rhamnoides* ssp. *turkestanica* and *H. salicifolia* growing in Lahaul-Spiti district of cold desert area of Himachal Pradesh. The fruit harvesting by tools and devices varied from 4.5 to 6.6 kg fruits/h, as compared to 2.8 kg/h by traditional method of hitting by a stick. Among all developed harvesters, the branch shaker and wire clip with handle were found more productive. The wire clip with handle is being provided to the farmers.

Keywords: *Seabuckthorn fruit harvesting tools and devices, Hippophae rhamnoides ssp. turkestanica and H. salicifolia, Cold desert area and Himachal Pradesh.*

INTRODUCTION

The seabuckthorn (*Hippophae rhamnoides* L.) belongs to the family *Elaeagnaceae*. The seabuckthorn is a medium deciduous shrub 2 to 4 m in height. The trees are dioecious meaning there are male and female plants. To ensure adequate pollination and suitable numbers of female trees, it is necessary for the orchard to have between 7 and 12 per cent of the trees as pollinators. The plants have tremendous potential for functional food and nutraceutical uses. The berries have very high vitamin C and antioxidant properties. Only female plants produce fruit. Flowers are small, yellow and appear before the leaves being produced on 3-year old wood. Seabuckthorn plants are very hardy and can withstand

Seabuckthorn (*Hippophae* L.): *A Multipurpose Wonder Plant, Vol. 4: Emerging Trends in Research and Technologies* (Virendra Singh, *Editor-in-Chief*, 2014), p. 584–591, Daya Publishing House, New Delhi, India.

Table 49.3: Technical Terms Adopted to Describe the Seed Surface Patterns

Term	*Explanation*
Blister	Having a expansible convex, each area outlined by line or curve being an interspace
Foveolatae	Marked with small shallow pits
Reticulatae	Having a raised network of narrow and sharply angled lines frequently presenting a geometric appreance, each area outlined by a reticulum being an interspace
Striatae	Marked with a series of fine narrow parallel bands wider than the lines of a lineate surface
Striate-blister	A type intermediate between striate and pimple types

Source: Stearn (1992); Font Quer (1993); Koul *et al.* (2000); Zeng *et al.* (2004).

RESULTS AND DISCUSSION

In the present study, based on the main external seabuckthorn seed morphology, seeds were traditionally distinguished into two groups: Non-flattened and flattened seeds. Flattened seeds were obtained from Urgup region (Figure 49.2), and non-flattened seeds from Ilgaz, Sivas and Trabzon regions (Figures 49.3–49.6). The observations made using scanning electron microscopy indicated that there are different types of seed surface sculpturing of *Hippophae rhamnoides* ssp. *caucasica* collected from different localities in Turkey.

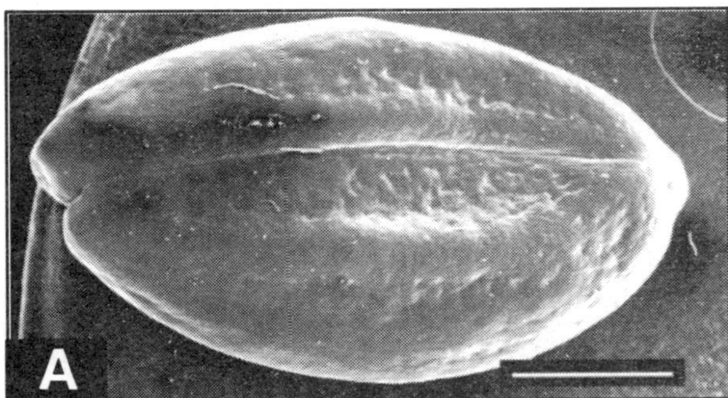

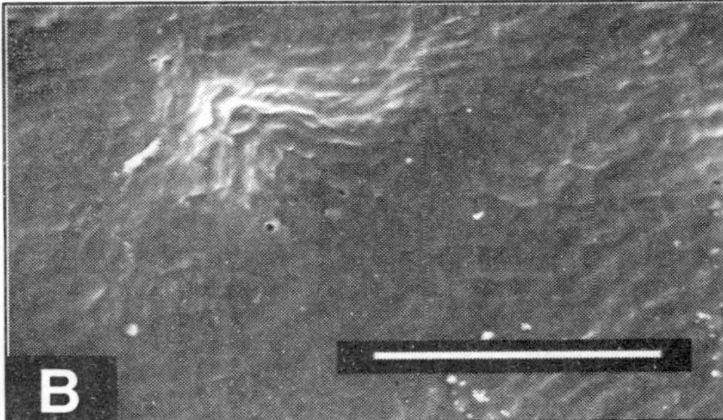

Figure 49.2: Complete Seed of Scale Bar 1 mm (A) and Seed Surface of Scale Bar is 100 µm (B) from Urgup Locality Showing Simple-Reticulate Seed Surface (Type I).

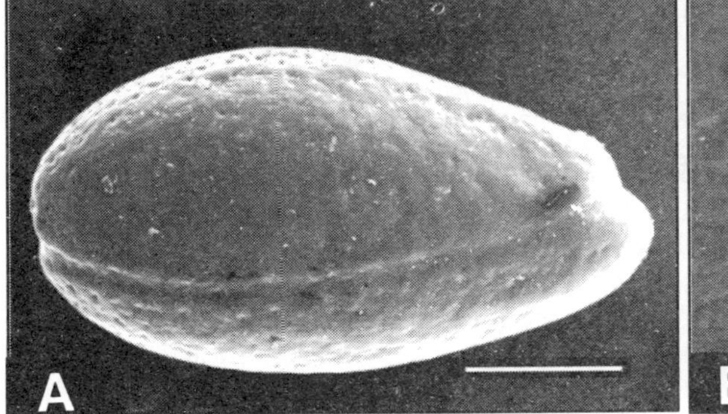

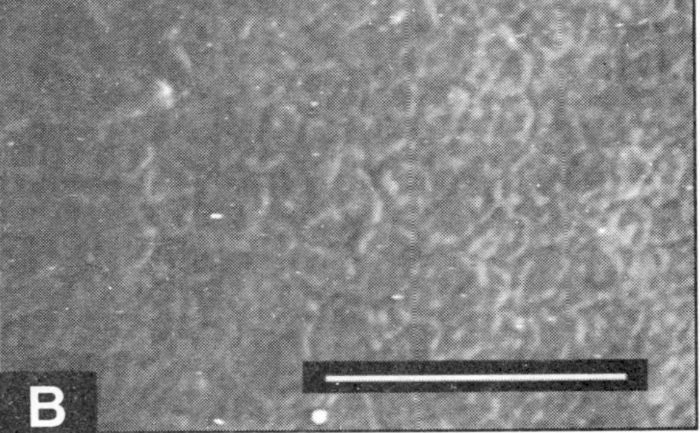

Figure 49.3: Complete Seed of Scale Bar 1 mm (A) and Seed Surface of Scale Bar is 100 µm (B) from Ilgaz Locality Showing Simple-Foveolate Seed Surface (Type II).

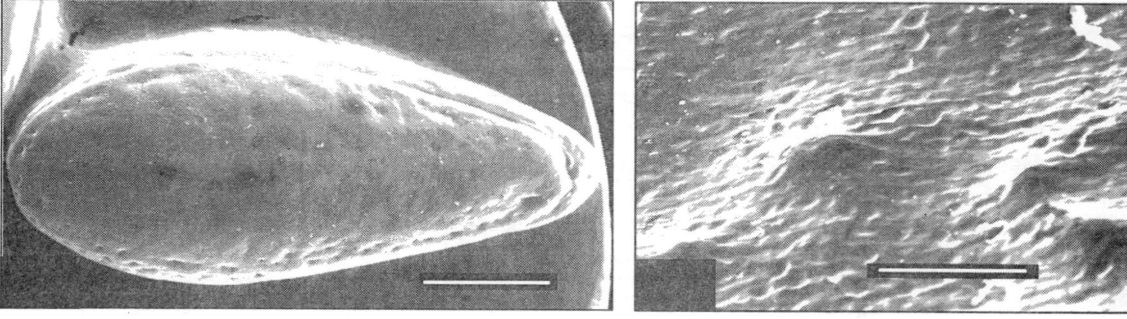

Figure 49.4: Complete Seed of Scale Bar 1 mm (A) and Seed Surface of Scale Bar is 100 µm (B) from Sivas (2) Locality Showing Striate-Blister Seed Surface (Type III).

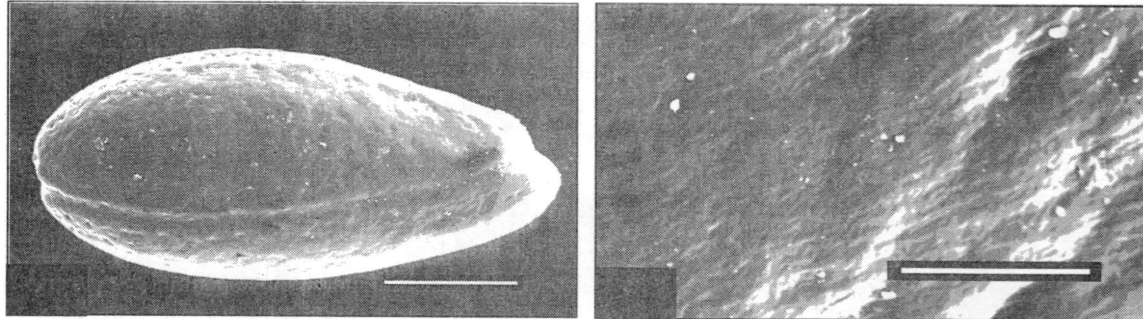

Figure 49.5: Complete Seed of Scale Bar 1 mm (A) and Seed Surface of Scale Bar is 100 µm (B) from Sivas (1) Locality Showing Blister Seed Surface (Type IV).

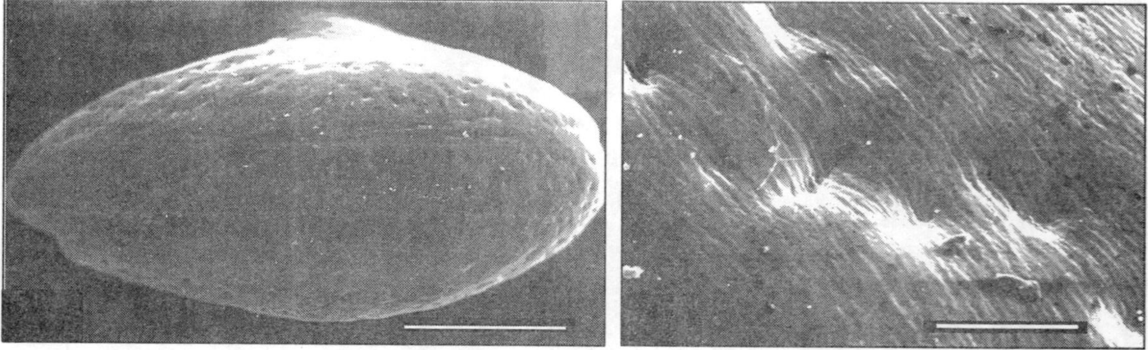

Figure 49.6: Complete Seed of Scale Bar 1 mm (A) and Seed Surface of Scale Bar is 100 µm (B) from Trabzon Locality Showing Blister Seed Surface (Type IV).

The variations observed in the seed coat pattern were specimen-specific. Four types of seed coat pattern were observed. Seed surface features of *H. rhamnoides* ssp. *caucasica*, growing in different localities in Turkey showed variability: Reticulatae, foveolate, striatae and striatae-pimple. Seed surface structure was discerned as striatae both of Sivas (1) and Trabzon, stiatae-blister in Sivas (2), foveolatae in Ilgaz and reticulatae in Urgup specimens. As a result, seed surface features of *H. rhamnoides* ssp. *caucasica*, growing in different localities in Turkey showed variability.

H. rhamnoides, which has a wide distribution, shows many differences in its morphological and anatomical structure due to adaptation within a broad ecological environment, and some morphological features can be very changeable. Rousi (1971) reported that some Turkish specimens have a certain resemblance to ssp. *turkestanica* and this resemblance probably results from adaptation to aridity.

All the previous studies (Aras-Tayhan 1995a; 1995b; 1997; Aras *et al.*, 2007) and present one on seed surface with SEM, showed that there is a taxonomic problem in *H. rhamnoides* ssp. *caucasica*, which known as one subspecies in Turkey. In this study, when climate types of these sites are considered, it is seen that seed surface features are not affected by climatic conditions.

Barthlott (1984) concluded that surface features of seed coats are little affected by the environmental conditions in which the plant grows and that differences in seed coats characters always reflect genetic-phylogenetic differences in the plants concerned. Also, results of many studies showed that seed surface is one of the most important features used in plant taxonomy (Heywood, 1971; Bouman, 1984; Foroughbakhch *et al.*, 2000; Buss *et al.*, 2001; Moro *et al.*, 2001; Fagundez and Izco, 2003a; 2003b; Zeng *et al.*, 2004).

ACKNOWLEDGEMENT

The present work was supported by the Research Fund of Istanbul University (Project No. UDP-580/12072005) and presented in the 2nd International Seabuckthorn Association Conference, August 26-29, 2005, Beijing, China.

REFERENCES

1. Aras, A., Akkemik, U. and Kaya, Z. 2007. *Hippophae rhamnoides* L.: Fruit and seed morphology and its taxonomic problems in Turkey. *Pakistan Journal of Botany* 39: 1907-1916.
2. Aras Tayhan, A. 1995a. Turkiye'nin *Hippophae rhamnoides* ssp. *caucasica* Rousi Polenleri. *Ulusal Palinoloji Kongresi, Aralýk 1995, I.U. Orman Fak, Proceedings*, p. 84-92, Aralýk, Istanbul.
3. Aras Tayhan, A. 1995b. *Hippophae rhamnoides* ssp. *caucasica* Rousi Tohumlarýnýn Morfolojisi. *I. U. Orm. Fak. Der.* 45 (A-1): 97-106.
4. Aras Tayhan, A. 1997. *Turkiye'nin Hippophae rhamnoides L.Taksonu Uzerinde Ksilolojik ve Palinolojik Arastýrmalar*. Ph.D. Thesis (Unpublished). Istanbul University, Turkey.
5. Barthlott, W. 1984. Microstructural features of seed surface. In: *Current concepts in Plant Taxonomy* (V.H. Heywood and D.M. Moore, Eds.), p. 95-105, Academic Press, London, 488p.
6. Bartish, I.V., Jeppsson, N. and Bartish, G.I. 2000. Inter-and intraspecific genetic variation in *Hippophae* (*Elaeagnaceae*) investigated by RAPD markers. *Plant Syst. Evol*. 225 (1-4): 85-101.
7. Bottema, S., Woldring, H. and Aytug, B. 1995. Late quaternary vegetation history of northern Turkey. *Palaeohistoria* 66: 13-72.
8. Bouman, F. 1984. The ovule. In: *Embryology of Angiosperms* (B.M. Johri, Ed.), p. 123-157, Springer-Verlag, Berlin, Germany.
9. Buss, C.C., Lammers, T.G. and Wise, R.R. 2001. Seed coat morphology and its systematic implication in *Cyanea* and other genera of *Lobelioideae* (*Campanulaceae*). *Am. J. Bot.* 88: 1301-1308.
10. Fagundez, J. and Izco, J. 2003a. Seed morphology of *Erica* L. sect. *Callicodon* Bentham. *Plant biosystems* 137: 111-116.

11. Fagundez, J. and Izco, J. 2003b. Seed morphology of *Erica* L. sect. *Chlorocodon* Bentham. *Acta Bot. Gallica* 150 (4): 401-410.

12. Font Quer, P. 1993. *Diccionario de Botanica*, Barcelona, Labor.

13. Foroughbakhch, R., Rodriguez, M., Maiti, R.K., Hernandez-Pinero, J.L., Hauad, L.A. and Cespedes, A.E. 2000. Seed morphology and ultrastructure of eight native shrubs and trees of northeastern Mexico. *Phyton-International Journal of Experimental Botany* 66: 25-31.

14. Gopinathan, M.C. and Babu, C.R. 1985. Structural diversity and its adaptive significance in seeds of *Vigna minima* (Roxb.) Ohwi and ohashi and its allies (*Leguminosae-Papilionoideae*). *Annals of Botany* 56: 723-732.

15. Heywood, V.H. 1971. *Flowering Plants of the World*. Oxford University Press, Oxford, London, 336p.

16. Koul, K.K., Nagpal, R. and Raina, S.N. 2000. Seed coat microsculpturing in *Brassica* and allied genera (subtribes *Brassicinae, raphaninae, Moricandiiae*). *Annals of Botany* 86: 385-397.

17. Lersten, N.R. 1979. A distinctive seed coat pattern in *Viceaceae* (*Papilionoideae, Leguminosae*). *Proceeding of Science* 86: 102-104.

18. Lian, Y. 1988. New discoveries of the genus *Hippophae* L. *Acta Phytotaxonomica Sinica* 26: 235-237.

19. Lian, Y.S., Chen, X.L., Sun, K. and Ma, R. 2003a. A new subspecies of *Hippophae* (*Elaeagnaceae*) from China. *Novon* 13 (2): 200-202.

20. Lian, Y.S., Chen, X.L. and Sun, K. 2003b. Clarification of the systematic position of *Hippophae goniocarpa* (*Elaeagnaceae*). *Bot. J. Linn Soc.* 142 (4): 425-430.

21. Liu, S.W. and He, T.N. 1978. The genus *Hippophae* from Qing-Zang Plateau. *Acta Phytotaxonomica Sinica* 16: 106-108. (In Chinese)

22. Mc Kean, D.R. 1982. *Flora of Turkey and the East Aegean Islands*. *Vol. 7*. (P.H. Davis, Ed.), p. 532-534, Edinburg, 947p.

23. Moro, F.V., Pintno, A.C.R., Dos Santos, J.M., and Damiao Filho, C.F. 2001. A scanning electron microscopy study of the seed and post-seminal development in *Angelonia salicariifolia* Bonpl. (*Scrophulariaceae*). *Annals of Botany* 88: 499-506.

24. Pan, R., Zhang, Z., Ma, Y., Sun, Z. and Deng, B. 1989. The distribution characters of seabuckthorn (*H. rhamnoides* L.) and its research progress in China. In: *Proceedings of the 1st International Symposium on Seabuckthorn* (*H. rhamnoides L.*), p.1-16, Xian, China.

25. Rejdali, M. 1990. Seed morphology and taxonomy of the North African species of *Sideritis* L. (*Lamiaceae*). *Botanical Journal of the Linnean Society* 103: 317-324.

26. Rousi, A. 1971. The genus *Hippophae* L.: A taxonomic study. *Ann. Bot. Fenn.* 8: 177-227.

27. Stearn, W.T. 1983. *Botanical Latin: History, Grammar, Syntax and Vocabulary* (3rd Ed.). David and Charles, London, 546p.

28. Stearn, W.T. 1992. *Botanical Latin: History, Grammar, Syntax, Terminology and Vocabulary* (4th Ed.) Newton Abbot: David and Charles, London, 546p.

29. Sun, K., Chen, X. and Ma, R. 2002. Molecular phylogenetics of *Hippophae* L. (*Elaeagnaceae*) based on the internal transcribed spacer (ITS) sequences of nrDNA. *Plant Syst. Evol.* 235 (1-4): 121-134.

30. Sun, K., Ma, R.J. and Chen, X.L. 2003. Hybrid origin of the diploid species *Hippophae goniocarpa* evidenced by the internal transcribed spacers (ITS) of nuclear rDNA. *Belg. J. Bot.* 136 (1): 91-96.

31. Zeng, C.H.L., Wang, J.B., Liu, A.H. and Wu, X.M. 2004. Seed coat microsculpturing changes during seed development in diploid and amphidiploid *Brassica* species. *Ann. Bot.* 93: 555-566.

Chapter 50

Pollen Morphology of Seabuckthorn (*Hippophae rhamnoides* L. subsp. *caucasica* Rousi) in Turkey

Aliye Aras Perk

Istanbul University, Faculty of Science, Botany Department, Istanbul-Turkey

SUMMARY

Pollen morphology of *Hippophae rhamnoides* from different locations in Turkey (Sivas, Trabzon, Ilgaz and Istanbul) has been examined with light microscope. The pollen grains were spheroidal and tricolporate and reticulate sculpture. The size of the pollen grains varied, ranging from 24.6 to 25.6 µm on the polar axis and from 24.62 to 26.24 µm on the equatorial axis. The results of the present study revealed that certain morphological characterstics of pollen grains from Trabzon region exhibited great variations as compared to other regions. According to current literature, only one taxon of *Hippophae i.e.* only *Hippophae rhamnoides* ssp. *caucasica* Rousi is found in Turkey. However, the results indicated that pollens from Trabzon region might be hybrid, suggesting a different taxon from *Hippophae rhamnoides* ssp. *caucasica* Rousi in Trabzon region. This taxon might be a sub-species or a separate variety.

Keywords: *Hippophae rhamnoides, Pollen morphology, Light microscopy and Turkey.*

INTRODUCTION

The genus *Hippophae*, a member of family *Elaeagnaceae*, has 7 species and 9 subspecies (Sun *et al.*, 2003). *Hippoppae rhamnoides* ssp. *caucasica* Rousi is the only known species growing in Turkey. This plant has been widely distributed all over Turkey in various floristic regions with different climatic conditions up to 2000 m asl. Fossil pollen records indicated that *Hippophae rhamnoides* was widespread during late quaternary in Anatolia (Bottema *et al.*, 1995), suggesting that *Hippophae* L. is the native plant of Anatolia.

Besides many molecular, taxonomic and phylogenetic studies (Liu and He, 1978; Lian, 1988; Yu *et al.*, 1989; Yao and Tigerstedt, 1994; Bartish *et al.*, 1999; 2000a; 2000b; Sun *et al.*, 2002; Lian *et al.*,

Seabuckthorn (*Hippophae* L.): *A Multipurpose Wonder Plant, Vol. 4: Emerging Trends in Research and Technologies* (Virendra Singh, *Editor-in-Chief*, 2014), p. 600–610, Daya Publishing House, New Delhi, India.

2003a; 2003b; Sun *et al.*, 2003) and investigations related to its agricultural, nutritional, medicinal and ornamental values (Kallio *et al.*, 2002; Narayanan *et al.*, 2005; Cupara *et al.*, 2011), few studies on pollen morphology of *Hippophae rhamnoides* L. were also carried out.

Erdtman (1952) reported that pollen grains from Switzerland were suboblate (24x29 µm); pouting at equatorial part of colpi, whereas grains from Sweden are oblate spheroidal (27x28 µm), occasionally 4-ruporate or irregular. Grains were spheroidal to suboblate, tricolpate, with narrow, tapering furrows, each with a single germ pore. Exine with reticulate texture, at germ pores, the ektexine is somewhat raised above the general surface of the grain. Hyde and Adams (1958) provided pollen grain pictures taken under LM in atlas of airborne pollen grains, and mentioned exine with thin reticulate texture.

Pollen morphology of *Hippophae* was also studied by Sorsa (1971) based on herbarium material supplied by Rousi (1971). The study was carried out on pollen grains of *Hippophae rhamnoides* ssp. *caucasica* Rousi collected from Turkey (Agri and Trabzon) and Iran. Pollen grains collected from Iran were smaller in size and having thicker exine as compared to those from Turkey (Sorsa, 1971).

Very little information is available on pollen morphology in this genus. Therefore, the aim of the present study was to provide detailed morphological and structural descriptions of pollens of *Hippophae rhamnoides* ssp. *caucasica* Rousi from Turkey using light microscope (LM).

MATERIALS AND METHODS

The samples of this study were collected from Sivas, Trabzon, Istanbul, and Ilgaz in Turkey. Information about the sampling sites and their locations are given in Table 50.1 and Figure 50.1.

Table 50.1: Some Characteristics of the Sampling Sites in Turkey

Sampling Sites	*Latitude/Longitude*	*Altitude (m asl)*	*Sampling Date*
Trabzon			
Esiroglu Beldesi Degirmendere basin	40°52'N 39°45'E	75	1995
Sivas			
Sincan stream	39°54'N 37°59'E	2000	1995
Ilgaz			
Bolu-Kastamonu	40°56'N 33°36'E	1000	1996
Istanbul			
I.U. Forestry Faculty Garden	41°09'N 29°03'E	129	1995

Pollen grains of the genus *Hippophae* were studied using LM. Pollen grains were first treated with 70 per cent alcohol to remove oily substances and then embedded in glycerine jelly stained with basic fuchsin, following the method of Wodehouse (1935). Additionally, the pollen slides were prepared by the acetolysis method as suggested by Erdtman (1969). Sculpture and structure of exine were determined by the method of L-O analysis.

All measurements were based on 30 pollen grains for each specimen. The polar and equatorial axis, colpus length (clg), colpus width (clt), pore length (plg), pore width (plt), apocolpium (t), exine,

Figure 50.1: Map of the Sampling Sites in Turkey.

intine, ectexine and endexine thickness, number of apertures were measured on the pollen grains (Figure 50.2). Also, the shape classification based on P/E ratios was taken following the method of Erdtman (1969).

The terminology used is in accordance with Erdtman (1952) and Faegri and Iverson (1975). The pollen slides were deposited in the Botany Department of Istanbul University in Turkey.

RESULTS AND DISCUSSION

The main palynological features of the species examined in this study are summarized in Table 50.2 and shown in Figures 50.3–50.7. Based on LM observations, pollen type was tricolporate, pollen shape was spheroidal/suboblate P/E: 0.96-0.98. The ornamentation (sculpture) of the pollen was reticulate.

Reticules were dense, fine, slightly distinctive, formless and smaller around the pores. Generally, pollen grains from all the four regions were of the same size (Table 50.2). Polar axes (P) of pollen grains from all the regions were larger than equatorial diameter (E), while polar axes (P) of pollen grains from Istanbul were slightly smaller than equatorial diameter (E). Broadly, the shape of the pollen was spheroidal for all the regions studied. However, pollens from Sivas region were spheroidal (1.04), similar to subprolate; pollens from Istanbul, Ilgaz and Trabzon regions were spheroidal (0.96 and 0.98), similar to suboblate.

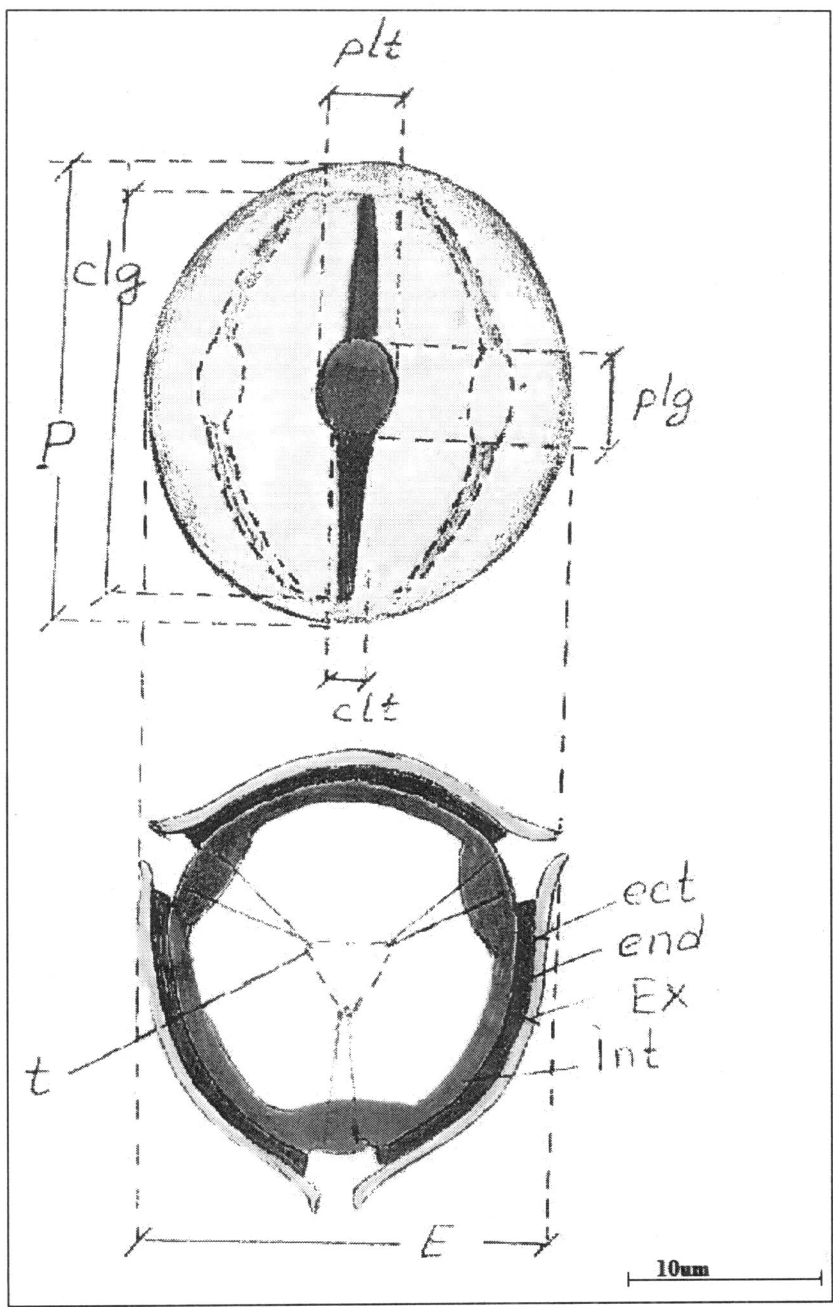

Figure 50.2: Measured Features of the Pollen Grains.

P: Polar axis; E: Equatorial axis; clg: Colpus length; clt: Colpus width; plg: Pore length; plt: Pore width; t: Apocolpium; ect: Ectexine; ent: Endexine; Ex: Exine; Int: Intine.

Table 50.2: Morphological Parameters of Pollen Grains of *Hippophae rhamnoides* ssp. *caucasica* Rousi

Parameters	*Sampling Sites*									
	Sivas		*Istanbul*		*Trabzon*		*Ilgaz*		*Agri-Trabzon (Sorsa, 1971)*	
	M	*Var.*	*M*	*Var.*	*M*	*Var.*	*M*	*Var.*	*M*	*Var.*
Polar axes (P)	25.57	24.30-29.16	24.67	21.06-27.54	25.43	19.44-29.16	25.69	21.06-30.78	26.1	25-28
Equatorial diameter (E)	24.62	22.60-27.54	25.69	21.06-27.54	26.03	17.82-32.40	26.24	22.68-32.40	24.3	23-27
P/E	1.04	–	0.96	–	0.98	–	0.98	–		
Ex	1.51	1.07-1.62	1.47	1.22-1.62	1.62	1.22-2.03	1.52	1.30-1.94	1.7	1.3-2.0
Ect/end	½	–	1/2	–	1/2	–	1/1.22	–	1/1	–
Ex/Int	1/1	–	1/1	–	1.5/1	–	1/1.6	–	–	–
clg	22.36	21.06-22.68	22.03	19.44-24.30	18.79	16.20-24.30	32.88	25.92-37.26	–	–
clt	0.89	0.53-1.07	1.34	0.81-1.62	1.04	0.81-1.62	1.28	0.81-2.03	–	–
plg	4.92	4.05-5.67	4.94	4.05-6.48	5.67	4.86-7.29	4.29	3.24-5.67	1.9	–
plt	4.54	4.05-5.67	4.37	4.05-5.67	5.35	4.05-8.10	4.71	4.05-4.86	4.7	–
t	5.62	4.05-8.10	5.31	4.86-6.48	7.13	4.86-8.10	2.51	1.62-3.24	5.3	5-6

M: Mean; Var.: Variation.

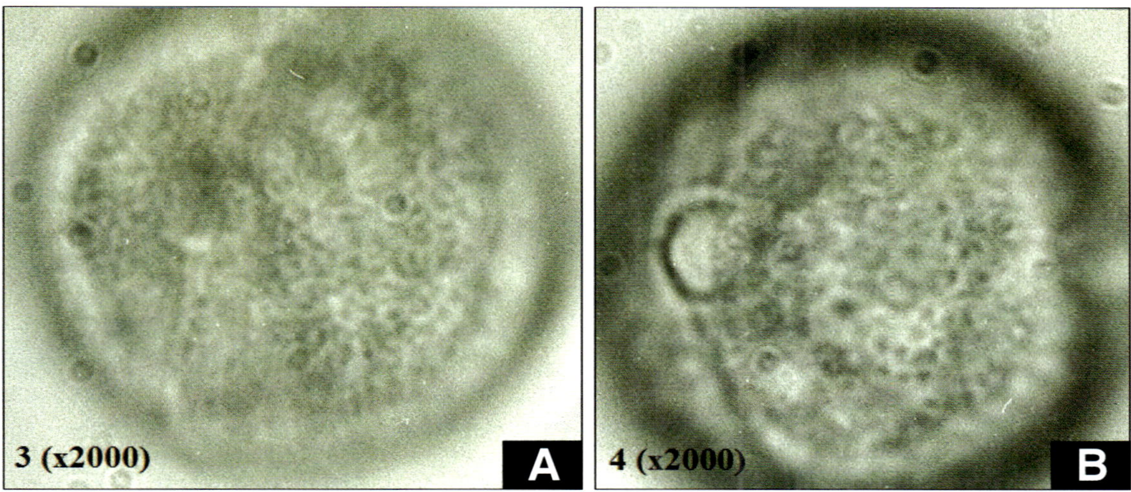

Figure 50.3: Reticulate Exine Surface from Ilgaz (A) and Istanbul (B) Samples.

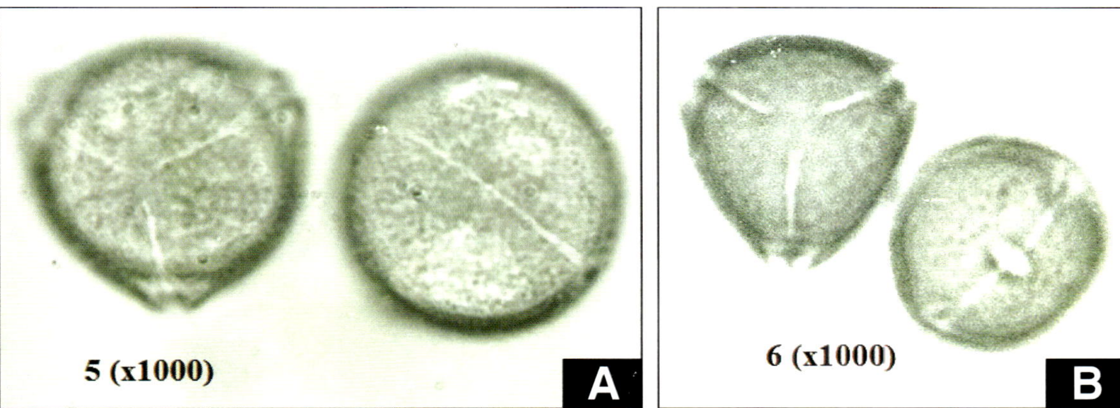

Figure 50.4: Apocolpium and Long Colpus from Ilgaz Samples (A) and Apocolpium and Short Colpus from Istanbul Samples (Acetolyzed) (B).

Thickness of exine of the pollen grains was also almost similar from all four regions. The ratio of ectexine thickness to endexine thickness was 1/2 for the pollens collected from Sivas, Istanbul and Trabzon, whereas it was 1/1.22 for pollen samples from Ilgaz. However, intine thickness under apertures was three fold to intine thickness within apertures. Colpus length was 22 µm each for Sivas and Istanbul pollens and 18.79 µm and 32.88 µm for Trabzon and Ilgaz pollens, respectively. Values of colpus width varied in pollens from different locations. Thus, 0.89, 1.34, 1.04 and 1.28 µm colpus width was observed in pollens from Sivas, Istanbul, Trabzon and Ilgaz regions. Length and width of pores were also nearly similar for Sivas, Ilgaz and Istanbul pollens, however these were larger in pollens from Trabzon region. The length of polar triangle (t) was found to be 5.62 µm and 5.31 µm for Sivas and Istanbul pollens, respectively, whereas it was much higher (7.13 µm) for Trabzon pollens and lower (2.51 µm) for Ilgaz pollens.

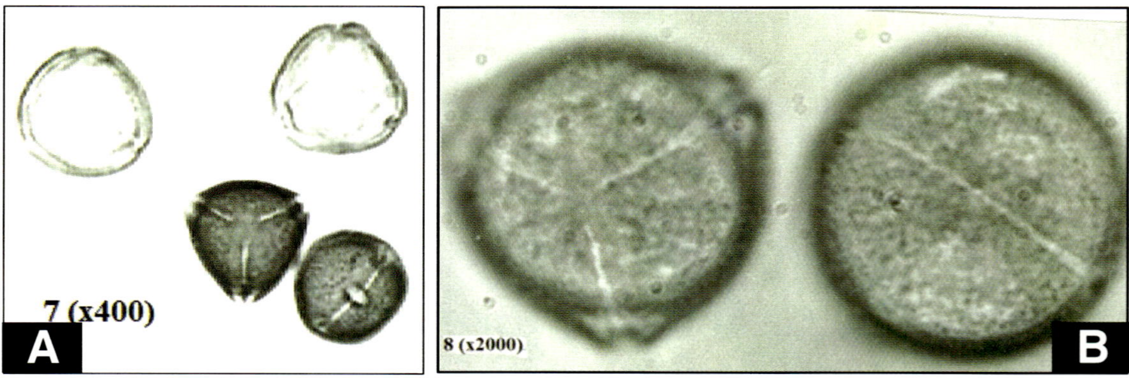

Figure 50.5: Acetolyzed Pollen Grains from Istanbul Samples (A) and Polar View, Ornamentation and Apocolpium from Ilgaz Samples (B).

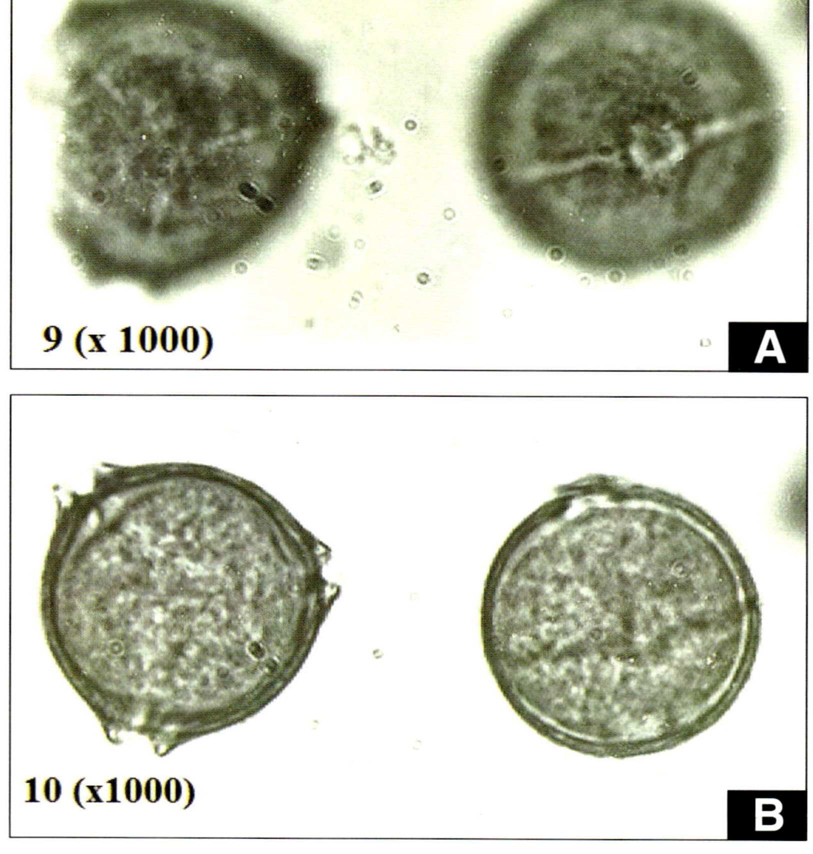

Figure 50.6: Pollen Grains from Sivas Samples (A and B).

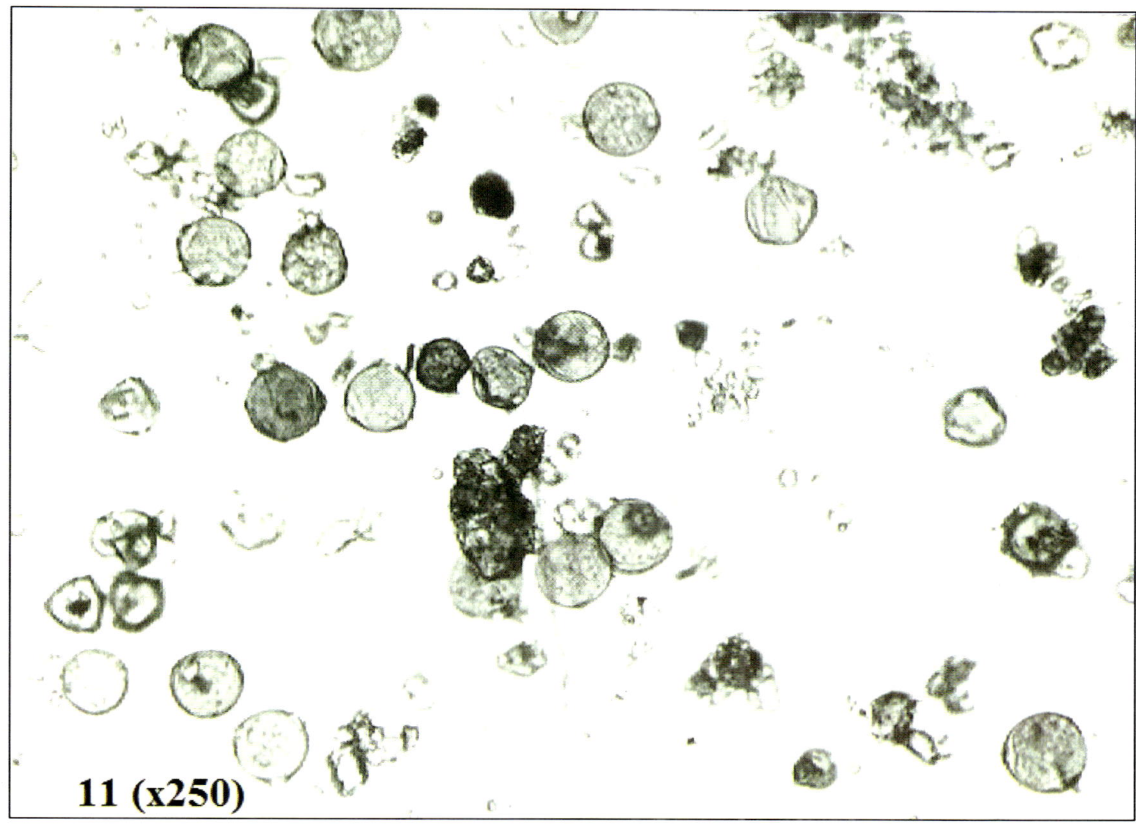

Figure 50.7: Pollen Grains from Trabzon Samples.

The results clearly indicated that the length of colpus (clg) and the size of polar triangle (t) were different in the pollens from Trabzon and Ilgaz regions as compared to Siva and Istanbul regions (Table 50.2).

These two features showed that pollens from Trabzon and Ilgaz regions were much different from the pollens of the other regions, and were tricolporate similar to syncolporatae (Figure 50.5).

Features of pollens from the studied regions showed that the pollens from Trabzon have distinctive characteristics in some of the morphological parameters (Table 50.2; Figures 50.8 and 50.9); P, E, clg, plg, plt and t values of Trabzon region were within vast variation limits, as compared to the pollens from the other regions. Exine ornamentation of the pollens from this region was also different. These features imply that these pollens might be hybrid.

The detailed work on *Hippophae rhamnoides* ssp. *caucasica* Rousi was only done by Sorsa (1971). The present study has been compared with that of Sorsa (1971) (Table 50.2). According to Sorsa (1971) the pollens of *Hippophae rhamnoides* ssp. *caucasica* Rousi have granular ornamentation on exine. Pollen type and form were not provided for this subspecies, but pollen type for *Hippophae* L. was stated to be

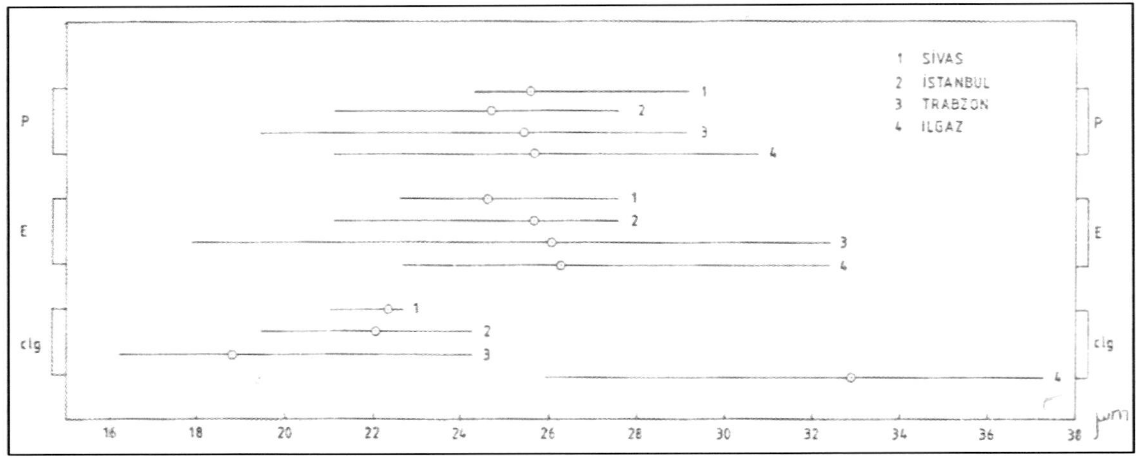

Figure 50.8: Graphical Comparison Showing Mean and Standard Deviations of Dimensions of some Pollen Features from Four different Regions.

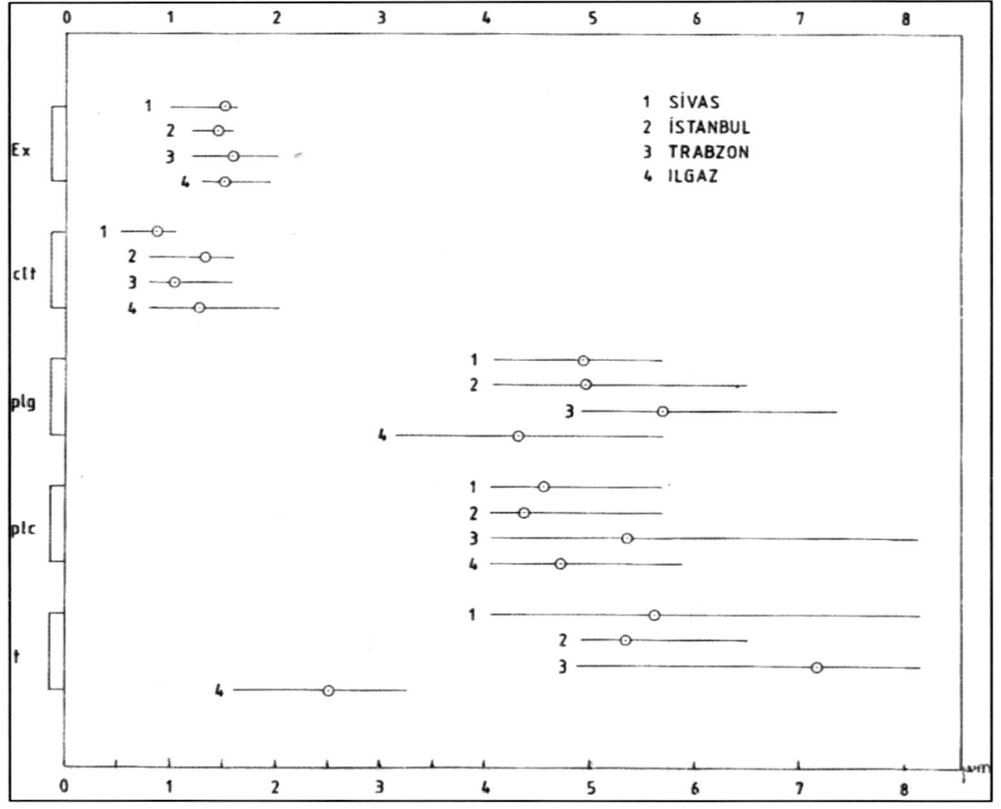

Figure 50.9: Graphical Comparison Showing Mean and Standard Deviations of Dimensions of some Pollen Features from Four different Regions.

"3-colporate, rarely 4 or 5-colporate" and pollen shape was stated to be prolate spheroidal and oblate spheroidal, rarely subprolate.

Sorsa did not give the values for length and width of colpus, and also did not mention intine. The plg was 1.9 µm which is much lower than the pollens from all the four regions. Since the exine thickness was 1.7 µm, it is in tune with the pollens we have measured. The polar triangle (t), which was measured 5.3 µm was equal to the pollens with the same features from regions of Istanbul and Sivas. Ectexine/entexine ratio was given 1/1, which is more or less closer to pollens from Ilgaz region. Information about the pollen structure is lacking. Concerning structure, the term "granulatae" was used, and related SEM figures were provided.

Ornemantation of *Hippophae rhamnoides* L. pollens was stated as "granulate-reticulate" (Erdtman, 1952) and "verrucate-spinule" (Nilsson *et al.*, 1977). While,Hyde and Adams (1958) have used the term finely reticulate for the ornamentation.

CONCLUSION

According to current literature, there is only *Hippophae rhamnoides* ssp. *caucasica* Rousi taxon growing in Turkey. However, the present data revealed that the pollen grains of *H. rhamnoides ssp caucasia* from Trabzon region might be hybrid, suggesting a taxon different from *Hippophae rhamnoides* ssp. *caucasica* Rousi in Trabzon region. This taxon might be a sub-species or separate variety. On the basis of these findings, it might be possible to trace out the parental plants through detailed comparative analysis of pollen characteristics of *Hippophae,* which may be close to or far from individuals from Trabzon region. In this case, it is needed to determine the distribution range of *Hippophae* L. taxa and to carry out a comprehensive taxonomic revision.

ACKNOWLEDGEMENT

This study is a part of Ph.D. thesis of the author.

REFERENCES

1. Bartish, I.V., Jeppsson, N. and Nybom, H. 1999. Population genetic structure in the dioecious pioneer plant species *Hippophae rhamnoides* investigated by random amplified polymorphic DNA (RAPD) markers. *Mol. Ecol.* 8 (5): 791-802.
2. Bartish, G.I., Jeppsson, N. and Bartish, I.V. 2000a. Assessment of genetic diversity using RAPD analysis in a germplasm collection of seabuckthorn. *Agric. Food Sci. Finland* 9 (4): 279-289.
3. Bartish, I.V., Jeppsson, N. and Bartish, G.I. 2000b. Inter-and intraspecific genetic variation in *Hippophae* (*Elaeagnaceae*) investigated by RAPD markers. *Plant Syst. Evol.* 225 (1-4): 85-101.
4. Bottema, S., Woldring, H. and Aytug, B. 1995. Late quaternary vegetation history of Northern Turkey. *Palaeohistoria* 35/36: 13-72.
5. Cupara, M.S., Ninkovic, B.M., Knezevic, G.M.,Vuckovic, M.I. and Jankovic, M.S. 2011. Wound healing potential of seabuckthorn oil topical emulsion. *Healthmed*. 5: 1218-1223.
6. Erdtman, G. 1952. *Pollen Morphology and Plant Taxonomy-Angiosperms: An Introduction to Palynology. Vol. I.* Almqvist and Wiksell, Stockholm, 539p.
7. Erdtman, G. 1969. *Handbook of Palynology, Morphology, Taxonomy and Ecology-An Introduction to the Study of Pollen Grains and Spores.* Munksgaard, Copenhagen, 486p.

8. Faegri, K. and Iverson, J. 1975. *Text Book of Pollen Analysis*. Munksgaard, Copenhagen, Denmark, 295p.
9. Hyde, H.A. and Adams, K.F. 1958. *An Atlas of Airborne Pollen Grains*. Macmillian Co. Ltd., London, 111p.
10. Kallio, H., Yang, B. and Peippo, P. 2002. Effect of different origins and harvesting time on vitamin C, tocopherols and tocotrienols in seabuckthorn (*Hippophae rhamnoides* L.) berries. *J. Agric. Food Chem.* 50: 6136-6142.
11. Lian, Y. 1988. New discoveries of the genus *Hippophae* L. *Acta Phytotaxonomica Sinica* 26: 235-237.
12. Lian, Y.S., Chen, X.L., Sun, K. and Ma, R. 2003a. A new subspecies of *Hippophae* (*Elaeagnaceae*) from China. *Novon* 13 (2): 200-202.
13. Lian, Y.S., Chen, X.L. and Sun, K. 2003b. Clarification of the systematic position of *Hippophae goniocarpa* (*Elaeagnaceae*). *Bot. J. Linn Soc.* 142 (4): 425-430.
14. Liu, S.W. and He, T.N. 1978. The genus *Hippophae* from Qing-Zang plateau. *Acta Phytotaxonomica Sinica* 16: 106-108. (In Chinese)
15. Narayanan, S., Ruma, D., Gitika, B., Sharma, S.K., Pauline, T., Sairam, M., Ilavazhagan, G., Sawhney, R.C., Kumar, D. and Banerjee, P.K. 2005. Antioxidant activities of seabuckthorn (*Hippophae rhamnoides*) during hypoxic induced oxidative stress in glial cells. *Mol. Cell. Biochem.* 278: 9-14.
16. Nilsson, S., Praglowski, J. and Nisson, L. 1977. *Atlas of Airborne Pollen Grains and Spores in Northern Europe*. Natur Och Kultur, Stockholm,159p.
17. Rousi, A. 1971. The genus *Hippophae* L.: A taxonomic study. *Ann. Bot. Fenn.* 8: 177-227.
18. Sorsa, P. 1971. Pollen morphological study of the genus *Hippophae* L. including the new taxa recognized by A. Rousi. *Ann. Bot. Fennici* 8: 228-236.
19. Sun, K., Chen, X. and Ma, R. 2002. Molecular phylogenetics of *Hippophae* L. (*Elaeagnaceae*) based on the internal transcribed spacer (ITS) sequences of nrDNA. *Plant Syst. Evol.* 235 (1-4): 121-134.
20. Sun, K., Ma, R.J. and Chen, X.L. 2003. Hybrid origin of the diploid species *Hippophae goniocarpa* evidenced by the internal transcribed spacers (ITS) of nuclear rDNA. *Belg. J. Bot.* 136 (1): 91-96.
21. Yao, Y. and Tigerstedt, P.M.A. 1994. Genetic diversity in *Hippophae* L. and its use in plant breeding. *Euphytica* 77: 165-169.
22. Yu, Z., Ao, F. and Lian, Y. 1989. Discussion on the problems of origin, classification, community and resource of seabuckthorn in China. In: *Proceedings of International Symposium on Seabuckthorn* (*H. rhamnoides* L.), p. 21-30, Xian, China.
23. Wodehouse, R.P. 1935. *Pollen Grains: Their Structure, Identification and Significance in Science and Medicine*. Hafner Publishing Company, New York, 574p.

Index

T